International Review of Nuclear Physics - Vol. 10

Relativistic Density Functional for Nuclear Structure

International Review of Nuclear Physics

ISSN: 1793-155X

Published:

Vol. 1	Quarks and Nuclei *edited by W Weise*
Vol. 3	Few-Body Problems *by E Hadjimichael and W Oelert*
Vol. 4	Cluster Models and Other Topics *by Y Akaishi, S A Chin, Horiuchi and K Ikeda*
Vol. 5 & 6	Relativistic Heavy Ion Physics — In 2 Volumes *edited by L P Csernai and D D Strottman*
Vol. 7	Electric and Magnetic Giant Resonances in Nucle *edited by J Speth*
Vol. 8	Nuclear Methods and the Nuclear Equation of State *edited by Marcello Baldo*
Vol. 9	Hadronic Physics from Lattice QCD *edited by Anthony M Green*
Vol. 10	Relativistic Density Functional for Nuclear Structure *edited by Jie Meng*

International Review of Nuclear Physics - Vol. 10

Relativistic Density Functional for Nuclear Structure

edited by

Jie Meng

Peking University, China

World Scientific

NEW JERSEY • LONDON • SINGAPORE • BEIJING • SHANGHAI • HONG KONG • TAIPEI • CHENNAI • TOKYO

Published by

World Scientific Publishing Co. Pte. Ltd.

5 Toh Tuck Link, Singapore 596224

USA office: 27 Warren Street, Suite 401-402, Hackensack, NJ 07601

UK office: 57 Shelton Street, Covent Garden, London WC2H 9HE

Library of Congress Cataloging-in-Publication Data

Names: Meng, Jie, 1966 January 25– author.

Title: Relativistic density functional for nuclear structure / Jie Meng (Peking University).

Other titles: International review of nuclear physics ; v. 10.

Description: Singapore ; Hackensack, NJ : World Scientific, [2015] | Series: International review of nuclear physics ; vol. 10 | Includes bibliographical references and index.

Identifiers: LCCN 2015033649| ISBN 9789814733250 (hardcover : alk. paper) | ISBN 9814733253 (hardcover : alk. paper)

Subjects: LCSH: Density functionals. | Many-body problem. | Nuclear structure.

Classification: LCC QC20.7.D43 M46 2015 | DDC 541/.28--dc23

LC record available at http://lccn.loc.gov/2015033649

British Library Cataloguing-in-Publication Data

A catalogue record for this book is available from the British Library.

Printed in Singapore

Preface

At the turn of the nineteenth and twentieth centuries, the door of nuclear science was opened with the discoveries of radioactivity by Becquerel and the Curies and the existence of a compact core at the center of an atom by Rutherford et al. These achievements conceived the birth of quantum mechanics, promoted the exploitation and utilization of nuclear energy and nuclear technology, and brought about a huge impact on human life. During the hundred years of studies, and in the development of nuclear physics itself, there emerged a number of significant milestones, including the discovery of the neutron by Chadwick which verified the composition of the nucleus as of protons and neutrons, the meson-exchange theory for the interaction between nucleons by Yukawa, the independent-particle shell model of the nucleus by Goeppert-Mayer and Jensen et al., and the collective Hamiltonian for nuclear rotations and vibrations by Bohr and Mottelson, etc.

However, since the 1950s, nuclear physics entered a more challenging stage. Although the independent-particle shell model could describe the single-particle motion in a nucleus with a phenomenological mean potential, it could not provide even a qualitative description of the nuclear bulk properties. On the other hand, a unified phenomenological description of nuclear vibrations and rotations could be achieved by a collective Hamiltonian but it was helpless in understanding the motion of a single nucleon. How to achieve microscopically and self-consistently a unified description of the single-nucleon and collective motions of the nucleus based on the theory of the strong interaction is a crucial question to be answered by nuclear scientists. In particular we should understand nuclear structure and reactions, the regular and simple patterns emerging in the structure of complex nuclei, and the evolution of nuclear properties with mass, isospin, angular momentum and temperature etc. in terms of the fundamental interactions between individual particles.

Much progress in this direction has been achieved in recent years considering the various scales. Having in mind quantum chromo-dynamics

(QCD), the basic theory of the strong interaction, and its symmetries, chiral symmetry is used for nuclear structure, i.e. in the region of low momenta. The non-relativistic form of the bare nucleon-nucleon forces is expressed in terms of chiral perturbation theory of various orders. At long and intermediate distances the pion, the Goldstone mode resulting from the breaking of chiral symmetry, plays the essential role in these processes. Only the short range part of the force, acting at a scale not influencing directly the details of nuclear physics, is written in terms of various renormalization parameters. At present these chiral forces still contain a large number of phenomenological parameters, which are adjusted to the experimental nucleon-nucleon scattering data. In the next step sophisticated numerical techniques have been developed to solve the nuclear many-body problem exactly in terms of these chiral forces. This is possible nowadays for very light nuclei. However, the results are disappointing and agreement with experimental binding energies, radii, and spectra of finite nuclei can be achieved only by introducing bare three-body forces containing additional phenomenological parameters adjusted to light nuclei.

For medium heavy nuclei, in particular for those near closed shells, approximate techniques based on these microscopic forces show considerable success, but most of the very successful shell model calculations that take into account the full mixing of all configurations in a limited space use effective interactions with additional phenomenological parameters. The wide range of deformed heavy and super-heavy nuclei, in particular those far from the valley of stability, are, however, described by density functional theory. It is based on universal energy density functionals. The essential idea of this theory, introduced originally in Coulombic systems, is an exact mapping of the complicated quantum-mechanical many-body problem onto an effective one-body problem, which is relatively easy to solve. In contrast to Coulombic systems, where the energy density functional can be derived from the bare Coulomb interaction, nuclear density functionals are more of less phenomenological. The ground state energy is expanded in terms of various densities, gradients and currents. Symmetries are used to restrict the numbers of phenomenological parameters, which are adjusted to experimental data in finite nuclei. For reasons of numerical simplicity, most of the applications of density functional theory in nuclei use a non-relativistic concept.

Relativistic or covariant density functional theory takes into account Lorentz invariance, one of the basic symmetries of QCD. Therefore it starts from a field theoretical Lagrangian. In the most popular version of covariant density functional theory, the nucleus is described as a system of Dirac nucleons that interact with each other via the exchange of mesons with various relativistic spin- and isospin quantum numbers: scalar-isoscalar, vector-isoscalar, scalar-isovector, and vector-isovector mesons build the minimal set of fields that, together with the electromagnetic field, are necessary for a description of the bulk and single-particle properties of nuclei. Moreover, the proper treatment of nuclear matter and surface properties requires a density dependence of the effective meson-nucleon vertices. It can be introduced in the Lagrangian by nonlinear meson self-interactions or by an explicit density dependence for the couplings parameters. Since the exchange of heavy mesons is associated with short-distance dynamics that cannot be resolved at low energies, covariant density functionals with point-coupling interactions have been proposed as an alternative. Here, in each channel (scalar-isoscalar, vector-isoscalar, scalar-isovector, and vector-isovector) the meson exchange is replaced by a corresponding local four-point interaction between the nucleons. Medium effects can be taken into account by including higher-order couplings or by assuming a density dependence of the strength parameters.

Compared with traditional non-relativistic theories of the nucleus, the meson-exchange picture provides a more fundamental and microscopic scheme for the effective nucleon-nucleon interaction. Relativistic density functional theory is of particular interest in nuclear physics as it exploits basic properties of QCD at low energies, such as symmetries and the separation of scales. Because of Lorentz covariance it naturally includes the spin-orbit component of the nuclear force and provides a new saturation mechanism for nuclear matter. It contains pseudospin as a relativistic symmetry for the nucleons and spin symmetry in the anti-nucleon spectrum. It also includes automatically nuclear magnetism, i.e. a consistent description of currents and time-odd fields, which play an important role in the description of nuclear magnetic moments and nuclear rotations. Furthermore, Lorentz symmetry puts stringent restrictions on the number of parameters in the corresponding functional without reducing the quality of the agreement with experimental data. Moreover, as in any mean field description, various kinds of correlations can be introduced via spontaneous symmetry breaking.

Based on covariant density functional theory, not only the description of the single-particle motion in nuclei but also the self-consistent description of nuclear collective modes, such as rotations, vibrations, and isospin excitations, can be achieved by restoring the symmetries and/or considering quantum fluctuations.

In the last decades, numerous efforts have been devoted to developing relativistic density functional theories including the finite-range meson-exchange and the zero-range point-coupling theories. Implemented with the newly developed or existing many-body techniques, the relativistic density functional theories have been successfully applied to describe a large variety of nuclear phenomena from finite nuclei to nuclear matter, from stable to extreme unstable nuclei near the neutron or proton drip-lines, from spherical shape to nuclei with novel shapes, from nuclear ground-state to excited-state properties, and from normal nuclear matter to hyper nuclear matter with strangeness degrees of freedom. Relativistic density functional theories have been widely used to investigate the equation of state for compact stars and provide important observables for nucleosynthesis in the astrophysical environment.

The current book, Relativistic Density Functional for Nuclear Structure, is one of the volumes in the series International Review of Nuclear Physics. It is intended to overview and to promote the development of relativistic density functional theories and their applications. All authors involved are experts and leading scientists in the related fields. The aim of the book is to make it helpful for students as well as research scientists interested in relativistic many-body theory or nuclear physics.

The book starts with some basic ideas and concepts of relativistic density functional theory. The theoretical framework is introduced including meson-exchange and point-coupling representations, with or without exchange terms. The hidden relativistic symmetries in the nuclear single-particle spectrum are discussed, in particular, pseudospin symmetry and spin symmetry, in connection with similarity renormalization group techniques and in a supersymmetric representation.

Relativistic density functional theories implemented with self-consistency and taking into account various correlations by spontaneously broken symmetries provide an excellent description for the ground-state properties, including the total energy and the other physical observables as the expectation values of local one-body operators. In order to describe open shell nuclei where pairing correlations play a major role, generalized density functional theories based on the Bardeen–Cooper–Schrieffer (BCS) or the Hartree–Fock–Bogoliubov (HFB) approach are included as well. Particular attention is focused on exotic nuclei with extreme neutron or proton numbers, where novel phenomena such as neutron halos may appear.

Static density functional theory includes calculations in a static external field, also a rotating field. Constraint relativistic density functional is therefore a powerful tool to investigate the ground state, shape isomers, shape-coexistence, and fission landscapes. Rotating relativistic density functional theory obtained by transforming the fields from the laboratory frame to the intrinsic frame has proven to be very successful in describing rotational spectra in nearly spherical, deformed, superdeformed, and triaxial nuclei.

In general, static density functional theory is not applicable for the description of excited states. Here, a proper method is time-dependent density functional theory. In analogy to the Hohenberg–Kohn theorem, there exists the Runge and Gross theorem which provides an exact mapping of the full time-dependent many-body problem onto a time-dependent single-particle problem. The corresponding single-particle field is not only time-dependent, but it also depends on the single-particle density with its full time dependence, i.e., it includes memory effects. Very little is known about the general properties of this new single-particle field, the self-energy. For vibrations with small amplitude, however, it can be expanded in the vicinity of the ground state and a connection to the well known static energy density functional can be found. In the adiabatic approximation, i.e. by neglecting the memory effects and the corresponding energy dependence in Fourier space, one ends up with the well-known random phase approximation (RPA) with a residual interaction derived, in a self-consistent way, as the second derivative of the static energy functional with respect to the density. This provides a very successful description of the mean energies of giant resonances in nuclei. However, RPA is not able to reproduce the decay width of these excitations which are produced to a large extent by the coupling to more complicated configurations. In order to describe such effects as well as the fragmentation of single-particle states, one has to go beyond mean field and consider the energy dependence of the self-energy. This can be done within the particle-vibrational coupling (PVC) model and is discussed in connection with the nuclear shell structure.

In transitional nuclei with very flat energy surfaces and several minima in deformation space, the small amplitude limit breaks down. To calculate excitation spectra and transition rates it is necessary to include correlations which arise from symmetry restoration and fluctuations around the mean-field equilibrium solution. The many-body energy takes the form of a functional of all transition density matrices between the various Slater determinants of different deformation and orientation. Mixing of these different configurations allows the restoration of symmetries and to take into account fluctuations around the mean-field equilibrium solution. In full space and with several deformation degrees of freedom, such calculations require considerable numerical efforts, at the limit of present computer capabilities. A considerable simplification is achieved by using the constraint relativistic mean field calculations for the derivation of a collective Hamiltonian in these degrees of freedom. It depends only on the underlying static energy density functional and does not contain any additional phenomenological parameters. The recent advances in beyond mean field relativistic density functional theory including angular-momentum and particle-number projection and mixing of various deformations, as well as the results of the corresponding microscopically determined collective Hamiltonians, are reviewed and compared. They are followed by illustrative applications including low-energy spectra of collective excitation and electromagnetic transition rates of nuclei characterized by quadrupole and/or octupole degrees of freedom.

With the important successes of the relativistic density functional theories, over the years, there has been growing interest to examine their applicability in understanding nucleosynthesis and the structure of neutron stars in astrophysics. The related topics covered here range from the nuclear physics inputs for rapid-neutron and rapid-proton capture processes, nucleosynthesis studies, nuclear cosmochronology, and the myriad of exotic states of matter believed to exist in a neutron star.

In the final chapter in the book, relativistic and non-relativistic density functionals are compared. Overall quality, predictive power, and correlations between observables are addressed using a statistical analysis on the basis of least squares fits. The possible directions for future developments of energy density functional theory are discussed.

Although, at present, covariant density functional theory is based on meson-exchange theory and on relativistic covariance, it is phenomenological to a large extent, and not explicitly derived from QCD, the basic theory of the strong interaction. As indicated by the successes achieved by the relativistic density functional, a unified description of nuclear ground state and excitation properties starting from a unified effective covariant density functional is feasible now. The aim of future investigations is to follow the successful concept of density functional theory in Coulombic systems and to try to derive this density functional from QCD. Of course, this has to be done in several steps. The first step would be the derivation of a relativistic form of the bare nucleon-nucleon interaction based on relativistic versions of chiral perturbation theory. Starting from such an interaction the homogeneous system should be solved exactly providing the basis for the energy density functional. Additional steps with constraining fields could allow the calculations of gradient terms, surface corrections and so on.

Jie Meng
September 20, 2015

Contents

Chapter 1

Concept of covariant density functional theory

Peter Ring

Physics Department, Technical University Munich, 85748 Garching, Germany

School of Physics, Peking University, Beijing 100871, China

ring@ph.tum.de

The concept of density functional theory is introduced. The history of nuclear density functional theory is briefly reviewed. The complexity and specialty of the density functional theories in nuclear system in connection with the degrees of freedom, covariance and pairing correlations are discussed in contrast to Coulombic systems. The existing problems and limits of the nuclear density functional theory are discussed. Perspective on the way to derive *ab initio* functionals is given.

Contents

1. Kohn–Sham density functional theory

Density Functional Theory (DFT) is one of the most popular and successful *ab initio* approaches to many quantum many-body systems (atoms, molecules, solids). The basic idea is that the ground-state energy of a stationary many-body system can be represented in terms of the ground-state density alone. Since the density is only a function with three spatial coordinates, rather than the N-body wave function with $3N$ coordinates,

DFT calculations are comparatively simple to implement yet often very accurate and computationally feasible even for systems with large particle numbers.

We start with DFT from the Hohenberg and Kohn (HK) theorem:[1] For an electromechanical many-body system in an external potential $v_{\text{ext}}(\boldsymbol{r})$ there exists an energy functional $E_v[\rho]$ of the density $\rho(\boldsymbol{r})$, such that the total energy is given by

$$E_v[\rho] = F_{\text{HK}}[\rho] + \int d^3\boldsymbol{r} v_{\text{ext}}(\boldsymbol{r})\rho(\boldsymbol{r}). \tag{1}$$

The functional F_{HK} does not depend on the external potential $v_{\text{ext}}(\boldsymbol{r})$ nor on the particle number of the system. It only depends on the form of the interaction between the particles. It is therefore identical for Coulomb systems, such as atoms, molecules and solids. In this sense, this energy density functional is supposed to be universal. Of course, nuclei governed by the strong interaction require a different functional, but it is again universal for the entire nuclear chart. For atoms and molecules, the external potential $v_{\text{ext}}(\boldsymbol{r})$ is usually the electrostatic potential from ions. The variational principle determines the exact ground-state density $\rho_{\text{gs}}(\boldsymbol{r})$ and the corresponding exact ground-state energy by minimizing $E_v[\rho]$ with respect to the density.

Starting with this form of the Hohenberg–Kohn theorem we immediately face three problems: (i) This theorem shows only the existence of the functional $F_{\text{HK}}[\rho]$. It gives a license to search for such energy density functional, but it offers no help in constructing the functional $F_{\text{HK}}[\rho]$. (ii) Even after more than half a century of intensive work nobody has been able to find a functional from which shell effects can be derived just from the variation of $E_v[\rho]$ with respect to the local density $\rho(\boldsymbol{r})$, which are an essential criterion for many quantum mechanical systems. (iii) The simple proof of Hohenberg and Kohn is based on the existence of a local external field $v_{\text{ext}}(\boldsymbol{r})$. This is usually no problem in Coulombic systems, where the ions provide such a field. Nuclei, however, are self-bound systems. We will come to this problem later on.

The problem of missing shell effects in the Hohenberg–Kohn theory has been solved by Kohn and Sham (KS) in a very elegant scheme[2,3] by mapping the exact local density $\rho(\boldsymbol{r})$ in a unique way to a local auxiliary single-particle potential $v_{\text{KS}}(\boldsymbol{r})$, defined in such a way that the exact ground-state density $\rho_{\text{gs}}(\boldsymbol{r})$ of the interacting system is the same as the ground-state density of the auxiliary non-interacting system,

$$\rho_{\rm gs}(\boldsymbol{r}) = \rho(\boldsymbol{r}) := \sum_i^N |\phi_i(\boldsymbol{r})|^2, \tag{2}$$

which is expressed in terms of the N lowest occupied single-particle orbital solutions of the Kohn–Sham equations

$$\left[-\frac{\nabla^2}{2m} + v_{\rm KS}(\boldsymbol{r})\right] \phi_i(\boldsymbol{r}) = \varepsilon_i \phi_i(\boldsymbol{r}). \tag{3}$$

To date, most practical applications of density functional theory use this Kohn–Sham (KS) scheme.

The exact energy density functional $F_{\rm HK}[\rho]$ can be decomposed into three separate terms:

$$F_{\rm HK}[\rho] = T_s[\rho] + E_H[\rho] + E_{xc}[\rho], \tag{4}$$

where T_s is the kinetic energy of the non-interacting A-nucleon system, E_H is the Hartree energy, and E_{xc} is defined as the exchange-correlation energy which, by definition, contains all other many-body effects which are not included in $T_s + E_H$.

It can be shown that the local Kohn–Sham potential $v_{\rm KS}(\boldsymbol{r})$ is given by

$$v_{\rm KS}(\boldsymbol{r}) = \frac{\delta(E_H + E_{xc})}{\delta\rho}, \tag{5}$$

and one can clearly see that the Kohn–Sham orbitals depend on the potential, which depends on the density, which depends on the orbitals. Therefore, this system has to be solved self-consistently, e.g., by iterating until convergence is achieved. Formally this procedure looks similar to the Hartree–Fock method. It has, however two essential advantages: (i) it is, in principle, exact and (ii) it avoids the complicated non-local Fock-term $V_{\rm F}(\boldsymbol{r}, \boldsymbol{r}')$ and deals only with local quantities. The complicated integro-differential equations of Hartree–Fock theory are replaced by simple differential equations (3).

It is clear that the usefulness of the Kohn–Sham scheme crucially depends on how accurate the universal exchange-correlation energy functional can be built. One possible way, which has been followed with great success in non-relativistic Coulombic systems,[4] is to develop the E_{xc} from first principles by incorporating known exact constraints. Another way is empirical, i.e., an empirical energy density function with the parameters optimized by fitting to a set of data. This way has been used in the early days of density functional theory in Coulombic systems and it is still the only way to use density functional in nuclear systems.

The Kohn–Sham concept can be written on local quantities. Without changing its basic idea, it can be generalized slightly, such that it can be used also for non-local problems, in particular in cases where other degrees besides the coordinates $\boldsymbol{r}$ play an essential role. The essential ingredient in this method are single particle wave functions $\phi_i(\boldsymbol{r})$ obtained from the diagonalization of a single particle Hamiltonian

$$\hat{h} = \left[-\frac{\nabla^2}{2m} + v_{\text{KS}}(\boldsymbol{r})\right] = \frac{\delta E[\hat{\rho}]}{\delta\hat{\rho}}, \tag{6}$$

which can be written as the derivative of the energy with respect to non-local density matrix

$$\hat{\rho}(\boldsymbol{r}, \boldsymbol{r}') = \sum_i^N \phi_i(\boldsymbol{r})\phi_i^*(\boldsymbol{r}'). \tag{7}$$

Obviously this density operator is not the exact single particle density operator, because it has single particle structure

$$\hat{\rho}^2 = \hat{\rho}. \tag{8}$$

It is only connected to the exact non-local single particle density operator by the fact that its local (diagonal) part is identical to the exact local density $\rho_{\text{gs}}(\boldsymbol{r}) = \hat{\rho}(\boldsymbol{r}, \boldsymbol{r})$. Within the concept of Kohn–Sham theory[2] there is a unique one-to-one correspondence between the exact density $\rho_{\text{gs}}(\boldsymbol{r})$ and the potential $v_{\text{KS}}(\boldsymbol{r})$. Therefore such a correspondence exists also between $\hat{\rho}$ and $\rho_{\text{gs}}(\boldsymbol{r})$ and the functional $E[\hat{\rho}]$ used in Eq. (6) is given by the Hohenberg–Kohn functional $E[\hat{\rho}] = F_{\text{HK}}[\rho_{\text{gs}}]$ in Eq. (4). It is clear that already the kinetic part of F_{HK} has a non-local structure. In the next section we will see that this concept of a non-local density operator with single particle structure is also very useful in nuclear physics, where we have to deal with other degrees of freedom besides the coordinates $\boldsymbol{r}$ and where the energy functional $E[\hat{\rho}]$ is given in terms of the single particle operator $\hat{\rho}$ instead of its local part only.

The extension of the Hohenberg and Kohn theorem[1] to relativistic systems was first formulated in Ref. 5 by utilizing a quantum electrodynamics (QED)-based Hamiltonian with the four-current. Within the no-sea approximation, which is also called no-pair approximation in atomic physics because it neglects all effects due to the creation of particle-antiparticle pairs, the total energy of the system can be expressed as a functional with respect to the four-current $j^\mu(\boldsymbol{r}) = (\rho(\boldsymbol{r}), \boldsymbol{j}(\boldsymbol{r}))$,[6,7] instead of only $\rho(r)$ in the non-relativistic case.

2. Historical remarks on nuclear density functional theory

Density functional theory was introduced in nuclear physics in the early seventies[8,9] and it has been used with great success since then.[10] It was introduced some years later than Kohn–Sham theory in Coulombic systems, not so much by the desire to have an exact theory of the nucleus but in order to understand the nuclear single particle model[11,12] in terms of an effective interaction. Since it was clear from the beginning that the bare nucleon-nucleon interaction was much too strong to be used within the framework of the Hartree–Fock approximation, several phenomenological effective two-body forces have been used within this framework, but none of these investigations could reproduce at the same time nuclear binding energies and nuclear radii. In order to understand the effective nuclear force from the bare nucleon-nucleon interaction, Brueckner et al[13] treated the scattering process between two nucleons in the nuclear medium in a similar way as in the vacuum by summing up the ladder diagrams in the Born series. They took into account the fact that in the intermediate states the nucleons could not scatter into orbits within the Fermi sea, which are occupied in the nuclear interior. This leads to a considerable reduction of the effective interaction, the so-called G-matrix, within the nuclear interior. From calculations in infinite nuclear matter it turns out this G-matrix, which can be used for Hartree–Fock calculations, is strongly density dependent. Based on this observation Vautherin and Brink[8] used the simple phenomenological force introduced by Skyrme[14] with a three-body term of zero range. This leads on the mean field level to an effective two-body force with a linear density dependence. In this way they were able by an appropriate fit to experimental data to reproduce not only experimental binding energies, but also experimental radii. Therefore at the beginning, and in many cases even today, nuclear density functional theory is not based on an energy density functional, but rather on a density dependent two-body interaction $V_{12}(\rho)$. The corresponding energy density functional $E[\hat{\rho}]$ is derived from it in Hartree–Fock approximation

$$E[\hat{\rho}] = \langle\Phi|T + V_{12}(\rho)|\Phi\rangle, \tag{9}$$

where $|\Phi\rangle$ is a Slater determinant. A variation with respect to the corresponding density matrix $\hat{\rho} = \langle\Phi|a^\dagger a|\Phi\rangle$ leads to the Hartree–Fock equations with rearrangement terms obtained from the variation of $V_{12}(\rho)$ with respect to ρ. It turned out that these rearrangement terms form the essential tool to reproduce at the same time nuclear binding energies and nuclear radii. They are also essential to guarantee that this type of theory obeys a variational principle and thermodynamic consistency.[15–17]

3. Degrees of freedom in nuclear density functional theory

In many applications of density functional theory in Coulombic systems only the coordinates of the particles play the essential role. Additional degrees of freedom, such as the spin, can often be neglected or treated separately. In all these cases it is enough to start with a density functional depending only on the local density, summed over the spin: $\rho(\boldsymbol{r}) = \sum_s \hat{\rho}(\boldsymbol{r}s, \boldsymbol{r}s)$. The spin degree of freedom shows up in such a case only in statistical factors.

In nuclei the situation is very different. We need in most cases several additional degrees of freedom with two discrete values. They behave similar to the spin and the corresponding 2×2 matrices can be expressed by the unit matrix and the three Pauli matrices.

Spin degree of freedom: The most obvious case is the spin degree of freedom. The spin-orbit force is very strong and the entire nuclear shell structure, in particular in medium heavy and heavy nuclei, can only be explained if the spin-orbit term is taken into account in the mean field from the beginning. It cannot be treated in any way by perturbation theory. Therefore single particle wave functions are spinors of dimension 2 and the local density is a 2×2 matrix, which can be decomposed into a scalar part with $S = 0$ and vector part with spin $S = 1$

$$\rho(\boldsymbol{r}) = \rho_0(\boldsymbol{r}) + \vec{\sigma}\cdot\vec{\rho}_1(\boldsymbol{r}). \tag{10}$$

Therefore, in general, the total energy depends on four densities $E[\rho_0, \vec{\rho}_1]$. Of course, because of rotational invariance the vector density $\vec{\rho}_1$ has to be coupled with other vectors to form finally a scalar. In order to describe the spin-orbit splitting in nuclei the Skyrme functionals contain also a two-body spin-orbit force of zero range.

Isospin degree of freedom: In addition we have in the nucleus two types of particles, neutrons and protons, which can be classified by the isospin. This leads to isoscalar densities and isovector densities. Since in most cases the Slater determinants have a fixed proton and neutron number, the third component of the isospin $T_3 = \frac{1}{2}(N-Z)$ is preserved and therefore only isospin-densities with $T_3 = 0$ are important, i.e. the total (isoscalar) density $\rho = \rho_+ = \rho_n + \rho_p$ and the isovector density $\rho_- = \rho_n - \rho_p$. Together with the isospin the single particle wave functions are therefore spinors of dimension 4.

Pairing degree of freedom: Most of the nuclei are open shell nuclei, where pairing correlations play a major role. Fortunately we can describe these correlations on a generalized mean field level within Bardeen–Cooper–Schrieffer (BCS)[18] or within Hartree–Fock–Bogoliubov (HFB) theory.[19–21] In this case, particle number is no longer a good quantum number. We have to deal with quasiparticles, which mix creation and annihilation operators

$$\alpha_k^\dagger = \sum_n U_{nk} c_n^\dagger + V_{nk} c_n. \tag{11}$$

The operators $(c, c^\dagger)$ induce an additional degree of freedom with the two discrete values 'annihilation' or 'creation'. In addition to the normal density matrix $\hat{\rho}$, we have now the anomalous density $\hat{\kappa}$

$$\rho_{nn'} = \langle\Phi|c_{n'}^\dagger c_n|\Phi\rangle, \qquad \kappa_{nn'} = \langle\Phi|c_{n'} c_n|\Phi\rangle, \tag{12}$$

where $|\Phi\rangle$ is the generalized Slater determinant (HFB-wave function), which describes the nuclear ground state and we can introduce the Valatin density matrix:[22]

$$\mathcal{R} = \begin{pmatrix} \hat{\rho} & \hat{\kappa} \\ -\hat{\kappa}^* & 1-\hat{\rho}^* \end{pmatrix}, \tag{13}$$

and the energy density functional depends on $\mathcal{R}$. This is usually written in a simplified way as

$$E[\mathcal{R}] = E_{\mathrm{MF}}[\hat{\rho}] + E_{\mathrm{pair}}[\hat{\kappa}]. \tag{14}$$

The mean field equations, which play the role of the Kohn–Sham equations in this case, are now HFB-equations

$$\begin{pmatrix} \hat{h} & \hat{\Delta} \\ -\Delta^* & -h^* \end{pmatrix} \begin{pmatrix} U \\ V \end{pmatrix}_k = \begin{pmatrix} U \\ V \end{pmatrix}_k E_k \tag{15}$$

with

$$\hat{h} = \frac{\delta E}{\delta \hat{\rho}}, \qquad \text{and} \qquad \hat{\Delta} = \frac{\delta E}{\delta \hat{\kappa}}\ , \tag{16}$$

and the quasiparticle wave functions $\begin{pmatrix} U(\boldsymbol{r}) \\ V(\boldsymbol{r}) \end{pmatrix}$ are now spinors of dimension 8.

Gogny introduced an effective interaction of finite range with a density dependent zero range part and a zero range two-body spin-orbit force. From this unique Hamiltonian $\hat{H}(\hat{\rho})$ he derived as well the mean field $\hat{h}$ as the pairing potential $\hat{\Delta}$. Because of the finite range of the density independent part of the two-body force he could avoid the ultraviolet divergence in the pairing channel, which usually needs an additional renormalization or a pairing window.

Relativistic degree of freedom: In principle nuclei are relativistic systems. The strong interaction has its origin in QCD, which is Lorentz invariant. Investigations using QCD sum rules[23,24] show that the scalar and the vector quark condensates induce in the nuclear medium two very large fields, an attractive scalar field S and a repulsive vector field V. All the non-relativistic nuclear density functionals are based on the fact that the velocities of the nucleons in the nucleus are small against their rest mass and therefore relativistic kinematics can be replaced by non-relativistic kinematics to a good approximation

$$\sqrt{c^2p^2 + (Mc^2)^2} \approx Mc^2 + \frac{p^2}{2M}. \tag{17}$$

In this way, one neglects, however, that the relativistic fields in the nucleus are relatively large. One has an attractive Lorenz scalar field $S \approx -400$ MeV and a repulsive Lorentz vector field $V \approx +350$ MeV. Only the sum of these fields seen by the large components in the Dirac equations is small $V + S \approx -50$ MeV. The small components feel the a very strong field $V - S \approx 750$ MeV. This leads to an extremely large spin-orbit splitting in nuclei, which is taken into account in non-relativistic models by a phenomenological parameter in the effective spin-orbit force used. Of course, it is more consistent to treat these large fields in the Dirac equation consistently in a self-consistent way and therefore relativistic density functional theories have certain advantages, which will be discussed in this volume in great detail. Of course, large and small components in the Dirac equation lead to a new discrete degree of freedom and all in all we have now to deal with spinors of dimension 16.

4. The complexity of the nuclear force

Coulombic systems are completely determined by the Coulomb force. It has a simple analytical structure and comes directly from QED. In nuclei the situation is very different. QCD is not analytically solvable at low energies. Highly non-linear processes in the gluon sector lead to color confinement and the effective residual interaction between nucleons is an induced interaction, relatively weak on the QCD scale. The situation is very similar to that of the forces between molecules caused by polarization effects of the various electron clouds. They are usually parameterized in a phenomenological way by the Lenard-Jones potential, but it is very hard to express them in an analytical way through the underlying Coulomb force. On the nuclear scale the forces between bare nucleons are relatively strong. At large distances the one-pion exchange dominates and they have an attractive tail of Yukawa shape, which decays relatively fast. At short distances they are strongly repulsive and in a phenomenological description by meson exchange they are dominated in this area by a strong repulsion caused by the exchange of vector mesons. In modern descriptions based on chiral perturbation theory this short range repulsion, which happens at scales outside the range of the validity of these theories, is often described by phenomenological zero-range forces and cut-off processes. At intermediate ranges the bare nucleon-nucleon force has a large tensor term, strongly off-diagonal in momentum space. It leads to the strong attraction of the nucleon-nucleon force at intermediate distances, which is the main reason for nuclear binding. It is caused to a large extent by two-pion exchange processes and, in phenomenological descriptions it is often described by the exchange of a scalar σ-meson.[25]

After more than fifty years of intensive experimental and theoretical investigations one has today various parameterizations of the bare nucleon-nucleon interaction which reproduce the nucleon-nucleon scattering data to a very high precision. They contain a large number of parameters. In older parameterizations they are phenomenological and fitted directly to the scattering phase shifts.[26] In more modern versions based on chiral perturbation theory[27] some of them, the low energy constants, are taken from the pion-nucleon Lagrangian adjusted to pion-scattering data. In recent years there are also attempts to derive the bare nucleon-nucleon force directly from QCD by large scale lattice calculations.[28] Most of these forces are non-relativistic, but there exist also relativistic versions, such as the relativistic one-boson-exchange potentials (OBEP) of the Bonn–Idaho group.[29,30]

One major problem with nuclear theory based on non-relativistic bare nucleon-nucleon interactions is the fact that it is not enough to determine the two-body interactions with high precisions. All the forces obtained in this way have problems reaching proper saturation in infinite symmetric nuclear matter. Depending on the strength of their tensor component the saturation points of all these non-relativistic forces lie roughly on a line in the E/A versus density ρ plane, the so-called *Coester line*.[31] Nowadays it is generally believed that the origin of this derivation are bare three-nucleon forces, not taken into account so far. In fact, recent investigations have shown that three-body forces play an important role in nuclei. The parameters of these forces are usually adjusted to properties of finite nuclei. The microscopic origin of these forces is still unknown to a large extent. The most famous case is the Fujita–Miyazawa diagram[32] involving a Δ-resonance as an intermediate state in one of the three nucleons, which is excited and absorbed by the exchange of two pions with the remaining nucleons. Of course there are also other intermediate configurations and other meson exchanges possible, such as for instance the virtual excitation of nucleon-anti-nucleon pairs (Z-diagram).[33] Recent non-relativistic Brueckner–Hartree–Fock calculations including such three-body forces[33] have shown that one can obtain the proper saturation properties in this way and that the Z-diagram plays a decisive role in this case. This is a strong hint that relativistic descriptions of the nuclear many-body problem, which include such effects from the beginning, can solve at least part of the problem with three-body forces. In fact, relativistic Brueckner–Hartree–Fock calculations[34–39] without three-body forces are able to shift the Coester line to the experimental area. Such investigations give us a better understanding of why covariant density functional theories are relatively successful.

5. Phenomenology in nuclear density functionals

In contrast to Coulombic systems, where one has very successful microscopic density functionals[4] which are completely derived from the Coulomb force and do not include any additional phenomenological parameters, the situation in nuclear physics is very different. At present all successful nuclear density functionals are based on phenomenological parameters. This is understandable because of the many difficulties in the nuclear case discussed before.

The form of the energy density functional in nuclear physics is determined by the principles of symmetries and simplicity. This is translational and rotational symmetry in the non-relativistic case and Poincare symmetry, including Lorentz invariance, in the relativistic case. In addition we have the well-known discrete symmetries such as parity and time-reversal. Isospin invariance is usually taken into account on the level of nuclear interactions, but it is broken by the Coulomb interaction. Only the simplest terms allowing for such symmetries are taken into account and the remaining parameters are fitted to experimental data in the many-nucleon system. The density dependence is taken into account in a very different way for relativistic and for non-relativistic functionals. In the conventional non-relativistic functionals one uses only an additional term with a power law $\sim \rho^{\alpha}$. The parameter α is in most cases 1, 1/3, or 1/6, and only rarely fitted to data. On the other side the density dependence of covariant functionals is rather sophisticated, either described by various non-linear terms in the meson fields or by a specific density dependence of the coupling constants. In all cases two to four additional parameters are used to optimize the density dependence in relativistic functionals. Most of the non-relativistic cases start from a density dependent Hamiltonian and derive from it the energy in the *ph*-channel including exchange terms, but only few of them derive also the pairing energy from the same Hamiltonian. Therefore in most non-relativistic and in all relativistic cases the total energy is not derived fully from a Hamiltonian or a Lagrangian. In particular, in most relativistic cases Fock-terms are not taken into account. Their effect is considered in an indirect way by the adjustment of the remaining parameters.

Neglecting pairing forces the number of adjustable parameters is roughly 10–12 for non-relativistic functionals and 4 for relativistic functionals. In addition there are the parameters for the density dependence, 1 in the non-relativistic case and 2–4 in the relativistic case. Various fitting strategies are used. Usually the data include binding energies and radii of a sample of nuclei distributed over the entire periodic table. Often basic properties of nuclear matter are taken into account too. By reasons of simplicity in most of the cases only spherical nuclei are considered, i.e. doubly magic and singly magic nuclei.[40–42] There are, however also new functionals adjusted only to the masses of many deformed nuclei,[43] because it is well known that the breaking of rotational symmetry allows us to take into account in deformed nuclei additional correlations and that, for this reason, deformed nuclei are much better candidates for the mean field approximation.

As discussed, tensor forces play an important role in the microscopic understanding of nuclear binding and in many details of shell model calculations with phenomenological forces.[44] There role in density functionals is still under debate. In non-relativistic functionals zero-range tensor terms can be easily included,[45] but it is not clear at all how to adjust their parameters, because they have little influence on binding energies and radii. They have, however, a strong influence on single particle energies. On the other hand, according to the concept of Kohn and Sham single particle energies are within this method purely artificial quantities, which should not be used in a fit. In fact single particle energies in nuclei are usually fragmented over several many-body states which should be described by particle-vibrational models in theories going beyond mean field (see Chapter 11 of this book). It is conceptually and technically difficult to take into account such effects in a fitting strategy. It is also not clear to what extent the zero-range tensor forces in Skyrme calculations can really describe the full effective tensor forces in the nuclear medium. On the other hand, in relativistic density functionals, one has only tensor forces of finite range.[46–48] They show up only in relativistic Hartree–Fock calculations, which require considerable numerical effort. Therefore many problems are left open for future investigations concerning tensor forces in nuclear density functionals.

6. On the way to derive *ab initio* functionals for nuclei

At present there are few attempts to derive nuclear density functionals from *ab initio* calculations. The way to proceed would be to follow concepts used in Coulombic systems, i.e. first solve the exact nuclear matter problem on the basis of bare nucleon-nucleon forces. Even this problem has not been solved so far with the necessary accuracy. In further steps one could try to derive exact results for gradient terms or surface terms etc. Following this concept Fayans[49,50] has proposed reducing the number of phenomenological parameters considerably by starting with the energy functional $E[\rho]$ of the homogeneous system (with constant ρ) derived from a state of the art many-body calculation and to add a few phenomenological surface parameters. The Barcelona–Catania–Paris–Madrid (BCPM) group used state of the art non-relativistic Brueckner results and derived in this way a successful non-relativistic density functional depending only on very few parameters.[51] A similar concept has been followed in Ref. 52 on the relativistic side. Here the density dependence of the coupling constants has been derived fully from microscopic *ab initio* calculations based on relativistic and non-relativistic state of the art Brueckner calculations and only 4 phenomenological parameters have been adjusted to nuclear binding energies of spherical nuclei all over the periodic table.

It is however to be expected that in nuclear physics we will not be able for a long time to exclude completely all phenomenological parameters. We have to consider that the required accuracy for binding energies in astrophysical reactions is of the order of 100 keV. For heavy nuclei this is an accuracy of a few times 10^{-5}. Considering the fact that the experimental phase shifts of nucleon-nucleon scattering have finite error bars and having in mind all the uncertainties in the nuclear many-body calculations, we will probably never be able to reach this accuracy without the fine tuning of a few parameters.

6.1. *Nuclei are self-bound systems*

There is an important difference between density functional theory in Coulombic systems and the nuclear case. The Hohenberg–Kohn theorem[1] was derived under the assumption that the system is confined by an external potential $v_{\text{ext}}(\boldsymbol{r})$. In atoms this is the central Coulomb field of the nucleus. This guarantees from the beginning, even neglecting the interaction between the electrons, that there exists a strong mean field and the interaction between the electrons provides only some refinement.

For nuclei, however, there exists no external potential. They are self-bound systems and they obey translational invariance. This means in particular that the exact density in the laboratory frame has to be a constant. The Hohenberg–Kohn theorem formulated with this constant density is certainly true, but it is relatively useless. On the other hand, it has been known since the pioneering work of Goeppert-Mayer and Jensen[11,53] that there exists shell structures in nuclei indicating the existence of an effective mean field, which could eventually be treated by Kohn–Sham theory. It has been shown[54,55] that the Hohenberg–Kohn theorem holds for the intrinsic density $\rho_I(\boldsymbol{r}) = \rho_I(\boldsymbol{r}+\boldsymbol{R})$. In non-relativistic systems with Galilean invariance this density is defined in an exact way, but it depends on the center of mass coordinate $\boldsymbol{R} = \frac{1}{A}\sum_i \boldsymbol{r}_i$ and it has therefore no longer single-particle character in the exact sense. In an expansion in terms of $1/A$, many-body effects show up in higher orders, for instance the well-known center-of-mass correction terms. Therefore it is still under debate[56] whether the Hohenberg–Kohn theorem holds in nuclei in an exact sense, or whether it is only true to a good approximation. In any case, at present, where we do not have the exact form of the nuclear energy density functional anyway, this question is somewhat academic.

7. Limits of a mean field description in nuclei

In principle the Kohn–Sham method provides a mapping of a complicated quantum-mechanical many-body problem onto a single particle problem, which can be treated relatively easily. Even if this mapping is in principle exact, we have to keep in mind that the information provided by this method is limited. In principle we can calculate only the exact local density $\rho(\boldsymbol{r})$ and all quantities which are directly connected to it, such as the total energy $E[\rho]$ or expectation values of local one-body operators

$$\langle \hat{F} \rangle = \int d^3r\, f(\boldsymbol{r})\rho(\boldsymbol{r}), \tag{18}$$

as for instance radii or quadrupole moments. Apart from the ground state energies of odd-mass nuclei, single particle energies and many other quantities should not be compared with experiment. Starting from an intrinsic density it is also clear that we first have to transform to the laboratory frame, which is connected with additional approximations. This can be done by angular momentum projection. In the limit of large deformations and a well defined orientation of the intrinsic system (needle approximation) we end up with the well-known relation of the Bohr model between the intrinsic matrix elements of the quadrupole operator and those in the laboratory frame.[57] In transitional nuclei with relatively flat energy surfaces in deformation space, one has to go beyond mean field and take fluctuations into account by describing the nuclear wave function as a superposition of many Slater determinants with different deformation. These methods go definitely beyond the mean field concept, but they have turned out to be very successful for the universal description of transitional nuclei and their low-lying spectra in terms of density functionals.[58] Therefore they will be discussed in detail in the following sections of the volume.

Of course, in principle, static density functional theory is only applicable to the ground states. There are, however extensions of density functional theories for the description of excited states. Many of them run under the concept of time-dependent density functional theory. In analogy to the Kohn–Sham theorem[2] there exists the Runge and Gross theorem[59] which provides an exact mapping of the full time-dependent many-body problem onto a time-dependent single particle problem. The corresponding auxiliary single particle field is not only time-dependent, it also depends on the full time-dependent single particle density, i.e. it includes memory effects. Therefore this theory is much more complicated than Kohn–Sham theory in the static case. Only in the vicinity of the ground state can one expand around the static density for vibrations with small amplitude and with additional approximations, where one neglects the memory effects[60] and the corresponding energy dependence in Fourier space, one ends up with the well-known random phase approximation (RPA)[61] or quasiparticle-RPA with a residual interaction derived as the second derivative of the energy functional with respect to the density. This is time-dependent mean field theory. It provides a very successful description of the mean energies of giant resonances in nuclei.[62] However, it is not able to reproduce the decay width of these excitations produced to a large extent by the coupling to more complicated configurations. In order to describe such effects as well as the fragmentation of single particle states one has to go beyond mean field and consider couplings to vibrational states as discussed in Chapter 11 of this book.

Acknowledgment

This work was supported by the DFG Cluster of Excellence "Origin and Structure of the Universe"(www.universecluster.de), and by the Overseas Distinguished Professor Project from Ministry of Education (MS2010BJDX001).

References

1. P. Hohenberg and W. Kohn, Inhomogeneous electron gas, *Phys. Rev.* **136**, B864–B871 (Nov, 1964). doi: 10.1103/PhysRev.136.B864. URL `http://link.aps.org/doi/10.1103/PhysRev.136.B864`.
2. W. Kohn and L. J. Sham, Self-consistent equations including exchange and correlation effects, *Phys. Rev.* **140**, A1133–A1138 (Nov, 1965). doi: 10.1103/PhysRev.140.A1133. URL `http://link.aps.org/doi/10.1103/PhysRev.140.A1133`.
3. W. Kohn, Nobel lecture: Electronic structure of matter-wave functions and density functionals, *Rev. Mod. Phys.* **71**, 1253–1266 (Oct, 1999). doi: 10.1103/RevModPhys.71.1253. URL `http://link.aps.org/doi/10.1103/RevModPhys.71.1253`.
4. J. P. Perdew and S. Kurth. Density functionals for non-relativistic Coulomb systrems in the new century. In eds. C. Fiolhais, F. Nogueira, and M. A. L. Marques, *A Primer in Density Functional Theory*, vol. 620, pp. 1–55. Springer Berlin Heidelberg, Berlin (2003). ISBN 978-3-540-03083-6. doi: 10.

1007/3-540-37072-2_1. URL http://dx.doi.org/10.1007/3-540-37072-2_1.
5. A. K. Rajagopal and J. Callaway, Inhomogeneous electron eas, *Phys. Rev. B.* **7**, 1912–1919 (Mar, 1973). doi: 10.1103/PhysRevB.7.1912. URL http://link.aps.org/doi/10.1103/PhysRevB.7.1912.
6. A. K. Rajagopal, Inhomogeneous relativistic electron gas, *J. Phys. C.* **11**, L943 (1978). URL http://iopscience.iop.org/0022-3719/11/24/002.
7. A. H. MacDonald and S. H. Vosko, A relativistic density functional formalism, *J. Phys. C.* **12**(15), 2977 (1979). doi: 10.1088/0022-3719/12/15/007.
8. D. Vautherin and D. M. Brink, Hartree–Fock calculations with Skyrme's interaction: I. spherical nuclei, *Phys. Rev. C.* **5**, 626–647 (Mar, 1972). doi: 10.1103/PhysRevC.5.626. URL http://link.aps.org/doi/10.1103/PhysRevC.5.626.
9. D. Vautherin, Hartree–Fock calculations with Skyrme's interaction: ii axially deformed nuclei, *Phys. Rev. C.* **7**, 296–316 (Jan, 1973). doi: 10.1103/PhysRevC.7.296. URL http://link.aps.org/doi/10.1103/PhysRevC.7.296.
10. M. Bender, P.-H. Heenen, and P.-G. Reinhard, Self-consistent mean-field models for nuclear structure, *Rev. Mod. Phys.* **75**, 121–180 (Jan, 2003). doi: 10.1103/RevModPhys.75.121. URL http://link.aps.org/doi/10.1103/RevModPhys.75.121.
11. O. Haxel, J. H. D. Jensen, and H. E. Suess, On the "magic numbers" in nuclear structure, *Phys. Rev.* **75**, 1766–1766 (Jun, 1949). doi: 10.1103/PhysRev.75.1766.2. URL http://link.aps.org/doi/10.1103/PhysRev.75.1766.2.
12. M. Goeppert-Mayer, On closed shells in nuclei. ii, *Phys. Rev.* **75**, 1969–1970 (Jun, 1949). doi: 10.1103/PhysRev.75.1969. URL http://link.aps.org/doi/10.1103/PhysRev.75.1969.
13. K. A. Brueckner, C. A. Levinson, and H. M. Mahmoud, Two-body forces and nuclear saturation. i. central forces, *Phys. Rev.* **95**, 217–228 (Jul, 1954). doi: 10.1103/PhysRev.95.217. URL http://link.aps.org/doi/10.1103/PhysRev.95.217.
14. T. H. R. Skyrme, The effective nuclear potential, *Nucl. Phys.* **9**(4), 615–634 (1959). ISSN 0029-5582. doi: 10.1016/0029-5582(58)90345-6. URL http://www.sciencedirect.com/science/article/pii/0029558258903456.
15. H. S. Koehler, Theory of finite nuclei, *Phys. Rev.* **137**, B1145–B1157 (Mar, 1965). doi: 10.1103/PhysRev.137.B1145. URL http://link.aps.org/doi/10.1103/PhysRev.137.B1145.
16. H. Müther, A. Faessler, and K. Goeke, Rearrangement in the Brueckner–Hartree–Fock approach, *Nucl. Phys. A.* **215**(1), 213–220 (1973). ISSN 0375-9474. doi: http://dx.doi.org/10.1016/0375-9474(73)90113-9. URL http://www.sciencedirect.com/science/article/pii/0375947473901139.
17. H. Lenske and C. Fuchs, Rearrangement in the density dependent relativistic field theory of nuclei, *Phys. Lett. B.* **345**(4), 355 (1995). ISSN 0370-2693. doi: 10.1016/0370-2693(94)01664-X. URL http://www.sciencedirect.com/science/article/pii/037026939401664X.

18. J. Bardeen, L. N. Cooper, and J. R. Schrieffer, Theory of superconductivity, *Phys. Rev.* **108**, 1175–1204 (Dec, 1957). doi: 10.1103/PhysRev.108.1175. URL `http://link.aps.org/doi/10.1103/PhysRev.108.1175`.
19. M. Baranger, Self-consistent field theory of nuclear shapes, *Phys. Rev.* **122**, 992–996 (May, 1961). doi: 10.1103/PhysRev.122.992. URL `http://link.aps.org/doi/10.1103/PhysRev.122.992`.
20. J. Dechargé and D. Gogny, Hartree–Fock–Bogolyubov calculations with the d1 effective interaction on spherical nuclei, *Phys. Rev. C.* **21**, 1568–1593 (Apr, 1980). doi: 10.1103/PhysRevC.21.1568. URL `http://link.aps.org/doi/10.1103/PhysRevC.21.1568`.
21. H. Kucharek and P. Ring, Relativistic field theory of superfluidity in nuclei, *Z. Phys. A.* **339**(1), 23–35 (Mar., 1991). doi: 10.1007/BF01282930. URL `http://dx.doi.org/10.1007/BF01282930`.
22. J. G. Valatin, Generalized Hartree–Fock method, *Phys. Rev.* **122**, 1012–1020 (May, 1961). doi: 10.1103/PhysRev.122.1012. URL `http://link.aps.org/doi/10.1103/PhysRev.122.1012`.
23. T. D. Cohen, R. J. Furnstahl, and D. K. Griegel, From QCD sum rules to relativistic nuclear physics, *Phys. Rev. Lett.* **67**, 961–964 (Aug, 1991). doi: 10.1103/PhysRevLett.67.961. URL `http://link.aps.org/doi/10.1103/PhysRevLett.67.961`.
24. R. J. Furnstahl, D. K. Griegel, and T. D. Cohen, QCD sum rules for nucleons in nuclear matter, *Phys. Rev. C.* **46**, 1507–1527 (Oct, 1992). doi: 10.1103/PhysRevC.46.1507. URL `http://link.aps.org/doi/10.1103/PhysRevC.46.1507`.
25. R. Machleidt, K. Holinde, and C. Elster, The Bonn meson-exchange model for the nucleon-nucleon interaction, *Phys. Rep.* **149**(1), 1–89 (1987). ISSN 0370-1573. doi: 10.1016/S0370-1573(87)80002-9. URL `http://www.sciencedirect.com/science/article/pii/S0370157387800029`.
26. R. B. Wiringa, V. G. Stoks, and R. Schiavilla, Accurate nucleon-nucleon potential with charge-independence breaking, *Phys. Rev. C.* **51**, 38–51 (Jan, 1995). doi: 10.1103/PhysRevC.51.38. URL `http://link.aps.org/doi/10.1103/PhysRevC.51.38`.
27. R. Machleidt and D. R. Entem, Chiral effective field theory and nuclear forces, *Phys. Rep.* **503**, 1 (2011). doi: doi:10.1016/j.physrep.2011.02.001.
28. S. Aoki, Hadron interactions in lattice QCD, *Prog. Part. Nucl. Phys.* **66**(4), 687–726 (2011). ISSN 0146-6410. doi: 10.1016/j.ppnp.2011.07.001. URL `http://www.sciencedirect.com/science/article/pii/S0146641011000949`.
29. R. Machleidt, The meson theory of nuclear forces and nuclear structure, *Adv. Nucl. Phys.* **19**, 189 (1989).
30. R. Brockmann and R. Machleidt, Relativistic nuclear structure. i. nuclear matter, *Phys. Rev. C.* **42**, 1965–1980 (Nov, 1990). doi: 10.1103/PhysRevC.42.1965. URL `http://link.aps.org/doi/10.1103/PhysRevC.42.1965`.
31. F. Coester, S. Cohen, B. Day, and C. M. Vincent, Variation in nuclear-matter binding energies with phase-shift-equivalent two-body potentials, *Phys. Rev. C.* **1**, 769–776 (Mar, 1970). doi: 10.1103/PhysRevC.1.769. URL `http://link.aps.org/doi/10.1103/PhysRevC.1.769`.

32. J. Fujita and H. Miyazawa, Pion theory of three-body forces, *Prog. Theor. Phys.* **17**, 360 (1957). doi: 10.1143/PTP.17.360. URL `http://ptp.ipap.jp/link?PTP/17/360/`.
33. W. Zuo, A. Lejeune, U. Lombardo, and J. Mathiot, Interplay of three-body interactions in the EOS of nuclear matter, *Nucl. Phys. A.* **706**(3–4), 418–430 (2002). ISSN 0375-9474. doi: http://dx.doi.org/10.1016/S0375-9474(02)00750-9. URL `http://www.sciencedirect.com/science/article/pii/S0375947402007509`.
34. M. R. Anastasio, L. S. Celenza, W. S. Pong, and C. M. Shakin, Relativistic nuclear structure physics, *Phys. Rep.* **100**(6), 327–392 (1983). ISSN 0370-1573. doi: 10.1016/0370-1573(83)90060-1. URL `http://www.sciencedirect.com/science/article/pii/0370157383900601`.
35. R. Brockmann and R. Machleidt, Nuclear saturation in a relativistic Brueckner–Hartree–Fock approach, *Phys. Lett. B.* **149**, 283–287 (1984). ISSN 0370-2693. doi: 10.1016/0370-2693(84)90407-6. URL `http://www.sciencedirect.com/science/article/pii/0370269384904076`.
36. B. ter Haar and R. Malfliet, Nucleons, mesons and deltas in nuclear matter a relativistic Dirac–Brueckner approach, *Phys. Rep.* **149**(4), 207–286 (1987). ISSN 0370-1573. doi: 10.1016/0370-1573(87)90085-8. URL `http://www.sciencedirect.com/science/article/pii/0370157387900858`.
37. H. Huber, F. Weber, and M. K. Weigel, Symmetric and asymmetric nuclear matter in the relativistic approach, *Phys. Rev. C.* **51**, 1790–1799 (Apr, 1995). doi: 10.1103/PhysRevC.51.1790. URL `http://link.aps.org/doi/10.1103/PhysRevC.51.1790`.
38. F. de Jong and H. Lenske, Relativistic Brueckner–Hartree–Fock calculations with explicit intermediate negative energy states, *Phys. Rev. C.* **58**, 890–899 (Aug, 1998). doi: 10.1103/PhysRevC.58.890. URL `http://link.aps.org/doi/10.1103/PhysRevC.58.890`.
39. E. N. E. van Dalen, C. Fuchs, and A. Faessler, Dirac–Brueckner–Hartree–Fock calculations for isospin asymmetric nuclear matter based on improved approximation schemes, *Eur. Phys. J. A.* **31**(1), 29–42 (2007). doi: 10.1140/epja/i2006-10165-x.
40. G. A. Lalazissis, J. König, and P. Ring, New parameterization for the lagrangian density of relativistic mean field theory, *Phys. Rev. C.* **55**(1), 540–543 (Jan., 1997). doi: 10.1103/PhysRevC.55.540. URL `http://link.aps.org/abstract/PRC/v55/p540`.
41. P. W. Zhao, Z. P. Li, J. M. Yao, and J. Meng, New parametrization for the nuclear covariant energy density functional with a point-coupling interaction, *Phys. Rev. C.* **82**, 054319 (Nov, 2010). doi: 10.1103/PhysRevC.82.054319. URL `http://link.aps.org/doi/10.1103/PhysRevC.82.054319`.
42. G. A. Lalazissis, T. Nikšić, D. Vretenar, and P. Ring, New relativistic mean field interaction with density dependent meson couplings, *Phys. Rev. C.* **71**, 024312 (2005). doi: 10.1103/PhysRevC.71.024312. URL `http://link.aps.org/doi/10.1103/PhysRevC.71.024312`.
43. T. Niksic, D. Vretenar, and P. Ring, Relativistic nuclear energy density functionals: adjusting parameters to binding energies., *Phys. Rev. C.* **78**,

034318 (2008). doi: 10.1103/PhysRevC.78.034318. URL http://link.aps.org/doi/10.1103/PhysRevC.78.034318.

44. T. Otsuka, T. Suzuki, R. Fujimoto, H. Grawe, and Y. Akaishi, Evolution of nuclear shells due to the tensor force, *Phys. Rev. Lett.* **95**, 232502 (Nov, 2005). doi: 10.1103/PhysRevLett.95.232502. URL http://link.aps.org/doi/10.1103/PhysRevLett.95.232502.
45. F. Stancu, D. M. Brink, and H. Flocard, The tensor part of Skyrme's interaction, *Phys. Lett. B.* **68**(2), 108–112 (1977). ISSN 0370-2693. doi: 10.1016/0370-2693(77)90178-2. URL http://www.sciencedirect.com/science/article/pii/0370269377901782.
46. W. H. Long, H. Sagawa, N. V. Giai, and J. Meng, Shell structure and rho-tensor correlations in density dependent relativistic Hartree–Fock theory, *Phys. Rev. C.* **76**, 034314 (Sep, 2007). doi: 10.1103/PhysRevC.76.034314. URL http://link.aps.org/doi/10.1103/PhysRevC.76.034314.
47. L. J. Jiang, Y. Shen, J. M. Dong, and W. H. Long, Self-consistent tensor effects on nuclear matter system under relativistic Hartree–Fock approach, *Phys. Rev. C.* **91**, 025802 (Feb, 2015). doi: 10.1103/PhysRevC.91.025802. URL http://link.aps.org/doi/10.1103/PhysRevC.91.025802.
48. G. A. Lalazissis, S. Karatzikos, M. Serra, T. Otsuka, and P. Ring, Covariant density functional theory: The role of the pion, *Phys. Rev. C.* **80**, 041301 (Oct, 2009). doi: 10.1103/PhysRevC.80.041301. URL http://link.aps.org/doi/10.1103/PhysRevC.80.041301.
49. S. Fayans, Towards a universal nuclear density functional, *JETP Lett.* **68**, 169–174 (1998). doi: 10.1134/1.567841.
50. S. Fayans and D. Zawischa, Local energy-density functional approach to many-body nuclear systems with s-wave pairing, *Int. J. Mod. Phys. B.* **15**, 1684 (2001). doi: 10.1142/S0217979201006203.
51. M. Baldo, P. Schuck, and X. Viñas, Kohn–Sham density functional inspired approach to nuclear binding, *Phys. Lett. B.* **663**(5), 390–394 (2008). ISSN 0370-2693. doi: 10.1016/j.physletb.2008.04.013. URL http://www.sciencedirect.com/science/article/pii/S0370269308004383.
52. X. Roca-Maza, X. Viñas, M. Centelles, P. Ring, and P. Schuck, Relativistic mean-field interaction with density-dependent meson-nucleon vertices based on microscopical calculations, *Phys. Rev. C.* **84**, 054309 (Nov, 2011). doi: 10.1103/PhysRevC.84.054309. URL http://link.aps.org/doi/10.1103/PhysRevC.84.054309.
53. M. Goeppert-Mayer, Nuclear configurations in the spin-orbit coupling model. i. empirical evidence, *Phys. Rev.* **78**, 16–21 (Apr, 1950). doi: 10.1103/PhysRev.78.16. URL http://link.aps.org/doi/10.1103/PhysRev.78.16.
54. J. Engel, Intrinsic-density functionals, *Phys. Rev. C.* **75**, 014306 (Jan, 2007). doi: 10.1103/PhysRevC.75.014306. URL http://link.aps.org/doi/10.1103/PhysRevC.75.014306.
55. B. G. Giraud, Density functionals in the laboratory frame, *Phys. Rev. C.* **77**, 014311 (Jan, 2008). doi: 10.1103/PhysRevC.77.014311. URL http://link.aps.org/doi/10.1103/PhysRevC.77.014311.
56. T. Nakatsukasa, Density functional approaches to collective phenomena

in nuclei: Time-dependent density functional theory for perturbative and non-perturbative nuclear dynamics, *Prog. Theor. Exp. Phys.* **01A207** (2012). doi: 10.1093/ptep/pts016. URL `http://ptep.oxfordjournals.org/content/2012/1/01A207.abstract`.

57. P. Ring, A. Hayashi, K. Hara, H. Emling, and E. Grosse, Electromagnetic transition rates from cranking model wave functions, *Phys. Lett. B.* **110**, 423–427 (1982). ISSN 0370-2693. doi: 10.1016/0370-2693(82)91030-9. URL `http://www.sciencedirect.com/science/article/pii/0370269382910309`.
58. T. Niksic, D. Vretenar, and P. Ring, Relativistic nuclear energy density functionals: Mean-field and beyond, *Prog. Part. Nucl. Phys.* **66**(3), 519–548 (2011). ISSN 0146-6410. doi: 10.1016/j.ppnp.2011.01.055. URL `http://www.sciencedirect.com/science/article/pii/S0146641011000561`.
59. E. Runge and E. K. U. Gross, Density-functional theory for time-dependent systems, *Phys. Rev. Lett.* **52**, 997–1000 (Mar, 1984). doi: 10.1103/PhysRevLett.52.997. URL `http://link.aps.org/doi/10.1103/PhysRevLett.52.997`.
60. M. A. L. Marques, C. A. Ullrich, F. Nogueira, A. Rubio, K. Burke, and E. K. U. G. (Eds.), *Time-Dependent Density Functional Theory*. Springer, Heidelberg (2006).
61. P. Ring, Z.-Y. Ma, N. Van Giai, D. Vretenar, A. Wandelt, and L.-G. Cao, The time-dependent relativistic mean field theory and the random phase approximation., *Nucl. Phys. A.* **694**, 249 (2001).
62. D. Vretenar, A. V. Afanasjev, G. A. Lalazissis, and P. Ring, Relativistic Hartree–Bogoliubov theory: static and dynamic aspects of exotic nuclear structure, *Phys. Rep.* **409**(3-4), 101–259 (Apr., 2005). doi: 10.1016/j.physrep.2004.10.001. URL `http://dx.doi.org/10.1016/j.physrep.2004.10.001`.

Chapter 2

Relativistic mean-field theory

Jie Meng
School of Physics, Peking University, Beijing 100871, P.R.China
mengj@pku.edu.cn

Peter Ring
Physics Department, Technical University Munich, 85748 Garching, Germany
School of Physics, Peking University, Beijing 100871, China
ring@ph.tum.de

Pengwei Zhao
School of Physics, Peking University, Beijing 100871, P.R.China
pwzhao@pku.edu.cn

In this chapter, the covariant energy density functional is constructed with both the meson-exchange and the point-coupling pictures. Several widely used functionals with either nonlinear or density-dependent effective interactions are introduced. The applications of covariant density functional theory are demonstrated for infinite nuclear matter and finite nuclei with spherical symmetry, axially symmetric quadrupole deformation, and triaxial quadrupole shapes. Finally, a relativistic description of the nuclear landscape has been discussed, which is not only important for nuclear structure, but also important for nuclear astrophysics, where we are facing the problem of a reliable extrapolation to the very neutron-rich nuclei.

Contents

1. Relativistic meson-exchange theory

In order to implement the Kohn–Sham concept[1] of density functional theory one needs in the first place a single particle equation with a single particle hamiltonian $\hat{h}$ derived from an energy functional $E[\rho]$. In the relativistic case the single particle equation is the Dirac equation and the energy density functional is most easily derived from a field theoretical Lagrangian obeying the necessary symmetries of Lorenz invariance, locality and causality. The Walecka model[2] provides an ideal starting point for such a scheme. As usual in a local field theory the interactions are provided by the exchange of mesons with the proper relativistic quantum numbers. Originally this model was introduced as a fully fledged renormalizable quantum field theory, as quantum hadron dynamics (QHD).[3] However, it was soon realized that renormalization is only possible on the tree level,[4] but it does not lead to reasonable results at higher orders.[5]

On the other side this model has turned out to be extremely useful for an implementation of a covariant density functional theory. For that purpose there are three basic assumptions necessary:

(i) On the mean field level, used in density functional theory, this is a classical field theory. The meson fields are, in a first step, not physical objects and they are not quantized. As a consequence the equations of motion do not contain Fock terms. On that level we therefore deal with a relativistic Hartree theory. The mesons have little to do with the mesons in free space. In particular the model contains no pions and the σ-meson, which plays the essential role for the large attractive scalar field in the nucleus, has never been observed. In short, the mesons serve only to introduce classical fields in the nucleus with the appropriate relativistic quantum numbers.

(ii) The effects of the Dirac sea are neglected, i.e., vacuum polarization is not taken into account. This is often called the "no-sea" approximation. In fact, it is a concept, which forms the basis of this model and not an approximation. It has been shown that, in principle, the Dirac sea can be taken into account on the tree level. This leads to a logarithmic divergence of the scalar density which can be renormalized by adding non-linear counter terms, cubic and quartic in the σ-field. Thus one finds changes of 30% in the results. However, since we deal here with a phenomenological model, one has to readjust the constants and after that there is practically no change.[4,6] This means, vacuum polarization is not neglected, but rather it is absorbed in a phenomenological way in the coupling constants of the model.

(iii) It was soon recognized[7] that the model with linear meson-nucleon couplings does not provide the proper density dependence of nuclear matter. The incompressibility is much too large and the nuclear surface properties are not described properly. Therefore an additional density dependence has to be introduced. This has been done in the literature in various ways, either by non-linear couplings in the meson sector[7–11] or by an explicit density dependence of the meson-nucleon coupling vertices (see below).

We discuss here the basic ingredients of the relativistic meson exchange models, as they are used today in the framework of covariant density functional theory.

The starting point of RMF theory is a Lagrangian density of the form

$$\begin{aligned}\mathcal{L} = \bar{\psi}\left[i\gamma^\mu\partial_\mu - M - g_\sigma\sigma - g_\omega\gamma^\mu\omega_\mu - g_\rho\gamma^\mu\vec{\tau}\cdot\vec{\rho}_\mu - e\gamma^\mu A_\mu\right]\psi \\ +\tfrac{1}{2}\partial^\mu\sigma\partial_\mu\sigma - U_\sigma(\sigma) - \tfrac{1}{4}\Omega^{\mu\nu}\Omega_{\mu\nu} + U_\omega(\omega_\mu) \\ -\tfrac{1}{4}\vec{R}^{\mu\nu}\cdot\vec{R}_{\mu\nu} + U_\rho(\vec{\rho}_\mu) - \tfrac{1}{4}F^{\mu\nu}F_{\mu\nu},\end{aligned} \tag{1}$$

where M and $m_m(g_m)$ $(m = \sigma, \omega, \rho)$ are the masses (coupling constants) of the nucleon and the mesons. e is the elementary charge for protons and it vanishes for neutrons. The field tensors are given by

$$\Omega^{\mu\nu} = \partial^\mu\omega^\nu - \partial^\nu\omega^\mu, \tag{2a}$$

$$\vec{R}^{\mu\nu} = \partial^\mu\vec{\rho}^{\,\nu} - \partial^\nu\vec{\rho}^{\,\mu}, \tag{2b}$$

$$F^{\mu\nu} = \partial^\mu A^\nu - \partial^\nu A^\mu, \tag{2c}$$

In the following, the vectors in isospin space are denoted by arrows and the space vectors by bold type. Greek indices μ and ν run over the Minkowski indices 0, 1, 2, 3 or $(x^\mu) = (t, x, y, z)$.

In this scheme, the isoscalar-scalar σ-meson provides the mid- and long-range attractive part of the nuclear interaction whereas the short-range repulsive part is provided by the isoscalar-vector ω-meson. The photon field $A^\mu(x)$ accounts for the Coulomb interaction while the isospin dependence of the nuclear force is described by the isovector-vector ρ-meson. The π-meson field is not included. Because of parity conservation it does not contribute at the Hartree level. Of course, the pion plays an essential role in a microscopic description of the nuclear force.[12–14] In particular the two-pion exchange induces the strong attraction at intermediate distances. This is taken into account in our model by the σ-meson in a phenomenological way.

Of course, in principle one can also introduce an isovector scalar meson, which is usually called δ-meson. It behaves similar as the scalar meson and leads to an isospin dependence of the Dirac mass. It plays a role in asymmetric nuclear matter at very high densities.[15] However, it can be shown[16] that at densities relevant for nuclear structure on earth, its effect can be totally absorbed by changing the coupling constant of the ρ-meson. In fact, present experimental data do not allow to determine the ρ- and δ-couplings separately. Therefore the δ-meson is neglected in most of the successful parameterizations.

The nonlinear self-coupling terms $U_\sigma(\sigma)$, $U_\omega(\omega_\mu)$, and $U_\rho(\vec{\rho}_\mu)$ for the σ-, ω-, and ρ-mesons in the Lagrangian density (1) have the following forms:

$$U_\sigma(\sigma) = \frac{1}{2}m_\sigma^2\sigma^2 + \frac{1}{3}g_2\sigma^3 + \frac{1}{4}g_3\sigma^4, \tag{3a}$$

$$U_\omega(\omega_\mu) = \frac{1}{2}m_\omega^2\omega^\mu\omega_\mu + \frac{1}{4}c_3\left(\omega^\mu\omega_\mu\right)^2, \tag{3b}$$

$$U_\rho(\vec{\rho}_\mu) = \frac{1}{2}m_\rho^2\vec{\rho}^{\,\mu}\cdot\vec{\rho}_\mu + \frac{1}{4}d_3\left(\vec{\rho}^{\,\mu}\cdot\vec{\rho}_\mu\right)^2. \tag{3c}$$

It should be mentioned that in the simplest RMF version, i.e., the σ-ω model,[2] the mesons do not interact among themselves, which leads to too large incompressibility in nuclear matter. Therefore, a nonlinear self-coupling of the σ-field was proposed.[7] In order to reproduce the density dependence of the vector and scalar potentials of Dirac–Brueckner calculations,[17] the nonlinear self-coupling of the ω-meson was introduced.[18] In addition a nonlinear self-coupling of the isovector ρ-meson[11] as well as a nonlinear couplings between ω and ρ mesons[19] have been considered to improve the density dependence in the isovector channel. Instead of the nonlinear meson couplings, one has also introduced RMF theories with an explicit density dependence of the meson-nucleon coupling constants.[11,17,20–24] In this case one can avoid instabilities at high densities.

From the Lagrangian density (1), the Hamiltonian operator can be obtained by the general Legendre transformation (for details see Ref. 25)

$$H = \int d^3r\, \mathcal{H} = \int d^3r \left[\sum_m \pi_m \frac{\partial \phi_m}{\partial t} - \mathcal{L}(\boldsymbol{r}) \right]. \tag{4}$$

The conjugate momenta of the field operators ϕ_m ($\phi_m = \psi, \sigma, \omega_\nu, \vec{\rho}_\nu, A_\nu$) are defined as

$$\pi_m = \frac{\partial \mathcal{L}}{\partial \left[\partial \phi_m / \partial t\right]}. \tag{5}$$

Then the Hamiltonian density of the system can be easily obtained as

$$\mathcal{H} = \mathcal{H}_\psi + \mathcal{H}_\sigma + \mathcal{H}_\omega + \mathcal{H}_\rho + \mathcal{H}_A + \mathcal{H}_{\text{int}} \tag{6}$$

with

$$\mathcal{H}_\psi = \psi^\dagger \left[\boldsymbol{\alpha} \cdot \boldsymbol{p} + \beta M\right] \psi, \tag{7}$$

$$\mathcal{H}_\sigma = -\frac{1}{2}\sigma \triangle \sigma + U_\sigma(\sigma), \qquad \mathcal{H}_\omega = \frac{1}{2}\omega_\mu \triangle \omega^\mu - U_\omega(\omega), \tag{8}$$

$$\mathcal{H}_\rho = \frac{1}{2}\vec{\rho}_\mu \triangle \vec{\rho}^\mu - U_\rho(\rho), \quad \mathcal{H}_A = \frac{1}{2} A_\mu \triangle A^\mu, \tag{9}$$

$$\mathcal{H}_{\text{int}} = g_\sigma \sigma \bar{\psi}\psi + g_\omega \omega_\mu \bar{\psi}\gamma^\mu \psi + g_\rho \vec{\rho}_\mu \cdot \bar{\psi}\gamma^\mu \vec{\tau}\psi + eA_\mu \bar{\psi}\gamma^\mu \psi. \tag{10}$$

The RMF theory is formulated on the basis of the above effective Lagrangian (1) under the mean-field and the no-sea approximations. The mean-field approximation removes all the quantum fluctuations of the meson fields and all the meson fields are treated as classical fields, which implies that the nucleons move independently in the meson fields. Therefore, the nucleon field operator can be expanded at all times in terms of the single-particle states with the index k as

$$\psi(x) = \sum_k \psi_k(x) c_k \quad \text{and} \quad \psi^\dagger(x) = \sum_k \psi_k^*(x) c_k^\dagger, \tag{11}$$

where c_k is the annihilation operator for a nucleon in the state k and ψ_k is the corresponding single-particle wave function. The operators c_k and its conjugate $c_k^\dagger$ satisfy the anticommutation rules

$$\{c_k, c_{k'}^\dagger\} = \delta_{kk'} \quad \text{and} \quad \{c_k, c_{k'}\} = \{c_k^\dagger, c_{k'}^\dagger\} = 0. \tag{12}$$

The ground state of the nucleus can be constructed as

$$|\Phi\rangle = \prod_i^A c_i^\dagger \,|0\rangle \quad \text{with} \quad \langle\Phi|\Phi\rangle = 1. \tag{13}$$

The product (13) is confined to the occupied single-particle states with positive energies. This is usually called *no-sea approximation*. To calculate the expectation value of the Hamiltonian (6) with respect to the nuclear ground state (13), one has to treat terms like $\langle\Phi|c^\dagger_{k'}c_k|\Phi\rangle$, $\langle\Phi|\phi c^\dagger_{k'}c_k|\Phi\rangle$, and $\langle\Phi|\phi|\Phi\rangle$, where ϕ is one of the meson fields σ, ω_μ, $\vec{\rho}_\mu$, or A_μ. Since ϕ are classical fields, one has

$$\begin{aligned}\langle\Phi|\phi|\Phi\rangle &= \phi,\\ \langle\Phi|\phi c^\dagger_{k'}c_k|\Phi\rangle &= \phi\langle\Phi|c^\dagger_{k'}c_k|\Phi\rangle,\end{aligned}$$

and one is left with the single particle density matrix

$$\hat{\rho}_{kk'} = \langle\Phi|c^\dagger_{k'}c_k|\Phi\rangle. \tag{14}$$

Therefore, it is straightforward to write down the expectation value of the Hamiltonian (6) with respect to $|\Phi\rangle$, i.e., the total energy of the system

$$E_{\rm RMF}(\hat{\rho},\phi) = \langle\Phi|\mathcal{H}|\Phi\rangle, \tag{15}$$

with

$$\begin{aligned}\langle\Phi|\mathcal{H}|\Phi\rangle = {}& \mathrm{Tr}\left[\hat{h}^0\hat{\rho}\right] + \mathrm{Tr}\left[\beta\left(g_\sigma\sigma + g_\omega\omega_\mu\gamma^\mu + g_\rho\vec{\tau}\cdot\vec{\rho}_\mu\gamma^\mu + eA_\mu\gamma^\mu\right)\hat{\rho}\right]\\ &+ \int d^3r\left\{-\frac{1}{2}\sigma\Delta\sigma + U_\sigma(\sigma) + \frac{1}{2}\omega_\mu\Delta\omega^\mu - U_\omega(\omega)\right.\\ &\left.+\frac{1}{2}\vec{\rho}_\mu\cdot\Delta\vec{\rho}^\mu - U_\rho(\rho) + \frac{1}{2}A_\mu\Delta A^\mu\right\},\end{aligned} \tag{16}$$

where the traces contain the integration in coordinate space and a summation over the Dirac index of the density matrix $\hat{\rho}$ and the single Hamiltonian

$$\hat{h}^0 = \boldsymbol{\alpha}\cdot\boldsymbol{p} + M. \tag{17}$$

For stationary states the time derivatives of the meson fields vanish and the nucleon single-particle wave function can be written as

$$\psi_a(x) = \psi_a(\boldsymbol{r})e^{-i\epsilon_a t}. \tag{18}$$

The density matrix can be written as

$$\hat{\rho}(\boldsymbol{r},\boldsymbol{r}') = \sum_i^A|\psi_i(\boldsymbol{r})\rangle\langle\psi_i^\dagger(\boldsymbol{r}')|. \tag{19}$$

Furthermore, we assume that in the present nuclear applications, the nucleon single-particle states do not mix isospin, i.e., the single-particle states are eigenstates of τ_3 and only the third component ρ_3^μ of $\vec{\rho}^\mu$ survives.

The equations of motion for the nucleons and the mesons can be obtained by requiring the energy functional (16) to be stationary with respect to variations of $\psi_k^\dagger$ and of ϕ,

$$\delta\left(E_{\mathrm{RMF}}(\hat{\rho},\phi) - \mathrm{Tr}(\Lambda\hat{\rho})\right) = 0, \tag{20}$$

where Λ is a diagonal matrix of Lagrange multipliers ϵ_k, which guarantee that the single particle states ψ_k are normalized. The index k runs over all levels in the configuration space under investigation. It turns out that the ϵ_k's are the single-particle energies introduced in Eq. (18).

Finally, one obtains for the nucleons the Kohn–Sham equation which has the form of a Dirac equation

$$\left[\boldsymbol{\alpha}\cdot(\boldsymbol{p}-\boldsymbol{V}) + V^0 + \beta(M+S)\right]\psi_k(\boldsymbol{r}) = \epsilon_k\psi_k(\boldsymbol{r}), \tag{21}$$

with the scalar potential S and vector potential V^μ:

$$S(\boldsymbol{r}) = g_\sigma\sigma(\boldsymbol{r}), \tag{22a}$$

$$V^\mu(\boldsymbol{r}) = g_\omega\omega^\mu(\boldsymbol{r}) + g_\rho\tau_3\rho_3^\mu(\boldsymbol{r}) + eA^\mu(\boldsymbol{r}). \tag{22b}$$

In equation (20) variation with respect to σ, ω^μ, ρ_3^μ, and A^μ leads to the Klein–Gordon equations for σ-, ω-, ρ-mesons, and the Poisson equation for the photon:

$$-\Delta\sigma + U_\sigma'(\sigma) = -g_\sigma\rho_s, \tag{23a}$$

$$-\Delta\omega^\mu + U_\omega^{'\mu}(\omega) = g_\omega j^\mu, \tag{23b}$$

$$-\Delta\rho_3^\mu + U_\rho^{'\mu}(\rho_3) = g_\rho j_3^\mu, \tag{23c}$$

$$-\Delta A^\mu = e j_c^\mu, \tag{23d}$$

where

$$U_\sigma' = \frac{\partial U_\sigma}{\partial\sigma}, \quad U_\omega^{'\mu} = \frac{\partial U_\omega}{\partial\omega_\mu}, \quad U_\rho^{'\mu} = \frac{\partial U_\rho}{\partial\rho_{3\mu}}. \tag{24}$$

The source terms in the Klein–Gordon equations are the scalar density ρ_s, the baryon current $j^\mu = (\rho_v, \boldsymbol{j})$, the isovector current $j_3^\mu = (\rho_3, \boldsymbol{j}_3)$ and the electromagnetic current $j_c^\mu = (\rho_c, \boldsymbol{j}_c)$ given by

$$\rho_s(\boldsymbol{r}) = \langle\Phi|\bar{\psi}\psi|\Phi\rangle = \sum_{i=1}^{A} \bar{\psi}_i(\boldsymbol{r})\psi_i(\boldsymbol{r}), \tag{25a}$$

$$j^\mu(\boldsymbol{r}) = \langle\Phi|\bar{\psi}\gamma^\mu\psi|\Phi\rangle = \sum_{i=1}^{A} \bar{\psi}_i(\boldsymbol{r})\gamma^\mu\psi_i(\boldsymbol{r}), \tag{25b}$$

$$j_3^\mu(\boldsymbol{r}) = \langle\Phi|\bar{\psi}\gamma^\mu\tau_3\psi|\Phi\rangle = \sum_{i=1}^{A} \bar{\psi}_i(\boldsymbol{r})\gamma^\mu\tau_3\psi_i(\boldsymbol{r}), \tag{25c}$$

$$j_c^\mu(\boldsymbol{r}) = \langle\Phi|\bar{\psi}\gamma^\mu\frac{1-\tau_3}{2}\psi|\Phi\rangle = \sum_{i=1}^{A} \bar{\psi}_i(\boldsymbol{r})\gamma^\mu\frac{1-\tau_3}{2}\psi_i(\boldsymbol{r}). \tag{25d}$$

The sum over the index i runs where the levels in the Fermi sea and A is the number of nucleons in the system.

The RMF energy of the system can be obtained from the energy functional (16),

$$E_{\text{RMF}} = E_{\text{kin}} + E_\sigma + E_\omega + E_\rho + E_{\text{e.m.}} + E_{\text{N.L.}}, \tag{26}$$

where E_{kin} denotes the kinetic energy, and E_σ, E_ω, and E_ρ represent the potential energies, more specifically,

$$\begin{aligned}
E_{\text{kin}} &= \int d^3r \sum_{i=1}^{A} \psi_i^\dagger(\boldsymbol{r})(\boldsymbol{\alpha}\cdot\boldsymbol{p} + \beta M)\psi_i(\boldsymbol{r}), && (27a)\\
E_\sigma &= \frac{1}{2}g_\sigma \int d^3r\, \rho_s(\boldsymbol{r})\sigma(\boldsymbol{r}), && (27b)\\
E_\omega &= \frac{1}{2}g_\omega \int d^3r\, j_\mu(\boldsymbol{r})\omega^\mu(\boldsymbol{r}), && (27c)\\
E_\rho &= \frac{1}{2}g_\rho \int d^3r\, j_{3\mu}(\boldsymbol{r})\rho_3^\mu(\boldsymbol{r}), && (27d)\\
E_{\text{e.m.}} &= \frac{1}{2}e \int d^3r\, j_{c\mu}(\boldsymbol{r})A^\mu(\boldsymbol{r}). && (27e)
\end{aligned}$$

The nonlinear energy $E_{\text{N.L.}}$ is defined as

$$E_{\text{N.L.}} = \int d^3r \left\{U_\sigma - \frac{1}{2}\sigma U'_\sigma - U_\omega + \frac{1}{2}\omega_\mu U'^\mu_\omega - U_\rho + \frac{1}{2}\rho_{3\mu}U'^\mu_\rho\right\}. \tag{28}$$

Models with density dependent coupling constants[11,17,20–23] do not contain any non-linear couplings in the meson sector. Usually it is assumed that the coupling constants $g_\sigma(\rho_v)$, $g_\omega(\rho_v)$, and $g_\rho(\rho_v)$ depend on the vector density ρ_v. This density dependence has to be taken into account in the variation.[20] This leads to additional terms in the Dirac equation

$$\left[\boldsymbol{\alpha} \cdot (\boldsymbol{p} - \boldsymbol{V}) + V^0(\boldsymbol{r}) + \Sigma_R^0(\boldsymbol{r}) + \beta\,(M + S(\boldsymbol{r}))\right] \psi_k(\boldsymbol{r}) = \varepsilon_k \psi_k(\boldsymbol{r}), \quad (29)$$

where Σ_R^0 is the rearrangement term

$$\Sigma_R^0 = \frac{\partial g_\sigma}{\partial \rho_v}\rho_s\sigma + \frac{\partial g_\omega}{\partial \rho_v}\rho_v\omega^0 + \frac{\partial g_\rho}{\partial \rho_v}\rho_3\rho_3^0. \quad (30)$$

Since translational symmetry is broken in the mean-field approximation, the center-of-mass (c.m.) correction should be made for the spurious c.m. motion. Nowadays, this is usually done by including the microscopic c.m. correction energy[26,27]

$$E_{\text{c.m.}}^{\text{mic}} = -\frac{1}{2MA}\langle \hat{\boldsymbol{P}}_{\text{c.m.}}^2 \rangle, \quad (31)$$

with A being the mass number and $\hat{\boldsymbol{P}}_{\text{c.m.}} = \sum_i^A \hat{\boldsymbol{p}}_i$ being the total momentum in the c.m. frame. Therefore, the total energy for the nuclear system becomes

$$E_{\text{tot}} = E_{\text{RMF}} + E_{\text{c.m.}}^{\text{mic}}. \quad (32)$$

2. Relativistic point-coupling theory

Relativistic point-coupling theory is another way to construct a relativistic energy density functional. In relativistic point-coupling theory,[28–31] the meson exchange in each channel (scalar-isoscalar, vector-isoscalar, scalar-isovector, and vector-isovector) is replaced by corresponding local (contact) interactions between the nucleons. As in non-relativistic Skyrme models, finite range effects are approximated by local derivative terms. The density dependence is taken into account either by density dependent coupling constants in the two-body interactions, or in analogy to non-linear meson couplings, by many-body contact terms. In recent years, the point-coupling theory has attracted more and more attention owing to the following advantages. Firstly, it is considerably simpler in numerical applications: it avoids the solution of the Klein–Gordon equations and, as in non-relativistic Skyrme models, it avoids complicated two-body matrix elements of finite range in applications going beyond static mean field theory.[32–36] Secondly, it is possible to study the role of naturalness[37,38] in effective theories for nuclear-structure-related problems. Thirdly, it is relatively easy to include the Fock terms,[39] and provides more opportunities to investigate its relationship to the non-relativistic approaches.[40] In the following, we present the theoretical framework of relativistic point-coupling theory.

The basic building blocks of covariant density functional theory with point-coupling vertices are

$$(\bar{\psi}\mathcal{O}\Gamma\psi), \qquad \mathcal{O} \in \{1, \vec{\tau}\}, \qquad \Gamma \in \{1, \gamma_\mu, \gamma_5, \gamma_5\gamma_\mu, \sigma_{\mu\nu}\}, \tag{33}$$

where ψ is Dirac spinor field of nucleon, $\vec{\tau}$ is the isospin Pauli matrix, and Γ generally denotes 4×4 Dirac matrices. There are ten such building blocks characterized by their transformation characteristics in isospin and Minkowski space.

A general effective Lagrangian can be written as a power series in the $\bar{\psi}\mathcal{O}\Gamma\psi$'s and their derivatives, with higher-order terms representing in-medium many-body correlations. In the actual application we start with the following Lagrangian density of the form:

$$\mathcal{L} = \mathcal{L}^{\text{free}} + \mathcal{L}^{\text{4f}} + \mathcal{L}^{\text{hot}} + \mathcal{L}^{\text{der}} + \mathcal{L}^{\text{em}} \tag{34}$$

which includes the Lagrangian density for free nucleons:

$$\mathcal{L}^{\text{free}} = \bar{\psi}(i\gamma_\mu\partial^\mu - M)\psi, \tag{35}$$

the four-fermion point-coupling terms:

$$\begin{aligned}\mathcal{L}^{\text{4f}} = & -\frac{1}{2}\alpha_S(\bar{\psi}\psi)(\bar{\psi}\psi) - \frac{1}{2}\alpha_V(\bar{\psi}\gamma_\mu\psi)(\bar{\psi}\gamma^\mu\psi) \\ & -\frac{1}{2}\alpha_{TS}(\bar{\psi}\vec{\tau}\psi)(\bar{\psi}\vec{\tau}\psi) - \frac{1}{2}\alpha_{TV}(\bar{\psi}\vec{\tau}\gamma_\mu\psi)(\bar{\psi}\vec{\tau}\gamma^\mu\psi),\end{aligned} \tag{36}$$

the higher order terms $\mathcal{L}^{\text{hot}}$ accounting for the medium effects, which lead in the mean-field approximation to density-dependent coupling constants with a density dependence of polynomial form:

$$\mathcal{L}^{\text{hot}} = -\frac{1}{3}\beta_S(\bar{\psi}\psi)^3 - \frac{1}{4}\gamma_S(\bar{\psi}\psi)^4 - \frac{1}{4}\gamma_V[(\bar{\psi}\gamma_\mu\psi)(\bar{\psi}\gamma^\mu\psi)]^2, \tag{37}$$

the derivative terms $\mathcal{L}^{\text{der}}$ to simulate the finite-range effects which are crucial for a quantitative description nuclear density distributions (e.g., nuclear radii):

$$\begin{aligned}\mathcal{L}^{\text{der}} = & -\frac{1}{2}\delta_S\partial_\nu(\bar{\psi}\psi)\partial^\nu(\bar{\psi}\psi) - \frac{1}{2}\delta_V\partial_\nu(\bar{\psi}\gamma_\mu\psi)\partial^\nu(\bar{\psi}\gamma^\mu\psi) \\ & -\frac{1}{2}\delta_{TS}\partial_\nu(\bar{\psi}\vec{\tau}\psi)\partial^\nu(\bar{\psi}\vec{\tau}\psi) - \frac{1}{2}\delta_{TV}\partial_\nu(\bar{\psi}\vec{\tau}\gamma_\mu\psi)\partial^\nu(\bar{\psi}\vec{\tau}\gamma_\mu\psi)\end{aligned} \tag{38}$$

and the electromagnetic interaction terms:

$$\mathcal{L}^{\mathrm{em}} = -\frac{1}{4}F^{\mu\nu}F_{\mu\nu} - \bar{\psi}eA_\mu\gamma^\mu\psi. \tag{39}$$

A_μ and $F_{\mu\nu}$ are the four-vector potential and field strength tensor of the electromagnetic field. e is the charge unit for protons, it vanishes for neutrons. There are in total 11 coupling constants, α_S, α_V, α_{TS}, α_{TV}, β_S, γ_S, γ_V, δ_S, δ_V, δ_{TS}, and δ_{TV}, in which α refers to the four-fermion terms, β and γ to the third- and fourth-order terms, and δ the derivative couplings. The subscripts S, V, and T indicate the symmetries of the couplings, i.e., S stands for scalar, V for vector, and T for isovector. As the δ-meson in the meson-coupling theories, the isovector-scalar channel including the terms α_{TS} and δ_{TS} in Eqs. (36) and (38) is usually neglected since a fit including the isovector-scalar interaction does not improve the description of nuclear ground-state properties.[29] Furthermore, the pseudoscalar γ_5 and pseudovector $\gamma_5\gamma_\mu$ channels are also neglected in Eq. (34) since they do not contribute at the Hartree level due to parity conservation in nuclei.

The Hamiltonian density, i.e., the 00 components of the energy-momentum tensor can be obtained by the Legendre transformation

$$\mathcal{H} = \frac{\partial\mathcal{L}}{\partial\dot{\phi}_i}\dot{\phi}_i - \mathcal{L}, \tag{40}$$

where ϕ_i represents the nucleon or photon field. Therefore, the total Hamiltonian reads

$$\begin{aligned} H &= \int d^3r\mathcal{H} \\ &= \int d^3r\, \{\psi^\dagger\left[\boldsymbol{\alpha}\cdot\boldsymbol{p} + M\right]\psi \\ &\quad + \frac{1}{2}\alpha_S(\bar{\psi}\psi)(\bar{\psi}\psi) + \frac{1}{2}\alpha_V(\bar{\psi}\gamma_\mu\psi)(\bar{\psi}\gamma^\mu\psi) + \frac{1}{2}\alpha_{TV}(\bar{\psi}\vec{\tau}\gamma_\mu\psi)(\bar{\psi}\vec{\tau}\gamma^\mu\psi) \\ &\quad + \frac{1}{3}\beta_S(\bar{\psi}\psi)^3 + \frac{1}{4}\gamma_S(\bar{\psi}\psi)^4 + \frac{1}{4}\gamma_V[(\bar{\psi}\gamma_\mu\psi)(\bar{\psi}\gamma^\mu\psi)]^2 \\ &\quad + \frac{1}{2}\delta_S(\bar{\psi}\psi)\triangle(\bar{\psi}\psi) + \frac{1}{2}\delta_V(\bar{\psi}\gamma_\mu\psi)\triangle(\bar{\psi}\gamma^\mu\psi) + \frac{1}{2}\delta_{TV}(\bar{\psi}\vec{\tau}\gamma_\mu\psi)\triangle(\bar{\psi}\vec{\tau}\gamma^\mu\psi) \\ &\quad + \bar{\psi}e\gamma^\mu A_\mu\psi + \frac{1}{2}A_\mu\triangle A^\mu\Big\}. \end{aligned} \tag{41}$$

Similar to the meson-exchange case and assuming that the nucleon single-particle states do not mix isospin, the energy density functional for the nuclear system can be represented as,

$$
\begin{aligned}
E_{\mathrm{RMF}} \equiv \langle\Phi|H|\Phi\rangle = \int d^3r \Bigg\{ & \sum_{i=1}^{A} \psi_i^\dagger(\boldsymbol{\alpha}\cdot\boldsymbol{p}+\beta M)\psi_i \\
& +\frac{1}{2}\alpha_S\rho_s^2 + \frac{1}{3}\beta_S\rho_s^3 + \frac{1}{4}\gamma_S\rho_s^4 + \frac{1}{2}\delta_S\rho_s\triangle\rho_s \\
& +\frac{1}{2}\alpha_V j_\mu j^\mu + \frac{1}{4}\gamma_V(j_\mu j^\mu)^2 + \frac{1}{2}\delta_V j_\mu\triangle j^\mu \\
& +\frac{1}{2}\alpha_{TV} j_{3\mu} j_3^\mu + \frac{1}{2}\delta_{TV} j_{3\mu}\triangle j_3^\mu \\
& +eA_\mu j_p^\mu + \frac{1}{2}A_\mu\triangle A^\mu \Bigg\},
\end{aligned} \tag{42}
$$

where the local densities and currents $\rho_s(\boldsymbol{r})$, $j^\mu(\boldsymbol{r})$, $j_3^\mu(\boldsymbol{r})$, and $j_c^\mu(\boldsymbol{r})$ are given in Eqs. (25),

By minimizing the energy density functional Eq. (42) with respect to $\bar{\psi}_i$, one obtains the Dirac equation for the nucleons

$$
[\boldsymbol{\alpha}\cdot(\boldsymbol{p}-\boldsymbol{V}) + V^0 + \beta(M+S)]\psi_k = \varepsilon_k\psi_k. \tag{43}
$$

Here, the local scalar $S(\boldsymbol{r})$ and vector $V^\mu(\boldsymbol{r})$ potentials read

$$
S(\boldsymbol{r}) = \Sigma_s, \quad V^\mu(\boldsymbol{r}) = \Sigma^\mu + \tau_3\Sigma_3^\mu, \tag{44}
$$

where the nucleon scalar-isoscalar Σ_s, vector-isoscalar Σ^μ, and vector-isovector Σ_3^μ self-energies are given in terms of the various densities,

$$
\Sigma_s = \alpha_S\rho_s + \beta_S\rho_s^2 + \gamma_S\rho_s^3 + \delta_S\triangle\rho_s, \tag{45a}
$$
$$
\Sigma^\mu = \alpha_V j^\mu + \gamma_V(j_\mu j^\mu)j^\mu + \delta_V\triangle j^\mu + eA^\mu, \tag{45b}
$$
$$
\Sigma_3^\mu = \alpha_{TV} j_3^\mu + \delta_{TV}\triangle j_3^\mu. \tag{45c}
$$

The Coulomb field A^μ which is determined by Poisson's equation (23d).

For a system with time reversal invariance, the space-like components of the currents $\boldsymbol{j}$ in Eq. (25) and the vector potential $\boldsymbol{V}$ in Eq. (44) vanish. The total energy is thus given by

$$
\begin{aligned}
E_{\mathrm{RMF}} = \int d^3x \Bigg\{ & \sum_{i=1}^{A} \psi_i^\dagger\left(\boldsymbol{\alpha}\cdot\boldsymbol{p}+\beta M\right)\psi_i + \frac{1}{2}\alpha_S\rho_s^2 + \frac{1}{2}\alpha_V\rho_v^2 + \frac{1}{2}\alpha_{TV}\rho_3^2 \\
& +\frac{1}{3}\beta_S\rho_s^3 + \frac{1}{4}\gamma_S\rho_s^4 + \frac{1}{4}\gamma_V\rho_v^4 + \frac{1}{2}\delta_S\rho_s\triangle\rho_s \\
& +\frac{1}{2}\delta_V\rho_v\triangle\rho_v + \frac{1}{2}\delta_{TV}\rho_3\triangle\rho_3 + \frac{1}{2}eA^0\rho_c \Bigg\}.
\end{aligned} \tag{46}
$$

Together with the center of mass correction in Eq. (31) we obtain

$$E_{\rm tot} = E_{\rm RMF} + E^{\rm mic}_{\rm c.m.}. \tag{47}$$

3. Effective interactions

It is an important issue to determine the coupling constants for the energy density functionals in Eqs. (16) and (42). In principle, they should be determined from more fundamental theories through *ab initio* calculations.[16,41,42] However, at present, such attempts are in their infancy. Therefore, the usual strategy to determine the energy density functional is by fitting the properties for selected finite nuclei and/or the equation of state (EOS) of infinite nuclear matter. These empirical density functionals, sometimes guided by microscopic nucleon self-energies in nuclear matter, gives us more freedom to investigate in detail the relationship between global properties of the nuclear matter equation of state (volume, surface, and asymmetry energies) and the corresponding predictions for properties of finite nuclei.

In Table 1 some widely used meson-exchange effective interactions are listed, including NLSH,[43] TM1,[18] NL3,[10] PK1,[11] TW99,[22] DD-ME2,[24] and PKDD.[11]

The effective interactions NLSH, TM1, NL3, and PK1 are nonlinear interactions, where g_2 and g_3 are the nonlinear coupling constants of the σ meson, and c_3 is that of the ω meson. Following the spirit of the NL3 parameter set, parametrization FSUGold is calibrated to the binding energies and charge radii of a variety of magic nuclei with two additional parameters to soften both the EOS of symmetric matter and the symmetry energy.[19]

The density dependent interaction TW99, DD-ME2, and PKDD are also listed in Table 1. Their density dependence is given by the Typel-Wolter ansatz[22] with the parameters given in Table 2. In this ansatz the density dependence of the coupling constants g_σ and g_ω are parameterized as

$$g_i(\rho_v) = g_i(\rho_{\rm sat}) f_i(x) \qquad \text{for } i = \sigma,\ \omega, \tag{48}$$

where

$$f_i(x) = a_i \frac{1 + b_i(x + d_i)^2}{1 + c_i(x + d_i)^2} \tag{49}$$

is a function of $x = \rho_v/\rho_{\rm sat}$, and $\rho_{\rm sat}$ denotes the baryonic saturation density of nuclear matter. The eight parameters for the density dependence of σ-N

Table 1. Some meson-exchange effective interactions including NLSH,[43] TM1,[18] NL3,[10] PK1,[11] TW99,[22] DD-ME2,[24] and PKDD.[11] The masses of nucleons and mesons are in MeV.

	NLSH	TM1	NL3	PK1	TW99	DD-ME2	PKDD
m_n	939.0000	938.0000	939.000	939.5731	939.00000	938.5000	939.5731
m_p	939.0000	938.0000	939.000	938.2796	939.00000	938.5000	938.2796
m_σ	526.0590	511.1980	508.194	514.0891	550.00000	550.1238	555.5112
m_ω	783.0000	783.0000	782.501	784.2540	783.00000	783.0000	783.0000
m_ρ	763.0000	770.0000	763.000	763.0000	763.00000	763.0000	763.0000
g_σ	10.4440	10.0289	10.217	10.3222	10.72854	10.5396	10.7385
g_ω	12.9450	12.6139	12.868	13.0131	13.29015	13.0189	13.1476
g_ρ	4.3830	4.6322	4.474	4.5297	3.66098	3.6836	4.2998
g_2	-6.9099	-7.2325	-10.431	-8.1688	—	—	—
g_3	-15.8337	0.6183	-28.885	-9.9976	—	—	—
c_3	—	71.3075	—	55.6360	—	—	—

Table 2. Density-dependent parameters of meson-nucleon couplings for several typical meson-exchange effective interactions including TW99,[22] DD-ME2,[24] and PKDD.[11]

	a_σ	b_σ	c_σ	d_σ	a_ω	b_ω	c_ω	d_ω	a_ρ
TW99	1.365469	0.226061	0.409704	0.901995	1.402488	0.172577	0.344293	0.983955	0.515
DD-ME2	1.3881	1.0943	1.7057	0.4421	1.3892	0.9240	1.4620	0.4775	0.5647
PKDD	1.327423	0.435126	0.691666	0.694210	1.342170	0.371167	0.611397	0.738376	0.183305

and ω-N coupling are not independent. Five constraints $f_i(1) = 1$, $f''_\sigma(1) = f''_\omega(1)$, and $f''_i(0) = 0$ reduce the number of independent parameters to three. For the ρ meson, an exponential dependence is suggested by Dirac–Brueckner calculations of asymmetric nuclear matter[44]

$$g_\rho = g_\rho(\rho_{\text{sat}}) \exp[-a_\rho(x-1)]. \tag{50}$$

Although relativistic point-coupling theory became popular only recently, there are already several successful parameter sets available by fitting the binding energies and charge radii of a variety of magic nuclei and / or infinite nuclear matter properties. Table 3 and Table 4 show the nonlinear point-coupling density functionals PC-LA,[28] PC-F1,[29] PC-PK1,[31] and DD-PC1.[30]

For the nonlinear point-coupling effective interactions PC-LA, PC-F1, and PC-PK1, β_S and γ_S are the nonlinear coupling constants in the scalar channel, and γ_V is that in the vector channel. For the density dependent interaction DD-PC1, guided by the microscopic density dependence of the vector and scalar self-energies, the following ansatz for the functional form

Table 3. Nonlinear point-coupling interactions PC-LA,[28] PC-F1,[29] and PC-PK1.[31]

Coupling Constant		PC-LA	PC-F1	PC-PK1
α_S	[MeV^{-2}]	-4.50765×10^{-4}	-3.83577×10^{-4}	-3.96291×10^{-4}
β_S	[MeV^{-5}]	1.10951×10^{-11}	7.68567×10^{-11}	8.6653×10^{-11}
γ_S	[MeV^{-8}]	5.73522×10^{-17}	-2.90443×10^{-17}	-3.80724×10^{-17}
δ_S	[MeV^{-4}]	-4.23944×10^{-10}	-4.18530×10^{-10}	-1.09108×10^{-10}
α_V	[MeV^{-2}]	3.42665×10^{-4}	2.59333×10^{-4}	2.6904×10^{-4}
γ_V	[MeV^{-8}]	-4.38858×10^{-17}	-3.87900×10^{-17}	-3.64219×10^{-18}
δ_V	[MeV^{-4}]	-1.14365×10^{-10}	-1.19210×10^{-10}	-4.32619×10^{-10}
α_{TV}	[MeV^{-2}]	3.25677×10^{-5}	3.46770×10^{-5}	2.95018×10^{-5}
δ_{TV}	[MeV^{-4}]	—	-4.20000×10^{-11}	-4.11112×10^{-10}
α_{TS}	[MeV^{-2}]	7.40291×10^{-7}	—	—

Table 4. The density-dependent point-coupling interaction DD-PC1.[30]

	a_S [fm^2]	b_S [fm^2]	c_S [fm^2]	d_S	a_V [fm^2]	b_V [fm^2]	d_V	b_{TV} [fm^2]	d_{TV}	δ_s [fm^4]
DD-PC1	−10.0462	−9.1504	−6.4273	1.3724	5.9195	8.8637	0.6584	1.8360	0.6400	−0.815

of the couplings is adopted:

$$\alpha_i(\rho_v) = a_i + (b_i + c_i x)e^{-d_i x} \quad (i = S, V, TV), \tag{51}$$

with $x = \rho_v/\rho_{\rm sat}$, and $\rho_{\rm sat}$ denotes the baryonic saturation density of nuclear matter. For the isovector channel, the two parameters a_{TV} and c_{TV} are set to zero.

4. Covariant density functional theory with pairing correlations

4.1. *Generalized mean field concepts in nuclei*

As we have seen in the last chapter, density functional theory provides a mapping of the correlated quantum mechanical many-body onto a single particle problem described by product states and single particle density matrices $\hat{\rho}$ with the property $\hat{\rho}^2 = \hat{\rho}$, whose local part is identical with the exact local density $\rho(\boldsymbol{r})$. Therefore density functional theory works best for systems where an effective mean field description is possible. As it is known from the early days of the shell model[45,46] and from the deformed shell model[47] nuclear physics is such a case. The basic idea of the mean field concept in nuclei is the fact, that correlations are taken into account by breaking symmetries. Apart from a few cases in very light nuclei, all nuclear mean field models break translational symmetry. When spherical and number symmetry is imposed, the nuclear shell model is applicable only for a few doubly magic nuclei, where the spherical shell gap is large enough such that admixtures of more complicated configurations can be neglected in a first approximation. For open shell nuclei we have to distinguish *ph*- and *pp*-correlations. For single magic nuclei, where the closed shell configuration of one type of particles guarantees to a large extend spherical shapes the most important correlations are *pp*-correlations. They are induced by effective short range interactions favoring $J = 0$ pairs and they can be taken into account within a generalized mean field concept by BCS[48,49] or Hartree–Bogoliubov[50–52] theory by breaking the gauge invariance connected with particle number. For nuclei with open proton and neutron shells the strong quadrupole-quadrupole parts of the effective proton-neutron force induce deformations, which are described very successfully in the deformed shell model of Nilsson.[47] Therefore the majority of nuclei in the nuclear chart can be described in a generalized mean field theory by a generalized density functional theory based on Hartree–Fock–Bogoliubov or Hartree–Bogoliubov theory. This theory breaks translational, as well as rotational, as well as gauche symmetry for protons and neutrons. Of course there exist also transitional nuclei with weak pairing correlations or small deformations and shape coexistence. They cannot be described well within present nuclear density functional theory based on the mean field level, but they are accessible to an extension of density functional theory beyond mean field.[36]

4.2. *Relativistic Hartree–Bogoliubov theory*

Pairing has often been taken into account in a very phenomenological way in the BCS model with the seniority force,[53] adjusted to the experimental odd-even mass differences. In many cases, however, this approach presents only a poor approximation. The physics of weakly-bound nuclei, in particular, necessitates a unified and self-consistent treatment of mean-field and pairing correlations. This has led to the formulation and development of the relativistic Hartree–Bogoliubov (RHB) model,[25,54–56] which represents a relativistic extension of the conventional Hartree–Fock–Bogoliubov framework. The RHB model provides a unified description of particle-hole (*ph*) and particle-particle (*pp*) correlations on a mean-field level by using two average potentials: the self-consistent mean field $\hat{h}$ that encloses all the long range *ph* correlations, and a pairing field $\hat{\Delta}$ which sums up the

pp-correlations. In analogy to Eq. (13) the ground state of the nucleus is described by a generalized Slater determinant $|\Phi\rangle$ that represents the vacuum with respect to independent quasiparticles. The quasiparticle operators are defined by the unitary Bogoliubov transformation of the single-nucleon creation and annihilation operators:

$$\alpha_k^+ = \sum_n U_{nk} c_n^+ + V_{nk} c_n \ , \tag{52}$$

where the index n refers to the original basis, e.g. an oscillator basis, or the coordinates $(\boldsymbol{r}, s, t)$ in space, spin and isospin. In addition, for the relativistic case the index $p = f, g$ will denote the large and small components of the corresponding Dirac spinor. U and V are the Hartree–Bogoliubov wave functions determined by the variational principle. In the presence of pairing the single-particle density matrix is generalized to two densities:[57] the normal density $\hat{\rho}$ and the pairing tensor $\hat{\kappa}$

$$\hat{\rho}_{nn'} = \langle\Phi|c_{n'}^\dagger c_n|\Phi\rangle, \qquad \hat{\kappa}_{nn'} = \langle\Phi|c_{n'} c_n|\Phi\rangle. \tag{53}$$

The RHB energy density functional thus depends on both densities

$$E_{\mathrm{RHB}}[\hat{\rho}, \hat{\kappa}] = E_{\mathrm{RMF}}[\hat{\rho}] + E_{\mathrm{pair}}[\hat{\kappa}] \ , \tag{54}$$

where $E_{\mathrm{RMF}}[\hat{\rho}]$ is the usual RMF-functional defined by Eqs. (42) or (46), and the pairing part of the RHB functional reads

$$E_{\mathrm{pair}}[\hat{\kappa}] = \frac{1}{4} \sum_{n_1 n_1'} \sum_{n_2 n_2'} \hat{\kappa}_{n_1 n_1'}^* \langle n_1 n_1'|V^{pp}|n_2 n_2'\rangle \hat{\kappa}_{n_2 n_2'}, \tag{55}$$

where $\langle n_1 n_1'|V^{pp}|n_2 n_2'\rangle$ are the matrix elements of the two-body pairing interaction. The RHB-coefficients U and V are obtained by the variational principle, which yields the RHB equations:

$$\begin{pmatrix} \hat{h}_D - \lambda & \hat{\Delta} \\ -\hat{\Delta}^* & -\hat{h}_D^* + \lambda \end{pmatrix} \begin{pmatrix} U_k \\ V_k \end{pmatrix} = E_k \begin{pmatrix} U_k \\ V_k \end{pmatrix} \ . \tag{56}$$

In the relativistic case the self-consistent mean-field corresponds to the single-nucleon Dirac Hamiltonian $\hat{h}_D$ of Eqs. (21) or (29). The chemical potential λ is determined by the particle number subsidiary condition such that the expectation value of the particle number operator in the ground state equals the number of nucleons. The pairing field $\hat{\Delta}$ reads

$$\hat{\Delta}_{n_1 n_1'} = \frac{1}{2} \sum_{n_2 n_2'} \langle n_1 n_1'|V^{pp}|n_2 n_2'\rangle \hat{\kappa}_{n_2 n_2'}. \tag{57}$$

The column vectors in the eigenvalue problem Eq. (56) denote the quasiparticle wave functions, and E_k are the quasiparticle energies. The dimension of the RHB matrix equation is two times the dimension of the corresponding Dirac matrix equation. Therefore, for each eigenvector (U_k, V_k) with positive quasiparticle energy $E_k > 0$, there exists an eigenvector (V_k^*, U_k^*) with negative quasiparticle energy $-E_k$. Since the baryon quasiparticle operators satisfy fermion commutation relations, the levels E_k and $-E_k$ cannot be occupied simultaneously, that is, one chooses either the positive or the negative eigenvalue and the corresponding eigenvector.[57] In the Hartree–Fock case, the choice of positive or negative value of the quasiparticle energy means that the level is either occupied or empty. For the non-relativistic HFB the ground state represents the minimum of the energy surface and, to form a vacuum with respect to all quasiparticles, one chooses only the positive quasiparticle energies

$$\alpha_k|\Phi\rangle = 0 \quad \text{for} \quad E_k > 0 \qquad \text{or} \qquad |\Phi\rangle = \prod_{E_k>0} \alpha_k|-\rangle. \tag{58}$$

$|\Phi\rangle$ denotes the quasiparticle vacuum, whereas $|-\rangle$ is the bare vacuum. $|\Phi\rangle$ corresponds to the occupation of states with lowest energy, because all the quasiparticle excitations have a positive excitation energy. The single-particle density and the pairing tensor Eq. (53) that correspond to this state can be expressed in terms of the quasi-particle wave functions:

$$\hat{\rho}_{nn'} = \sum_{E_k>0} V_{nk}^* V_{n'k},, \qquad \hat{\kappa}_{nn'} = \sum_{E_k>0} V_{nk}^* U_{n'k}. \tag{59}$$

In the relativistic case one finds solutions in the Dirac sea (usually called negative energy solutions and denoted by the index a), and solutions above the Dirac sea (usually called positive energy solutions and denoted by the index p). In the Dirac equation without pairing they can easily be distinguished by the sign of the corresponding eigenvalues. For the RHB equations (56) this is no longer the case but, because of the large gap between the Dirac and the Fermi sea ($\approx$ 1200 MeV), one can easily distinguish the levels in the Dirac sea $|E_a| > 1200$ MeV, from those above the Dirac sea (note that for the Hamiltonian $\hat{h} - M$ the positive energy continuum corresponds to zero energy and the negative energy continuum corresponds to $-2M$). The *no-sea approximation* means that we have to choose solutions with positive quasiparticle energies $E_p > 0$ for the states above the Dirac sea, and solutions with negative quasiparticle energies $E_a < 0$ for all the levels in the Dirac sea. Therefore, the RHB ground state is given by[58]

$$|\Phi\rangle = \prod_{E_p>0} \alpha_p \prod_{E_a<0} \alpha_a|-\rangle. \tag{60}$$

Finally, we find for this wave function the following expressions for the single-particle density and the pairing tensor:

$$\hat{\rho}_{nn'} = \sum_{E_p>0} V^*_{np} V_{n'p} + \sum_{E_a<0} V^*_{na} V_{n'a}, \tag{61a}$$

$$\hat{\kappa}_{nn'} = \sum_{E_p>0} V^*_{np} U_{n'p} + \sum_{E_a<0} V^*_{na} U_{n'a}. \tag{61b}$$

It should be noted that the coefficients V_{na} are negligibly small because of the large Dirac gap and, therefore, in many of the practical applications, the second term in Eqs. (61a) and (61b) is neglected. In this case, however, the pairing tensor $\hat{\kappa}$ is only approximately antisymmetric.

Pairing correlations in nuclei are restricted to an energy window of a few MeV around the Fermi level, and their scale is well separated from the scale of binding energies, that are in the range from several hundred to thousand MeV. There is no empirical evidence for any relativistic effect in the nuclear pairing field $\hat{\Delta}$ and, therefore, a hybrid RHB model[54] with a non-relativistic pairing interaction can be formulated. For a general two-body interaction, the matrix elements of the relativistic pairing field read

$$\hat{\Delta}_{n_1p_1,n_1'p_1'} = \frac{1}{2} \sum_{n_2p_2,n_2'p_2'} \langle n_1p_1, n_1'p_1'|V^{pp}|n_2p_2, n_2'p_2'\rangle \hat{\kappa}_{n_2p_2,n_2'p_2'}. \tag{62}$$

The indices $(p_1, p_2, p_3, p_4 \equiv f, g)$ refer to the large and small components of the quasiparticle Dirac spinors:

$$U_k = \begin{pmatrix} f_k^{(U)} \\ ig_k^{(U)} \end{pmatrix} \qquad V_k = \begin{pmatrix} f_k^{(V)} \\ ig_k^{(V)} \end{pmatrix}. \tag{63}$$

In practical applications of the RHB model only the large components of the spinors U_k and V_k are used to build the non-relativistic pairing tensor $\hat{\kappa}$ in Eq. (59). The resulting pairing field reads

$$\hat{\Delta}_{n_1f,n_1'f} = \frac{1}{2} \sum_{n_2n_2'} \langle n_1f, n_1'f|V^{pp}|n_2f, n_2'f\rangle_a \; \hat{\kappa}_{n_2f,n_2'f}. \tag{64}$$

The other components: $\hat{\Delta}_{fg}$, $\hat{\Delta}_{gf}$, and $\hat{\Delta}_{gg}$ can be safely omitted.[56]

The RHB equations (56) are mostly solved in a discrete basis, either in the oscillator basis[9,58] or in a Saxon-Woods basis[59–61] or in coordinate space.[62–64]

4.3. *Various paring forces*

There have been several types of effective pairing forces V^{pp} used in the literature. The seniority force of Kerman[53] with constant pairing matrix elements G takes only into account Cooper pairs coupled to angular momentum $J = 0$. For problems with time-reversal symmetry the corresponding pairing matrix $\hat{\Delta}$ in Eq. (64) is proportional to unity for this force and RHB theory is equivalent to RMF + BCS. This force is widely used, but it has many limitations, e.g. correlations in pairs with higher angular momentum are neglected, the scattering between pairs with different shells is constant, which is not true for realistic pairing forces, the coupling to the continuum is not properly taken into account,[65] and the predictive power is limited. Nonetheless this method is used in the constant gap approximation in most of the large scale calculations for adjusting the RMF-parameter sets, in particular also for the sets DD-ME2[24] and DD-PC1.[30] For each nucleus in the fit, the gap parameter is determined directly from odd-even mass differences of neighboring nuclei. In this case the occupation numbers v_k^2 in the neighborhood of the Fermi surface which depend crucially on the gap parameter have rather reasonable values and in this way all quantities depending only on the v_k^2's are not influenced further by the value of G nor by the pairing window. Of course, the pairing energy (55) depends on the constant G and on the pairing window. The actual value of G producing this experimental gap parameter is determined after the self-consistent solution of the BCS equations and depends on the nucleus under consideration and on the pairing window. However, for a reasonable pairing window the total change in binding energy caused by pairing, which is the difference between the gain in binding due to the pairing energy (55) and the loss in binding due to the reoccupation of the single particle levels, is rather small, only a few MeV. Therefore there is a clear separation of scales between the total binding energy, which is for heavy nuclei in the order of 1000 MeV and more and the additional binding of a few MeV caused by pairing. By this reason the conventional procedure to adjust the parameters of the Lagrangian in the constant gap approximation by RMF+BCS calculations and to use for all further calculations a more realistic pairing force is very successful. In this way all the problems of the monopole pairing force are avoided.

For reasons of simplicity in many RHB calculations,[62] in particular for those which serve as a basis for investigations beyond mean field[36,66] a zero-range pairing force has been chosen. However, this force shares with the monopole force with constant G the problem of an ultraviolet divergence and requires the use of a pairing window. Finite range forces avoid this. Therefore the pairing part of the Gogny force D1S[67] has been used with great success for many conventional RHB calculations for nuclear ground state properties,[54] for the study of rotational bands in the rotating frame,[68] and for investigations of giant resonances in the framework of the relativistic quasiparticle random phase approximation (RQRPA).[69] On the mean field level such calculations are possible, but they require, in particular for axial and triaxial deformations an extreme numerical effort. For investigations beyond mean field using the generator coordinate method (GCM) or projection they are practically impossible.

Therefore, Tian et al have introduced a separable version of the Gogny pairing force.[70–72] It can be represented as a sum of a finite number of separable terms in the harmonic oscillator basis

$$\langle n_1 n_2 | V^{pp} | n_1' n_2' \rangle = G \sum_{N}^{N_{\max}} W_{n_1 n_2}^{N*} W_{n_1' n_2'}^{N} \,. \tag{65}$$

In this case the pairing field $\hat{\Delta}$ takes the form

$$\hat{\Delta}_{n_1 n_2} = G \sum_{N}^{N_{\max}} P_N \, W_{n_1 n_2}^{N*} \qquad \text{with} \quad P_N = \frac{1}{2}\mathrm{Tr}(W^N \hat{\kappa}) \tag{66}$$

and the pairing energy in the nuclear ground state is given by:

$$E_{\text{pair}} = -G \sum_{N}^{N_{\max}} P_N^* P_N. \tag{67}$$

5. Infinite Nuclear Matter

In the following, both the meson-exchange and the point-coupling models will be used to describe infinite nuclear matter which is defined as a static, homogeneous infinite system.

5.1. *Nonlinear meson-exchange model*

For infinite nuclear matter, the Coulomb field is neglected. Since this system is invariant under translation, the nucleon wave function has good linear momentum **k**. In addition, the spatial derivatives of the meson fields and densities are zero due to the homogeneity.

Therefore, one can simplify the Dirac equation Eq. (21) as

$$\left(\boldsymbol{\alpha} \cdot \boldsymbol{p} + V^0 + \beta M^*\right) \psi_k(\boldsymbol{r}) = \epsilon_k \psi_k(\boldsymbol{r}). \tag{68}$$

with the effective Dirac mass $M^* = M + S$ and the constant fields $S = g_\sigma \sigma$ and $V^0 = g_\omega \omega^0 + g_\rho \tau_3 \rho_3^0$. The eigenstates are characterized by the momentum k, the spin s and the sign of the energy

$$\psi_{ks}^{(+)}(\boldsymbol{r}) = u(k,s)e^{i\boldsymbol{k}\cdot\boldsymbol{r}} \quad \text{with} \quad \varepsilon_k = V^0 + E_k, \tag{69a}$$

$$\psi_{ks}^{(-)}(\boldsymbol{r}) = v(k,s)e^{-i\boldsymbol{k}\cdot\boldsymbol{r}} \quad \text{with} \quad \varepsilon_k = V^0 - E_k, \tag{69b}$$

where $E_k = \sqrt{k^2 + M^{*2}}$ and $u(k,s)$, $v(k,s)$ are the well known[73] plane wave solutions of the Dirac equation with positive and negative energy

$$u(k,s) = \mathcal{N}\begin{pmatrix} \chi(k,s) \\ \frac{\boldsymbol{k\sigma}}{E_k + M^*}\chi(k,s) \end{pmatrix}, \quad v(k,s) = \mathcal{N}\begin{pmatrix} \frac{\boldsymbol{k\sigma}}{E_k + M^*}\phi(k,s) \\ \phi(k,s) \end{pmatrix}, \tag{70}$$

Here χ and ϕ are two-component spinors and

$$\mathcal{N} = \sqrt{\frac{E_k + M^*}{2M^*}} \tag{71}$$

is a normalization constant, such that

$$\bar{u}(k,s)u(k,s') = \frac{M^*}{E_k}u^\dagger(k,s)u(k,s') = \delta_{ss'}, \tag{72a}$$

$$\bar{v}(k,s)v(k,s') = -\frac{M^*}{E_k}v^\dagger(k,s)v(k,s') = -\delta_{ss'}. \tag{72b}$$

In contrast to the non-relativistic case, where in nuclear matter the wave functions do not depend on the mean potential U nor on the density, here the wave functions depend on the effective Dirac mass M^*, i.e. on the scalar potential S and on the density.

The Klein–Gordon equations (23a–23d) have the form

$$m_\sigma^2 \sigma + g_2 \sigma^2 + g_3 \sigma^3 = -g_\sigma \rho_s, \tag{73a}$$

$$m_\omega^2 \omega^0 + c_3 (\omega^0)^3 = g_\omega \rho_v, \tag{73b}$$

$$m_\rho^2 \rho_3^0 + d_3 (\rho_3^0)^3 = g_\rho (\rho_v^n - \rho_v^p), \tag{73c}$$

with the vector and scalar densities

$$\rho_v = 4\int_0^{k_F} \frac{d^3k}{(2\pi)^3} = \frac{2}{3\pi^2}k_F^3, \tag{74a}$$

$$\rho_s = 4\int_0^{k_F} \frac{d^3k}{(2\pi)^3}\frac{M^*}{E_k} = \frac{M^*}{\pi^2}\left\{\frac{k_F}{M^*}\frac{E_F}{M^*} - \ln\left(\frac{k_F}{M^*} + \frac{E_F}{M^*}\right)\right\}, \tag{74b}$$

where k_F is the Fermi momentum and $E_F = \sqrt{k_F^2 + M^{*2}}$.

Finally the energy density $\mathcal{H}$ can be obtained as

$$\mathcal{H} = \frac{1}{4}(3\rho_v E_F + M^*\rho_s) + \frac{1}{2}m_\sigma^2\sigma^2 + \frac{1}{2}m_\omega^2(\omega^0)^2 + \frac{1}{2}m_\rho^2(\rho_3^0)^2 + \frac{1}{3}g_2\sigma^3 + \frac{1}{4}g_3\sigma^4 + \frac{3}{4}c_3(\omega^0)^4 + \frac{3}{4}d_3(\rho_3^0)^4. \quad (75)$$

5.2. *Density-dependent meson-exchange model*

In the density-dependent meson-exchange model, the Dirac equation reads

$$\left(\boldsymbol{\alpha}\cdot\boldsymbol{p} + \beta(M + g_\sigma\sigma) + g_\omega\omega^0 + g_\rho\tau_3\rho_3^0 + \Sigma_R^0\right)\psi_a(\boldsymbol{r}) = \epsilon_a\psi_a(\boldsymbol{r}), \quad (76)$$

where Σ_R^0 is the rearrangement term,

$$\Sigma_R^0 = \frac{\partial g_\sigma}{\partial\rho_v}\rho_s\sigma + \frac{\partial g_\omega}{\partial\rho_v}\rho_v\omega^0 + \frac{\partial g_\rho}{\partial\rho_v}(\rho_v^n - \rho_v^p)\rho_3^0, \quad (77)$$

which plays an essential role in its applications.

Accordingly, the meson equations read

$$m_\sigma^2\sigma = -g_\sigma\rho_s, \quad (78a)$$

$$m_\omega^2\omega^0 = g_\omega\rho_v, \quad (78b)$$

$$m_\rho^2\rho_3^0 = g_\rho(\rho_v^n - \rho_v^p), \quad (78c)$$

with the same scalar and vector densities as in Eqs. (74b) and (74a).

By solving these equations, the energy density can be derived as,

$$\mathcal{H} = \frac{1}{4}(3\rho_v E_F + M^*\rho_s) + \frac{1}{2}m_\sigma^2\sigma^2 + \frac{1}{2}m_\omega^2(\omega^0)^2 + \frac{1}{2}m_\rho^2(\rho_3^0)^2. \quad (79)$$

5.3. *Nonlinear point-coupling model*

In the nonlinear point-coupling model, the Dirac equation for infinite nuclear matter can be written as

$$[\boldsymbol{\alpha}\cdot\boldsymbol{p} + V^0 + \beta(M + S)]\psi_k = \varepsilon_k\psi_k, \quad (80)$$

which is coupled with the scalar $S(\boldsymbol{r})$ and vector $V^\mu(\boldsymbol{r})$ potentials

$$S(\boldsymbol{r}) = \alpha_S\rho_s + \beta_S\rho_s^2 + \gamma_S\rho_s^3, \quad (81a)$$

$$V^0(\boldsymbol{r}) = \alpha_V\rho_v + \gamma_V\rho_v^3 + \alpha_{TV}\rho_3. \quad (81b)$$

The scalar and vector densities are given by Eqs. (74b) and (74a) and the energy density is,

$$\mathcal{H} = \frac{1}{4}(3\rho_v E_F + M^*\rho_s) + \frac{1}{2}\alpha_S\rho_s^2 + \frac{1}{2}\alpha_V\rho_v^2 + \frac{1}{2}\alpha_{TV}(\rho_v^n - \rho_v^p)^2 + \frac{2}{3}\beta_S\rho_s^3 + \frac{3}{4}\gamma_S\rho_s^4 + \frac{1}{4}\gamma_V\rho_v^4. \quad (82)$$

5.4. *Density-dependent point-coupling model*

In the density-dependent point-coupling model, the Dirac equation for infinite nuclear matter can be written as

$$[\boldsymbol{\alpha}\cdot\boldsymbol{p}+V^0+\beta(M+S)]\psi_k=\varepsilon_k\psi_k, \tag{83}$$

with the scalar $S(\boldsymbol{r})$ and vector $V^\mu(\boldsymbol{r})$ potentials

$$S(\boldsymbol{r})=\alpha_S\rho_s, \tag{84a}$$

$$V^0(\boldsymbol{r})=\alpha_V\rho_v+\alpha_{TV}\rho_3+\Sigma_R^0. \tag{84b}$$

Here, Σ_R^0 is the rearrangement term

$$\Sigma_R^0=\frac{1}{2}\left(\frac{\partial\alpha_S}{\partial\rho_v}\rho_s^2+\frac{\partial g_\omega}{\partial\rho_v}\rho_v^2+\frac{\partial\alpha_{TV}}{\partial\rho_v}(\rho_v^n-\rho_v^p)^2\right), \tag{85}$$

with the scalar and vector densities given in Eqs. (74b) and (74a).

After solving these equations, the energy density can be obtained by

$$\mathcal{H}=\frac{1}{4}(3\rho_vE_F+M^*\rho_s)+\frac{1}{2}\alpha_S\rho_s^2+\frac{1}{2}\alpha_V\rho_v^2+\frac{1}{2}\alpha_{TV}(\rho_v^n-\rho_v^p)^2. \tag{86}$$

5.5. *Saturation properties for symmetric nuclear matter*

In Table 5, the saturation properties for nuclear matter, including the binding energy per nucleon E/A, saturation density ρ_0, incompressibility K_0, nucleon effective mass M_D^* and M_L^*, symmetry energy $E_{\rm sym}$, as well as the characteristics L and $K_{\rm asy}$ for the density dependence of $E_{\rm sym}$, predicted by different covariant density functionals are listed.

There are several definitions for the nucleon effective mass.[74,75] Here we mainly focus on the Dirac mass and Landau mass. The Dirac mass M_D^* is defined through the nucleon scalar self-energy in the Dirac equation, i.e., $M_D^*=M+S$. It is directly related to the spin-orbit potential in finite nuclei and is thus a genuine relativistic quantity without non-relativistic correspondence. While the Landau mass $M_L^*=pdp/dE$ is related to the density of state both in relativistic and non-relativistic models. In relativistic models, the relation between the Dirac mass and the Landau mass is $M_L^*=\sqrt{p_F^2+M_D^{*2}}$, where p_F is the Fermi momentum.

The density dependence of the nuclear symmetry energy is very important to understand the properties of exotic nuclei with extreme isospin values,[76] in particular the slope $L\equiv 3\rho_0(dE_{\rm sym}/d\rho)_{\rho=\rho_0}$ and curvature $K_{\rm sym}\equiv 9\rho_0^2(d^2E_{\rm sym}/d^2\rho)_{\rho=\rho_0}$ of the symmetry energy at the saturation density ρ_0. In Refs. 77, 78, the isospin-dependent part, $K_{\rm asy}\approx K_{\rm sym}-6L$,

Table 5. The saturation properties for nuclear matter in CDFT with different covariant density functionals.

	NLSH	TM1	NL3	PK1	TW99	DD-ME2	PKDD	PC-PK1	PC-F1	PC-LA	DD-PC1
ρ_0 (fm^{-3})	0.146	0.145	0.148	0.148	0.153	0.152	0.150	0.154	0.151	0.148	0.152
E/A(MeV)	−16.35	−16.26	−16.25	−16.27	−16.25	−16.14	−16.27	−16.12	−16.17	−16.13	−16.06
M_D^*/M	0.60	0.63	0.59	0.60	0.55	0.57	0.57	0.59	0.61	0.58	0.58
M_L^*/M	0.66	0.69	0.65	0.66	0.62	0.63	0.63	0.65	0.67	0.64	0.64
K_0(MeV)	355	281	272	283	240	251	262	238	255	264	230
$E_{\rm sym}$(MeV)	36.1	36.9	37.4	37.6	32.8	32.3	36.8	35.6	37.8	37.2	33
L(MeV)	113	111	119	116	55	51	90	113	117	108	70
$K_{\rm asy}$(MeV)	−602	−631	−611	−641	−457	−393	−622	−583	−627	−709	−528

in the isobaric incompressibility $K(\delta) = K_0 + K_{\text{asy}}\delta^2$ (with $\delta \equiv (\rho_n - \rho_p)/\rho$), is used to characterize the density dependence of the symmetry energy as both L and K_{asy} can be extracted from the experiment empirically (see Ref. 79 and references therein).

For the effective masses, all the functionals give values between 0.55 and 0.60 for the Dirac mass M_D^*/M as required by the spin-orbit splitting data in finite nuclei,[80] but smaller Landau masses M_L^*/M compared with the empirical constraint 0.8±0.1,[81] implying that they would give a small single-particle level density at the Fermi energy in finite nuclei in comparison with the data.

The symmetry energies in the calculations with the nonlinear effective interactions shown here are always larger than those given by the density-dependent interactions. This leads to the prediction of larger neutron skin thicknesses in finite nuclei. This can be understood by the fact that most of the non-linear meson models shown here have no density dependence in the isovector channel. Only PK1 has a density dependence in the isovector channel induced by a quartic coupling in the ρ-meson. It is however very week leading also to a relatively large value for the symmetry energy. On the other side the non-linear parameter set FSU-Gold,[19] with a relatively strong non-linear ω-ρ coupling has a smaller symmetry energy $E_{\text{sym}} = 32.59$ MeV.[82]

6. Finite nuclei

In the following, covariant density functional theory (CDFT) will be applied to describe ground state properties of finite nuclei. In CDFT, the relativistic (covariant) energy density functional is constructed as a functional of the one-body nucleon density matrix that corresponds to a product of single-particle or single-quasiparticle states. Since symmetries simplify the calculations considerably, in practical application, the equations of motion are usually solved by assuming some symmetries. In the following, we will mainly discuss the solution of the CDFT for even-even nuclei with spherical symmetry, axially symmetric quadrupole deformation, and triaxial quadrupole shapes, respectively. In addition, the results thus obtained for nuclear bulk properties including nuclear masses and charge radii will be discussed.

For nuclei with spherical symmetry, the relativistic equations of motion in coordinate space are reduced to a simple set of radial integro-differential equations.[62–64] For deformed nuclei, however, the solution of integro-differential equations in coordinate space is numerically very time-consuming. As an efficient implementation, the relativistic equations of motion solved in the configuration space of harmonic oscillator wave functions with appropriate symmetry is introduced, whereas the densities are computed in coordinate space. This method can be applied to spherical, axially and non-axially deformed nuclei, and the corresponding numerical code is also available.[58,83] However, it should be noted that for the nuclei far away from the β-stability line, i.e., the so-called exotic nuclei, the usage of the harmonic oscillator basis cannot provide an adequate asymptotic behavior at large distances for the wave function in the weakly bound states and continuum (the details will be discussed in Chapter 3).

For the relativistic Hartree–Bogoliubov (RHB) equation in Eq. (56),

$$\begin{pmatrix} \hat{h}_D - \lambda & \hat{\Delta} \\ -\hat{\Delta}^* & -\hat{h}_D^* + \lambda \end{pmatrix} \begin{pmatrix} U_k \\ V_k \end{pmatrix} = E_k \begin{pmatrix} U_k \\ V_k \end{pmatrix}, \tag{87}$$

with the Dirac Hamiltonian

$$\hat{h}_D = \boldsymbol{\alpha} \cdot \mathbf{p} + \beta(M + S) + V^0. \tag{88}$$

The corresponding RHB eigenvalue problem in configuration space reads,

$$\begin{pmatrix} \mathcal{A} - \lambda & \mathcal{B} & \Delta_{ff} & 0 \\ \mathcal{B}^T & \mathcal{C} - \lambda & 0 & 0 \\ -\Delta_{ff} & 0 & -\mathcal{A} + \lambda & -\mathcal{B} \\ 0 & 0 & -\mathcal{B}^T & -\mathcal{C} + \lambda \end{pmatrix} \begin{pmatrix} f^U \\ g^U \\ f^V \\ g^V \end{pmatrix} = E \begin{pmatrix} f^U \\ g^U \\ f^V \\ g^V \end{pmatrix}. \tag{89}$$

The diagonalization of the RHB matrix equation yields the wave functions in configuration space. The resulting density matrix is computed in configuration space

$$\begin{pmatrix} \rho_{nn'} & \rho_{n\tilde{n}'} \\ \rho_{\tilde{n}n'} & \rho_{\tilde{n}\tilde{n}'} \end{pmatrix} = \begin{pmatrix} \sum f_n^{V*} f_{n'}^V & i \sum f_n^{V*} g_{\tilde{n}'}^V \\ -i \sum g_{\tilde{n}}^{V*} f_{n'}^V & \sum g_{\tilde{n}}^{V*} g_{\tilde{n}'}^V \end{pmatrix}, \tag{90}$$

where n and $\tilde{n}$ denote the indices of an expansion of the large and small components of the Dirac spinor in the oscillator basis. The density matrix is then transformed to the coordinate space, and the resulting vector and scalar densities are used to calculate the potentials.

For the nuclei near the β-stability line, the BCS method is usually much simpler than solving the RHB equation with the corresponding quasiparticle operators that are defined by a special Bogoliubov transformation (see Chapter 1). In the BCS method, the mean-field part and the pairing part are not necessarily diagonalized simultaneously, and therefore the dimension of corresponding Hamiltonian matrix is reduced to half.

6.1. *Spherical nuclei*

6.1.1. *Numerical implementation*

Most single or doubly magic nuclei are spherical. Therefore, we use spherical coordinates (r, θ, ϕ), and the nucleon densities depend only on the radial coordinate r. The spinor is labeled by the nucleon angular momentum j, its projection on the z axis m, parity π, and the isospin projection $t = \pm 1/2$ for neutrons and protons, respectively

$$\psi_k(\boldsymbol{r}, s, t) = \begin{pmatrix} f_k(r)\Phi_{l_k j_k m_k}(\theta, \phi, s) \\ i g_k(r)\Phi_{\tilde{l}_k j_k m_k}(\theta, \phi, s) \end{pmatrix} \chi_{t_k}(t). \tag{91}$$

Here, Φ_{ljm} is a two-dimensional spinor with the angular momentum quantum numbers ljm

$$\Phi_{ljm}(\theta, \phi, s) = [\chi_{\frac{1}{2}}(s) \otimes Y_l(\theta, \phi)]_{jm}. \tag{92}$$

The orbital angular momenta corresponding to large (l) and small $(\tilde{l})$ components of the spinor are determined by the total angular momentum j and parity π

$$l = j \pm \frac{1}{2}, \quad \tilde{l} = j \mp \frac{1}{2}, \quad \pi = (-)^l, \quad \kappa = \pm\left(j + \frac{1}{2}\right). \tag{93}$$

The corresponding radial Dirac equation reads,

$$\begin{pmatrix} M + S(r) + V(r) & \frac{\partial}{\partial_r} - \frac{\kappa - 1}{r} \\ -\frac{\partial}{\partial r} - \frac{\kappa+1}{r} & -M - S(r) + V(r) \end{pmatrix} \begin{pmatrix} f_k \\ g_k \end{pmatrix} = \epsilon_k \begin{pmatrix} f_k \\ g_k \end{pmatrix}. \tag{94}$$

The scalar potential $S(r)$ and the vector potential $V(r)$ depend on the scalar and vector densities.

The large and small components of the Dirac spinors can be expanded in terms of the radial functions $R_{nl}(r, b_0)$ of a spherical harmonic oscillator potential with the corresponding oscillator length b_0

$$f_k(r) = \sum_n^{n_{\max}} f_n^{(i)} R_{nl_k}(r, b_0), \qquad g_k(r) = \sum_{\tilde{n}}^{\tilde{n}_{\max}} g_{\tilde{n}}^{(i)} R_{\tilde{n}\tilde{l}_k}(r, b_0). \tag{95}$$

The radial oscillator wave functions read

$$R_{nl}(r, b_0) = b_0^{-3/2} R_{nl}(\xi) = b_0^{-3/2} N_{nl} \xi^l L_n^{l+1/2}(\xi^2) e^{-\xi^2/2}, \tag{96}$$

where $\xi = r/b_0$ corresponds to the radial distance in units of the oscillator length. The quantum number $n = 0, 1, 2, \ldots$ counts the number of radial nodes, and $L_n^m(\xi^2)$ denote the associated Laguerre polynomials. The normalization factor is

$$N_{nl} = \sqrt{\frac{2n!}{(l+n+1/2)!}}. \tag{97}$$

The upper limits $n_{\max}$ and $\tilde{n}_{\max}$ are determined by the corresponding major shell quantum numbers $N_{\max} = 2n_{\max} + l_{\max}$ and $\tilde{N}_{\max} = 2\tilde{n}_{\max} + \tilde{l}_{\max}$. In order to avoid spurious contributions to the solution of the Dirac equation,[9] the small components are expanded up to $\tilde{N}_{\max} = N_{\max} + 1$.

If we use $|\alpha\rangle = |nljm\rangle$ for the expansion of the upper, and $|\tilde{\alpha}\rangle = |\tilde{n}\tilde{l}jm\rangle$ for the expansion of the lower components of the Dirac spinor, the matrix elements of the Dirac Hamiltonian in Eq. (89) for the case of spherical symmetry are

$$\mathcal{A}_{\alpha\alpha'} = \int_0^\infty r^2 dr R_{nl}(r)\left[M + S(r) + V(r)\right] R_{n'l}(r), \tag{98}$$

$$\mathcal{C}_{\tilde{\alpha}\tilde{\alpha}'} = \int_0^\infty r^2 dr R_{\tilde{n}\tilde{l}}(r)\left[-M - S(r) + V(r)\right] R_{\tilde{n}'\tilde{l}}(r), \tag{99}$$

$$\mathcal{B}_{\alpha\tilde{\alpha}'} = \int_0^\infty r^2 dr R_{nl}(r)\left(\frac{\partial}{\partial r} - \frac{\kappa - 1}{r}\right) R_{\tilde{n}'\tilde{l}}(r). \tag{100}$$

The potential for protons includes the direct Coulomb field,

$$V_C(\boldsymbol{r}) = \frac{e^2}{4\pi}\int d^3\boldsymbol{r}'\,\frac{\rho_p(\boldsymbol{r}')}{|\boldsymbol{r} - \boldsymbol{r}'|}. \tag{101}$$

However, the integrand has a logarithmic singularity, which makes the numerical calculation difficult. By using the identity,

$$\triangle_{\boldsymbol{r}'}|\boldsymbol{r} - \boldsymbol{r}'| = \frac{2}{|\boldsymbol{r} - \boldsymbol{r}'|}, \tag{102}$$

the Coulomb potential can be rewritten as,

$$V_C(\boldsymbol{r}) = \frac{e^2}{8\pi}\int d^3\boldsymbol{r}'|\boldsymbol{r} - \boldsymbol{r}'|\triangle_{\boldsymbol{r}'}\rho_p(\boldsymbol{r}'). \tag{103}$$

After integrating the angular part analytically, the remaining radial part reads

$$V_C(r) = \frac{e^2}{4}\int_0^\infty dr' r'^2\left(3r + \frac{r'^2}{r}\right)\frac{d^2\rho_p(r')}{dr'^2}. \tag{104}$$

For the meson-exchange models, one also has to solve the Klein–Gordon equations for different mesons. For spherical symmetry, the Klein–Gordon equations for the meson fields $\phi = \sigma, \omega, \rho$ read

$$\left(-\frac{\partial^2}{\partial r^2} - \frac{2}{r}\frac{\partial}{\partial r} + m_\phi^2\right)\phi(r) = S_\phi(r). \tag{105}$$

One can expand the solution of this equation in terms of a complete set of basis states

$$\phi(r) = \sum_{n=0}^{n_b} \phi_n R_{n0}(r, b_0), \qquad S_\phi(r) = \sum_{n=0}^{n_b} s_n^\phi R_{n0}(r, b_0). \tag{106}$$

The maximal radial quantum number n_b in the expansion is determined by a cut-off parameter $N_B = 2n_b$. Therefore, one can rewrite the Klein–Gordon equation into a set of inhomogeneous linear equations,

$$\sum_{n'}^{n_b} \mathcal{H}_{nn'}\phi_{n'} = s_n^\phi \tag{107}$$

with the matrix elements

$$\begin{aligned}\mathcal{H}_{nn'} &= -b_0^{-2}\delta_{nn'}(2n + 3/2) + b_0^{-2}\delta_{nn'+1}\sqrt{(n+1)(n+3/2)} \\ &\quad + b_0^{-2}\delta_{n'n+1}\sqrt{(n'+1)(n'+3/2)}.\end{aligned} \tag{108}$$

Inverting the matrix $\mathcal{H}$, this set of equations can be easily solved.

The pairing matrix elements are antisymmetric and can be obtained similarly based on the specific pairing interaction. Note that the pairing properties are determined by correlations in an energy window of a few MeV around the Fermi surface, and therefore the matrix elements associated with the small components could be neglected safely.

6.1.2. *Binding energy*

The binding energies for the Ca, Ni, Sn, and Pb isotopes are calculated with the covariant density functional PC-PK1,[31] DD-PC1,[30] PC-F1,[29] PC-LA,[28] and NL3*,[84] and their deviations from the data[85] are shown in Fig. 1 in comparison with those calculated with DD-PC1, PC-F1, PC-LA, and NL3*.

The experimental data are reproduced in general quite well by the calculations. In particular, the calculations with PC-PK1 reproduce the experimental binding energies within 1 MeV for the Ca isotopes and 2 MeV for both Sn and Pb isotopes. For Ni isotopes, there is an underestimation of several MeV for the even-even $^{58-64}$Ni. In fact, in some former investigations, e.g., in Ref. 86, it has been shown that these nuclei are soft against deformation. Therefore, the dynamic correlation energies gained by restoration of rotational symmetry and configuration mixing are expected

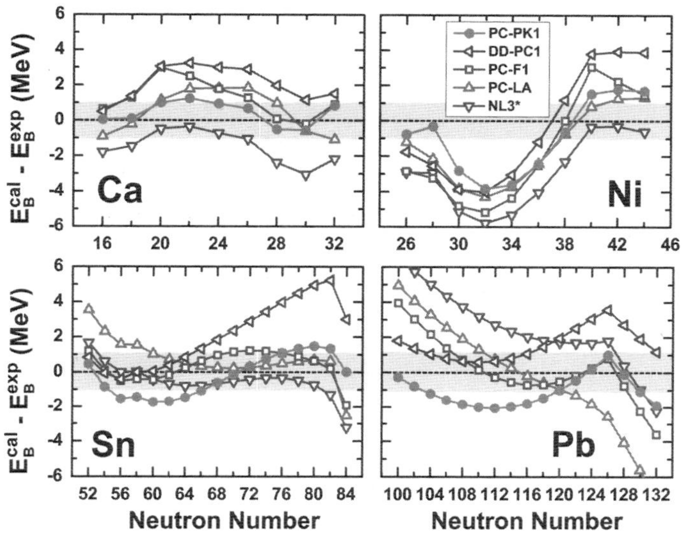

Fig. 1. (Color online) Deviations of the calculated binding energies with the covariant density functional PC-PK1,[31] DD-PC1,[30] PC-F1,[29] PC-LA,[28] and NL3*[84] for Ca, Ni, Sn, and Pb isotopes from the data.[85] Taken from Ref. 31.

to reduce these deviations. Similarly, the underestimations of the binding energies for neutron-deficient Pb isotopes can also be improved by configuration mixing, as demonstrated in the non-relativistic calculations for the even-even $^{182-194}$Pb.[87] For the spherical Ca and Sn isotopes, on the other hand, the energies gained from the restoration of rotational symmetry and configuration mixing are expected to be much smaller as illustrated in the systematic beyond mean-field studies.[88,89] Therefore, the inclusion of these energies will not change significantly the discrepancy between the mean-field results and the corresponding data for such spherical isotopes.

6.1.3. *Charge radii*

The root-mean-square (rms) charge radii are obtained by using the simple formula:

$$r_c = \sqrt{r_p^2 + 0.64} \qquad \text{(fm)}, \tag{109}$$

where r_p denotes the rms radius of the proton density distribution, and the term 0.64 fm^2 accounts for the finite size of the proton. In Fig. 2, the charge radii for Sn and Pb isotopes predicted by the CDFT with PC-PK1,[31] DD-PC1,[30] PC-F1,[29] PC-LA,[28] and NL3*[84] are shown in comparison with the

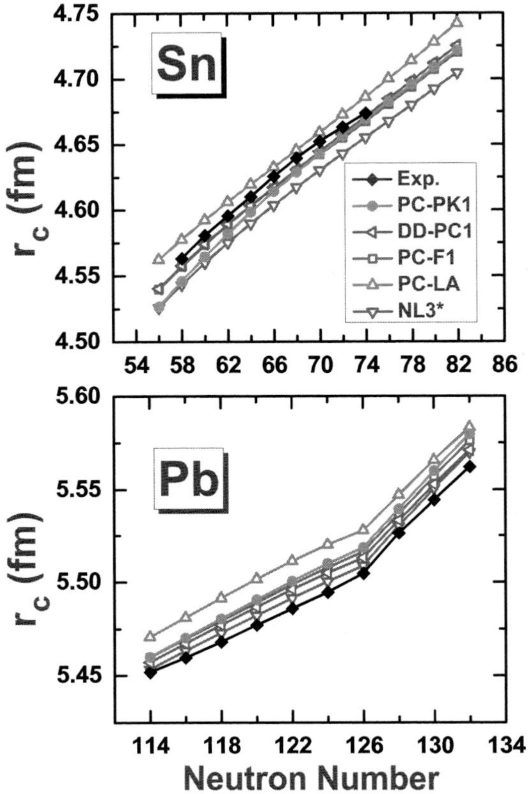

Fig. 2. (Color online) The calculated charge radii with the covariant density functional PC-PK1,[31] DD-PC1,[30] PC-F1,[29] PC-LA,[28] and NL3*[84] for Sn and Pb isotopes in comparison with the data.[90,91] Taken from Ref. 31.

data.[90,91] It is seen that the charge radii from all density functionals follow closely the observed ones for Sn isotopes. For the Pb isotopes, the kink in the charge radii has been excellently reproduced by all the density functionals. This kink is connected with the kink shown in the famous isotope shifts in the Pb-region, which have been measured with extreme accuracy by atomic beam laser spectroscopy.[92] As in Refs. 93, 55, this anomalous kink of the charge radii of the isotopic chain of Pb nuclei has been discussed intensively in the CDFT. It shows that the origin for this kink is the isospin dependence of the spin-orbit term. Different from the phenomenological spin-orbit terms in the non-relativistic Skyrme approaches, in the covariant density functional theory, the spin-orbit interaction originates naturally

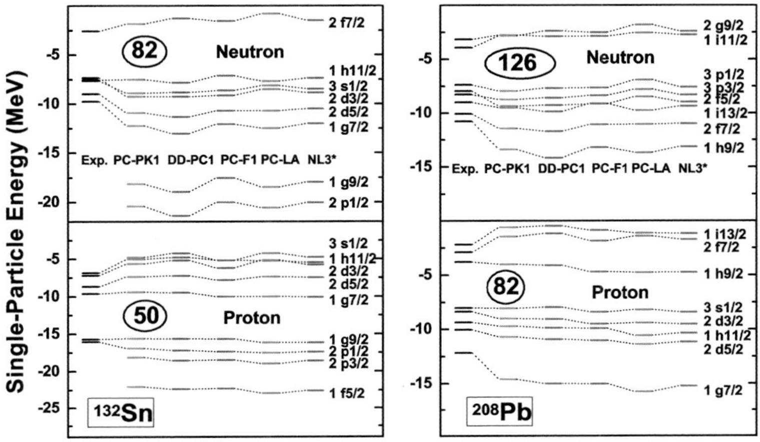

Fig. 3. (Color online) The calculated single-particle energies with the covariant density functional PC-PK1,[31] DD-PC1,[30] PC-F1,[29] PC-LA,[28] and NL3*[84] for ^{132}Sn and ^{208}Pb in comparison with data.[94] Taken from Ref. 31.

from the coupling of scalar and vector potentials. The spin-orbit splitting is responsible for putting different orbitals in space and thus determining the structure of a nucleus.

6.1.4. *Single-particle spectrum*

In Fig. 3, the calculated single-particle energies with the covariant density functional PC-PK1,[31] DD-PC1,[30] PC-F1,[29] PC-LA,[28] and NL3*[84] are shown in the vicinity of the Fermi surface for ^{132}Sn and ^{208}Pb and in comparison with the data.[94] The experimental values are extracted from the single-nucleon separation energies or excitation energies.[94] It is clearly shown that the energy gap between the last occupied and the first unoccupied level is more or less reproduced. It is observed, however, that the experimental spectra are generally more compressed as compared to that of the theory. This is related to the relatively small effective masses for the present covariant density functionals. Moreover, the theoretical single-particle energies here are only the eigenvalues of the Dirac equation for nucleon, and the particle-vibration coupling effects which would increase the single-particle level density are neglected,[33] see the following chapters. In addition, spurious shells are found at $Z = 58$ (^{132}Sn) and $Z = 92$ (^{208}Pb) for all the density functionals here. The problem of these spurious shells could be improved by the inclusion of ρ-tensor couplings.[95]

6.2. *Deformed nuclei*

6.2.1. *Axially deformed nuclei*

Most nuclei in the nuclide chart are deformed, and it turns out that the quadrupole deformation is the most common deformation for nuclear system. In the case of quadrupole deformation with axial symmetry, the parity and the third component j_z of the total angular momentum is conserved, which define the quantum number π_k and K_k that label the nucleon Dirac spinor

$$\psi_k(\boldsymbol{r}, s, t) = \begin{pmatrix} f_k^+(r_\perp, z)e^{i\Lambda_- \phi} \\ f_k^-(r_\perp, z)e^{i\Lambda_+ \phi} \\ ig_k^+(r_\perp, z)e^{i\Lambda_- \phi} \\ ig_k^-(r_\perp, z)e^{i\Lambda_+ \phi} \end{pmatrix} \chi_{t_k}(t), \tag{110}$$

where $\Lambda_\pm = K_k \pm 1/2$, and $(r_\perp, z, \phi)$ are the standard cylindrical coordinates. It is convenient to expand the nucleon wave functions in a basis of eigenfunctions of a single-particle Hamiltonian for an axially deformed harmonic oscillator potential,[9,58]

$$V_{\rm osc}(z, r_\perp) = \frac{1}{2}m\omega_z^2 z^2 + \frac{1}{2}m\omega_\perp^2 r_\perp^2. \tag{111}$$

Imposing volume conservation, the two oscillator frequencies $\hbar\omega_\perp$ and $\hbar\omega_z$ can be expressed in terms of a deformation parameter β_0,

$$\hbar\omega_z = \hbar\omega_0 e^{-\sqrt{\frac{5}{4\pi}}\beta_0}, \quad \hbar\omega_\perp = \hbar\omega_0 e^{\frac{1}{2}\sqrt{\frac{5}{4\pi}}\beta_0}. \tag{112}$$

The corresponding oscillator length parameters are

$$b_z = \sqrt{\frac{\hbar}{m\omega_z}}, \quad b_\perp = \sqrt{\frac{\hbar}{m\omega_\perp}}, \tag{113}$$

and $b_\perp^2 b_z = b_0^3$ because of volume conservation. The basis is now determined by the two constants $\hbar\omega_0$ and β_0. The eigenfunctions of the deformed harmonic oscillator potential are labeled by the set of quantum numbers

$$|\alpha\rangle = |n_z n_r \Lambda m_s\rangle, \tag{114}$$

where n_z and n_r are the number of nodes in the z- and $r_\perp$-directions, respectively. Λ and m_s are the components of the orbital angular momentum and the spin on the intrinsic z-axis, respectively. The eigenvalue of j_z which is the z projection of the total single-nucleon angular momentum and the parity are determined by

$$K = \Lambda + m_s, \quad \pi = (-1)^{n_z+\Lambda}. \tag{115}$$

By introducing two dimensionless variables $\xi = z/b_z$ and $\eta = r_\perp^2/b_\perp^2$, the harmonic oscillator eigenfunctions read

$$\Phi_\alpha(\boldsymbol{r}, s) = \varphi_{n_z}(z, b_z)\varphi_{n_r}^\Lambda(r_\perp, b_\perp)\frac{e^{i\Lambda\phi}}{\sqrt{2\pi}}\chi_{m_s}(s), \tag{116}$$

with

$$\varphi_{n_z}(z, b_z) = \frac{N_{n_z}}{\sqrt{b_z}}H_{n_z}(\xi)e^{-\xi^2/2}, \tag{117}$$

$$\varphi_{n_r}^\Lambda(r_\perp, b_\perp) = \frac{N_{n_r}^\Lambda}{b_\perp}\sqrt{2}\eta^{|\Lambda|/2}L_{n_r}^{|\Lambda|}(\eta)e^{-\eta/2}, \tag{118}$$

where $H_{n_z}(\xi)$ and $L_{n_r}^{|\Lambda|}(\eta)$ denote the Hermite and associated Laguerre polynomials, respectively. The normalization factors are

$$N_{n_z} = \frac{1}{\sqrt{(\sqrt{\pi}2^{n_z}n_z!)}}, \qquad N_{n_r}^\Lambda = \sqrt{\frac{n_r!}{(n_r + |\Lambda|)!}}. \tag{119}$$

The large and small components of a Dirac spinor are expanded independently in terms of the oscillator eigenfunctions

$$f_k(\boldsymbol{r}, s, t) = \frac{1}{\sqrt{2\pi}}\begin{pmatrix} f_k^+(r_\perp, z)e^{i\Lambda_-\phi} \\ f_k^-(r_\perp, z)e^{i\Lambda_+\phi}\end{pmatrix} = \sum_\alpha^{\alpha_{max}} f_\alpha^{(i)}\Phi_\alpha(\boldsymbol{r}, s)\chi_{t_k}(t), \tag{120}$$

$$g_k(\boldsymbol{r}, s, t) = \frac{1}{\sqrt{2\pi}}\begin{pmatrix} g_k^+(r_\perp, z)e^{i\Lambda_-\phi} \\ g_k^-(r_\perp, z)e^{i\Lambda_+\phi}\end{pmatrix} = \sum_{\tilde\alpha}^{\tilde\alpha_{max}} g_{\tilde\alpha}^{(i)}\Phi_{\tilde\alpha}(\boldsymbol{r}, s)\chi_{t_k}(t), \tag{121}$$

To avoid the appearance of spurious states, the quantum numbers α_{max} and $\tilde\alpha_{max}$ are chosen in such a way that the corresponding major quantum numbers $N = 2n_r + |\Lambda| + n_z$ are not larger than N_{max} and $N_{max} + 1$ for the large and small components, respectively.

By means of the expansions in Eqs. (120) and (121), the matrix elements of the Dirac Hamiltonian are given by

$$\begin{aligned}\mathcal{A}_{\alpha\alpha'} &= \delta_{\Lambda\Lambda'}\delta_{m_s m_s'}\int_{-\infty}^{\infty} dz\varphi_{n_z}(z)\varphi_{n_z'}(z)\int_0^{\infty} dr_\perp r_\perp \varphi_{n_r}^{\Lambda}(r_\perp)\varphi_{n_r'}^{\Lambda}(r_\perp)\\ &\quad\times[M+S(r_\perp,z)+V(r_\perp,z)], \qquad (122)\\ \mathcal{C}_{\tilde\alpha\tilde\alpha'} &= \delta_{\tilde\Lambda\tilde\Lambda'}\delta_{\tilde m_s\tilde m_s'}\int_{-\infty}^{\infty} dz\varphi_{\tilde n_z}(z)\varphi_{\tilde n_z'}(z)\int_0^{\infty} dr_\perp r_\perp \varphi_{\tilde n_r}^{\tilde\Lambda}(r_\perp)\varphi_{\tilde n_r'}^{\tilde\Lambda}(r_\perp)\\ &\quad\times[-M-S(r_\perp,z)+V(r_\perp,z)], \qquad (123)\\ \mathcal{B}_{\alpha\tilde\alpha} &= \delta_{\Lambda\tilde\Lambda}\delta_{m_s\tilde m_s}\delta_{n_r\tilde n_r}(-1)^{1/2-m_s}\frac{1}{b_z}\left(\delta_{n_z\tilde n_z-1}\sqrt{\frac{\tilde n_z}{2}}-\delta_{n_z\tilde n_z+1}\sqrt{\frac{n_z}{2}}\right)\\ &\quad+\delta_{m_s\tilde m_s-1}\delta_{n_z\tilde n_z}\\ &\quad\times\int_0^{\infty} dr_\perp r_\perp\varphi_{n_r}^{\Lambda}(r_\perp)\partial_{r_\perp}\varphi_{\tilde n_r}^{\tilde\Lambda}(r_\perp)-\tilde\Lambda\int_0^{\infty}\varphi_{n_r}^{\Lambda}(r_\perp)\varphi_{\tilde n_r}^{\tilde\Lambda}(r_\perp)\\ &\quad+\delta_{m_s\tilde m_s+1}\delta_{n_z\tilde n_z}\\ &\quad\times\int_0^{\infty} dr_\perp r_\perp\varphi_{n_r}^{\Lambda}(r_\perp)\partial_{r_\perp}\varphi_{\tilde n_r}^{\tilde\Lambda}(r_\perp)+\tilde\Lambda\int_0^{\infty}\varphi_{n_r}^{\Lambda}(r_\perp)\varphi_{\tilde n_r}^{\tilde\Lambda}(r_\perp). \qquad (124)\end{aligned}$$

Similar to the spherical case, the Coulomb potential is given by

$$V_C(\boldsymbol{r}) = \frac{e^2}{8\pi}\int d^3\boldsymbol{r}'|\boldsymbol{r}-\boldsymbol{r}'|\triangle_{\boldsymbol{r}'}\rho_p(\boldsymbol{r}'). \qquad (125)$$

After integrating over the azimuthal angle ϕ, one obtains the following expression

$$V_C(r_\perp,z) = \frac{e^2}{2\pi}\int_0^{\infty} dr_\perp' r_\perp'\int_{-\infty}^{\infty} dz' d(r_\perp,z)E\left(\frac{4r_\perp r_\perp'}{d(r_\perp,z)}\right)\triangle\rho_p(r_\perp',z') \qquad (126)$$

with $d(r_\perp,z)=\sqrt{(z-z')^2+(r_\perp+r_\perp')^2}$. The complete elliptic integral of the second kind is approximated by the standard polynomial formula.[96]

For the meson-exchange model, one also need to solve the Klein–Gordon equations for different mesons. In the case of axial symmetry, the Klein–Gordon equations for the meson fields $\phi=\sigma,\omega,\rho$ read

$$\left(-\frac{\partial^2}{\partial r_\perp^2}-\frac{1}{r_\perp}\frac{\partial}{\partial r_\perp}-\frac{\partial^2}{\partial z^2}+m_\phi^2\right)\phi(r_\perp,z) = S_\phi(r_\perp,z). \qquad (127)$$

One can expand the solution of this equation in terms of a complete set of basis states

$$\phi(r_\perp,z) = \sum_{n_z n_r}^{n_b}\phi_{n_z n_r}\varphi_{n_z}(z,b_z)\varphi_{n_r}^0(r_\perp,b_\perp). \qquad (128)$$

It is convenient to use the same deformation parameter β_0 and oscillator frequency $\hbar\omega_0$ as for the nucleon wave functions. Inserting the expansion into the Klein–Gordon equation, one can rewrite the Klein–Gordon equation into a set of inhomogeneous linear equations,

$$\sum_{n'_z n'_r}^{n_b} \mathcal{H}_{n_z n_r n'_z n'_r} \phi_{n'_z n'_r} = s^{\phi}_{n_z n_r}, \tag{129}$$

with the matrix elements

$$\begin{aligned}\mathcal{H}_{n_z n_r n'_z n'_r} &= \delta_{n_r n'_r}\delta_{n_z n'_z}\left[b_z^{-2}(n_z+1/2) + b_\perp^{-2}(2n_r+1/2) + m_\phi^2\right]\\ &\quad -\delta_{n_r n'_r}\frac{1}{2}b_z^{-2}\left(\sqrt{(n_z+1)n'_z}\delta_{n_z n'_z-2} + \sqrt{n_z(n'_z+1)}\delta_{n_z n'_z+2}\right)\\ &\quad +\delta_{n_z n'_z} b_\perp^{-2}(n'_r\delta_{n_r n'_r-1} + n_r\delta_{n_r n'_r+1}).\end{aligned} \tag{130}$$

By inverting the matrix $\mathcal{H}$, this set of equations can be easily solved.

6.2.2. *Triaxially deformed nuclei*

The search for conclusive evidence for the existence of stable triaxial nuclear shapes has been a central theme of recent studies in nuclear structure. In particular, nuclear wobbling[97] and chirality[98–100] are considered to be direct and natural consequences of the deviation from axial symmetry in the nuclear mean field. Although triaxiality itself is not directly measurable, it is an essential prerequisite for the manifestation of chirality and wobbling and, thus, the exact characterization of these modes provides robust evidence for stable triaxial nuclear shapes.

For triaxially deformed nuclei, the single-particle Dirac spinors are expanded in a basis of eigenfunctions of a three-dimensional harmonic oscillator in Cartesian coordinates:[101,102]

$$V(x,y,z) = \frac{1}{2}m\omega_x^2 x^2 + \frac{1}{2}m\omega_y^2 y^2 + \frac{1}{2}m\omega_z^2 z^2. \tag{131}$$

According to the condition of volume conservation, the three oscillator frequencies can be expressed in terms of the basis deformation parameters β_0 and γ_0,

$$\hbar\omega_x = \hbar\omega_0 e^{-\sqrt{\frac{5}{4\pi}}\beta_0\cos(\gamma_0-2\pi/3)}, \tag{132}$$

$$\hbar\omega_y = \hbar\omega_0 e^{-\sqrt{\frac{5}{4\pi}}\beta_0\cos(\gamma_0+2\pi/3)}, \tag{133}$$

$$\hbar\omega_z = \hbar\omega_0 e^{-\sqrt{\frac{5}{4\pi}}\beta_0\cos\gamma_0}. \tag{134}$$

The corresponding oscillator lengths are $b_k = \sqrt{\hbar/m\omega_k}, k = x,y,z$, and because of volume conservation, $b_x b_y b_z = b_0^3$. The basis is completely determined by the three constants $\hbar\omega_0$, β_0, and γ_0. The basis states are a combination of the spatial part and spin part of the wave functions. The spin part is chosen in such a way that the basis state is an eigenfunction of the x-simplex operator $\hat{S}_x = \hat{P}e^{-i\pi J_x}$, where $\hat{P}$ denotes the parity operator. Therefore, the basis states with positive and negative x-simplex read

$$\begin{cases} \Phi_{\underline{\alpha}}(\boldsymbol{r}, s) & = & \varphi_{n_x}(x)\varphi_{n_y}(y)\varphi_{n_z}(z)\dfrac{i^{n_y}}{\sqrt{2}}\begin{pmatrix} 1 \\ (-1)^{n_x+1} \end{pmatrix} \\ & & \text{with } \underline{\alpha} = |n_x, n_y, n_z, +i\rangle, \\ \Phi_{\bar{\underline{\alpha}}}(\boldsymbol{r}, s) & = & \varphi_{\tilde{n}_x}(x)\varphi_{\tilde{n}_y}(y)\varphi_{\tilde{n}_z}(z)\dfrac{i^{\tilde{n}_y}}{\sqrt{2}}(-1)^{\tilde{n}_x+\tilde{n}_y+1}\begin{pmatrix} 1 \\ (-1)^{\tilde{n}_x} \end{pmatrix} \\ & & \text{with } \bar{\underline{\alpha}} = |\tilde{n}_x, \tilde{n}_y, \tilde{n}_z, -i\rangle, \end{cases} \tag{135}$$

where $\varphi_{n_k}(x_k)$ are the one-dimensional harmonic oscillator wave functions in k direction ($x_k = x, y, z$),

$$\varphi_{n_k}(x_k) = \frac{N_{n_k}}{\sqrt{b_k}} H_{n_k}\left(\frac{x_k}{b_k}\right) \exp\left[-\frac{1}{2}\left(\frac{x_k}{b_k}\right)^2\right] \tag{136}$$

with $N_n = (\sqrt{\pi}2^n n!)^{-1/2}$ being the normalization factor, and $H_n(\xi) = (-1)^n e^{\xi^2} \frac{d^n}{d\xi^n} e^{-\xi^2}$ is the Hermite polynomial $H_n(\xi)$ is defined in Ref. 96.

For a Dirac spinor with a positive simplex eigenvalue, the large component corresponds to positive eigenvalues and the small component to negative eigenvalues. The large and small components are expanded in terms of the basis states

$$\begin{cases} f_k(\boldsymbol{r}, \pm) = \displaystyle\sum_{\alpha}^{\alpha_{max}} f_k^{\alpha} \Phi_{\alpha}(\boldsymbol{r}, \pm), \\ g_k(\boldsymbol{r}, \pm) = \displaystyle\sum_{\tilde{\alpha}}^{\tilde{\alpha}_{max}} g_k^{\tilde{\alpha}} \Phi_{\tilde{\alpha}}(\boldsymbol{r}, \pm). \end{cases} \tag{137}$$

Note that usually the parity is also conserved, thus allowing a further reduction of the basis to four simplex-parity blocks. To avoid the appearance of spurious states, the quantum numbers α_{max} and $\tilde{\alpha}_{max}$ are chosen in such a way that the corresponding major quantum number $N = n_x + n_y + n_z$ does not exceed N_{max} and $\tilde{N}_{max} = N_{max} + 1$ for large and small components, respectively.

In addition to x-simplex and parity, one assumes that the densities and fields are reflection-symmetric with respect to the yz, xz, and xy planes

$$V(-x, y, z) = V(x, y, z), \tag{138}$$

$$V(x, -y, z) = V(x, y, z), \tag{139}$$

$$V(x, y, -z) = V(x, y, z). \tag{140}$$

The matrix elements of the Dirac Hamiltonian in Cartesian coordinates are thus given by

$$\begin{aligned}
\mathcal{A}_{\alpha\alpha'} &= \delta_{(n_x+n'_x),\text{even}}\delta_{(n_y+n'_y),\text{even}}\delta_{(n_z+n'_z),\text{even}}(-1)^{\frac{n_y+n'_y}{2}} \\
&\quad\times \int dxdydz\varphi_{n_x}\varphi_{n_y}\varphi_{n_z}(M+S+V)\varphi_{n'_x}\varphi_{n'_y}\varphi_{n'_z}, \qquad (141)\\
\mathcal{C}_{\tilde{\alpha}\tilde{\alpha}'} &= \delta_{(\tilde{n}_x+\tilde{n}'_x),\text{even}}\delta_{(\tilde{n}_y+\tilde{n}'_y),\text{even}}\delta_{(\tilde{n}_z+\tilde{n}'_z),\text{even}}(-1)^{\frac{\tilde{n}_y+\tilde{n}'_y}{2}} \\
&\quad\times \int dxdydz\varphi_{\tilde{n}_x}\varphi_{\tilde{n}_y}\varphi_{\tilde{n}_z}(-M-S+V)\varphi_{\tilde{n}'_x}\varphi_{\tilde{n}'_y}\varphi_{\tilde{n}'_z}, \qquad (142)\\
\mathcal{B}_{\alpha\tilde{\alpha}} &= \frac{1}{\sqrt{2}b_x}\delta_{n_y\tilde{n}_y}\delta_{n_z\tilde{n}_z}(-1)^{n_y}\left(\sqrt{n_x}\delta_{n_x\tilde{n}_x+1} - \sqrt{\tilde{n}_x}\delta_{n_x\tilde{n}_x-1}\right) \\
&\quad+\frac{1}{\sqrt{2}b_y}\delta_{n_x\tilde{n}_x}\delta_{n_z\tilde{n}_z}(-1)^{n_y}\left(\sqrt{n_y}\delta_{n_y\tilde{n}_y+1} + \sqrt{\tilde{n}_y}\delta_{n_y\tilde{n}_y-1}\right) \\
&\quad+\frac{1}{\sqrt{2}b_z}\delta_{n_x\tilde{n}_x}\delta_{n_y\tilde{n}_y}(-1)^{n_x+n_y}\left(\sqrt{n_z}\delta_{n_z\tilde{n}_z+1} - \sqrt{\tilde{n}_z}\delta_{n_z\tilde{n}_z-1}\right).
\end{aligned}$$

$$(143)$$

Similar to the spherical and axial case, the direct Coulomb potential can be obtained with a direct integration of the Poisson equation. It should be noted that in the triaxial case, the number of mesh points on the three dimensional grid can easily become prohibitively large. In this case the Poisson equation can be solved using the conjugate gradient method,[103] or by replacing the Coulomb interaction with an integral over Gaussian interactions,[104,105] or using the method based on the Coulomb Green function as in Refs. 106, 58.

For the meson-exchange model, one also need to solve the Klein–Gordon equations for different mesons. In the triaxial case, the Klein–Gordon equations for the meson fields $\phi = \sigma, \omega, \rho$ read

$$\left(-\frac{\partial^2}{\partial x^2} - \frac{\partial^2}{\partial y^2} - \frac{\partial^2}{\partial z^2} + m_\phi^2\right)\phi(x,y,z) = S_\phi(x,y,z). \qquad (144)$$

One can expand the solution of this equation in terms of a complete set of basis states

$$\phi(x,y,z) = \sum_{n_x n_y n_z}^{n_b} \phi_{n_x n_y n_z}\varphi_{n_x}\varphi_{n_y}\varphi_{n_z}. \qquad (145)$$

Inserting the expansion into the Klein–Gordon equation, one can rewrite the Klein–Gordon equation into a set of inhomogeneous linear equations,

$$\sum_{n'_x n'_y n'_z}^{n_b} \mathcal{H}_{n_x n_y n_z n'_x n'_y n'_z} \phi_{n'_x n'_y n'_z} = s^{\phi}_{n_x n_y n_z}, \tag{146}$$

with the matrix elements

$$\begin{aligned}
&\mathcal{H}_{n_x n_y n_z n'_x n'_y n'_z} \\
&= -\delta_{n_y n'_y}\delta_{n_z n'_z}\frac{1}{b_x^2}\left(\sqrt{(n_x+1)n'_x\delta_{n_x n'_x-2}} + \sqrt{(n'_x+1)n_x\delta_{n'_x n_x-2}}\right) \\
&\quad -\delta_{n_x n'_x}\delta_{n_z n'_z}\frac{1}{b_y^2}\left(\sqrt{(n_y+1)n'_y\delta_{n_y n'_y-2}} + \sqrt{(n'_y+1)n_y\delta_{n'_y n_y-2}}\right) \\
&\quad -\delta_{n_x n'_x}\delta_{n_y n'_y}\frac{1}{b_z^2}\left(\sqrt{(n_z+1)n'_z\delta_{n_z n'_z-2}} + \sqrt{(n'_z+1)n_z\delta_{n'_z n_z-2}}\right) \\
&\quad +\delta_{n_x n'_x}\delta_{n_y n'_y}\delta_{n_z n'_z}\left(\frac{n_x+1/2}{b_x^2} + \frac{n_y+1/2}{b_y^2} + \frac{n_z+1/2}{b_z^2} + m_\phi^2\right).
\end{aligned} \tag{147}$$

It is worthwhile to mention that in Refs. 107, 108, the authors developed the so-called multidimensionally-constrained covariant density functional theory by expanding the Dirac equation with a set of axially deformed harmonic oscillator basis. In this multidimensional CDFT, the quadrupole deformation with triaxiality and the octupole deformation can be considered simultaneously.

6.2.3. *Binding energy and deformation*

By taking the well deformed Yb and U isotopes as examples, the ground-state binding energies and quadrupole deformations were calculated in covariant density functional theory with the axial and reflection symmetry imposed. In the upper panels of Fig. 4, the deviations of the calculated binding energies with the energy functionals PC-PK1,[31] DD-PC1,[30] and PC-F1[29] from the data[85] are shown as circles, triangles, and squares respectively.

It shows that the binding energies are systematically underestimated by around 3 MeV for both Yb and U isotopes for PC-PK1, while for the PC-F1 results, the differences between the calculated results and the data exhibit very strong isospin dependence. The predicted binding energies by DD-PC1 are in very good agreement with the data (within 1 MeV) since almost all the nuclei shown here are used to adjust the DD-PC1 functional.

However, for deformed nuclei, since the rotational symmetry of the ground state is broken, one needs to take into account the energy cor-

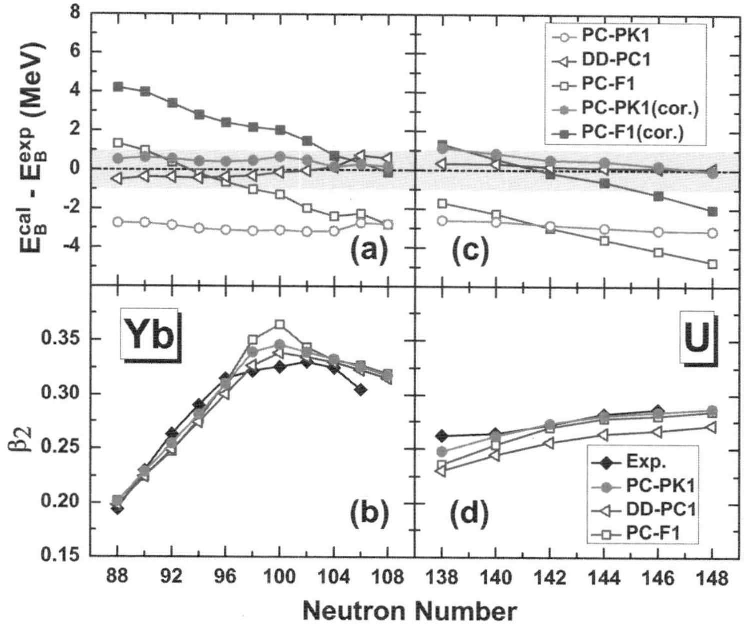

Fig. 4. (Color online) Deviations of the calculated binding energies with the covariant density functional PC-PK1,[31] DD-PC1,[30] and PC-F1[29] for Yb and U isotopes from the data[85] (upper panels). Axial and reflection symmetry is imposed in the calculations, and the corresponding calculated ground-state deformations are in comparison with the data[109] (lower panels). The filled circles and squares in the upper panels correspond to the rotational corrected ones given by PC-PK1 and PC-F1 respectively. Taken from Ref. 31.

rection resulting from the restoration of the rotational symmetry. After including the rotation correction energy, although the differences between the corrected binding energies given by PC-F1 and data are still large, the predicted energies given by PC-PK1 reproduces the data quite well for both Yb and U isotopes, and the deviations are within 1 MeV. Note that the rotational correction energy is not considered in the DD-PC1 calculations, since the functional DD-PC1 is adjusted to the binding energies of 64 well-deformed nuclei, and we presume the rotational correction effects have been effectively included.

Generally, the calculated quadrupole deformation is related to the quadrupole moments Q_{20} and Q_{22}, which are given by

$$Q_{20} = \sqrt{\frac{5}{16\pi}}\langle 2z^2 - x^2 - y^2\rangle, \tag{148}$$

$$Q_{22} = \sqrt{\frac{5}{32\pi}}\langle x^2 - y^2\rangle. \tag{149}$$

The relations which express the quadrupole moments in terms of deformation parameters a_{20} and a_{22}

$$Q_{20} = \frac{3A}{4\pi}R_0^2 a_{20}, \quad Q_{22} = \frac{3A}{4\pi}R_0^2 a_{22}, \tag{150}$$

determine the deformations parameters β and γ

$$\beta = \sqrt{a_{20}^2 + 2a_{22}^2}, \quad \gamma = \arctan\left(\sqrt{2}\frac{a_{22}}{a_{20}}\right), \tag{151}$$

with $R_0 = 1.2A^{1/3}$ (fm). For the case of axial symmetry, the deformation parameters β is determined by the quadrupole moment Q_{20}:

$$Q_{20} = \sqrt{\frac{9}{5\pi}}AR_0^2\beta. \tag{152}$$

Positive β values correspond to prolate shapes and negative values to the oblate shapes.

In the lower panels of Fig. 4, the calculated quadrupole deformations for the ground states by PC-PK1, DD-PC1, and PC-F1 are given in comparison with the corresponding data.[109] It shows that the deformations and their corresponding evolutions with neutron number for both Yb and U isotopes are reproduced quite well.

6.2.4. *Fission barrier*

Nuclear fission is one of the most important topics in the field of nuclear physics. Since its first interpretation with the barrier penetration, it has been a difficult task to describe this phenomenon theoretically. Accurate information about the fission barrier is very crucial to study the fission problem. For example, a 1-MeV difference in the fission barrier could result in several orders of magnitude difference in the fission half-life. Moreover, the accurate predictions of fission barriers for superheavy nuclei are also very important to explore the island of stability of the superheavy nuclei. For details, readers are referred to Chapter 5.

Here, by taking the actinide nucleus ^{240}Pu as an example, the potential energy curves from an oblate shape with β_{20} about -0.2 to the fission configuration with β_{20} around 2.0 are shown in Fig. 5. As discussed in Ref. 110, these potential energy curves are obtained from the CDFT calculations with different self-consistent symmetries imposed: the axial or triaxial symmetries combined with reflection-symmetric or -asymmetric cases. It is clearly

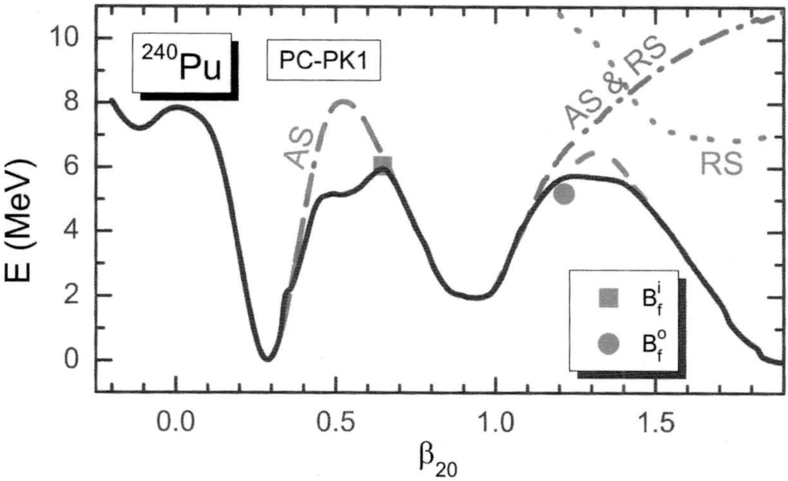

Fig. 5. (Color online) Calculated potential energy curves of ^{240}Pu with the covariant density functional PC-PK1 under various symmetries imposed. The solid black curve represents the results with V_4 symmetry imposed, the red dashed curve represents that with only axial symmetry (AS) imposed, the green dotted curve represents that with only reflection symmetry (RS) imposed, the violet dotted-dashed line represents that with both AS and RS imposed. The empirical inner (outer) barrier height is denoted by the gray square (circle). The energy is normalized with respect to the binding energy of the ground state. Taken from Ref. 107.

seen that the triaxial deformation reduces the inner barrier height by more than 2 MeV and results in a better agreement with the empirical datum, and the reflection-asymmetric shape is favored beyond the fission isomer and lowers very much the outer fission barrier. The importance of the triaxial deformation and the octupole deformation discussed here has also been stressed in other investigations, e.g., Refs. 104, 111, 112. Furthermore, since the authors broke the axial and reflection symmetries simultaneously in Ref. 107, it was also found that the outer barrier is also considerably lowered by about 1 MeV when the triaxial deformation is allowed, and thus achieves a better reproduction of the empirical barrier height.

7. Nuclear landscape

To date, the nuclear masses of approximately 3000 nuclei are known from experiment.[113] This number is much smaller than the number expected in most theoretical models, although different models and different parametrizations show rather large variations in predictions of the neutron drip line. From experiments, the proton drip line has been delineated up to protactinium ($Z = 91$), while the position of the neutron drip line beyond $Z = 8$ is still undetermined. Moreover, due to the experimental limitations, it will be only possible to define the location of the neutron drip line for the majority of elements in theoretical calculations in the foreseeable future.

The theoretical description of ground state properties of nuclei is not only important for our understanding of their structure, but also important for nuclear astrophysics, where we are facing the problem of an extrapolation to the very neutron-rich nuclei. In the current situation that many of such nuclei will not be studied experimentally even with the next generation of facilities, it is important to answer two questions from the theoretical side. First, how well do the existing nuclear theoretical models describe the experimental data available? Second, how well can they be extrapolated to the region of unknown nuclei?

The theoretical determination of nuclear masses can be traced to the von Weizsäcker mass formula which was proposed based on the famous liquid drop model (LDM).[114] Since then, a tremendous effort has been made in pursuing different possible extensions of the LDM which are known as the macroscopic-microscopic models, e.g., the finite-range droplet model (FRDM),[115] the extended Thomas–Fermi plus Strutinsky integral with shell-quenching (ETFSI-Q) model,[116] and the Weizsäcker–Skyrme (WS) model.[117] It turns out that these macroscopic-microscopic models work pretty well in the description of known nuclei, but their predictions show a large deviation for very neutron-rich nuclei. In addition, the local mass relations such as the isobaric multiplet mass equation (IMME),[118] the Garvey–Kelson (GK) relations,[119] and the residual proton-neutron interactions[120] are also used to give predictions of unmeasured masses.

In principle, an ideal mass formula would be the one in which the masses of all nuclei are derived from the basic nucleonic interactions. In this regard, density functional theories (DFT) are extremely useful for the microscopic description of nuclear systems. They have been applied with great success for many years in Coulombic systems,[1] where the functional can be derived directly from the Coulomb interaction. For nuclear system, the situation is much more complicated. The bare nuclear force is usually adjusted to scattering data. This requires additional assumptions and is connected with additional uncertainties such as the off-shell behavior. As compared to the Coulomb force, the two-body part of the nuclear force is extremely strong at short distances and has a relatively short range. There are convincing indications that it contains, on the non-relativistic level, an important three-body part.

Despite all these restrictions, in the past decades, both non-relativistic and relativistic (covariant) DFT have been developed and successfully applied to the description of a variety of nuclear phenomena with great success.[69,121,122] In particular, a series of investigations for nuclear landscape have been performed based on the non-relativistic density functional theory, see e.g., Refs. 123–125.

Apart from the non-relativistic DFT, the CDFT has also been applied for the investigation of nuclear landscape. The first relativistic mass table was reported in Ref. 126 for about 2000 even-even nuclei with $8 \leq Z \leq 120$ up to the proton and neutron drip lines. However, the pairing correlations are not considered in these calculations. Later on, by including the pairing correlations with BCS method, the ground-state properties of 1315 even-even nuclei with $10 \leq Z \leq 98$ were calculated in Ref. 127. In 2005, the first systematic study of the ground-state properties of over 7000 nuclei ranging from the proton drip line to the neutron drip line was performed.[128] The obtained nuclear masses have also been used in the rapid neutron capture (r-) process nucleosynthesis calculations.[129,130] In Ref. 131, the rotational correction energies were considered though with a simple formula, and the binding energies even-even nuclei are calculated in CDFT with the pairing correlation being treated with the constant gap approximation.

Very recently, within the framework of relativistic Hartree–Bogoliubov model, in Ref. 132, the nuclear ground-properties have been calculated with the covariant density functionals NL3*,[84] DD-ME2,[24] DD-MEδ,[133] and DD-PC1,[30] and a detailed analysis of two-neutron drip lines in covariant and non-relativistic DFT has also been performed in Ref. 134. In addition, in Ref. 135, the binding energies for even-even nuclei with proton numbers ranging from $Z = 8$ to $Z = 108$ were calculated with the covariant density functional PC-PK1, and the dynamical correlation energies include the rotational correction energies obtained with the cranking approximation and the quadrupole vibrational correction energies were investigated.

Thus, so far there have been already many systematic CDFT calculations for the binding energies of nuclei which are experimentally available. Although the set of nuclei considered in different works are slightly different from each other, in comparison with the experimental data, the root-mean-square (rms) errors for the nuclear masses found for most of the present CDFT calculations are above 2 MeV. For example, in Fig. 6, the energy differences between the calculated binding energies for even-even nuclei with

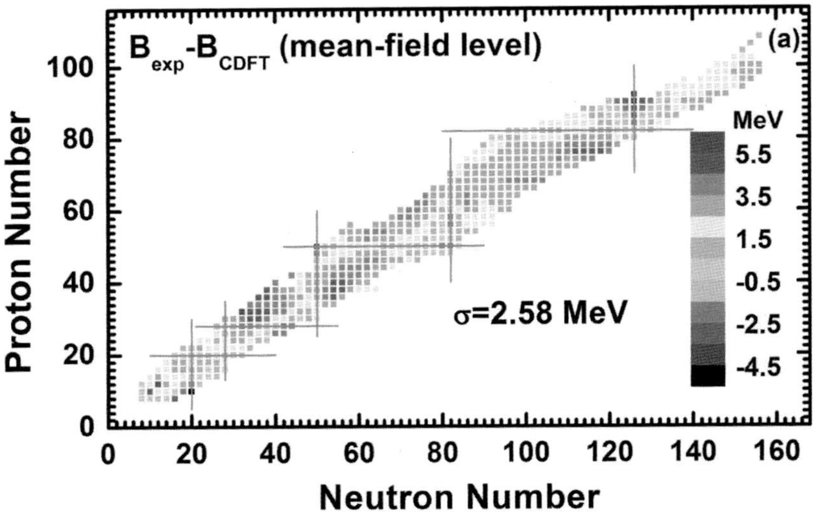

Fig. 6. (Color online) Energy differences between the calculated binding energies for even-even nuclei given by CDFT with PC-PK1 and the data available.[85] Taken from Ref. 135.

PC-PK1 functional and the data available[85] are shown. It should be noted that this is achieved with a relatively small number of model parameters fitted to a rather modest set of data on several tens of finite nuclei. Actually, in contrast to the non-relativistic DFT's where mass tables based on an extensive use of experimental data were generated in the Skyrme[123] and the Gogny[124] functionals, no similar attempt, up to date, has been carried out in CDFT by fitting to the full set of available experimental masses. However, one should be kept in mind that the number of free parameters in such fits could be considerably increased by including some more or less phenomenological terms e.g., the Wigner energy for $N \simeq Z$ nuclei.

One should also note that the beyond mean-field effects are usually very important for the description of nuclear masses. This is in particular clearly visible by comparing Figs. 6 and 7, the calculated binding energies with or without the rotational correction energies for even-even nuclei are compared with the data available.[85] The rotational correction energies were evaluated with the cranking approximation, and the details were discussed in Ref. 135. One can clearly see that in Fig. 6, the CDFT calculation with PC-PK1 systematically underestimate the binding energies for most nuclei by about 3 MeV. However, by including the rotational correction energies,

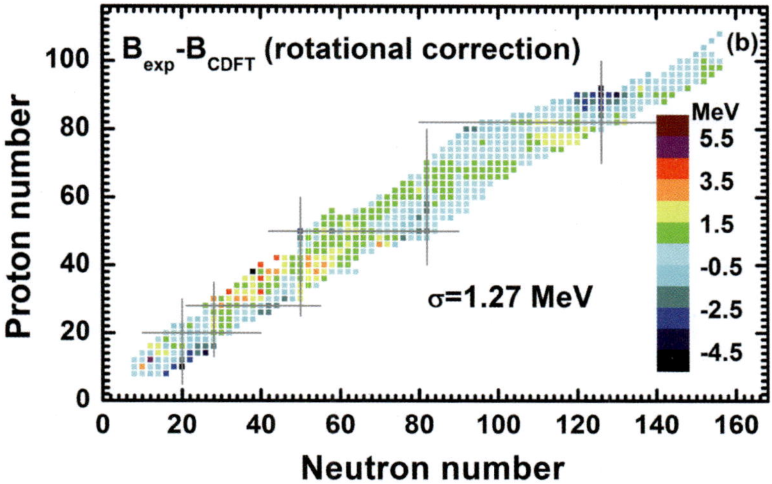

Fig. 7. Energy differences between the calculated binding energies including the rotational corrections for even-even nuclei given by CDFT with PC-PK1 and the data available.[85] Taken from Ref. 135.

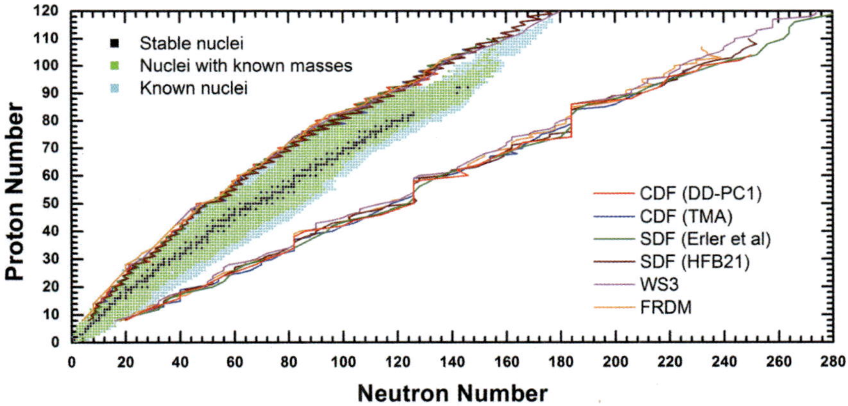

Fig. 8. Nuclear landscape as a function of the proton number Z and the neutron number N. The nuclei known experimentally are shown as black squares for stable nuclei, cyan squares for the observed nuclei,[143] and green squares for the nuclei with masses measured.[113] The proton and neutron drip lines are shown as solid lines, and they are obtained from calculations with the covariant density functionals,[128,132] the Skyrme density functionals,[125,144] and the macroscopic-microscopic mass models WS3[145] and FRDM.[115]

the rms error for the nuclear masses is significantly reduced from 2.58 MeV to 1.27 MeV. Similar results have also been reported in Ref. 136, where the masses of all the nuclei including even-even, odd-A, and odd-odd nuclei are compared with the data, and the rms error is found to be around 1.4 MeV.

One can see that the deviations are especially pronounced in the region of light nuclei, where other effects including the configuration mixing effects,[137–139] the clusterization effects,[140] and the Wigner term[141,142] for the $N \simeq Z$ nuclei are important and are not taken into account in the density functionals.

Next, we turn to discuss the nucleon drip lines which would answer the question how many protons and neutrons can be bound in a nucleus. In Fig. 8, the nuclear landscape is shown as a functional of the proton and neutron number. There are only around 300 of the several thousand nuclides that are either stable or practically stable, which forms the so-called valley of stability. Away from this valley, the nuclei tend to short-lived and radioactive. Nuclear existence ends at the drip lines, where there is no longer enough binding energy to prevent the last nucleons from escaping the nucleus. As indicated in Fig. 8, the proton drip line has been experimentally delineated up to proton number $Z = 91$, and the proton drip lines given by different models are very close to each other. However, for the neutron drip lines, there are still a lot of nuclei unobserved in the neutron-rich side, and the predicted neutron drip lines given by different models are also quite different, in particular in the superheavy region.

Concerning these discrepancies, firstly, the isovector properties of energy density functionals are poorly defined so far. In particular, in the light of the model limitations and the relatively narrow isospin range measured in experiment, it still remains an open question whether the isovector properties of energy density functionals can be accurately defined from masses alone. Secondly, the inevitable inaccuracies in the DFT description of the single-particle energies leads to different underlying shell structures. For better single-particle properties, one may need to improve the density functionals with additional terms, such as tensor terms which cannot be firmly constrained by only nuclear-matter properties and the nuclear masses and radii. So far, no clear correlations have been found between the location of the neutron drip line and the nuclear-matter properties.

It is worthwhile to mention that recently the continuum effects on the nuclear landscape have been investigated in Ref. 146 by using the relativistic continuum Hartree–Bogoliubov (RCHB) theory, which provides an appropriate treatment of pairing correlations in the presence of the continuum.[122] It has been known that the continuum effects play crucial roles in the description of nuclear halo phenomena.[62,147] In the investigation for nuclei from $Z = 8$ to $Z = 22$,[146] it was found that the continuum couplings could extend the neutron drip line towards the neutron-rich site. At present, such investigation has been carried out only in this light nuclei region and the spherical symmetry was imposed in the calculations. It will definitely be interesting to extend such calculations to the heavier nuclei mass region and also to the deformed case.

Acknowledgment

This work was supported in part by the National Key Basic Research Program of China (Grant No. 2013CB834400), the Natural Science Foundation of China (Grants No. 10975008, No. 11175002, No. 11105005, No. 11105010, No. 11035007, No. 11128510, No. 11121403, No. 11120101005, No. 11211120152, No. 11235002, No. 11275248, No. 11205004, and No. 11335002) the Research Fund for the Doctoral Program of Higher Education (Grant No. 20110001110087), the Chinese Academy of Sciences (Grant No. KJCX2-EW-N01), the DFG Cluster of Excellence "Origin and Structure of the Universe"(www.universecluster.de), and the Overseas Distinguished Professor Project from Ministry of Education (MS2010BJDX001).

References

1. W. Kohn and L. J. Sham, Self-consistent equations including exchange and correlation effects, *Phys. Rev.* **140**, A1133–A1138 (Nov, 1965). doi: 10.1103/PhysRev.140.A1133. URL `http://link.aps.org/doi/10.1103/PhysRev.140.A1133`.
2. J. D. Walecka, A theory of highly condensed matter, *Ann. Phys. (N.Y.)*. **83**(2), 491–529 (1974). ISSN 0003-4916. doi: 10.1016/0003-4916(74)90208-5. URL `http://www.sciencedirect.com/science/article/pii/0003491674902085`.
3. B. D. Serot and J. D. Walecka, The relativistic nuclear many-body problem, *Adv. Nucl. Phys.* **16**, 1–327 (1986).
4. S. A. Chin, A relativistic many-body theory of high density matter, *Ann. Phys. (N.Y.)*. **108**(2), 301–367 (1977). ISSN 0003-4916. doi: http://dx.doi.org/10.1016/0003-4916(77)90016-1. URL `http://www.sciencedirect.com/science/article/pii/0003491677900161`.
5. R. J. Furnstahl, R. J. Perry, and B. D. Serot, Two-loop corrections for nuclear matter in the Walecka model, *Phys. Rev. C.* **40**, 321–353 (Jul, 1989). doi: 10.1103/PhysRevC.40.321. URL `http://link.aps.org/doi/10.1103/PhysRevC.40.321`.

6. Z. Zhu, H. J. Mang, and P. Ring, Vacuum polarization in a relativistic description of open shell nuclei, *Phys. Lett. B.* **254**(3–4), 325–331 (1991). ISSN 0370-2693. doi: 10.1016/0370-2693(91)91163-P. URL `http://www.sciencedirect.com/science/article/pii/037026939191163P`.
7. J. Boguta and A. R. Bodmer, Relativistic calculation of nuclear matter and the nuclear surface, *Nucl. Phys. A.* **292**(3), 413–428 (Dec., 1977). ISSN 0375-9474. doi: 10.1016/0375-9474(77)90626-1. URL `http://dx.doi.org/10.1016/0375-9474(77)90626-1`.
8. P.-G. Reinhard, M. Rufa, J. Maruhn, W. Greiner, and J. Friedrich, Nuclear ground state properties in a relativistic meson field model, *Z. Phys. A.* **323**, 13 (1986). doi: 10.1007/BF01294551.
9. Y. K. Gambhir, P. Ring, and A. Thimet, Relativistic mean field theory for finite nuclei, *Ann. Phys. (N.Y.).* **198**, 132–179 (1990). ISSN 0003-4916. doi: 10.1016/0003-4916(90)90330-Q. URL `http://www.sciencedirect.com/science/article/pii/000349169090330Q`.
10. G. A. Lalazissis, J. König, and P. Ring, New parameterization for the Lagrangian density of relativistic mean field theory., *Phys. Rev. C.* **55**(1), 540–543 (Jan., 1997). doi: 10.1103/PhysRevC.55.540. URL `http://link.aps.org/abstract/PRC/v55/p540`.
11. W. Long, J. Meng, N. V. Giai, and S.-G. Zhou, New effective interactions in relativistic mean field theory with nonlinear terms and density-dependent meson-nucleon coupling, *Phys. Rev. C.* **69**(3), 034319–15 (mar, 2004). doi: 10.1103/PhysRevC.69.034319. URL `http://link.aps.org/abstract/PRC/v69/e034319`.
12. R. Machleidt, K. Holinde, and C. Elster, The bonn meson-exchange model for the nucleon-nucleon interaction, *Phys. Rep.* **149**(1), 1–89 (1987). ISSN 0370-1573. doi: 10.1016/S0370-1573(87)80002-9. URL `http://www.sciencedirect.com/science/article/pii/S0370157387800029`.
13. R. Machleidt, The meson theory of nuclear forces and nuclear structure, *Adv. Nucl. Phys.* **19**, 189 (1989).
14. D. R. Entem and R. Machleidt, Accurate charge-dependent nucleon-nucleon potential at fourth order of chiral perturbation theory, *Phys. Rev. C.* **68**, 041001 (Oct, 2003). doi: 10.1103/PhysRevC.68.041001. URL `http://link.aps.org/doi/10.1103/PhysRevC.68.041001`.
15. B. Liu, V. Greco, V. Baran, M. Colonna, and M. DiToro, Asymmetric nuclear matter: The role of the isovector scalar channel, *Phys. Rev. C.* **65**, 045201 (Mar, 2002). doi: 10.1103/PhysRevC.65.045201. URL `http://link.aps.org/doi/10.1103/PhysRevC.65.045201`.
16. X. Roca-Maza, X. Viñas, M. Centelles, P. Ring, and P. Schuck, Relativistic mean-field interaction with density-dependent meson-nucleon vertices based on microscopical calculations, *Phys. Rev. C.* **84**, 054309 (Nov, 2011). doi: 10.1103/PhysRevC.84.054309. URL `http://link.aps.org/doi/10.1103/PhysRevC.84.054309`.
17. R. Brockmann and H. Toki, Relativistic density-dependent Hartree approach for finite nuclei, *Phys. Rev. Lett.* **68**, 3408 (Jun, 1992). doi: 10.1103/PhysRevLett.68.3408. URL `http://link.aps.org/doi/10.1103/PhysRevLett.68.3408`.

18. Y. Sugahara and H. Toki, Relativistic mean-field theory for unstable nuclei with non-linear σ and ω terms, *Nucl. Phys. A.* **579**, 557–572 (1994). ISSN 0375-9474. doi: 10.1016/0375-9474(94)90923-7. URL `http://www.sciencedirect.com/science/article/pii/0375947494909237`.
19. B. G. Todd-Rutel and J. Piekarewicz, Neutron-rich nuclei and neutron stars: A new accurately calibrated interaction for the study of neutron-rich matter, *Phys. Rev. Lett.* **95**, 122501 (Sep, 2005). doi: 10.1103/PhysRevLett.95.122501. URL `http://link.aps.org/doi/10.1103/PhysRevLett.95.122501`.
20. H. Lenske and C. Fuchs, Rearrangement in the density dependent relativistic field theory of nuclei, *Phys. Lett. B.* **345**(4), 355 (1995). ISSN 0370-2693. doi: 10.1016/0370-2693(94)01664-X. URL `http://www.sciencedirect.com/science/article/pii/037026939401664X`.
21. C. Fuchs, H. Lenske, and H. H. Wolter, Density dependent hadron field theory, *Phys. Rev. C.* **52**, 3043–3060 (Dec, 1995). doi: 10.1103/PhysRevC.52.3043. URL `http://link.aps.org/doi/10.1103/PhysRevC.52.3043`.
22. S. Typel and H. H. Wolter, Relativistic mean field calculations with density-dependent meson-nucleon coupling, *Nucl. Phys. A.* **656**(3–4), 331–364 (1999). ISSN 0375-9474. doi: 10.1016/S0375-9474(99)00310-3. URL `http://www.sciencedirect.com/science/article/pii/S0375947499003103`.
23. T. Niksic, D. Vretenar, P. Finelli, and P. Ring, Relativistic Hartree–Bogoliubov model with density dependent meson-nucleon couplings., *Phys. Rev. C.* **66**, 024306 (Aug, 2002). doi: 10.1103/PhysRevC.66.024306. URL `http://link.aps.org/doi/10.1103/PhysRevC.66.024306`.
24. G. A. Lalazissis, T. Nikšić, D. Vretenar, and P. Ring, New relativistic mean field interaction with density dependent meson couplings, *Phys. Rev. C.* **71**, 024312 (2005). doi: 10.1103/PhysRevC.71.024312. URL `http://link.aps.org/doi/10.1103/PhysRevC.71.024312`.
25. H. Kucharek and P. Ring, Relativistic field theory of superfluidity in nuclei, *Z. Phys. A.* **339**(1), 23–35 (Mar., 1991). doi: 10.1007/BF01282930. URL `http://dx.doi.org/10.1007/BF01282930`.
26. M. Bender, K. Rutz, P.-G. Reinhard, and J. A. Maruhn, Consequences of the center-of-mass correction in nuclear mean-field models, *Eur. Phys. J. A.* **7**(4), 467–478 (Apr., 2000). doi: 10.1007/PL00013645. URL `http://dx.doi.org/10.1007/PL00013645`.
27. P.-W. Zhao, B.-Y. Sun, and J. Meng, Deformation effect on the center-of-mass correction energy in nuclei ranging from oxygen to calcium, *Chin. Phys. Lett.* **26**(11), 112102– (nov, 2009). ISSN 0256-307X. URL `http://stacks.iop.org/0256-307X/26/i=11/a=112102`.
28. A. A. Nikolaus, T. Hoch, and D. Madland, Nuclear ground state properties in a relativistic point coupling model, *Phys. Rev. C.* **46**, 1757–1781 (Nov, 1992). doi: 10.1103/PhysRevC.46.1757. URL `http://link.aps.org/doi/10.1103/PhysRevC.46.1757`.
29. T. Bürvenich, D. G. Madland, J. A. Maruhn, and P.-G. Reinhard, Nuclear ground state observables and QCD scaling in a refined relativistic point coupling model, *Phys. Rev. C.* **65**, 044308 (Mar, 2002). doi:

10.1103/PhysRevC.65.044308. URL `http://link.aps.org/doi/10.1103/PhysRevC.65.044308`.

30. T. Niksic, D. Vretenar, and P. Ring, Relativistic nuclear energy density functionals: adjusting parameters to binding energies, *Phys. Rev. C.* **78**, 034318 (2008). doi: 10.1103/PhysRevC.78.034318. URL `http://link.aps.org/doi/10.1103/PhysRevC.78.034318`.
31. P. W. Zhao, Z. P. Li, J. M. Yao, and J. Meng, New parametrization for the nuclear covariant energy density functional with a point-coupling interaction, *Phys. Rev. C.* **82**, 054319 (Nov, 2010). doi: 10.1103/PhysRevC.82.054319. URL `http://link.aps.org/doi/10.1103/PhysRevC.82.054319`.
32. P. Ring, Z.-Y. Ma, N. Van Giai, D. Vretenar, A. Wandelt, and L.-G. Cao, The time-dependent relativistic mean field theory and the random phase approximation, *Nucl. Phys. A.* **694**, 249 (2001).
33. E. Litvinova and P. Ring, Covariant theory of particle-vibrational coupling and its effect on the single-particle spectrum., *Phys. Rev. C.* **73**, 044328 (2006). doi: 10.1103/PhysRevC.73.044328.
34. J. M. Yao, J. Meng, P. Ring, and D. P. Arteaga, Three-dimensional angular momentum projection in relativistic mean-field theory, *Phys. Rev. C.* **79**, 044312 (Apr, 2009). doi: 10.1103/PhysRevC.79.044312. URL `http://link.aps.org/doi/10.1103/PhysRevC.79.044312`.
35. P. Ring and E. Litvinova, Particle vibrational coupling in covariant density functional theory, *Phys. At. Nucl.* **72**, 1285–1304 (2009). Translated from Yadernaya Fisika **72** (2009) No. 8, 1.
36. T. Niksic, D. Vretenar, and P. Ring, Relativistic nuclear energy density functionals: Mean-field and beyond, *Prog. Part. Nucl. Phys.* **66**(3), 519–548 (2011). ISSN 0146-6410. doi: 10.1016/j.ppnp.2011.01.055. URL `http://www.sciencedirect.com/science/article/pii/S0146641011000561`.
37. J. L. Friar, D. G. Madland, and B. W. Lynn, QCD scales in finite nuclei, *Phys. Rev. C.* **53**, 3085–3087 (Jun, 1996). doi: 10.1103/PhysRevC.53.3085. URL `http://link.aps.org/doi/10.1103/PhysRevC.53.3085`.
38. A. Manohar and H. Georgi, Chiral quarks and the non-relativistic quark model, *Nucl. Phys. B.* **234**(1), 189–212 (1984). doi: 10.1016/0550-3213(84)90231-1. URL `http://www.sciencedirect.com/science/article/pii/0550321384902311`.
39. H. Liang, P. Zhao, P. Ring, X. Roca-Maza, and J. Meng, Localized form of Fock terms in nuclear covariant density functional theory, *Phys. Rev. C.* **86**, 021302 (Aug, 2012). doi: 10.1103/PhysRevC.86.021302. URL `http://link.aps.org/doi/10.1103/PhysRevC.86.021302`.
40. A. Sulaksono, T. Buervenich, J. Maruhn, P.-G. Reinhard, and W. Greiner, Mapping exchange in relativistic Hartree Fock, *Ann. Phys. (N.Y.).* **306**(1), 36–57 (2003). ISSN 0003-4916. doi: 10.1016/S0003-4916(03)00073-3. URL `http://www.sciencedirect.com/science/article/pii/S0003491603000733`.
41. J. Drut, R. Furnstahl, and L. Platter, Toward *ab initio* density functional theory for nuclei, *Prog. Part. Nucl. Phys.* **64**(1), 120–168 (2010). ISSN 0146-6410. doi: 10.1016/j.ppnp.2009.09.001. URL `http://www.sciencedirect.com/science/article/pii/S0146641009000660`.

42. P. Ring, On the way to a microscopic derivation of covariant density functional theory, *J. Phys. Conf. Ser.* **580**(1), 012005 (2015). doi: 10.1088/1742-6596/580/1/012005. URL `http://stacks.iop.org/1742-6596/580/i=1/a=012005`.
43. M. Sharma, M. Nagarajan, and P. Ring, Rho meson coupling in the relativistic mean field theory and description of exotic nuclei, *Phys. Lett. B.* **312**(4), 377–381 (1993). ISSN 0370-2693. doi: 10.1016/0370-2693(93)90970-S. URL `http://www.sciencedirect.com/science/article/pii/037026939390970S`.
44. F. de Jong and H. Lenske, Asymmetric nuclear matter in the relativistic Brueckner–Hartree–Fock approach, *Phys. Rev. C.* **57**, 3099–3107 (Jun, 1998). doi: 10.1103/PhysRevC.57.3099. URL `http://link.aps.org/doi/10.1103/PhysRevC.57.3099`.
45. M. Goeppert-Mayer, On closed shells in nuclei. ii, *Phys. Rev.* **75**, 1969–1970 (Jun, 1949). doi: 10.1103/PhysRev.75.1969. URL `http://link.aps.org/doi/10.1103/PhysRev.75.1969`.
46. O. Haxel, J. H. D. Jensen, and H. E. Suess, On the "magic numbers" in nuclear structure, *Phys. Rev.* **75**, 1766–1766 (Jun, 1949). doi: 10.1103/PhysRev.75.1766.2. URL `http://link.aps.org/doi/10.1103/PhysRev.75.1766.2`.
47. S. G. Nilsson, *Mat. Fys. Medd. Dan. Vid. Selsk.* **29**, No. 16 (1955).
48. J. Bardeen, L. N. Cooper, and J. R. Schrieffer, Theory of superconductivity, *Phys. Rev.* **108**, 1175–1204 (Dec, 1957). doi: 10.1103/PhysRev.108.1175. URL `http://link.aps.org/doi/10.1103/PhysRev.108.1175`.
49. A. Bohr, B. R. Mottelson, and D. Pines, Possible analogy between the excitation spectra of nuclei and those of the superconducting metallic state, *Phys. Rev.* **110**, 936–938 (May, 1958). doi: 10.1103/PhysRev.110.936. URL `http://link.aps.org/doi/10.1103/PhysRev.110.936`.
50. L. P. Gor'kov, On the energy spectrum of superconductors, *Sov. Phys. JETP.* **34**(3), 505–508 (1958).
51. M. Baranger, Extension of the shell model for heavy spherical nuclei, *Phys. Rev.* **120**, 957–968 (Nov, 1960). doi: 10.1103/PhysRev.120.957. URL `http://link.aps.org/doi/10.1103/PhysRev.120.957`.
52. J. G. Valatin, Generalized Hartree–Fock method, *Phys. Rev.* **122**, 1012–1020 (May, 1961). doi: 10.1103/PhysRev.122.1012. URL `http://link.aps.org/doi/10.1103/PhysRev.122.1012`.
53. A. K. Kerman, Pairing forces and nuclear collective motion, *Ann. Phys. (N.Y.).* **12**(2), 300–329 (1961). ISSN 0003-4916. doi: http://dx.doi.org/10.1016/0003-4916(61)90008-2. URL `http://www.sciencedirect.com/science/article/pii/0003491661900082`.
54. T. Gonzalez-Llarena, J. Egido, G. Lalazissis, and P. Ring, Relativistic Hartree–Bogoliubov calculations with finite range pairing forces, *Phys. Lett. B.* **379**, 13–19 (1996). ISSN 0370-2693. doi: 10.1016/0370-2693(96)00461-3. URL `http://www.sciencedirect.com/science/article/pii/0370269396004613`.
55. P. Ring, Relativistic mean field theory in finite nuclei, *Prog. Part. Nucl.*

Phys. **37**, 193–263 (1996). ISSN 0146-6410. doi: 10.1016/0146-6410(96)00054-3. URL http://dx.doi.org/10.1016/0146-6410(96)00054-3.
56. M. Serra and P. Ring, Relativistic Hartree–Bogoliubov theory for finite nuclei, *Phys. Rev. C.* **65**(6), 064324–9 (June, 2002). doi: 10.1103/PhysRevC.65.064324. URL http://link.aps.org/abstract/PRC/v65/e064324.
57. P. Ring and P. Schuck, *The Nuclear Many-Body Problem*. Springer-Verlag, Berlin (1980).
58. T. Niksic, N. Paar, D. Vretenar, and P. Ring, Dirhb - a relativistic self-consistent mean-field framework for atomic nuclei, *Comp. Phys. Comm.* **185**(6), 1808–1821 (2014). ISSN 0010-4655. doi: 10.1016/j.cpc.2014.02.027. URL http://www.sciencedirect.com/science/article/pii/S0010465514000836.
59. S.-G. Zhou, J. Meng, and P. Ring, Spherical relativistic Hartree theory in a Woods–Saxon basis, *Phys. Rev. C.* **68**(3), 034323–12 (Sept., 2003). doi: 10.1103/PhysRevC.68.034323. URL http://link.aps.org/abstract/PRC/v68/e034323.
60. W. H. Long, P. Ring, N. V. Giai, and J. Meng, Relativistic Hartree–Fock–Bogoliubov theory with density dependent meson-nucleon couplings, *Phys. Rev. C.* **81**(2), 024308–10 (Feb., 2010). doi: 10.1103/PhysRevC.81.024308. URL http://link.aps.org/doi/10.1103/PhysRevC.81.024308.
61. L. Li, J. Meng, P. Ring, E.-G. Zhao, and S.-G. Zhou, Deformed relativistic Hartree–Bogoliubov theory in continuum, *Phys. Rev. C.* **85**, 024312 (Feb, 2012). doi: 10.1103/PhysRevC.85.024312. URL http://link.aps.org/doi/10.1103/PhysRevC.85.024312.
62. J. Meng and P. Ring, Relativistic Hartree–Bogoliubov description of the neutron halo in ^{11}Li, *Phys. Rev. Lett.* **77**(19), 3963–3966 (Nov, 1996). doi: 10.1103/PhysRevLett.77.3963. URL http://link.aps.org/abstract/PRL/v77/p3963.
63. J. Meng, W. Pöschl, and P. Ring, Relativistic Hartree–Bogoliubov description of lithium isotopes., *Z. Phys. A.* **358**, 123 (1997). doi: 10.1007/s002180050285. URL http://www.springerlink.com/content/0939-7922/358/2/.
64. J. Meng, Relativistic continuum Hartree–Bogoliubov theory with both zero range and finite range Gogny force and their application, *Nucl. Phys. A.* **635**(1–2), 3–42 (May, 1998). ISSN 0375-9474. doi: 10.1016/S0375-9474(98)00178-X. URL http://dx.doi.org/10.1016/S0375-9474(98)00178-X.
65. J. Dobaczewski, H. Flocard, and J. Treiner, Hartree–Fock–Bogolyubov description of nuclei near the neutron-drip line, *Nucl. Phys. A.* **422**(1), 103–139 (June, 1984). ISSN 0375-9474. doi: 10.1016/0375-9474(84)90433-0. URL http://dx.doi.org/10.1016/0375-9474(84)90433-0.
66. J. M. Yao, K. Hagino, Z. P. Li, J. Meng, and P. Ring, A microscopic benchmark-study of triaxiality in low-lying states of ^{76}Kr, *Phys. Rev. C.* **89**, 054306 (May, 2014). doi: 10.1103/PhysRevC.89.054306. URL http://link.aps.org/doi/10.1103/PhysRevC.89.054306.
67. J. F. Berger, M. Girod, and D. Gogny, Time-dependent quantum collective dynamics applied to nuclear fission, *Comp. Phys. Comm.* **63**, 365–374

(1991). ISSN 0010-4655. doi: 10.1016/0010-4655(91)90263-K. URL `http://www.sciencedirect.com/science/article/pii/001046559190263K`.

68. A. V. Afanasjev, J. König, and P. Ring, Cranked relativistic Hartree–Bogoliubov theory: Superdeformed bands in the $A \sim 190$ region, *Phys. Rev. C.* **60**, 051303R (Oct, 1999). doi: 10.1103/PhysRevC.60.051303. URL `http://link.aps.org/doi/10.1103/PhysRevC.60.051303`.
69. D. Vretenar, A. V. Afanasjev, G. A. Lalazissis, and P. Ring, Relativistic Hartree–Bogoliubov theory: static and dynamic aspects of exotic nuclear structure, *Phys. Rep.* **409**(3–4), 101–259 (Apr., 2005). doi: 10.1016/j.physrep.2004.10.001. URL `http://dx.doi.org/10.1016/j.physrep.2004.10.001`.
70. Y. Tian, Z. Y. Ma, and P. Ring, A finite range pairing force for density functional theory in superfluid nuclei, *Phys. Lett. B.* **676**, 44 (2009). doi: 10.1016/j.physletb.2009.04.067.
71. Y. Tian, Z. Y. Ma, and P. Ring, Axially deformed relativistic Hartree Bogoliubov with a separable pairing force, *Phys. Rev. C.* **80**, 024313 (2009). doi: 10.1103/PhysRevC.80.024313.
72. T. Niksic, P. Ring, D. Vretenar, Y. Tian, and Z. Y. Ma, 3d relativistic Hartree–Bogoliubov model with a separable pairing interaction: Triaxial ground-state shapes, *Phys. Rev. C.* **81**(5), 054318 (May, 2010). doi: 10.1103/PhysRevC.81.054318. URL `http://link.aps.org/doi/10.1103/PhysRevC.81.054318`.
73. J. D. Bjorken and S. D. Drell, *Relativistic Quantum Fields.* McGraw-Hill, New York (1965).
74. M. Jaminon and C. Mahaux, Effective masses in relativistic approaches to the nucleon-nucleus mean field, *Phys. Rev. C.* **40**, 354–367 (Jul, 1989). doi: 10.1103/PhysRevC.40.354. URL `http://link.aps.org/doi/10.1103/PhysRevC.40.354`.
75. E. N. E. van Dalen, C. Fuchs, and A. Faessler, Effective nucleon masses in symmetric and asymmetric nuclear matter, *Phys. Rev. Lett.* **95**, 022302 (Jul, 2005). doi: 10.1103/PhysRevLett.95.022302. URL `http://link.aps.org/doi/10.1103/PhysRevLett.95.022302`.
76. B.-A. Li, L.-W. Chen, and C.-M. Ko, Recent progress and new challenges in isospin physics with heavy-ion reactions, *Phys. Rep.* **464**(4–6), 113–281 (2008). ISSN 0370-1573. doi: http://dx.doi.org/10.1016/j.physrep.2008.04.005. URL `http://www.sciencedirect.com/science/article/pii/S0370157308001269`.
77. M. Prakash and K. S. Bedell, Incompressibility of neutron-rich nuclear matter, *Phys. Rev. C.* **32**(3), 1118–1121 (Sep, 1985). doi: 10.1103/PhysRevC.32.1118. URL `http://link.aps.org/doi/10.1103/PhysRevC.32.1118`.
78. V. Baran, M. Colonna, M. D. Toro, V. Greco, M. Zielinska-Pfabé, and H. H. Wolter, Isospin effects in nuclear fragmentation, *Nucl. Phys. A.* **703**(3–4), 603–632 (2002). doi: 10.1016/S0375-9474(01)01671-2.
79. L.-W. Chen, C. M. Ko, and B.-A. Li, Isospin-dependent properties of asymmetric nuclear matter in relativistic mean field models, *Phys. Rev. C.* **76**, 054316 (Nov, 2007). doi: 10.1103/PhysRevC.76.054316. URL `http://link.aps.org/doi/10.1103/PhysRevC.76.054316`.

80. T. Marketin, D. Vretenar, and P. Ring, Calculation of beta-decay rates in a relativistic model with momentum-dependent self-energies, *Phys. Rev. C.* **75**, 024304 (Feb, 2007). doi: 10.1103/PhysRevC.75.024304. URL http://link.aps.org/doi/10.1103/PhysRevC.75.024304.
81. P.-G. Reinhard, Skyrme forces and giant resonances in exotic nuclei, *Nucl. Phys. A.* **649**(1–4), 305–314 (1999). ISSN 0375-9474. doi: 10.1016/S0375-9474(99)00076-7. URL http://www.sciencedirect.com/science/article/pii/S0375947499000767.
82. F. J. Fattoyev and J. Piekarewicz, Has a thick neutron skin in ^{208}Pb been ruled out?, *Phys. Rev. Lett.* **111**, 162501 (Oct, 2013). doi: 10.1103/PhysRevLett.111.162501. URL http://link.aps.org/doi/10.1103/PhysRevLett.111.162501.
83. P. Ring, Y. Gambhir, and G. Lalazissis, Computer program for the relativistic mean field description of the ground state properties of even-even axially deformed nuclei, *Comp. Phys. Comm.* **105**(1), 77–97 (1997). ISSN 0010-4655. doi: 10.1016/S0010-4655(97)00022-2. URL http://www.sciencedirect.com/science/article/pii/S0010465597000222.
84. G. A. Lalazissis, S. Karatzikos, R. Fossion, D. Peña Arteaga, A. V. Afanasjev, and P. Ring, The effective force NL3 revisited., *Phys. Lett. B.* **671**, 36 (2009). doi: 10.1016/j.physletb.2008.11.070.
85. G. Audi, A. H. Wapstra, and C. Thibault, The 2003 atomic mass evaluation: (ii). tables, graphs and references, *Nucl. Phys. A.* **729**(1), 337–676 (Dec., 2003). doi: 10.1016/j.nuclphysa.2003.11.003. URL http://dx.doi.org/10.1016/j.nuclphysa.2003.11.003.
86. S. Hilaire and M. Girod, Large-scale mean-field calculations from proton to neutron drip lines using the D1S Gogny force, *Eur. Phys. J. A.* **33**, 237–241 (2007). doi: DOI10.1140/epja/i2007-10450-2.
87. M. Bender, P. Bonche, T. Duguet, and P.-H. Heenen, Configuration mixing of angular momentum projected self-consistent mean-field states for neutron-deficient Pb isotopes, *Phys. Rev. C.* **69**, 064303 (Jun, 2004). doi: 10.1103/PhysRevC.69.064303. URL http://link.aps.org/doi/10.1103/PhysRevC.69.064303.
88. M. Bender, G. F. Bertsch, and P. H. Heenen, Global study of quadrupole correlation effects, *Phys. Rev. C.* **73**, 034322 (Mar, 2006). doi: 10.1103/PhysRevC.73.034322. URL http://link.aps.org/doi/10.1103/PhysRevC.73.034322.
89. J. M. Yao, H. Mei, H. Chen, J. Meng, P. Ring, and D. Vretenar, Configuration mixing of angular-momentum-projected triaxial relativistic mean-field wave functions. II. microscopic analysis of low-lying states in magnesium isotopes, *Phys. Rev. C.* **83**(1), 014308–11 (Jan., 2011). doi: 10.1103/PhysRevC.83.014308. URL http://link.aps.org/doi/10.1103/PhysRevC.83.014308.
90. H. De Vries, C. W. De Jager, and C. De Vries, Nuclear charge-density-distribution parameters from elastic electron scattering, *At. Data Nucl. Data Tables.* **36**(3), 495–536 (May, 1987). ISSN 0092-640X. doi: 10.1016/0092-640X(87)90013-1. URL http://www.sciencedirect.com/science/article/pii/0092640X87900131.

91. E. G. Nadjakov, K. P. Marinova, and Y. P. Gangrsky, Systematics of nuclear charge radii, *At. Data Nucl. Data Tables.* **56**(1), 133 (Jan., 1994). doi: 10.1006/adnd.1994.1004.
92. E. W. Otten. Nuclear radii and moments of unstable isotopes. In ed. D. A. Bromley, *Treatise on Heavy Ion Science*, vol. 8, pp. 517–638. Springer US (1989). ISBN 978-1-4612-8045-3. doi: 10.1007/978-1-4613-0713-6_7. URL `http://dx.doi.org/10.1007/978-1-4613-0713-6_7`.
93. M. Sharma, G. Lalazissis, and P. Ring, Anomaly in the charge radii of Pb isotopes, *Phys. Lett. B.* **317**(1–2), 9–13 (1993). ISSN 0370-2693. doi: 10.1016/0370-2693(93)91561-Z. URL `http://www.sciencedirect.com/science/article/pii/037026939391561Z`.
94. V. I. Isakov, K. I. Erokhina, H. Mach, M. Sanchez-Vega, and B. Fogelberg, On the difference between proton and neutron spin-orbit splittings in nuclei, *Eur. Phys. J. A.* **14**(1), 29–36 (May, 2002). ISSN 1434-6001. doi: 10.1140/epja/iepja1393. URL `http://dx.doi.org/10.1140/epja/iepja1393`.
95. W. H. Long, H. Sagawa, N. V. Giai, and J. Meng, Shell structure and rho-tensor correlations in density dependent relativistic Hartree–Fock theory, *Phys. Rev. C.* **76**, 034314 (Sep, 2007). doi: 10.1103/PhysRevC.76.034314. URL `http://link.aps.org/doi/10.1103/PhysRevC.76.034314`.
96. M. Abramowitz and I. A. Stegun, *Handbook of Mathematical Functions*. Dover, New York (1965).
97. A. Bohr and B. R. Mottelson, *Nuclear Structure Volume II: Nuclear Deformation*. W. A. Benjamin, Inc., Reading, Mass. (1975).
98. S. Frauendorf and J. Meng, Tilted rotation of triaxial nuclei, *Nucl. Phys. A.* **617**(2), 131–147 (1997). ISSN 0375-9474. doi: 10.1016/S0375-9474(97)00004-3. URL `http://www.sciencedirect.com/science/article/pii/S0375947497000043`.
99. S. Frauendorf, Spontaneous symmetry breaking in rotating nuclei, *Rev. Mod. Phys.* **73**, 463–514 (Jun, 2001). doi: 10.1103/RevModPhys.73.463. URL `http://link.aps.org/doi/10.1103/RevModPhys.73.463`.
100. J. Meng and S. Zhang, Open problems in understanding the nuclear chirality, *J. Phys. G.* **37**, 064025 (2010).
101. W. Koepf and P. Ring, A relativistic description of rotating nuclei: The yrast line of ^{20}Ne, *Nucl. Phys. A.* **493**(1), 61–82 (1989). ISSN 0375-9474. doi: 10.1016/0375-9474(89)90532-0. URL `http://www.sciencedirect.com/science/article/pii/0375947489905320`.
102. J. Peng, J. Meng, P. Ring, and S. Q. Zhang, Covariant density functional theory for magnetic rotation, *Phys. Rev. C.* **78**, 024313 (Aug, 2008). doi: 10.1103/PhysRevC.78.024313. URL `http://link.aps.org/doi/10.1103/PhysRevC.78.024313`.
103. P. Bonche, H. Flocard, and P.-H. Heenen, Solution of the Skyrme HF-BCS equation on a 3D mesh, *Comp. Phys. Comm.* **171**(1), 49–62 (2005). ISSN 0010-4655. doi: 10.1016/j.cpc.2005.05.001. URL `http://www.sciencedirect.com/science/article/pii/S0010465505002821`.
104. M. Girod and B. Grammaticos, Triaxial Hartree–Fock–Bogolyubov calculations with d1 effective interaction, *Phys. Rev. C.* **27**, 2317–2339 (May,

1983). doi: 10.1103/PhysRevC.27.2317. URL `http://link.aps.org/doi/10.1103/PhysRevC.27.2317`.

105. M. Stoitsov, J. Dobaczewski, W. Nazarewicz, and P. Ring, Axially deformed solution of the Skyrme–Hartree–Fock–Bogolyubov equations using the transformed harmonic oscillator basis. the program hfbtho (v1.66p), *Comp. Phys. Comm.* **167**(1), 43–63 (2005). ISSN 0010-4655. doi: 10.1016/j.cpc.2005.01.001. URL `http://www.sciencedirect.com/science/article/pii/S0010465505000305`.
106. J. Dobaczewski and J. Dudek, Solution of the Skyrme–Hartree–Fock equations in the Cartesian deformed harmonic oscillator basis. (i) the method, *Comp. Phys. Comm.* **102**, 166–182 (1997). doi: 10.1016/S0010-4655(97)00004-0.
107. B.-N. Lu, E.-G. Zhao, and S.-G. Zhou, Potential energy surfaces of actinide nuclei from a multidimensional constrained covariant density functional theory: Barrier heights and saddle point shapes, *Phys. Rev. C.* **85**, 011301 (Jan, 2012). doi: 10.1103/PhysRevC.85.011301. URL `http://link.aps.org/doi/10.1103/PhysRevC.85.011301`.
108. B.-N. Lu, J. Zhao, E.-G. Zhao, and S.-G. Zhou, Multi-dimensional constraint relativistic mean field models and potential energy surfaces of actinide nuclei, *Phys. Rev. C.* **89**, 014323 (Jan, 2014). doi: 10.1103/PhysRevC.89.014323. URL `http://link.aps.org/doi/10.1103/PhysRevC.89.014323`.
109. S. Raman, C. W. Nestor, and P. Tikkanen, Transition probability from the ground to the first-excited 2+ state of even-even nuclides, *At. Data Nucl. Data Tables.* **78**(1), 1–128 (May, 2001). doi: http://dx.doi.org/10.1006/adnd.2001.0858. URL `http://www.sciencedirect.com/science/article/B6WBB-45BT4PT-7/2/2704f8783712265744f9f940d01ff835`.
110. B.-N. Lu, E.-G. Zhao, and S.-G. Zhou, Pseudospin symmetry in single particle resonant states, *Phys. Rev. Lett.* **109**(7), 072501–5 (Aug., 2012). doi: 10.1103/PhysRevLett.109.072501. URL `http://link.aps.org/doi/10.1103/PhysRevLett.109.072501`.
111. K. Rutz, J. A. Maruhn, P.-G. Reinhard, and W. Greiner, Fission barriers and asymmetric ground states in the relativistic mean-field theory, *Nucl. Phys. A.* **590**, 680–702 (1995). ISSN 0375-9474. doi: 10.1016/0375-9474(95)00192-4. URL `http://www.sciencedirect.com/science/article/pii/0375947495001924`.
112. H. Abusara, A. V. Afanasjev, and P. Ring, Fission barriers in actinides in covariant density functional theory: The role of triaxiality, *Phys. Rev. C.* **82**, 044303 (2010). doi: 10.1103/PhysRevC.82.044303. URL `http://link.aps.org/doi/10.1103/PhysRevC.82.044303`.
113. M. Wang, G. Audi, A. Wapstra, F. Kondev, M. MacCormick, X. Xu, and B. Pfeiffer, The AME2012 atomic mass evaluation, *Chin. Phys. C.* **36**(12), 1603 (2012). doi: 10.1088/1674-1137/36/12/003. URL `http://stacks.iop.org/1674-1137/36/i=12/a=003`.
114. C. F. von Weizsäcker, Zur theorie der kernmassen, *Z. Phys.* **96**, 431 (1935).
115. P. Möller, J. R. Nix, W. D. Myers, and W. J. Swiatecki, Nuclear ground-

state masses and deformations, *At. Data Nucl. Data Tables.* **59**(2), 185–381 (Mar., 1995). doi: 10.1006/adnd.1995.1002. URL `http://dx.doi.org/10.1006/adnd.1995.1002`.

116. J. M. Pearson, R. C. Nayak, and S. Goriely, Nuclear mass formula with Bogolyubov-enhanced shell-quenching: application to r-process, *Phys. Lett. B.* **387**(3), 455–459 (1996). doi: 10.1016/0370-2693(96)01071-4. URL `http://www.sciencedirect.com/science/article/pii/0370269396010714`.
117. N. Wang, Z. Liang, M. Liu, and X. Wu, Mirror nuclei constraint in nuclear mass formula, *Phys. Rev. C.* **82**, 044304 (Oct, 2010). doi: 10.1103/PhysRevC.82.044304. URL `http://link.aps.org/doi/10.1103/PhysRevC.82.044304`.
118. W. E. Ormand, Mapping the proton drip line up to $A = 70$, *Phys. Rev. C.* **55**, 2407–2417 (May, 1997). doi: 10.1103/PhysRevC.55.2407. URL `http://link.aps.org/doi/10.1103/PhysRevC.55.2407`.
119. J. Barea, A. Frank, J. G. Hirsch, P. V. Isacker, S. Pittel, and V. Velázquez, Garvey–Kelson relations and the new nuclear mass tables, *Phys. Rev. C.* **77**, 041304 (Apr, 2008). doi: 10.1103/PhysRevC.77.041304. URL `http://link.aps.org/doi/10.1103/PhysRevC.77.041304`.
120. G. J. Fu, Y. Lei, H. Jiang, Y. M. Zhao, B. Sun, and A. Arima, Description and evaluation of nuclear masses based on residual proton-neutron interactions, *Phys. Rev. C.* **84**, 034311 (Sep, 2011). doi: 10.1103/PhysRevC.84.034311. URL `http://link.aps.org/doi/10.1103/PhysRevC.84.034311`.
121. M. Bender, P.-H. Heenen, and P.-G. Reinhard, Self-consistent mean-field models for nuclear structure, *Rev. Mod. Phys.* **75**, 121–180 (Jan, 2003). doi: 10.1103/RevModPhys.75.121. URL `http://link.aps.org/doi/10.1103/RevModPhys.75.121`.
122. J. Meng, H. Toki, S. G. Zhou, S. Q. Zhang, W. H. Long, and L. S. Geng, Relativistic continuum Hartree Bogoliubov theory for ground-state properties of exotic nuclei, *Prog. Part. Nucl. Phys.* **57**(2), 470–563 (Oct., 2006). doi: 10.1016/j.ppnp.2005.06.001. URL `http://dx.doi.org/10.1016/j.ppnp.2005.06.001`.
123. S. Goriely, N. Chamel, and J. M. Pearson, Skyrme–Hartree–Fock–Bogoliubov nuclear mass formulas: Crossing the 0.6 MeV accuracy threshold with microscopically deduced pairing, *Phys. Rev. Lett.* **102**, 152503 (Apr, 2009). doi: 10.1103/PhysRevLett.102.152503. URL `http://link.aps.org/doi/10.1103/PhysRevLett.102.152503`.
124. S. Goriely, S. Hilaire, M. Girod, and S. Péru, First Gogny–Hartree–Fock–Bogoliubov nuclear mass model, *Phys. Rev. Lett.* **102**, 242501 (Jun, 2009). doi: 10.1103/PhysRevLett.102.242501. URL `http://link.aps.org/doi/10.1103/PhysRevLett.102.242501`.
125. J. Erler, N. Birge, M. Kortelainen, W. Nazarewicz, E. Olsen, A. M. Perhac, and M. Stoitsov, The limits of the nuclear landscape, *Nature.* **486**, 509–512 (2012). doi: 10.1038/nature11188.
126. D. Hirata, K. Sumiyoshi, I. Tanihata, Y. Sugahara, T. Tachibana, and H. Toki, A systematic study of even-even nuclei up to the drip lines within the relativistic mean field framework, *Nucl. Phys. A.* **616**(1–

2), 438–445 (1997). ISSN 0375-9474. doi: http://dx.doi.org/10.1016/S0375-9474(97)00115-2. URL `http://www.sciencedirect.com/science/article/pii/S0375947497001152`.

127. G. A. Lalazissis, S. Raman, and P. Ring, Ground-state properties of even-even nuclei in the relativistic mean-field theory., *At. Data Nucl. Data Tables.* **71**, 1–40 (1999). ISSN 0092-640X. doi: 10.1006/adnd.1998.0795. URL `http://www.sciencedirect.com/science/article/pii/S0092640X98907951`.
128. L. S. Geng, H. Toki, and J. Meng, Masses, deformations and charge radii–nuclear ground-state properties in the relativistic mean field model, *Prog. Theor. Phys.* **113**(4), 785–800 (2005). doi: 10.1143/PTP.113.785. URL `http://ptp.ipap.jp/link?PTP/113/785/`.
129. B. Sun, F. Montes, L. S. Geng, H. Geissel, Y. A. Litvinov, and J. Meng, Application of the relativistic mean-field mass model to the r-process and the influence of mass uncertainties, *Phys. Rev. C.* **78**(2), 025806 (Aug, 2008). doi: 10.1103/PhysRevC.78.025806.
130. Z. M. Niu, B. H. Sun, and J. Meng, Influence of nuclear physics inputs and astrophysical conditions on the Th/U chronometer, *Phys. Rev. C.* **80**(6), 065806 (Dec, 2009). doi: 10.1103/PhysRevC.80.065806.
131. P.-G. Reinhard and B. K. Agrawal, Energy systematics of heavy nuclei — mean field models in comparison, *Int. J. Mod. Phys. E.* **20**, 1379–1390 (2011). doi: 10.1142/S0218301311018472.
132. S. E. Agbemava, A. V. Afanasjev, D. Ray, and P. Ring, Global performance of covariant energy density functionals: ground state observables of even-even nuclei and the estimate of theoretical uncertainties, *Phys. Rev. C.* **89**, 054320 (May, 2014). doi: 10.1103/PhysRevC.89.054320. URL `http://link.aps.org/doi/10.1103/PhysRevC.89.054320`.
133. X. Roca-Maza, X. Viñas, M. Centelles, P. Ring, and P. Schuck, Relativistic mean-field interaction with density-dependent meson-nucleon vertices based on microscopical calculations, *Phys. Rev. C.* **84**, 054309 (Nov, 2011). doi: 10.1103/PhysRevC.84.054309.
134. A. V. Afanasjev, S. E. Agbemava, D. Ray, and P. Ring, Nuclear landscape in covariant density functional theory, *Phys. Lett. B.* **726**(4–5), 680–684 (2013). ISSN 0370-2693. doi: http://dx.doi.org/10.1016/j.physletb.2013.09.017. URL `http://www.sciencedirect.com/science/article/pii/S037026931300734X`.
135. Q.-S. Zhang, Z.-M. Niu, Z.-P. Li, J.-M. Yao, and J. Meng, Global dynamical correlation energies in covariant density functional theory: Cranking approximation, *Front. Phys.* pp. 1–8 (march, 2014). ISSN 2095-0462. doi: 10.1007/s11467-014-0413-5. URL `http://dx.doi.org/10.1007/s11467-014-0413-5`.
136. J. Meng, J. Peng, S. Q. Zhang, and P. W. Zhao, Progress on tilted axis cranking covariant density functional theory for nuclear magnetic and antimagnetic rotation, *Front. Phys.* **8**, 55–59 (2013). doi: 10.1007/s11467-013-0287-y. URL `http://dx.doi.org/10.1007/s11467-013-0287-y`.
137. M. Bender, P. Bonche, P.-H. Heenen, M. Bender, P. Bonche, and P.-

H. Heenen, Shape coexistence in neutron-deficient Kr isotopes: Constraints on the single-particle spectrum of self-consistent mean-field models from collective excitations, *Phys. Rev. C.* **74**, 024312 (Aug, 2006). doi: 10.1103/PhysRevC.74.024312. URL `http://link.aps.org/doi/10.1103/PhysRevC.74.024312`.

138. N. Hinohara, T. Nakatsukasa, M. Matsuo, and K. Matsuyanagi, Microscopic description of oblate-prolate shape mixing in proton-rich Se isotopes, *Phys. Rev. C.* **80**, 014305 (Jul, 2009). doi: 10.1103/PhysRevC.80.014305. URL `http://link.aps.org/doi/10.1103/PhysRevC.80.014305`.
139. Y. Fu, H. Mei, J. Xiang, Z. P. Li, J. M. Yao, and J. Meng, Beyond relativistic mean-field studies of low-lying states in neutron-deficient krypton isotopes, *Phys. Rev. C.* **87**, 054305 (May, 2013). doi: 10.1103/PhysRevC.87.054305. URL `http://link.aps.org/doi/10.1103/PhysRevC.87.054305`.
140. W. Von Oertzen, M. Freer, and Y. Kanada-En'yo, Nuclear clusters and nuclear molecules, *Phys. Rep.* **432**(2), 43–113 (Sep, 2006). ISSN 0370-1573. doi: 10.1016/j.physrep.2006.07.001.
141. E. Wigner, On the structure of nuclei beyond oxygen, *Phys. Rev.* **51**, 947–958 (Jun, 1937). doi: 10.1103/PhysRev.51.947. URL `http://link.aps.org/doi/10.1103/PhysRev.51.947`.
142. W. D. Myers and W. J. Świątecki, The congruence energy: a contribution to nuclear masses, deformation energies and fission barriers, *Nucl. Phys. A.* **612**(2), 249–261 (1997). ISSN 0375-9474. doi: 10.1016/S0375-9474(97)80014-0. URL `http://www.sciencedirect.com/science/article/pii/S0375947497800140`.
143. M. Thoennessen, Current status and future potential of nuclide discoveries, *Rep. Prog. Phys.* **76**, 056301 (2013). doi: 10.1088/0034-4885/76/5/056301.
144. S. Goriely, N. Chamel, and J. M. Pearson, Further explorations of Skyrme–Hartree–Fock–Bogoliubov mass formulas. xii. stiffness and stability of neutron-star matter, *Phys. Rev. C.* **82**(3), 035804–18 (Sept., 2010). doi: 10.1103/PhysRevC.82.035804. URL `http://link.aps.org/doi/10.1103/PhysRevC.82.035804`.
145. M. Liu, N. Wang, Y. Deng, and X. Wu, Further improvements on a global nuclear mass model, *Phys. Rev. C.* **84**, 014333 (Jul, 2011). doi: 10.1103/PhysRevC.84.014333. URL `http://link.aps.org/doi/10.1103/PhysRevC.84.014333`.
146. X. Qu, Y. Chen, S. Zhang, P. Zhao, I. J. Shin, Y. Lim, Y. Kim, and J. Meng, Extending the nuclear chart by continuum: From oxygen to titanium, *Sci. China Phys. Mech. Astron.* **56**(11), 2031–2036 (Oct, 2013). doi: 10.1007/s11433-013-5329-5.
147. J. Meng and P. Ring, Giant halo at the neutron drip line, *Phys. Rev. Lett.* **80**(3), 460–463 (Jan, 1998). doi: 10.1103/PhysRevLett.80.460. URL `http://link.aps.org/doi/10.1103/PhysRevLett.80.460`.

Chapter 3

Relativistic mean field description of exotic nuclei

Jie Meng

School of Physics, Peking University, Beijing 100871, P.R.China
mengj@pku.edu.cn

Peter Ring

Physics Department, Technical University Munich, 85748 Garching, Germany
School of Physics, Peking University, Beijing 100871, China
ring@ph.tum.de

Pengwei Zhao

School of Physics, Peking University, Beijing 100871, P.R.China
pwzhao@pku.edu.cn

Shan-Gui Zhou

State Key Laboratory of Theoretical Physics, Institute of Theoretical Physics, Chinese Academy of Sciences, Beijing 100190, China
Center of Theoretical Nuclear Physics, National Laboratory of Heavy Ion Accelerator, Lanzhou 730000, China
sgzhou@itp.ac.cn

In this chapter, we will present relativistic mean field (RMF) models with pairing treated by the Bardeen–Cooper–Schrieffer (BCS) and the relativistic Hartree–Bogoliubov (RHB) approaches and applications for exotic nuclear phenomena including nuclear halos, the position of the proton drip line and proton radioactivity, the surface diffuseness and its relation to nuclear exotic phenomena, and the effects of pairing correlations on the nuclear size.

Contents

1. Introduction

The exploration of the nuclear landscape is always one of the most fascinating topics in nuclear physics. One can study the limits of the existence of nuclei from several different aspects, e.g., nuclei far from the β-stability line, superheavy nuclei (Chapter 5), hypernuclei (Chapter 7), fast rotating nuclei (Chapters 8 and 9), nuclei with high temperature, and so on. In this chapter, we will focus on the study of nuclei far from the β-stability line, the so-called "exotic nuclei".

In nature, there exists less than 300 types of nuclei which are stable or at least have half-lives comparable with or longer than the expected life of the Solar System. These nuclei form the valley of β-stability in the nuclear chart. In laboratories, one can produce nuclei which do not fall on this line and nowadays around 3,000 nuclides have been produced in this way.[1–4] In nuclei far away from the β-stability line, particularly in those close to or beyond the proton or neutron drip lines, one finds many nuclear phenomena which are exotic in the sense that they do not follow the predictions of available nuclear models, e.g., nuclear halos,[5] the changes of nuclear magicity,[6] one-proton radioactivity[7–9] or two-proton radioactivity,[10,11] etc. Therefore these phenomena are called exotic nuclear phenomena and those nuclei are called exotic nuclei.

The nuclear chart will be extended further with a new generation radioactive ion beam (RIB) facilities developed around the world, including the RIKEN Radioactive Ion Beam Factory (RIBF) in Japan,[12] the Facility for Antiproton and Ion Research (FAIR) in Germany,[13] the Second Generation System On-Line Production of Radioactive Ions (SPIRAL2) at GANIL in France,[14] the Facility for Rare Isotope Beams (FRIB) in USA,[15,16] the Rare Isotope Science Project (RISP) in Korea,[17] the Cooler Storage Ring (CSR) at Heavy Ion Research Facility in Lanzhou (HIRFL) in China,[18,19] etc.

Covariant density functional theories (CDFTs) can describe well not only properties of stable nuclei but also those of exotic nuclei.[20–29] For the study of exotic nuclei, CDFTs have been used not only to investigate observed exotic nuclear phenomena, but also to predict new ones.[30–33] In this chapter, we will present the formalism of relativistic mean field (RMF) models with pairing in the continuum (Section 2) and the applications of CDFTs on the nuclear halo phenomena (Section 3), the position of the proton drip line and proton radioactivity (Section 6), the surface diffuseness and its relation to nuclear exotic phenomena (Section 4), and the effects of the pairing on nuclear size (Section 5).

2. Relativistic models with pairing in the continuum

Pairing correlations are crucial in the description of open-shell nuclei. The general concept of the description of pairing in relativistic models has been discussed in Chapter 2. In this Section, we present the proper treatment of the continuum in connection with the pairing correlations in relativistic density functional theory. For the concept of CDFT and the formalism of RMF models, please refer to Chapter 1 and Chapter 2, respectively. The details of the relativistic Hartree–Fock–Bogoliubov theory are presented in Chapter 4.

2.1. *Relativistic Hartree–Bogoliubov theory in the continuum*

For the description of pairing correlations the simple BCS approach[34] is used widely. In nuclear physics it was introduced[35] first for cases with fixed mean field potentials and for nuclei in the valley of stability, where the Fermi surface is relatively far from the continuum limit, such that the

BCS-occupation probabilities for orbits in the continuum vanish. In these cases it works very well, in particular if it is combined with a self-consistent description of the mean field within the HF+BCS or RMF+BCS models,[22,23] where in each step of the iteration the BCS equations are solved for the corresponding mean field.

For exotic nuclei, in particular those close to the neutron drip line, the Fermi surface comes close to the continuum threshold. In this case one has finite occupation numbers in the continuum and a gas of unbound neutrons develops extending up to infinity. Therefore the conventional BCS approach presents only a poor approximation for exotic bound nuclei close to the drip lines. Using specific techniques to remove the influence of this neutron gas one has used the resonant-BCS (rBCS) approach,[36–38] to include the contribution of resonant states.

A consistent treatment of the physics of weakly-bound nuclei, however, requires a unified and self-consistent treatment of the continuum, the mean-field potentials and the pairing correlations. In the non-relativistic case it has been shown in Refs. 39, 40 that Hartree–Fock–Bogoliubov (HFB) theory fulfills these conditions. Relativistic Hartree–Bogoliubov (RHB) theory[41,42] has the same properties. By using two average potentials, these theories provide a unified description of the correlations in the particle-hole (ph) and in the particle-particle (pp) channel on the mean-field level. These two potentials are the self-consistent mean field $\hat{h}$ that includes all the long range ph correlations, and the pairing field $\hat{\Delta}$ which contains the pp correlations. Details are given in Chapter 2.

These potentials and the corresponding wave functions $U_k(\boldsymbol{r})$ and $V_k(\boldsymbol{r})$ are determined by the self-consistent solution of the Bogoliubov equations[a]

$$\begin{pmatrix} \hat{h}-\lambda & \hat{\Delta} \\ -\hat{\Delta}^* & -\hat{h}^*+\lambda \end{pmatrix} \begin{pmatrix} U_k(\boldsymbol{r}) \\ V_k(\boldsymbol{r}) \end{pmatrix} = E_k \begin{pmatrix} U_k(\boldsymbol{r}) \\ V_k(\boldsymbol{r}) \end{pmatrix} . \tag{1}$$

In the simple BCS-model approximation the operator $\hat{h}$ is diagonalized

$$\hat{h}\varphi_k(\boldsymbol{r}) = \varepsilon_k \varphi_k(\boldsymbol{r}) \tag{2}$$

and, in this basis, only the diagonal elements $\Delta_k = \Delta_{k\bar{k}}$ are considered, all other matrix elements are neglected. Then one ends up with a set of 2×2 problems

$$\begin{pmatrix} \varepsilon_k-\lambda & \Delta_k \\ \Delta_k & -\varepsilon_k+\lambda \end{pmatrix} \begin{pmatrix} u_k \\ v_k \end{pmatrix} = E_k \begin{pmatrix} u_k \\ v_k \end{pmatrix} , \tag{3}$$

[a]for simplicity we neglect here and in the following the spin (s) and isospin (t) coordinates and replace $(\boldsymbol{r}, s, t)$ by $(\boldsymbol{r})$.

with the well known solution for the BCS occupation probabilities

$$\left.\begin{array}{l} u_k^2 \\ v_k^2 \end{array}\right\} = \frac{1}{2}\left\{1 \pm \frac{\varepsilon_k - \lambda}{\sqrt{(\varepsilon_k - \lambda)^2 + \Delta_k^2}}\right\} \tag{4}$$

and the approximate wave functions of RMF+BCS theory

$$U_k(\boldsymbol{r}) = u_k \varphi_k(\boldsymbol{r}) \quad \text{and} \quad V_k(\boldsymbol{r}) = v_k \varphi_k(\boldsymbol{r}). \tag{5}$$

Here U_k and V_k have the same radial dependence, they differ only in amplitude. This leads to the essential difference between full RHB theory as compared to RMF+BCS theory. In RMF+BCS theory the asymptotic behavior of the coefficients $U_k(\boldsymbol{r})$ and $V_k(\boldsymbol{r})$ are identical, whereas in RHB theory their asymptotic behavior can be different. For bound states with pairing in the continuum all the wave functions $V_k(\boldsymbol{r})$ decay exponentially for $r \to \infty$, only the wave functions $U_k(\boldsymbol{r})$ in the continuum have oscillating character. Therefore the density

$$\rho(\boldsymbol{r}) = \sum_k V_k^2(\boldsymbol{r}) \tag{6}$$

decays exponentially and is localized. This is impossible in the simple BCS model, where we have for levels in the continuum $U_k(\boldsymbol{r}) \sim V_k(\boldsymbol{r}) \sim \varphi(\boldsymbol{r})$. As a consequence only full HFB or RHB theory allows a proper treatment of the continuum in the presence of pairing correlations.[40,42]

These useful properties of full HFB or RHB theory are not restricted to the solution of the corresponding equations (1) in a r-space extending to infinity. As long as one finds convergence, it is also valid for solutions in r-space with a finite box radius R or for solutions in a finite oscillator or Woods–Saxon basis. Convergence means in this case convergence with respect to the box size R or/and the number of basis states. This means in particular, that the oscillator basis is usually not well suited for the description of halo phenomena. The oscillator functions decay for large r-values with Gaussian shape and therefore one needs a forbiddingly large number of basis states to reproduce the exponential decay of realistic halo phenomena. One possibility to avoid this problem is the transformed harmonic oscillator (THO) basis[43,44] or the Woods–Saxon basis. On the other hand, halo phenomena are restricted to nuclei with low ℓ orbits with small centrifugal barriers in the neighborhood of the Fermi surface. Therefore all the other exotic nuclei, even close to the drip lines, can be very well described by the solutions of the HFB or the RHB equation in a spherical or a deformed oscillator basis, even if there is an essential coupling to the continuum.

HFB and RHB theory have another important property. According to the Bloch–Messiah theorem[45] any HFB or RHB wave function, i.e. any generalized Slater determinant $|\Phi\rangle$ can be represented in a specific single particle basis $|\mu\rangle$ as a BCS state

$$|\Phi\rangle = \prod_{\mu>0} (u_\mu + v_\mu a^\dagger_\mu a^\dagger_{\bar{\mu}})|-\rangle. \tag{7}$$

This is the so-called *canonical* basis with the single particle wave functions $\phi_\mu(\boldsymbol{r}) = \langle \boldsymbol{r}|a^\dagger_\mu|-\rangle$. They are not defined as eigenstates of a specific Hamiltonian, but, for systems with even particle number, they are found as doubly degenerate eigenstates of the single particle density matrix

$$\int d^3r' \rho(\boldsymbol{r},\boldsymbol{r}')\, \phi_\mu(\boldsymbol{r}') = v^2_\mu\, \phi_\mu(\boldsymbol{r}) \tag{8}$$

i.e. for $\mu > 0$ as pairs of *canonical conjugate* eigenstates $|\mu\rangle$ and $|\bar{\mu}\rangle$ with the same eigenvalue v^2_μ. In most of the cases the kernel $\rho(\boldsymbol{r},\boldsymbol{r}')$ of the integral equation (8) is of Hilbert Schmidt type[46] and therefore the spectrum v^2_μ is discrete with an accumulation point $v^2_\mu \to 0$. This means that the so-called *canonical states* $|\mu\rangle$ are discrete states, eigenstates of $\hat{\rho}$ and not of $\hat{h}$. Their energies, the so-called *canonical energies* ϵ_μ, are expectation values of $\hat{h}$ and, in cases without time-reversal symmetry, they are given by

$$\epsilon_\mu = \frac{1}{2}\left(\langle\mu|\hat{h}|\mu\rangle + \langle\bar{\mu}|\hat{h}|\bar{\mu}\rangle\right). \tag{9}$$

It can be shown[47] that the occupation probabilities in the canonical basis have the form

$$\left.\begin{matrix} u^2_\mu \\ v^2_\mu \end{matrix}\right\} = \frac{1}{2}\left\{1 \pm \frac{\epsilon_\mu - \lambda}{\sqrt{(\epsilon_\mu - \lambda)^2 + \Delta^2_\mu}}\right\}, \tag{10}$$

where $\Delta_\mu = \Delta_{\mu\bar{\mu}}$. Of course, the BCS occupation amplitudes u_μ and v_μ (10), defined for any HFB or RHB function, should be distinguished from the corresponding quantities u_k and v_k (4) in the HF+BCS or RMF+BCS approximation.

It has to be emphasized that, in general, most of the eigenvalues v^2_μ of the eigenvalue problem (8) are zero or very close to zero. This is particularly the case for the relativistic problem, where, because of the no-sea approximation, all the levels in the Dirac sea are empty. As a consequence these eigenvalues are highly degenerate and the canonical states $|\mu\rangle$ are not well defined. This leads to arbitrary spurious mixtures of states in the empty continuum and in the Dirac sea. The corresponding canonical energies ϵ_μ in Eq. (9) are completely arbitrary and depend on small numerical uncertainties during the diagonalization process. The problem can be solved by the fact, that, in the subspace with vanishing occupation probabilities $v_\mu^2 = 0$, the Hamiltonian $\hat{h}$ commutes with the density matrix $\hat{\rho}$ and therefore the canonical energies in this subspace can be determined uniquely by diagonalizing $\hat{h}$ in this subspace. In this case one can avoid the spurious admixtures discussed above.

In the relativistic case the single particle Hamiltonian $\hat{h} = \hat{h}_D$ corresponds to the Dirac Hamiltonian and the RHB coefficients U_k and V_k are Dirac spinors with large and small components

$$U_k = \begin{pmatrix} g_k^U \\ if_k^U \end{pmatrix} \qquad V_k = \begin{pmatrix} g_k^V \\ if_k^V \end{pmatrix} . \tag{11}$$

We therefore introduce in the original basis, besides the non-relativistic quantum number n the relativistic quantum number $p = (g, f)$ characterizing large and small components. For a general effective two-body *pp*-interaction V^{pp}, the relativistic pairing field reads

$$\hat{\Delta}_{n_1p_1,n_1'p_1'} = \frac{1}{2} \sum_{n_2p_2,n_2'p_2'} \langle n_1p_1, n_1'p_1'|V^{pp}|n_2p_2, n_2'p_2'\rangle \kappa_{n_2p_2,n_2'p_2'}. \tag{12}$$

However, the energy scale of pairing correlations in nuclei is restricted to a few MeV around the Fermi level. It is well separated from the scale of binding energies, which are in the range from several hundreds to thousands MeV. There is no empirical evidence[48,49] for any relativistic effect in the nuclear pairing field $\hat{\Delta}$. Therefore, in most practical applications of the RHB model only the large components of the spinors U_k and V_k are used to build the non-relativistic pairing tensor $\hat{\kappa}$. The resulting pairing field reads

$$\hat{\Delta}_{n_1g,n_1'g} = \frac{1}{2} \sum_{n_2n_2'} \langle n_1g, n_1'g|V^{pp}|n_2g, n_2'g\rangle_a \; \kappa_{n_2g,n_2'g}. \tag{13}$$

The other components: $\hat{\Delta}_{gf}$, $\hat{\Delta}_{fg}$, and $\hat{\Delta}_{ff}$ can be safely omitted.[49]

For practical applications of RHB theory several types of effective pairing interactions have been used in the literature.[50] Early attempts to use the same force in the *pp*-channel and in the *ph*-channel, i.e. to use a relativistic meson-exchange force, have failed completely.[41] The strong repulsion introduced by the exchange of ω-mesons leads to completely unrealistic values of the gap parameter at the Fermi surface. Therefore, in all relativistic applications of RMF+BCS or RHB theory the effective *pp*-force is different from the effective *ph*-force. The simplest form of V^{pp} is the seniority force of Kerman[51] also called monopole pairing with the constant G. In this case the pairing field operator $\hat{\Delta}$ is a multiple of unity and therefore the Hamiltonian $\hat{h}$ is diagonal in the canonical basis and RHB theory coincides with RMF+BCS theory. Therefore this force has been used in many applications of covariant density functional theory with pairing correlations for stable nuclei far from the drip lines. However, as discussed before, this pairing force does not allow a proper treatment of the continuum. Another simple pairing force often used in HFB and RHB calculations at the drip lines[42,52] is the zero-range pairing force

$$V(\boldsymbol{r}, \boldsymbol{r}') = V_0 \delta(\boldsymbol{r} - \boldsymbol{r}') \frac{1}{4} (1 - \boldsymbol{\sigma} \cdot \boldsymbol{\sigma}') \left(1 - \eta \frac{\rho(r)}{\rho_0}\right), \tag{14}$$

with an adjusted strength parameter V_0. It has the advantage of being local and therefore relatively simple to apply for the solution of the HFB and RHB equations in r-space. There are two versions: for $\eta = 0$ it is called volume pairing, because it depends on the density, and for $\eta = 1$ it is called surface pairing, because it vanishes in the nuclear interior and has its largest contribution in the region of the surface. However, zero range forces share with the monopole force with constant G the problem of an ultraviolet divergence and require the use of a rather arbitrary pairing window and its strength V_0 has to be renormalized accordingly.[50]

Finite range pairing forces do not show the ultraviolet divergence, because the finite range introduces a natural cut-off in momentum space. In order to avoid the ambiguities of the size of the pairing window, which reduce the predictive power of such calculations in areas where the experimental gap parameters cannot be deduced from odd-even mass differences, the pairing part of the finite-range Gogny force[53]

$$V(\boldsymbol{r}, \boldsymbol{r}') = \sum_{i=1,2} e^{((r-r')/\mu_i)^2} (W_i + B_i P^\sigma - H_i P^\tau - M_i P^\sigma P^\tau), \tag{15}$$

has been introduced in a hybride model[54] with relativistic forces in the *ph*-channel and non-relativistic Gogny-forces in the *pp*-channel. In such calculations one uses usually with great success the parameters μ_i, W_i, B_i, H_i and M_i of the Gogny force D1S.[55] It has been shown that after proper renormalization, the above density-dependent force of zero range and the finite range Gogny force produce essentially the same results.[56] Recently, a separable form of the Gogny force[57] has been introduced for RHB calculations in oscillator space which reduces considerably the computational cost, in particular for nuclei with axial and triaxial deformations.[58–60]

2.2. *The rBCS approach*

For weakly bound nuclei, the contribution of continuum states cannot be included properly in the conventional BCS approach. One way to avoid this problem is the rBCS method. Here one first locates resonant states in the continuum and determines the resonance parameters with various methods.[61–67] Then by using the resonant-BCS (rBCS) approach,[36] one includes the contribution of these resonant states and studies the halo phenomena.[37,38,68,69] The main physics is the following. The anomalous density matrix would diverge whenever a local pairing interaction is used (see Ref. 39 and references therein); therefore, a certain form of cutoff is needed to simulate the finite-range nature of the pairing interaction,[36,39] which is sensitive only to the inner part of the resonant wave functions. That is, the wave functions of resonant states have to be taken to be zero beyond a cutoff radius.[36] Without this cutoff, an unnatural neutron gas will appear outside the nucleus.[36,40] This procedure is justified because in the case of resonant states, which are well localized inside the nucleus, there is a large region outside the nucleus where the resonant wave functions have values close to zero before they start oscillating. Therefore, the HF+rBCS and RMF+rBCS results do not depend sensitively on the cutoff radius chosen in such a region.[36]

The extended BCS equations for a general (finite range) pairing interaction including the contribution of the resonant continuum read:[36,37]

$$\Delta_i = \sum_j V_{i\bar{i}j\bar{j}} u_j v_j + \sum_\nu V_{i\bar{i},\nu\epsilon_\nu\overline{\nu\epsilon_\nu}} \int_{I_\nu} g_\nu(\epsilon) u_\nu(\epsilon) v_\nu(\epsilon) d\epsilon \ , \tag{16}$$

$$\Delta_\nu \equiv \sum_j V_{\nu\epsilon_\nu\overline{\nu\epsilon_\nu},j\bar{j}} u_j v_j + \sum_{\nu'} V_{\nu\epsilon_\nu\overline{\nu\epsilon_\nu},\nu'\epsilon_{\nu'}\overline{\nu'\epsilon_{\nu'}}} \int_{I_{\nu'}} g_{\nu'}(\epsilon') u_{\nu'}(\epsilon') v_{\nu'}(\epsilon') d\epsilon' \ , \tag{17}$$

$$N = \sum_i v_i^2 + \sum_\nu \int_{I_\nu} g_\nu^c(\epsilon) v_\nu^2(\epsilon) d\epsilon \ . \tag{18}$$

For details of the notation see Refs. 36, 37. Here Δ_i are the gaps for the bound states and Δ_ν, ϵ_ν are the averaged gaps and energies for the resonant states. The quantity $g_\nu^c(\epsilon) = \frac{2j_\nu+1}{\pi}\frac{d\delta_\nu}{d\epsilon}$ is the total level density and δ_ν is the phase shift of angular momentum $(l_\nu j_\nu)$. The factor $g_\nu^c(\epsilon)$ takes into account the variation of the localisation of scattering states in the energy region of a resonance (i.e. the width effect) and goes to a delta function in the limit of a very narrow resonance. The interaction matrix elements are calculated with the scattering wave functions at the resonance energies and normalized inside a volume where the pairing interaction is active.

In the RMF+rBCS approach, one should first solve the RMF equations with the scattering-type boundary conditions or any other methods like the analytical continuation of coupling constant (ACCC) method, the real stabilization method, the complex scaling method, etc.[61–67,70] and find the energies, widths and wave functions for the resonant states. Then rBCS equations (16,17,18) can be solved by including both bound states and resonant states.

2.3. *Relativistic continuum Hartree–Bogoliubov theory for spherical nuclei in r-space*

For spherical nuclei, one can solve the RHB equations in coordinate (r) space,

$$\int d\boldsymbol{r}'\mathcal{H}(\boldsymbol{r},\boldsymbol{r}')\begin{pmatrix} U_k(\boldsymbol{r}') \\ V_k(\boldsymbol{r}') \end{pmatrix} = E_k \begin{pmatrix} U_k(\boldsymbol{r}) \\ V_k(\boldsymbol{r}) \end{pmatrix}, \tag{19}$$

where $\mathcal{H}$ is the RHB matrix

$$\mathcal{H}(\boldsymbol{r},\boldsymbol{r}') = \begin{pmatrix} h_D(\boldsymbol{r},\boldsymbol{r}') - \lambda\delta(r-r') & \Delta(\boldsymbol{r},\boldsymbol{r}') \\ \Delta(\boldsymbol{r},\boldsymbol{r}') & -h_D(\boldsymbol{r},\boldsymbol{r}') + \lambda\delta(r-r') \end{pmatrix}. \tag{20}$$

The pairing potential can be written as,

$$\Delta_\alpha(\boldsymbol{r},\boldsymbol{r}') = -\frac{1}{2}\sum_\beta V^{pp}_{\alpha\beta}(\boldsymbol{r},\boldsymbol{r}')\,\kappa_\beta(\boldsymbol{r},\boldsymbol{r}'), \tag{21}$$

with the pairing tensor κ

$$\kappa_\alpha(\boldsymbol{r},\boldsymbol{r}') = V_\alpha(\boldsymbol{r})^* U^T_\alpha(\boldsymbol{r}'). \tag{22}$$

The spherical quasiparticle Dirac spinors read,

$$U_k = \frac{1}{r}\begin{pmatrix} \mathrm{i}G^k_U(r)\mathcal{Y}^l_{jm}(\theta,\phi) \\ -F^k_U(r)\mathcal{Y}^{l'}_{jm}(\theta,\phi) \end{pmatrix}, \tag{23}$$

$$V_k = \frac{1}{r}\begin{pmatrix} \mathrm{i}G^k_V(r)\mathcal{Y}^l_{jm}(\theta,\phi) \\ -F^k_V(r)\mathcal{Y}^{l'}_{jm}(\theta,\phi) \end{pmatrix}. \tag{24}$$

The angular part of the RHB equation can be integrated out and the radial RHB equations are derived as

$$
\begin{aligned}
&\frac{dG_U(r)}{dr} + \frac{\kappa}{r}G_U(r) - (E + \lambda - \Sigma_0(r) + \Sigma_S(r))F_U(r) \\
&\qquad\qquad + r\int r'dr'\Delta_F(r,r')F_V(r') = 0, \\
&\frac{dF_U(r)}{dr} - \frac{\kappa}{r}F_U(r) + (E + \lambda - \Sigma_0(r) - \Sigma_S(r))G_U(r) \\
&\qquad\qquad + r\int r'dr'\Delta_G(r,r')G_V(r') = 0, \\
&\frac{dG_V(r)}{dr} + \frac{\kappa}{r}G_V(r) + (E - \lambda + \Sigma_0(r) - \Sigma_S(r))F_V(r) \\
&\qquad\qquad + r\int r'dr'\Delta_F(r,r')F_U(r') = 0, \\
&\frac{dF_V(r)}{dr} - \frac{\kappa}{r}F_V(r) - (E - \lambda + \Sigma_0(r) + \Sigma_S(r))G_V(r) \\
&\qquad\qquad + r\int r'dr'\Delta_G(r,r')G_U(r') = 0.
\end{aligned}
\tag{25}
$$

Here $\Sigma_S(r) = M + S(r)$ and $\Sigma_0(r) = V(r)$ are the scalar and vector potentials discussed in Chapter 2 and the pairing potentials are calculated as

$$
\begin{aligned}
\Delta_G^\kappa(r,r') &= \tfrac{1}{4}\textstyle\sum_{\kappa'} V_{\kappa\kappa'}^{J=0}(r,r')g_{\kappa'}\sum_{E_k>0}[G_U^k(r)G_V^k(r') + G_U^k(r')G_V^k(r)], \\
\Delta_F^\kappa(r,r') &= \tfrac{1}{4}\textstyle\sum_{\kappa'} V_{\kappa\kappa'}^{J=0}(r,r')g_{\kappa'}\sum_{E_k>0}[F_U^k(r)F_V^k(r') + F_U^k(r')F_V^k(r)],
\end{aligned}
\tag{26}
$$

in which, in comparison with Eq. (13), the components $\hat{\Delta}_{ff}$ is included as well. If the zero-range pairing force is used, the above coupled integro-differential equations are reduced to differential ones, which may be directly solved in coordinate space.[71]

2.4. *Deformed relativistic continuum Hartree–Bogoliubov theory*

For deformed nuclei the solution of HFB or RHB equations in r space is numerically a very demanding task. Considerable efforts have been made to develop mean field models either in r space or in a basis with an improved asymptotic behavior at large distances.[72–78] In particular, an expansion in a Woods–Saxon (WS) basis was proposed as a reconciler between the HO basis and the integration in coordinate space.[75] Woods–Saxon wave functions have a much more realistic asymptotic behavior at large r than the harmonic oscillator wave functions. A discrete set of Woods–Saxon wave functions is obtained by using box boundary conditions to discretize the continuum. It has been shown in Ref. 75 for spherical systems that the solution of the relativistic Hartree equations in a Woods–Saxon basis is equivalent to the solution in coordinate space. The Woods–Saxon basis has been used recently for both non-relativistic[79] and relativistic Hartree–Fock–Bogoliubov theories with finite range pairing force (Chapter 4).

Over the past years, lots of efforts have been made to develop a deformed Relativistic Hartree (RH) theory[80] and a deformed relativistic Hartree–Bogoliubov theory in the continuum (DRHBc) theory.[81] As a first application, halo phenomena in deformed nuclei have been investigated within the DRHBc theory.[32] The detailed theoretical framework is available in Ref. 33.

For axially deformed nuclei with spatial reflection symmetry, the potentials $\Sigma_S(\boldsymbol{r})$ and $\Sigma_0(\boldsymbol{r})$ and various densities can be expanded in terms of the Legendre polynomials,[82]

$$f(\boldsymbol{r}) = \sum_\lambda f_\lambda(r) P_\lambda(\cos\theta), \ \lambda = 0, 2, 4, \cdots, \tag{27}$$

with

$$f_\lambda(r) = \frac{2\lambda+1}{2}\int_{-1}^{1} d(\cos\theta) f(\boldsymbol{r}) P_\lambda(\cos\theta). \tag{28}$$

The quasiparticle wave functions U_k and V_k are expanded in terms of spherical Dirac spinors $\varphi_{n\kappa m}(\boldsymbol{r}sp)$ with the eigenvalues $\epsilon_{n\kappa}$ obtained from the solution of a Dirac equation $h_D^{(\lambda=0)}$ containing spherical potentials $S^{(0)}(r)$ and $V^{(0)}(r)$ of Woods–Saxon shape:[75,83]

$$U_k(\boldsymbol{r}sp) = \sum_{n\kappa} u_{k,(n\kappa)}^{(m)} \varphi_{n\kappa m}(\boldsymbol{r}sp), \tag{29}$$

$$V_k(\boldsymbol{r}sp) = \sum_{n\kappa} v_{k,(n\kappa)}^{(m)} \bar{\varphi}_{n\kappa m}(\boldsymbol{r}sp). \tag{30}$$

The basis wave functions read

$$\varphi_{n\kappa m}(\boldsymbol{r}s) = \frac{1}{r}\begin{pmatrix} iG_{n\kappa}(r)\mathcal{Y}_{jm}^{l}(\Omega s) \\ -F_{n\kappa}(r)\mathcal{Y}_{jm}^{\tilde{l}}(\Omega s)\end{pmatrix}, \tag{31}$$

where $G_{n\kappa}(r)/r$ and $F_{n\kappa}(r)/r$ are the radial wave functions for the upper and lower components. The state $\bar{\varphi}_{n\kappa m}(\boldsymbol{r}sp)$ is the time reversal state of $\varphi_{n\kappa m}(\boldsymbol{r}sp)$. The spherical spinor $\varphi_{n\kappa m}$ is characterized by the radial quantum number n, angular momentum j, and the parity π; the latter two are combined to the relativistic quantum number $\kappa = \pi(-1)^{j+1/2}(j+1/2)$ which runs over positive and negative integers $\kappa = \pm1, \pm2, \cdots$. For the spinor spherical harmonics $\mathcal{Y}_{jm}^{l}$ and $\mathcal{Y}_{jm}^{\tilde{l}}$, $l = j + \frac{1}{2}\text{sign}(\kappa)$ and $\tilde{l} = j -$

$\frac{1}{2}\text{sign}(\kappa)$. These states form a complete spherical and discrete basis in Dirac space (see Ref. 75 for details). Because of the axial symmetry the z-component m of the angular momentum j is a conserved quantum number and the RHB Hamiltonian can be decomposed into blocks characterized by m and parity π. For each $(m\pi)$-block, solving the RHB equation is equivalent to the diagonalization of the matrix

$$\begin{pmatrix} \mathcal{A}-\lambda & \mathcal{B} \\ \mathcal{B}^\dagger & -\mathcal{A}^*+\lambda \end{pmatrix} \begin{pmatrix} \mathcal{U}_k \\ \mathcal{V}_k \end{pmatrix} = E_k \begin{pmatrix} \mathcal{U}_k \\ \mathcal{V}_k \end{pmatrix}, \tag{32}$$

where

$$\mathcal{U}_k = \left(u_{k,(n\kappa)}^{(m)}\right), \ \mathcal{V}_k = \left(v_{k,(n\kappa)}^{(m)}\right), \tag{33}$$

and

$$\mathcal{A} = \left(h_{D(n\kappa)(n'\kappa')}^{(m)}\right) = \left(\langle n\kappa m|h_D|n'\kappa', m\rangle\right), \tag{34}$$

$$\mathcal{B} = \left(\Delta_{(n\kappa)(n'\kappa)}^{(m)}\right) = \left(\langle n\kappa m|\Delta|\overline{n'\kappa', m}\rangle\right). \tag{35}$$

Further details are given in Ref. 33.

3. Halos, giant halos and deformed halos

Halo has become one of the most interesting exotic nuclear phenomena since it was discovered in ^{11}Li.[5] The RMF+BCS models have been used to study nuclear halo phenomena since the early 1990s.[84–101] By incorporating the Bogoliubov transformation in the relativistic Hartree theory, the relativistic continuum Hartree–Bogoliubov (RCHB) theory has been developed and extensively used to study the halo phenomenon in spherical nuclei.[30,42,71,102,103] The RCHB theory was also extended to include the hyperon and used to study neutron halos in Λ-hypernuclei.[104–108] The relativistic Hartree–Fock–Bogoliubov theory in the continuum (RHFBc) has been developed and used to study the influence of exchange terms on the formation of neutron halos.[109] Over the past years, lots of efforts have been made to develop a deformed relativistic Hartree–Bogoliubov theory in the continuum (DRHBc).[81] Halo phenomena in deformed Ne and Mg isotopes have been investigated with the DRHBc theory.[32,33,110] In this Section, we present recent progress on the application of the RMF+rBCS, RCHB and DRHBc theories for the neutron halo phenomena in spherical and deformed nuclei. More details can be found in Ref. 29.

3.1. *Neutron halo in* ^{11}Li

The ground state properties of Li isotopes have been investigated using the RCHB theory.[42] A satisfactory agreement with experimental values is found for the binding energies and the radii of $^{6-11}$Li. The matter radius, showing a considerable increase from the nucleus ^{9}Li to ^{11}Li, has been reproduced by the RCHB theory. In contrast to the earlier mean field calculations,[86,111,112] the results given in Ref. 42 were obtained without any artificial modifications of the potential.

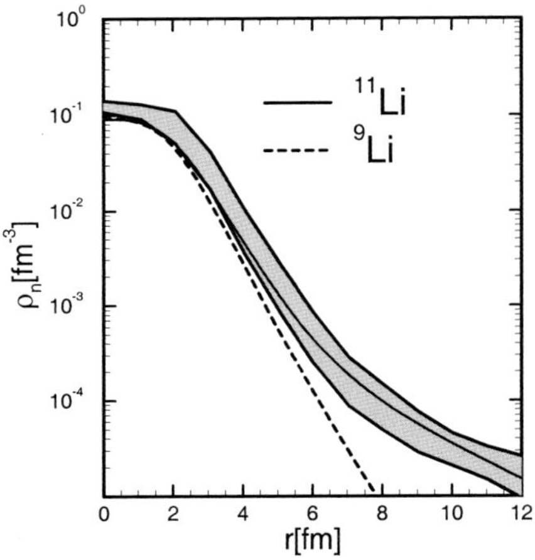

Fig. 1. Calculated and experimental density distributions in ^{11}Li and ^{9}Li. The solid line shows the result for ^{11}Li while the dashed line corresponds to the calculation for ^{9}Li. The shaded area gives the experimental results with error bars. Taken from Ref. 42.

In Fig. 1 the neutron density distributions of ^{9}Li and ^{11}Li are shown. It is clearly seen that the increase of the matter radius is caused by the large neutron density distribution in ^{11}Li. Its density distribution is in very good agreement with the experimental density shown with its error bars by the shaded area.

By examining the mean fields for protons and neutrons and the single particle levels in the canonical basis, it has been shown that the halo in ^{11}Li is formed by Cooper-pairs scattered in the two levels $1p_{1/2}$ and $2s_{1/2}$ with the former below the Fermi surface and the latter in the continuum; the pairing interaction couples those levels below the Fermi surface with the continuum. The RCHB theory provides a much more general mechanism for halo: a halo appears if the Fermi level is close to the continuum and around the Fermi level there are single particle levels with low orbital angular momenta and correspondingly low centrifugal barriers.

3.2. *Giant halos*

The halo nuclei observed so far are composed of one or two halo nucleons only. The RCHB theory has been used to study the influence of correlations and many-body effects in nuclei with a larger number of neutrons distributed in the halo channel. A giant neutron halo has been predicted in Zr isotopes close to the neutron drip line.[30] It is formed by up to six neutrons scattered among the positive energy states outside the ^{122}Zr core with the magic neutron number $N = 82$.

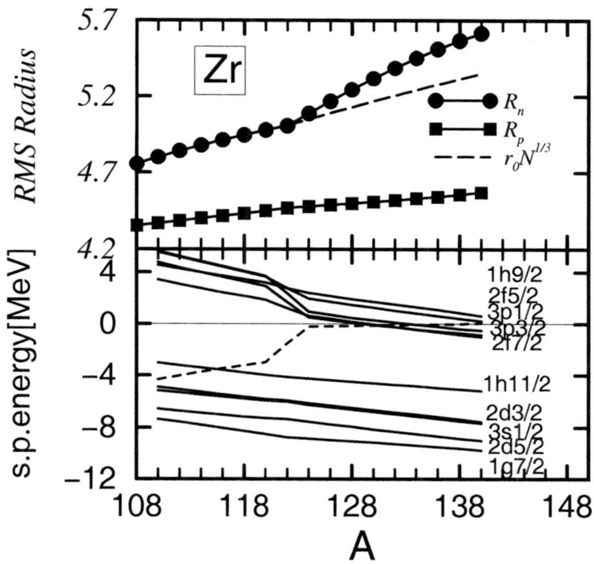

Fig. 2. Upper part: Root mean square radii for neutrons and protons in Zr isotopes close to the neutron drip line as a function of the mass number A. Lower part: single particle energies for neutrons in the canonical basis. The dashed line indicates the chemical potential. Taken from Ref. 30.

In the upper panel of Fig. 2, the rms radii of the protons and neutrons are presented for the Zr isotopes. There is a clear kink for the neutron rms radius at the magic neutron number $N = 82$. This kink can be understood by examining the single particle levels in the canonical basis which is shown in the lower panel of Fig. 2. Going from $N = 70$ to $N = 100$, there is a big gap above the $1h_{11/2}$ orbital. For $N > 82$, the neutrons are filled to the levels in the continuum or weakly bound states in the order of $3p_{3/2}$, $2f_{7/2}$, $3p_{1/2}$, $2f_{5/2}$ and $1h_{9/2}$. The neutron chemical potential is given by a dashed line. It approaches rapidly the continuum threshold already shortly after the magic neutron number $N = 82$ and crosses the continuum at $N = 100$ for the nucleus ^{140}Zr. In this region the chemical potential keeps a very small but negative value for a large A range. This means that the additional neutrons are added with a very small, nearly vanishing binding energy at the edge of the continuum provided by pairing correlations. The total binding energies E for the isotopes above ^{122}Zr are therefore almost identical. This has been recognized in Ref. 85. However, the RMF calculations with the BCS approximation expanded on an oscillator basis is definitely not appropriate for exotic nuclei close to the continuum limit.

From the occupation probabilities in the canonical basis, one can get the number of halo neutrons. It was found that there are 2 valence neutrons in ^{124}Zr, 4 in ^{126}Zr, 6 in ^{128}Zr, ..., roughly 5 in ^{138}Zr, and roughly 6 in ^{140}Zr where the neutron drip line is reached. With the very large neutron rms radii of these systems, one can estimate the number of valence neutrons which fill in the same volume outside the ^{122}Zr core if packed with normal neutron density. This number is 24 for ^{134}Zr and 34 for ^{140}Zr. This phenomenon is therefore clearly a neutron halo and was called a Giant Halo because of the large number of particles in the halo region.[30]

It is very difficult for experimentalists to study giant halos in Zr isotopes because they are too heavy to be synthesized by the RIB facilities at present. Thus it is useful to investigate the giant halo phenomena in lighter nuclei which are easily accessible with available facilities. In Refs. 31,113, ground state properties of all the even-even O, Ca, Ni, Zr, Sn, Pb isotopes ranging from the proton drip line to the neutron drip line were investigated by using the RCHB theory with the effective interaction NLSH.[114] Recently, a systematic calculation for nuclei above oxygen has also been performed with the effective interaction PC-PK1,[115] aiming to investigate the global impact of the continuum for the nuclear boundary; the results ranging from O to Ti isotopes have been published in Ref. 116. In such a systematic study, it is found that the appearance of giant halos, the contribution of continuum and boundaries of nuclear chart are tightly connected. We note that the functional PC-PK1 has proven to be very successful in describing the isospin dependence of the nuclear masses,[117] the Coulomb displacement energies between mirror nuclei,[118] fission barriers[119,120] and nuclear rotations,[121–124] etc.

For nuclei far from the valley of stability and with small nucleon separation energy, the Fermi surface is very close to the continuum threshold and the valence nucleons may extend over quite a wide space to form low density nuclear matter. Therefore the nucleon separation energies are sensitive quantities to test theory and examine the halo phenomena. The two-neutron separation energies for O, Ca, Ni, Zr, Sn, and Pb isotopes have been investigated to examine their connection with systematic behavior of the nuclear size.[31,113,125]

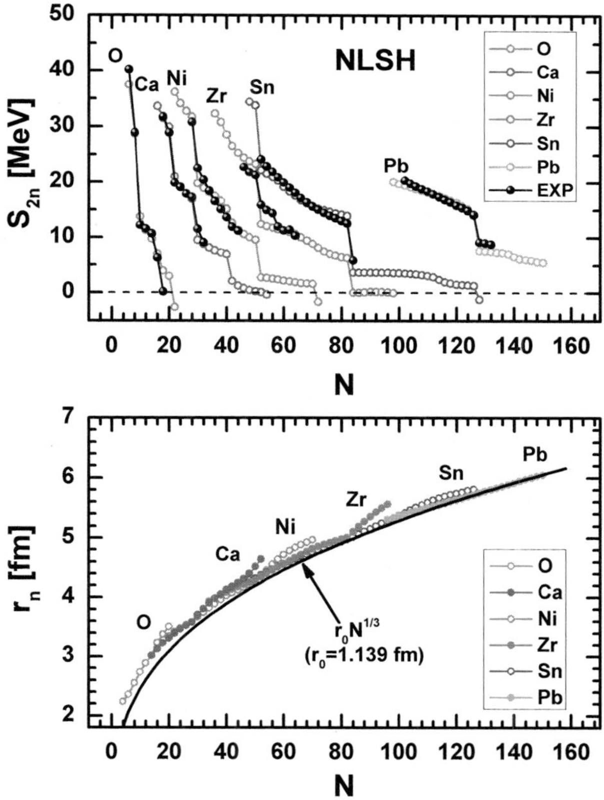

Fig. 3. (Color online) (a) The two-neutron separation energies S_{2n} calculated by RCHB theory[71] with NLSH[114] (Open symbols) in comparison to available data (solid ones) for the even O, Ca, Ni, Zr, Sn, and Pb isotopes against the neutron number N. (b) The root mean square (rms) neutron radii r_n for even O, Ca, Ni, Zr, Sn, and Pb isotopes by RCHB calculations against the neutron number N. The curve $r_0N^{1/3}$ with $r_0 = 1.139$ fm has been included to guide the eye. Taken from Ref. 31 with modified color.

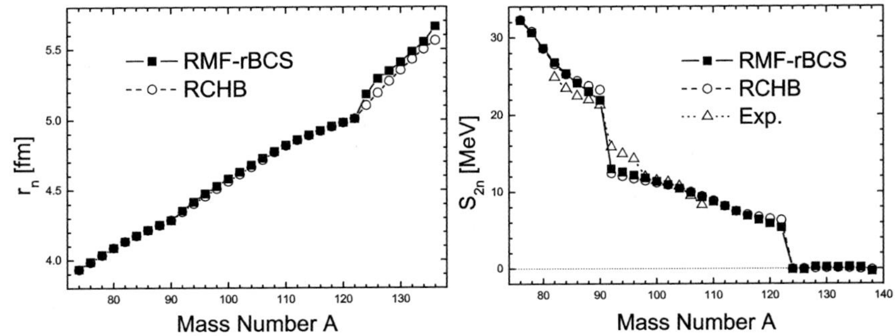

Fig. 4. The rms neutron radii (left panel) and the two-neutron separation energies (right panel) of even Zr isotopes as a function of the mass number A. Taken from Ref. 37.

In Fig. 3, similar to Ref. 31, the two-neutron separation energies S_{2n} and the neutron radii r_n from the RCHB calculation for the even-even nuclei of the O, Ca, Ni, Zr, Sn, and Pb isotope chains have been shown. The predicted r_n curve using the simple empirical equation $r_n = r_0 N^{1/3}$ with $r_0 = 1.139$ fm normalizing to ^{208}Pb is also represented in the figure. This simple formula for r_n agrees with the calculated neutron radii with exceptions in neutron-rich Ca and Zr isotopes.

The good agreement between experimental and calculated S_{2n} is clearly shown in Fig. 3. The experimental magic or submagic numbers $N = 20, 28,$ and 40 are reproduced. The S_{2n} values for several nuclei in exotic Ca and Zr isotopes are extremely close to zero. As discussed in Refs. 30,31, if taking ^{60}Ca or ^{122}Zr respectively as a core, the valence neutrons will gradually occupy the loosely bound states and the continuum above the sub-shell of $N = 40$ in Ca or closed shell of $N = 82$ in Zr isotopes. Such a typical behavior of S_{2n} can be taken as evidence of the occurrence of giant halos in Ca chain and Zr chain.[30]

In Fig. 3, it is very interesting to see that r_n follows the $N^{1/3}$ systematics well for stable nuclei although their proton numbers are quite different. Near the drip line, distinct abnormal behaviors appear at $N = 40$ in Ca isotopes and at $N = 82$ in Zr isotopes. As pointed out in Ref. 31, nuclei having the abnormal r_n increase correspond to those having small S_{2n}, which provides another piece of evidence to the emergence of giant halos. The increase of r_n in exotic Ni, Sn and Pb nuclei is not as rapid as those in Ca and Zr chains. The regions of the abnormal increases of the neutron radii are just the same as those for S_{2n}. Both behaviors are connected with the formation of giant halo.[30,31,113,125]

The RMF+rBCS method has also been applied to study neutron-rich Zr isotopes.[37] It was found that the sudden increase of neutron radii close to the neutron drip line depends on a few resonant states close to the continuum threshold. Including into the RMF-BCS calculations only these resonant states one gets for the neutron radii and neutron separation energies practically the same results as those obtained in the RCHB calculations (see Fig. 4).

3.3. *Halos in hyper nuclei*

Motivated by knowledge of Λ-N interaction and understanding of giant halo,[30,31,113] it is interesting to study the possible appearances of halos in exotic hyper nuclei.

By examining two neutron separation energies S_{2n} for normal nuclei, single-Λ hyper nuclei and double-Λ hyper nuclei of Ca isotopes, labelled by $\Lambda = 0$, $\Lambda = 1$ and $\Lambda = 2$, respectively, from the proton drip line to neutron drip line were investigated.[106] It was found that one or two Λ hyperons lower the Fermi level a little bit but keep the neutron shell structure unchanged. Therefore, the neutron drip line is pushed outside from $N = 52$ in Ca isotope chain to $N = 54$ in hyper isotope chain. Meanwhile, giant halo due to pairing correlation and the contribution from the continuum still exist in Ca hyper nuclei similar to that in Ca isotopes.[31] This is a slight but rewarding step for exploring the limit of drip line nuclei on the basis of the giant halo. For more details, see Ref. 106.

Apart from neutron halos in hyper nuclei, as Λ hyperon is less bound than the corresponding nucleon in nuclei, it is worth investigating the existence of hyperon halos.

In Refs. 126,127, a Λ halo was tentatively suggested with the RMF+BCS model. However, since the conventional BCS method will have an unphysical solution involving a baryon gas,[71] such a prediction needs further study. In Ref. 105, hyper carbon isotopes are studied by using the RCHB theory. By investigating the baryon density distributions in hyper nuclei $^{13}_{\Lambda}$C, $^{14}_{2\Lambda}$C, and $^{15}_{3\Lambda}$C, it was found that for up to two Λ hyperons added to the core ^{12}C, the nucleon density distributions remain the same and hyperon density distributions at the tail are comparable with those of the nucleons. An intriguing phenomenon appears in $^{15}_{3\Lambda}$C where the hyperon density distribution has a long tail extended far outside of its core $^{14}_{2\Lambda}$C. This is a signature of hyperon halo due to the weakly bound state $1p^{\Lambda}_{3/2}$ in $^{15}_{3\Lambda}$C which has a small hyperon separation energy and a density distribution with long tail.[105]

3.4. *Deformed halo and shape decoupling*

Most open shell nuclei are deformed. The interplay between deformation and weak binding raises interesting questions, such as whether or not there exist halos in deformed nuclei and, if yes, what their new features are.[128–133] The DRHBc theory has been used to study Mg isotopes and answer these questions concerning halos in deformed nuclei.

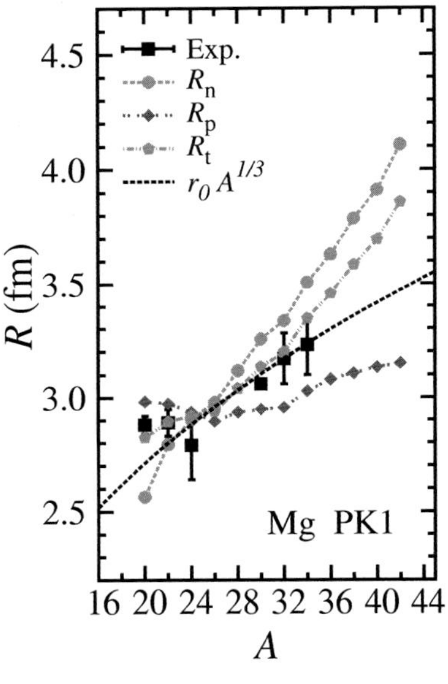

Fig. 5. (Color online) The root mean square (rms) radii for Mg isotopes calculated in DRHBc theory plotted as functions of the mass number. The neutron radii $R_{\rm n}$, proton radii $R_{\rm p}$, matter radii $R_{\rm t}$, the $r_0 A^{1/3}$ curve, and the available data for $R_{\rm t}$[134,135] are given. The neutron radius $R_{\rm n}$, proton radius $R_{\rm p}$ and total radius $R_{\rm t}$ of Mg isotopes are calculated with PK1. The $r_0 A^{1/3}$ curve is included to guide the eye. The data for matter radii labeled by "Exp." are taken from Refs. 134,135. Taken from Ref. 33.

In Fig. 5, the rms radii for Mg isotopes are plotted as functions of the mass number, including the neutron radii $R_{\rm n}$, proton radii $R_{\rm p}$, matter radii $R_{\rm t}$, the $r_0 A^{1/3}$ curve, and the available data for $R_{\rm t}$.[134,135] The parameter $r_0 = 1.26$ fm in the $r_0 A^{1/3}$ curve is determined by the neutron radius of ^{24}Mg. It is shown that the deformed RHB results agree well with experiment for matter radii. The proton radii are almost constant with a slow increase with A due to the neutron-proton coupling included in the mean field. The neutron radii $R_{\rm n}$ follows more or less the $r_0 A^{1/3}$ curve up to

$N = 20$ with two exceptions. One is ^{32}Mg for which the neutron radius $R_{\rm n}$ is a bit below the $r_0 A^{1/3}$ curve. This is due to the strong shell effect at $N = 20$ in the calculations. Another exception is after ^{32}Mg which shows a quick increase in the neutron radius. This indicates a halo structure.

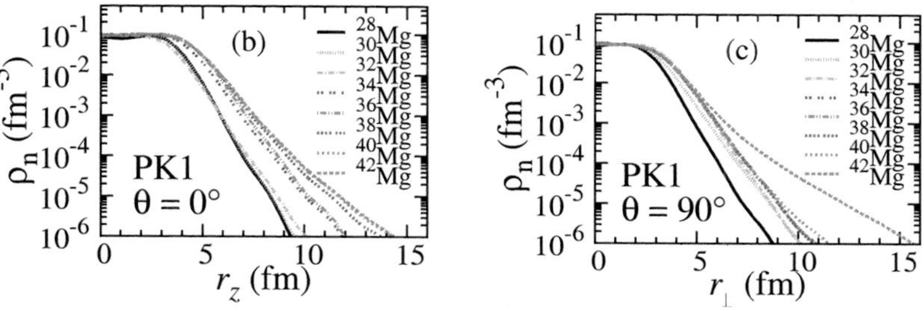

Fig. 6. (Color online) Neutron density profiles of even-even Mg isotopes with $A \geq 28$ calculated in DRHBc theory with the parameter set PK1. The left panel refers to the density distribution $\rho_{\rm n}(z, r_\perp = 0)$ along the symmetry axis z ($\theta = 0°$) with $r_\perp = \sqrt{x^2 + y^2}$ and the right panel refers to $\rho_{\rm n}(z = 0, r_\perp)$ that is perpendicular to the symmetry axis z ($\theta = 90°$). Taken from Ref. 33.

Figure 6 shows neutron density profiles of even-even Mg isotopes with $A \geq 28$ calculated in DRHBc theory with the parameter set PK1.[33] The density distributions along the symmetry axis z ($\theta = 0°$) and the axis perpendicular to the symmetry axis z ($\theta = 90°$) are respectively represented by $\rho_{\rm n}(z, r_\perp = 0)$ and $\rho_{\rm n}(z = 0, r_\perp)$ with $r_\perp = \sqrt{x^2 + y^2}$.

The density distribution along the symmetry axis $\rho_{\rm n}(z, r_\perp = 0)$ changes abruptly from ^{32}Mg to ^{34}Mg. This can be understood easily from the shape change in these two nuclei: ^{32}Mg is spherical but ^{34}Mg is prolate which means that it is elongated in the z axis. In the direction perpendicular to the symmetry axis, the neutron density $\rho_{\rm n}(z = 0, r_\perp)$ of ^{42}Mg extends very far away from the center of the nucleus and a long tail emerges, revealing the formation of a halo. By comparing $\rho_{\rm n}(z, r_\perp = 0)$ and $\rho_{\rm n}(z = 0, r_\perp)$ for ^{42}Mg, it was found that in the tail part, the neutron density distributes more along the direction perpendicular to the symmetry axis. Since this nucleus as a whole is prolate, it indicates that the tail part has a different shape from the nucleus itself. This is the shape decoupling which was first predicted in Ref. 32.

The deformed relativistic Hartree–Bogoliubov theory in continuum with the parameter set PK1 predicted the last bound neutron rich nucleus for Mg isotopes as ^{42}Mg.[33] There are both prolate and oblate stable solutions for

^{42}Mg. The ground state of ^{42}Mg is prolate. With the triaxially deformed relativistic mean field calculations,[136] it is shown that the obtained oblate minimum is not an isomeric state but a saddle point in the potential energy surface. It is found that the ground state of ^{42}Mg is well deformed with $\beta \approx 0.41$ and two neutron separation energy $S_{2n} \approx 0.22$ MeV.[33] The density distribution of such a weakly bound nucleus has a very long tail in the direction perpendicular to the symmetry axis, see Fig. 6, which indicates that the prolate ^{42}Mg has an oblate halo.

The density distribution is decomposed into contributions of the oblate "halo" and of the prolate "core" in Fig. 7. The density distribution of this weakly bound nucleus has a very long tail in the direction perpendicular to the symmetry axis which indicates the prolate nucleus ^{42}Mg has an oblate halo and the deformations of the core and the halo are quite different. This is the decoupling between the deformations of the core and the halo.[32] It should be noted that in the calculations with NL3, ^{46}Mg is the last bound nucleus.[32]

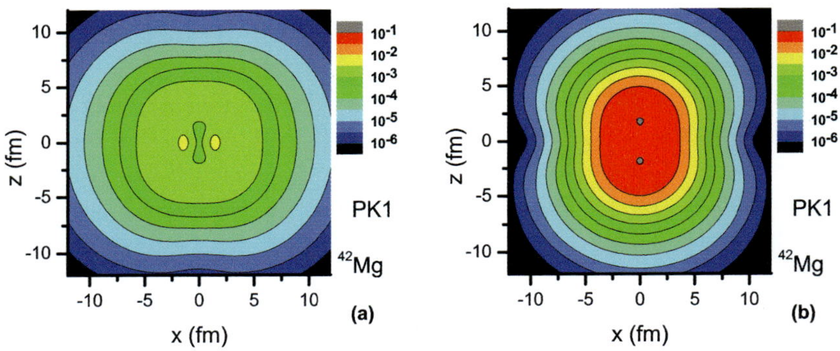

Fig. 7. Density distributions of the ground state of ^{42}Mg with the z axis as the symmetry axis: (a) the neutron halo, and (b) the neutron core. Taken from Ref. 33.

The single particle spectrum around the Fermi level for the ground state of ^{42}Mg is shown in Fig. 8.[33] The good quantum numbers of each single particle state are also shown. The occupation probabilities v^2 in the canonical basis have BCS-form[47] and are given by the length of the horizontal lines in Fig. 8. The levels close to the threshold are labeled by the number i according to their energies, and their conserved quantum number Ω^π as well as the main spherical components are given at the right hand side. The neutron Fermi level is within the pf shell and most of the single particle levels have negative parities. Since the chemical potential λ_n is close to the continuum, orbitals above the threshold have noticeable occupations due to the pairing correlations. The single neutron levels of ^{42}Mg can be divided into two parts, the deeply bound levels ($\varepsilon_{\rm can} < -2$ MeV) corresponding to the "core", and the remaining weakly bound levels close to the threshold ($\varepsilon_{\rm can} > -0.3$ MeV) and in the continuum corresponding to the "halo".

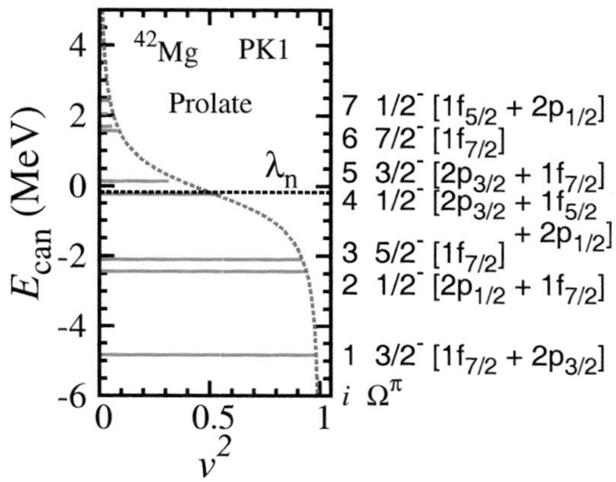

Fig. 8. (Color online) Single neutron levels of ground state of ^{42}Mg in the canonical basis as a function of the occupation probability v^2. The blue dashed line corresponds to the BCS-formula with an average pairing gap. Taken from Ref. 33.

As discussed in Refs. 32, 137, 33, the shape of the halo originates from the intrinsic structure of the weakly bound or continuum orbitals. By examining the neutron density distribution, it was found that for the ground state of ^{42}Mg, the halo is mainly formed by level 4 and level 5. A detailed analysis of the intrinsic properties of these orbits revealed that in ^{42}Mg the shape of the halo is slightly oblate and decouples from the prolate core. We note that this kind of shape decoupling was later also found in non-relativistic HFB calculations.[138] An exotic "egg"-like structure consisting of a spherical core plus a prolate halo was predicted in ^{38}Ne, in which the near-threshold nonresonant continuum plays an essential role.

4. Surface diffuseness

In atomic nuclei, the surface diffuse properties are intimately related to the spin-orbit splitting,[139] the evolution of single particle shell structure,[30] the pseudospin symmetry[140,141] and spin symmetry,[142] and the nuclear surface energy and thus nuclear masses.[143] In this section, we will present recent progress on nuclear surface diffuseness and the corresponding impact on nuclear shell structure and size.

4.1. *Spin-orbit splitting and shell structure*

The RCHB theory has been used to systematically study the isospin dependence of the mean potential for the Sn isotopes from proton drip line to neutron drip line[139] and the spin-orbit splitting for the whole isotope chain was examined in detail. As an example, the spin-orbit splitting $E_{ls} = (E_{lj=l-1/2} - E_{lj=l+1/2})/(2l+1)$ versus the binding energy $E = [(l+1)E_{lj=l-1/2} + lE_{lj=l+1/2}]/(2l+1)$ in ^{A}Sn with $A = 110, 120, \cdots, 170$ are given in Fig. 9 for the neutron spin-orbit partners $(1d_{3/2}, 1d_{5/2})$, $(1g_{7/2}, 1g_{9/2})$, $(1i_{11/2}, 1i_{13/2})$, $(1p_{1/2}, 1p_{3/2})$, $(1f_{5/2}, 1f_{7/2})$ and $(1h_{9/2}, 1h_{11/2})$, and the proton spin-orbit partners $(1d_{3/2}, 1d_{5/2})$ and $(1f_{5/2}, 1f_{7/2})$.

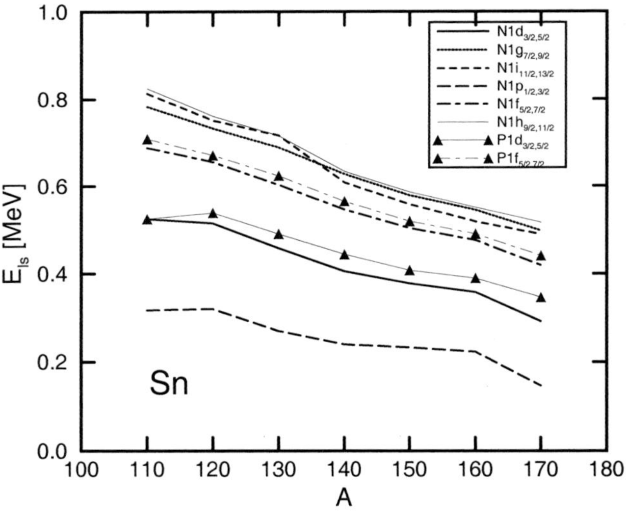

Fig. 9. The neutron spin-orbit splitting $E_{ls} = (E_{lj=l-1/2} - E_{lj=l+1/2})/(2l+1)$ in Sn isotopes for neutron $(1d_{3/2}, 1d_{5/2})$, $(1g_{7/2}, 1g_{9/2})$, $(1i_{11/2}, 1i_{13/2})$, $(1p_{1/2}, 1p_{3/2})$, $(1f_{5/2}, 1f_{7/2})$, and $(1h_{9/2}, 1h_{11/2})$ orbitals and proton $(1d_{3/2}, 1d_{5/2})$, and $(1f_{5/2}, 1f_{7/2})$ orbitals, respectively. Taken from Ref. 139.

One can find that 1) the spin-orbit splitting for the neutron and proton is very close to each other, at least for $(1d_{3/2}, 1d_{5/2})$ and $(1f_{5/2}, 1f_{7/2})$ cases and 2) the splitting decreases monotonically from the proton drip line to the neutron drip line. These behaviors were understood by examining the origin of the spin-orbit splitting in the Dirac equation.[139] For both proton and neutron, the derivatives of the potentials $\Sigma_0(r) - \Sigma_S(r)$ are almost the same. As the potential $\Sigma_0 - \Sigma_S$ is a big quantity ($\sim$ 700 MeV), the isospin dependence in the spin-orbit potential could be neglected. That is the reason why the spin-orbit splitting for the neutron and proton is very close to each other in Fig. 9. From ^{110}Sn to ^{170}Sn, the amplitude of the derivative of $\Sigma_0 - \Sigma_S$ decreases monotonically due to the surface diffuseness.

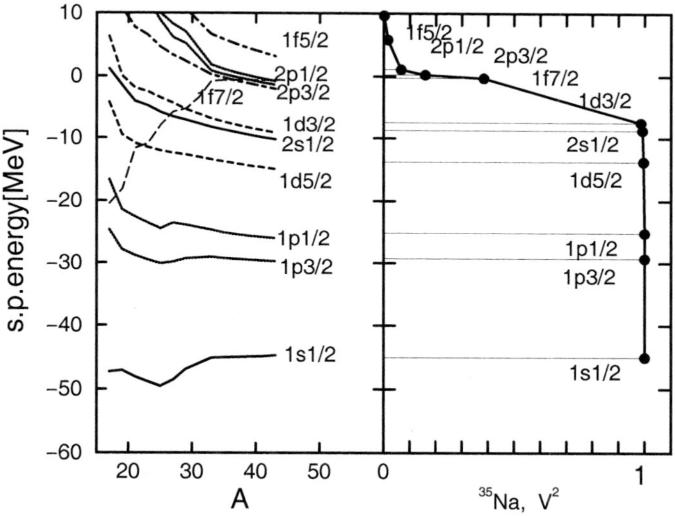

Fig. 10. Left part: Single particle energies for neutrons in the canonical basis as a function of the mass number for Na isotopes. The dashed line indicates the chemical potential. Right part: The occupation probabilities in the canonical basis for ^{35}Na. Taken from Ref. 103.

Similar RCHB calculations have been carried out for all the nuclei in Na isotopes with mass number A ranging from 17 to 45 with effective interaction NLSH.[103] Apart from the diffuseness of the potential, the shell evolution with neutron number has been examined there as well. In Fig. 10, the microscopic structure of the single particle energies in the canonical basis[47,71] is given. In the left panel of Fig. 10, the single particle levels in the canonical basis for the Na isotopes with an even neutron number are shown. The level density becomes dense with the neutron number due to the decrease of the spin-orbit potential or diffuseness of the potential. Going from $A = 19$ to $A = 45$, a big gap above the $N = 8$, $N = 20$ major shell, and $N = 14$ sub-shell can be observed. The $N = 28$ shell for stable nuclei fails to appear, as the $2p_{3/2}$ and $2p_{1/2}$ come so close to $1f_{7/2}$. When $N \geq 20$, the neutrons are filled to the levels in the continuum or weakly bound states in the order of $1f_{7/2}$, $2p_{3/2}$, $2p_{1/2}$, and $1f_{5/2}$. In the right part, the occupation probabilities in the canonical basis of all the neutron levels below $E = 10$ MeV have been given for ^{35}Na to show how the levels are filled in nuclei near the drip line. The importance of careful treatment of the pairing correlation and the scattering of particle pairs to higher lying levels are noted in the figure.

4.2. *Interaction cross section*

The neutron and proton distributions and radii are fundamental quantities to define the halo phenomena. However, the radius for unstable nuclei is normally extracted from the interaction cross sections.[5] Definitely in order to avoid model dependence for extracting the matter distribution, it is better to use the density distribution from the self-consistent models and calculate the interaction cross sections directly from the Glauber model.

Such systematic calculations have been carried out for all Na isotopes with mass number A ranging from 17 to 45 with the RCHB theory and the effective interaction NLSH.[30] The calculated binding energies E and the interaction cross sections with the Glauber model are shown in Fig. 11. The calculated binding energies E are in a good agreement with the empirical values.[144] The resonance states of ^{17}Na and ^{18}Na (with a positive Fermi energy) are exactly reproduced. The nucleus ^{19}Na is bound but unstable against proton emission, reproducing the experimental observation. The neutron drip line nucleus has been predicted to be ^{45}Na. The difference between the calculations and the empirical values for the stable isotopes is from the deformation, which has been neglected.[30]

As the interaction cross sections σ_I is well reproduced, the density distributions of the whole isotopes from the RCHB theory have been examined and the relation between the development of halo and shell effect has been studied in detail.[30]

Similar calculations have been performed for C, N, O and F isotopes up to the neutron drip line by the self-consistent RCHB theory.[145] The calculated one neutron separation energies S_n are in good agreement with the available experimental values for most cases. A Glauber model calculation for the total charge-changing cross section has been carried out with the density distribution obtained from the RCHB theory. A good agreement was obtained with the measured cross sections for studied nuclei with

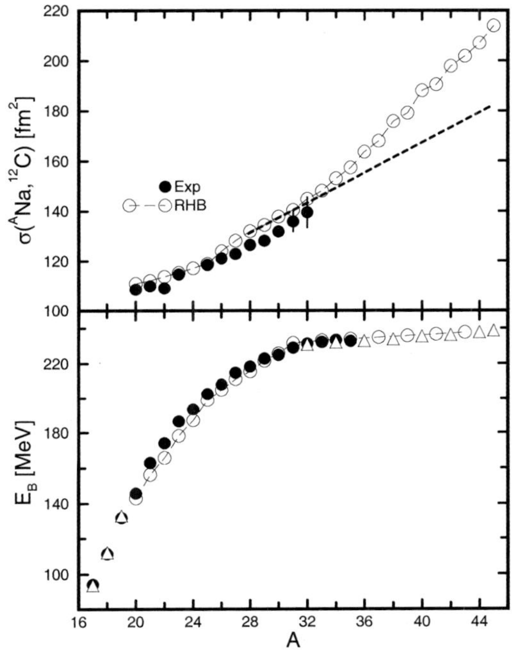

Fig. 11. Upper part: The interaction cross sections σ_I of ^{A}Na isotopes on a carbon target at 950A MeV: the open circles are the result of RCHB calculation and the available experimental data ($A = 20 - 23, 25 - 32$) are given by solid circles with their error-bars. The dashed line is a simple extrapolation based on the RCHB calculation for $^{28-31}$Na. Lower part: Binding energies for Na isotopes, the convention is the same as the upper part, but the RCHB result for particle unstable isotopes are indicated by triangles. Taken from Ref. 30.

^{12}C as a target. An important conclusion was that, contrary to the usual expectation, the proton density distribution is not very sensitive to the neutron number along the isotope chain. Instead it is almost unchanged from stability to the neutron drip line. Such calculations are quite useful in extracting both the proton and neutron distributions inside the nucleus and define the neutron/proton skin or halo.

4.3. *Surface diffuseness correction in global mass formula*

For finite nuclei, the diffuseness of nuclear surface is an important degree-of-freedom in the calculations of nuclear masses.[143] The density distributions of most stable nuclei are of the "neutron skin-type", with a typical value around 0.5 fm for the surface diffuseness. For nuclei near the neutron drip line, halos and giant halos may develop and the neutrons may distribute more extensive spatially than the protons, which implies the enhanced neutron surface diffuseness for these extremely neutron-rich nuclei. However, in global mass calculations, the surface diffuseness of exotic nuclei near the drip lines has not been properly considered. Recently, the surface diffuseness from the RCHB theory and its impact on the symmetry energy and shell correction have been studied in Ref. 143 and the accuracy of the global mass formula has been considerably improved.

Inspired by the Skyrme energy-density functional, a macroscopic-microscopic mass formula, the Weizsäcker–Skyrme (WS) formula,[146–148] was proposed with a rms deviation around 336 keV with respect to the 2149 measured masses[149] in 2003 Atomic Mass Evaluation (AME). The WS models provide the best accuracy for nuclear masses in several mass regions[150,151] and has been used in r-process simulations.[152,153]

In the WS formula, an axially deformed Woods–Saxon potential, with a constant surface diffuseness parameter a for all nuclei, was used to obtain the single particle levels of nuclei. In Ref. 143, by taking into account the surface diffuseness correction for unstable nuclei, a new global mass formula, WS4, has been proposed. The surface diffuseness a of the Woods–Saxon potential was given by $a = a_0\,(1 + 2\varepsilon\delta_q)$, with $\varepsilon = (I - I_0)^2 - I^4$ the correction factor for surface diffuseness, a_0 the diffuseness of the Woods–Saxon potential, $I_0 = 0.4A/(A + 200)$ the isospin asymmetry of the nuclei along the β-stability line described by Green's formula, and $\delta_q = 1$ for neutrons (protons) in the nuclei with $I > I_0$ ($I < I_0$) and $\delta_q = 0$ for other cases. This means that the surface diffuseness of neutron distribution is larger than that of protons at the neutron-rich side and smaller than that of protons at the proton-rich side. The rms deviation with the available mass data[2] falls to 298 keV, crossing for the first time the 0.3 MeV accuracy threshold for mass formulas (models) within the mean field framework.

In Fig. 12(a), the deviations of the calculated masses from the experimental values are shown. For all the 2353 nuclei with Z and $N \geq 8$, the deviation is within 1.23 MeV. In Fig. 12(b), the difference between the WS3 (constant surface diffuseness a_0) and WS4 calculated masses is shown. For most nuclei, the results of WS3 and WS4 are consistent (with deviations smaller than one MeV). For nuclei near the neutron drip line, the masses given by WS4 are larger than the results of WS3 by several MeV. This is due to the enhancement of the nuclear symmetry energy coming from the surface diffuseness effect in extremely neutron-rich nuclei.

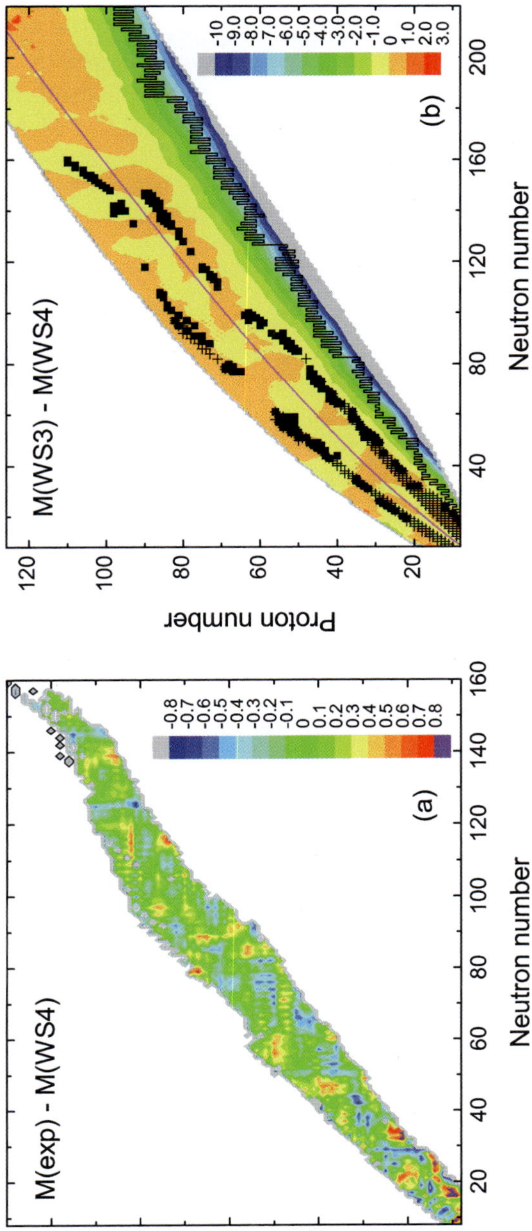

Fig. 12. (a) Difference between measured and the WS4 calculated masses. (b) Difference between WS3 and WS4 calculated masses. The squares and crosses denote the nuclei first appearing in AME2012 and the nuclei with $|I - I_0| > 0.1$, respectively. The smooth and the zigzag curves denote the β-stability line from the Green's formula and the neutron drip line from the WS4 formula, respectively. Taken from Ref. 143.

It is worthwhile to note that with an accuracy of 258 keV for all the available neutron separation energies and of 237 keV for the alpha-decay Q-values of superheavy nuclei, the proposed mass formula will be important not only for the reliable description of the r-process of nucleosynthesis but also for the study of the synthesis of superheavy nuclei.

5. Pairing correlations and nuclear size

The mean square radius of a system deduced from the asymptotic Hartree–Fock (HF) solution is

$$\langle r^2 \rangle_{\mathrm{HF}} \propto \frac{\hbar^2}{2m|\varepsilon_k|}, \tag{36}$$

which diverges in the limit $\varepsilon_k \to 0$, while the mean square radius in a superfluid system deduced from the asymptotic Hartree–Fock–Bogoliubov (HFB) density is

$$\langle r^2 \rangle_{\mathrm{HFB}} \propto \frac{\hbar^2}{2m(E_k - \lambda)} \tag{37}$$

with the quasiparticle energy $E_k = \sqrt{(\varepsilon_k - \lambda)^2 + \Delta_k^2}$. If the pairing gap Δ_k is finite, the radius will never diverge in the limit of small separation energy $\varepsilon_k \simeq \lambda \to 0$.

In Ref. 154, the asymptotic HF and HFB densities characterized by $l = 0$ orbitals were compared and it was emphasized that pairing correlations reduce the nuclear size and an extreme halo with infinite radius cannot be formed in superfluid nuclear systems. Here pairing correlations act against the formation of an infinite radius. Therefore this effect was called "pairing anti-halo effect" in Ref. 154.

In fact, the mechanism leading to a halo in Eq. (36), i.e., the valence nucleons occupy the weakly-bound orbital below and very close to the continuum threshold, has been used in early interpretation for nuclear halo phenomena.[86,111,112] The limiting condition $\varepsilon_k \to 0$ and the radius deduced from this $l = 0$ orbital alone correspond to an extremely ideal situation, which is difficult to find in real nuclei.

In Ref. 155, the influences of pairing correlations on the nuclear size and on the development of a nuclear halo were studied in detail by the self-consistent RCHB theory.[71] In order to simplify the problem, neutron-rich nuclei with the neutron Fermi surface below, between and above two weakly-bound $2p$ levels are investigated with a fixed Wood–Saxon potential, which is followed by the self-consistent calculation on the well studied neutron halos in ^{11}Li and ^{32}Ne with low-l orbitals in the continuum.

5.1. *Fermi surface and nuclear size*

In order to investigate the impact of nuclear binding on nuclear size for varying pairing potentials, a spherical Woods-Saxon shape mean field potential fitted to the corresponding neutron potentials resulting from a self-consistent calculation for ^{42}Mg is adopted.[155]

The total neutron radius of the nucleus is given by

$$R_N = \frac{1}{N}\sum_{nlj}(2j+1)\,\langle r\rangle_{nlj}\,v^2_{nlj}, \tag{38}$$

which is determined by the rms radii of the orbitals $\langle r\rangle_{nlj}$ and the corresponding occupation probabilities v^2_{nlj}.

In Fig. 13, the single neutron levels ε_k in the canonical basis,[47] the rms radii $\langle r\rangle_{nlj}$, the occupation probabilities v^2_{nlj}, the corresponding contributions $v^2_{nlj}\times\langle r\rangle_{nlj}$ to the neutron rms radius for the orbitals near the Fermi surface ($2p$ and $1f$), and the total neutron rms radii R_N for four neutron-rich Mg isotopes with $N = 28,\ 30,\ 32$, and 34 are plotted as functions of the pairing gap

$$\Delta = \frac{1}{N}\sum_{nlj}(2j+1)\Delta^{(lj)}_n v^2_{nlj}. \tag{39}$$

Here $\Delta^{(lj)}_n$ are the diagonal matrix elements of the pairing field in the canonical basis.

The single neutron levels in canonical basis are obtained by solving the corresponding RCHB equations for various pairing strengths. In principle, the energy levels $\varepsilon_{nlj} = h^{(lj)}_{nn}$ depend on Δ. However, since this solution was obtained for fixed WS potentials, as shown in the upper panels of Fig. 13, these states remain almost unchanged with increasing pairing correlations.

For Mg isotopes with $N = 28,\ 30,\ 32$, and 34, the neutron Fermi surface (shown as dashed line in upper panels of Fig. 13) is located just below, between and above the two weakly-bound $2p$ levels. As the pairing strength increases, more neutrons are scattered from occupied levels in the Fermi sea to empty levels above the Fermi sea, and therefore the Fermi surface is rising.

In Fig. 13, the rms radii of the $2p$ orbitals are much larger than those of the $1f$ orbitals due to the lower centrifugal barrier. With increasing

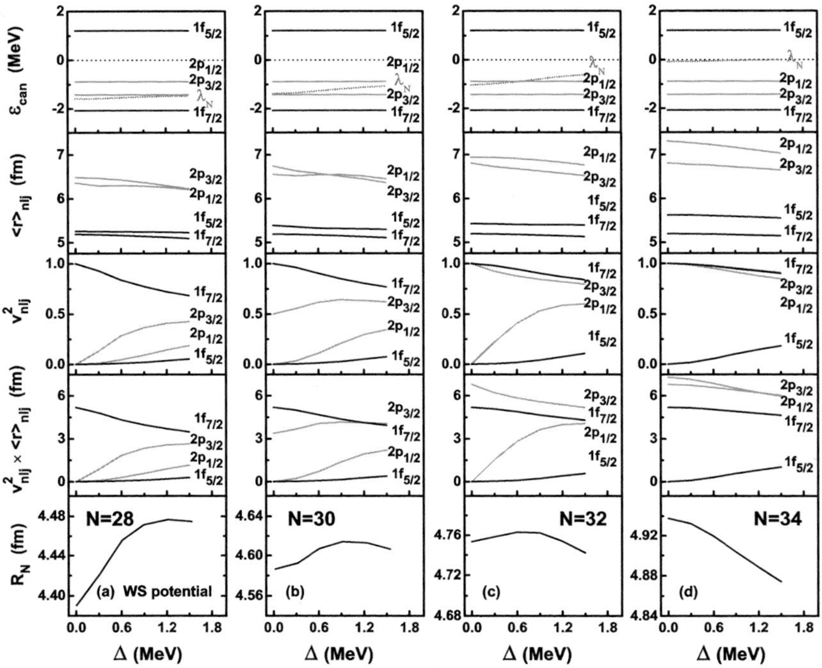

Fig. 13. (Color online) Single neutron levels $2p$ and $1f$ near the Fermi surface in the canonical basis, the corresponding rms radii $\langle r\rangle_{nlj}$, occupation probabilities v^2_{nlj}, contributions $v^2_{nlj} \times \langle r\rangle_{nlj}$ to the neutron rms radii, and the total neutron rms radii R_N for $N = 28$, 30, 32, and 34 as a function of average pairing gap Δ. The Fermi surfaces for the neutron numbers are plotted as dashed lines. Taken from Ref. 155.

pairing correlations, the rms radius decreases for $2p$ and $1f$ levels. This is the so-called "pairing anti-halo effect" discussed in Ref. 154. However, this effect concerns only the radii $\langle r\rangle_{nlj}$ of the individual orbitals. The total neutron radius of the nucleus in Eq. (38) is determined by the rms radii of the orbitals and the corresponding occupation probabilities, which depend strongly on the pairing correlations.

As shown in Fig. 13, for $N = 28$ the neutron Fermi surface is just below the two weakly-bound $2p$ levels. Without pairing correlations, the occupation probability is 1.0 for the $1f_{7/2}$ orbital, and it vanishes for the $2p$ and the $1f_{5/2}$ orbitals. As the pairing strength increases, the neutrons on $1f_{7/2}$ orbital are scattered to $2p$ and $1f_{5/2}$ orbitals which have much larger rms radii. Therefore, the contributions to the total neutron rms radius from $2p$ and $1f_{5/2}$ states grow more than the contribution from $1f_{7/2}$ state decreases. As a result, the total neutron rms radius increases monotonically, where the contributions from $2p$ orbitals play a dominant role.

For $N = 30$ in Fig. 13, the $2p_{1/2}$ and the $1f_{5/2}$ orbitals are empty for zero pairing, while the $1f_{7/2}$ orbital is fully occupied and the $2p_{3/2}$ orbital is half occupied. With increasing pairing strength, the neutrons in the $1f_{7/2}$ orbital are scattered to the $2p$ orbitals which contribute strongly to the neutron rms radius. Therefore, the total neutron rms radius R_N increases with the growing pairing strength up to $\Delta = 0.9$ MeV. Increasing the pairing strength further, the neutrons continue to be scattered from the $1f_{7/2}$ to the $2p_{1/2}$ state, while the neutron number in the $2p_{3/2}$ orbital decreases slightly and the resonance state $1f_{5/2}$ begins to be occupied. Together with the monotonic decreasing rms radii of the individual orbitals, the increase resulting from the $2p_{1/2}$ and $1f_{5/2}$ states is smaller than the decreases of the contributions from the $2p_{3/2}$ and $1f_{7/2}$ states. Therefore, for very strong pairing correlations, the total neutron rms radius finally decreases. As it is clearly seen this decrease is not caused by the pairing anti-halo effect, because the radius of the $1f_{7/2}$ state stays completely constant. It has its origin in the reoccupation caused by pairing.

For $N = 32$ in Fig. 13, the $1f_{7/2}$ and $2p_{3/2}$ orbitals are fully occupied without pairing correlations, while the $2p_{1/2}$ and $1f_{5/2}$ orbitals are empty. As the pairing strength increases, the neutrons in the $1f_{7/2}$ and $2p_{3/2}$ orbitals are scattered to the $2p_{1/2}$ orbital, whose rms radius is larger than the ones of $1f_{7/2}$ and $2p_{3/2}$ levels. The contribution of the $2p_{1/2}$ orbitals causes an increase in the total neutron rms radius up to a pairing gap $\Delta = 0.9$ MeV. As the pairing strength continues to grow beyond this value, the neutrons in the $2p_{3/2}$ state begin to be scattered to the $1f_{5/2}$ state which provides a smaller contribution than the $2p_{3/2}$ state, while the occupation probability remains almost constant for the $1p_{1/2}$ state. Together with the decreasing rms radius of the individual orbitals, the total neutron rms radius finally decreases for very large pairing.

For $N = 34$, the two weakly-bound $2p$ levels are fully occupied for zero pairing, and the Fermi surface is just above the weakly-bound $2p$ levels. With increasing pairing strength, the neutrons in the $1f_{7/2}$ and $2p$ orbitals are scattered to the $1f_{5/2}$ orbital with a much smaller rms radius than the $2p$ orbitals. The contributions to the total neutron rms radius from the $1f_{7/2}$ and $2p$ orbitals decrease more than the increase of the contribution from $1f_{5/2}$ orbital, and therefore the total neutron rms radius finally decreases monotonically. Here the two effects, decreasing rms radius of the individual orbitals and changes of the occupation probabilities, act in the same direction. The decrease of the total neutron radius R_N with increasing pairing correlations comes not only from the decreasing rms radii of the individual orbitals but also from the change of the occupation probabilities.

It is clear that pairing correlations can change the rms radii of the individual weakly-bound orbitals and simultaneously the corresponding occupation probabilities. As a result, they have an influence on the nuclear rms radius and the total nuclear size. Furthermore, the contributions of the weakly-bound $2p$ orbitals to the nuclear rms radius play an important role.

5.2. *Crucial low l orbitals*

Following the investigation on the impact of nuclear binding on nuclear size, the next question is the impact of orbital binding on the nuclear size as a function of the pairing potential. In Ref. 155, this problem was investigated for the orbitals $2p$ as well-bound, weakly-bound, around the continuum threshold, and in the continuum cases.

In Fig. 14(a), the single neutron levels $1d_{3/2}$, $1f_{7/2}$, $2p_{3/2}$, $2p_{1/2}$, and $1f_{5/2}$ for $N = 28$ in WS potentials with varying depth such that $\varepsilon_{2p_{1/2}} = -5.0$, -0.5, 0.0, and $+0.5$ MeV are shown. In Fig. 14(b), for each $\varepsilon_{2p_{1/2}}$, the change of the total neutron rms radius with respect to that of zero pairing, $\delta R_N = R_N(\Delta) - R_N(\Delta = 0)$, is plotted as a function of the pairing gap.

For $\varepsilon_{2p_{1/2}} = -5.0$ MeV which corresponds to a well bound nucleus, the neutron rms radius $R_N(\Delta)$ increases slightly with the pairing gap increasing from 0.0 to 1.5 MeV. For $\varepsilon_{2p_{1/2}} = -0.5,\ 0.0,\ +0.5$ MeV which correspond to a weakly-bound nucleus, increases by more than 0.1 fm are observed. It has to be noticed that for $\varepsilon_{2p_{1/2}} = +0.5$ MeV, both levels $2p_{1/2}$ and $2p_{3/2}$ are in the continuum. If the pairing correlation is switched on, the system is basically not bound. The eigenvalue problem is solved in a finite box and the solution depends on the box size. As an illustration, when the box size changes from 20 to 25 fm, the neutron rms radius calculated with a box radius of 25 is about 0.05 fm larger than that obtained with a box radius of 20 fm for $\Delta = 1.5$ MeV as shown in the shadowed region in Fig. 14(b).

For the discussion on the influence of pairing correlations on the nuclear size so far, the mean fields are fixed in the form of spherical WS potentials, and only the pairing strength and the Fermi level for $N = 28,\ 30,\ 32,\ 34$ are considered. Here, with the changing pairing strength and the Fermi

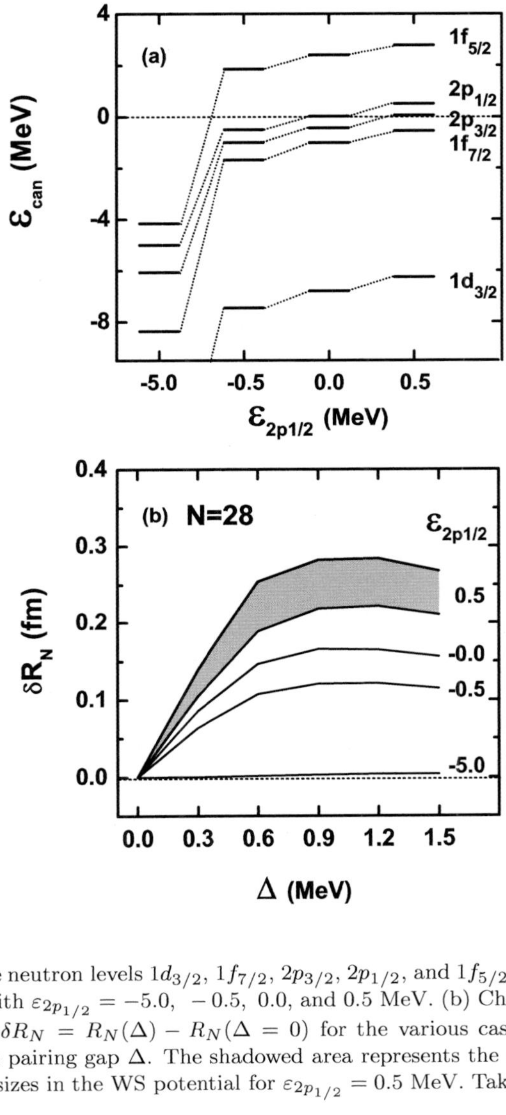

Fig. 14. (a) Single neutron levels $1d_{3/2}$, $1f_{7/2}$, $2p_{3/2}$, $2p_{1/2}$, and $1f_{5/2}$ in WS potentials of varying depth with $\varepsilon_{2p_{1/2}} = -5.0$, -0.5, 0.0, and 0.5 MeV. (b) Changes of the total neutron rms radii $\delta R_N = R_N(\Delta) - R_N(\Delta = 0)$ for the various cases of $\varepsilon_{2p_{1/2}}$ as a function of average pairing gap Δ. The shadowed area represents the results calculated with different box sizes in the WS potential for $\varepsilon_{2p_{1/2}} = 0.5$ MeV. Taken from Ref. 155.

level, the single neutron levels stay almost constant except for a slight modification by the pairing field. In Ref. 155, fully self-consistent RCHB calculation for neutron-rich nuclei $^{40,\,42,\,44,\,46}$Mg was performed, it should be emphasized that qualitatively the effect of pairing correlations on the nuclear size agrees with that found for fixed WS potentials.

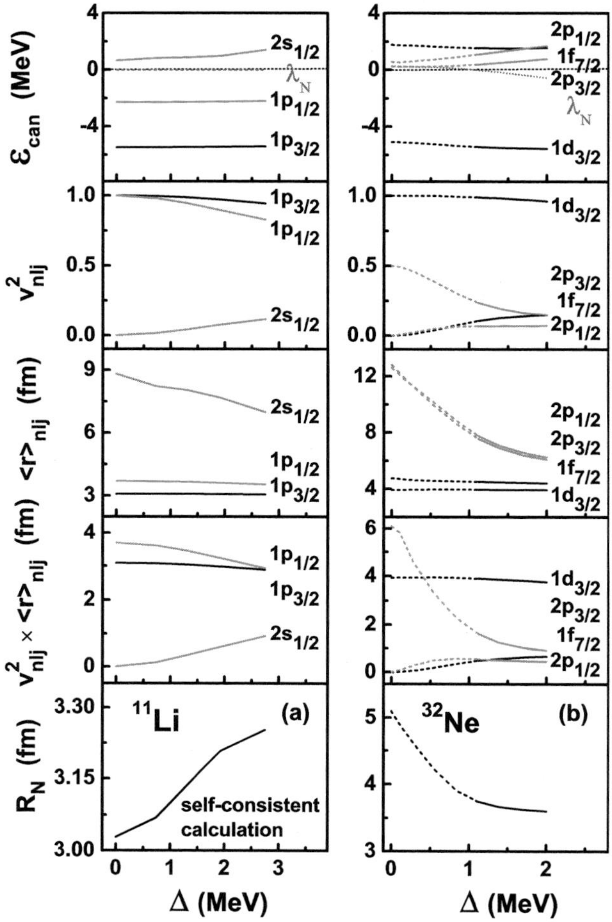

Fig. 15. (Color online) Same as Fig. 13, but for ^{11}Li and ^{32}Ne. Taken from Ref. 155.

5.3. *Pairing correlations and halos*

In order to investigate the effect of pairing correlations on the development of a nuclear halo, self-consistent calculations are performed with different pairing strength for the well-known neutron halo nuclei ^{11}Li[42] and ^{32}Ne,[102] where the low-l orbitals are in the continuum.[155] Single neutron levels in the canonical basis, the corresponding rms radii $\langle r\rangle_{nlj}$, the occupation probabilities v^2_{nlj}, the contributions to the neutron rms radius $v^2_{nlj} \times \langle r\rangle_{nlj}$, and the total neutron rms radii are plotted in Fig. 15 as a function of the average pairing gap for the nuclei ^{11}Li (panel a) and ^{32}Ne (panel b).

For ^{11}Li in Fig. 15(a), a halo is developed by scattering the neutrons to the $2s_{1/2}$ orbitals in continuum and the increasing pairing correlations promote the development of this halo.

For ^{32}Ne in Fig. 15(b), the $2p$ and $1f_{7/2}$ orbitals are in the continuum. Without pairing correlations, the last two neutrons occupy the $2p_{3/2}$ orbital and therefore this nucleus is unbound and the rms radius depends on the box size. Increasing the pairing correlations, the Fermi surface comes down and becomes negative when the pairing gap gets larger than 1.0 MeV. It shows that for the halo nuclei such as ^{32}Ne, where the low-l orbitals are in the continuum and unoccupied without pairing correlations, increasing pairing correlations will decrease the size of the nuclear halo, but nonetheless pairing plays an essential role. Without pairing the nucleus would stay unbound.

It was concluded that the weakly-bounded orbitals with low orbital angular momenta l in the neighborhood of the threshold to the continuum play an important role.[155] The pairing correlations have a two-fold influence on the total nuclear rms radius and the nuclear size. First, they can change the rms radii of individual weakly bound orbitals and, second, they can change their occupation probabilities. With increasing pairing strength, the individual rms radii are reduced (pairing anti-halo effect). Meanwhile pairing changes also the occupation probabilities. The total nuclear size is determined by the competition between this two effects.

For low-l orbitals in the continuum, it was found that when the low-l orbitals are empty without pairing, the pairing correlations can give rise to a possible halo by scattering the neutrons up to the low-l orbitals in continuum, and can promote the development of this halo with increasing pairing correlations, even though the increasing pairing correlations decrease the rms radii of the individual single particle levels. Conversely, when the low-l orbitals are in the continuum and partially occupied for some pairing, growing pairing correlations will decrease the rms radii and increase the occupation probabilities of the low-l states and which effect dominates depends on the details. Pairing plays an essential role also in cases where the nucleus is not bound without pairing correlations. In Ref. 155, the strength of the pairing correlations was an external variable parameter. In realistic nuclei, the strength of pairing depends on the level density in the vicinity of the Fermi surface. Therefore, even orbitals with high l-values and large centrifugal barriers can contribute indirectly to changes of the nuclear size.[102] If they come close to the Fermi surface they enhance pairing correlations influencing at the same time the occupation of low-l orbitals close to the continuum limit and therefore the nuclear size.

6. Proton drip line and proton radioactivity

Due to the Coulomb barrier, some proton rich nuclei beyond the proton drip line can exist for a certain time and decay via emission of a single proton. Since the first proton emitter was discovered in an isomeric state of ^{53}Co* in 1970,[7,156] more than 30 proton emitters have been observed.[157,158] To understand the physics beyond the drip line, comprehensive theoretical efforts have also been made.[159–173] In this Section, we will highlight some theoretical progresses made with CDFT.

6.1. *Spherical proton emitters*

Spherical proton emitters have been systematically studied with or based on RMF models. In Ref. 171, the proton radioactivity was described by a fully self-consistent calculation based on relativistic density functionals derived from meson exchange and point coupling models. The experimental data were well reproduced and a strong evidence of the effects of configuration mixing was pointed out for nuclei away from the $N = 82$ magic number. In Ref. 172, the half-life of proton radioactivity of spherical proton emitters was studied within the scheme of CDFT. The authors also extracted, for the first time, the potential barrier that prevents proton emission by using the similarity renormalization group method. The estimated half-lives are in good agreement with the experimental values.

Quite recently, spherical proton emitters in the region from I ($Z = 53$) to Bi ($Z = 83$) isotopes have been studied using the relativistic continuum Hartree–Bogoliubov theory.[173] In Fig. 16 the nuclear chart from the I ($Z = 53$) to Bi ($Z = 83$) isotopes is shown, where the experimentally known proton emitters[157] are highlighted with symbols $\diamond$. The proton drip-line nuclei given by the RCHB calculations[116] with the relativistic energy density functional PC-PK1[115] and pairing strength $V_0 = 685$ MeV fm^{-3} for the density-dependent δ pairing interaction are compared with the last bound or last mass-measured nuclei in AME 2012.[1–3] The proton emitters should be located outside the proton drip line with positive Q values for emitting one or two protons. Thus the criteria from the calculation point of view are: $S_p > 0$, $S_{2p} > 0$ and $\lambda_p < 0$ where S_p and S_{2p} are the one- and two-proton separation energies, respectively, and λ_p is the Fermi energy. For most of the proton emitters which have been discovered in this mass region, the self-consistent RCHB calculations fulfill the above criteria with several exceptions for nuclei with $A > 167$.

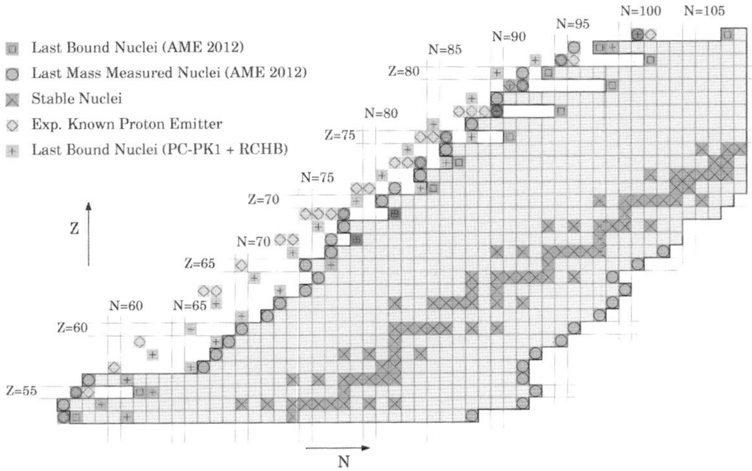

Fig. 16. (Color online) Nuclear chart from I ($Z = 53$) to Bi ($Z = 83$) isotopes calculated by using the RCHB theory with PC-PK1. The experimentally known proton emitters[157] are denoted with the symbol ◇, while the proton drip-line nuclei given by the spherical RCHB calculations are shown with the symbol +. The stable nuclei as well as the last mass-measured nuclei and the last bound nuclei in AME 2012[1–3] are also shown to guide the eye. Taken from Ref. 173.

In Ref. 173, the authors also calculated the half-lives of the proton emitters using the WKB method. In Fig. 17, the ratios of the calculated half-lives $t_{1/2}^{\rm th}$ to experimental values $t_{1/2}^{\rm exp}$ are shown as a function of mass number A. The blue solid and red open symbols represent the proton emissions from the ground states and isomeric states, respectively. The error bars indicate the uncertainties of the experiment values, $\delta t_{1/2}^{\rm exp}/t_{1/2}^{\rm exp}$. It is seen that the proton emission half-lives range from a few microseconds to a few seconds, which are mainly determined by the orbital angular momentum and Q value. For all the cases covering six orders of magnitude, the theoretical results by the relativistic energy density functional PC-PK1 are in good agreement with experiment.

6.2. *Deformed proton emitters*

The relativistic Hartree–Bogoliubov (RHB) model has been used to study ground-state properties of proton-rich odd-Z nuclei with $53 \leq Z \leq 69$.[163,164] With the NL3 effective interaction used for the mean-field Lagrangian and the finite range Gogny interaction D1S for pairing

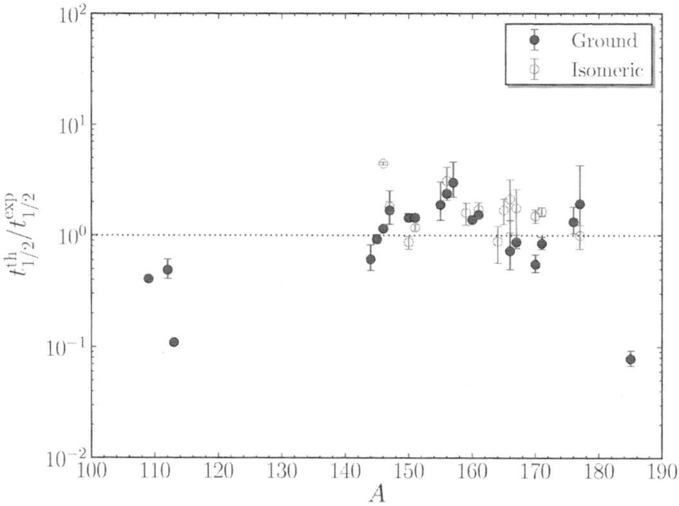

Fig. 17. (Color online) Ratios of the calculated half-lives $t^{th}_{1/2}$ to experimental half-lives $t^{exp}_{1/2}$[157,174,175] as a function of mass number A, where the error bars indicate the uncertainties of the experiment values, $\delta t^{exp}_{1/2}/t^{exp}_{1/2}$. The proton emissions from the ground states and isomeric states are denoted with blue solid and red open symbols, respectively. Taken from Ref. 173.

correlations, the RHB model predicted the location of the proton drip-line, the ground-state quadrupole deformations and one-proton separation energies at and beyond the drip-line, the deformed single-particle orbitals occupied by the odd valence proton, and information about the deformed single-particle orbitals occupied by the odd valence proton.

In Fig. 18 the one-proton separation energies S_p for the odd-Z nuclei $59 \leq Z \leq 69$ are presented as function of the number of neutrons.[164] From this figure one can find that the following nuclei are beyond the proton drip line: ^{124}Pr, ^{129}Pm, ^{134}Eu, ^{139}Tb, ^{146}Ho and ^{152}Tm. In Ref. 164, the calculated S_p values have been compared with experimental data on proton radioactivity from ^{131}Eu, ^{141}Ho, ^{147}Tm and ^{145}Tm and good agreement was found for the first three nuclei. A detailed analysis of single proton levels, including spectroscopic factors, was performed in the canonical basis and predictions for the deformed single-particle orbitals occupied by the valence odd protons are consistent with experimental half-lives for proton transitions.

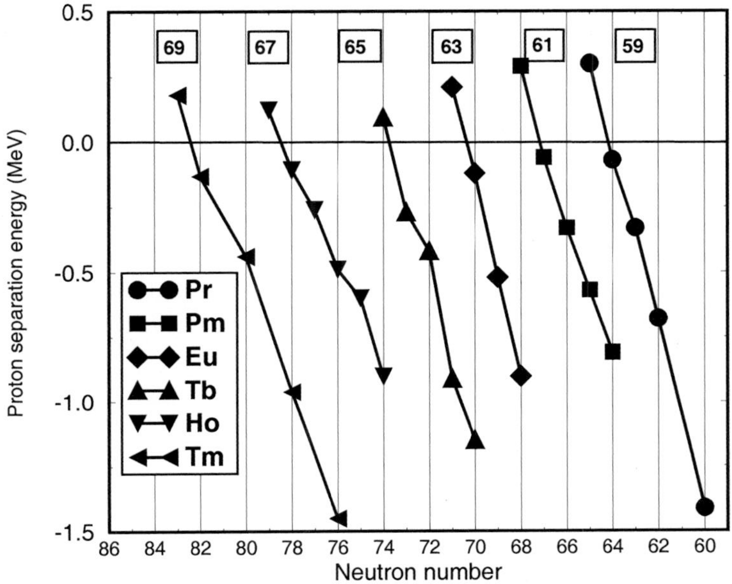

Fig. 18. One-proton separation energies for odd-Z nuclei with $59 \leq Z \leq 69$ at and beyond the drip-line calculated with the RHB model. Taken from Ref. 164.

7. Summary

In this chapter, we have reviewed relativistic mean field (RMF) models with pairing treated by the BCS and Bogoliubov approaches and the applications of CDFTs to exotic nuclear phenomena including the nuclear halos, the position of the proton drip line and proton radioactivity, the surface diffuseness and its relation to nuclear exotic phenomena, and the effects of the pairing on the nuclear size.

In the mean field study of nuclear halos, a proper description of nuclear pairing and the continuum contribution is very important. The RMF+rBCS approach, the relativistic continuum Hartree–Bogoliubov (RCHB) theory for spherical nuclei and deformed relativistic Hartree–Bogoliubov theory in the continuum have been developed. Now we summarize this Chapter.

From the RCHB theory, a self-consistent description of the neutron halo in ^{11}Li has been achieved and giant neutron halos, consisting of up to six halo neutrons, have been predicted in Zr and Ca isotopes close to the neutron drip line. The RCHB theory has also been extended to include hyperons and even-A Ca isotopes ranging from the proton drip line to the neutron drip line have been studied systematically. Halo phenomena in deformed nuclei were investigated within the DRHBc theory. The generic conditions for the occurrence of this decoupling effects have been given and a shape decoupling of the halo orbitals from the deformation of the core was predicted.

Proton emitters sitting outside the proton drip line present themselves as interesting open systems for quantum mechanical tunneling as the emitted proton goes through the barrier of Coulomb and centrifugal potentials. The RMF models have been used to study proton emitters including the Q values, the spectroscopic information, and the lifetimes. Satisfactory agreement with experiment achieved.

The RCHB theory has been also used to describe even-even Sn isotopes from the proton drip line to the neutron drip line. The change of the potential with the neutron number was investigated and the connection between the diffuseness of the potential and the spin-orbit splitting was analyzed. The relation between the shell effects and the diffuseness and the tail of the matter distribution has been also examined by studying Na isotopes.

Finally, we discussed the influence of pairing correlations on the size of the nucleus and the development of nuclear halos. It was revealed that the contributions of weakly bounded orbits in the continuum with low orbital angular momenta l play an essential role. Pairing correlations have a two-fold influence on the density distribution of the neutrons and therefore on the total nuclear size. First, they can change the root-mean-square radius of the individual weakly bound orbits and, second, they can change the occupation probabilities of these orbits in the nuclear system. Both effects are important and the total radius is dominated by their competition.

Acknowledgment

We would like to thank Y. Chen, L. S. Geng, N. V. Giai, Y. Kim, L. L. Li, H. Z. Liang, M. Liu, W. H. Long, H. F. Lü, H. Sagawa, I. Tanihata, J. Terasaki, H. Toki, N. Wang, X. Z. Wu, X. W. Xia, S. Yamaji, J. Y. Zeng, S. Q. Zhang, and E. G. Zhao for fruitful collaborations. This work was supported in part by the National Key Basic Research Program of China (Grant No. 2013CB834400), the Natural Science Foundation of China (Grants No. 10975008, No. 11175002, No. 11105005, No. 11105010, No. 11035007, No. 11128510, No. 11121403, No. 11120101005, No. 11211120152, No. 11235002,

No. 11275248, No. 11205004, No. 11335002 and No. 11525524) the Research Fund for the Doctoral Program of Higher Education (Grant No. 20110001110087), the Chinese Academy of Sciences (Grant No. KJCX2-EW-N01), the DFG Cluster of Excellence "Origin and Structure of the Universe" (www.universecluster.de), and the Overseas Distinguished Professor Project from Ministry of Education (MS2010BJDX001).

References

1. G. Audi, F. G. Kondev, M. Wang, B. Pfeiffer, X. Sun, J. Blachot, and M. MacCormick, The NUBASE2012 evaluation of nuclear properties, *Chin. Phys. C.* **36**(12), 1157–1286 (Dec., 2012). URL `http://amdc.impcas.ac.cn/`.
2. G. Audi, M. Wang, A. H. Wapstra, F. G. Kondev, M. MacCormick, X. Xu, and B. Pfeiffer, The NUBASE2012 evaluation of nuclear properties (I). Evaluation of input data, adjustment procedures, *Chin. Phys. C.* **36**(12), 1287–1602 (Dec., 2012). URL `http://amdc.impcas.ac.cn/`.
3. M. Wang, G. Audi, A. H. Wapstra, F. G. Kondev, M. MacCormick, X. Xu, and B. Pfeiffer, The NUBASE2012 evaluation of nuclear properties (II). Tables, graphs and references, *Chin. Phys. C.* **36**(12), 1603–2014 (Dec., 2012). URL `http://amdc.impcas.ac.cn/`.
4. M. Thoennessen, Current status and future potential of nuclide discoveries, *Rep. Prog. Phys.* **76**(5), 056301–14 (May, 2013). ISSN 0034-4885. doi: 10.1088/0034-4885/76/5/056301. URL `http://stacks.iop.org/0034-4885/76/i=5/a=056301`. See also `http://www.nscl.msu.edu/~thoennes/isotopes/`.
5. I. Tanihata, H. Hamagaki, O. Hashimoto, Y. Shida, N. Yoshikawa, K. Sugimoto, O. Yamakawa, T. Kobayashi, and N. Takahashi, Measurements of interaction cross sections and nuclear radii in the light p-shell region, *Phys. Rev. Lett.* **55**(24), 2676–2679 (Dec., 1985). doi: 10.1103/PhysRevLett.55.2676. URL `http://link.aps.org/abstract/PRL/v55/p2676`.
6. A. Ozawa, T. Kobayashi, T. Suzuki, K. Yoshida, and I. Tanihata, New magic number, n=16, near the neutron drip line, *Phys. Rev. Lett.* **84**(24), 5493–5495 (June, 2000). doi: 10.1103/PhysRevLett.84.5493. URL `http://link.aps.org/doi/10.1103/PhysRevLett.84.5493`.
7. K. P. Jackson, C. U. Cardinal, H. C. Evans, N. A. Jelley, and J. Cerny, ^{53}Co^{m}: A proton-unstable isomer, *Phys. Lett. B.* **33**(4), 281–283 (Oct., 1970). ISSN 0370-2693. doi: 10.1016/0370-2693(70)90269-8. URL `http://www.sciencedirect.com/science/article/pii/0370269370902698`.
8. S. Hofmann, W. Reisdorf, G. Münzenberg, F. P. Heßberger, J. R. H. Schneider, and P. Armbruster, Proton radioactivity of ^{151}Lu, *Z. Phys. A.* **305**(2), 111–123 (1982). ISSN 0939-7922. doi: 10.1007/BF01415018. URL `http://dx.doi.org/10.1007/BF01415018`.
9. O. Klepper, T. Batsch, S. Hofmann, R. Kirchner, W. Kurcewicz, W. Reis-

dorf, E. Roeckl, D. Schardt, and G. Nyman, Direct and beta-delayed proton decay of very neutron-deficient rare-earth isotopes produced in the reaction ^{58}Ni+^{92}Mo, *Z. Phys. A.* **305**(2), 125–130 (1982). ISSN 0939-7922. doi: 10.1007/BF01415019. URL `http://dx.doi.org/10.1007/BF01415019`.

10. J. Giovinazzo, B. Blank, M. Chartier, S. Czajkowski, A. Fleury, M. J. Lopez Jimenez, M. S. Pravikoff, J.-C. Thomas, F. de Oliveira Santos, M. Lewitowicz, V. Maslov, M. Stanoiu, R. Grzywacz, M. Pfützner, C. Borcea, and B. A. Brown, Two-proton radioactivity of ^{45}Fe, *Phys. Rev. Lett.* **89**(10), 102501– (Aug., 2002). doi: 10.1103/PhysRevLett.89.102501. URL `http://link.aps.org/doi/10.1103/PhysRevLett.89.102501`.
11. M. Pfützner, E. Badura, C. Bingham, B. Blank, M. Chartier, H. Geissel, J. Giovinazzo, L. V. Grigorenko, R. Grzywacz, M. Hellström, Z. Janas, J. Kurcewicz, A. S. Lalleman, C. Mazzocchi, I. Mukha, G. Münzenberg, C. Plettner, E. Roeckl, K. P. Rykaczewski, K. Schmidt, R. S. Simon, M. Stanoiu, and J.-C. Thomas, First evidence for the two-proton decay of ^{45}Fe, *Euro. Phys. J. A.* **14**(3), 279–285 (2002). ISSN 1434-6001. doi: 10.1140/epja/i2002-10033-9. URL `http://dx.doi.org/10.1140/epja/i2002-10033-9`.
12. T. Motobayashi, RIKEN RI beam factory — recent results and perspectives, *Nucl. Phys. A.* **834**(1-4), 707c–712c (Mar., 2010). ISSN 0375-9474. doi: 10.1016/j.nuclphysa.2010.01.128. URL `http://dx.doi.org/10.1016/j.nuclphysa.2010.01.128`.
13. C. Sturm, B. Sharkov, and H. Stöcker, 1, 2, 3 ... FAIR!, *Nucl. Phys. A.* **834**(1-4), 682c–687c (Mar., 2010). ISSN 0375-9474. doi: 10.1016/j.nuclphysa.2010.01.124. URL `http://dx.doi.org/10.1016/j.nuclphysa.2010.01.124`.
14. S. Gales, SPIRAL2 at GANIL: Next generation of ISOL facility for intense secondary radioactive ion beams, *Nucl. Phys. A.* **834**(1-4), 717c–723c (Mar., 2010). ISSN 0375-9474. doi: 10.1016/j.nuclphysa.2010.01.130. URL `http://dx.doi.org/10.1016/j.nuclphysa.2010.01.130`.
15. M. Thoennessen, Plans for the facility for rare isotope beams, *Nucl. Phys. A.* **834**(1-4), 688c–693c (Mar., 2010). ISSN 0375-9474. doi: 10.1016/j.nuclphysa.2010.01.125. URL `http://dx.doi.org/10.1016/j.nuclphysa.2010.01.125`.
16. P. D. Cottle and K. W. Kemper, A walk along the dripline, *Phys.* **5**, 49– (Apr., 2012). doi: 10.1103/Physics.5.49. URL `http://link.aps.org/doi/10.1103/Physics.5.49`.
17. S. Choi. KoRIA project — RI accelerator in Korea. URL `http://indico.ihep.ac.cn/conferenceTimeTable.py?confId=1714` Invited talk given in International Symposium on Nuclear Physics in Asia, 14–15 October, 2010, Beihang University, Beijing.
18. J. W. Xia, W. L. Zhan, B. W. Wei, Y. J. Yuan, M. T. Song, W. Z. Zhang, X. D. Yang, P. Yuan, D. Q. Gao, H. W. Zhao, X. T. Yang, G. Q. Xiao, K. T. Man, J. R. Dang, X. H. Cai, Y. F. Wang, J. Y. Tang, W. M. Qiao, Y. N. Rao, Y. He, L. Z. Mao, and Z. Z. Zhou, The heavy ion cooler-storage-ring project (HIRFL-CSR) at Lanzhou, *Nucl. Instrum. Methods Phys. Res. A.* **488**(1-2),

11–25 (Aug., 2002). ISSN 0168-9002. doi: 10.1016/S0168-9002(02)00475-8. URL `http://dx.doi.org/10.1016/S0168-9002(02)00475-8`.

19. W. L. Zhan, H. S. Xu, G. Q. Xiao, J. W. Xia, H. W. Zhao, and Y. J. Yuan, Progress in HIRFL-CSR, *Nucl. Phys. A.* **834**(1-4), 694c–700c (Mar., 2010). ISSN 0375-9474. doi: 10.1016/j.nuclphysa.2010.01.126. URL `http://dx.doi.org/10.1016/j.nuclphysa.2010.01.126`.
20. B. D. Serot and J. D. Walecka, The relativistic nuclear many-body problem, *Adv. Nucl. Phys.* **16**, 1–327 (1986).
21. P. G. Reinhard, The relativistic mean-field description of nuclei and nuclear dynamics, *Rep. Prog. Phys.* **52**(4), 439–514 (Apr., 1989). ISSN 0034-4885. doi: 10.1088/0034-4885/52/4/002. URL `http://dx.doi.org/10.1088/0034-4885/52/4/002`.
22. Y. K. Gambhir, P. Ring, and A. Thimet, Relativistic mean field theory for finite nuclei, *Ann. Phys. (NY).* **198**(1), 132–179 (Feb., 1990). doi: 10.1016/0003-4916(90)90330-Q. URL `http://dx.doi.org/10.1016/0003-4916(90)90330-Q`.
23. P. Ring, Relativistic mean field theory in finite nuclei, *Prog. Part. Nucl. Phys.* **37**, 193–263 (1996). ISSN 0146-6410. doi: 10.1016/0146-6410(96)00054-3. URL `http://dx.doi.org/10.1016/0146-6410(96)00054-3`.
24. B. D. Serot and J. D. Walecka, Recent progress in quantum hadrodynamics, *Int. J. Mod. Phys. E.* **6**(4), 515–631 (dec, 1997). doi: 10.1142/S0218301397000299. URL `http://www.worldscientific.com/doi/abs/10.1142/S0218301397000299`.
25. A. Afanasjev, D. Fossan, G. Lane, and I. Ragnarsson, Termination of rotational bands: disappearance of quantum many-body collectivity, *Phys. Rep.* **322**(1-2), 1–124 (Dec, 1999). ISSN 0370-1573. doi: 10.1016/s0370-1573(99)00035-6. URL `http://dx.doi.org/10.1016/S0370-1573(99)00035-6`.
26. D. Vretenar, A. V. Afanasjev, G. A. Lalazissis, and P. Ring, Relativistic Hartree–Bogoliubov theory: static and dynamic aspects of exotic nuclear structure, *Phys. Rep.* **409**(3-4), 101–259 (Apr., 2005). doi: 10.1016/j.physrep.2004.10.001. URL `http://dx.doi.org/10.1016/j.physrep.2004.10.001`.
27. J. Meng, H. Toki, S. G. Zhou, S. Q. Zhang, W. H. Long, and L. S. Geng, Relativistic continuum Hartree Bogoliubov theory for ground-state properties of exotic nuclei, *Prog. Part. Nucl. Phys.* **57**(2), 470–563 (Oct., 2006). doi: 10.1016/j.ppnp.2005.06.001. URL `http://dx.doi.org/10.1016/j.ppnp.2005.06.001`.
28. T. Nikšić, D. Vretenar, and P. Ring, Relativistic nuclear energy density functionals: Mean-field and beyond, *Prog. Part. Nucl. Phys.* **66**(3), 519–548 (July, 2011). ISSN 0146-6410. doi: 10.1016/j.ppnp.2011.01.055. URL `http://dx.doi.org/10.1016/j.ppnp.2011.01.055`.
29. J. Meng and S. G. Zhou, Halos in medium-heavy and heavy nuclei with covariant density functional theory in continuum, *J. Phys. G: Nucl. Part. Phys.* **42**(9), 093101–52 (Jul, 2015). ISSN 1361-6471. doi: 10.1088/0954-3899/42/9/093101. URL `http://dx.doi.org/10.1088/0954-3899/42/9/093101`.

30. J. Meng and P. Ring, Giant halo at the neutron drip line, *Phys. Rev. Lett.* **80**(3), 460–463 (Jan., 1998). doi: 10.1103/PhysRevLett.80.460. URL http://link.aps.org/abstract/PRL/v80/p460.
31. J. Meng, H. Toki, J. Y. Zeng, S. Q. Zhang, and S.-G. Zhou, Giant halo at the neutron drip line in Ca isotopes in relativistic continuum Hartree–Bogoliubov theory, *Phys. Rev. C.* **65**(4), 041302(R)–4 (Mar., 2002). doi: 10.1103/PhysRevC.65.041302. URL http://link.aps.org/abstract/PRC/v65/e041302.
32. S.-G. Zhou, J. Meng, P. Ring, and E.-G. Zhao, Neutron halo in deformed nuclei, *Phys. Rev. C.* **82**(1), 011301(R)–5 (July, 2010). doi: 10.1103/PhysRevC.82.011301. URL http://link.aps.org/doi/10.1103/PhysRevC.82.011301.
33. L. Li, J. Meng, P. Ring, E.-G. Zhao, and S.-G. Zhou, Deformed relativistic Hartree–Bogoliubov theory in continuum, *Phys. Rev. C.* **85**(2), 024312–17 (Feb., 2012). doi: 10.1103/PhysRevC.85.024312. URL http://link.aps.org/doi/10.1103/PhysRevC.85.024312.
34. J. Bardeen, L. N. Cooper, and J. R. Schrieffer, Theory of superconductivity, *Phys. Rev.* **108**(5), 1175–1204 (Dec, 1957). ISSN 0031-899X. doi: 10.1103/physrev.108.1175. URL http://dx.doi.org/10.1103/PhysRev.108.1175.
35. A. Bohr, B. R. Mottelson, and D. Pines, Possible analogy between the excitation spectra of nuclei and those of the superconducting metallic state, *Phys. Rev.* **110**(4), 936–938 (May, 1958). doi: 10.1103/PhysRev.110.936. URL http://link.aps.org/doi/10.1103/PhysRev.110.936.
36. N. Sandulescu, N. Van Giai, and R. J. Liotta, Resonant continuum in the Hartree–Fock+BCS approximation, *Phys. Rev. C.* **61**(6), 061301(R)–5 (May, 2000). doi: 10.1103/PhysRevC.61.061301. URL http://link.aps.org/abstract/PRC/v61/e061301.
37. N. Sandulescu, L. S. Geng, H. Toki, and G. C. Hillhouse, Pairing correlations and resonant states in the relativistic mean field theory, *Phys. Rev. C.* **68**(5), 054323–6 (Nov., 2003). doi: 10.1103/PhysRevC.68.054323. URL http://link.aps.org/doi/10.1103/PhysRevC.68.054323.
38. S.-S. Zhang, E.-G. Zhao, and S.-G. Zhou, Theoretical study of the two-proton halo candidate ^{17}Ne including contributions from resonant continuum and pairing correlations, *Euro. Phys. J. A.* **49**(6), 77 (June, 2013). ISSN 1434-6001. doi: 10.1140/epja/i2013-13077-8. URL http://dx.doi.org/10.1140/epja/i2013-13077-8.
39. A. Bulgac. Hartree–Fock–Bogoliubov approximation for finite systems. IPNE FT-194-1980, Bucharest (arXiv: nucl-th/9907088). URL http://arxiv.org/abs/nucl-th/9907088 (1980).
40. J. Dobaczewski, H. Flocard, and J. Treiner, Hartree–Fock–Bogolyubov description of nuclei near the neutron-drip line, *Nucl. Phys. A.* **422**(1), 103–139 (June, 1984). ISSN 0375-9474. doi: 10.1016/0375-9474(84)90433-0. URL http://dx.doi.org/10.1016/0375-9474(84)90433-0.
41. H. Kucharek and P. Ring, Relativistic field theory of superfluidity in nuclei, *Z. Phys. A.* **339**(1), 23–35 (Mar., 1991). doi: 10.1007/BF01282930. URL http://dx.doi.org/10.1007/BF01282930.

42. J. Meng and P. Ring, Relativistic Hartree–Bogoliubov description of the neutron halo in ^{11}Li, *Phys. Rev. Lett.* **77**(19), 3963–3966 (Nov., 1996). doi: 10.1103/PhysRevLett.77.3963. URL `http://link.aps.org/abstract/PRL/v77/p3963`.
43. M. Stoitsov, P. Ring, D. Vretenar, and G. A. Lalazissis, Solution of relativistic Hartree–Bogoliubov equations in configurational representation: Spherical neutron halo nuclei, *Phys. Rev. C.* **58**(4), 2086–2091 (Oct., 1998). doi: 10.1103/PhysRevC.58.2086. URL `http://link.aps.org/abstract/PRC/v58/p2086`.
44. M. Stoitsov, J. Dobaczewski, W. Nazarewicz, and P. Ring, Axially deformed solution of the Skyrme–Hartree–Fock–Bogolyubov equations using the transformed harmonic oscillator basis. the program HFBTHO (v1.66p), *Comput. Phys. Commun.* **167**(1), 43–63 (Apr., 2005). URL `http://www.sciencedirect.com/science/article/B6TJ5-4FJGWB5-2/2/4c9351240204a50e09853221c938dbf3`.
45. C. Bloch and A. Messiah, The canonical form of an antisymmetric tensor and its application to the theory of superconductivity, *Nucl. Phys.* **39**, 95–106 (Dec, 1962). ISSN 0029-5582. doi: 10.1016/0029-5582(62)90377-2. URL `http://dx.doi.org/10.1016/0029-5582(62)90377-2`.
46. M. Reed and B. Simon, *Methods of Modern Mathematical Physics*. Academic Press, New York (1972).
47. P. Ring and P. Schuck, *The nuclear many-body problem*. Springer-Verlag, New York (1980).
48. M. Serra, A. Rummel, and P. Ring, Relativistic theory of pairing in infinite nuclear matter, *Phys. Rev. C.* **65**(1), 014304 (Dec, 2001). ISSN 1089-490X. doi: 10.1103/physrevc.65.014304. URL `http://dx.doi.org/10.1103/PhysRevC.65.014304`.
49. M. Serra and P. Ring, Relativistic Hartree–Bogoliubov theory for finite nuclei, *Phys. Rev. C.* **65**(6), 064324–9 (June, 2002). doi: 10.1103/PhysRevC.65.064324. URL `http://link.aps.org/abstract/PRC/v65/e064324`.
50. S. Karatzikos, A. V. Afanasjev, G. A. Lalazissis, and P. Ring, The fission barriers in Actinides and superheavy nuclei in covariant density functional theory, *Phys. Lett. B.* **689**(2-3), 72–81 (May, 2010). ISSN 0370-2693. doi: 10.1016/j.physletb.2010.04.045. URL `http://dx.doi.org/10.1016/j.physletb.2010.04.045`.
51. A. K. Kerman, Pairing forces and nuclear collective motion, *Ann. Phys.* **12**(2), 300–329 (Feb, 1961). ISSN 0003-4916. doi: 10.1016/0003-4916(61)90008-2. URL `http://dx.doi.org/10.1016/0003-4916(61)90008-2`.
52. G. F. Bertsch and H. Esbensen, Pair correlations near the neutron drip line, *Ann. Phys.* **209**(2), 327–363 (Aug, 1991). ISSN 0003-4916. doi: 10.1016/0003-4916(91)90033-5. URL `http://dx.doi.org/10.1016/0003-4916(91)90033-5`.
53. J. Decharge and D. Gogny, Hartree–Fock–Bogolyubov calculations with the D1 effective interaction on spherical nuclei, *Phys. Rev. C.* **21**(4), 1568–1593 (Apr., 1980). doi: 10.1103/PhysRevC.21.1568. URL `http://link.aps.org/doi/10.1103/PhysRevC.21.1568`.

54. T. Gonzalez-Llarena, J. L. Egido, G. A. Lalazissis, and P. Ring, Relativistic Hartree–Bogoliubov calculations with finite range pairing forces, *Phys. Lett. B.* **379**(1-4), 13–19 (Jun, 1996). ISSN 0370-2693. doi: 10.1016/0370-2693(96)00461-3. URL `http://dx.doi.org/10.1016/0370-2693(96)00461-3`.
55. J. F. Berger, M. Girod, and D. Gogny, Microscopic analysis of collective dynamics in low energy fission, *Nucl. Phys. A.* **428**, 23–36 (Oct., 1984). ISSN 0375-9474. doi: 10.1016/0375-9474(84)90240-9. URL `http://dx.doi.org/10.1016/0375-9474(84)90240-9`.
56. J. Meng, Pairing interaction in exotic nuclei: Finite range or zero range?, *Phys. Rev. C.* **57**(3), 1229–1232 (Mar, 1998). ISSN 1089-490X. doi: 10.1103/physrevc.57.1229. URL `http://dx.doi.org/10.1103/PhysRevC.57.1229`.
57. Y. Tian, Z. Ma, and P. Ring, A finite range pairing force for density functional theory in superfluid nuclei, *Phys. Lett. B.* **676**(1-3), 44–50 (June, 2009). ISSN 0370-2693. doi: 10.1016/j.physletb.2009.04.067. URL `http://dx.doi.org/10.1016/j.physletb.2009.04.067`.
58. Y. Tian, Z.-y. Ma, and P. Ring, Axially deformed relativistic Hartree Bogoliubov theory with a separable pairing force, *Phys. Rev. C.* **80**(2), 024313–7 (Aug., 2009). doi: 10.1103/PhysRevC.80.024313. URL `http://link.aps.org/doi/10.1103/PhysRevC.80.024313`.
59. T. Nikšić, P. Ring, D. Vretenar, Y. Tian, and Z.-y. Ma, 3d relativistic Hartree–Bogoliubov model with a separable pairing interaction: Triaxial ground-state shapes, *Phys. Rev. C.* **81**(5), 054318–12 (May, 2010). doi: 10.1103/PhysRevC.81.054318. URL `http://link.aps.org/doi/10.1103/PhysRevC.81.054318`.
60. T. Niksic, N. Paar, D. Vretenar, and P. Ring, DIRHB — A relativistic self-consistent mean-field framework for atomic nuclei, *Comput. Phys. Commun.* **185**(6), 1808–1821 (Jun, 2014). ISSN 0010-4655. doi: 10.1016/j.cpc.2014.02.027. URL `http://dx.doi.org/10.1016/j.cpc.2014.02.027`.
61. S.-C. Yang, J. Meng, and S.-G. Zhou, Exploration of unbound states by analytical continuation in the coupling constant method within relativistic mean field theory, *Chin. Phys. Lett.* **18**(2), 196–198 (Feb., 2001). doi: 10.1088/0256-307X/18/2/314. URL `http://stacks.iop.org/0256-307X/18/196`.
62. L.-G. Cao and Z.-Y. Ma, Exploration of resonant continuum and giant resonance in the relativistic approach, *Phys. Rev. C.* **66**(2), 024311–5 (Aug., 2002). doi: 10.1103/PhysRevC.66.024311. URL `http://link.aps.org/doi/10.1103/PhysRevC.66.024311`.
63. S. S. Zhang, J. Meng, S. G. Zhou, and G. C. Hillhouse, Analytic continuation of single-particle resonance energy and wave function in relativistic mean field theory, *Phys. Rev. C.* **70**(3), 034308–8 (Sept., 2004). doi: 10.1103/PhysRevC.70.034308. URL `http://link.aps.org/abstract/PRC/v70/e034308`.
64. L. Zhang, S.-G. Zhou, J. Meng, and E.-G. Zhao, Real stabilization method for nuclear single-particle resonances, *Phys. Rev. C.* **77**(1), 014312–6 (Jan., 2008). doi: 10.1103/PhysRevC.77.014312. URL `http://link.aps.org/doi/10.1103/PhysRevC.77.014312`.

65. S.-G. Zhou, J. Meng, and E.-G. Zhao, A spherical-box approach for resonances in the presence of the Coulomb interaction, *J. Phys. B: At. Mol. Phys.* **42**(24), 245001–5 (Dec., 2009). ISSN 0953-4075. doi: 10.1088/0953-4075/42/24/245001. URL `http://dx.doi.org/10.1088/0953-4075/42/24/245001`.
66. Z. P. Li, J. Meng, Y. Zhang, S. G. Zhou, and L. N. Savushkin, Single-particle resonances in a deformed Dirac equation, *Phys. Rev. C.* **81**(3), 034311–6 (Mar., 2010). doi: 10.1103/PhysRevC.81.034311. URL `http://link.aps.org/doi/10.1103/PhysRevC.81.034311`.
67. J.-Y. Guo, X.-Z. Fang, P. Jiao, J. Wang, and B.-M. Yao, Application of the complex scaling method in relativistic mean-field theory, *Phys. Rev. C.* **82**(3), 034318–8 (Sept., 2010). doi: 10.1103/PhysRevC.82.034318. URL `http://link.aps.org/doi/10.1103/PhysRevC.82.034318`.
68. H. L. Yadav, M. Kaushik, and H. Toki, Description of drip-line nuclei within the relativistic mean field plus BCS approach, *Int. J. Mod. Phys. E.* **13**(3), 647–696 (June, 2004). doi: 10.1142/S0218301304002375. URL `http://dx.doi.org/10.1142/S0218301304002375`.
69. S.-S. Zhang, M. S. Smith, Z.-S. Kang, and J. Zhao, Microscopic self-consistent study of neon halos with resonant contributions, *Phys. Lett. B.* **730**, 30–35 (Mar, 2014). doi: 10.1016/j.physletb.2014.01.023. URL `http://www.sciencedirect.com/science/article/pii/S0370269314000355`.
70. M. Shi, Q. Liu, Z.-M. Niu, and J.-Y. Guo, Relativistic extension of the complex scaling method for resonant states in deformed nuclei, *Phys. Rev. C.* **90**(3), 034319 (Sep, 2014). ISSN 1089-490X. doi: 10.1103/physrevc.90.034319. URL `http://dx.doi.org/10.1103/PhysRevC.90.034319`.
71. J. Meng, Relativistic continuum Hartree–Bogoliubov theory with both zero range and finite range Gogny force and their application, *Nucl. Phys. A.* **635**(1-2), 3–42 (May, 1998). ISSN 0375-9474. doi: 10.1016/S0375-9474(98)00178-X. URL `http://dx.doi.org/10.1016/S0375-9474(98)00178-X`.
72. J. Terasaki, P. H. Heenen, H. Flocard, and P. Bonche, 3D solution of Hartree–Fock–Bogoliubov equations for drip-line nuclei, *Nucl. Phys. A.* **600**(3), 371–386 (Apr., 1996). ISSN 0375-9474. doi: 10.1016/0375-9474(96)00036-X. URL `http://dx.doi.org/10.1016/0375-9474(96)00036-X`.
73. M. V. Stoitsov, J. Dobaczewski, P. Ring, and S. Pittel, Quadrupole deformations of neutron-drip-line nuclei studied within the Skyrme Hartree–Fock–Bogoliubov approach, *Phys. Rev. C.* **61**(3), 034311–14 (Feb., 2000). doi: 10.1103/PhysRevC.61.034311. URL `http://link.aps.org/abstract/PRC/v61/e034311`.
74. E. Teran, V. E. Oberacker, and A. S. Umar, Axially symmetric Hartree–Fock–Bogoliubov calculations for nuclei near the drip lines, *Phys. Rev. C.* **67**(6), 064314–13 (June, 2003). doi: 10.1103/PhysRevC.67.064314. URL `http://link.aps.org/abstract/PRC/v67/e064314`.
75. S.-G. Zhou, J. Meng, and P. Ring, Spherical relativistic Hartree theory in a Woods–Saxon basis, *Phys. Rev. C.* **68**(3), 034323–12 (Sept., 2003). doi: 10.1103/PhysRevC.68.034323. URL `http://link.aps.org/abstract/PRC/v68/e034323`.

76. N. Tajima, Canonical-basis solution of the Hartree–Fock–Bogoliubov equation on a three-dimensional cartesian mesh, *Phys. Rev. C.* **69**(3), 034305–22 (Mar., 2004). doi: 10.1103/PhysRevC.69.034305. URL http://link.aps.org/abstract/PRC/v69/e034305.
77. M. Stoitsov, N. Michel, and K. Matsuyanagi, New efficient method for performing Hartree–Fock–Bogoliubov calculations for weakly bound nuclei, *Phys. Rev. C.* **77**(5), 054301–12 (May, 2008). doi: 10.1103/PhysRevC.77.054301. URL http://link.aps.org/abstract/PRC/v77/e054301.
78. H. Nakada, Application of Gaussian expansion method to nuclear mean-field calculations with deformation, *Nucl. Phys. A.* **808**(1-4), 47–59 (Aug., 2008). ISSN 0375-9474. doi: 10.1016/j.nuclphysa.2008.05.011. URL http://www.sciencedirect.com/science/article/B6TVB-4SMNXRD-1/2/5b3d912dab1e2b213bc161417b54afd3.
79. N. Schunck and J. L. Egido, Nuclear halos and drip lines in symmetry-conserving continuum Hartree–Fock–Bogoliubov theory, *Phys. Rev. C.* **78**(6), 064305–14 (Dec., 2008). doi: 10.1103/PhysRevC.78.064305. URL http://link.aps.org/abstract/PRC/v78/e064305.
80. S.-G. Zhou, J. Meng, and P. Ring, Toward a deformed relativistic Hartree–Bogoliubov model for exotic nuclei, *AIP Conf. Proc.* **865**(1), 90–95 (Nov., 2006). doi: 10.1063/1.2398833. URL http://link.aip.org/link/?APC/865/90/1.
81. S.-G. Zhou, J. Meng, and P. Ring. Deformed relativistic Hartree–Bogoliubov model for exotic nuclei. In eds. D. T. Khoa, P. Egelhof, S. Gales, N. Van Giai, and T. Motobayashi, *Physics of Unstable Nuclei*, pp. 402–408, World Scientific (2008). doi: 10.1142/9789812776150_0059. URL http://www.worldscibooks.com/physics/6649.html. Proceedings of the International Symposium on Physics of Unstable Nuclei, July 3-7, 2007, Hoi An, Vietnam.
82. C. E. Price and G. E. Walker, Self-consistent Hartree description of deformed nuclei in a relativistic quantum field theory, *Phys. Rev. C.* **36**(1), 354–364 (July, 1987). doi: 10.1103/PhysRevC.36.354. URL http://link.aps.org/abstract/PRC/v36/p354.
83. W. Koepf and P. Ring, The spin-orbit field in superdeformed nuclei: a relativistic investigation, *Z. Phys. A.* **339**(1), 81–90 (Mar., 1991). doi: 10.1007/BF01282936. URL http://dx.doi.org/10.1007/BF01282936.
84. W. Koepf, Y. K. Gambhir, P. Ring, and M. M. Sharma, Neutron halo in lithium nuclei: A relativistic mean-field approach, *Z. Phys. A.* **340**(2), 119–124 (June, 1991). ISSN 0939-7922. doi: 10.1007/BF01303823. URL http://dx.doi.org/10.1007/BF01303823.
85. M. M. Sharma, G. A. Lalazissis, W. Hillebrandt, and P. Ring, Shell effects in nuclei near the neutron-drip line, *Phys. Rev. Lett.* **72**(10), 1431–1434 (Mar., 1994). doi: 10.1103/PhysRevLett.72.1431. URL http://link.aps.org/doi/10.1103/PhysRevLett.72.1431.
86. Z. Y. Zhu, W. Q. Shen, Y. H. Cai, and Y. G. Ma, Study of halo nuclei with phenomenological relativistic mean field approach, *Phys. Lett. B.* **328**(1-2), 1–4 (May, 1994). ISSN 0370-2693. doi: 10.1016/0370-2693(94)90418-9. URL http://dx.doi.org/10.1016/0370-2693(94)90418-9.

87. Z. Ren, W. Mittig, B. Chen, and Z. Ma, Neutron halos in O isotopes, *Phys. Rev. C.* **52**(1), R20–R22 (July, 1995). doi: 10.1103/PhysRevC.52.R20. URL http://link.aps.org/doi/10.1103/PhysRevC.52.R20.
88. Z. Ren, G. Xu, B. Chen, Z. Ma, and W. Mittig, Structure of halo nuclei ^{14}Be and ^{32}Ne, *Phys. Lett. B.* **351**(1-3), 11–17 (May, 1995). doi: 10.1016/0370-2693(95)00364-Q. URL http://dx.doi.org/10.1016/0370-2693(95)00364-Q.
89. Z. Ren, B. Chen, Z. Ma, and G. Xu, Relativistic mean-field study of odd-A N and F isotopes, *J. Phys. G: Nucl. Part. Phys.* **22**(1), L1–L5 (Jan, 1996). ISSN 1361-6471. doi: 10.1088/0954-3899/22/1/001. URL http://dx.doi.org/10.1088/0954-3899/22/1/001.
90. Z. Ren, B. Chen, Z. Ma, Z. Zhu, and G. Xu, One-neutron halos in N-22, O-23 and F-24 and three-neutron halo in F-26, *J. Phys. G: Nucl. Part. Phys.* **22**(4), L1–L5 (Jan, 1996). doi: 10.1088/0954-3899/22/4/013. URL http://dx.doi.org/10.1088/0954-3899/22/4/013.
91. Z. Ren, Z. Y. Zhu, Y. H. Cai, and G. Xu, Relativistic mean-field study of exotic carbon nuclei, *Nucl. Phys. A.* **605**(1), 75–86 (July, 1996). ISSN 0375-9474. doi: 10.1016/0375-9474(96)00186-8. URL http://www.sciencedirect.com/science/article/pii/0375947496001868.
92. Z. Ren, B. Chen, Z. Ma, and G. Xu, One-proton halo in ^{26}P and two-proton halo in ^{27}S, *Phys. Rev. C.* **53**(2), R572–575 (Feb., 1996). doi: 10.1103/PhysRevC.53.R572. URL http://link.aps.org/abstract/PRC/v53/pR572.
93. Z. Z. Ren, A. Faessler, and A. Bobyk, Relativistic mean-field description of a proton halo in the first excited $(1/2)^+$ state of ^{17}F, *Phys. Rev. C.* **57**, 2752–2755 (1998). doi: 10.1103/PhysRevC.57.2752. URL http://link.aps.org/doi/10.1103/PhysRevC.57.2752.
94. G. Gangopadhyay and S. Roy, Relativistic mean field study of neutron-rich even-even C and Be isotopes, *J. Phys. G: Nucl. Phys.* **31**(9), 1111–1122 (Sept., 2005). ISSN 0954-3899. doi: 10.1088/0954-3899/31/9/012. URL http://www.iop.org/EJ/abstract/0954-3899/31/9/012/.
95. Y.-j. Liang, Z.-h. Liu, and H.-y. Zhou, Investigations on the nuclear halo structures, *Int. J. Mod. Phys. E.* **17**(09), 1729–1738 (Oct, 2008). ISSN 1793-6608. doi: 10.1142/s0218301308010738. URL http://dx.doi.org/10.1142/S0218301308010738.
96. Y.-J. Liang, Y.-S. Li, Z.-H. Liu, and H.-Y. Zhou, Relativistic mean field study on halo structures of mirror nuclei, *Chinese Phys. Lett.* **26**(3), 032102 (Feb, 2009). ISSN 1741-3540. doi: 10.1088/0256-307x/26/3/032102. URL http://dx.doi.org/10.1088/0256-307X/26/3/032102.
97. J. G. Chen, X. Z. Cai, Y. Shi, Y. G. Ma, Z. Z. Ren, D. Q. Fang, W. Guo, H. W. Wang, W. D. Tian, and W. Q. Shen, Nuclear halo and its scaling laws in the excited states of nuclei near the beta-stability line, *Int. J. Mod. Phys. E.* **17** (Supp01), 50–65 (Dec, 2008). ISSN 1793-6608. doi: 10.1142/s0218301308011756. URL http://dx.doi.org/10.1142/S0218301308011756.
98. G. Saxena, D. Singh, and M. Kaushik, Sn isotopes: a study within

RMF+BCS approach, *Rom. Rep. Phys.* **65**(4), 1301–1313 (2013). URL `http://www.rrp.infim.ro/2013_65_4.html`.

99. G. Saxena and D. Singh, Study of neutron magic drip-line nuclei within relativistic mean-field plus BCS approach, *Int. J. Mod. Phys. E.* **22**(05), 1350025 (2013). doi: 10.1142/S0218301313500250. URL `http://www.worldscientific.com/doi/abs/10.1142/S0218301313500250`.
100. G. Saxena and D. Singh, Shell closures, loosely bound structures, and halos in exotic nuclei, *J. Exp. Theor. Phys.* **116**(4), 567–578 (2013). ISSN 1063-7761. doi: 10.1134/S1063776113040092. URL `http://dx.doi.org/10.1134/S1063776113040092`.
101. G. Saxena, D. Singh, M. Kaushik, and S. S. Singh, RMF+BCS approach for drip-line isotopes of Si, *Can. J. Phys.* **92**(3), 253–258 (Mar, 2014). ISSN 1208-6045. doi: 10.1139/cjp-2013-0358. URL `http://dx.doi.org/10.1139/cjp-2013-0358`.
102. W. Pöschl, D. Vretenar, G. A. Lalazissis, and P. Ring, Relativistic Hartree–Bogoliubov theory with finite range pairing forces in coordinate space: Neutron halo in light nuclei, *Phys. Rev. Lett.* **79**(20), 3841–3844 (Nov., 1997). doi: 10.1103/PhysRevLett.79.3841. URL `http://link.aps.org/abstract/PRL/v79/p3841`.
103. J. Meng, I. Tanihata, and S. Yamaji, The proton and neutron distributions in Na isotopes: the development of halo and shell structure, *Phys. Lett. B.* **419**(1-4), 1–6 (Feb., 1998). doi: 10.1016/S0370-2693(97)01386-5. URL `http://www.sciencedirect.com/science/article/B6TVN-3VGPWM4-44/1/f83ff8aa3662905a3a6f21e8b7464573`.
104. D. Vretenar, W. Poschl, G. A. Lalazissis, and P. Ring, Relativistic mean-field description of light Lambda hypernuclei with large neutron excess, *Phys. Rev. C.* **57**(3), R1060–R1063 (Mar., 1998). doi: 10.1103/PhysRevC.57.R1060. URL `http://link.aps.org/doi/10.1103/PhysRevC.57.R1060`.
105. H.-F. Lü and J. Meng, Hyperon haloes in λ hypernuclei in the relativistic continuum Hartree–Bogoliubov theory, *Chin. Phys. Lett.* **19**(12), 1775–1778 (Dec., 2002). ISSN 0256-307X. doi: 10.1088/0256-307X/19/12/310. URL `http://stacks.iop.org/0256-307X/19/i=12/a=310`.
106. H. F. Lü, J. Meng, S. Q. Zhang, and S. G. Zhou, Neutron halos in hypernuclei, *Eur. Phys. J. A.* **17**(1), 19–24 (May, 2003). doi: 10.1140/epja/i2002-10136-3. URL `http://dx.doi.org/10.1140/epja/i2002-10136-3`.
107. H. F. Lu, J. Meng, and S. Q. Zhang, Giant neutron halos in Λ-hypernuclei ${}^{A}{}_{\Lambda}$Ca and ${}^{A}{}_{\Lambda\Lambda}$Ca in the relativistic continuum Hartree–Nogoliubov theory, *High Ener. Phys. Nucl. Phys.* **27**(5), 411–415 (May, 2003). URL `http://cpc-hepnp.ihep.ac.cn/qikan/Cpaper/zhaiyao.asp?bsid=3493`. (in Chinese).
108. H.-F. Lü, Extreme exotic calcium lambda hypernuclei in the relativistic continuum Hartree–Bogoliubov theory, *Chin. Phys. Lett.* **25**(10), 3613–3616 (Oct., 2008). ISSN 0256-307X. doi: 10.1088/0256-307X/25/10/025. URL `http://stacks.iop.org/0256-307X/25/i=10/a=025`.
109. W. H. Long, P. Ring, J. Meng, N. Van Giai, and C. A. Bertulani, Nuclear halo structure and pseudospin symmetry, *Phys. Rev. C.* **81**(3), 031302(R)–

5 (Mar., 2010). URL http://link.aps.org/doi/10.1103/PhysRevC.81.031302.

110. Y. Chen, L. Li, H. Liang, and J. Meng, Density-dependent deformed relativistic Hartree–Bogoliubov theory in continuum, *Phys. Rev. C.* **85**(6), 067301–5 (June, 2012). doi: 10.1103/PhysRevC.85.067301. URL http://link.aps.org/doi/10.1103/PhysRevC.85.067301.

111. G. F. Bertsch, B. A. Brown, and H. Sagawa, High-energy reaction cross sections of light nuclei, *Phys. Rev. C.* **39**(3), 1154–1157 (Mar., 1989). doi: 10.1103/PhysRevC.39.1154. URL http://link.aps.org/doi/10.1103/PhysRevC.39.1154.

112. H. Sagawa, Density distributions of halo nuclei, *Phys. Lett. B.* **286**(1-2), 7–12 (Jul, 1992). ISSN 0370-2693. doi: 10.1016/0370-2693(92)90150-3. URL http://dx.doi.org/10.1016/0370-2693(92)90150-3.

113. S.-Q. Zhang, J. Meng, and S.-G. Zhou, Proton magic even-even isotopes and giant halos of Ca isotopes with relativistic continuum Hartree–Bogoliubov theory, *Sci. China G.* **46**(6), 632–658 (Dec., 2003). doi: 10.1360/03yw0140. URL http://www.scichina.com:8083/sciGe/EN/abstract/abstract408047.shtml.

114. M. M. Sharma, M. A. Nagarajan, and P. Ring, Rho meson coupling in the relativistic mean field theory and description of exotic nuclei, *Phys. Lett. B.* **312**(4), 377–381 (Aug., 1993). ISSN 0370-2693. URL http://www.sciencedirect.com/science/article/B6TVN-470G13S-69/2/b3e2b625f3a9be50fc271b3986c87fda.

115. P. W. Zhao, Z. P. Li, J. M. Yao, and J. Meng, New parametrization for the nuclear covariant energy density functional with a point-coupling interaction, *Phys. Rev. C.* **82**(5), 054319–14 (Nov., 2010). doi: 10.1103/PhysRevC.82.054319. URL http://link.aps.org/doi/10.1103/PhysRevC.82.054319.

116. X. Qu, Y. Chen, S. Zhang, P. Zhao, I. Shin, Y. Lim, Y. Kim, and J. Meng, Extending the nuclear chart by continuum: From oxygen to titanium, *Sci. China-Phys. Mech. Astron.* **56**(11), 2031–2036 (Nov., 2013). ISSN 1674-7348. doi: 10.1007/s11433-013-5329-5. URL http://dx.doi.org/10.1007/s11433-013-5329-5.

117. P. W. Zhao, L. S. Song, B. Sun, H. Geissel, and J. Meng, Crucial test for covariant density functional theory with new and accurate mass measurements from Sn to Pa, *Phys. Rev. C.* **86**(6), 064324–6 (Dec., 2012). doi: 10.1103/PhysRevC.86.064324. URL http://link.aps.org/doi/10.1103/PhysRevC.86.064324.

118. B. Sun, P. Zhao, and J. Meng, Mass prediction of proton-rich nuclides with the coulomb displacement energies in the relativistic point-coupling model, *Sci. China-Phys. Mech. Astron.* **54**(2), 210–214 (Jan, 2011). ISSN 1869-1927. doi: 10.1007/s11433-010-4222-8. URL http://dx.doi.org/10.1007/s11433-010-4222-8.

119. B.-N. Lu, E.-G. Zhao, and S.-G. Zhou, Potential energy surfaces of actinide nuclei from a multidimensional constrained covariant density functional theory: Barrier heights and saddle point shapes, *Phys. Rev. C.*

85(1), 011301(R)–5 (Jan., 2012). doi: 10.1103/PhysRevC.85.011301. URL http://link.aps.org/doi/10.1103/PhysRevC.85.011301.

120. B.-N. Lu, J. Zhao, E.-G. Zhao, and S.-G. Zhou, Multidimensionally-constrained relativistic mean-field models and potential-energy surfaces of actinide nuclei, *Phys. Rev. C.* **89**(1), 014323–15 (Jan., 2014). doi: 10.1103/PhysRevC.89.014323. URL http://link.aps.org/doi/10.1103/PhysRevC.89.014323.
121. P. W. Zhao, J. Peng, H. Z. Liang, P. Ring, and J. Meng, Antimagnetic rotation band in nuclei: A microscopic description, *Phys. Rev. Lett.* **107**(12), 122501–5 (Sept., 2011). doi: 10.1103/PhysRevLett.107.122501. URL http://link.aps.org/doi/10.1103/PhysRevLett.107.122501.
122. P. W. Zhao, S. Q. Zhang, J. Peng, H. Z. Liang, P. Ring, and J. Meng, Novel structure for magnetic rotation bands in ^{60}Ni, *Phys. Lett. B.* **699**(3), 181–186 (May, 2011). ISSN 0370-2693. doi: 10.1016/j.physletb.2011.03.068. URL http://dx.doi.org/10.1016/j.physletb.2011.03.068.
123. P. W. Zhao, J. Peng, H. Z. Liang, P. Ring, and J. Meng, Covariant density functional theory for antimagnetic rotation, *Phys. Rev. C.* **85**(5), 054310–14 (May, 2012). doi: 10.1103/PhysRevC.85.054310. URL http://link.aps.org/doi/10.1103/PhysRevC.85.054310.
124. J. Meng, J. Peng, S.-Q. Zhang, and P.-W. Zhao, Progress on tilted axis cranking covariant density functional theory for nuclear magnetic and antimagnetic rotation, *Front. Phys.* **8**(1), 55–79 (Jan., 2013). ISSN 2095-0462. doi: 10.1007/s11467-013-0287-y. URL http://dx.doi.org/10.1007/s11467-013-0287-y.
125. S.-Q. Zhang, J. Meng, S.-G. Zhou, and J.-Y. Zeng, Giant neutron halo in exotic calcium nuclei, *Chin. Phys. Lett.* **19**(3), 312–314 (Mar., 2002). ISSN 0256-307X. doi: 10.1088/0256-307X/19/3/308. URL http://dx.doi.org/10.1088/0256-307X/19/3/308.
126. M. Rufa, J. Schaffner, J. Maruhn, H. Stoecker, W. Greiner, and P.-G. Reinhard, Multi-lambda hypernuclei and the equation of state of hypermatter, *Phys. Rev. C.* **42**(6), 2469–2478 (Dec., 1990). doi: 10.1103/PhysRevC.42.2469. URL http://link.aps.org/doi/10.1103/PhysRevC.42.2469.
127. M. Rufa, J. Schaffner, J. Maruhn, H. Stoecker, W. Greiner, and P.-G. Reinhard, Erratum: Multi-lambda hypernuclei and the equation of state of hypermatter, *Phys. Rev. C.* **43**(4), 2020–2020 (Apr., 1991). doi: 10.1103/PhysRevC.43.2020. URL http://link.aps.org/doi/10.1103/PhysRevC.43.2020.
128. X. Li and P.-H. Heenen, Self-consistent calculations of Be isotopes, *Phys. Rev. C.* **54**(4), 1617–1621 (Oct., 1996). doi: 10.1103/PhysRevC.54.1617. URL http://link.aps.org/abstract/PRC/v54/p1617.
129. T. Misu, W. Nazarewicz, and S. Åberg, Deformed nuclear halos, *Nucl. Phys. A.* **614**(1), 44–70 (Feb., 1997). ISSN 0375-9474. doi: 10.1016/S0375-9474(96)00458-7. URL http://dx.doi.org/10.1016/S0375-9474(96)00458-7.
130. L. Guo, E.-G. Zhao, and F. Sakata, Ground-state properties of C, O, and Ne isotopes in Hartree–Fock–Bogoliubov calculation with Gogny interaction,

Commun. Theor. Phys. **40**(5), 573–576 (Nov., 2003). URL http://ctp.itp.cn/qikan/Epaper/zhaiyao.asp?bsid=1081.

131. J. Meng, H. F. Lü, S. Q. Zhang, and S. G. Zhou, Giant, hyperon, and deformed halos near the particle drip line, *Nucl. Phys. A.* **722**, 366c–371c (July, 2003). ISSN 0375-9474. doi: 10.1016/S0375-9474(03)01391-5. URL http://dx.doi.org/10.1016/S0375-9474(03)01391-5.
132. F. M. Nunes, Valence pairing, core deformation and the development of two-neutron halos, *Nucl. Phys. A.* **757**(3-4), 349–359 (Aug., 2005). ISSN 0375-9474. doi: 10.1016/j.nuclphysa.2005.04.005. URL http://dx.doi.org/10.1016/j.nuclphysa.2005.04.005.
133. J. C. Pei, F. R. Xu, and P. D. Stevenson, Deformation effects on the structures of $n = 7$ halo nuclei, *Nucl. Phys. A.* **765**(1-2), 29–38 (Jan., 2006). ISSN 0375-9474. doi: 10.1016/j.nuclphysa.2005.10.004. URL http://dx.doi.org/10.1016/j.nuclphysa.2005.10.004.
134. T. Suzuki, H. Geissel, O. Bochkarev, L. Chulkov, M. Golovkov, N. Fukunishi, D. Hirata, H. Irnich, Z. Janas, H. Keller, T. Kobayashi, G. Kraus, G. Müzenberg, S. Neumaier, F. Nickel, A. Ozawa, A. Piechaczeck, E. Roeckl, W. Schwab, K. Sümerer, K. Yoshida, and I. Tanihata, Nuclear radii of Na and Mg isotopes, *Nucl. Phys. A.* **630**(3-4), 661–677 (1998). ISSN 0375-9474. doi: 10.1016/S0375-9474(98)00799-4. URL http://dx.doi.org/10.1016/S0375-9474(98)00799-4.
135. R. Kanungo, A. Prochazka, W. Horiuchi, C. Nociforo, T. Aumann, D. Boutin, D. Cortina-Gil, B. Davids, M. Diakaki, F. Farinon, H. Geissel, R. Gernhaeuser, J. Gerl, R. Janik, B. Jonson, B. Kindler, R. Knoebel, R. Kruecken, M. Lantz, H. Lenske, Y. Litvinov, B. Lommel, K. Mahata, P. Maierbeck, A. Musumarra, T. Nilsson, C. Perro, C. Scheidenberger, B. Sitar, P. Strmen, B. Sun, Y. Suzuki, I. Szarka, I. Tanihata, Y. Utsuno, H. Weick, and M. Winkler, Matter radii of $^{32-35}$Mg, *Phys. Rev. C.* **83**(2), 021302(R)–4 (Feb., 2011). doi: 10.1103/PhysRevC.83.021302. URL http://link.aps.org/doi/10.1103/PhysRevC.83.021302.
136. B.-N. Lu, E.-G. Zhao, and S.-G. Zhou, Quadrupole deformation (β,γ) of light λ hypernuclei in a constrained relativistic mean field model: Shape evolution and shape polarization effect of the λ hyperon, *Phys. Rev. C.* **84**(1), 014328–10 (July, 2011). doi: 10.1103/PhysRevC.84.014328. URL http://link.aps.org/doi/10.1103/PhysRevC.84.014328.
137. S.-G. Zhou, J. Meng, P. Ring, and E.-G. Zhao, Neutron halo in deformed nuclei from a relativistic Hartree–Bogoliubov model in a Woods–Saxon basis, *J. Phys: Conf. Ser.* **312**(9), 092067–7 (Sept., 2011). ISSN 1742-6596. doi: 10.1088/1742-6596/312/9/092067. URL http://stacks.iop.org/1742-6596/312/i=9/a=092067. arXiv:1101.3158v1 [nucl-th].
138. J. C. Pei, Y. N. Zhang, and F. R. Xu, Evolution of surface deformations of weakly bound nuclei in the continuum, *Phys. Rev. C.* **87**(5), 051302(R)–5 (May, 2013). doi: 10.1103/PhysRevC.87.051302. URL http://link.aps.org/doi/10.1103/PhysRevC.87.051302.
139. J. Meng and I. Tanihata, The surface diffuseness and the spin-orbital splitting in relativistic continuum Hartree–Bogoliubov the-

ory, *Nucl. Phys. A.* **650**(2), 176–196 (Apr., 1999). doi: 10.1016/S0375-9474(99)00104-9. URL http://www.sciencedirect.com/science/article/B6TVB-3WSV1PV-3/1/a568ac0e55bc594c2a148681ea97c9e4.

140. J. Meng, K. Sugawara-Tanabe, S. Yamaji, P. Ring, and A. Arima, Pseudospin symmetry in relativistic mean field theory, *Phys. Rev. C.* **58**(2), R628–631 (Aug., 1998). doi: 10.1103/PhysRevC.58.R628. URL http://link.aps.org/abstract/PRC/v58/pR628.
141. J. Meng, K. Sugawara-Tanabe, S. Yamaji, and A. Arima, Pseudospin symmetry in Zr and Sn isotopes from the proton drip line to the neutron drip line, *Phys. Rev. C.* **59**(1), 154–163 (Jan., 1999). doi: 10.1103/PhysRevC.59.154. URL http://link.aps.org/abstract/PRC/v59/p154.
142. S.-G. Zhou, J. Meng, and P. Ring, Spin symmetry in the antinucleon spectrum, *Phys. Rev. Lett.* **91**(26), 262501–4 (Dec., 2003). doi: 10.1103/PhysRevLett.91.262501. URL http://link.aps.org/abstract/PRL/v91/e262501.
143. N. Wang, M. Liu, X. Wu, and J. Meng, Surface diffuseness correction in global mass formula, *Phys. Lett. B.* **734**(0), 215–219 (June, 2014). ISSN 0370-2693. doi: 10.1016/j.physletb.2014.05.049. URL http://www.sciencedirect.com/science/article/pii/S037026931400358X.
144. G. Audi and A. H. Wapstra, The 1995 update to the atomic mass evaluation, *Nucl. Phys. A.* **595**(4), 409–480 (Dec., 1995). doi: 10.1016/0375-9474(95)00445-9. URL http://www.sciencedirect.com/science/article/B6TVB-3YYTR61-1/1/4d118297fb99a6b35c8527554b90e57d.
145. J. Meng, S. G. Zhou, and I. Tanihata, The relativistic continuum Hartree–Bogoliubov description of charge-changing cross section for C, N, O and F isotopes, *Phys. Lett. B.* **532**(3-4), 209–214 (Apr., 2002). ISSN 0370-2693. doi: 10.1016/S0370-2693(02)01574-5. URL http://dx.doi.org/10.1016/S0370-2693(02)01574-5.
146. N. Wang, M. Liu, and X. Wu, Modification of nuclear mass formula by considering isospin effects, *Phys. Rev. C.* **81**(4), 044322–8 (Apr., 2010). doi: 10.1103/PhysRevC.81.044322. URL http://link.aps.org/doi/10.1103/PhysRevC.81.044322.
147. N. Wang, Z. Liang, M. Liu, and X. Wu, Mirror nuclei constraint in nuclear mass formula, *Phys. Rev. C.* **82**(4), 044304–6 (Oct., 2010). doi: 10.1103/PhysRevC.82.044304. URL http://link.aps.org/doi/10.1103/PhysRevC.82.044304.
148. M. Liu, N. Wang, Y. Deng, and X. Wu, Further improvements on a global nuclear mass model, *Phys. Rev. C.* **84**(1), 014333–8 (July, 2011). doi: 10.1103/PhysRevC.84.014333. URL http://link.aps.org/doi/10.1103/PhysRevC.84.014333.
149. G. Audi, A. H. Wapstra, and C. Thibault, The 2003 atomic mass evaluation: (II). tables, graphs and references, *Nucl. Phys. A.* **729**(1), 337–676 (Dec., 2003). doi: 10.1016/j.nuclphysa.2003.11.003. URL http://dx.doi.org/10.1016/j.nuclphysa.2003.11.003.
150. A. Sobiczewski and Y. A. Litvinov, Accuracy of theoretical descriptions of nuclear masses, *Phys. Rev. C.* **89**(2), 024311–4 (Feb, 2014). doi:

10.1103/PhysRevC.89.024311. URL http://link.aps.org/doi/10.1103/PhysRevC.89.024311.

151. A. Sobiczewski and Y. A. Litvinov, Predictive power of nuclear-mass models, *Phys. Rev. C.* **90**(1), 017302–5 (July, 2014). doi: 10.1103/PhysRevC.90.017302. URL http://link.aps.org/doi/10.1103/PhysRevC.90.017302.
152. Z. Li, Z.-M. Niu, B.-H. Sun, N. Wang, and J. Meng, WLW mass model in nuclear r-process calculations, *Acta Phys. Sin.* **61**(7), 072601–6 (2012). doi: 10.7498/aps.61.072601. URL http://dx.doi.org/10.7498/aps.61.072601. (in Chinese).
153. J. d. J. Mendoza-Temis, G. Martinez-Pinedo, K. Langanke, A. Bauswein, and H. T. Janka. On the robustness of the r-process in neutron-star mergers. arXiv:1409.6135 [astro-ph.HE]. URL http://arxiv.org/abs/1409.6135 (Sept., 2014).
154. K. Bennaceur, J. Dobaczewski, and M. Ploszajczak, Pairing anti-halo effect, *Phys. Lett. B.* **496**(3-4), 154–160 (Dec., 2000). ISSN 0370-2693. doi: 10.1016/S0370-2693(00)01292-2. URL http://www.sciencedirect.com/science/article/B6TVN-41XM9PX-4/2/d261a458d12e29f0df079cabe8081520.
155. Y. Chen, P. Ring, and J. Meng, Influence of pairing correlations on the size of the nucleus in relativistic continuum Hartree–Bogoliubov theory, *Phys. Rev. C.* **89**(1), 014312–9 (Jan., 2014). doi: 10.1103/PhysRevC.89.014312. URL http://link.aps.org/doi/10.1103/PhysRevC.89.014312.
156. J. Cerny, J. Esterl, R. Gough, and R. Sextro, Confirmed proton radioactivity of $^{53}\mathrm{Co}^m$, *Phys. Lett. B.* **33**(4), 284–286 (Oct, 1970). ISSN 0370-2693. doi: 10.1016/0370-2693(70)90270-4. URL http://dx.doi.org/10.1016/0370-2693(70)90270-4.
157. B. Blank and M. J. G. Borge, Nuclear structure at the proton drip line: Advances with nuclear decay studies, *Prog. Part. Nucl. Phys.* **60**(2), 403–483 (Apr, 2008). ISSN 0146-6410. doi: 10.1016/j.ppnp.2007.12.001. URL http://dx.doi.org/10.1016/j.ppnp.2007.12.001.
158. M. Pfutzner, M. Karny, L. V. Grigorenko, and K. Riisager, Radioactive decays at limits of nuclear stability, *Rev. Mod. Phys.* **84**(2), 567–619 (Apr., 2012). doi: 10.1103/RevModPhys.84.567. URL http://link.aps.org/doi/10.1103/RevModPhys.84.567.
159. S. Aberg, P. B. Semmes, and W. Nazarewicz, Spherical proton emitters, *Phys. Rev. C.* **56**(4), 1762–1773 (Oct, 1997). ISSN 1089-490X. doi: 10.1103/physrevc.56.1762. URL http://dx.doi.org/10.1103/PhysRevC.56.1762.
160. E. Maglione, L. S. Ferreira, and R. J. Liotta, Nucleon decay from deformed nuclei, *Phys. Rev. Lett.* **81**(3), 538–541 (Jul, 1998). ISSN 1079-7114. doi: 10.1103/physrevlett.81.538. URL http://dx.doi.org/10.1103/PhysRevLett.81.538.
161. D. Vretenar, G. A. Lalazissis, and P. Ring, Proton drip-line nuclei in relativistic Hartree–Bogoliubov theory, *Phys. Rev. C.* **57**(6), 3071–3078 (Jun, 1998). ISSN 1089-490X. doi: 10.1103/physrevc.57.3071. URL http://dx.doi.org/10.1103/PhysRevC.57.3071.
162. G. A. Lalazissis and S. Raman, Proton drip-line nuclei in relativistic mean-

field theory, *Phys. Rev. C.* **58**(3), 1467–1472 (Sep, 1998). ISSN 1089-490X. doi: 10.1103/physrevc.58.1467. URL `http://dx.doi.org/10.1103/PhysRevC.58.1467`.

163. G. A. Lalazissis, D. Vretenar, and P. Ring, Ground-state properties of deformed proton emitters in the relativistic Hartree–Bogoliubov model, *Nucl. Phys. A.* **650**(2), 133–156 (Apr., 1999). ISSN 0375-9474. doi: 10.1016/S0375-9474(99)00121-9. URL `http://dx.doi.org/10.1016/S0375-9474(99)00121-9`.
164. D. Vretenar, G. A. Lalazissis, and P. Ring, Relativistic Hartree–Bogoliubov description of the deformed ground-state proton emitters, *Phys. Rev. Lett.* **82**(23), 4595–4598 (June, 1999). doi: 10.1103/PhysRevLett.82.4595. URL `http://link.aps.org/abstract/PRL/v82/p4595`.
165. G. A. Lalazissis, D. Vretenar, and P. Ring, Transitional Lu and spherical Ta ground-state proton emitters in the relativistic Hartree–Bogoliubov model, *Phys. Rev. C.* **60**(5), 051302(R) (Oct, 1999). ISSN 1089-490X. doi: 10.1103/physrevc.60.051302. URL `http://dx.doi.org/10.1103/PhysRevC.60.051302`.
166. H. Esbensen and C. N. Davids, Coupled-channels treatment of deformed proton emitters, *Phys. Rev. C.* **63**(1), 014315– (Dec, 2000). ISSN 1089-490X. doi: 10.1103/physrevc.63.014315. URL `http://dx.doi.org/10.1103/PhysRevC.63.014315`.
167. G. A. Lalazissis, D. Vretenar, and P. Ring, Mapping the proton drip line from $Z = 31$ to $Z = 49$, *Nucl. Phys. A.* **679**(3-4), 481–493 (Jan, 2001). ISSN 0375-9474. doi: 10.1016/s0375-9474(00)00375-4. URL `http://dx.doi.org/10.1016/S0375-9474(00)00375-4`.
168. G. Fiorin, E. Maglione, and L. S. Ferreira, Theoretical description of deformed proton emitters: Nonadiabatic quasiparticle method, *Phys. Rev. C.* **67**(5), 054302–4 (May, 2003). doi: 10.1103/PhysRevC.67.054302. URL `http://link.aps.org/abstract/PRC/v67/e054302`.
169. G. A. Lalazissis, D. Vretenar, and P. Ring, Mapping the proton drip line, *Nucl. Phys. A.* **719**, C209–C212 (May, 2003). ISSN 0375-9474. doi: 10.1016/s0375-9474(03)00919-9. URL `http://dx.doi.org/10.1016/S0375-9474(03)00919-9`.
170. G. A. Lalazissis, D. Vretenar, and P. Ring, Mapping the proton drip line in the suburanium region and for superheavy elements, *Physical Review C.* **69**(1), 017301 (Jan, 2004). ISSN 1089-490X. doi: 10.1103/physrevc.69.017301. URL `http://dx.doi.org/10.1103/physrevc.69.017301`.
171. L. Ferreira, E. Maglione, and P. Ring, Self-consistent description of proton radioactivity, *Phys. Lett. B.* **701**(4), 508–511 (July, 2011). ISSN 0370-2693. doi: 10.1016/j.physletb.2011.06.026. URL `http://dx.doi.org/10.1016/j.physletb.2011.06.026`.
172. Q. Zhao, J. M. Dong, J. L. Song, and W. H. Long, Proton radioactivity described by covariant density functional theory with the similarity renormalization group method, *Phys. Rev. C.* **90**(5), 054326– (Nov., 2014). doi: 10.1103/PhysRevC.90.054326. URL `http://link.aps.org/doi/10.1103/PhysRevC.90.054326`.

173. Y. Lim, X. Xia, and Y. Kim. Proton radioactivity in relativistic continuum Hartree–Bogoliubov theory (2015). Submitted to Phys. Rev. C.
174. I. G. Darby, R. D. Page, D. T. Joss, L. Bianco, T. Grahn, D. S. Judson, J. Simpson, S. Eeckhaudt, P. T. Greenlees, P. M. Jones, and et al., Precision measurements of proton emission from the ground states of Ta-156 and Re-160, *Phys. Rev. C.* **83**(6), 064320–4 (Jun, 2011). ISSN 1089-490X. doi: 10.1103/physrevc.83.064320. URL `http://dx.doi.org/10.1103/PhysRevC.83.064320`.
175. C. Qi, D. S. Delion, R. J. Liotta, and R. Wyss, Effects of formation properties in one-proton radioactivity, *Phys. Rev. C.* **85**(1), 011303(R)–5 (Jan., 2012). doi: 10.1103/PhysRevC.85.011303. URL `http://link.aps.org/doi/10.1103/PhysRevC.85.011303`.

Chapter 4

Relativistic Hartree–Fock–Bogoliubov theory: ground states and excitations

Wen Hui Long

School of Nuclear Science and Technology, Lanzhou University, Lanzhou 730000, China
longwh@lzu.edu.cn

Jie Meng

School of Physics, Peking University, Beijing 100871, China
mengj@pku.edu.cn

Nguyen Van Giai

Institut de Physique Nucléaire, IN2P3-CNRS, Université Paris-Sud, F-91406 Orsay, France
nguyen@ipno.in2p3.fr

The covariant density functional (CDF) theory with the Fock diagrams, the indivisible part of the effective nuclear interaction, is introduced, including both the relativistic Hartree–Fock and its extension — the relativistic Hartree–Fock–Bogoliubov methods. The specific roles played by Fock diagrams, particularly for the new degrees of freedom associated with the π and ρ-tensor fields and the non-local mean fields, are discussed in determining the nuclear energy functional, the shell structure and the evolution, and nuclear isospin excitations. The existing problems and limits of the CDF theory with Fock terms are also discussed, and the perspective on a new algorithm of dealing with the non-local Fock terms is given.

Contents

1. Introduction

As mentioned in previous parts of the book, the relativistic mean field (RMF) theory,[1,2] namely the relativistic Hartree (RH) approach with no-sea approximation, has achieved great success in describing numerous nuclear phenomena including nuclear ground state and excitation properties over the whole nuclear chart.[3–8] The advantage of the simple RMF framework is mainly due to the covariant formulation of the model which maintains the distinction between Lorentz scalars and vectors (more precisely, the time components of four-vectors). Such models with large scalar and vector fields, of the order of a few hundred MeV, provide a simpler and more efficient description of nuclear systems than the non-relativistic picture. The most remarkable feature is the existence of the spin-orbit potential which comes out as a natural consequence of the covariant structure of the RMF scheme.

However, the defects of the RMF model are also evident. At the level of a mean field approach, both direct (Hartree) and exchange (Fock) diagrams are the indivisible parts of the nuclear dynamics. Among the mesons mediating the nuclear interactions, the pion would be one of the most important. Because of its negative parity, the pion contributes to the mean field only via the Fock diagram. Another important degree of freedom associated with the tensor ρ field is also present in most of the Fock contributions. Furthermore, the RMF approach cannot take into account other effective spin-spin interactions such as the nuclear tensor force. Except for simplicity, there is no robust physical reason to exclude the Fock diagrams.

At this point, we need to use a more precise wording to discuss the so-called "tensor interactions". In non-relativistic approaches the name of tensor force is commonly used to designate that part of the interaction which behaves like an irreducible spherical tensor of rank 2,[9] and we shall refer to it as the non-relativistic (NR) tensor force. In covariant theory, the nucleon-nucleon interactions can be separated into various components (Lorentz scalar, pseudo-scalar, etc.) among which is the second-rank Lorentz tensor $\sigma^{\mu\nu}$,[10] and we call it the Lorentz tensor interaction component.

Although the Fock diagrams increase greatly the complexity of the model, many attempts have been made by several groups in developing and applying the relativistic Hartree–Fock (RHF) approach to describe nuclear systems since the 1980s.[11–14] In the 1980's the first systematic investigations within the RHF framework were performed by Bouyssy and collaborators for nuclear matter and finite nuclei.[12] The importance of isovector-meson contributions was recognized in determining nuclear matter properties.[11] They gave a satisfactory description on the density profiles of doubly magic nuclei, but failed on the value of nuclear incompressibility as well as the binding energy of finite nuclei, due to the fact that the nuclear in-medium effects were not treated properly, e.g., by allowing for non-linear self-couplings of meson fields or density-dependent meson-nucleon couplings. Other attempts were devoted to the nuclear in-medium effects by introducing the non-linear self-coupling of the σ-field[13] or the non-linear self-interaction of the scalar field in the zero-range limit.[14] Although these studies brought distinct improvements on both the nuclear incompressibility and binding energies, they were still a long way from the RMF theory, either in the quantitative description of nuclear systems or their extrapolation to the exotic regions. Compared with the wide range of applications of the RMF theory, the disparity seemed quite large.

In recent years a new RHF method, namely the density-dependent relativistic Hartree–Fock (DDRHF) theory was proposed, where density-dependent meson-nucleon couplings were introduced to account for the nuclear in-medium effects.[15,16] Later on, it was extended to the relativistic Hartree–Fock–Bogoliubov (RHFB) theory[17] to explore the unstable exotic nuclei.[18] For the first time, it provided an appropriate RHF description of nuclear matter and finite nuclei, with a quantitative precision comparable with the RMF models. Due to the non-local mean fields coming from the Fock diagrams, the DDRHF theory improves the isospin- and energy-dependence of nuclear effective masses, leading to consistent results with the Brueckner–Hartree–Fock predictions.[16] In contrast with the RMF approach, the new degrees of freedom associated with the pion and Lorentz tensor ρ fields bring significant improvements to the description of nuclear shell structure[19] and its evolution,[20] respectively because of the new balance of nuclear attraction and repulsion achieved with fairly strong Lorentz tensor ρ attraction and the NR tensor force components carried by the pion exchange potential. Besides the pion and Lorentz tensor ρ fields, the Fock diagrams of the isoscalar meson fields (σ and ω) also present substantial effects in determining the density-dependent behavior of the symmetry energy,[21] as well as in reproducing the enhancement of the proton sub-shell $Z = 64$ along the isotonic chain of $N = 82$.[22]

Besides nuclear ground state properties, the Fock diagrams have also substantial impact on nuclear excitations and decay modes. Combined with the random phase approximation (RPA), the RHF+RPA model for the first time provides full self-consistent description of nuclear Gamow–Teller, and spin-dipole (SD) excitations,[23] where the Fock diagrams of the isoscalar σ and ω meson fields were found to play an essential role. Furthermore, such effects due to σ and ω Fock terms are also important for a good description of the fine structure of the SD excitation in ^{16}O.[24] In contrast to the RMF+RPA model, a unified description of the nuclear mean fields and particle-hole residual interactions is achieved by the RHF+RPA scheme with the same RHF functional,[23] thus demonstrating the advantage of the method. Based on the spherical RHFB framework, a fully self-consistent proton-neutron quasiparticle random phase approximation (QRPA) for spherical nuclei was developed[25] for the study of the β-decay half-lives of neutron-rich nuclei with $20 \leqslant Z \leqslant 50$, and the available data are reproduced by including an isospin-dependent proton-neutron pairing interaction in the isoscalar channel of the RHFB+QRPA model.[25]

It should be also noted that, in infinite nuclear matter one could re-express the Fock diagrams in terms of direct interactions by means of the Fierz transformation whereas this cannot be done in finite nuclei. Thus, it is unavoidable to have full-fledged RHF and RHFB theories for nuclear systems. The RHF and RHFB methods for spherical nuclei will be introduced in Sections 2–5. In Section 6 the role of Fock diagrams in describing nuclear ground state and excitation properties will be discussed, specifically the essential roles played by the pion and Lorentz tensor ρ fields and the non-local mean fields originated from the isoscalar channels. Finally the conclusions and perspectives are presented in Section 7.

2. General Lagrangian and Hamiltonian for RHFB

Based on the meson-exchange picture of the nuclear force suggested by Yukawa in 1935,[26] the two-body interaction between two nucleons can be decomposed into two types of diagrams (see Fig. 1): the direct (Hartree) and exchange (Fock) processes. Following such criterion, the starting point of the covariant density functional (CDF) theory, the Lagrangian density can be constructed by including the degrees of freedom associated with the nucleon (ψ), the mesons (σ, ω and ρ) and the photon (A) fields as introduced in Chapter 2. Due to the limitations of the Hartree approach, significant degrees of freedom related to the NR tensor force components — the π- and Lorentz tensor ρ-fields — are missing, and these missing components will emerge naturally with the presence of Fock diagrams of the nucleon-nucleon (NN) interaction (see Fig. 1).

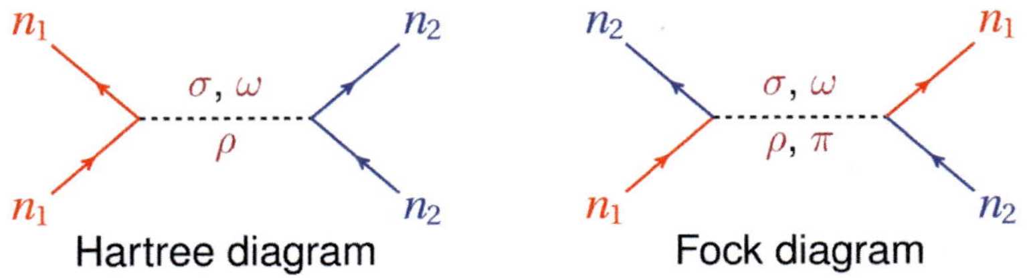

Fig. 1. The Feynman diagrams for the Hartree and Fock contributions to the nucleon-nucleon interaction mediated by mesons.

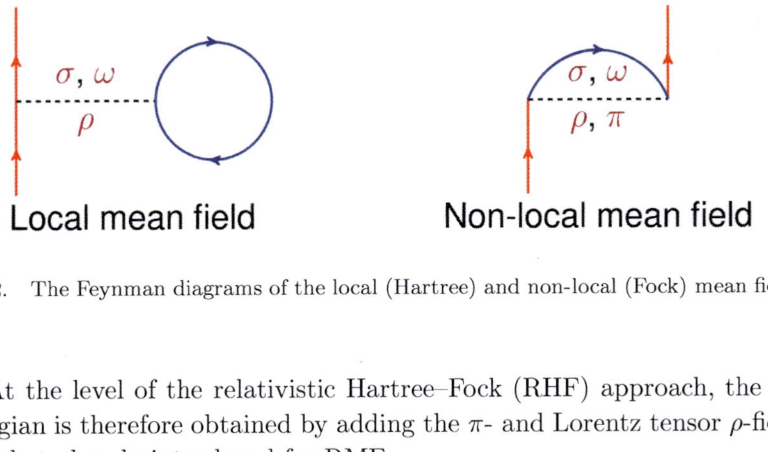

Fig. 2. The Feynman diagrams of the local (Hartree) and non-local (Fock) mean fields.

At the level of the relativistic Hartree–Fock (RHF) approach, the Lagrangian is therefore obtained by adding the π- and Lorentz tensor ρ-fields into that already introduced for RMF,

$$\mathscr{L}_\pi = + \frac{1}{2}\partial_\mu \vec{\pi} \cdot \partial^\mu \vec{\pi} - \frac{1}{2} m_\pi^2 \vec{\pi} \cdot \vec{\pi} - \frac{f_\pi}{m_\pi} \bar{\psi} \gamma_5 \gamma_\mu \partial^\mu \vec{\pi} \cdot \vec{\tau} \psi, \tag{1}$$

$$\mathscr{L}_\rho = - \frac{1}{4} \vec{R}_{\mu\nu} \cdot \vec{R}^{\mu\nu} + \frac{1}{2} m_\rho^2 \vec{\rho}^\mu \cdot \vec{\rho}_\mu - \bar{\psi}\Big(g_\rho \gamma^\mu \vec{\rho}_\mu \cdot \vec{\tau} + \frac{f_\rho}{2M} \sigma^{\mu\nu} \partial_\nu \vec{\rho}_\mu \cdot \vec{\tau}\Big)\psi, \tag{2}$$

where f_π (m_π) and f_ρ (m_ρ) are the coupling strengths (masses) of the π- and Lorentz tensor ρ-fields, respectively. In principle, for the π-mediated

NN interaction one can choose two possible forms, the pseudo-scalar or pseudo-vector form, to obtain the one-pion exchange potentials. From general arguments (chiral symmetry) and from the construction of the NN potential (with pair suppression mechanism), the pseudo-vector (PV) coupling has to be used to get reasonable results.[12]

From the Lagrangian densities (1, 2), the variations with respect to π- and ρ-meson fields lead to the following field equations, namely the inhomogeneous Klein–Gordon equations,

$$(\Box + m_\pi^2)\vec{\pi} = + \partial^\mu \frac{f_\pi}{m_\pi} \vec{j}_{5\mu}, \tag{3}$$

$$(\Box + m_\rho^2)\vec{\rho}^\nu = + g_\rho \vec{j}^\nu - \partial_\mu \frac{f_\rho}{2M} \vec{j}^{\mu\nu}, \tag{4}$$

where the d'Alembert operator $\Box = \partial_\mu \partial^\mu$ and the nucleon currents read as,

$$\vec{j}^{5\nu} \equiv \bar{\psi}\gamma^5\gamma^\nu\vec{\tau}\psi, \qquad \vec{j}^{\mu\nu} = \bar{\psi}\sigma^{\mu\nu}\vec{\tau}\psi. \tag{5}$$

Notice that the current conservation condition $\partial^\nu \vec{\rho}_\nu = 0$ has been used to abbreviate the field equation of the ρ-meson. With the retarded Green's functions $D_\pi(x, x')$ and $D_\rho(x, x')$, the π- and ρ-meson fields can be analytically expressed as:

$$\vec{\pi}(x) = - \int d^4x' \frac{f_\pi}{m_\pi} \vec{j}_{5\mu}(x')\partial^\mu(x') D_\pi(x, x'), \tag{6}$$

$$\vec{\rho}_\nu(x) = \int d^4x' \left[g_\rho \vec{j}_\nu(x') + \frac{f_\rho}{2M} \vec{j}_{\mu\nu}(x')\partial^\mu(x') \right] D_\rho(x, x'). \tag{7}$$

At the mean field level, one can make the simplifying assumption of neglecting the time component of the four-momentum carried by the mesons, which means that the meson fields are time independent. This assumption has no consequence on the Hartree diagrams, while for the Fock diagrams it amounts to neglecting the retardation effects. Thus, the retarded Green's functions correspond to the Yukawa propagator,

$$D_\phi(\boldsymbol{x}, \boldsymbol{x}') = \frac{1}{4\pi} \frac{e^{-m_\phi |\boldsymbol{x} - \boldsymbol{x}'|}}{|\boldsymbol{x} - \boldsymbol{x}'|}, \tag{8}$$

where the coordinate variables can be separated conveniently, e.g., in terms of Bessel functions and harmonic polynomials if one assumes spherical symmetry.[12]

Eventually, the Hamiltonian can be derived as the functional of the nucleon field operator ψ,

$$\begin{aligned} H = & \int d^3x \Big[\bar{\psi}\big(- i\gamma\cdot\boldsymbol{\nabla} + M\big)\psi\Big] \\ & + \frac{1}{2}\int d^3x d^3x' \sum_{\phi} \bar{\psi}(x)\bar{\psi}(x')\Gamma_{\phi}(x,x')D_{\phi}(x,x')\psi(x')\psi(x), \end{aligned} \tag{9}$$

where the interaction vertices Γ_ϕ with the corresponding coupling constants are,

$$\Gamma_{\sigma\text{-S}} \equiv - \big[g_\sigma\big]_x \big[g_\sigma\big]_{x'}, \tag{10a}$$

$$\Gamma_{\omega\text{-V}} \equiv + \big[g_\omega\gamma_\mu\big]_x \big[g_\omega\gamma^\mu\big]_{x'}, \tag{10b}$$

$$\Gamma_{A\text{-V}} \equiv + \Big[e\gamma_\mu\frac{1-\tau_3}{2}\Big]_x \Big[e\gamma^\mu\frac{1-\tau_3}{2}\Big]_{x'}, \tag{10c}$$

$$\Gamma_{\pi\text{-PV}} \equiv - \Big[\frac{f_\pi}{m_\pi}\vec{\tau}\gamma_5\gamma_\mu\partial^\mu\Big]_x \cdot \Big[\frac{f_\pi}{m_\pi}\vec{\tau}\gamma_5\gamma_\nu\partial^\nu\Big]_{x'}, \tag{10d}$$

$$\Gamma_{\rho\text{-V}} \equiv + \big[g_\rho\gamma_\mu\vec{\tau}\big]_x \cdot \big[g_\rho\gamma^\mu\vec{\tau}\big]_{x'}, \tag{10e}$$

$$\Gamma_{\rho\text{-T}} \equiv + \Big[\frac{f_\rho}{2M}\sigma_{\nu k}\vec{\tau}\partial^k\Big]_x \cdot \Big[\frac{f_\rho}{2M}\sigma^{\nu l}\vec{\tau}\partial_l\Big]_{x'}, \tag{10f}$$

$$\Gamma_{\rho\text{-VT}} \equiv + \Big[\frac{f_\rho}{2M}\sigma^{k\nu}\vec{\tau}\partial_k\Big]_x \cdot \big(g_\rho\gamma_\nu\vec{\tau}\big)_{x'} + \big(g_\rho\gamma_\nu\vec{\tau}\big)_x \cdot \Big[\frac{f_\rho}{2M}\sigma^{k\nu}\vec{\tau}\partial_k\Big]_{x'}. \tag{10g}$$

In Eq. (9), ϕ stands for the meson (photon) channels, namely the σ-scalar (σ-S), ω-vector (ω-V), photon-vector (A-V), π-pseudo-vector (π-PV), ρ-vector (ρ-V) and Lorentz tensor (ρ-T), ρ-vector-tensor (ρ-VT) coupling channels.

3. RHF Energy functional

To derive the RHF energy functional, namely the expectation value of the Hamiltonian (9), one may expand the nucleon field operator ψ on a complete and orthogonal basis defined by the stationary solutions of the Dirac equation.[12] In terms of the annihilation and creation operators defined by the solutions of the Dirac equation, i.e., $\{c_k, c_k^\dagger\}$ and $\{d_l, d_l^\dagger\}$ which correspond to the positive and negative energy states, respectively, the field operator ψ is expanded as:

$$\psi = \sum_k \psi_k(\boldsymbol{x})e^{-i\varepsilon_k t}c_k + \sum_l \psi_l(\boldsymbol{x})e^{-i\varepsilon_l t}d_l^\dagger, \tag{11a}$$

$$\psi^\dagger = \sum_k \psi_k^\dagger(\boldsymbol{x})e^{+i\varepsilon_k t}c_k^\dagger + \sum_l \psi_l^\dagger(\boldsymbol{x})e^{+i\varepsilon_l t}d_l, \tag{11b}$$

where ψ_k, ψ_l are the Dirac spinors.

In keeping with the mean field approach, the contributions from the negative energy states, i.e., the d_l and $d_l^\dagger$ terms in Eq. (11), are dropped, which amounts to neglecting the Dirac sea: this is the so-called no-sea approximation.[2] In fact, the neglected terms may lead to divergence and require a cumbersome renormalization procedure. With the no-sea approximation, the trial ground state can be constructed as:

$$|\Phi_0\rangle = \prod_i c_i^\dagger |0\rangle, \tag{12}$$

where i denotes the positive energy states in the Fermi sea and $|0\rangle$ is the vacuum state.[3] With respect to the trial ground state $|\Phi_0\rangle$, the RHF energy functional is determined from the expectation value of the Hamiltonian (9),

$$E = \langle\Phi_0| H |\Phi_0\rangle = \langle\Phi_0| T |\Phi_0\rangle + \sum_\phi \langle\Phi_0| V_\phi |\Phi_0\rangle, \tag{13}$$

where T and V_ϕ correspond to the kinetic and potential energy parts, respectively:

$$T = \int d\boldsymbol{x} \sum_{\alpha,\beta} \bar{\psi}_\alpha(\boldsymbol{x})(-i\boldsymbol{\gamma}\cdot\boldsymbol{\nabla} + M)\psi_\beta(\boldsymbol{x})c_\alpha^\dagger c_\beta, \tag{14}$$

$$V_\phi = \frac{1}{2}\int d\boldsymbol{x}d\boldsymbol{x}' \sum_{\alpha\beta;\alpha'\beta'} c_\alpha^\dagger c_\beta^\dagger c_{\beta'} c_{\alpha'} \bar{\psi}_\alpha(\boldsymbol{x})\bar{\psi}_\beta(\boldsymbol{x}')\Gamma_\phi D_\phi \psi_{\beta'}(\boldsymbol{x}')\psi_{\alpha'}(\boldsymbol{x}). \tag{15}$$

At the mean field level, the expectation value of the two-body operator in Eq. (15), i.e., $\langle\Phi_0| c_\alpha^\dagger c_\beta^\dagger c_{\beta'} c_{\alpha'} |\Phi_0\rangle$, may lead to two types of contributions, the direct (Hartree: $\beta = \beta'$ and $\alpha = \alpha'$) and exchange (Fock: $\beta = \alpha'$ and $\alpha = \beta'$) terms. In Fig. 2 are shown the Feynman diagrams of the Hartree and Fock contributions to single-particle energies coming from the σ-, ω-, ρ- and π-meson mean fields.

If one considers only the Hartree diagrams (see the left panels of Figs. 1 and 2), the expression (13) leads to the energy functional of the RMF model. On the other hand, with the presence of both Hartree and Fock diagrams, i.e., in the RHF picture, it leads to:

$$E = E_k + \sum_\phi \left(E_\phi^D + E_\phi^E\right), \tag{16}$$

where E_k stands for the kinetic energy, and E_ϕ^D and E_ϕ^E are the Hartree and Fock potential energies in the coupling channel ϕ, respectively. The variations of E_ϕ^D and E_ϕ^E lead to two types of mean fields, the Hartree and

Fock potentials, and the relevant Feynman diagrams are shown in Fig. 2. It can be easily seen from Fig. 2 that the Hartree mean field is local whereas the one arising from the Fock diagram is non-local in coordinate space.

4. Relativistic Hartree–Fock method

In the π-PV and ρ-T coupling channels, there exist repulsive contact components that can be derived from the momentum representation of $\boldsymbol{\nabla}\boldsymbol{\nabla}' D_\phi(\boldsymbol{r}, \boldsymbol{r}')$,

$$\frac{\boldsymbol{q}^2}{m_\phi^2 + \boldsymbol{q}^2} = 1 - \frac{m_\phi^2}{m_\phi^2 + \boldsymbol{q}^2}, \tag{17}$$

where ϕ stands for π- and ρ-mesons and $\boldsymbol{q} = \boldsymbol{p} - \boldsymbol{p}'$ is the momentum transfer. These contact parts will be counterbalanced by the strong repulsion from the heavy ω-meson exchanges, and their contributions to the energy functional will be cancelled by adding the following terms,

$$H_\pi^\delta = \frac{1}{3} \cdot \frac{1}{2} \int d\boldsymbol{r} d\boldsymbol{r}' \Big[\frac{f_\pi}{m_\pi} \bar{\psi}\gamma^5\gamma^k \vec{\tau}\psi\Big]_{\boldsymbol{r}} \cdot \Big[\frac{f_\pi}{m_\pi} \bar{\psi}\gamma_5\gamma_k \vec{\tau}\psi\Big]_{\boldsymbol{r}'} \delta[\boldsymbol{r} - \boldsymbol{r}'], \tag{18}$$

$$H_{\rho\text{-T}}^\delta = \frac{1}{3} \cdot \frac{1}{2} \int d\boldsymbol{r} d\boldsymbol{r}' \Big[\frac{f_\rho}{2M} \bar{\psi}\sigma_{\mu i} \vec{\tau}\psi\Big]_{\boldsymbol{r}} \cdot \Big[\frac{f_\rho}{2M} \bar{\psi}\sigma^{\mu i} \vec{\tau}\psi\Big]_{\boldsymbol{r}'} \delta[\boldsymbol{r} - \boldsymbol{r}']. \tag{19}$$

The detailed derivation of the RHF functional (16) can be found in Refs. 12, 15, and we will not repeat it here.

To perform a realistic calculation, it is in general convenient and efficient to assume a specific geometrical symmetry, for instance the spherical symmetry which is well fulfilled by many nuclei in the nuclear chart. From the RHF energy functional (16), the spherical RHF equation, i.e., the integro-differential Dirac equation, can be obtained by the variation with respect to the Dirac spinors ψ_α (for details, see Ref. 15) as,

$$\int d\boldsymbol{r}' h(\boldsymbol{r}, \boldsymbol{r}')\psi_\alpha(\boldsymbol{r}') = \varepsilon_\alpha \psi_\alpha(\boldsymbol{r}), \tag{20}$$

where ε_α denotes the single-particle energy and $h(\boldsymbol{r}, \boldsymbol{r}')$ is the single-particle Dirac Hamiltonian,

$$h = h^{\text{kin.}} + h^D + h^E, \tag{21}$$

with the kinetic ($h^{\text{kin.}}$), direct (Hartree: h^D) and exchange (Fock: h^E) terms,

$$
\begin{aligned}
h^{\text{kin.}}(\boldsymbol{r},\boldsymbol{r}') =& \big(\boldsymbol{\alpha}\cdot\boldsymbol{p} + \beta M\big)\delta(\boldsymbol{r}-\boldsymbol{r}'), && (22)\\
h^{D}(\boldsymbol{r},\boldsymbol{r}') =& \Big[\Sigma_0(\boldsymbol{r}) + \beta\Sigma_S(\boldsymbol{r}) + \Sigma_T(\boldsymbol{r})\gamma_5\Big]\delta(\boldsymbol{r}-\boldsymbol{r}'), && (23)\\
h^{E}(\boldsymbol{r},\boldsymbol{r}') =& \begin{pmatrix} Y_G(\boldsymbol{r},\boldsymbol{r}') & Y_F(\boldsymbol{r},\boldsymbol{r}') \\ X_G(\boldsymbol{r},\boldsymbol{r}') & X_F(\boldsymbol{r},\boldsymbol{r}') \end{pmatrix}. && (24)
\end{aligned}
$$

In the direct component h^D, Σ_S, Σ_0 and Σ_T are the scalar, vector (the time component of four-vector) and Lorentz tensor local potentials:

$$
\begin{aligned}
\Sigma_S =& g_\sigma\sigma, && (25)\\
\Sigma_0 =& g_\omega\omega + g_\rho\left(\rho^V + \rho^{TV}\right)\tau_3 + e\frac{1-\tau_3}{2}A + \Sigma_R, && (26)\\
\Sigma_T =& \frac{f_\rho}{2M}\left(\rho^{VT} + \rho^T\right)\tau_3, && (27)
\end{aligned}
$$

where the Hartree mean fields σ, ω, ρ^V, ρ^T, ρ^{TV} and ρ^{VT}, A are determined by the variations of the Hartree potential energies $E^D_{\sigma\text{-S}}$, $E^D_{\omega\text{-V}}$, $E^D_{\rho\text{-V}}$, $E^D_{\rho\text{-T}}$, $E^D_{\rho\text{-VT}}$, $E^D_{\text{A-V}}$, respectively, and Σ_R is the rearrangement term due to the density dependence of the meson-nucleon couplings.[15] For spherical nuclei, the non-local potentials X_G, X_F, Y_G and Y_F in h^E can be derived from the variations of the Fock potential energy E^E_ϕ:

$$
\begin{aligned}
X^{(\phi)}_{G_a}(r,r') =& \sum_b \mathscr{T}^\phi_{ab}(2j_b+1)\big(g_\phi F_b\big)_r \mathscr{R}^{X_G}_{ab}(m_\phi;r,r')\big(g_\phi G_b\big)_{r'}, && (28)\\
X^{(\phi)}_{F_a}(r,r') =& \sum_b \mathscr{T}^\phi_{ab}(2j_b+1)\big(g_\phi F_b\big)_r \mathscr{R}^{X_F}_{ab}(m_\phi;r,r')\big(g_\phi F_b\big)_{r'}, && (29)\\
Y^{(\phi)}_{G_a}(r,r') =& \sum_b \mathscr{T}^\phi_{ab}(2j_b+1)\big(g_\phi G_b\big)_r \mathscr{R}^{Y_G}_{ab}(m_\phi;r,r')\big(g_\phi G_b\big)_{r'}, && (30)\\
Y^{(\phi)}_{F_a}(r,r') =& \sum_b \mathscr{T}^\phi_{ab}(2j_b+1)\big(g_\phi G_b\big)_r \mathscr{R}^{Y_F}_{ab}(m_\phi;r,r')\big(g_\phi F_b\big)_{r'}, && (31)
\end{aligned}
$$

where ϕ (resp. g_ϕ) stands for the meson-nucleon coupling channels (resp. constants), the isospin factor $\mathscr{T}^{(\phi)}_{ab}$ being $\delta_{\tau_a\tau_b}$ for the isoscalar channels and $2-\delta_{\tau_a\tau_b}$ for the isovector ones.

In contrast to the radial differential Dirac equation of the RMF case, the RHF equation (20) is a set of coupled integro-differential equations which are difficult to solve directly. The major complication is due to the Fock contributions that give rise to the integral term h^E in the single-particle Hamiltonian. In practice it is advantageous to rewrite the integral terms $X(r)$ and $Y(r)$ as a completely equivalent local form,[12] e.g.,

$$\begin{aligned} X_a(r) =& \frac{G_a(r)X_a(r)}{G_a^2(r)+F_a^2(r)}G_a(r) + \frac{F_a(r)X_a(r)}{G_a^2(r)+F_a^2(r)}F_a(r) \\ \equiv& \bar{X}_{G_a}(r)G_a(r) + \bar{X}_{F_a}(r)F_a(r), \qquad (32) \\ Y_a(r) \equiv& \bar{Y}_{G_a}(r)G_a(r) + \bar{Y}_{F_a}(r)F_a(r), \qquad (33) \end{aligned}$$

where

$$\begin{pmatrix} Y_a(r) \\ X_a(r) \end{pmatrix} = \int dr' \begin{pmatrix} Y_{G_a}(r,r') & Y_{F_a}(r,r') \\ X_{G_a}(r,r') & X_{F_a}(r,r') \end{pmatrix} \begin{pmatrix} G_a(r') \\ F_a(r') \end{pmatrix}. \qquad (34)$$

With the above decomposition of the integral terms $X(r)$ and $Y(r)$, the integro-differential equation (20) is transformed into a set of coupled differential ones that can be solved by the Runge–Kutta algorithm, as in the RMF scheme. It is illustrated by realistic applications[12,15,16,19] to nuclei with spherical symmetry. However, if one needs to extend the RHF scheme to more complex situations, such as introducing the Bogoliubov pairing or the deformation, this decomposition becomes too cumbersome. It is then better to solve the relevant equations by expanding the Dirac spinors on an appropriate basis, such as the harmonic oscillator basis[27] or Dirac–Woods–Saxon basis.[17]

5. Relativistic Hartree–Fock–Bogoliubov method

For open-shell nuclei, the pairing correlations become significant in determining the nuclear properties. In many cases, the BCS method can be used to deal with the pairing correlations, and this does not bring major changes to the theoretical scheme itself. While if one approaches the vicinity of neutron/proton drip lines, namely the unstable exotic nuclei, the nucleon separation energies become comparable with the pairing gap energy so that the Cooper pairs can be scattered into the continuum by the pairing interaction. The role of the pairing correlations therefore becomes more important and delicate in these weakly bound exotic nuclei and one needs a reliable treatment of the continuum as well as of the bound states.[6]

In dealing with the pairing effects in exotic nuclei, the simplest BCS method meets serious difficulties. Particularly if there are low angular momentum states around the continuum threshold, which is tightly related to the occurrence of the exotic structure mode — the halos, the BCS treatment may lead to the unphysical neutron gas. It is therefore necessary to treat the pairing correlations with the Bogoliubov method[28] as introduced in Chapter 3. The advantage of this method is that the mean field and pairing field can be unified into the Bogoliubov scheme and the continuum effects can be treated properly.[29] When combined with the RHF approach introduced in the previous section, this leads to the relativistic Hartree–Fock–Bogoliubov (RHFB) theory.[17] The variation of the RHFB energy functional[30] gives the RHFB equation:

$$\int d\boldsymbol{r}' \begin{pmatrix} h(\boldsymbol{r},\boldsymbol{r}') & \Delta(\boldsymbol{r},\boldsymbol{r}') \\ -\Delta(\boldsymbol{r},\boldsymbol{r}') & h(\boldsymbol{r},\boldsymbol{r}') \end{pmatrix} \begin{pmatrix} \psi_U(\boldsymbol{r}') \\ \psi_V(\boldsymbol{r}') \end{pmatrix} = \begin{pmatrix} \lambda + E_q & 0 \\ 0 & \lambda - E_q \end{pmatrix} \begin{pmatrix} \psi_U(\boldsymbol{r}) \\ \psi_V(\boldsymbol{r}) \end{pmatrix}, \tag{35}$$

where E_q stands for the quasi-particle energies, the chemical potential λ is introduced to preserve the particle number on the average, and h and Δ are respectively the RHF single-particle Hamiltonian [see Eq. (20)] and pairing potential. The pairing potential is:

$$\Delta_\alpha(\boldsymbol{r},\boldsymbol{r}') = -\frac{1}{2}\sum_\beta V^{pp}_{\alpha\beta}(\boldsymbol{r},\boldsymbol{r}')\kappa_\beta(\boldsymbol{r},\boldsymbol{r}'), \tag{36}$$

where the pairing tensor κ is:

$$\kappa_\alpha(\boldsymbol{r},\boldsymbol{r}') = \psi^*_{V_\alpha}(\boldsymbol{r})\psi_{U_\alpha}(\boldsymbol{r}'). \tag{37}$$

For the pairing force V^{pp}, in principle it should also be deduced from the meson exchange scheme, just as the NN interaction. Unfortunately, this leads generally to too strong pairing effects, e.g., the deduced pairing gaps are two or three times larger than the empirical values for uniform nuclear matter. In practice, a phenomenological pairing force is adopted in the particle-particle channel. Typical choices are the zero-range δ forces or finite range Gogny-type forces.[31]

Notice that the RHFB equation (35) is a set of coupled integro-differential equations, where the integral terms come from the RHF mean field as well as from the pairing potential if one uses a finite-range pairing force. Within spherical symmetry, the quasi-particle spinors ψ_U and ψ_V have a form similar to the Dirac spinor ψ_α. Then, the RHFB equation (35) can be reduced into its radial form as,

$$\left(\frac{d}{dr} + \frac{\kappa_a}{r} + \Sigma_T\right)G_{U_a} - (E_a + \lambda - \Sigma_-)F_{U_a} + X_{U_a} + r\int r'dr'\Delta(r,r')F_{V_a}(r') = 0, \tag{38a}$$

$$\left(\frac{d}{dr} - \frac{\kappa_a}{r} - \Sigma_T\right)F_{U_a} + (E_a + \lambda - \Sigma_+)G_{U_a} - Y_{U_a} + r\int r'dr'\Delta(r,r')G_{V_a}(r') = 0, \tag{38b}$$

$$\left(\frac{d}{dr} + \frac{\kappa_a}{r} + \Sigma_T\right)G_{V_a} + (E_a - \lambda + \Sigma_-)F_{V_a} + X_{V_a} + r\int r'dr'\Delta(r,r')F_{U_a}(r') = 0, \tag{38c}$$

$$\left(\frac{d}{dr} - \frac{\kappa_a}{r} - \Sigma_T\right)F_{V_a} - (E_a - \lambda + \Sigma_+)G_{V_a} - Y_{V_a} + r\int r'dr'\Delta(r,r')G_{U_a}(r') = 0, \tag{38d}$$

where $\Sigma_+ \equiv \Sigma_0 + \Sigma_S$, $\Sigma_- \equiv \Sigma_0 - \Sigma_S - 2M$, and the integral terms X_U, Y_U, X_V and Y_V contain the non-local mean field that originates from the Fock diagrams,

$$\begin{pmatrix} Y_{U \text{ or } V}(r) \\ X_{U \text{ or } V}(r) \end{pmatrix} = \int dr' \begin{pmatrix} Y_G(r,r') & Y_F(r,r') \\ X_G(r,r') & X_F(r,r') \end{pmatrix} \begin{pmatrix} G_{U \text{ or } V}(r') \\ F_{U \text{ or } V}(r') \end{pmatrix}. \tag{39}$$

To solve such complicated integro-differential equations (38), an intuitive choice is to localize the above integral terms by a decomposition similar to that introduced in solving the RHF equations (20). A drawback is that this may lead to divergences.

To solve the integro-differential RHFB equations (38), one may expand the quasi-particle spinors on some basis, such as the harmonic oscillator basis or the Dirac–Woods–Saxon (DWS) basis. The former basis works well for ordinary nuclei.[32] When treating exotic nuclei, especially those with dilute matter distributions, the harmonic oscillator basis cannot provide an appropriate asymptotic behavior for the density tail. Although one cannot obtain analytical solutions of the Dirac or Schrödinger equations with a Woods–Saxon potential, the numerical solutions still form a complete and orthogonal basis for the expansion of the quasi-particle spinors. In particular, it can provide a proper asymptotic behavior for the radial wave functions.[33]

Compared to the Woods–Saxon basis composed of the solutions of Schrödinger equations with a Woods–Saxon potential, the DWS basis allows for a simpler and more straightforward expansion. With the spherical symmetry, the quasi-particle spinors can be expanded with respect to the principal quantum number for a fixed κ value, since the Dirac spinors of the DWS basis with different κ are orthogonal.[17] Thus, the integro-differential RHFB equation (38) is transformed into an eigenvalue problem (for the details, see Ref. 17). In solving the RHFB equations, several parameters of the basis have to be chosen carefully: the radial cut-off $R_{\rm m}$, and the numbers of positive and negative energy states N_F and N_D. For most applications, it is sufficient to choose $R_{\rm m}$ =20 fm, N_F =28, N_D =12. For weakly bound nuclei, especially those with dilute matter distributions, one needs to choose a larger $R_{\rm m}$ and the value of N_F should be enlarged simultaneously to keep a consistent energy cut-off whereas N_D remains stable with respect to $R_{\rm m}$.[17]

Notice that the DWS basis is only introduced to solve the integro-differential RHFB equation while the RHF mean fields and pairing potentials are still calculated in the coordinate space. This will not bring too much additional numerical complexity so that the RHFB method can be naturally extended, e.g., to the deformed nuclei with axial symmetry. Nowadays there already exist some RHF calculations of the deformed nuclei with axial symmetry using the harmonic oscillator basis.[27] Due to the complexity induced by the expansion of non-local Fock potentials, the applications are so far limited to light nuclei. On the other hand, one can also expand the quasi-particle spinors of deformed nuclei on the spherical DWS basis.[34] Compared to the harmonic oscillator basis, the DWS basis has the innate advantage that the basis wave functions have appropriate asymptotic behaviors suitable for the applications to weakly bound nuclei. Potentially, the DWS basis could be used also for the extensions of RHFB theory to the deformed cases.

6. The Role of Fock diagrams

In the previous sections, the RHF and RHFB methods were introduced briefly for ordinary and exotic nuclei. Compared to the CDF models which drop the Fock diagrams, the numerical complexity increases greatly. In this respect, the question why the Fock diagrams are necessary in describing the nuclear systems may emerge naturally. Conceptually there is no doubt that the Fock diagrams have to be taken into account because they are the indivisible parts of the two-body interactions at the mean field level. Although one may hope that the Fock contributions can be mocked up by an appropriate parametrization of the CDF functional, many important aspects contained in the Fock diagrams cannot be described in the Hartree limit, such as the one-pion exchange potential, Lorentz tensor ρ-couplings, the non-local mean field effects, etc.

Table 1. Detailed contributions (in MeV) to the total energy in ^{208}Pb and ^{16}O calculated with the RHF functional PKA1.[19] $E_{\text{c.m.}}$ stands for the center-of-mass correction.[35]

	^{208}Pb			^{16}O		
	Neutron	Proton	Total	Neutron	Proton	Total
E_k	1596.57	907.90	2504.46	96.21	94.67	190.88
$E^D_{\sigma\text{-S}}$	−14320.86	−10022.49	−24343.35	−667.25	−659.85	−1327.10
$E^D_{\omega\text{-V}}$	11520.81	7962.50	19483.32	534.17	527.83	1062.00
$E^D_{A\text{-V}}$	0.00	827.64	827.64	0.00	16.10	16.10
$E^D_{\rho\text{-V}}$	98.42	−65.12	33.30	0.13	−0.12	0.00
$E^D_{\rho\text{-VT}}$	−1.65	1.08	−0.57	−0.01	0.01	−0.00
$E^D_{\rho\text{-T}}$	−0.31	0.21	−0.10	−0.00	0.00	−0.00
E^D	−1107.02	−388.28	−1495.30	−36.75	−21.36	−58.12
$E^E_{\sigma\text{-S}}$	3503.24	1774.28	5277.53	152.29	149.09	301.38
$E^E_{\omega\text{-V}}$	−2451.22	−1265.20	−3716.43	−113.44	−111.08	−224.51
$E^E_{A\text{-V}}$	0.00	−29.02	−29.02	0.00	−2.83	−2.83
$E^E_{\rho\text{-V}}$	−266.20	−210.10	−476.30	−17.40	−17.31	−34.70
$E^E_{\rho\text{-VT}}$	122.51	89.16	211.66	5.66	5.60	11.25
$E^E_{\rho\text{-T}}$	−687.90	−531.94	−1219.85	−46.00	−45.79	−91.79
$E^E_{\pi\text{-PV}}$	−103.26	−79.75	−183.01	−9.14	−9.13	−18.26
E^E	117.17	−252.57	−135.42	−28.03	−31.45	−59.46
$E_{\text{c.m.}}$	−3.83	−2.43	−6.26	−4.75	−4.664	−9.41
E	−993.67	−643.30	−1636.97	−69.51	−57.48	−126.99

The presence of Fock terms in the mean field will essentially change the balance between strong attractions and repulsions which originate mainly from the scalar σ- and vector ω-meson exchanges. Taking the RHF functional PKA1[19] as an example, Table 1 shows the detailed contributions from different channels in the doubly magic nuclei ^{208}Pb and ^{16}O which correspond to spherical ground states. It is found that, in the isoscalar channels (σ-S and ω-V) the Fock contributions cancel about 20%∼25% of the Hartree contributions whereas for the isovector ones (ρ-V, ρ-T, ρ-VT and π-PV) the contributions are dominated by the Fock diagrams rather than the Hartree ones. Specifically, a fairly strong attraction is provided by the Fock diagram of ρ-T coupling, which is even comparable with the cancellation between the strong σ-S attraction and ω-V repulsion. This indicates that the ρ-T coupling may essentially affect the balance between nuclear attraction and repulsion. This could be the reason why the RHF functional PKA1 can cure the common disease of the CDF models which give artificial shell closures at $N/Z = 58$ and 92.[19,36] Compared to the other channels, the Coulomb exchange term presents relatively tiny contributions to the energy functional while it is of special significance in describing some excitation modes, such as the charge-exchange excitations.[37]

6.1. *Roles of π and ρ-T couplings*

We have seen that the PV π- and Lorentz tensor ρ-couplings only or mostly contribute to the Fock diagrams. Both have substantial effects on the nuclear structure properties. For the π contribution, it was recognized to be composed of the central and NR tensor force components,[12,20] and it is also treated as the origin of the nuclear NR tensor force.[38] Because of the negative parity carried by the π-meson, it does not contribute to the Hartree diagrams so understanding the role of the pion cannot be done in RMF.

In recent years, it was found that the one-pion exchange potential, especially its NR tensor component, plays an essential role in the self-consistent description of nuclear shell evolutions along the chains of $N = 82$ isotones and $Z = 50$ isotopes.[20] As shown in the left plot of Fig. 3, the energy gap ΔE between the proton orbits $\pi 1h_{11/2}$ and $\pi 1g_{7/2}$ presents characteristic isospin dependences with respect to the neutron excess $N - Z$ along the isotopic chain of $Z = 50$.[39] Theoretically, such a behavior can be partially interpreted only by the RHF functionals PKO1[16] and PKO3[20] which contain the π-PV couplings. From the right plot of Fig. 3, it is also seen that the one-pion exchange potential in PKO1 (also PKO3), dominated by the NR tensor part, plays an essential role to reproduce the isospin dependence of ΔE, particularly for the experimental kink around $N - Z = 12$. In contrast, the RHF functional PKO2[20] and RMF functional PKDD,[35] in which the degree of freedom associated with π-meson is missing, just provide a monotonous isospin dependence for the energy gap ΔE.

For the ρ-T coupling, which brings very tiny contributions to the Hartree diagrams, it provides another possible source of NR tensor force with opposite effects as compared to π-PV coupling. In realistic applications, it is

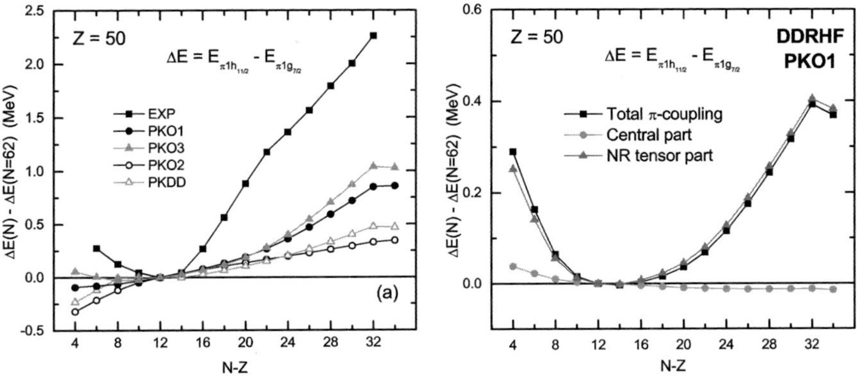

Fig. 3. (Color online) Left plot: Energy difference $\Delta E(N) = E_{\pi 1h_{11/2}} - E_{\pi 1g_{7/2}}$ in $Z = 50$ isotopes as a function of neutron excess $N - Z$. Right plot: Contributions to the energy difference ΔE from the one-pion exchange potentials. The results are extracted from the CDF calculations with PKO1,[16] PKO2, PKO3[20] and PKDD,[35] in comparison with the data.[39] The figures are taken from Ref. 20.

found that the ρ-T coupling gives rise to a fairly strong attraction, mainly by its central part, thus influencing the balance between nuclear attraction and repulsion as shown in Table 1. As a result, the common weakness of CDF calculations, i.e., the artificial shell closures $N/Z = 58$ and 92, is cured by the presence of the degree of freedom associated with Lorentz tensor ρ-field.[19] Such a result can also be interpreted by the self-consistent relation between the pseudo-spin symmetry and spin-orbit effects.[40,41] For the artificial shell closures, they are determined by the splitting between the pseudo-spin partner states, e.g., the $2f_{7/2}$ and $1h_{9/2}$ states for N or $Z = 92$ as shown in Fig. 4. Compared to other functionals, e.g., PKO1, PK1 and DD-ME2 as shown in Fig. 4, the RHF functional PKA1[19] presents enhanced spin-orbit effects such that the unphysically large gaps at N or $Z = 58$ and 92 vanish with the presence of the ρ-T coupling. In addition, it also brings some improvement to the ordering of single-particle levels, e.g., for the neutron orbits $1i_{11/2}$ and $2g_{9/2}$ beyond the $N = 126$ magic shell in ^{208}Pb (see right plot of Fig. 4).

6.2. *Non-local mean field effects*

Besides the new degrees of freedom associated with π and Lorentz tensor-ρ fields, the Fock diagrams from the other meson channels also bring significant influence on the properties of nuclear systems. As typical examples,

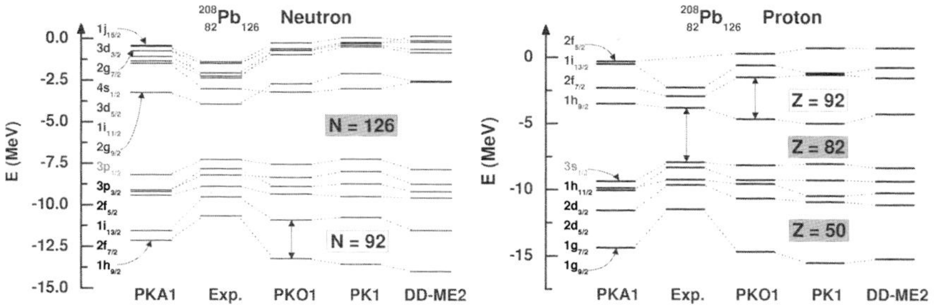

Fig. 4. (Color online) Neutron (left plot) and proton (right plot) spectra determined by the CDF models with the functionals PKA1,[19] PKO1,[16] PK1[35] and DD-ME2,[42] in comparison with the data.[43] The figure is taken from Ref. 19.

essential effects are found in determining the behaviors of nucleon effective masses[16] and symmetry energy.[21] It was found that, with the inclusion of Fock diagrams the isospin and energy dependences of nucleon effective masses are improved by the RHF method and the results are consistent with the predictions of Brueckner–Hartree–Fock models at low density and energy regions.[16,44,45] Unexpectedly the Fock diagrams of the isoscalar σ-S and ω-V couplings present substantial contributions to the nuclear symmetry energy compared to the isovector channels,[21] thus indicating that not only the Hartree diagram of the isovector ρ-field but also the Fock diagrams of both isovector and isoscalar ones do influence nuclear isovector properties.

Beyond nuclear matter bulk properties, significant improvements are also found in nuclear shell structure due to the non-locality of the Fock mean fields. Experimentally, the sub-shell closure $Z = 64$ has been identified and the enhancement of the shell gap towards $Z = 64$ was also discovered, as well as the corresponding spin-orbit splitting.[46] In addition, high-spin isomer studies along the isotonic chain of $N = 83$ also indicate similar enhancements for $Z = 64$ sub-shell gap, which increases from 2.0 MeV to 2.4 MeV when the proton number increases from $Z = 61$ to 65.[47] Within the CDF or Skyrme–Hartree–Fock (SHF) schemes, such a trend cannot be properly reproduced. The CDF models which exclude the ρ-T couplings even fail to reproduce the sub-shell closure $Z = 64$ that is compressed by the neighboring unphysical shell closure $Z = 58$.[19,36]

With the presence of the degrees of freedom associated with the Lorentz tensor ρ-field, the common drawback of unphysical shell closures $N/Z = 58$ and 92 was cured eventually by the density-dependent relativistic Hartree–

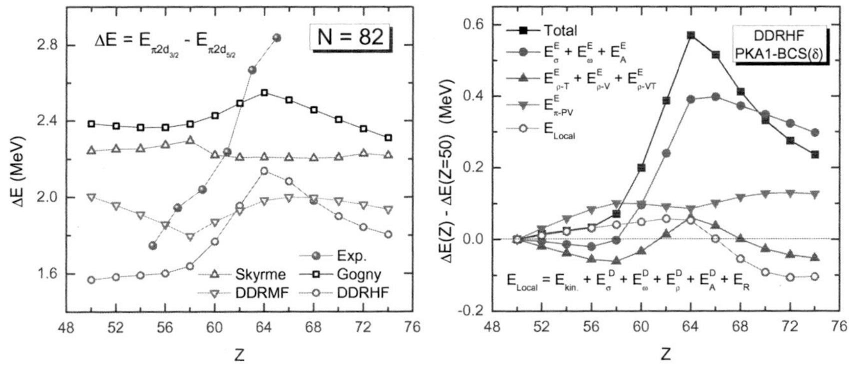

Fig. 5. (Color online) Left plot: Spin-orbit splittings ΔE_{2d} of proton $\pi 2d$ states in $N = 82$ isotones calculated by DDRHF with PKA1,[19] Gogny-HFB with D1S,[48] SHF with SLy4[49] and RMF with PKDD,[35] as compared to the data.[46] Right plot: Detailed contributions to ΔE_{2d} from different channels given by the DDRHF calculations with PKA1. The figure is taken from Ref. 22.

Fock (DDRHF) theory[16] with the RHF functional PKA1[19] and consequently the sub-shell closure $Z = 64$ was also recovered properly.[22] Based on this improvement, it was shown that the DDRHF model with PKA1 can provide a self-consistent interpretation of the enhancement of the sub-shell $Z = 64$ that is essentially determined by the spin-orbit splitting (ΔE) of proton spin partners $\pi 2d$. In Fig. 5 are shown the values of ΔE and detailed contributions from different channels along the isotonic chain of $N = 82$. To clarify the hidden mechanism, the comparison is performed between DDRHF with PKA1[19] and other three selected models, namely Gogny-HFB with D1S,[48] SHF with SLy4[49] and RMF with PKDD.[35] Compared to the data,[46] it is found from the left plot that the DDRHF calculations present a stronger enhancement (∼0.6MeV) than the Gogny and DDRMF results (∼0.2MeV). In contrast, the SHF model gives a nearly constant spin-orbit splitting, even slightly quenched from $Z = 58$ to 64.

Specifically, the results in the right plot of Fig. 5 show that the enhancement of ΔE toward $Z = 64$ is mainly determined by the non-local Fock mean field whereas very little effects come from the local one, i.e., $E_{\text{Local}} = E_k + E_\sigma^{\text{D}} + E_\omega^{\text{D}} + E_\rho^{\text{D}} + E_A^{\text{D}} + E_R$. Among the Fock terms, the isoscalar σ-S and ω-V couplings together with the Coulomb field, i.e., $E_{\sigma+\omega+A}^{\text{E}} = E_\sigma^{\text{E}} + E_\omega^{\text{E}} + E_A^{\text{E}}$, play a dominant role in the enhancement of ΔE towards $Z = 64$. In contrast, the Fock diagrams of the isovector couplings (ρ and π) present small effects. As shown in the right plot of Fig. 5

there exist distinct cancellations between π and ρ (mainly Lorentz tensor ρ) contributions along the isotonic chain $N = 82$, because of the opposite non-relativistic tensor effects embedded in π and ρ mean fields.[20] Therefore, the non-local mean field, mainly from the Fock diagrams of the isoscalar meson-nucleon couplings, becomes especially important for reproducing the enhancement of spin-orbit splittings of the $\pi 2d$ states in the $N = 82$ isotones.

Among the four models, the central and spin-orbit potentials of SHF are local thus giving nearly constant spin-orbit splittings. In contrast, DDRHF has non-local Fock terms and the integro-differential RHF equation also leads to a non-local spin-orbit potential after non-relativistic reduction. Due to the non-locality in both central and spin-orbit potentials, the DDRHF-PKA1 model can properly reproduce the enhancement of the spin-orbit splitting along the isotonic chain $N = 82$. As for the Gogny HFB model, it presents a much weaker enhancement because it has a non-local mean central potential, whereas the one-body spin-orbit potential is local.

6.3. *Fully self-consistent description of nuclear isospin excitations*

The effects of Fock diagrams are important not only for nuclear ground state properties, and they have also evident impacts on nuclear excitations, such as the Gamow-Teller (GT) transitions, spin-dipole (SD) excitations,[23,37] as well as the decay modes, e.g. β-decay.[25] Based on the relativistic Hartree–Fock (RHF) approach,[12,16] a fully self-consistent charge-exchange relativistic RPA model was established to explore the nuclear excitation modes.[23] Two major advantages of this RHF+RPA approach are that the pion is included in both the ground state description and the particle-hole (p-h) residual interaction, and the zero-range pionic counter-term with $g' = 1/3$ is maintained self-consistently. Without any adjusted p-h residual interaction or refitting process, this RHF+RPA approach provides a reliable description of nuclear excitations.[23]

Taking the fine structure of the SD excitation in ^{16}O as an example, one can illustrate the fact that the Fock diagrams play an essential role in the self-consistent description of the excitation modes.[37] Figure 6 shows the strength distributions of the SD excitations in ^{16}O for the T_- channel and the comparison is performed between the RHF+RPA with PKO1[16] (left plot) and RMF+RPA with DD-ME2[42] (right plot). From the left plot, it can be seen that the fine structure of the SD excitations can be

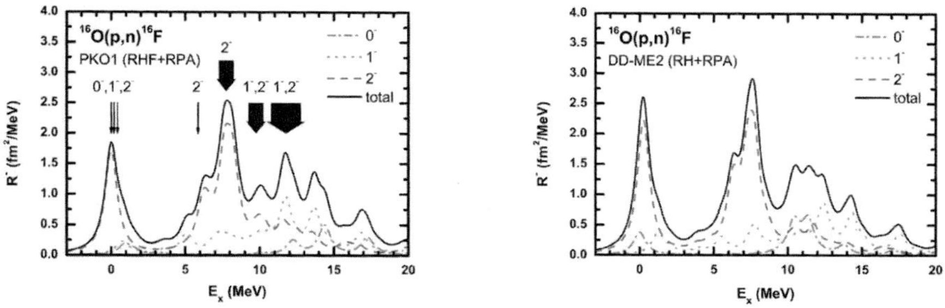

Fig. 6. (Color online) Strength distributions of the SD excitations in ^{16}O for the T_- channel. The $J^\pi = 0^-, 1^-$, and 2^- contributions calculated by RHF+RPA with PKO1[16] (left plot) and RMF+RPA with DD-ME2[42] (right plot) are shown as the dash-dotted, dotted, and dashed lines, respectively, while their sums are shown as the solid lines. The energy of the lowest RPA state is taken as reference and a Lorentzian smearing parameter $\Gamma = 1$ MeV is used. The experimental data[50,51] are shown with arrows, whose widths illustrate the widths of the corresponding resonances. The figure is taken from Ref. 37.

described fairly well by the RHF+RPA without any readjustment. For the RMF+RPA calculation with DD-ME2, it should be noticed that the fraction of pionic zero-range counter term is readjusted to 0.52 according to Ref. 52. If one looks at the general profile in the region $E_x \lesssim 8$ MeV, there is no evident difference between the two models. However, beyond $E_x = 8$ MeV substantial discrepancies emerge. As seen in the right plot, the mixture of the $J^\pi = 1^-$ and 2^- excitations at $E_x \simeq 9.5$ MeV is missing and the $J^\pi = 1^-$ resonances determined by RMF+RPA with DD-ME2 are located at too high energy as compared to the data.

To further understand the discrepancies and evaluate the significance of the self-consistency of the model, Figure 7 shows the contributions to the p-h residual interactions from the σ and ω mesons, the ρ meson, the pseudo-vector π-N coupling and its zero-range counter term, and the total. For RHF+RPA, it is found that the total strength of the p-h residual interaction is essentially determined by the delicate balance between σ and ω mesons via the exchange terms, namely the Fock diagrams of the isoscalar mesons. The ρ meson seems to be important in the natural-parity channel, while the pseudovector π-N coupling and its zero-range counter term play minor roles due to the strong suppression of the coupling strength f_π of PKO1[16] in nuclear medium. As the σ and ω mesons are well calibrated by the central and spin-orbit potentials for the nuclear ground-state properties, the proper description of the SD excitations provides a stringent and critical test of the theoretical model. It is noted that, in contrast to the Skyrme HF+RPA calculations,[53] here the explicit NR tensor interactions

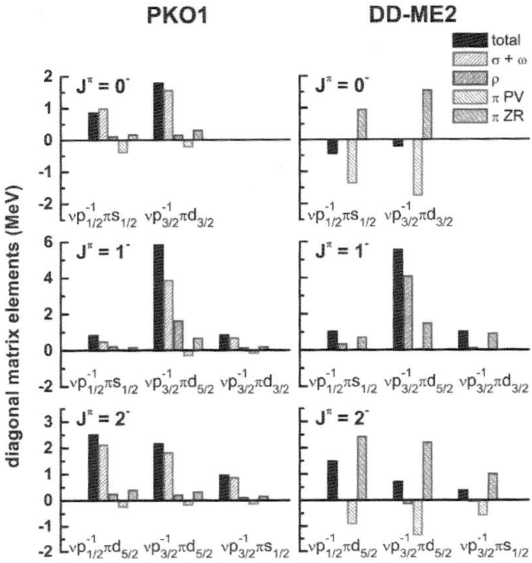

Fig. 7. (Color online) Diagonal matrix elements of the particle-hole residual interactions for the $J^\pi = 0^-, 1^-$, and 2^- excitations. The total contributions are decomposed into those from the σ and ω mesons ($\sigma + \omega$), the ρ meson, the pseudo-vector π-N coupling (π PV) and its zero-range counter term (π ZR). The configurations are indicated as neutron-hole proton-particle. The figure is taken from Ref. 37.

are not needed to reproduce the data. For RMF+RPA, the p-h residual interactions are essentially determined by the ρ meson for the natural-parity channel and by the pseudovector π-N coupling and its zero-range counter term for the unnatural parity channels. Since the pseudovector π-N coupling and its zero-range counter term are absent in describing the nuclear ground-state properties, the adjusted fraction of pionic zero-range counter term introduces an extra free parameter,[52] thus bringing some inconsistency between the mean field and p-h residual interaction.

7. Conclusion and perspectives

In this chapter, the RHF and RHFB methods were briefly introduced and the role of Fock diagrams in determining nuclear structure properties was discussed. Although the theoretical and numerical complexities are increased very much with the presence of Fock diagrams, significant improvements due to the exchange (Fock) diagrams are also found in describing nuclear ground states and excitations. The typical example is the fully self-consistent treatment of the nuclear excitation modes achieved by the RHF+RPA model, as well as the self-consistent description of nuclear shell structure and its evolution due to the new degrees of freedom associated with the pion and Lorentz tensor-ρ fields. In fact, common successes in nuclear excitations and shell evolutions were also achieved by the Skyrme–Hartree–Fock (SHF) plus the NR nuclear tensor force components at zero-range limit,[53–56] as well as in nuclear β-decay,[25,57] where the NR tensor force components were found to play a crucial role. On the other hand, the Fock terms after non-relativistic reduction also contain the tensor force components, which may explain the consensus between the SHF+Tensor and RHF models.[58] In prospect, the RHF scheme based on the meson-exchange picture may improve our understanding of the nature of nuclear force, as inspired by the existence of the NR tensor force components associated with the Fock diagrams.[58,59]

Currently, the applications of the RHF and RHFB models are mainly concentrated on nuclear matter and spherical nuclei. Based on the harmonic oscillator basis, the RHF method has been extended to describe the deformed nuclei with axial symmetry.[27] Due to the numerical complexity induced by the non-local Fock mean fields, the applications are still limited to the light nuclei. As a perspective, a new algorithm for dealing with the non-local mean fields would be needed for the extension of the RHFB method to heavier deformed systems.

Acknowledgment

This work is supported in part by the National Key Basic Research Program of China (Grant No. 2013CB834400), the Natural Science Foundation of China (Grants No. 11075066, No. 11175002, No. 11375076, and No. 11335002), and the Specialized Research Fund for the Doctoral Program of Higher Education (Grant No. 20110001110087, and No. 20130211110005).

References

1. J. D. Walecka, A theory of highly condensed matter, *Ann. Phys.* **83**(2), 491–529 (1974). doi: 10.1016/0003-4916(74)90208-5.
2. B. D. Serot and J. D. Walecka, The relativistic nuclear many-body problem, *Adv. Nucl. Phys.* **16**, 1–327 (1986).
3. P.-G. Reinhard, The relativistic mean-field description of nuclei and nuclear-

dynamics, *Rep. Prog. Phys.* **52**(4), 439–514 (1989). doi: 10.1088/0034-4885/52/4/002.

4. P. Ring, Relativistic mean field theory in finite nuclei, *Prog. Part. Nucl. Phys.* **37**, 193–263 (1996). doi: 10.1016/0146-6410(96)00054-3.
5. M. Bender, P.-H. Heenen, and P.-G. Reinhard, Self-consistent mean-field models for nuclear structure, *Rev. Mod. Phys.* **75**(1), 121–180 (Jan, 2003). doi: 10.1103/RevModPhys.75.121.
6. J. Meng, H. Toki, S. Zhou, S. Zhang, W. Long, and L. Geng, Relativistic continuum Hartree–Bogoliubov theory for ground-state properties of exotic nuclei, *Prog. Part. Nucl. Phys.* **57**(2), 470–563 (2006). doi: 10.1016/j.ppnp.2005.06.001.
7. T. Nikšić, D. Vretenar, and P. Ring, Relativistic nuclear energy density functionals: Mean-field and beyond, *Prog. Part. Nucl. Phys.* **66**(3), 519–548 (2011). ISSN 0146-6410. doi: 10.1016/j.ppnp.2011.01.055.
8. H. Liang, J. Meng, and S.-G. Zhou, Hidden pseudospin and spin symmetries and their origins in atomic nuclei, *Physics Reports.* **570**, 1–84 (2015). ISSN 0370-1573. doi: http://dx.doi.org/10.1016/j.physrep.2014.12.005. URL `http://www.sciencedirect.com/science/article/pii/S0370157315000502`.
9. D. M. Brink and G. R. Satchler, *Angular momentum.* Oxford University Press (1968).
10. J. D. Bjorken and S. D. Drell, *Relativistic Quantum Mechanics.* McGraw-Hill, New York (1964).
11. A. Bouyssy, S. Marcos, J. F. Mathiot, and N. Van Giai, Isovector-meson contributions in the Dirac–Hartree–Fock approach to nuclear matter, *Phys. Rev. Lett.* **55**(17), 1731–1733 (Oct., 1985). doi: 10.1103/PhysRevLett.55.1731.
12. A. Bouyssy, J.-F. Mathiot, N. Van Giai, and S. Marcos, Relativistic description of nuclear systems in the Hartree–Fock approximation, *Phys. Rev. C.* **36**(1), 380–401 (July, 1987). doi: 10.1103/PhysRevC.36.380.
13. P. Bernardos, V. Fomenko, N. Van Giai, M. Quelle, S. Marcos, R. Niembro, and L. Savushkin, Relativistic Hartree–Fock approximation in a nonlinear model for nuclear matter and finite nuclei, *Phys. Rev. C.* **48**(6), 2665–2672 (Dec., 1993). doi: 10.1103/PhysRevC.48.2665.
14. S. Marcos, L. Savushkin, V. Fomenko, M. López-Quelle, and R. Niembro, Description of nuclear systems within the relativistic Hartree–Fock method with zero-range self-interactions of the scalar field, *J. Phys. G.* **30**(6), 703–721 (2004). doi: 10.1088/0954-3899/30/6/002.
15. W. H. Long. *Density-dependent relativistic Hartree–Fock theory.* PhD thesis, University Paris-Sud XI (2005). in English.
16. W. H. Long, N. Van Giai, and J. Meng, Density-dependent relativistic Hartree–Fock approach, *Phys. Lett. B.* **640**(4), 150–154 (2006). doi: 10.1016/j.physletb.2006.07.064.
17. W. H. Long, P. Ring, N. Van Giai, and J. Meng, Relativistic Hartree–Fock–Bogoliubov theory with density dependent meson-nucleon couplings, *Phys. Rev. C.* **81**(2), 024308 (2010). doi: 10.1103/PhysRevC.81.024308.

18. W. H. Long, P. Ring, J. Meng, N. Van Giai, and C. A. Bertulani, Nuclear halo structure and pseudospin symmetry, *Phys. Rev. C.* **81**(3), 031302 (Mar., 2010). doi: 10.1103/PhysRevC.81.031302.
19. W. H. Long, H. Sagawa, N. Van Giai, and J. Meng, Shell structure and rho-tensor correlations in density dependent relativistic Hartree–Fock theory, *Phys. Rev. C.* **76**(3), 034314 (Sep, 2007). doi: 10.1103/PhysRevC.76.034314.
20. W. H. Long, H. Sagawa, J. Meng, and N. Van Giai, Evolution of nuclear shell structure due to the pion exchange potential, *Europhys. Lett.* **82**(1), 12001 (2008). doi: 10.1209/0295-5075/82/12001.
21. B. Sun, F. Montes, L. S. Geng, H. Geissel, Y. A. Litvinov, and J. Meng, Application of the relativistic mean-field mass model to the r-process and the influence of mass uncertainties, *Phys. Rev. C.* **78**(2), 025806 (Aug, 2008). doi: 10.1103/PhysRevC.78.025806.
22. W. H. Long, T. Nakatsukasa, H. Sagawa, J. Meng, H. Nakada, and Y. Zhang, Non-local mean field effect on nuclei near $Z = 64$ sub-shell, *Phys. Lett. B.* **680**(5), 428–431 (2009). doi: 10.1016/j.physletb.2009.09.034.
23. H. Liang, N. Van Giai, and J. Meng, Spin-isospin resonances: A self-consistent covariant description, *Phys. Rev. Lett.* **101**(12), 122502 (2008). doi: 10.1103/PhysRevLett.101.122502.
24. H. Liang, P. Zhao, P. Ring, X. Roca-Maza, and J. Meng, Localized form of Fock terms in nuclear covariant density functional theory, *Phys. Rev. C.* **86**, 021302 (Aug, 2012). doi: 10.1103/PhysRevC.86.021302.
25. Z. M. Niu, Y. F. Niu, H. Z. Liang, W. H. Long, T. Nikšić, D. Vretenar, and J. Meng, β-decay half-lives of neutron-rich nuclei and matter flow in the r-process, *Phys. Lett.* **B 723**, 172–176 (2013).
26. H. Yukawa, On the interaction of elementary particles, *Proc. Phys.-Math. Soc. Japan.* **Ser. 3, 17**, 48–57 (1935).
27. J. P. Ebran, E. Khan, D. Peña Artega, and D. Vretenar, Relativistic Hartree–Fock–Bogoliubov model for deformed nuclei, *Phys. Rev. C.* **83**, 064323 (2011).
28. P. Ring and P. Schuck, *The nuclear many-body problem.* Springer-Verlag, New York (1980).
29. J. Meng, Relativistic continuum Hartree–Bogoliubov theory with both zero range and finite range gogny force and their application, *Nucl. Phys. A.* **635** (1-2), 3–42 (1998). doi: 10.1016/S0375-9474(98)00178-X.
30. H. Kucharek and P. Ring, Relativistic field theory of superfluidity in nuclei, *Z. Phys. A.* **339**(1), 23–35 (1991). doi: 10.1007/BF01282930.
31. J. Meng, Pairing interaction in exotic nuclei: Finite range or zero range?, *Phys. Rev. C.* **57**, 1229–1232 (Mar, 1998). doi: 10.1103/PhysRevC.57.1229. URL `http://link.aps.org/doi/10.1103/PhysRevC.57.1229`.
32. Y. K. Gambhir, P. Ring, and A. Thimet, Relativistic mean field theory for finite nuclei, *Ann. Phys.* **198**, 132 (1990). doi: 10.1016/0003-4916(90)90330-Q.
33. S.-G. Zhou, J. Meng, and P. Ring, Spherical relativistic Hartree theory in a Woods–Saxon basis, *Phys. Rev. C.* **68**(3), 034323 (2003). doi: 10.1103/PhysRevC.68.034323.
34. S.-G. Zhou, J. Meng, P. Ring, and E.-G. Zhao, Neutron halo in deformed nuclei, *Phys. Rev. C.* **82**(1), 011301(R) (2010). doi: 10.1103/PhysRevC.82.011301.

35. W. Long, J. Meng, N. Van Giai, and S.-G. Zhou, New effective interactions in relativistic mean field theory with nonlinear terms and density-dependent meson-nucleon coupling, *Phys. Rev. C.* **69**(3), 034319 (Mar, 2004). doi: 10.1103/PhysRevC.69.034319.
36. L.-S. Geng, J. Meng, H. Toki, W.-H. Long, and G. Shen, Spurious shell closures in the relativistic mean field model, *Chin. Phys. Lett.* **23**(5), 1139–1141 (2006). doi: 10.1088/0256-307X/23/5/021.
37. H. Liang, P. Zhao, and J. Meng, Fine structure of charge-exchange spin-dipole excitations in ^{16}O, *Phys. Rev. C.* **85**, 064302 (2012).
38. T. Otsuka, T. Suzuki, R. Fujimoto, H. Grawe, and Y. Akaishi, Evolution of nuclear shells due to the tensor force, *Phys. Rev. Lett.* **95**, 232502 (Nov, 2005). doi: 10.1103/PhysRevLett.95.232502.
39. J. P. Schiffer, S. J. Freeman, J. A. Caggiano, C. Deibel, A. Heinz, C.-L. Jiang, R. Lewis, A. Parikh, P. D. Parker, K. E. Rehm, S. Sinha, and J. S. Thomas, Is the nuclear spin-orbit interaction changing with neutron excess?, *Phys. Rev. Lett.* **92**(16), 162501 (2004). doi: 10.1103/PhysRevLett.92.162501.
40. H. Liang, S. Shen, P. Zhao, and J. Meng, Pseudospin symmetry in supersymmetric quantum mechanics: Schrödinger equations, *Phys. Rev. C.* **87**, 014334 (Jan, 2013). doi: 10.1103/PhysRevC.87.014334.
41. S. Shen, H. Liang, P. Zhao, S. Zhang, and J. Meng, Pseudospin symmetry in supersymmetric quantum mechanics. ii. spin-orbit effects, *Phys. Rev. C.* **88**, 024311 (Aug, 2013). doi: 10.1103/PhysRevC.88.024311.
42. G. A. Lalazissis, T. Nikšić, D. Vretenar, and P. Ring, New relativistic mean-field interaction with density-dependent meson-nucleon couplings, *Phys. Rev. C.* **71**(2), 024312 (Feb, 2005). doi: 10.1103/PhysRevC.71.024312.
43. A.-M. Oros, Study of the coupling between collective and single-particle excitation modes in medium heavy spherical nuclei, *Ph. D. thesis, University of Köln* (1996).
44. I. Bombaci and U. Lombardo, Asymetric nuclear matter equation of state, *Phys. Rev.* **C 44**, 1892–1900 (1991).
45. E. N. E. van Dalen, C. Fuchs, and A. Faessler, Effective nucleon masses in symmetric and asymmetric nuclear matter, *Phys. Rev. Lett.* **95**(2), 022302 (Jul, 2005). doi: 10.1103/PhysRevLett.95.022302.
46. Y. Nagai, J. Styczen, M. Piiparinen, P. Kleinheinz, D. Bazzacco, P. V. Brentano, K. O. Zell, and J. Blomqvist, Proton single-particle states above $Z = 64$, *Phys. Rev. Lett.* **47**(18), 1259–1262 (1981). doi: 10.1103/PhysRevLett.47.1259.
47. A. Odahara, Y. Gono, S. Mitarai, T. Morikawa, T. Shizuma, M. Kidera, M. Shibata, T. Kishida, E. Ideguchi, K. Morita, A. Yoshida, H. Kumagai, Y. Zhang, A. Ferragut, T. Murakami, M. Oshima, H. Iimura, M. Shibata, S. Hamada, H. Kusakari, M. Sugawara, M. Ogawa, M. Nakajima, B. Min, J. Kim, S. Chae, and H. Sagawa, High-spin isomer and level structure of ^{145}Sm, *Nucl. Phys. A.* **620**, 363–384 (1997).
48. J. F. Berger, M. Girod, and D. Gogny, Microscopic analysis of collective dynamics in low energy fission, *Nucl. Phys. A.* **428**, 23–36 (1984). doi: 10.1016/0375-9474(84)90240-9.

49. E. Chabanat, P. Bonche, P. Haensel, J. Meyer, and R. Schaeffer, A Skyrme parametrization from subnuclear to neutron star densities part II. Nuclei far from stabilities, *Nucl. Phys. A.* **635**(1-2), 231–256 (1998). doi: 10.1016/S0375-9474(98)00180-8.
50. T. Wakasa, M. Okamoto, M. Takaki, M. Dozono, K. Hatanaka, M. Ichimura, T. Noro, H. Okamura, and Y. Sakemi, Complete set of polarization transfer observables for the ^{16}O(p,n)^{16}F reaction at 296 MeV and 0 degrees, *Phys. Rev.* **C 84**, 014614 (2011).
51. D. R. Tilley, H. R. Weller, and C. M. Cheves, Energy levels of light nuclei $A = 16$–17, *Nucl. Phys.* **A 564**, 1–183 (1993).
52. N. Paar, D. Vretenar, T. Marketin, and P. Ring, Inclusive charged-current neutrino-nucleus reactions calculated with the relativistic quasiparticle random-phase approximation, *Phys. Rev. C.* **77**(2), 024608 (2008). doi: 10.1103/PhysRevC.77.024608.
53. C. L. Bai, H. Sagawa, G. Colò, H. Q. Zhang, and X. Z. Zhang, Spin-dipole excitations in ^{16}O and tensor correlations, *Phys. Rev. C.* **84**(4), 044329 (2011).
54. G. Colò, H. Sagawa, S. Fracasso, and P. F. Bortignon, Spin–orbit splitting and the tensor component of the Skyrme interaction, *Phys. Lett.* **B 646**(5), 227–231 (2007).
55. C. L. Bai, H. Q. Zhang, H. Sagawa, X. Z. Zhang, G. Colò, and F. R. Xu, Effect of the tensor force on the charge exchange spin-dipole excitations of ^{208}Pb, *Phys. Rev. Lett.* **105**(7), 072501 (2010).
56. C. L. Bai, H. Q. Zhang, H. Sagawa, X. Z. Zhang, G. Colò, and F. R. Xu, Spin-isospin excitations as quantitative constraints for the tensor force, *Phys. Rev. C.* **83**(5), 054316 (2011).
57. F. Minato and C. L. Bai, Impact of tensor force on β decay of magic and semimagic nuclei, *Phys. Rev. Lett.* **110**(12), 122501 (2013).
58. L. J. Jiang, Y. Shen, J. M. Dong, and W. H. Long, Self-consistent tensor effects on nuclear matter system under relativistic Hartree–Fock approach, *Phys. Rev. C.* **91**, 025802 (Feb, 2015). doi: 10.1103/PhysRevC.91.025802. URL `http://link.aps.org/doi/10.1103/PhysRevC.91.025802`.
59. L. J. Jiang, S. Yang, B. Y. Sun, W. H. Long, and H. Q. Gu, Nuclear tensor interaction in a covariant energy density functional, *Phys. Rev. C.* **91**, 034326 (Mar, 2015). doi: 10.1103/PhysRevC.91.034326. URL `http://link.aps.org/doi/10.1103/PhysRevC.91.034326`.

Chapter 5

Superheavy nuclei and fission barriers

Bing-Nan Lu

State Key Laboratory of Theoretical Physics, Institute of Theoretical Physics, Chinese Academy of Sciences, Beijing 100190, China
Institut für Kernphysik (IKP-3) and Jülich Center for Hadron Physics, Forschungszentrum Jülich, D-52425 Jülich, Germany
b.lu@fz-juelich.de

Jie Zhao

State Key Laboratory of Theoretical Physics, Institute of Theoretical Physics, Chinese Academy of Sciences, Beijing 100190, China
Physics Department, Faculty of Science, University of Zagreb, Bijenicka 32, Zagreb 10000, Croatia
zhaojie@itp.ac.cn

En-Guang Zhao

State Key Laboratory of Theoretical Physics, Institute of Theoretical Physics, Chinese Academy of Sciences, Beijing 100190, China
Center of Theoretical Nuclear Physics, National Laboratory of Heavy Ion Accelerator, Lanzhou 730000, China
egzhao@mail.itp.ac.cn

Shan-Gui Zhou

State Key Laboratory of Theoretical Physics, Institute of Theoretical Physics, Chinese Academy of Sciences, Beijing 100190, China
Center of Theoretical Nuclear Physics, National Laboratory of Heavy Ion Accelerator, Lanzhou 730000, China
sgzhou@itp.ac.cn

In this chapter, we will present relativistic mean field (RMF) description of heavy and superheavy nuclei (SHN). We will discuss the shell structure and magic numbers in the mass region of SHN, binding energies and

α decay Q values, shapes of ground states and potential energy surfaces and fission barriers. We particularly focus on the multidimensionally-constrained covariant density functional theories (CDFT) and the applications of CDFT to the study of exotic nuclear shapes and fission barriers.

Contents

1. Introduction

The nuclear landscape can be extended along various directions with each of them representing an important forefront of modern nuclear physics study. In this book, many of such extensions have been discussed, including exotic nuclei close to or beyond the drip lines (Chapter 3), hypernuclei containing strangeness degrees of freedom (Chapter 7), fast rotating nuclei with high spins (Chapters 8 and 9), and so on. In this chapter, we will focus on the study of heavy and superheavy nuclei (SHN).

If one treats atomic nuclei as charged liquid drops, the existence limit of nuclear charge and mass is mainly constrained by the subtle balance between the attractive nuclear force among all nucleons and the repulsive Coulomb force among protons. Then nuclei with $Z > 104$ will disintegrate immediately after their formation due to the large static electric forces. However, atomic nuclei are quantum many-body systems and quantum effects play an important role in determining their stability. One striking feature is that there appears shell structure in nuclear single particle spectrum and nuclei with neutrons or protons exactly filling a major shell, corresponding to the "magic numbers" $Z/N = 2, 8, 20, 28, 50, 82$ and $N = 126$, are considerably more stable compared to their neighbours. Therefore, soon after the nuclear shell model was established, the possibility of the existence of long-lived nuclides around $Z = 126$ was discussed.[1–3] Based on an assumed shell gap corresponding to $Z = 126$, one can study semiquantitatively the stability of SHN[4] and there were also some controversial conclusions.[5] The shell correction energy can be evaluated more reliably by using the Strutinsky approach[6] based on single particle spectrum determined by some realistic mean field potentials. In 1960s, several groups made a lot of efforts along this direction and predicted the so-called "Island of Stability " of SHN centering at $Z = 114$ and $N = 184$.[7–10]

So far, much progress has been made in producing superheavy elements (SHE).[11] In GSI, Germany, new elements with $Z = 107$–112 were synthesized.[12] The production of Z=113 element via the cold fusion reaction was achieved in RIKEN, Japan.[13,14] Meanwhile, the hot fusion technique with ^{48}Ca as the projectile was applied to synthesize new elements with $Z = 113$–118 at the Joint Institute for Nuclear Research, Russia.[15,16] Note that in Ref. 17, one α-decay chain was observed for ^{271}Ds using the gas-filled recoil separator at Heavy Ion Research Facility in Lanzhou (HIRFL), China and the measured decay properties are consistent with those reported previously.

Nowadays, one confronts many problems to synthesize even longer-lived or even heavier SHN by using heavy ion fusion reactions, e.g., the tiny production cross section, only neutron deficient SHN being produced with stable nuclei as projectiles and targets, and the difficulty in finding proper target material.[11] There are also many open problems on the theoretical side concerning the structure and the synthesis mechanism of SHN: the shell structure and magicity which determines the location of the island of stability, the shape and isomeric states, the potential energy surface (PES) and fission barrier, and the fusion mechanism, etc.[18]

The relativistic mean field (RMF) models[19–24] have been used to study various properties of SHN. On the one hand, the single particle structure, shapes and binding energies of ground states, the α decay Q values, and the neutron/proton separation energies are the most important quantities that determine the location of the island of stability and the lifetimes of the SHN against the particle emissions. On the other hand, we need to know the potential energy surfaces and fission barriers because fission is one of the dominant decay channels of SHN. All these properties have been investigated with the RMF models.

In this chapter we will present the RMF description of the magic numbers in the SHN region predicted by the RMF models (Section 2), the α-decay Q values (Section 3), and shapes, fission barriers and potential energy surfaces in heavy nuclei and SHN (Section 4).

2. Magic numbers in the SHN mass region

The existence of SHN is mainly due to the quantum shell effect, thus the location of the island of stability is mainly determined by the single particle shell structure of SHN. For example, the splitting of $\pi 2f_{5/2,7/2}$ determines whether the next proton shell closure appears at $Z = 114$ or 120. Experimentally one cannot study directly the single particle level structure of SHN with $Z \geq 110$ though there are experimental investigations of lighter nuclei with $Z \approx 100$ and $N \approx 152$ which can reveal indirectly features of single particle orbitals which are important in determining the shell closures in SHN.[25–31]

On the theoretial side, the magicity in the SHE region has been investigated with various modern microscopic models and parameter sets. Different models often predict different proton magic numbers after 82; even with the same model, the predictions may depend on the parameters.[32] The Skyrme–Hartree–Fock (SHF) method with effective interactions SkP and SLy7 predicts magic numbers at $Z = 126$ and $N = 184$, and also predicts a deformed shell closure at N=162.[33] In Ref. 34 the doubly magic spherical nuclei $^{184}114$, $^{172}120$, and $^{184}126$ were predicted in the calculations with non-relativistic SHF effective interactions SkM*, SkP, SLy6, SkI1, SkI3, and SkI4 and RMF effective interactions PL-40, NLSH, NL-Z, and TM1. In Ref. 35, shell corrections to the nuclear binding energy in SHN were also studied with both SHF and RMF models. It was shown that for the vast majority of Skyrme interactions, the strongest shell stabilization appears for $Z = 124$ and 126 and for $N = 184$. On the other hand, in the relativistic approaches the strongest spherical shell effect appears systematically for $Z = 120$ and $N = 172$. Systematic calculations in the framework of relativistic Hartree–Bogoliubov (RHB) theory with effective interaction NLSH in the *ph* channel and Gogny interaction in the *pp* channel indicated strong magicity for $Z = 114$ and $N = 160$, $N = 166$, and $N = 184$.[36] In Ref. 37, the authors made predictions for the magic numbers by examining the two-nucleon separation energies S_{2p} and S_{2n}, the two-nucleon gaps δ_{2p} and δ_{2n}, the shell correction energies E_{shell}, the pairing energies E_{pair}, and the pairing gap Δ_p and Δ_n with the relativistic continuum Hartree–Bogoliubov (RCHB) theory (see Chapter 3). In this section, we will present some results given in Ref. 37.

The nuclei among more than 1200 even-even nuclei in the nuclear chart with proton number $Z = 100$–140 and neutron number $N = (Z + 30) - (2Z+32)$ (i.e., $N = 130$–312) by the RCHB calculation with the effective interactions NL1,[38] NL3,[39] NLSH,[40] TM1,[41] TW99,[42] DD-ME1,[43] and PK1 and PK1R.[44] As discussed in other chapters, e.g., in Chapter 3, the RCHB theory has been very successful in describing not only stable nuclei but also the nuclei near the drip lines.[45–49] In principle, as will be discussed in Section 4, a calculation in a multidimensional deformation space may be more appropriate to define ground state shapes of SHN. However, since SHN are expected to be spherical and are located on the nuclear chart around a spherical doubly magic nucleus next to $^{208}_{126}$Pb, the RCHB theory with the assumption of spherical shape can be applied to a preliminary scan for magic numbers. In Ref. 37, the changes of the two-nucleon separation energies were calculated as the second difference of the binding energies, i.e., the two-nucleon gaps:

$$\begin{aligned}\delta_{2n}(N,Z) &= 2E_B(N,Z) - E_B(N+2,Z) - E_B(N-2,Z)\\ &= S_{2n}(N,Z) - S_{2n}(N+2,Z),\\ \delta_{2p}(N,Z) &= 2E_B(N,Z) - E_B(N,Z+2) - E_B(N,Z-2)\\ &= S_{2p}(N,Z) - S_{2p}(N,Z+2).\end{aligned}$$

A peak in the two-nucleon gap indicates drastic changes in the two-nucleon separation energies, thus being used as an evidence of the magic number.

The two-proton gaps δ_{2p} from the RCHB calculation for even-even nuclei with $Z = 102$–138 as a function of Z with the eight effective interactions, NL1, NL3, NL-SH, TM1, TW-99, DD-ME1, PK1, and PK1R, are shown in Fig. 1. The two-proton gaps δ_{2p} with the same N are connected as a curve. A peak at a certain Z in the curve suggests the existence of a magic proton number. The sharpness of the peaks represent the robustness of the corresponding closed shell while the quenching effects are associated with the bundle of the curves at the certain Z. It was observed that common magic proton numbers $Z = 120$, 132, and 138 exist for all effective interactions while $Z = 106$ is observed only for NL3, NLSH, TW99, DD-ME1, PK1, and PK1R. Furthermore, the peak at $Z = 114$ for NLSH and TW99 and the peak at $Z = 126$ for NL1 were also noticed, though they are not as obvious as that at $Z = 120$. Moreover, small peaks are evident for $N = 154$ (NLSH and TW99), $N = 164$ (NL3, NLSH, TW99, DD-ME1, PK1, and PK1R), and $N = 216$ (NLSH, TW99, PK1, and PK1R). In general, based on two-nucleon gaps δ_{2p} and δ_{2n}, the magic numbers $Z = 120$, 132, and 138, and $N = 138$, 172, 184, 198, 228, 238, 258, and 274 can be observed for RCHB calculation with all effective interactions. Apart from $Z = 106$ and $N = 164$ and 216, additional 10 magic numbers $Z = 114$ and 126,

$N = 154$ appear in the two-nucleon gaps δ_{2p} and δ_{2n} in Figs. 1 and 2 for some effective interactions. These magic nuclei are presented as shaded area in Fig. 3.

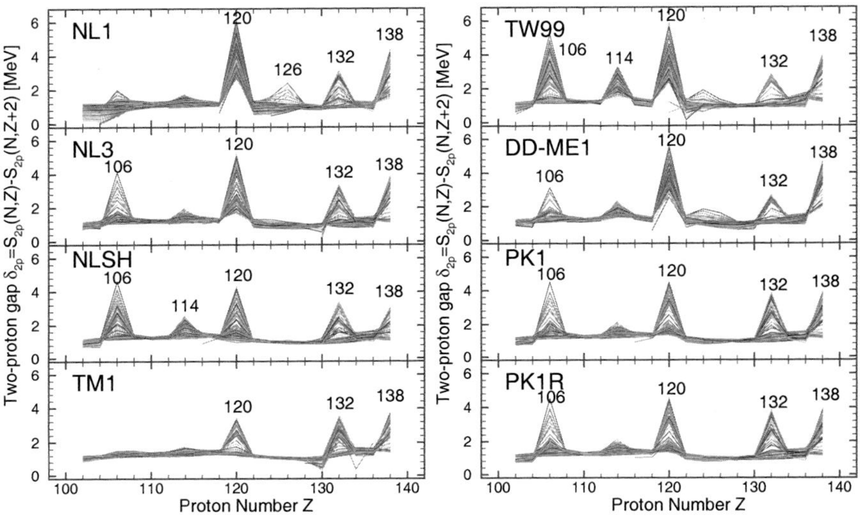

Fig. 1. (Color online) The two-proton gaps $\delta_{2p}(N, Z) = S_{2p}(N, Z) - S_{2p}(N, Z + 2)$ as a function of proton number obtained by RCHB calculation with effective interactions NL1, NL3, NL-SH, TM1, TW-99, DD-ME1, PK1, and PK1R, respectively. Taken from Ref. 37.

In Ref. 37, besides two-nucleon gaps (δ_{2p} and δ_{2n}) discussed above, other quantities such as two-nucleon separation energies (S_{2p} and S_{2n}), shell correction energies (E^p_{shell} and E^n_{shell}), pairing energies (E^p_{pair} and E^n_{pair}), and the effective pairing gaps Δ_p and Δ_n were analyzed in detail and based on these analyses, new magic numbers in the SHN mass region were predicted. The doubly magic superheavy nuclei can be the combination of magic proton numbers $Z = 120$, 132, and 138 and magic neutron numbers $N = 172$, 184, 198, 228, 238, and 258. Such magic numbers are presented as gray lines in Fig. 3. It should be noted that the intersections of lines in Fig. 3 do not necessarily mean doubly magic nuclei as there are also shell quenching effects. However, these crosses include all the possible doubly magic nuclei in the state-of-the-art relativistic approach.

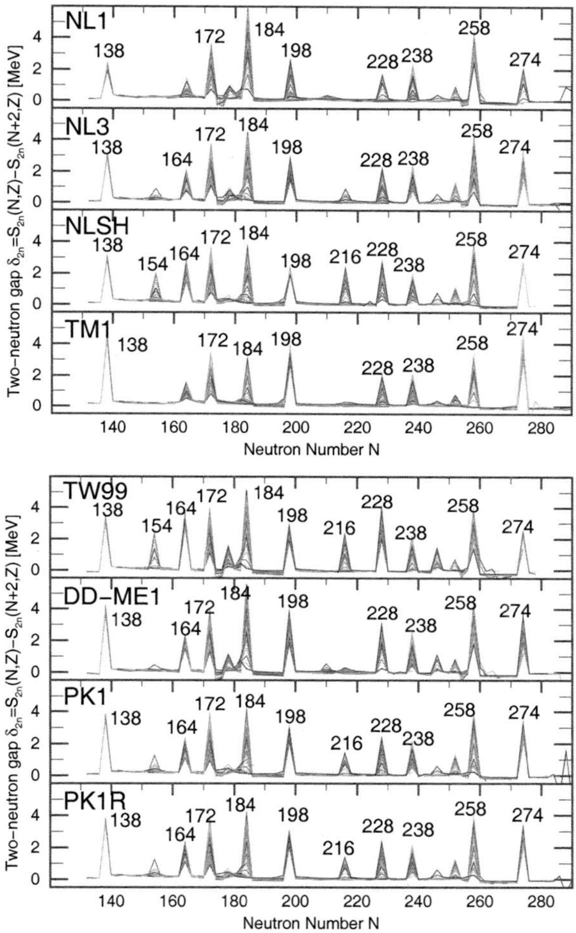

Fig. 2. (Color online) The two-neutron gaps $\delta_{2n}(N, Z) = S_{2n}(N, Z) - S_{2n}(N+2, Z)$ as a function of proton number obtained by RCHB calculation with effective interactions NL1, NL3, NL-SH, TM1, TW-99, DD-ME1, PK1, and PK1R, respectively. Taken from Ref. 37.

3. Binding energies and α-decay Q values of SHN

The binding energy is one of the most important properties of atomic nuclei. In the previous section, it has been shown that with binding energies, one can evaluate various quantities, such as separation energies and two-nucleon gaps, and then study the magicity in the SHN mass region. For SHN, the

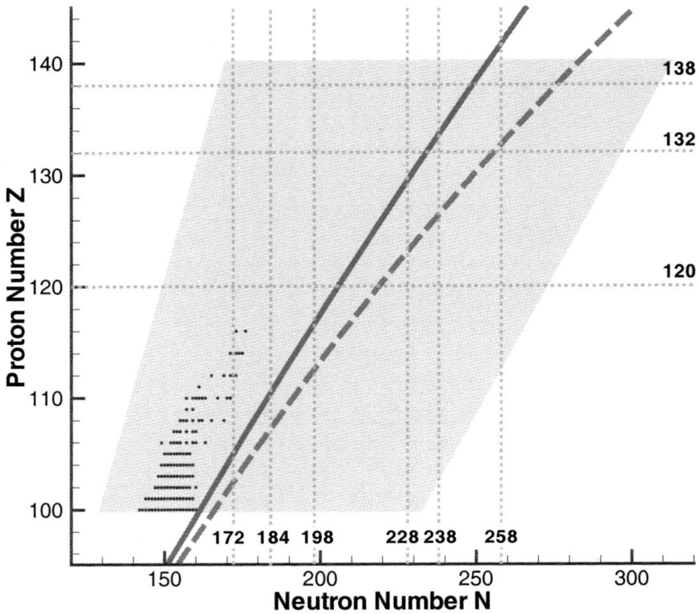

Fig. 3. (Color online) Nuclear chart in the superheavy region. The long-dashed and solid lines represent the β-stability lines defined by two empirical formulas. The dots lying at the lower-left corner represent the superheavy nuclei experimentally observed. Taken from Ref. 37.

binding energies are particularly important because one can study the α-decay properties of SHN. α-decay is one of the most important decay modes of SHN. Nowadays, the identification of a synthesized SHN is primarily to identify the successive α-decays of the evaporation residues of compound nuclei to known nuclei by the method of α-α correlations.[50,51] The α-decay half-lives are sensitive to the α-decay Q_α values which can be obtained from the binding energy difference of the parent nucleus and the daughter nucleus

$$Q_\alpha(Z,N) = B(Z-2,N-2) + B_\alpha - B(Z,N), \tag{1}$$

where $B_\alpha = 28.3$ MeV is the binding energy of an α particle. The binding energies of a SHN and its daughter nuclei in the α-deay chain can be calculated with the RMF models. Then α-decay energies can be obtained and α-decay half-lives can be evaluated with some empirical formulas. Along this line, lots of efforts have been made with the RMF models.[52–64] Note that although many of calculations appear after some experiments, e.g., in the synthesis and theoretical study of ^{259}Db,[54,65] it also often happens that theoretical predictions are made beforehand, and thus are very useful for experiments, e.g., in the study of ^{265}Bh.[57,66] In this section, we present some results of α-decay Q values for some $Z = 105$, 115 and 117 nuclides.

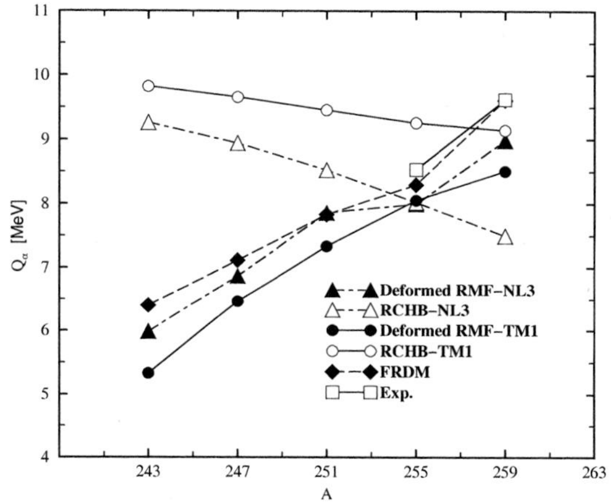

Fig. 4. The Q value of ^{259}Db and its α-decay daughter nuclei calculated with the RCHB theory and the deformed RMF+BCS model using NL3 and TM1, respectively, compared with the data and the results from the finite range droplet model (FRDM). Taken from Ref. 54.

In Ref. 54, the structure of then newly discovered nuclide, ^{259}Db ($Z =$ 105), and its α decay chain[65] was systematically studied in the RMF model with effective interactions NL3 and TM1. It was found that both the pairing correlation and deformation are essential to reproduce the experiment Q_α values. There are many possible channels for the α decay of odd–A nuclei or odd–odd nuclei.[67] Only the Q values of the ground state to ground state α decay were given in Ref. 54 which are shown in Fig. 4 with comparison to the data[65] and the results from the finite range droplet model (FRDM).[68] The deformed RMF+BCS calculations reproduce the experimental Q_α values well. In addition, both the deformed RMF model and the FRDM give the same trend as the data, i.e., the Q values become smaller along the α decay chain. In contrast, the Q values from RCHB theory become larger along this α decay chain. This implies the importance of the deformation effects in the studied nuclei.

Soon after the element 115 was synthesized in the ^{48}Ca + ^{243}Am reaction at Dubna, Russia,[69,70] the deformed RMF+BCS model with a density-independent δ function interaction in the pairing channel was applied to

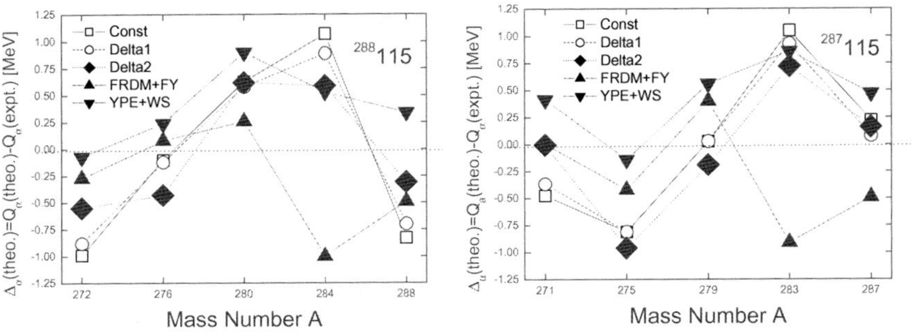

Fig. 5. The difference between calculated Q_α(theo.) and experimental Q_α(expt.), Δ_α(theo.) = Q_α(theo.) − Q_α(expt.), for the $^{288}115$ and $^{287}115$ α-decay chains. Taken from Ref. 58.

the analysis of these synthesized superheavy nuclei.[58] In Fig. 5 the difference between the predicted Q_α values by using the RMF+BCS model with constant pairing gap ("Const") and two versions of contact pairing force ("Delta1" and "Delta2"), those from the FRDM with folded-Yukawa single-particle potentials (FRDM+FY)[71] or the Yukawa-plus-exponential model with Woods–Saxon single-particle potentials (YPE+WS)[72] and the experimental values, Δ_α(theo.) = Q_α(theo.) − Q_α(expt.) for the $^{288}115$ chain and the $^{287}115$ chain are presented, respectively. It can be seen that the calculated α-decay energies agree well with the experimental values and with those of the FRDM.

In Ref. 63, the RMF model with the point-coupling density functional PC-PK1[73] and the meson-exchange density functional PK1[44] were used to describe the shapes and α-decay Q_α values and half lives of nuclei in the α-decay chains of $^{293,294}117$. Both the normally deformed minima (NDM) and superdeformed minima (SDM) were considered to calculate the Q_α values. It was found the experimental Q_α of $^{293,294}117$ lie in between the two theoretical values (Q_α values between NDM and those between SDM). Note that the rotational correction has been calculated in Ref. 63. The rotational correction energy does not have much influence on the absolute value of Q_α, as is seen in Fig. 6, but might be sensitive to the refined fluctuation of Q_α as a function of nucleon number.

In Fig. 6, the Q_α values for the α-decay chains of $^{293,294}117$, based on their NDM (not necessarily the global minimum) in PES, were compared to the available data.[74,75] In Fig. 6, it can be seen that the general trend of Q_α is well reproduced by both PC-PK1 and PK1 density functionals. The rms

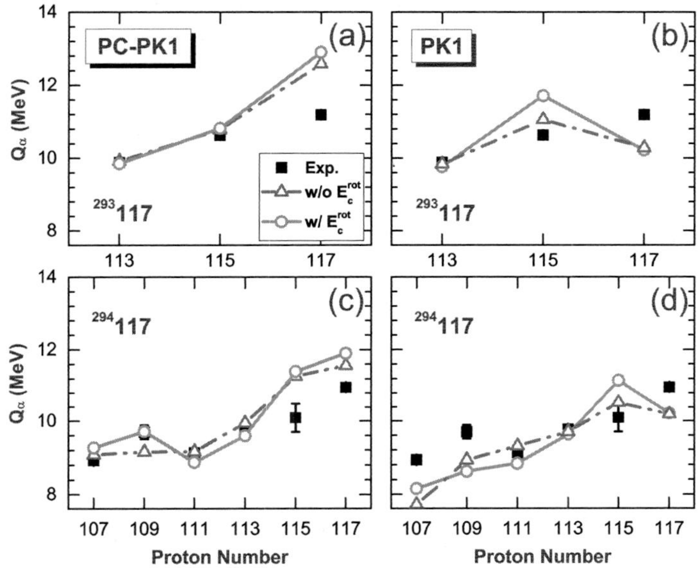

Fig. 6. (Color online) The Q_α values for the decay chains of $^{293,294}117$. The theoretical values were calculated as the difference between the normal deformed minima (not necessarily the global minimum) of the parent and daughter nuclei. The results without and with rotational correction energies are denoted by the open triangles and circles, respectively. Taken from Ref. 63.

deviation between the theoretical results and the data is about 0.65 MeV. The average difference between the Q_α values calculated with and without $E_c^{\rm rot}$ is about 0.28 MeV, and the largest difference (0.65 MeV and 0.62 MeV) appears in $^{289,290}115$ with PK1. It is shown in Fig. 6 that the consideration of rotational correction energy improves the agreement of the calculated Q_α for PC-PK1 with experiment. In Ref. 63, this has been attributed to the fact that the PC-PK1 was proposed by fitting to observables of selected spherical nuclei,[73] where the contamination from the long-range rotational correlation was excluded. The Q_α data in the α-decay chains of $^{293,294}117$ are reproduced by PC-PK1, except for the nuclei $^{293,294}117$ and $^{290}115$. By taking the rotational correction energy into account, the small jump of Q_α at $Z = 109$ could be well reproduced with PC-PK1. Whether this jump is caused by the rotational correction or by other mechanisms needs further investigations.

4. Shapes and fission barriers of heavy and superheavy nuclei

The occurrence of spontaneous symmetry breaking leads to nuclear shapes with a variety of symmetries[76,77] and a lot of nuclear phenomena are connected with the nuclear deformation.[78,79] The shape of a nucleus can be described by the parametrization of the nuclear surface or the nucleon density distribution by, e.g., making a multipole expansion,

$$R(\theta,\varphi) = R_0 \left[1 + \beta_{00} + \sum_{\lambda=1}^{\infty} \sum_{\mu=-\lambda}^{\lambda} \beta^*_{\lambda\mu} Y_{\lambda\mu}(\theta,\varphi) \right], \tag{2}$$

where $\beta_{\lambda\mu}$'s are deformation parameters. The majority of observed nuclear shapes is of the spheroidal form which can be described by the axial-quadrupole deformation parameter β_{20}. The non-axial-quadrupole (triaxial) deformation β_{22} (or, equivalently, γ) has been predicted in atomic nuclei for quite a long time and it may play important roles in superheavy nuclei.[80] The octupole shapes with $\lambda = 3$ are predicted to exist in nuclei in several mass regions.[81] In addition, the influence of the non-axial octupole β_{32} deformation on the low-lying spectra have been traced out theoretically and experimentally.[82] For SHN, due to the considerably large repulsion among protons and the complexity of the shell structure, the ground state shape may become very exotic: (1) New types of topology such as bubble and toroidal nuclei have been proposed since the 1970s.[83–85] (2) The superdeformed prolate shape may become the ground state for some SHN.[53,55,86,87] (3) Tetrahedral shapes due to the strong nonaxial-octupole coupling may persist up to element 108[88] and play a crucial role in determining the shell stability in even heavier nuclei.[89]

The shape degrees of freedom are important not only for the ground states or small amplitude collective motions, but also for large amplitude collective motions such as fission. To explore the island of stability of SHN, it is more and more desirable to have accurate predictions of fission barriers of SHN. β_{20}, which describes the elongation of a fissile nucleus, is certainly the most important deformation for fission. Besides β_{20} and β_{40} (relevant to the size of a neck), many other shape degrees of freedom are also crucial for determining the shape of fission barriers and the fission path. For example, actinide nuclei are characterized by a two-humped fission barrier[90] and there may also exist a third minimum. It has long been known from the MM model calculations that the inner fission barrier is lowered by the non-axial-quadrupole deformation[91–93] and the outer barrier by the reflection asymmetric (RA) shape.[94] Later, the important roles played by the non-axial-quadrupole deformation and octupole deformation were confirmed in the non-relativistic[95] and relativistic[96] density functional calculations, respectively. Therefore it is desirable to include simultaneously all these important shape degrees of freedom in model calculations for unknown nuclei, in particular, for SHN.

The RMF models have been extensively used to study both the ground states and fission barriers of heavy and superheavy nuclei.[22,61,87,97–108] In recent years, we developed multidimensionally-constrained covariant density functional theory (MDC-CDFT) by breaking the reflection and the axial symmetries simultaneously.[98,105,109] The MDC-CDFT has been used to study ground state shapes, potential energy surfaces, and fission barriers of heavy nuclei and SHN. In this section we will present the formalism of the MDC-CDFT and some applications.

4.1. *The multidimensionally-constrained covariant density functional theory*

Since the general framework of the RMF model has been given in Chapter 2, here we only present briefly the formalism which is specific in the MDC-CDFT. It is customary to solve the deformed RMF equations by expanding the auxiliary single-particle wave functions in a complete basis, e.g., the axially deformed harmonic oscillator (ADHO) basis[110,111] or the Woods–Saxon (WS) basis,[112–116] as is discussed in Chapters 2 and 3. By using a basis in a two-centre harmonic oscillator potential, a reflection asymmetric relativistic mean field (RAS-RMF) approach has been developed.[117] In the MDC-CDFT, the single-particle wave functions are expanded in the ADHO basis. These basis states are eigenstates of the z component of the angular momentum $\hat{j}_z$ with eigenvalues $K_\alpha = m_l + m_s$. For $\Phi_\alpha(\boldsymbol{r}\sigma)$, the time-reversal state is defined as $\Phi_{\bar{\alpha}}(\boldsymbol{r}\sigma) = \mathcal{T}\Phi_\alpha(\boldsymbol{r}\sigma)$, where $\mathcal{T} = i\sigma_y\hat{K}$ is the time-reversal operator and $\hat{K}$ is the complex conjugation operator. Obviously, we have $K_{\bar{\alpha}} = -K_\alpha$ and $\pi_{\bar{\alpha}} = \pi_\alpha$, where $\pi_\alpha = \pm 1$ is the parity.

These basis states form a complete set for expanding any two-component spinors. In practical calculations a truncation has to be introduced. Following Ref. 118, for the large component of the Dirac wave function, the states satisfying $[n_z/Q_z + (2n_\rho + |m_l|)/Q_\rho] \leq N_f$ are included in the expansion, where $Q_z = \max(1, b_z/b_0)$ and $Q_\rho = \max(1, b_\rho/b_0)$ are constants related to the oscillator lengths $b_0 \equiv 1/\sqrt{M\omega_0}$, b_z, and b_ρ. For the expansion of the small component, the truncation is made up to $N_g = N_f + 1$ major shells in order to avoid the spurious states.[110]

If a nucleus is invariant under rotation around the symmetry z axis and spatial reflection, the angular momentum projection on the z-axis and the parity are conserved. Then one can diagonalize the Dirac equation in blocks characterized by (K_α, π_α) with K the projection of the angular momentum and π the parity. If the axial symmetry and the spatial reflection symmetry are both broken, components with different K and π are mixed together; thus, we must diagonalize a larger single-particle Hamiltonian matrix with non-zero matrix elements between two basis states with different K and π. There remains one symmetry operator that makes the Hamiltonian matrix block-diagonal. Due to the axial symmetry of the basis, it is convenient to introduce a simplex operator $\hat{S} = ie^{-i\pi\hat{j}_z}$. For a fermionic system with a half-integer spin, $\hat{S}$ is a Hermitian operator and $\hat{S}^2 = 1$. This operator corresponds to the rotation by π around the z axis, thus leaving the nuclear mean field invariant. The eigenvalue of $\hat{S}$, S, is also a good quantum number for the basis, $\hat{S}\Phi_\alpha = S\Phi_\alpha = (-1)^{K_\alpha - 1/2}\Phi_\alpha$, which means that the basis Φ_α with $K_\alpha = +1/2, -3/2, +5/2, -7/2, \ldots$ span the subspace with $S = 1$, while their time-reversal states span the one with $S = -1$. Therefore the blocks with $K = +1/2, -3/2, +5/2, -7/2, \ldots$ are mixed. For a nucleus with time-reversal symmetry, only the basis with $S = 1$ are used in the expansion. The basis states with $S_\alpha = -1$ are obtained by simply applying $\mathcal{T}$ on those with $S_\alpha = 1$. Furthermore, for systems with time-reversal symmetry, it is only necessary to diagonalize the matrix with $S = 1$ and the other half is obtained by a time-reversal operation on the obtained single-particle wave functions.

For deformed nuclei with the V_4 symmetry, the potentials $V(\boldsymbol{r})$, $S(\boldsymbol{r})$, and the densities can be expanded in terms of the Fourier series,

$$f(\rho, \varphi, z) = \sum_{\mu=-\infty}^{\infty} f_\mu(\rho, z) \frac{1}{\sqrt{2\pi}} \exp(i\mu\varphi). \tag{3}$$

Applying the symmetry conditions, one has $f_\mu = f_\mu^* = f_{\bar{\mu}}$ and $f_n = 0$ for odd n. Thus the expansion (3) can be simplified as

$$f(\rho, \varphi, z) = f_0(\rho, z)\frac{1}{\sqrt{2\pi}} + \sum_{n=1}^{\infty} f_n(\rho, z)\frac{1}{\sqrt{\pi}} \cos(2n\varphi), \tag{4}$$

where

$$f_0(\rho,z) = \frac{1}{\sqrt{2\pi}} \int_0^{2\pi} d\varphi f(\rho,\varphi,z), \tag{5}$$

$$f_n(\rho,z) = \frac{1}{\sqrt{\pi}} \int_0^{2\pi} d\varphi f(\rho,\varphi,z) \cos(2n\varphi), \tag{6}$$

are real functions of ρ and z. The details for calculating the matrix elements of the Dirac Hamiltonian and various densities and their derivatives can be found in Ref. 105.

For open shell nuclei the pairing interaction becomes crucial and must be included. It has been shown that fission barriers depend very much on the form and strength of the effective pairing interactions.[119] Several methods have been developed to treat the pairing effects, e.g., the BCS approach, the Bogoliubov transformation, and the particle number conserving method;[120–126] all of them have been used in the study of PES's and fission barriers. Both the BCS approach and the Bogoliubov transformation were incorporated in the MDC-CDFT and for the pairing force, either the δ-force or the finite-range separable force[127–129] is implemented.

To obtain a PES one can perform a constraint calculation which is equivalent to adding an external potential during the iteration.[77] The quadratic constraint method is usually used,

$$E' = E_{\mathrm{RMF}} + \sum_{\lambda\mu} \frac{1}{2} C_{\lambda\mu} \left(Q_{\lambda\mu} - m_{\lambda\mu}\right)^2, \tag{7}$$

where $C_{\lambda\mu}$ is the spring constant, $Q_{\lambda\mu}$ is the intrinsic multipole moment, and $m_{\lambda\mu}$ is a desired moment. With this method the calculation always converges to a deformation point on the PES other than the desired one. To overcome this shortcoming and to get a PES with equally distributed points, we use a modified linear constraint method. The Routhian reads,

$$E' = E_{\mathrm{RMF}} + \sum_{\lambda\mu} \frac{1}{2} C_{\lambda\mu} Q_{\lambda\mu}, \tag{8}$$

where the variables $C_{\lambda\mu}$'s change their values during the iteration through the following relation,

$$C_{\lambda\mu}^{(n+1)} = C_{\lambda\mu}^{(n)} + k_{\lambda\mu} \left(\beta_{\lambda\mu}^{(n)} - \beta_{\lambda\mu}\right), \tag{9}$$

where $\beta_{\lambda\mu}$ is the desired deformation, $k_{\lambda\mu}$ is a constant, and $C_{\lambda\mu}^{(n)}$ is the value at the nth step. This constraint method works well in our multidimensionally-constrained calculations.

In the MDC-CDFT, the RMF functional can be one of the following four forms: the meson exchange or point-coupling nucleon interactions combined with the nonlinear or density-dependent couplings. The pairing is treated with either the BCS approach or the Bogoliubov transformation. We name the MDC-CDFT with the BCS approach for the pairing as the MDC-RMF models[98,105] and those with the Bogoliubov transformation as the MDC-RHB models.[109] The MDC-CDFT has recently been applied to the study of PES's and fission barriers of actinides,[98,100,105–107] the nonaxial-octupole Y_{32} correlations in $N = 150$ isotones,[101] the third minima in light actinides,[108] and shapes of hypernuclei.[130,131]

4.2. *Non-axial octupole shapes*

It has been anticipated that β_{32} deformation occurs in the ground states of some nuclei with special combinations of the neutron and proton numbers.[82,132,133] Recently, lots of theoretical studies focus on this type of nuclear shape, either from the T_d^D-symmetric single particle spectra[132–135] or from various nuclear models including the macroscopic-microscopic model,[133,135–137] the Skyrme Hartree–Fock (SHF), SHF+BCS, or Skyrme Hartree–Fock–Bogoliubov models,[135–142] the generator coordinate method,[140,142] and the reflection asymmetric shell model (RASM).[88,89,143] For heavy nuclei, one of the relevant and interesting topics is how to explain the low-lying 2^- states in some $N = 150$ even-even nuclei.[144,145] In Ref. 88, it was proposed that the non-axial octupole Y_{32}-correlation results in the experimentally observed low-energy 2^- bands in the $N = 150$ isotones.

A microscopic and self-consistent study of the Y_{32} effects in the $N = 150$ isotones has been carried out with the MDC-RMF model in Ref. 101. In Fig. 7, potential energy curves $E \sim \beta_{32}$ are shown for these $N = 150$ isotones ^{246}Cm, ^{248}Cf, ^{250}Fm, and ^{252}No. For ^{246}Cm, the ground state has a nonzero value of the nonaxial octupole deformation parameter, $\beta_{32} = 0.020$. The potential pocket is rather shallow and also flat around the minimum. The energy difference between the ground state and the point with $\beta_{32} = 0$ was introduced, E_{depth}, to measure the energy gain with respect to the β_{32} distortion. For ^{246}Cm, E_{depth} is only 34 keV. For ^{248}Cf, ^{250}Fm and ^{252}No, the minima are located at $\beta_{32} = 0.037$, 0.034, and 0.025, respectively. The corresponding energy gain $E_{\text{depth}} = 0.351$, 0.328, and 0.104 MeV. These calculations indicate a strong Y_{32}-correlation in these nuclei. Both the non-axial octupole parameter β_{32} and the energy gain E_{depth} reach maximal

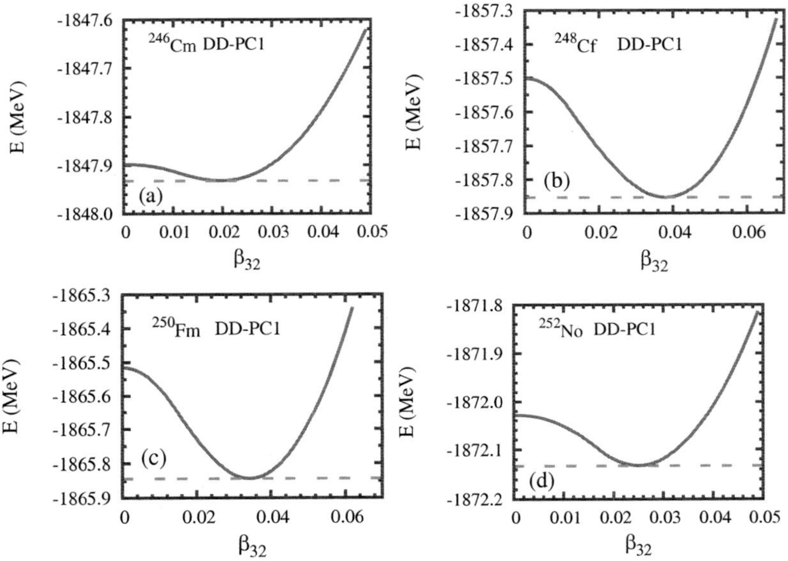

Fig. 7. (Color online) The binding energy E for $N = 150$ isotones ^{246}Cm, ^{248}Cf, ^{250}Fm, and ^{252}No as a function of the non-axial octupole deformation parameter β_{32}. Adapted from Ref. 101.

values at ^{248}Cf in these $N = 150$ isotones. This is consistent with the analysis given in Refs. 88, 145 and the experimental observation that in ^{248}Cf, the 2^- state is the lowest among these nuclei.

In Ref. 101, we analyzed the origin of these strong Y_{32}-correlations. The nonaxial octupole couplings occur between pairs of single-particle orbits with $\Delta j = \Delta l = 3$ and $\Delta K = 2$ where j and l are the total and orbital angular momenta of single particles respectively and K the projection of j on the z axis. If the Fermi surface of a nucleus lies close to a pair of such orbitals and these two orbitals are nearly degenerate, a strong nonaxial octupole β_{32} correlation is expected. In Fig. 8, the proton and neutron single-particle levels near the Fermi surface for ^{248}Cf is shown as functions of quadrupole deformation β_{20} on the left side and of β_{32} with β_{20} fixed at 0.3 on the right side.

In the upper panel of Fig. 8, one finds a strong spherical shell closure at $Z = 92$. As discussed in Ref. 146, this shell closure is a spurious one and it is commonly predicted in relativistic mean field calculations. The spherical proton orbitals $\pi 2f_{7/2}$ and $\pi 1i_{13/2}$ are very close to each other and this near degeneracy results in octupole correlations. The two proton levels, [521]3/2 originating from $2f_{7/2}$ and [633]7/2 originating from $1i_{13/2}$,

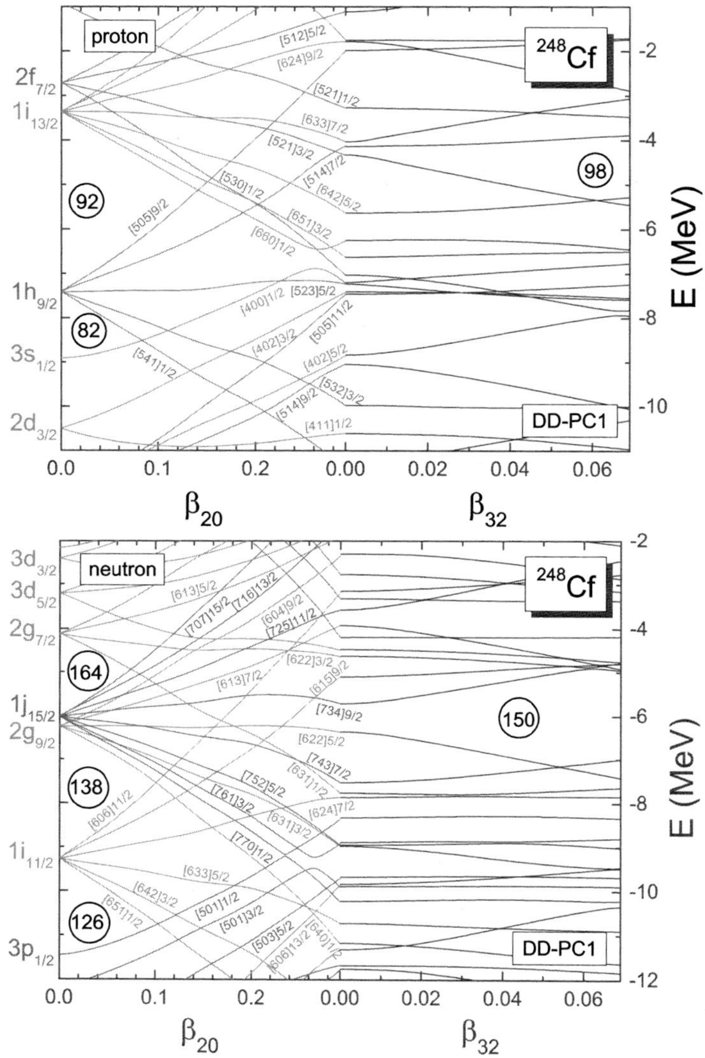

Fig. 8. (Color online) The single-particle levels near the Fermi surface for protons (upper panel) and neutrons (lower panel) of ^{248}Cf as functions of quadrupole deformation β_{20} (left side) and of β_{32} with β_{20} fixed at 0.3 (right side). Taken from Ref. 101.

satisfying the $\Delta j = \Delta l = 3$ and $\Delta K = 2$ condition, are very close to each other when β_{20} varies from 0 to 0.3. Therefore the non-axial octupole Y_{32} correlation develops and with β_{32} increasing from zero, an energy gap appears at $Z = 98$. The appearance of the spurious shell closure at $Z = 92$

is mainly due to the lowering of $\pi 1h_{9/2}$ orbital which also results in the proton [514]7/2 level lying in between [521]3/2 and [633]7/2, thus making the gap at $Z = 98$ smaller and weakening the Y_{32} correlation. Similarly, the spherical neutron orbitals $\nu 2g_{9/2}$ and $\nu 1j_{15/2}$ are very close to each other, as seen in the lower panel of Fig. 8. The two neutron levels, [734]9/2 originating from $1j_{15/2}$ and [622]5/2 originating from $2g_{9/2}$, are also close to each other and they just lie above and below the Fermi surface. This leads to the development of a gap at $N = 150$ with β_{32} increasing. Therefore the Y_{32} correlation in $N = 150$ isotones is from both protons and neutrons and for ^{248}Cf the correlation is the most pronounced. The gap around $N = 150$ is larger than that around $Z = 98$, which may indicate that the non-axial octupole effect originating from neutrons is larger than that from protons. If there is no spurious shell gap at $Z = 92$, one can expect more pronounced non-axial octupole correlations from protons.

For SHN, the β_{32} deformation may also play an important role. From the RASM calculations, it was predicted that the strong nonaxial-octupole effect may persist up to element 108[88] and play a crucial role in determining the shell stability in even heavier nuclei.[89] For example, the PES for the ground state of ^{310}126 shows a well established minimum at $\epsilon_2 = 0.02$ and $\epsilon_{32} = 0.3$.[89] This result strongly supports the tetrahedral symmetry of nucleus ^{310}126 in its ground state, where the quadrupole deformation is vanishingly small and the tetrahedral deformation is large. Thus, the investigation of tetrahedral SHN with MDC-CDFT would be an interesting topic.

4.3. *Third minima and hyperdeformed states*

The PES's of actinide nuclei are characterized by a two-humped barrier structure along the static fission path. In the 1970s, macroscopic-microscopic model calculations predicted the occurrence of shallow third minima beyond the second barrier.[147–149] But contradictory results were obtained later by using different models. The PES's of nuclei in the Ra–Th region were computed using the macroscopic-microscopic model with a modified oscillator potential, and in many nuclei a third minimum was found to be very deep.[150] This model, with shell corrections calculated by adopting an axially-deformed Woods–Saxon potential, was later used to systematically study the PES's of even-even Rn, Ra, Th, and U isotopes.[151] Very deep minima, or even two hyperdeformed minima, were predicted in many of these nuclei.

In the macroscopic-microscopic model based on the Woods–Saxon potential, calculations with additional shape degrees of freedom were preformed in Refs. 152, 153. It was found that the third barrier is lowered substantially by including the β_1 deformation. Consequently, the third minimum disappeared in many nuclei, except for ^{230,232}Th, in which only a shallow third minimum with a depth of less than 400 keV was found. Furthermore, in Ref. 154 an analysis of PES's computed with the finite-range liquid-drop model[155] revealed that only a few nuclei accessible to experiment exhibit third minima in their PES's and the depth of the third well is less than 1 MeV for the light Th and U isotopes. Results of constrained Hartree–Fock or Hartree–Fock–Bogoliubov calculations with Skyrme forces[156–158] and the Gogny force[159–161] did not show deep third minima in actinide nuclei. In Ref. 96 a shallow third minimum with a depth of 1–2 MeV in ^{232}Th was found in axially-deformed relativistic mean-field calculations with the effective interactions PL-40, NL1, and NL-SH.

In Ref. 108, the occurrence and properties of the third minima on the PES's of light actinides were studied by using the MDC-RMF model with the nonlinear point-coupling functional PC-PK1 and the density-dependent meson exchange functional DD-ME2 in the particle-hole channel. Pairing correlations are treated in the BCS approximation with a separable pairing force of finite range.

By examining the two-dimensional PES's, it was found that the third minima and barriers appear for ^{226,228,230,232}Th and ^{232,234,238}U. For these nuclei the energy curves along the lowest static fission path, calculated in the ADHO basis with $N_f = 20$ shells, are shown in Fig. 9. In the vicinity of the second saddle point, MDC-RMF calculations are performed with and without the inclusion of triaxial quadrupole shapes and both results are displayed for comparison. As reported in Ref. 98, the inclusion of triaxial configurations, in addition to the axial octupole deformation, modifies the shape and height of the second fission barrier. From Fig. 9, it can be seen that the functional DD-ME2 predicts the occurrence of a third barrier in all Th nuclei and ^{238}U. The third minima in ^{230,232}Th are very shallow, whereas those in ^{226,228}Th and ^{238}U are quite prominent. With the functional PC-PK1 a third barrier is found only in ^{226,228,230}Th. This tells us that the prediction of a third barrier on the PES's of actinide nuclei depends on the effective interaction used in MDC-RMF calculations: More pronounced minima are predicted by the DD-ME2 functional, as compared to the functional PC-PK1. In Fig. 9, one also finds that the depth of the third well in Th isotopes decreases with increasing neutron number. The

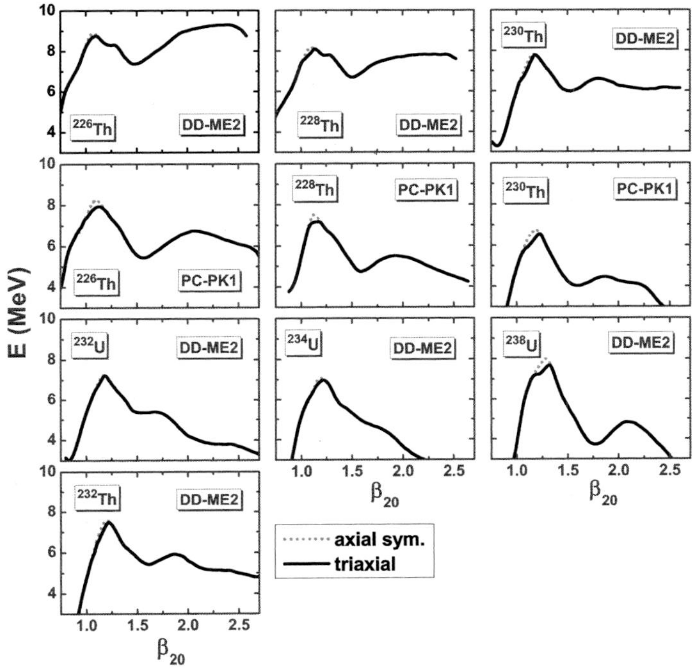

Fig. 9. (Color online) MDC-RMF energy curves of ^{226,228,230,232}Th and ^{232,234,238}U. For each nucleus the energy is normalized with respect to the binding energy at the absolute minimum. Taken from Ref. 108.

energies of the third minima, the heights of the second and third barriers, and the depths of the third wells $\Delta E \equiv B_{\rm III} - E_{\rm III}$, together with empirical parameters were given in Table I of Ref. 108.

Similar to the study of Y_{32}-correlations in $N = 150$ isotones, the origin of the third minimum can also be traced by examining the single particle levels around the Fermi surface. This has been done for ^{226}Th in Ref. 108 and it was found that the formation of the third minimum is mainly due to the $Z = 90$ proton energy gap at $\beta_{20} \approx 1.5$ and $\beta_{30} \approx 0.7$.

4.4. *Fission barriers of actinide nuclei*

The double-humped fission barriers of actinide nuclei are usually used as a benchmark for theoretical models. Many shape degrees of freedom play important and different roles in determining the heights of the inner and outer barriers. It has long been known that the inner barrier is lowered when the triaxial deformation is allowed,[91–93] while for the outer barrier, the reflection asymmetric shape is favored.[94] We have shown that the reflection asymmetric outer barrier may be further lowered by including the nonaxial shape degrees of freedom.[98] In this subsection we show some results of multi-dimensional PES's for actinides. We will first present the one-, two, and three-dimensional PES of ^{240}Pu, then display systematic results of the actinide nuclei.[98,105]

In Fig. 10 we show the potential energy curves which are obtained with different self-consistent symmetries imposed in the calculations: the axial (AS) or triaxial (TS) symmetries combined with reflection symmetric (RS) or asymmetric (TA) cases. The effect of the triaxial deformation on the inner barrier and that of the octupole deformation on the outer barrier can be clearly seen: The triaxial deformation lowers the inner barrier by more than 2 MeV;[97] the RA shape is favored beyond the fission isomer and lowers very much the outer fission barrier.[96] It was found for the first time that the outer barrier is also considerably lowered by about 1 MeV by the triaxial deformation and a better reproduction of the empirical barrier height is achieved for the outer barrier. This lowering effect can be revealed only when the axial and reflection symmetries are both broken.

How the PES of ^{240}Pu evolves with the triaxial distortion can be seen much more clearly in Fig. 11 in which we show two-dimensional PES's from calculations without and with the triaxial deformation. When the triaxial deformation is allowed, the binding energy of ^{240}Pu assumes its lowest possible value at each (β_{20}, β_{30}) point. At some points we get non-zero β_{22} values. That is, non-axial solutions are favored at these points compared to the axial ones. The triaxial deformation appears mainly in two regions in Fig. 11. One region starts from the first saddle point and extends roughly along the direction of the β_{30} axis up to a very asymmetric shape with $\beta_{30} \sim 1.0$. In this region the values of β_{22} are about $0.06 \sim 0.12$ corresponding to $\gamma \sim 10^\circ$. The energy, especially the inner barrier height, is lowered by about 2 MeV. The other region is around the outer barrier and the β_{22} values are about $0.02 \sim 0.03$ corresponding to $\gamma \sim 2^\circ$, about 1 MeV is gained for the binding energy at the second saddle point due to the triaxiality. In other regions, e.g., in the ground state and fission isomer valleys, only axially symmetric solutions are obtained.

In Figs. 12 and 13 the three-dimensional PES of ^{240}Pu are shown. In these calculations, both non-axial and reflection asymmetric shapes are allowed. In each sub-figure we fix the value of β_{20} and display the energy as a function of β_{22} and β_{30}. Note that in three-dimensional constraints

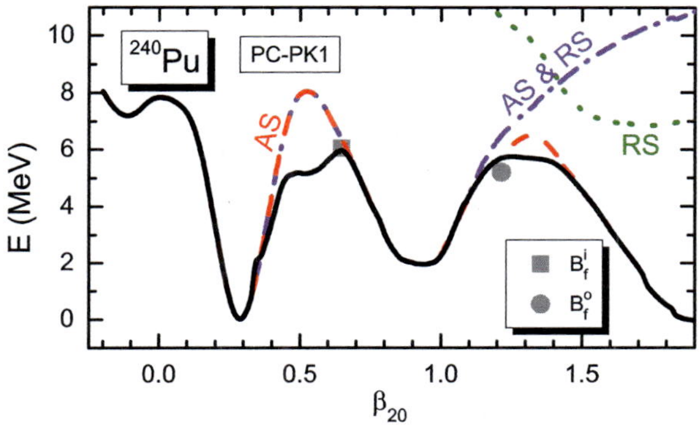

Fig. 10. Potential energy curves of ^{240}Pu with various self-consistent symmetries imposed: The black solid, red dashed, green dotted, and violet dot-dashed curves represents the lowest fission path with V_4, axial symmetry (AS), reflection symmetry (RS), and AS & RS symmetries imposed. The empirical inner (outer) barrier height $B_{\rm emp}$ is denoted by the grey square (circle). The energy is normalized with respect to the binding energy of the ground state. Taken from Ref. 98.

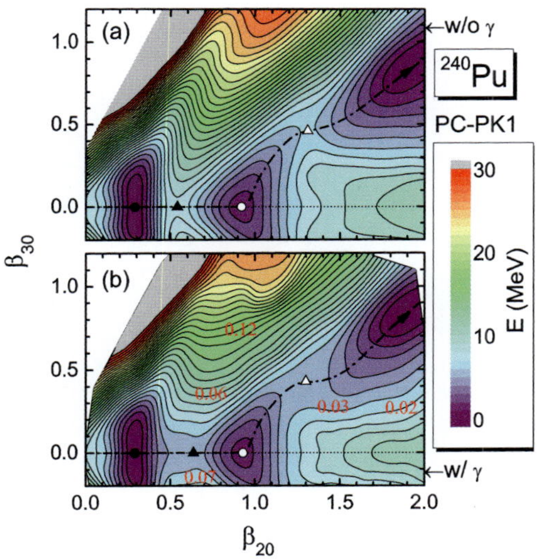

Fig. 11. Potential energy surfaces of ^{240}Pu in the (β_{20}, β_{30}) plane from calculations (a) without and (b) with the triaxial deformation included. The energy is normalized with respect to the binding energy of the ground state. The numbers in (b) show the values of β_{22} at these points. The fission path is represented by a dash-dotted line. The ground state and fission isomer are denoted by full and open circles. The first and second saddle points are denoted by full and open triangles. The contour interval is 1 MeV. Taken from Ref. 98.

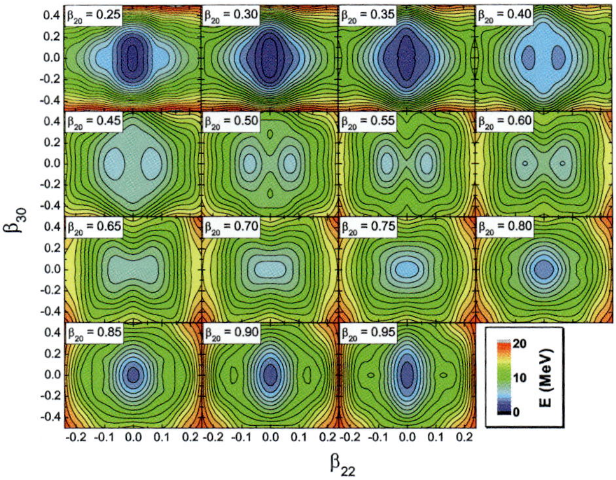

Fig. 12. Sections of the three-dimensional potential energy surface, $E = E(\beta_{20}, \beta_{22}, \beta_{30})$, of ^{240}Pu around the ground state, the inner barrier, and the fission isomer from MDC-RMF calculations. In each sub-figure the energy is shown as a function of the deformation parameters β_{22} and β_{30} when β_{20} is fixed at a certain value. The energies are normalized with respect to the binding energy of the ground state. The contour interval is 1 MeV. Taken from Ref. 105.

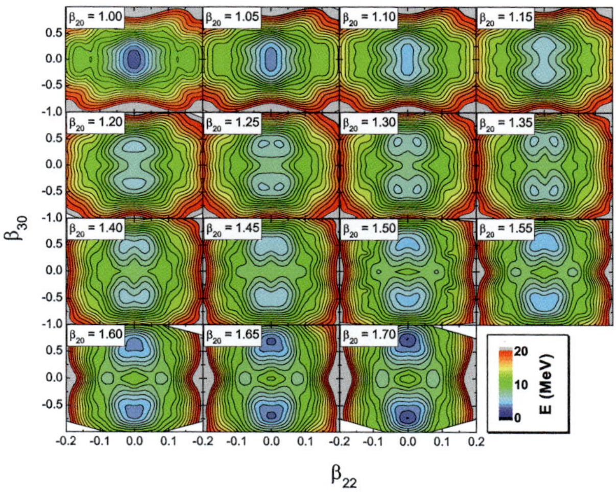

Fig. 13. Sections of the three-dimensional potential energy surface, $E = E(\beta_{20}, \beta_{22}, \beta_{30})$, of ^{240}Pu around the fission isomer and the outer barrier from MDC-RMF calculations. In each sub-figure the energy is shown as a function of the deformation parameters β_{22} and β_{30} when β_{20} is fixed at a certain value. The energies are normalized with respect to the binding energy of the ground state. The contour interval is 1 MeV. Taken from Ref. 105.

on the corresponding multipole moments, other shape degrees of freedom $\beta_{\lambda\mu}$ with even μ, e.g., β_{32}, β_{40}, β_{42}, β_{44}, β_{50}, ..., are also included in the calculations self-consistently.

From the first three sub-figures with $\beta_{20} = 0.25$, 0.30, and 0.35, it can be seen that the ground state with $\beta_{20} \sim 0.3$ is both axially symmetric and reflection symmetric, though it is a little soft against the octupole distortion. When the nucleus is stretched by the quadrupole constraining potential, it becomes softer against the triaxial as well as the octupole distortions. In the sub-figures with $\beta_{20} = 0.40 \sim 0.65$, two symmetric minima with non-zero triaxial deformation β_{22} appear and the corresponding fission paths are much more favored than the axially symmetric one. Although $\beta_{30} = 0$ for all minima in these sub-figures, the softness against the octupole distortion changes as the nucleus is elongated. The inner fission barrier is located near the deformation $\beta_{20} \sim 0.60$. From the last six sub-figures in Fig. 12 with $\beta_{20} = 0.70 \sim 0.95$, one finds that the nucleus becomes axially symmetric and reflection symmetric again.

As the deformation β_{20} becomes larger, the second minimum and the second saddle point of the PES appear. From the last three sub-figures of Fig. 12 and the first three sub-figures of Fig. 13 it is seen that, the nucleus stays reflection symmetric near the fission isomeric state with $\beta_{20} \sim 0.95$, but it becomes softer against the β_{30} distortion. At $\beta_{20} = 1.15$, two minima corresponding to reflection asymmetric shapes appear. Here the effect of the nonaxial deformation is not apparent, but along the β_{22} direction the PES becomes softer. At $\beta_{20} = 1.2$ the energy of the reflection asymmetric shape is lower by about 1 MeV than that of the reflection symmetric one. When β_{20} increases further, around the top of the second barrier, each reflection asymmetric minimum splits into two minima with non-zero β_{22}. This is the lowering effects of triaxiality on the outer barrier. Around the second barrier, the largest energy gain due to the triaxial distortion is about 1 MeV. The nucleus becomes axially symmetric again when $\beta_{20} > 1.6$.

From the above discussions, the following conclusions for ^{240}Pu have been drawn in Ref. 105: (1) Both the ground state and the fission isomeric state are axial and reflection symmetric; (2) Around the first fission barrier it assumes triaxial and reflection symmetric shapes; (3) Around the second fission barrier both triaxial and octupole deformations are important.

In Ref. 105, a systematic study of even-even actinide nuclei was carried out with the MDC-RMF model. Since around the inner barrier an actinide nucleus assumes triaxial but reflection symmetric shapes, we have made a one-dimensional constraint calculation with the triaxial deformation

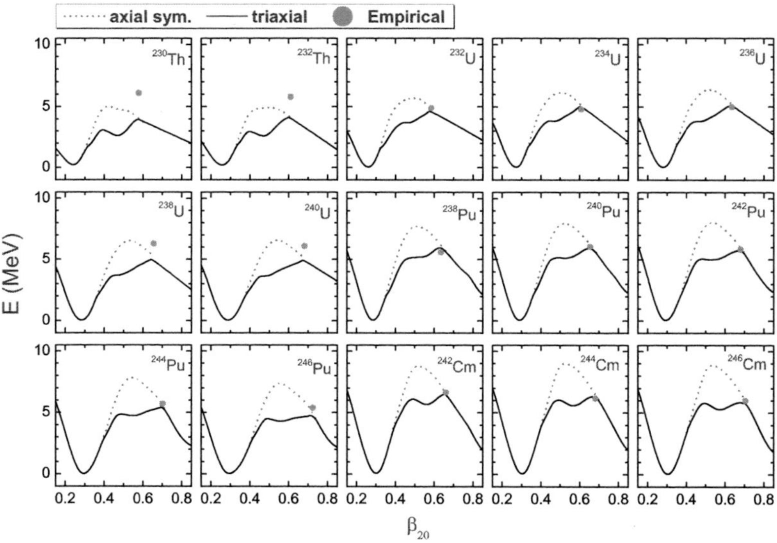

Fig. 14. (Color online) Potential energy curves of even-even actinide nuclei around the ground states and the first fission barriers from MDC-RMF calculations. The axially symmetric results are displayed by dotted curves, while those from the triaxial calculations are shown by solid curves. The binding energy is normalized with respect to that of the ground state of each nucleus. The empirical values of fission barriers are taken from Ref. 162 and shown by red dots. Taken from Ref. 105.

allowed and the reflection symmetry imposed. In Fig. 14 we present potential energy curves of even-even actinide nuclei around the ground state and the first fission barrier from MDC-RMF calculations. It is clearly seen that the triaxial deformation lowers considerably the inner barrier of these actinide nuclei.

Potential energy curves of even-even actinide nuclei around the fission isomer and the second fission barrier from MDC-RMF calculations are shown in Fig. 15. From Fig. 15 one finds that for most of the nuclei investigated here, the triaxiality lowers the second barrier by about $0.5 \sim 1$ MeV (about $10 \sim 20\%$ of the barrier height). The calculation results with the triaxiality included agrees well with the empirical values for all actinide nuclei shown in Fig. 15.

4.5. *Fission barriers of SHN*

Due to the heavy computational burden, full MDC-CDFT calculations for SHN with both axial and reflection symmetries broken are still in progress.

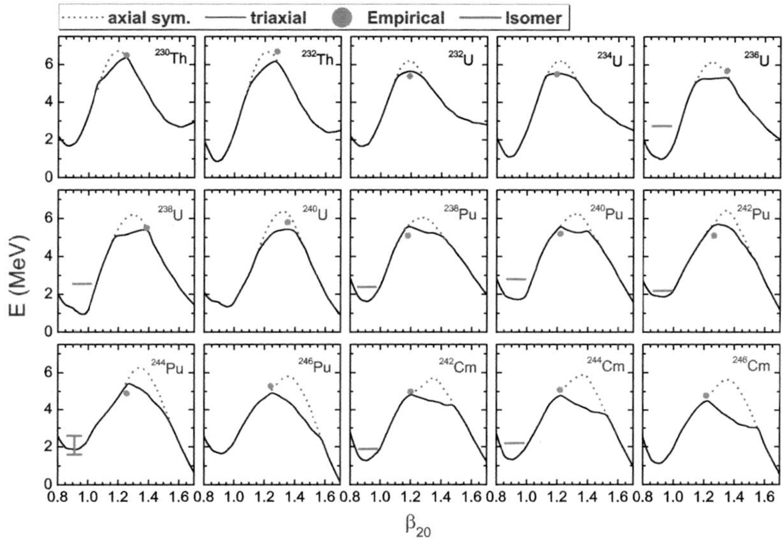

Fig. 15. (Color online) Potential energy curves of even-even actinide nuclei around the fission isomers and the second fission barriers from MDC-RMF calculations. The reflection asymmetric shapes are allowed in the calculations. The axially symmetric results are displayed by dotted curves, while those from the triaxial calculations are shown by solid curves. The binding energy is normalized with respect to that of the ground state of each nucleus. The empirical values of fission barriers are taken from Ref. 162 and shown by red dots. Experimental values of energies of fission isomers are taken from Ref. 163 and shown by green horizontal lines. For ^{244}Pu, only an energy range was given for the isomer. Taken from Ref. 105.

In Ref. 87, Abusara et al. have performed a systematic study of fission barriers for even-even superheavy nuclei with $Z = 112$–120 by using the RMF model with triaxial or octupole shapes included separately. It was found that the triaxiality has a considerable impact on the shape and the height of outer fission barrier which is lowered by ~ 3 MeV in $^{292}120$ nucleus, as is seen in Fig. 16. Interestingly, it was also seen that for this nucleus, the lowering of outer fission barrier due to the octupole deformation is substantially smaller than the lowering effect due to the triaxiality. Therefore in RMF calculations, the lowering of the triaxial deformation on the second fission barrier exists not only for actinide nuclei but also for SHN. However, in a recent work within the MM model, it was found that the influence of the triaxiality on the second fission barriers is negligible.[164] This raises an open problem: What are the main reasons for the different effects of the triaxiality on the second fission barriers predicted by the MM and RMF

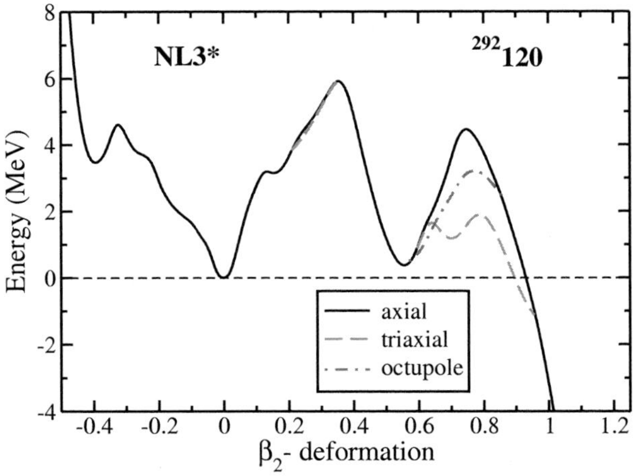

Fig. 16. (Color online) Deformation energy curves for the $Z = 120$, $N = 172$ nucleus obtained with the NL3* parametrization. The black solid, red dashed and blue dot-dashed lines display the deformation energy curves for the axially symmetric, triaxial and axial octupole deformed solutions. The deformation energy curves are shown for the last two solutions only in the range of β_2 values where it is lower in energy than the deformation energy curve of the axially symmetric solution. Taken from Ref. 87.

models? It would be interesting to make more investigations with different models, especially, with the non-relativistic DFTs.

5. Summary

In this chapter, we have given a brief review of RMF descriptions of heavy and superheavy nuclei. We mainly discussed the shell structure and magic numbers in the mass region of SHN, binding energies and α decay Q values, shapes of ground states and potential energy surfaces, and fission barriers.

The magicity in the SHE region is one of the most important problems in the study of SHN. Predictions for the magic numbers next to ^{208}Pb have been made by examining the two-nucleon separation energies S_{2p} and S_{2n}, the two-nucleon gaps δ_{2p} and δ_{2n}, the shell correction energies E_{shell}, the pairing energies E_{pair}, and the pairing gap Δ_p and Δ_n with the relativistic continuum Hartree–Bogoliubov (RCHB) theory. The doubly magic SHN can be the combination of magic proton numbers $Z = 120$, 132, and 138 and magic neutron numbers $N = 172$, 184, 198, 228, 238, and 258.

The binding energy is one of the most important properties of atomic nuclei. In particular, for SHN, the binding energies are very important because one can study the α-decay properties of SHN. The binding energies of an SHN and its daughter nuclei in the α-deay chain can be calculated with the RMF models. Then α-decay energies can be obtained and α-decay half-lives can be evaluated with some empirical formulas. Along this line, lots of efforts have been made with the RMF models. In this chapter, we have presented some results of α-decay Q values for some $Z = 105$, 115 and 117 nuclides.

The shape degrees of freedom are important not only for the ground states or small amplitude collective motions, but also for large amplitude collective motions such as fission. For SHN, due to the large repulsion among protons and the complexity of the shell structure, the ground state shapes may become very exotic, e.g., bubble and toroidal shapes, superdeformed shapes, and tetrahedral shapes. Many shape degrees of freedom, such as the triaxial and octupole shapes, are also important for determining the fission path and fission barriers. To include these deformations simultaneously, the MDC-CDFT was developed in which all shape degrees of freedom $\beta_{\lambda\mu}$ with even μ are included self-consistently. The MDC-CDFT has been used to study ground state shapes, potential energy surfaces, and fission barriers of heavy nuclei and SHN.

The nonaxial octupole correlations in some $N = 150$ isotones are investigated with the MDC-CDFT. For the ground states of ^{248}Cf and ^{250}Fm, the nonaxial octupole deformation parameter $\beta_{32} > 0.03$ and the energy gain due to the β_{32} distortion is larger than 300 keV. In ^{246}Cm and ^{252}No, shallow β_{32} minima are found. The occurrence of the non-axial octupole β_{32} correlations is mainly from a pair of neutron orbitals [734]9/2 ($\nu j_{15/2}$) and [622]5/2 ($\nu g_{9/2}$) which are close to the neutron Fermi surface, and a pair of proton orbitals [521]3/2 ($\pi f_{7/2}$) and [633]7/2 ($\pi i_{13/2}$) which are close to the proton Fermi surface.

Triple-humped barriers in actinide nuclei have been investigated with the MDC-RMF model. Two-dimensional PES's of ^{226,228,230,232}Th and ^{232,234,236,238}U were mapped and the third minima on these surfaces were located. It was found that the possible occurrence of a third barrier on the PES's of actinide nuclei depends on the effective interaction used in multidimensional CDFT calculations. More pronounced minima are predicted by the DD-ME2 functional, as compared to the functional PC-PK1. The depth of the third well in Th isotopes decreases with increasing neutron number. The third minimum is caused by the proton $Z = 90$ shell gap at relevant deformations.

The potential energy surfaces of actinide nuclei have been studied with the MDC-RMF model. It was found that besides the octupole deformation, the triaxiality also plays an important role upon the second fission barriers. Both the outer and the inner barriers are lowered by the triaxial deformation compared with axially symmetric results. This lowering effect for the reflection-asymmetric outer barrier is 0.5–1 MeV, accounting for 10%–20% of the barrier height. With the inclusion of the triaxial deformation, a good agreement with empirical heights for the outer barriers of actinide nuclei was achieved.

Using the RMF model with triaxial or octupole shapes included separately, a recent study of fission of even-even SHN with $Z = 112$–120 revealed that both the triaxiality and the octupole deformation lower the height of the outer fission barrier of SHN. For the nucleus $^{292}120$, the outer barrier is lowered by the triaxial deformation by about 3 MeV. It was also predicted that for this nucleus, the lowering of outer fission barrier due to the octupole deformation is substantially smaller than the lowering due to the triaxiality. It is very desirable to carry out MDC-CDFT calculations for SHN with the axial and reflection symmetries simultaneously broken.

Acknowledgment

We thank G. Adamian, A. V. Afanasjev, N. Antonenko, R. Capote, Y. S. Chen, L. S. Geng, R. Jolos, M. Kowal, L. L. Li, Z. P. Li, H. Liang, W. H. Long, A. Lopez-Martens, U.-G. Meissner, J. Meng, P. Ring, N. Schunck, T. Shneidman, J. Skalski, H. Toki, D. Vretenar, K. Wen, X. Z. Wu, S. Q. Zhang, W. Zhang, and Z. H. Zhang for useful discussions and/or their contributions to the work reviewed in this chapter. This work was supported in part by the National Key Basic Research Program of China (Grant No. 2013CB834400), the National Natural Science Foundation of China (Grants No. 11121403, No. 11120101005, No. 11211120152, No. 11275248, and No. 11475115, and No. 11525524), and the Knowledge Innovation Project of the Chinese Academy of Sciences (Grant No. KJCX2-EW-N01). The results described in this paper are obtained on the High-performance Computing Cluster of SKLTP/ITP-CAS and the ScGrid of the Supercomputing Center, Computer Network Information Center of the Chinese Academy of Sciences.

References

1. J. A. Wheeler. Nuclear fission and nuclear stability. In eds. W. Pauli, L. Rosenfeld, and V. Weisskopf, *Niels Bohr and the Development of Physics:*

Essays Dedicated to Niels Bohr on the Occasion of His Seventieth Birthday, pp. 163–184. McGraw-Hill (1955). URL http://babel.hathitrust.org/cgi/pt?id=mdp.39015001324709.

2. G. Scharff-Goldhaber, Nuclear physics, *Nukeonika (Nucleonics)*. **15**(9), 122–124 (Sept., 1957).
3. F. G. Werner and J. A. Wheeler, Superheavy nuclei, *Phys. Rev.* **109**(1), 126–144 (Jan., 1958). doi: 10.1103/PhysRev.109.126. URL http://link.aps.org/abstract/PR/v109/p126.
4. W. D. Myers and W. J. Swiatecki, Nuclear masses and deformations, *Nucl. Phys.* **81**(1), 1–60 (June, 1966). doi: 10.1016/0029-5582(66)90639-0. URL http://dx.doi.org/10.1016/0029-5582(66)90639-0.
5. P. J. Siemens and H. A. Bethe, Shape of heavy nuclei, *Phys. Rev. Lett.* **18** (17), 704–706 (Apr., 1967). doi: 10.1103/PhysRevLett.18.704. URL http://link.aps.org/doi/10.1103/PhysRevLett.18.704.
6. V. M. Strutinsky, "Shells" in deformed nuclei, *Nucl. Phys. A.* **122**(1), 1–33 (Dec., 1968). ISSN 0375-9474. doi: 10.1016/0375-9474(68)90699-4. URL http://dx.doi.org/10.1016/0375-9474(68)90699-4.
7. A. Sobiczewski, F. Gareev, and B. Kalinkin, Closed shells for $Z > 82$ and $N > 126$ in a diffuse potential well, *Phys. Lett.* **22**(4), 500–502 (Sept., 1966). doi: 10.1016/0031-9163(66)91243-1. URL http://dx.doi.org/10.1016/0031-9163(66)91243-1.
8. H. Meldner, Predictions of new magic regions and masses for super-heavy nuclei from calculations with realistic shell model single particle Hamiltonian, *Arkiv Fysik.* **36**(2), 593–598 (1967). Proceedings of the Lysekil Symposium: Nuclides far off the Stability Line, Aug. 21–27, 1966, Sweden.
9. S. G. Nilsson, C. F. Tsang, A. Sobiczewski, Z. Szymanski, S. Wycech, C. Gustafson, I.-L. Lamm, P. Möller, and B. Nilsson, On the nuclear structure and stability of heavy and superheavy elements, *Nucl. Phys. A.* **131**(1), 1–66 (June, 1969). doi: 10.1016/0375-9474(69)90809-4. URL http://dx.doi.org/10.1016/0375-9474(69)90809-4.
10. U. Mosel and W. Greiner, On the stability of superheavy nuclei against fission, *Z. Phys. A.* **222**(3), 261–282 (June, 1969). doi: 10.1007/BF01392125. URL http://dx.doi.org/10.1007/BF01392125.
11. J. Hamilton, S. Hofmann, and Y. Oganessian, Search for superheavy nuclei, *Annu. Rev. Nucl. Part. Sci.* **63**(1), 383–405 (2013). doi: 10.1146/annurev-nucl-102912-144535. URL http://www.annualreviews.org/doi/abs/10.1146/annurev-nucl-102912-144535.
12. S. Hofmann and G. Münzenberg, The discovery of the heaviest elements, *Rev. Mod. Phys.* **72**(3), 733–767 (July, 2000). doi: 10.1103/RevModPhys.72.733. URL http://link.aps.org/abstract/RMP/v72/p733.
13. K. Morita, K. Morimoto, D. Kaji, T. Akiyama, S.-i. Goto, H. Haba, E. Ideguchi, R. Kanungo, K. Katori, H. Koura, H. Kudo, T. Ohnishi, A. Ozawa, T. Suda, K. Sueki, H.-S. Xu, T. Yamaguchi, A. Yoneda, A. Yoshida, and Y.-L. Zhao, Experiment on the synthesis of element 113 in the reaction ^{209}Bi(^{70}Zn,n)^{278}113, *J. Phys. Soc. Jpn.* **73**(10), 2593–2596

(2004). ISSN 0375-9474. doi: 10.1143/JPSJ.73.2593. URL http://jpsj.ipap.jp/link?JPSJ/73/2593/.
14. K. Morita, K. Morimoto, D. Kaji, H. Haba, K. Ozeki, Y. Kudou, T. Sumita, Y. Wakabayashi, A. Yoneda, K. Tanaka, S. Yamaki, R. Sakai, T. Akiyama, S.-i. Goto, H. Hasebe, M. Huang, T. Huang, E. Ideguchi, Y. Kasamatsu, K. Katori, Y. Kariya, H. Kikunaga, H. Koura, H. Kudo, A. Mashiko, K. Mayama, S.-i. Mitsuoka, T. Moriya, M. Murakami, H. Murayama, S. Namai, A. Ozawa, N. Sato, K. Sueki, M. Takeyama, F. Tokanai, T. Yamaguchi, and A. Yoshida, New result in the production and decay of an isotope, $^{278}113$, of the 113th element, *J. Phys. Soc. Jpn.* **81**(10), 103201–4 (2012). doi: 10.1143/JPSJ.81.103201. URL http://jpsj.ipap.jp/link?JPSJ/81/103201/.
15. Y. Oganessian, Heaviest nuclei from ^{48}Ca-induced reactions, *J. Phys. G: Nucl. Phys.* **34**(4), R165–R242 (Apr., 2007). ISSN 0954-3899. doi: 10.1088/0954-3899/34/4/R01. URL http://dx.doi.org/10.1088/0954-3899/34/4/R01.
16. Y. T. Oganessian, F. S. Abdullin, P. D. Bailey, D. E. Benker, M. E. Bennett, S. N. Dmitriev, J. G. Ezold, J. H. Hamilton, R. A. Henderson, M. G. Itkis, Y. V. Lobanov, A. N. Mezentsev, K. J. Moody, S. L. Nelson, A. N. Polyakov, C. E. Porter, A. V. Ramayya, F. D. Riley, J. B. Roberto, M. A. Ryabinin, K. P. Rykaczewski, R. N. Sagaidak, D. A. Shaughnessy, I. V. Shirokovsky, M. A. Stoyer, V. G. Subbotin, R. Sudowe, A. M. Sukhov, Y. S. Tsyganov, V. K. Utyonkov, A. A. Voinov, G. K. Vostokin, and P. A. Wilk, Synthesis of a new element with atomic number $Z = 117$, *Phys. Rev. Lett.* **104**(14), 142502–4 (Apr., 2010). doi: 10.1103/PhysRevLett.104.142502. URL http://link.aps.org/doi/10.1103/PhysRevLett.104.142502.
17. Z.-Y. Zhang, Z.-G. Gan, L. Ma, M.-H. Huang, T.-H. Huang, X.-L. Wu, G.-B. Jia, G.-S. Li, Z.-Z. Ren, S.-G. Zhou, X.-H. Zhou, H.-S. Xu, H.-Q. Zhang, and G.-Q. Xiao, Observation of superheavy nuclide ^{271}Ds, *Chin. Phys. Lett.* **29**(1), 012502–3 (Jan., 2012). doi: 10.1088/0256-307X/28/1/012502. URL http://dx.doi.org/10.1088/0256-307X/29/1/012502.
18. S.-G. Zhou. Theoretical study of structure and synthesis mechanism of superheavy nuclei. In *Nuclear Structure in China 2014: Proceedings of the 15th National Conference on Nuclear Structure in China (NSC2014)*, World Scientific, Singapore (2015).
19. B. D. Serot and J. D. Walecka, The relativistic nuclear many-body problem, *Adv. Nucl. Phys.* **16**, 1–327 (1986).
20. P. G. Reinhard, The relativistic mean-field description of nuclei and nuclear dynamics, *Rep. Prog. Phys.* **52**(4), 439–514 (Apr., 1989). ISSN 0034-4885. doi: 10.1088/0034-4885/52/4/002. URL http://dx.doi.org/10.1088/0034-4885/52/4/002.
21. P. Ring, Relativistic mean field theory in finite nuclei, *Prog. Part. Nucl. Phys.* **37**, 193–263 (1996). ISSN 0146-6410. doi: 10.1016/0146-6410(96)00054-3. URL http://dx.doi.org/10.1016/0146-6410(96)00054-3.
22. M. Bender, P.-H. Heenen, and P.-G. Reinhard, Self-consistent mean-field models for nuclear structure, *Rev. Mod. Phys.* **75**(1), 121–180 (Jan., 2003).

doi: 10.1103/RevModPhys.75.121. URL `http://link.aps.org/doi/10.1103/RevModPhys.75.121`.

23. D. Vretenar, A. V. Afanasjev, G. A. Lalazissis, and P. Ring, Relativistic Hartree–Bogoliubov theory: static and dynamic aspects of exotic nuclear structure, *Phys. Rep.* **409**(3-4), 101–259 (Apr., 2005). doi: 10.1016/j.physrep.2004.10.001. URL `http://dx.doi.org/10.1016/j.physrep.2004.10.001`.
24. J. Meng, H. Toki, S. G. Zhou, S. Q. Zhang, W. H. Long, and L. S. Geng, Relativistic continuum Hartree–Bogoliubov theory for ground-state properties of exotic nuclei, *Prog. Part. Nucl. Phys.* **57**(2), 470–563 (Oct., 2006). doi: 10.1016/j.ppnp.2005.06.001. URL `http://dx.doi.org/10.1016/j.ppnp.2005.06.001`.
25. M. Leino and F. P. Hessberger, The nuclear structure of heavy-actinide and trans actinide nuclei, *Annu. Rev. Nucl. Part. Sci.* **54**, 175–215 (Dec., 2004). doi: 10.1146/annurev.nucl.53.041002.110332. URL `http://dx.doi.org/10.1146/annurev.nucl.53.041002.110332`.
26. R.-D. Herzberg and P. Greenlees, In-beam and decay spectroscopy of transfermium nuclei, *Prog. Part. Nucl. Phys.* **61**(2), 674–720 (Oct., 2008). ISSN 0146-6410. doi: 10.1016/j.ppnp.2008.05.003. URL `http://dx.doi.org/10.1016/j.ppnp.2008.05.003`.
27. S. K. Tandel, P. Chowdhury, S. Lakshmi, U. S. Tandel, I. Ahmad, M. P. Carpenter, S. Gros, R. V. F. Janssens, T. L. Khoo, F. G. Kondev, J. P. Greene, D. J. Hartley, T. Lauritsen, C. J. Lister, D. Peterson, A. Robinson, D. Seweryniak, and S. Zhu, Rotational bands in odd-A Cm and Cf isotopes: Exploring the highest neutron orbitals, *Phys. Rev. C.* **82**(4), 041301R–5 (Oct., 2010). doi: 10.1103/PhysRevC.82.041301. URL `http://link.aps.org/doi/10.1103/PhysRevC.82.041301`.
28. D. Ackermann, Superheavy elements at GSI — investigating exotic nuclear matter, *Acta Phys. Pol. B.* **42**(3-4), 577–586 (Mar., 2011). URL `http://www.actaphys.uj.edu.pl/vol42/abs/v42p0577`.
29. J. Piot, B. J.-P. Gall, O. Dorvaux, P. T. Greenlees, N. Rowley, L. L. Andersson, D. M. Cox, F. Dechery, T. Grahn, K. Hauschild, G. Henning, A. Herzan, R.-D. Herzberg, F. P. Hessberger, U. Jakobsson, P. Jones, R. Julin, S. Juutinen, S. Ketelhut, T.-L. Khoo, M. Leino, J. Ljungvall, A. Lopez-Martens, P. Nieminen, J. Pakarinen, P. Papadakis, E. Parr, P. Peura, P. Rahkila, S. Rinta-Antila, J. Rubert, P. Ruotsalainen, M. Sandzelius, J. Saren, C. Scholey, D. Seweryniak, J. Sorri, B. Sulignano, and J. Uusitalo, In-beam spectroscopy with intense ion beams: Evidence for a rotational structure in ^{246}Fm, *Phys. Rev. C.* **85**(4), 041301(R)–6 (Apr., 2012). doi: 10.1103/PhysRevC.85.041301. URL `http://link.aps.org/doi/10.1103/PhysRevC.85.041301`.
30. E. M. Ramirez, D. Ackermann, K. Blaum, M. Block, C. Droese, C. E. Dullmann, M. Dworschak, M. Eibach, S. Eliseev, E. Haettner, F. Herfurth, F. P. Hessberger, S. Hofmann, J. Ketelaer, G. Marx, M. Mazzocco, D. Nesterenko, Y. N. Novikov, W. R. Plass, D. Rodriguez, C. Scheidenberger, L. Schweikhard, P. G. Thirolf, and C. Weber, Direct mapping of

nuclear shell effects in the heaviest elements, *Science.* **337**(6099), 1207–1210 (Sept., 2012). doi: 10.1126/science.1225636. URL http://www.sciencemag.org/content/337/6099/1207.abstract.

31. P. T. Greenlees, J. Rubert, J. Piot, B. J. P. Gall, L. L. Andersson, M. Asai, Z. Asfari, D. M. Cox, F. Dechery, O. Dorvaux, T. Grahn, K. Hauschild, G. Henning, A. Herzan, R.-D. Herzberg, F. P. Hessberger, U. Jakobsson, P. Jones, R. Julin, S. Juutinen, S. Ketelhut, T.-L. Khoo, M. Leino, J. Ljungvall, A. Lopez-Martens, R. Lozeva, P. Nieminen, J. Pakarinen, P. Papadakis, E. Parr, P. Peura, P. Rahkila, S. Rinta-Antila, P. Ruotsalainen, M. Sandzelius, J. Saren, C. Scholey, D. Seweryniak, J. Sorri, B. Sulignano, C. Theisen, J. Uusitalo, and M. Venhart, Shell-structure and pairing interaction in superheavy nuclei: Rotational properties of the $Z = 104$ nucleus ^{256}Rf, *Phys. Rev. Lett.* **109**(1), 012501–5 (July, 2012). doi: 10.1103/PhysRevLett.109.012501. URL http://link.aps.org/doi/10.1103/PhysRevLett.109.012501.
32. A. Sobiczewski and K. Pomorski, Description of structure and properties of superheavy nuclei, *Prog. Part. Nucl. Phys.* **58**(1), 292–349 (Jan., 2007). ISSN 0146-6410. doi: 10.1016/j.ppnp.2006.05.001. URL http://dx.doi.org/10.1016/j.ppnp.2006.05.001.
33. S. Ćwiok, J. Dobaczewski, P. H. Heenen, P. Magierski, and W. Nazarewicz, Shell structure of the superheavy elements, *Nucl. Phys. A.* **611**(2-3), 211–246 (Dec., 1996). doi: 10.1016/S0375-9474(96)00337-5. URL http://dx.doi.org/10.1016/S0375-9474(96)00337-5.
34. K. Rutz, M. Bender, T. Burvenich, T. Schilling, P.-G. Reinhard, J. A. Maruhn, and W. Greiner, Superheavy nuclei in self-consistent nuclear calculations, *Phys. Rev. C.* **56**(1), 238–243 (July, 1997). doi: 10.1103/PhysRevC.56.238. URL http://link.aps.org/abstract/PRC/v56/p238.
35. A. T. Kruppa, M. Bender, W. Nazarewicz, P.-G. Reinhard, T. Vertse, and S. Ćwiok, Shell corrections of superheavy nuclei in self-consistent calculations, *Phys. Rev. C.* **61**(3), 034313–13 (Feb., 2000). doi: 10.1103/PhysRevC.61.034313. URL http://link.aps.org/doi/10.1103/PhysRevC.61.034313.
36. G. A. Lalazissis, M. M. Sharma, P. Ring, and Y. K. Gambhir, Superheavy nuclei in the relativistic mean-field theory, *Nucl. Phys. A.* **608**(2), 202–226 (Oct., 1996). ISSN 0375-9474. doi: 10.1016/0375-9474(96)00273-4. URL http://dx.doi.org/10.1016/0375-9474(96)00273-4.
37. W. Zhang, J. Meng, S. Zhang, L. Geng, and H. Toki, Magic numbers for superheavy nuclei in relativistic continuum Hartree–Bogoliubov theory, *Nucl. Phys. A.* **753**(1-2), 106–135 (May, 2005). ISSN 0375-9474. doi: 10.1016/j.nuclphysa.2005.02.086. URL http://dx.doi.org/10.1016/j.nuclphysa.2005.02.086.
38. S.-J. Lee, J. Fink, A. Balantekin, M. Strayer, A. Umar, P. Reinhard, J. Maruhn, and W. Greiner, Relativistic Hartree calculations for axially deformed nuclei, *Phys. Rev. Lett.* **57**(23), 2916–2919 (Dec, 1986). ISSN 0031-9007. doi: 10.1103/physrevlett.57.2916. URL http://dx.doi.org/10.1103/PhysRevLett.57.2916.

39. G. A. Lalazissis, J. Konig, and P. Ring, New parametrization for the Lagrangian density of relativistic mean field theory, *Phys. Rev. C.* **55**(1), 540–543 (Jan., 1997). doi: 10.1103/PhysRevC.55.540. URL `http://link.aps.org/abstract/PRC/v55/p540`.
40. M. M. Sharma, M. A. Nagarajan, and P. Ring, Rho meson coupling in the relativistic mean field theory and description of exotic nuclei, *Phys. Lett. B.* **312**(4), 377–381 (Aug., 1993). ISSN 0370-2693. doi: 10.1016/0370-2693(93) 90970-S. URL `http://dx.doi.org/10.1016/0370-2693(93)90970-S`.
41. Y. Sugahara and H. Toki, Relativistic mean-field theory for unstable nuclei with non-linear sigma and omega terms, *Nucl. Phys. A.* **579**(3-4), 557–572 (Oct., 1994). ISSN 0375-9474. doi: 10.1016/0375-9474(94)90923-7. URL `http://dx.doi.org/10.1016/0375-9474(94)90923-7`.
42. S. Typel and H. H. Wolter, Relativistic mean field calculations with density-dependent meson-nucleon coupling, *Nucl. Phys. A.* **656**(3-4), 331–364 (Sep, 1999). ISSN 0375-9474. doi: 10.1016/s0375-9474(99)00310-3. URL `http://dx.doi.org/10.1016/S0375-9474(99)00310-3`.
43. T. Nikšić, D. Vretenar, P. Finelli, and P. Ring, Relativistic Hartree–Bogoliubov model with density-dependent meson-nucleon couplings, *Phys. Rev. C.* **66**(2), 024306–15 (Aug., 2002). doi: 10.1103/PhysRevC.66.024306. URL `http://link.aps.org/doi/10.1103/PhysRevC.66.024306`.
44. W. Long, J. Meng, N. V. Giai, and S.-G. Zhou, New effective interactions in relativistic mean field theory with nonlinear terms and density-dependent meson-nucleon coupling, *Phys. Rev. C.* **69**(3), 034319–15 (Mar., 2004). doi: 10.1103/PhysRevC.69.034319. URL `http://link.aps.org/abstract/PRC/v69/e034319`.
45. J. Meng and P. Ring, Relativistic Hartree–Bogoliubov description of the neutron halo in ^{11}Li, *Phys. Rev. Lett.* **77**(19), 3963–3966 (Nov., 1996). doi: 10.1103/PhysRevLett.77.3963. URL `http://link.aps.org/abstract/PRL/v77/p3963`.
46. J. Meng and P. Ring, Giant halo at the neutron drip line, *Phys. Rev. Lett.* **80**(3), 460–463 (Jan., 1998). doi: 10.1103/PhysRevLett.80.460. URL `http://link.aps.org/abstract/PRL/v80/p460`.
47. J. Meng, I. Tanihata, and S. Yamaji, The proton and neutron distributions in Na isotopes: The development of halo and shell structure, *Phys. Lett. B.* **419**(1-4), 1–6 (Feb., 1998). doi: 10.1016/S0370-2693(97)01386-5. URL `http://www.sciencedirect.com/science/article/B6TVN-3VGPWM4-44/1/f83ff8aa3662905a3a6f21e8b7464573`.
48. J. Meng, S. G. Zhou, and I. Tanihata, The relativistic continuum Hartree–Bogoliubov description of charge-changing cross section for C, N, O and F isotopes, *Phys. Lett. B.* **532**(3-4), 209–214 (Apr., 2002). ISSN 0370-2693. doi: 10.1016/S0370-2693(02)01574-5. URL `http://dx.doi.org/10.1016/S0370-2693(02)01574-5`.
49. J. Meng, H. Toki, J. Y. Zeng, S. Q. Zhang, and S.-G. Zhou, Giant halo at the neutron drip line in Ca isotopes in relativistic continuum Hartree–Bogoliubov theory, *Phys. Rev. C.* **65**(4), 041302(R)–4 (Mar., 2002). doi:

10.1103/PhysRevC.65.041302. URL http://link.aps.org/abstract/PRC/v65/e041302.
50. E. D. Donets, V. A. Shchegolev, and V. A. Ermakov, The properties of the isotope 102^{254}, *Soviet Atomic Energy.* **20**(3), 257–263 (1966). ISSN 0038-531X. URL http://dx.doi.org/10.1007/BF01127385.
51. A. Ghiorso, J. M. Nitschke, J. R. Alonso, C. T. Alonso, M. Nurmia, G. T. Seaborg, E. K. Hulet, and R. W. Lougheed, Element 106, *Phys. Rev. Lett.* **33**(25), 1490–1493 (Dec., 1974). doi: 10.1103/PhysRevLett.33.1490. URL http://link.aps.org/doi/10.1103/PhysRevLett.33.1490.
52. J. Meng and N. Takigawa, Structure of superheavy elements suggested in the reaction of ^{86}Kr with ^{208}Pb, *Phys. Rev. C.* **61**(6), 064319–7 (May, 2000). doi: 10.1103/PhysRevC.61.064319. URL http://link.aps.org/abstract/PRC/v61/e064319.
53. Z. Ren and H. Toki, Superdeformation in the newly discovered superheavy elements, *Nucl. Phys. A.* **689**(3-4), 691–706 (July, 2001). ISSN 0375-9474. doi: 10.1016/S0375-9474(00)00689-8. URL http://dx.doi.org/10.1016/S0375-9474(00)00689-8.
54. W. Long, J. Meng, and S.-G. Zhou, Structure of the new nuclide ^{259}Db and its alpha-decay daughter nuclei, *Phys. Rev. C.* **65**(4), 047306–4 (Apr., 2002). doi: 10.1103/PhysRevC.65.047306. URL http://link.aps.org/abstract/PRC/v65/e047306.
55. Z. Ren, Shape coexistence in even-even superheavy nuclei, *Phys. Rev. C.* **65**(5) (Apr, 2002). ISSN 1089-490X. doi: 10.1103/physrevc.65.051304. URL http://dx.doi.org/10.1103/PhysRevC.65.051304.
56. Z. Ren, F. Tai, and D.-H. Chen, Systematic calculations of the ground state properties of superheavy nuclei, *Phys. Rev. C.* **66**(6), 064306– (Dec., 2002). doi: 10.1103/PhysRevC.66.064306. URL http://link.aps.org/doi/10.1103/PhysRevC.66.064306.
57. Z. Ren, D.-H. Chen, F. Tai, H. Y. Zhang, and W. Q. Shen, Ground state properties of odd-Z superheavy nuclei, *Phys. Rev. C.* **67**(6), 064302–12 (June, 2003). doi: 10.1103/PhysRevC.67.064302. URL http://link.aps.org/abstract/PRC/v67/e064302.
58. L. S. Geng, H. Toki, and J. Meng, α-decay chains of $^{288}_{173}115$ and $^{287}_{172}115$ in the relativistic mean field theory, *Phys. Rev. C.* **68**(6), 061303R–5 (Dec., 2003). doi: 10.1103/PhysRevC.68.061303. URL http://link.aps.org/abstract/PRC/v68/e061303.
59. Y. K. Gambhir, A. Bhagwat, and M. Gupta, Microscopic description of superheavy nuclei, *Ann. Phys. (N.Y.).* **320**(2), 429–452 (Dec., 2005). ISSN 0003-4916. doi: 10.1016/j.aop.2005.05.009. URL http://dx.doi.org/10.1016/j.aop.2005.05.009.
60. M. Bhuyan, S. K. Patra, and R. K. Gupta, Relativistic mean-field study of the properties of $Z = 117$ nuclei and the decay chains of the $^{293,294}117$ isotopes, *Phys. Rev. C.* **84**(1), 014317–7 (July, 2011). doi: 10.1103/PhysRevC.84.014317. URL http://link.aps.org/doi/10.1103/PhysRevC.84.014317.
61. V. Prassa, T. Nikšić, G. A. Lalazissis, and D. Vretenar, Relativistic en-

ergy density functional description of shape transitions in superheavy nuclei, *Phys. Rev. C.* **86**(2), 024317–12 (Aug., 2012). doi: 10.1103/PhysRevC.86.024317. URL `http://link.aps.org/doi/10.1103/PhysRevC.86.024317`.

62. S. Ahmad, M. Bhuyan, and S. K. Patra, Properties of $Z = 120$ nuclei and the alpha-decay chains of the $^{292,304}120$ isotopes using relativistic and non-relativistic formalisms, *Int. J. Mod. Phys. E.* **21**(11), 1250092– (Oct., 2012). ISSN 0218-3013. doi: 10.1142/S0218301312500929. URL `http://dx.doi.org/10.1142/S0218301312500929`.
63. W. Zhang, Z. P. Li, and S. Q. Zhang, Description of alpha-decay chains for $^{293,294}117$ within covariant density functional theory, *Phys. Rev. C.* **88**(5), 054324–6 (Nov., 2013). doi: 10.1103/PhysRevC.88.054324. URL `http://link.aps.org/doi/10.1103/PhysRevC.88.054324`.
64. A. V. Afanasjev and S. E. Agbemava, Nuclear structure theory of the heaviest nuclei, *Acta Phys. Pol. B.* **46**(3), 405–417 (2015). ISSN 1509-5770. doi: 10.5506/aphyspolb.46.405. URL `http://dx.doi.org/10.5506/APhysPolB.46.405`.
65. Z. G. Gan, Z. Qin, H. M. Fan, X. G. Lei, Y. B. Xu, J. J. He, H. Y. Liu, X. L. Wu, J. S. Guo, X. H. Zhou, S. G. Yuan, and G. M. Jin, A new alpha-particle-emitting isotope ^{259}Db, *Eur. Phys. J. A.* **10**(1), 21–25 (Feb., 2001). ISSN 1434-6001. doi: 10.1007/s100500170140. URL `http://dx.doi.org/10.1007/s100500170140`.
66. Z. G. Gan, J. S. Guo, X. L. Wu, Z. Qin, H. M. Fan, X. G. Lei, H. Y. Liu, B. Guo, H. G. Xu, R. F. Chen, C. F. Dong, F. M. Zhang, H. L. Wang, C. Y. Xie, Z. Q. Feng, Y. Zhen, L. T. Song, P. Luo, H. S. Xu, X. H. Zhou, G. M. Jin, and Z. Ren, New isotope ^{265}Bh, *Eur. Phys. J. A.* **20**(3), 385–387 (June, 2004). doi: 10.1140/epja/i2004-10020-2. URL `http://dx.doi.org/10.1140/epja/i2004-10020-2`.
67. A. N. Bezbakh, V. G. Kartavenko, G. G. Adamian, N. V. Antonenko, R. V. Jolos, and V. O. Nesterenko, Quasiparticle structure of superheavy nuclei along the α-decay chain of 115-288, *Phys. Rev. C.* **92**(1), 014329–11 (Jul, 2015). ISSN 1089-490X. doi: 10.1103/physrevc.92.014329. URL `http://dx.doi.org/10.1103/PhysRevC.92.014329`.
68. P. Möller, J. R. Nix, W. D. Myers, and W. J. Swiatecki, Nuclear ground-state masses and deformations, *At. Data Nucl. Data Tables.* **59**(2), 185–381 (Mar., 1995). doi: 10.1006/adnd.1995.1002. URL `http://dx.doi.org/10.1006/adnd.1995.1002`.
69. Y. T. Oganessian, V. K. Utyonkoy, Y. V. Lobanov, F. S. Abdullin, A. N. Polyakov, I. V. Shirokovsky, Y. S. Tsyganov, G. G. Gulbekian, S. L. Bogomolov, A. N. Mezentsev, S. Iliev, V. G. Subbotin, A. M. Sukhov, A. A. Voinov, G. V. Buklanov, K. Subotic, V. I. Zagrebaev, M. G. Itkis, J. B. Patin, K. J. Moody, J. F. Wild, M. A. Stoyer, N. J. Stoyer, D. A. Shaughnessy, J. M. Kenneally, and R. W. Lougheed. Experiments on the synthesis of element 115 in the reaction ^{243}Am(^{48}Ca,xn)$^{291-x}$115. JINR Communication E7-2003-178. URL `http://www1.jinr.ru/Preprints/2003/178(E7-2003-178).pdf` (sep, 2003).
70. Y. T. Oganessian, V. K. Utyonkoy, Y. V. Lobanov, F. S. Abdullin, A. N.

Polyakov, I. V. Shirokovsky, Y. S. Tsyganov, G. G. Gulbekian, S. L. Bogomolov, A. N. Mezentsev, S. Iliev, V. G. Subbotin, A. M. Sukhov, A. A. Voinov, G. V. Buklanov, K. Subotic, V. I. Zagrebaev, M. G. Itkis, J. B. Patin, K. J. Moody, J. F. Wild, M. A. Stoyer, N. J. Stoyer, D. A. Shaughnessy, J. M. Kenneally, and R. W. Lougheed, Experiments on the synthesis of element 115 in the reaction ^{243}Am(^{48}Ca,xn)$^{291-x}$115, *Phys. Rev. C.* **69**(2), 021601(R)–5 (Feb., 2004). doi: 10.1103/PhysRevC.69.021601. URL http://link.aps.org/abstract/PRC/v69/e021601.

71. P. Möller, J. R. Nix, and K. L. Kratz, Nuclear properties for astrophysical and radioactive-ion-beam applications, *At. Data Nucl. Data Tables.* **66**(2), 131–343 (July, 1997). ISSN 0092-640X. doi: 10.1006/adnd.1997.0746. URL http://dx.doi.org/10.1006/adnd.1997.0746.
72. I. Muntian, Z. Patyk, and A. Sobiczewski, Calculated masses of heaviest nuclei, *Phys. At. Nucl.* **66**(6), 1015–1019 (June, 2003). ISSN 1063-7788. doi: 10.1134/1.1586412. URL http://dx.doi.org/10.1134/1.1586412.
73. P. W. Zhao, Z. P. Li, J. M. Yao, and J. Meng, New parametrization for the nuclear covariant energy density functional with a point-coupling interaction, *Phys. Rev. C.* **82**(5), 054319–14 (Nov., 2010). doi: 10.1103/PhysRevC.82.054319. URL http://link.aps.org/doi/10.1103/PhysRevC.82.054319.
74. Y. T. Oganessian, F. S. Abdullin, P. D. Bailey, D. E. Benker, M. E. Bennett, S. N. Dmitriev, J. G. Ezold, J. H. Hamilton, R. A. Henderson, M. G. Itkis, Y. V. Lobanov, A. N. Mezentsev, K. J. Moody, S. L. Nelson, A. N. Polyakov, C. E. Porter, A. V. Ramayya, F. D. Riley, J. B. Roberto, M. A. Ryabinin, K. P. Rykaczewski, R. N. Sagaidak, D. A. Shaughnessy, I. V. Shirokovsky, M. A. Stoyer, V. G. Subbotin, R. Sudowe, A. M. Sukhov, R. Taylor, Y. S. Tsyganov, V. K. Utyonkov, A. A. Voinov, G. K. Vostokin, and P. A. Wilk, Eleven new heaviest isotopes of elements $Z = 105$ to $Z = 117$ identified among the products of ^{249}Bk+^{48}Ca reactions, *Phys. Rev. C.* **83**(5), 054315–14 (May, 2011). doi: 10.1103/PhysRevC.83.054315. URL http://link.aps.org/doi/10.1103/PhysRevC.83.054315.
75. Y. T. Oganessian, F. S. Abdullin, C. Alexander, J. Binder, R. A. Boll, S. N. Dmitriev, J. Ezold, K. Felker, J. M. Gostic, R. K. Grzywacz, J. H. Hamilton, R. A. Henderson, M. G. Itkis, K. Miernik, D. Miller, K. J. Moody, A. N. Polyakov, A. V. Ramayya, J. B. Roberto, M. A. Ryabinin, K. P. Rykaczewski, R. N. Sagaidak, D. A. Shaughnessy, I. V. Shirokovsky, M. V. Shumeiko, M. A. Stoyer, N. J. Stoyer, V. G. Subbotin, A. M. Sukhov, Y. S. Tsyganov, V. K. Utyonkov, A. A. Voinov, and G. K. Vostokin, Production and decay of the heaviest nuclei ^{293,294}117 and ^{294}118, *Phys. Rev. Lett.* **109** (16), 162501–5 (Oct., 2012). doi: 10.1103/PhysRevLett.109.162501. URL http://link.aps.org/doi/10.1103/PhysRevLett.109.162501.
76. A. Bohr and B. R. Mottelson, *Nuclear Structure.* vol. I, World Scientific, Singapore (1998).
77. P. Ring and P. Schuck, *The nuclear many-body problem.* Springer-Verlag, New York (1980).
78. S. Frauendorf, Spontaneous symmetry breaking in rotating nuclei, *Rev.*

Mod. Phys. **73**(2), 463–514 (June, 2001). doi: 10.1103/RevModPhys.73.463. URL http://link.aps.org/abstract/RMP/v73/p463.
79. W. Nazarewicz. The nuclear collective motion. In eds. J. Arias and M. Lozano, *Lecture Notes in Physics*, vol. 581, pp. 102–140. Springer, Berlin/Heidelberg (2001). doi: 10.1007/3-540-44620-6_4. URL http://dx.doi.org/10.1007/3-540-44620-6_4.
80. S. Ćwiok, P.-H. Heenen, and W. Nazarewicz, Shape coexistence and triaxiality in the superheavy nuclei, *Nature.* **433**(7027), 705–709 (Feb., 2005). ISSN 0028-0836. doi: 10.1038/nature03336. URL http://dx.doi.org/10.1038/nature03336.
81. P. A. Butler and W. Nazarewicz, Intrinsic reflection asymmetry in atomic nuclei, *Rev. Mod. Phys.* **68**(2), 349–421 (Apr., 1996). doi: 10.1103/RevModPhys.68.349. URL http://link.aps.org/doi/10.1103/RevModPhys.68.349.
82. J. Dudek, A. Gozdz, K. Mazurek, and H. Molique, Mean-field theory of nuclear stability and exotic point-group symmetries, *J. Phys. G: Nucl. Phys.* **37**(6), 064032–16 (June, 2010). ISSN 0954-3899. doi: 10.1088/0954-3899/37/6/064032. URL http://stacks.iop.org/0954-3899/37/i=6/a=064032.
83. C. Y. Wong, Toroidal and spherical bubble nuclei, *Ann. Phys.* **77**(1–2), 279–353 (May, 1973). ISSN 0003-4916. doi: 10.1016/0003-4916(73)90420-X. URL http://dx.doi.org/10.1016/0003-4916(73)90420-X.
84. K. Dietrich and K. Pomorski, Stability of bubble nuclei through shell effects, *Phys. Rev. Lett.* **80**(1), 37– (Jan, 1998). ISSN 1079-7114. doi: 10.1103/physrevlett.80.37. URL http://dx.doi.org/10.1103/PhysRevLett.80.37.
85. J. C. Pei, F. R. Xu, and P. D. Stevenson, Density distributions of superheavy nuclei, *Phys. Rev. C.* **71**(3), 034302–7 (Mar, 2005). ISSN 1089-490X. doi: 10.1103/physrevc.71.034302. URL http://dx.doi.org/10.1103/PhysRevC.71.034302.
86. S. K. Patra, M. Bhuyan, M. S. Mehta, and R. K. Gupta, Superdeformed and hyperdeformed states in $Z = 122$ isotopes, *Phys. Rev. C.* **80**(3), 034312–8 (Sept., 2009). doi: 10.1103/PhysRevC.80.034312. URL http://link.aps.org/doi/10.1103/PhysRevC.80.034312.
87. H. Abusara, A. V. Afanasjev, and P. Ring, Fission barriers in covariant density functional theory: Extrapolation to superheavy nuclei, *Phys. Rev. C.* **85**(2), 024314–17 (Feb., 2012). doi: 10.1103/PhysRevC.85.024314. URL http://link.aps.org/doi/10.1103/PhysRevC.85.024314.
88. Y.-S. Chen, Y. Sun, and Z.-C. Gao, Nonaxial-octupole effect in superheavy nuclei, *Phys. Rev. C.* **77**(6), 061305(R)–5 (June, 2008). doi: 10.1103/PhysRevC.77.061305. URL http://link.aps.org/doi/10.1103/PhysRevC.77.061305.
89. Y. Chen and Z. Gao, Tetrahedral symmetry in superheavy nuclei, *Nucl. Phys. Rev.* **30**(3), 278–283 (Sept., 2013). doi: 10.11804/NuclPhysRev.30.03.278. URL http://www.npr.ac.cn/qikan/Cpaper/zhaiyao.asp?bsid=16379.
90. M. Brack, J. Damgaard, A. S. Jensen, H. C. Pauli, V. M. Strutinsky, and C. Y. Wong, Funny hills: The shell-correction approach to nuclear

shell effects and its applications to the fission process, *Rev. Mod. Phys.* **44** (2), 320–405 (Apr., 1972). doi: 10.1103/RevModPhys.44.320. URL http://link.aps.org/doi/10.1103/RevModPhys.44.320.

91. V. V. Pashkevich, The energy of non-axial deformation of heavy nuclei, *Nucl. Phys. A.* **133**(2), 400–404 (Aug., 1969). ISSN 0375-9474. doi: 10.1016/0375-9474(69)90641-1. URL http://dx.doi.org/10.1016/0375-9474(69)90641-1.
92. P. Möller and S. G. Nilsson, The fission barrier and odd-multipole shape distortions, *Phys. Lett. B.* **31**(5), 283–286 (Mar., 1970). ISSN 0370-2693. doi: 10.1016/0370-2693(70)90171-1. URL http://dx.doi.org/10.1016/0370-2693(70)90171-1.
93. J. Randrup, S. E. Larsson, P. Moller, S. G. Nilsson, K. Pomorski, and A. Sobiczewski, Spontaneous-fission half-lives for even nuclei with $Z \geq 92$, *Phys. Rev. C.* **13**(1), 229–239 (Jan., 1976). doi: 10.1103/PhysRevC.13.229. URL http://link.aps.org/abstract/PRC/v13/p229.
94. T. Ledergerber and H.-C. Pauli, On the dynamics of fission: The role of reflection asymmetry in the nuclear shape, *Nucl. Phys. A.* **207**(1), 1–32 (June, 1973). ISSN 0375-9474. doi: 10.1016/0375-9474(73)90022-5. URL http://www.sciencedirect.com/science/article/pii/0375947473900225.
95. M. Girod and B. Grammaticos, Triaxial Hartree–Fock–Bogolyubov calculations with D1 effective interaction, *Phys. Rev. C.* **27**(5), 2317–2339 (May, 1983). doi: 10.1103/PhysRevC.27.2317. URL http://link.aps.org/doi/10.1103/PhysRevC.27.2317.
96. K. Rutz, J. A. Maruhn, P. G. Reinhard, and W. Greiner, Fission barriers and asymmetric ground states in the relativistic mean-field theory, *Nucl. Phys. A.* **590**(3-4), 680–702 (July, 1995). ISSN 0375-9474. doi: 10.1016/0375-9474(95)00192-4. URL http://dx.doi.org/10.1016/0375-9474(95)00192-4.
97. H. Abusara, A. V. Afanasjev, and P. Ring, Fission barriers in actinides in covariant density functional theory: The role of triaxiality, *Phys. Rev. C.* **82**(4), 044303–11 (Oct., 2010). doi: 10.1103/PhysRevC.82.044303. URL http://link.aps.org/doi/10.1103/PhysRevC.82.044303.
98. B.-N. Lu, E.-G. Zhao, and S.-G. Zhou, Potential energy surfaces of actinide nuclei from a multidimensional constrained covariant density functional theory: Barrier heights and saddle point shapes, *Phys. Rev. C.* **85** (1), 011301(R)–5 (Jan., 2012). doi: 10.1103/PhysRevC.85.011301. URL http://link.aps.org/doi/10.1103/PhysRevC.85.011301.
99. B.-N. Lu. *Multi-dimensional constrained relativistic mean field theory and the potential energy surfaces and fission barriers of actinide nuclei.* PhD thesis, Institute of Theoretical Physics, Chinese Academy of Sciences (2012). in Chinese.
100. B.-N. Lu, J. Zhao, E.-G. Zhao, and S.-G. Zhou, Potential energy surfaces of actinide and transfermium nuclei from multi-dimensional constraint covariant density functional theories, *EPJ Web Conf.* **38**, 05003 (Dec., 2012). doi: 10.1051/epjconf/20123805003. URL http://dx.doi.org/10.1051/epjconf/20123805003.

101. J. Zhao, B.-N. Lu, E.-G. Zhao, and S.-G. Zhou, Nonaxial-octupole Y_{32} correlations in N=150 isotones from multidimensional constrained covariant density functional theories, *Phys. Rev. C.* **86**(5), 057304–4 (Nov., 2012). doi: 10.1103/PhysRevC.86.057304. URL `http://link.aps.org/doi/10.1103/PhysRevC.86.057304`.
102. A. V. Afanasiev, H. Abusara, and P. Ring, Recent progress in the study of fission barriers in covariant density functional theory, *Int. J. Mod. Phys. E.* **31**(5), 1250025–10 (May, 2012). doi: 10.1142/S0218301312500255. URL `http://dx.doi.org/10.1142/S0218301312500255`.
103. A. V. Afanasjev. Fission of actinides and superheavy nuclei: covariant density functional perspective. In eds. J. H. Hamilton and A. V. Ramayya, *Fission and Properties of Neutron-Rich Nuclei*, Proceedings of the Fifth International Conference on Fission and Properties of Neutron-Rich Nuclei (ICFN5), Nov. 4-10, 2012, Sanibel Island, Florida, USA, pp. 303–310, World Scientific, Singapore (Oct, 2013). ISBN http://id.crossref.org/isbn/978-981-4525-43-5. doi: 10.1142/9789814525435_0033. URL `http://dx.doi.org/10.1142/9789814525435_0033`.
104. A. V. Afanasjev, H. Abusara, and P. Ring, Nuclear fission in covariant density functional theory, *EPJ Web of Conferences.* **62**, 03003 (2013). ISSN 2100-014X. doi: 10.1051/epjconf/20136203003. URL `http://dx.doi.org/10.1051/epjconf/20136203003`.
105. B.-N. Lu, J. Zhao, E.-G. Zhao, and S.-G. Zhou, Multidimensionally-constrained relativistic mean-field models and potential-energy surfaces of actinide nuclei, *Phys. Rev. C.* **89**(1), 014323–15 (Jan., 2014). doi: 10.1103/PhysRevC.89.014323. URL `http://link.aps.org/doi/10.1103/PhysRevC.89.014323`.
106. B.-N. Lu, J. Zhao, E.-G. Zhao, and S.-G. Zhou, Multidimensionally constrained relativistic mean field model and applications in actinide and transfermium nuclei, *Phys. Scr.* **89**(5), 054028–5 (2014). ISSN 1402-4896. doi: 10.1088/0031-8949/89/5/054028. URL `http://stacks.iop.org/1402-4896/89/i=5/a=054028`.
107. B.-N. Lu, J. Zhao, E.-G. Zhao, and S.-G. Zhou, Multi-dimensional potential energy surfaces and non-axial octupole correlations in actinide and transfermium nuclei from relativistic mean field models, *J. Phys: Conf. Ser.* **492**(1), 012014–8 (2014). ISSN 1742-6596. doi: 10.1088/1742-6596/492/1/012014. URL `http://stacks.iop.org/1742-6596/492/i=1/a=012014`.
108. J. Zhao, B.-N. Lu, D. Vretenar, E.-G. Zhao, and S.-G. Zhou, Multidimensionally constrained relativistic mean-field study of triple-humped barriers in actinides, *Phys. Rev. C.* **91**(1), 014321–9 (Jan., 2015). doi: 10.1103/PhysRevC.91.014321. URL `http://link.aps.org/doi/10.1103/PhysRevC.91.014321`.
109. B.-N. Lu et al. to be published.
110. Y. K. Gambhir, P. Ring, and A. Thimet, Relativistic mean field theory for finite nuclei, *Ann. Phys. (NY).* **198**(1), 132–179 (Feb., 1990). doi: 10.1016/0003-4916(90)90330-Q. URL `http://dx.doi.org/10.1016/0003-4916(90)90330-Q`.

111. P. Ring, Y. K. Gambhir, and G. A. Lalazissis, Computer program for the relativistic mean field description of the ground state properties of even-even axially deformed nuclei, *Comput. Phys. Commun.* **105**(1), 77–97 (Sept., 1997). ISSN 0010-4655. doi: 10.1016/S0010-4655(97)00022-2. URL http://dx.doi.org/10.1016/S0010-4655(97)00022-2.
112. S.-G. Zhou, J. Meng, and P. Ring, Spherical relativistic Hartree theory in a Woods–Saxon basis, *Phys. Rev. C.* **68**(3), 034323–12 (Sept., 2003). doi: 10.1103/PhysRevC.68.034323. URL http://link.aps.org/abstract/PRC/v68/e034323.
113. S.-G. Zhou, J. Meng, and P. Ring, Toward a deformed relativistic Hartree Bogoliubov model for exotic nuclei, *AIP Conf. Proc.* **865**(1), 90–95 (Nov., 2006). doi: 10.1063/1.2398833. URL http://link.aip.org/link/?APC/865/90/1.
114. S.-G. Zhou, J. Meng, P. Ring, and E.-G. Zhao, Neutron halo in deformed nuclei, *Phys. Rev. C.* **82**(1), 011301(R)–5 (July, 2010). doi: 10.1103/PhysRevC.82.011301. URL http://link.aps.org/doi/10.1103/PhysRevC.82.011301.
115. L. Li, J. Meng, P. Ring, E.-G. Zhao, and S.-G. Zhou, Deformed relativistic Hartree–Bogoliubov theory in continuum, *Phys. Rev. C.* **85**(2), 024312–17 (Feb., 2012). doi: 10.1103/PhysRevC.85.024312. URL http://link.aps.org/doi/10.1103/PhysRevC.85.024312.
116. Y. Chen, L. Li, H. Liang, and J. Meng, Density-dependent deformed relativistic Hartree–Bogoliubov theory in continuum, *Phys. Rev. C.* **85**(6), 067301–5 (June, 2012). doi: 10.1103/PhysRevC.85.067301. URL http://link.aps.org/doi/10.1103/PhysRevC.85.067301.
117. L.-S. Geng, J. Meng, and H. Toki, Reflection asymmetric relativistic mean field approach and its application to the octupole deformed nucleus ^{226}Ra, *Chin. Phys. Lett.* **24**(7), 1865–1868 (July, 2007). ISSN 0256-307X. doi: 10.1088/0256-307X/24/7/021. URL http://dx.doi.org/10.1088/0256-307X/24/7/021.
118. M. Warda, J. L. Egido, L. M. Robledo, and K. Pomorski, Self-consistent calculations of fission barriers in the Fm region, *Phys. Rev. C.* **66**(1), 014310–11 (July, 2002). doi: 10.1103/PhysRevC.66.014310. URL http://link.aps.org/doi/10.1103/PhysRevC.66.014310.
119. S. Karatzikos, A. V. Afanasjev, G. A. Lalazissis, and P. Ring, The fission barriers in Actinides and superheavy nuclei in covariant density functional theory, *Phys. Lett. B.* **689**(2-3), 72–81 (May, 2010). ISSN 0370-2693. doi: 10.1016/j.physletb.2010.04.045. URL http://dx.doi.org/10.1016/j.physletb.2010.04.045.
120. J. Y. Zeng and T. S. Cheng, Particle-number-conserving method for treating the nuclear pairing correlation, *Nucl. Phys. A.* **405**(1), 1–28 (Aug., 1983). ISSN 0375-9474. doi: 10.1016/0375-9474(83)90320-2. URL http://dx.doi.org/10.1016/0375-9474(83)90320-2.
121. H. Molique and J. Dudek, Fock-space diagonalization of the state-dependent pairing Hamiltonian with the Woods–Saxon mean field, *Phys. Rev. C.* **56**(4), 1795–1813 (Oct., 1997). doi: 10.1103/PhysRevC.56.1795. URL http://link.aps.org/doi/10.1103/PhysRevC.56.1795.

122. J. Meng, J.-Y. Guo, L. Liu, and S.-Q. Zhang, Shell-model-like approach (slap) for the nuclear properties in relativistic mean field theory, *Frontiers Phys. China.* **1**(1), 38–46 (Jan., 2006). ISSN 1673-3487. doi: 10.1007/s11467-005-0013-5. URL `http://dx.doi.org/10.1007/s11467-005-0013-5`.
123. N. Pillet, P. Quentin, and J. Libert, Pairing correlations in an explicitly particle-number conserving approach, *Nucl. Phys. A.* **697**(1-2), 141–163 (Jan., 2002). ISSN 0375-9474. doi: 10.1016/S0375-9474(01)01240-4. URL `http://www.sciencedirect.com/science/article/pii/S0375947401012404`.
124. T. V. Nhan Hao, P. Quentin, and L. Bonneau, Parity restoration in the highly truncated diagonalization approach: Application to the outer fission barrier of ^{240}Pu, *Phys. Rev. C.* **86**(6), 064307–8 (Dec., 2012). doi: 10.1103/PhysRevC.86.064307. URL `http://link.aps.org/doi/10.1103/PhysRevC.86.064307`.
125. Z.-H. Zhang, J.-Y. Zeng, E.-G. Zhao, and S.-G. Zhou, Particle-number conserving analysis of rotational bands in ^{247,249}Cm and ^{249}Cf, *Phys. Rev. C.* **83**(1), 011304(R)–5 (Jan., 2011). doi: 10.1103/PhysRevC.83.011304. URL `http://link.aps.org/doi/10.1103/PhysRevC.83.011304`.
126. Z.-H. Zhang, X.-T. He, J.-Y. Zeng, E.-G. Zhao, and S.-G. Zhou, Systematic investigation of the rotational bands in nuclei with $Z \approx 100$ using a particle-number conserving method based on a cranked shell model, *Phys. Rev. C.* **85**(1), 014324–17 (Jan., 2012). doi: 10.1103/PhysRevC.85.014324. URL `http://link.aps.org/doi/10.1103/PhysRevC.85.014324`.
127. Y. Tian and Z.-Y. Ma, A separable pairing force in nuclear matter, *Chin. Phys. Lett.* **23**(12), 3226–3229 (Dec., 2006). ISSN 0256-307X. doi: 10.1088/0256-307X/23/12/029. URL `http://stacks.iop.org/0256-307X/23/i=12/a=029`.
128. Y. Tian, Z. Ma, and P. Ring, A finite range pairing force for density functional theory in superfluid nuclei, *Phys. Lett. B.* **676**(1-3), 44–50 (June, 2009). ISSN 0370-2693. doi: 10.1016/j.physletb.2009.04.067. URL `http://dx.doi.org/10.1016/j.physletb.2009.04.067`.
129. Y. Tian, Z.-y. Ma, and P. Ring, Separable pairing force for relativistic quasiparticle random-phase approximation, *Phys. Rev. C.* **79**(6), 064301–7 (June, 2009). doi: 10.1103/PhysRevC.79.064301. URL `http://link.aps.org/doi/10.1103/PhysRevC.79.064301`.
130. B.-N. Lu, E.-G. Zhao, and S.-G. Zhou, Quadrupole deformation (β, γ) of light Λ hypernuclei in a constrained relativistic mean field model: Shape evolution and shape polarization effect of the Λ hyperon, *Phys. Rev. C.* **84**(1), 014328–10 (July, 2011). doi: 10.1103/PhysRevC.84.014328. URL `http://link.aps.org/doi/10.1103/PhysRevC.84.014328`.
131. B.-N. Lu, E. Hiyama, H. Sagawa, and S.-G. Zhou, Superdeformed Λ hypernuclei within relativistic mean field models, *Phys. Rev. C.* **89**(04), 044307–7 (Apr, 2014). doi: 10.1103/PhysRevC.89.044307. URL `http://link.aps.org/doi/10.1103/PhysRevC.89.044307`.
132. X. Li and J. Dudek, Deformed atomic nuclei with degeneracies of the nu-

cleonic levels higher than 2, *Phys. Rev. C.* **49**(3), R1250–R1252 (Mar., 1994). doi: 10.1103/PhysRevC.49.R1250. URL `http://link.aps.org/doi/10.1103/PhysRevC.49.R1250`.

133. J. Dudek, A. Gozdz, N. Schunck, and M. Miskiewicz, Nuclear tetrahedral symmetry: Possibly present throughout the periodic table, *Phys. Rev. Lett.* **88**(25), 252502–4 (June, 2002). doi: 10.1103/PhysRevLett.88.252502. URL `http://link.aps.org/doi/10.1103/PhysRevLett.88.252502`.
134. J. Dudek, A. Gozdz, and N. Schunck, Atomic nuclei with tetrahedral and octahedral symmetries, *Acta Phys. Pol. B.* **34**(4), 2491– (Apr., 2003). URL `http://th-www.if.uj.edu.pl/acta/vol34/abs/v34p2491.htm`.
135. J. Dudek, J. Dobaczewski, N. Dubray, A. Gozdz, V. Pangon, and N. Schunck, Nuclei with tetrahedral symmetry, *Int. J. Mod. Phys. E.* **16** (2), 516–532 (2007). doi: 10.1142/S0218301307005958. URL `http://dx.doi.org/10.1142/S0218301307005958`.
136. N. Schunck, J. Dudek, A. Gozdz, and P. H. Regan, Tetrahedral symmetry in ground and low-lying states of exotic $A \sim 110$ nuclei, *Phys. Rev. C.* **69**(6), 061305(R)–5 (June, 2004). doi: 10.1103/PhysRevC.69.061305. URL `http://link.aps.org/doi/10.1103/PhysRevC.69.061305`.
137. J. Dudek, D. Curien, N. Dubray, J. Dobaczewski, V. Pangon, P. Olbratowski, and N. Schunck, Island of rare earth nuclei with tetrahedral and octahedral symmetries: Possible experimental evidence, *Phys. Rev. Lett.* **97**(7), 072501–4 (Aug., 2006). doi: 10.1103/PhysRevLett.97.072501. URL `http://link.aps.org/doi/10.1103/PhysRevLett.97.072501`.
138. M. Yamagami, K. Matsuyanagi, and M. Matsuo, Symmetry-unrestricted Skyrme–Hartree–Fock–Bogoliubov calculations for exotic shapes in $N = Z$ nuclei from ^{64}Ge to ^{84}Mo, *Nucl. Phys. A.* **693**(3-4), 579–602 (Oct., 2001). ISSN 0375-9474. doi: 10.1016/S0375-9474(01)00918-6. URL `http://dx.doi.org/10.1016/S0375-9474(01)00918-6`.
139. P. Olbratowski, J. Dobaczewski, P. Powalowski, M. Sadziak, and K. Zberecki, Skyrme–Hartree–Fock and Hartree–Fock–Bogolyubov calculations for nuclei with tetrahedral deformation, *Int. J. Mod. Phys. E.* **15** (2), 333–338 (2006). doi: 10.1142/S021830130600417X. URL `http://dx.doi.org/10.1142/S021830130600417X`.
140. K. Zberecki, P.-H. Heenen, and P. Magierski, Nuclear tetrahedral configurations at spin zero, *Phys. Rev. C.* **79**(1), 014319–7 (Jan., 2009). doi: 10.1103/PhysRevC.79.014319. URL `http://link.aps.org/doi/10.1103/PhysRevC.79.014319`.
141. S. Takami, K. Yabana, and M. Matsuo, Tetrahedral and triangular deformations of $Z = N$ nuclei in mass region $A \sim 60 - 80$, *Phys. Lett. B.* **431**(3–4), 242–248 (July, 1998). ISSN 0370-2693. doi: 10.1016/S0370-2693(98)00545-0. URL `http://www.sciencedirect.com/science/article/pii/S0370269398005450`.
142. K. Zberecki, P. Magierski, P.-H. Heenen, and N. Schunck, Tetrahedral correlations in ^{80}Zr and ^{98}Zr, *Phys. Rev. C.* **74**(5), 051302(R)–5 (Nov., 2006). doi: 10.1103/PhysRevC.74.051302. URL `http://link.aps.org/doi/10.1103/PhysRevC.74.051302`.

143. Z.-C. Gao, Y.-S. Chen, and J. Meng, Rotational band structures of non-axial octupole deformed shapes, *Chin. Phys. Lett.* **21**(5), 806–809 (May, 2004). ISSN 0256-307X. doi: 10.1088/0256-307X/21/5/012. URL http://stacks.iop.org/0256-307X/21/i=5/a=012.
144. A. P. Robinson, T. L. Khoo, I. Ahmad, S. K. Tandel, F. G. Kondev, T. Nakatsukasa, D. Seweryniak, M. Asai, B. B. Back, M. P. Carpenter, P. Chowdhury, C. N. Davids, S. Eeckhaudt, J. P. Greene, P. T. Greenlees, S. Gros, A. Heinz, R.-D. Herzberg, R. V. F. Janssens, G. D. Jones, T. Lauritsen, C. J. Lister, D. Peterson, J. Qian, U. S. Tandel, X. Wang, and S. Zhu, $K^\pi = 8^-$ isomers and $K^\pi = 2^-$ octupole vibrations in $N = 150$ shell-stabilized isotones, *Phys. Rev. C.* **78**(3), 034308–6 (Sept., 2008). doi: 10.1103/PhysRevC.78.034308. URL http://link.aps.org/doi/10.1103/PhysRevC.78.034308.
145. R. V. Jolos, L. A. Malov, N. Y. Shirikova, and A. V. Sushkov, Structure of some low lying two-quasiparticle and collective states in nuclei with Z $\sim$ 100 considered in the quasiparticle phonon model, *J. Phys. G: Nucl. Part. Phys.* **38**(11), 115103– (2011). ISSN 0954-3899. doi: 10.1088/0954-3899/38/11/115103. URL http://stacks.iop.org/0954-3899/38/i=11/a=115103.
146. L.-S. Geng, J. Meng, H. Toki, W.-H. Long, and G. Shen, Spurious shell closures in the relativistic mean field model, *Chin. Phys. Lett.* **23**(5), 1139– (2006). ISSN 0256-307X. URL http://stacks.iop.org/0256-307X/23/i=5/a=021.
147. P. Möller, Odd-multipole shape distortions and the fission barriers of elements in the region $84 \leq Z \leq 120$, *Nucl. Phys. A.* **192**(3), 529–580 (Sept., 1972). ISSN 0375-9474. doi: 10.1016/0375-9474(72)90090-5. URL http://www.sciencedirect.com/science/article/pii/0375947472900905.
148. P. Möller and J. R. Nix. Calculation of fission barriers. In *Proceedings of the Third IAEA Symposium on Physics and Chemistry of Fission, Rochester, New York, 13-17 August 1973*, vol. 1, pp. 103–140, International Atomic Energy Agency, Vienna (1974). URL http://www-nds.iaea.org/publications/item.php?key=INDCSpecial1.
149. P. Möller and J. Nix, Calculation of fission barriers with the droplet model and folded Yukawa single-particle potential, *Nucl. Phys. A.* **229**(2), 269–291 (Sept., 1974). ISSN 0375-9474. doi: 10.1016/0375-9474(74)90788-X. URL http://www.sciencedirect.com/science/article/pii/037594747490788X.
150. R. Bengtsson, I. Ragnarsson, S. Aberg, A. Gyurkovich, A. Sobiczewski, and K. Pomorski, Properties of nuclei at the third-minimum deformation, *Nucl. Phys. A.* **473**(1), 77–110 (Oct., 1987). ISSN 0375-9474. doi: 10.1016/0375-9474(87)90156-4. URL http://www.sciencedirect.com/science/article/pii/0375947487901564.
151. S. Ćwiok, W. Nazarewicz, J. Saladin, W. Plociennik, and A. Johnson, Hyperdeformations and clustering in the actinide nuclei, *Phys. Lett. B.* **322**(4), 304–310 (Feb., 1994). ISSN 0370-2693. doi: 10.1016/0370-2693(94)91157-6. URL http://www.sciencedirect.com/science/article/pii/0370269394911576.

152. M. Kowal and J. Skalski, Examination of the existence of third, hyper-deformed minima in actinide nuclei, *Phys. Rev. C.* **85**(6), 061302(R)–4 (June, 2012). doi: 10.1103/PhysRevC.85.061302. URL `http://link.aps.org/doi/10.1103/PhysRevC.85.061302`.
153. P. Jachimowicz, M. Kowal, and J. Skalski, Eight-dimensional calculations of the third barrier in ^{232}Th, *Phys. Rev. C.* **87**(4), 044308–4 (Apr., 2013). doi: 10.1103/PhysRevC.87.044308. URL `http://link.aps.org/doi/10.1103/PhysRevC.87.044308`.
154. T. Ichikawa, P. Möller, and A. J. Sierk, Character and prevalence of third minima in actinide fission barriers, *Phys. Rev. C.* **87**(5), 054326–4 (May, 2013). doi: 10.1103/PhysRevC.87.054326. URL `http://link.aps.org/doi/10.1103/PhysRevC.87.054326`.
155. P. Möller, A. J. Sierk, T. Ichikawa, A. Iwamoto, R. Bengtsson, H. Uhrenholt, and S. Aberg, Heavy-element fission barriers, *Phys. Rev. C.* **79**(6), 064304–38 (June, 2009). doi: 10.1103/PhysRevC.79.064304. URL `http://link.aps.org/abstract/PRC/v79/e064304`.
156. L. Bonneau, P. Quentin, and D. Samsoen, Fission barriers of heavy nuclei within a microscopic approach, *Eur. Phys. J. A.* **21**(3), 391–406 (Sept., 2004). ISSN 1434-6001. doi: 10.1140/epja/i2003-10224-x. URL `http://dx.doi.org/10.1140/epja/i2003-10224-x`.
157. M. Samyn, S. Goriely, and J. M. Pearson, Further explorations of Skyrme–Hartree–Fock–Bogoliubov mass formulas. V. extension to fission barriers, *Phys. Rev. C.* **72**(4), 044316–24 (Oct., 2005). doi: 10.1103/PhysRevC.72.044316. URL `http://link.aps.org/doi/10.1103/PhysRevC.72.044316`.
158. J. D. McDonnell, W. Nazarewicz, and J. A. Sheikh, Third minima in thorium and uranium isotopes in a self-consistent theory, *Phys. Rev. C.* **87**(5), 054327–8 (May, 2013). doi: 10.1103/PhysRevC.87.054327. URL `http://link.aps.org/doi/10.1103/PhysRevC.87.054327`.
159. J. Berger, M. Girod, and D. Gogny, Constrained Hartree–Fock and beyond, *Nucl. Phys. A.* **502**(0), 85–104 (Oct., 1989). ISSN 0375-9474. doi: 10.1016/0375-9474(89)90656-8. URL `http://www.sciencedirect.com/science/article/pii/0375947489906568`.
160. J.-P. Delaroche, M. Girod, H. Goutte, and J. Libert, Structure properties of even-even actinides at normal and super deformed shapes analysed using the Gogny force, *Nucl. Phys. A.* **771**, 103–168 (May, 2006). ISSN 0375-9474. doi: 10.1016/j.nuclphysa.2006.03.004. URL `http://dx.doi.org/10.1016/j.nuclphysa.2006.03.004`.
161. N. Dubray, H. Goutte, and J.-P. Delaroche, Structure properties of ^{226}Th and ^{256,258,260}Fm fission fragments: Mean-field analysis with the Gogny force, *Phys. Rev. C.* **77**(1), 014310–13 (Jan., 2008). doi: 10.1103/PhysRevC.77.014310. URL `http://link.aps.org/doi/10.1103/PhysRevC.77.014310`.
162. R. Capote, M. Herman, P. Oblozinsky, P. Young, S. Goriely, T. Belgya, A. Ignatyuk, A. Koning, S. Hilaire, V. Plujko, M. Avrigeanu, O. Bersillon, M. Chadwick, T. Fukahori, Z. Ge, Y. Han, S. Kailas, J. Kopecky, V. Maslov, G. Reffo, M. Sin, E. Soukhovitskii, and P. Talou, RIPL reference input pa-

rameter library for calculation of nuclear reactions and nuclear data evaluations, *Nucl. Data Sheets.* **110**(12), 3107–3214 (Dec., 2009). ISSN 0090-3752. doi: 10.1016/j.nds.2009.10.004. URL `http://dx.doi.org/10.1016/j.nds.2009.10.004`.

163. B. Singh, R. Zywina, and R. B. Firestone, Table of superdeformed nuclear bands and fission isomers: Third edition (october 2002), *Nucl. Data Sheets.* **97**(2), 241–592 (Oct., 2002). ISSN 0090-3752. doi: 10.1006/ndsh.2002.0018. URL `http://www.sciencedirect.com/science/article/pii/S0090375202900189`.

164. P. Jachimowicz, M. Kowal, and J. Skalski, Secondary fission barriers in even-even actinide nuclei, *Phys. Rev. C.* **85**(3), 034305–6 (Mar., 2012). doi: 10.1103/PhysRevC.85.034305. URL `http://link.aps.org/doi/10.1103/PhysRevC.85.034305`.

Chapter 6

Relativistic symmetries in nuclear single-particle spectra

Jian-You Guo

School of Physics and Material Science, Anhui University,
Hefei 230601, China
jianyou@ahu.edu.cn

Hao Zhao Liang

RIKEN Nishina Center, Wako 351-0198, Japan
haozhao.liang@riken.jp

Jie Meng

School of Physics, Peking University, Beijing 100871, China
mengj@pku.edu.cn

Shan-Gui Zhou

State Key Laboratory of Theoretical Physics,
Institute of Theoretical Physics,
Chinese Academy of Sciences, Beijing 100190, China
Center of Theoretical Nuclear Physics,
National Laboratory of Heavy Ion Accelerator,
Lanzhou 730000, China
sgzhou@itp.ac.cn

Symmetry is a fundamental concept in quantum physics. The quasi-degeneracy between single-particle orbitals $(n, l, j = l + 1/2)$ and $(n - 1, l + 2, j = l + 3/2)$ indicates a hidden symmetry in atomic nuclei, the so-called pseudospin symmetry. Since the pseudospin symmetry was recognized as a relativistic symmetry in 1990s, many special features, including the spin symmetry for anti-nucleons, and many new concepts have been introduced. In this Chapter, we will illustrate the schematic picture of spin and pseudospin symmetries, derive the basic formalism, highlight the recent progress from several different aspects, and discuss selected open issues in this topic.

Contents

1. Introduction

As early as the first half of the 20th century, nuclear physicists and chemists found that the nuclei with nucleon number 2, 8, 20, 28, 50, or 82 are extraordinarily stable. That such so-called magic numbers exist was first remarked by Elsasser in 1933.[1] Besides the stability, the magic numbers manifest themselves as a sudden jump in the plots of two-nucleon separation energies, α-decay half-lives, neutron-capture cross sections, and isotope shifts as functions of nucleon number. The magic numbers also appear as peaks in the abundance pattern of isotopes in the solar system in astrophysics.

Based on these properties, it was conjectured that protons and neutrons in a nucleus form shell structures and the magic numbers correspond to the occupation numbers in a full shell, like that of electrons in an atom. In order to understand these magic numbers, the Schrödinger equations with some special potentials, such as the square-well and harmonic-oscillator potentials, were solved to reproduce the shell structure but without success. In 1949, Haxel, Jensen, and Suess[2] and Mayer[3] introduced a strong spin-orbit (SO) potential to the usual mean field, e.g., the harmonic-oscillator, square-well, or Woods–Saxon potential. The solutions of the corresponding Schrödinger equations reproduced all traditional magic numbers in nuclei excellently. From then on, the shell model with strong spin-orbit potential became the strongest candidate of the standard nuclear model, which not only reproduces the magic numbers, but also provides satisfactory descriptions of nuclear ground-state and some low-lying excited-state properties. In other words, the substantial spin symmetry (SS) breaking between the spin doublets $(n, l, j = l \pm 1/2)$ is one of the most important concepts in nuclear structure. Even nowadays, searching for new magic numbers and investigating shell-structure evolution for unstable nuclei are still the key topics for the radioactive ion beam facilities worldwide.[4]

The success of the shell model made the spin-orbit coupling attract tremendous attention in nuclear physics as well as other fields. The spin symmetry is broken with the introduction of the spin-orbit coupling, but a new symmetry comes into being in nuclear single-particle spectra. In 1969, Hecht and Adler[5] and Arima, Harvey, and Shimizu[6] re-examined the single-particle spectra in heavy nuclei, and they discovered that pairs of single-particle states with the quantum numbers $(n, l, j = l + 1/2)$ and $(n-1, l+2, j = l+3/2)$ are always nearly degenerate. In order to understand such near degeneracy, they introduced the so-called pseudospin symmetry (PSS) by defining new quantum numbers $(\tilde{n} = n - 1, \tilde{l} = l + 1, j = \tilde{l} \pm 1/2)$ as pseudospin doublets. The introduction of pseudospin concept has led to successful interpretations explicitly or implicitly a large number of phenomena in nuclei, such as nuclear superdeformed configurations,[7] identical bands,[8,9] quantized alignment,[10] pseudospin partner bands,[11,12] and so on. In addition, the relevance of PSS in the structure of halo nuclei[13] and superheavy nuclei[14,15] was pointed out. Due to these reasons, probing the origin of the spin and pseudospin symmetries and their breaking mechanisms has been a hot topic. More details can be found in reviews by Ginocchio[16] and by Liang, Meng, and Zhou.[17]

In this Chapter, we will first illustrate the schematic picture of spin and pseudospin symmetries in Section 2. The typical Dirac equation used in nuclear physics will be presented together with its Schrödinger-like equations in Section 3. The PSS in realistic nuclei will be checked in Section 4. Section 5 will be devoted to highlighting the progress during the last decade from several different aspects, and also to discussing the selected open issues. Summary and perspectives will be given in Section 6.

2. Schematic picture of spin and pseudospin symmetries

In order to see clearly the spin and pseudospin symmetries, we recall the description of shell model for the single-particle levels.[2,3] According to the idea of shell model, each nucleon is assumed to move independently in a

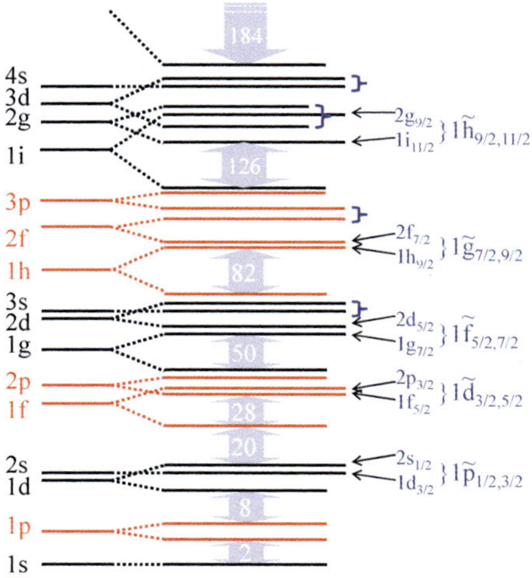

Fig. 1. Schematic nuclear single-particle spectrum. On the one hand, the strong spin-orbit splittings between the spin doublets $(n, l, j = l \pm 1/2)$ lead to the traditional magic numbers. On the other hand, pairs of single-particle states in braces, $(n, l, j = l + 1/2)$ and $(n - 1, l + 2, j = l + 3/2)$, are quasi-degenerate. They are defined as the pseudospin doublets $(\tilde{n} = n - 1, \tilde{l} = l + 1, j = \tilde{l} \pm 1/2)$, and the pseudospin symmetry is introduced for such near degeneracy. Taken from Ref. 17.

mean field, e.g., harmonic-oscillator, square-well, or Woods–Saxon potential, with a strong spin-orbit potential. The solutions of the corresponding equation present the single-particle spectrum with a clear shell structure, as illustrated in Fig. 1 for a schematic example. It is clearly seen that the remarkable energy splittings between the spin doublets $(n, l, j = l \pm 1/2)$ lead to the traditional magic numbers. On the other hand, pairs of single-particle states with quantum numbers $(n, l, j = l+1/2)$ and $(n-1, l+2, j = l+3/2)$ are found to be nearly degenerate. These pairs of states are defined as the pseudospin doublets $(\tilde{n} = n - 1, \tilde{l} = l + 1, j = \tilde{l} \pm 1/2)$, and the PSS is introduced for such near degeneracy.

In the above notation, the spherical shape of nuclei is assumed; however, it is well known that most nuclei in the nuclide chart are deformed. In 1969, Nilsson *et al.*[18] extended the idea of spherical shell model to the deformed nuclei, and calculated the corresponding single-particle spectra. Assuming axial symmetry, the parity π and the projection of angular momentum on the third axis Ω are good quantum numbers. The single-particle orbitals can be labeled as $\Omega[Nn_3\Lambda]$ in the asymptotic Nilsson quantum numbers. In Ref. 19, Bohr, Hamamoto, and Mottelson investigated the SS and PSS for the axially deformed nuclei, and pointed out that the single-particle states with the Nilsson quantum numbers ($\Omega = \Lambda + 1/2[N, n_3, \Lambda]$) and ($\Omega = \Lambda + 3/2[N, n_3, \Lambda + 2]$) are quasi-degenerate and they can be regarded as the pseudospin doublets ($\Omega = \tilde{\Lambda} \pm 1/2[\tilde{N} = N - 1, \tilde{n}_3 = n_3, \tilde{\Lambda} = \Lambda + 1]$). For deformed nuclei, the existence of the quasi-degenerate pseudospin doublets near the Fermi surface was verified in Ref. 20, and the PSS of Dirac eigenfunctions from the relativistic mean-field (RMF) calculations was examined extensively in Ref. 21.

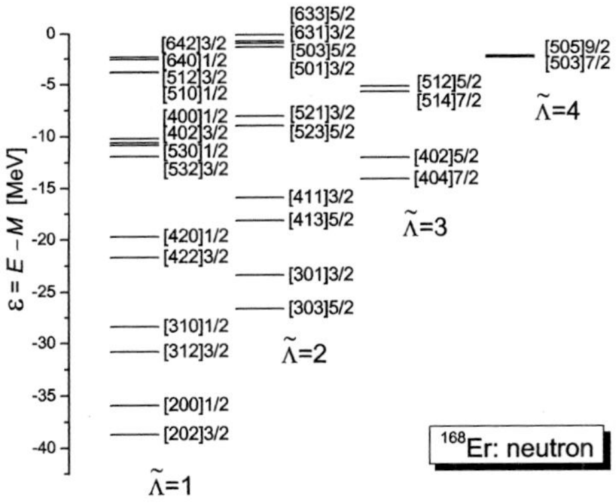

Fig. 2. Single-neutron spectra for the pseudospin doublets in the deformed nucleus ^{168}Er. Taken from Ref. 21.

The single-neutron spectra for the axially deformed nucleus ^{168}Er is plotted in Fig. 2. From left to right the states with the pseudo-orbital angular momentum projection $\tilde{\Lambda} = 1, 2, 3$, and 4, are shown respectively. The energy splittings between the pseudospin doublets decrease as the single-particle binding energies decrease. For the pseudospin doublets with binding energy larger than 5 MeV, the pseudospin-down (spin-up) state is higher than the pseudospin-up (spin-down) one. In contrast, the opposite is observed for the bound doublets with binding energy less than 5 MeV. This indicates that the PSS for deformed nuclei depends on the single-particle energy. With this intuitive knowledge, in the following we will present the basic theoretical formalism on the spin and pseudospin symmetries to deepen our understanding of these symmetries.

3. Theoretical formalism

Since the introduction of the PSS concept in atomic nuclei, there have been comprehensive efforts to understand its origin. Apart from the formal relabelling of quantum numbers, various explicit transformations from the normal scheme to the pseudospin scheme have been proposed.[19,22,23] Although these transformations have provided us with an intuitional understanding of the PSS, its origin and breaking mechanism are still ambiguous. In 1997, Ginocchio[24] pointed out that the PSS is essentially a relativistic symmetry of the Dirac Hamiltonian and the pseudo-orbital angular momentum $\tilde{l}$ is nothing but the orbital angular momentum of the lower component of the Dirac spinor. Following this idea, we show the basic formalism in the following.

3.1. *Dirac equation*

Based on the covariant density functional theory (CDFT),[25] the Dirac equation for the motion of nucleons can be written as

$$H\psi(\mathbf{r}) = \epsilon\psi(\mathbf{r}) \,. \tag{1}$$

Here ϵ is the single-particle energy including the rest mass of nucleon M, and H is the Dirac Hamiltonian in the form of

$$H = \boldsymbol{\alpha} \cdot \mathbf{p} + \beta[M + S(\mathbf{r})] + V(\mathbf{r}) \,, \tag{2}$$

where $\boldsymbol{\alpha}$ and β are the Dirac matrices, $S(\mathbf{r})$ and $V(\mathbf{r})$ are the scalar and vector potentials, respectively. In this Chapter, natural units $\hbar = c = 1$ is adopted.

When spherical symmetry is imposed, the single-particle eigenstates are specified by a set of quantum numbers (n, l, j, m), and the single-particle wave functions can be factorized as

$$\psi(\mathbf{r}) = \frac{1}{r}\begin{pmatrix} iG(r) \\ F(r)\hat{\boldsymbol{\sigma}} \cdot \hat{\mathbf{r}} \end{pmatrix} Y^l_{jm}(\hat{\mathbf{r}}) \,, \tag{3}$$

with the spherical harmonics spinor $Y^l_{jm}(\hat{\mathbf{r}})$ for the angular and spin parts.[26]

For the lower component of the Dirac spinor (3), one has

$$\hat{\boldsymbol{\sigma}} \cdot \hat{\mathbf{r}} Y^l_{jm}(\hat{\mathbf{r}}) = -Y^{\tilde{l}}_{jm}(\hat{\mathbf{r}}) \,, \tag{4}$$

with

$$\tilde{l} = 2j - l \,. \tag{5}$$

Thus, the single-particle wave functions can also be expressed as

$$\psi(\mathbf{r}) = \frac{1}{r}\begin{pmatrix} iG(r)Y^{l}_{jm}(\hat{\mathbf{r}}) \\ -F(r)Y^{\tilde{l}}_{jm}(\hat{\mathbf{r}}) \end{pmatrix}. \tag{6}$$

In such a way, the pseudo-orbital angular momentum $\tilde{l}$ is found to be the orbital angular momentum of the lower component of the Dirac spinor.[24]

The corresponding radial Dirac equation reads

$$\begin{pmatrix} M+\Sigma(r) & -\frac{d}{dr}+\frac{\kappa}{r} \\ \frac{d}{dr}+\frac{\kappa}{r} & -M+\Delta(r) \end{pmatrix}\begin{pmatrix} G(r) \\ F(r) \end{pmatrix} = \epsilon \begin{pmatrix} G(r) \\ F(r) \end{pmatrix}, \tag{7}$$

where $\Sigma(r) = V(r)+S(r)$ and $\Delta(r) = V(r)-S(r)$ denote the combinations of the vector and scalar potentials, and κ is a good quantum number defined as $\kappa = \mp(j+1/2)$ for the $j = l \pm 1/2$ orbitals. In Eq. (7), it is not straightforward to observe the symmetries of the Dirac Hamiltonian. In order to uncover the symmetries hidden in the Dirac Hamiltonian, one can transform the Dirac equation into a Schrödinger-like form.

3.2. *Schrödinger-like equations*

Equation (7) is a set of coupling equations, and it can be decoupled to the Schrödinger-like equation for the upper component $G(r)$ or lower component $F(r)$. By substituting the lower component $F(r)$ with

$$F(r) = \frac{1}{M-\Delta(r)+\epsilon}\left(\frac{d}{dr}+\frac{\kappa}{r}\right)G(r), \tag{8}$$

in Eq. (7), one can derive the Schrödinger-like equation for the upper component $G(r)$ as

$$\left\{-\frac{1}{M_+}\frac{d^2}{dr^2}+\frac{1}{M_+^2}\frac{dM_+}{dr}\frac{d}{dr}+\left[(M+\Sigma)+\frac{1}{M_+}\frac{\kappa(\kappa+1)}{r^2}+\frac{1}{M_+^2}\frac{dM_+}{dr}\frac{\kappa}{r}\right]\right\}G = \epsilon G, \tag{9}$$

with the energy-dependent effective mass $M_+(r) = M-\Delta(r)+\epsilon$. Similarly, one can derive the Schrödinger-like equation for the lower component $F(r)$ by using

$$G(r) = \frac{1}{-M-\Sigma(r)+\epsilon}\left(-\frac{d}{dr}+\frac{\kappa}{r}\right)F(r), \tag{10}$$

and obtain

$$\left\{-\frac{1}{M_-}\frac{d^2}{dr^2}+\frac{1}{M_-^2}\frac{dM_-}{dr}\frac{d}{dr}+\left[(-M+\Delta)+\frac{1}{M_-}\frac{\kappa(\kappa-1)}{r^2}-\frac{1}{M_-^2}\frac{dM_-}{dr}\frac{\kappa}{r}\right]\right\}F=\epsilon F, \tag{11}$$

with the energy-dependent effective mass $M_-(r) = -M - \Sigma(r) + \epsilon$.

For Eq. (9), in analogy with the Schrödinger equation, $\Sigma(r)$ is the central potential in which particles move; the term proportional to $l(l+1) = \kappa(\kappa+1)$ corresponds to the centrifugal barrier; and the last term corresponds to the SO potential, which leads to the substantial SO splittings in nuclear single-particle spectra. Namely,

$$V_{\rm CB}(r) = \frac{1}{M_+(r)}\frac{\kappa(\kappa+1)}{r^2} \qquad \text{and} \qquad V_{\rm SO}(r) = \frac{1}{M_+^2(r)}\frac{dM_+(r)}{dr}\frac{\kappa}{r}. \tag{12}$$

It is clear that there is no SO splitting if $V_{\rm SO}$ vanishes. This indicates the condition,

$$\frac{dM_+(r)}{dr} = \frac{d\Delta(r)}{dr} = 0\,, \tag{13}$$

is the SS limit.

Analogically, if one uses the Schrödinger-like equation (11) for the lower component, such a term proportional to $\tilde{l}(\tilde{l}+1) = \kappa(\kappa-1)$ can be regarded as the pseudo-centrifugal barrier and the last term can be regarded as the pseudospin-orbit (PSO) potential, i.e.,

$$V_{\rm PCB}(r) = \frac{1}{M_-(r)}\frac{\kappa(\kappa-1)}{r^2} \quad \text{and} \quad V_{\rm PSO}(r) = -\frac{1}{M_-^2(r)}\frac{dM_-(r)}{dr}\frac{\kappa}{r}, \tag{14}$$

although $\Delta(r)$ in Eq. (11) does not stand for the potential in which particles move. As pointed out in Ref. 27, if $V_{\rm PSO}(r)$ vanishes, i.e.,

$$\frac{dM_-(r)}{dr} = \frac{d\Sigma(r)}{dr} = 0\,, \tag{15}$$

each pair of pseudospin doublets will be degenerate and the PSS will be exactly conserved. This is called the PSS limit. From the physical point of view, $\Sigma(r) = 0$ or constants is never fulfilled in realistic nuclei as it implies there exist no bound states for nucleons,[28] but $d\Sigma(r)/dr \sim 0$ can be approximately satisfied in exotic nuclei with highly diffuse potentials,[29] as we will see in the next Section.

4. Pseudospin symmetry in realistic nuclei

For realistic nuclei, both the scalar $S(r)$ and vector $V(r)$ potentials in Eq. (2) vanish at large distance from the center. In such a case, no bound state can exist in the Fermi sea at the PSS limit. In other words, the PSS must be broken in real nuclei. Hence, much attention is focused on the extent of the PSS breaking in realistic nuclei and what the symmetry-breaking mechanism is.

4.1. Pseudospin symmetry in single-nucleon spectra

To check the extent of the PSS breaking, the neutron-rich doubly magic nucleus ^{132}Sn was chosen as an example in Ref. 30. The potentials $\Sigma(r)$ and $\Delta(r)$ were adopted from the RMF calculations with the effective interaction PK1.[31] The corresponding depths of potentials are $\Sigma(r) \sim 70$ MeV and $\Delta(r) \sim 700$ MeV, respectively.

The single-particle energies of the neutron bound states are shown in Fig. 3, where $E = \epsilon - M$ excluding the rest mass of nucleon. In order to show the SO and PSO splittings and to see their energy dependence clearly, the reduced SO splittings $\Delta E_{\rm SO} = (E_{j_<} - E_{j_>})/(2l+1)$ and the reduced PSO splittings $\Delta E_{\rm PSO} = (E_{j_<} - E_{j_>})/(2\tilde{l}+1)$ versus their average single-particle energies $E_{\rm av} = (E_{j_<} + E_{j_>})/2$ are plotted in the left and right panels of Fig. 4, respectively. For simplicity, $j_<$ ($j_>$) denotes the states with $j = l - 1/2$ ($j = l + 1/2$) for the spin doublets and the states with $j = \tilde{l} - 1/2$ ($j = \tilde{l} + 1/2$) for the pseudospin doublets.

From Fig. 4, it can be seen that the energy dependence is obvious for the reduced PSO splittings, and less obvious for the reduced SO splittings. For example, the reduced PSO splitting of the $1\tilde{p}$ doublets is 1.291 MeV, whereas that of the $2\tilde{p}$ doublets is 0.138 MeV, roughly smaller than the former by a factor of 10. This indicates the PSS is better for the doublets closer to the Fermi surface, which is in agreement with the experimental observations. Furthermore, with the levels approaching the single-particle threshold, the PSO splittings become smaller and even reversed, e.g., $\Delta E_{\rm PSO} = -0.068$ MeV for the $1\tilde{g}$ doublets. In contrast to the PSO splittings, the reduced SO splittings are almost a constant, as seen for the $1h$, $1g$, and $1f$ doublets around the Fermi surface.

To understand the reason why the energy splittings between the pseudospin doublets change with the single-particle energies, Meng *et al.*[27,29] compared the contributions of the PSO potential $V_{\rm PSO}$ and the pseudo-centrifugal barrier $V_{\rm PCB}$ to the single-particle energy E. The calculations were performed by the integrals with the lower component of the Dirac spinor as follows,

$$E_i = \frac{\int F^*(r) V_i(r) F(r) dr}{\int F^*(r) F(r) dr} . \tag{16}$$

By comparing the contributions from the different components in the above equation, the condition of the approximate PSS was suggested as

$$|V_{\rm PSO}(r)| \ll |V_{\rm PCB}(r)| . \tag{17}$$

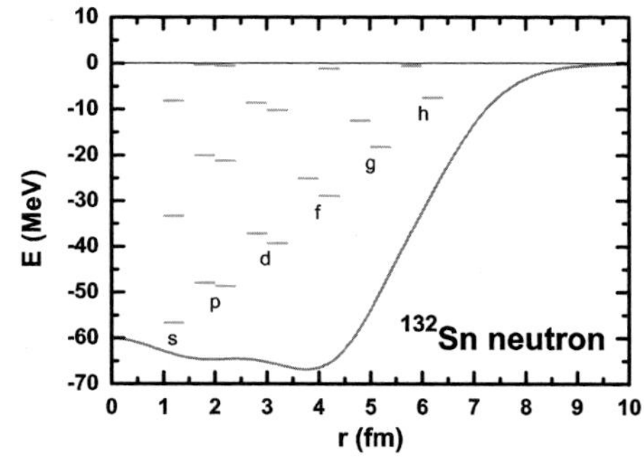

Fig. 3. (Color online) Neutron single-particle spectrum in ^{132}Sn calculated by the RMF theory with PK1. For each pair of spin doublets, the left levels are those with $j_< = l-1/2$ and the right ones with $j_> = l+1/2$. Potential $\Sigma(r)$ is shown as the solid line. Taken from Ref. 17.

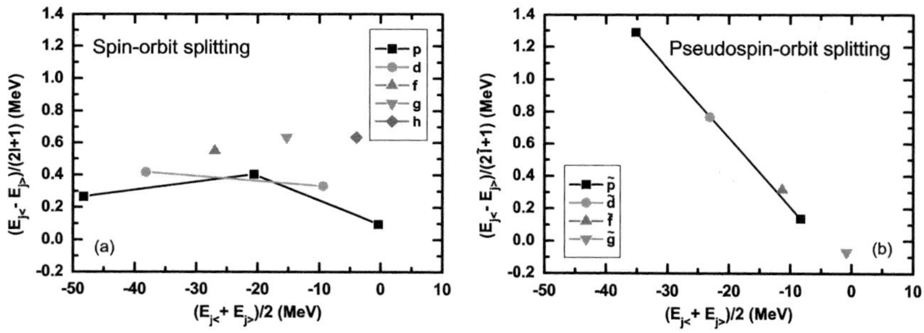

Fig. 4. (Color online) Reduced spin-orbit splittings $(E_{j<}-E_{j>})/(2l+1)$ (the left panel) and reduced pseudospin-orbit splittings $(E_{j<}-E_{j>})/(2\tilde{l}+1)$ (the right panel) versus their average single-particle energies $E_{\rm av} = (E_{j<}+E_{j>})/2$ in neutron single-particle spectrum of ^{132}Sn calculated by the RMF theory with PK1. Taken from Ref. 17.

Unfortunately, it is difficult to check the condition because of the singularity appearing in $M_-(r)$, where $M_-(r) = 0$ at $r = r_0$. In order to avoid this singularity, Meng *et al.* introduced the effective pseudo-centrifugal barrier, $M_-\kappa(\kappa-1)/r^2$, and the effective PSO potential, $(-dM_-/dr)(\kappa/r)$, which correspond respectively to the pseudo-centrifugal barrier and the PSO potential multiplied by a common factor M_-^2.

The effective PSO potential depends on the angular momentum and parity, but does not depend on the single-particle energy, while the effective pseudo-centrifugal barrier depends on the energy. In Fig. 5, the effective PSO potential and the effective pseudo-centrifugal barrier are shown for the

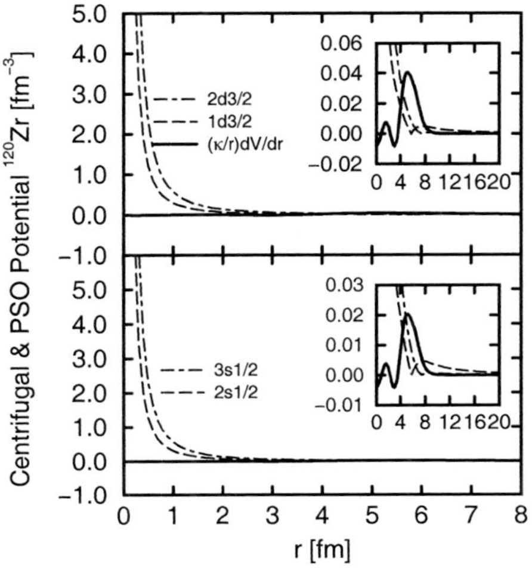

Fig. 5. Comparison of the effective pseudo-centrifugal barrier (dashed and dot-dashed lines) and the effective PSO potential (solid line) in arbitrary scale for $d_{3/2}$ (the upper panel) and $s_{1/2}$ (the lower panel) states in ^{120}Zr. The dashed lines are for $1d_{3/2}$ and $2s_{1/2}$ states, and the dot-dashed lines are for $2d_{3/2}$ and $3s_{1/2}$ states. The insets show the behaviors of the effective pseudo-centrifugal barrier and the effective PSO potential near the nuclear surface. Taken from Ref. 29.

$s_{1/2}$ and $d_{3/2}$ orbitals in ^{120}Zr with arbitrary scale. Their behaviors near the nuclear surface are shown in the insets. It can be seen that the PSS is conserved much better for the less-bound pseudospin partners, because the effective pseudo-centrifugal barrier is smaller for more deeply bound states. This conclusion is in agreement with the previous judgement.

4.2. *Pseudospin symmetry in single-nucleon wave functions*

Since the concept of PSS was introduced in nuclei, more attention has been focused on the approximate degeneracy in the single-particle spectra. Until the relativistic origin was recognized, a lot of attention was paid to the wave functions of PSS.[32,33] Within the PSS limit shown in Eq. (15), the Schrödinger-like equation for the lower component $F(r)$ of the Dirac spinor is expressed as

$$\left[\frac{d^2}{dr^2} - \frac{\kappa(\kappa-1)}{r^2} + (\epsilon - M - \Sigma_0)(\epsilon + M - \Delta(r))\right] F(r) = 0\,. \qquad (18)$$

It is clear that this equation is identical for the pseudospin doublets a and b with $\kappa_a(\kappa_a - 1) = \kappa_b(\kappa_b - 1)$, i.e., $\kappa_a + \kappa_b = 1$. Therefore, the eigenfunctions

$F_a(r)$ and $F_b(r)$ are also exactly the same,[32]

$$F_a(r) = F_b(r) \,. \tag{19}$$

As a step further, together with Eq. (8), the first-order differential relation for their upper components can be obtained,[33]

$$\left(\frac{d}{dr} + \frac{\kappa_a}{r}\right) G_a(r) = \left(\frac{d}{dr} + \frac{\kappa_b}{r}\right) G_b(r) \,. \tag{20}$$

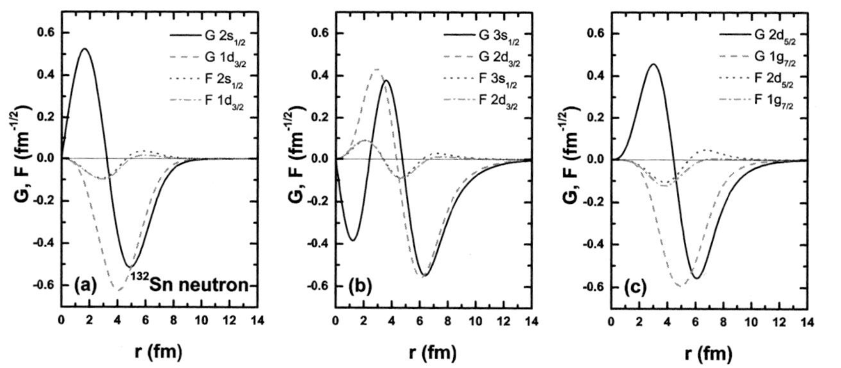

Fig. 6. (Color online) Neutron single-particle wave functions of the $1\tilde{p}$, $2\tilde{p}$, and $1\tilde{f}$ pseudospin doublets in ^{132}Sn calculated by the RMF theory with the effective interaction PC-PK1. Taken from Ref. 17.

To test the relation in Eq. (19), the single-particle wave functions for the neutrons in ^{132}Sn are calculated by the self-consistent point-coupling RMF theory with the effective interaction PC-PK1.[34] In panels (a), (b), and (c) of Fig. 6 are shown the wave functions of the $1\tilde{p}$, $2\tilde{p}$, and $1\tilde{f}$ pseudospin doublets, respectively. For each pair of pseudospin doublets, their upper components G have different number of nodes and radial shape, however, their lower components F are similar to each other except on the nuclear surface. By comparing these three panels, it is found that for the same pseudo-orbital angular momentum $\tilde{l}$, the relation (19) is better satisfied as the single-particle energy increases. Meanwhile, for roughly the same single-particle energy, this relation holds better for smaller $\tilde{l}$. These features are in agreement with the behaviors of the PSO splittings shown in Fig. 4.

5. Highlights on recent progress and open issues

Since the introduction of the spin and pseudospin symmetries, their origin and breaking mechanism as well as their applications in nuclear structure studies have attracted wide attention. In particular, more attention in recent years are devoted to the descriptions of exotic phenomena, including the SS in anti-nucleon spectra, the PSS in the resonant states, and so on. Furthermore, these symmetries are further investigated by various new methods, including the similarity renormalization group (SRG), the supersymmetric (SUSY) quantum mechanics, and their combinations.

5.1. *Spin symmetry in anti-nucleon spectra*

The SS and PSS symmetries were initially found in the single-nucleon spectra, and little attention was paid to the anti-nucleon spectra in addition to being mentioned in Ref. 35. In 2003, Zhou, Meng, and Ring[36] adopted the eigenstates of the Dirac Hamiltonian to form a basis in the RMF calculations. Certainly the negative-energy states in the Dirac sea must be included for the completeness. By examining the single-particle spectra of negative-energy states, Zhou *et al.* found unexpectedly that the PSS of those negative-energy states in the Dirac sea, or equivalently, the SS in the anti-nucleon spectra is conserved very well.[37] In this Section, we will introduce the SS in the single-anti-nucleon spectra and wave functions.

5.1.1. *Spin symmetry in single-anti-nucleon spectra*

Following Ref. 37, we label nucleon states with "N" and anti-nucleon states with "A". The Dirac equation for nucleons reads

$$[\boldsymbol{\alpha} \cdot \mathbf{p} + V_{\rm N}(\mathbf{r}) + \beta(M + S_{\rm N}(\mathbf{r}))]\, \psi_{\rm N}(\mathbf{r}) = \epsilon_{\rm N}\psi_{\rm N}(\mathbf{r})\,, \tag{21}$$

where $V_{\rm N}(\mathbf{r}) = V(\mathbf{r})$ and $S_{\rm N}(\mathbf{r}) = S(\mathbf{r})$. For a spherical system, the Dirac spinor $\psi_{\rm N}$ has the form [cf. Eq. (6)]

$$\psi_{\rm N}(\mathbf{r}) = \frac{1}{r}\begin{pmatrix} iG_{n\kappa}(r)Y^{l}_{jm}(\hat{\mathbf{r}}) \\ -F_{\tilde{n}\kappa}(r)Y^{\tilde{l}}_{jm}(\hat{\mathbf{r}}) \end{pmatrix}, \qquad j = l \pm \frac{1}{2}\,. \tag{22}$$

The charge conjugation leaves the scalar potential $S_{\rm N}(\mathbf{r})$ invariant, while it changes the sign of the vector potential $V_{\rm N}(\mathbf{r})$. That is, for anti-nucleons, $V_{\rm A}(\mathbf{r}) = -V_{\rm N}(\mathbf{r}) = -V(\mathbf{r})$ and $S_{\rm A}(\mathbf{r}) = S_{\rm N}(\mathbf{r}) = S(\mathbf{r})$. Thus, the charge conjugation of Eq. (22) gives the Dirac spinor for an anti-nucleon as

$$\psi_{\mathrm{A}}(\mathbf{r}) = \frac{1}{r}\begin{pmatrix} -F_{\tilde{n}\tilde{\kappa}}(r)Y^{\tilde{l}}_{jm}(\hat{\mathbf{r}}) \\ iG_{n\tilde{\kappa}}(r)Y^{l}_{jm}(\hat{\mathbf{r}}) \end{pmatrix}, \qquad j = l \pm \frac{1}{2}, \tag{23}$$

with $\tilde{\kappa} = -\kappa$.

For particles there are positive- and negative-energy solutions of the Dirac equation, and the same is true for anti-particles. For the positive-energy states, the normal quantum numbers follow the upper component that is dominant. A particle state is labelled by $\{nl\kappa m\}$, while its pseudo-quantum numbers are $\{\tilde{n}\tilde{l}\tilde{\kappa}m\}$: $\tilde{n} = n+1$ for $\kappa > 0$; $\tilde{n} = n$ for $\kappa < 0$.[28] An anti-particle state is labeled by $\{\tilde{n}\tilde{l}\tilde{\kappa}m\}$ and its pseudo-quantum numbers are $\{nl\kappa m\}$. In analogy to Ref. 28, the following relation holds for the anti-nucleon states,

$$n = \tilde{n} + 1 \quad \text{for} \quad \tilde{\kappa} > 0\,; \qquad n = \tilde{n} \quad \text{for} \quad \tilde{\kappa} < 0\,. \tag{24}$$

With $\kappa(\kappa+1) = \tilde{\kappa}(\tilde{\kappa}-1) = l(l+1)$ and $\kappa(\kappa-1) = \tilde{\kappa}(\tilde{\kappa}+1) = \tilde{l}(\tilde{l}+1)$ in mind, one derives the Schrödinger-like equations for the upper and the lower components [cf. Eqs. (9) and (11)],

$$\left\{-\frac{1}{M_+}\frac{d^2}{dr^2} + \frac{1}{M_+^2}\frac{dM_+}{dr}\frac{d}{dr} + \left[(M+\Sigma) + \frac{1}{M_+}\frac{l(l+1)}{r^2} + \frac{1}{M_+^2}\frac{dM_+}{dr}\frac{\kappa}{r}\right]\right\}G(r)$$
$$= \begin{cases} +\epsilon_{\mathrm{N}}G(r)\,, \\ -\epsilon_{\mathrm{A}}G(r)\,, \end{cases} \tag{25}$$

and

$$\left\{-\frac{1}{M_-}\frac{d^2}{dr^2} + \frac{1}{M_-^2}\frac{dM_-}{dr}\frac{d}{dr} + \left[(-M+\Delta) + \frac{1}{M_-}\frac{\tilde{l}(\tilde{l}+1)}{r^2} + \frac{1}{M_-^2}\frac{dM_-}{dr}\frac{\tilde{\kappa}}{r}\right]\right\}F(r)$$
$$= \begin{cases} +\epsilon_{\mathrm{N}}F(r)\,, \\ -\epsilon_{\mathrm{A}}F(r)\,, \end{cases} \tag{26}$$

where $M_+(r) = M - \Delta(r) + \epsilon$ and $M_-(r) = -M - \Sigma(r) + \epsilon$ with $\epsilon = +\epsilon_{\mathrm{N}}$ for the particle states or $-\epsilon_{\mathrm{A}}$ for the anti-particle states. Both equations are fully equivalent to the original Dirac equation with the full spectrum of particle and anti-particle states.

The relation between SS or PSS and the external fields is given in Table 1. If $d\Delta/dr = 0$, there is an exact SS in the particle spectrum and exact PSS in the anti-particle spectrum, because states with the same l but different κ are degenerate in Eq. (25). Note that l is the orbital angular momentum of particle states and pseudo-orbital angular momentum of anti-particle states. When $d\Delta/dr \neq 0$, the symmetries are broken, but

Table 1. Relation between symmetries and external fields. Taken from Ref. 37.

	Particle	Anti-particle
$d\Delta/dr = 0$	Spin symmetry	Pseudospin symmetry
$d\Sigma/dr = 0$	Pseudospin symmetry	Spin symmetry

if $d\Delta/dr$ is so small that the SO term is much smaller than the centrifugal barrier, there will be approximate symmetries.

Similarly, when $d\Sigma/dr = 0$ in Eq. (26), there is an exact PSS in the particle spectra. On the other hand, for the anti-particle states, the SS is exactly conserved because now $\tilde{l}$ is the orbital angular momentum. If $d\Sigma/dr \neq 0$ but is small, there are approximate PSS in the particle spectra and approximate SS in the anti-particle spectra. This implies that the SS in the anti-particle spectra has the same origin as the PSS in the particle spectra.

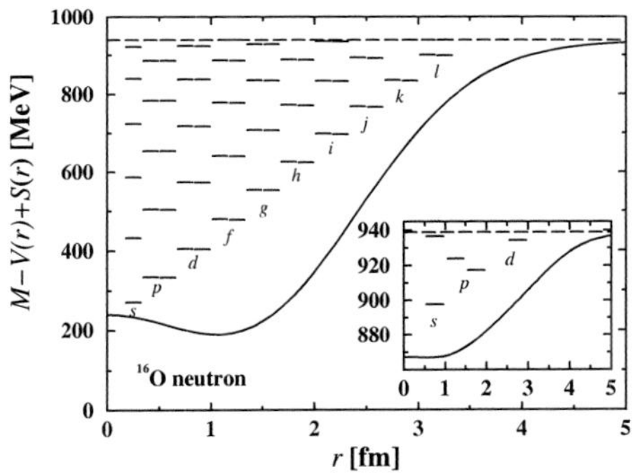

Fig. 7. Single-anti-neutron potential $M - V(r) + S(r)$ and spectrum of ^{16}O. For each pair of spin doublets, the left level is with $\tilde{\kappa} < 0$ and the right one with $\tilde{\kappa} > 0$. The inset gives single-neutron potential $M + V(r) + S(r)$ and spectrum. Taken from Ref. 37.

As revealed in Ref. 37, there is an essential difference in the degree to which the symmetry is broken: The factor $1/M_-^2 = 1/(\epsilon - \Sigma - M)^2$ is two orders of magnitude smaller for the anti-nucleon states than that for the nucleon states. The SS for the anti-particle states is therefore much less broken than the PSS for the particle states,[37] as shown in Fig. 7 by taking the neutron and anti-neutron spectra in ^{16}O as example.

5.1.2. *Spin symmetry in single-anti-nucleon wave functions*

Since the spin-orbit term in Eq. (26) is so small, for a pair of spin doublets in the anti-nucleon spectrum, the radial wave functions of dominant components should be very similar to each other in realistic nuclei, and be the same under the SS limit, i.e.,

$$F_{j_<}(r) = F_{j_>}(r) \qquad \text{if} \quad d\Sigma/dr = 0\,. \tag{27}$$

This relation has been tested in Ref. 37: The dominant components $F(r)$ are nearly identical for the spin doublets; on the other hand, their small components $G(r)$ show dramatic deviations from each other. However, the small components should satisfy a certain relation under the SS limit,[38]

$$\left(\frac{d}{dr} + \frac{\tilde{l}+1}{r}\right) G_{j_>}(r) = \left(\frac{d}{dr} - \frac{\tilde{l}}{r}\right) G_{j_<}(r)\,. \tag{28}$$

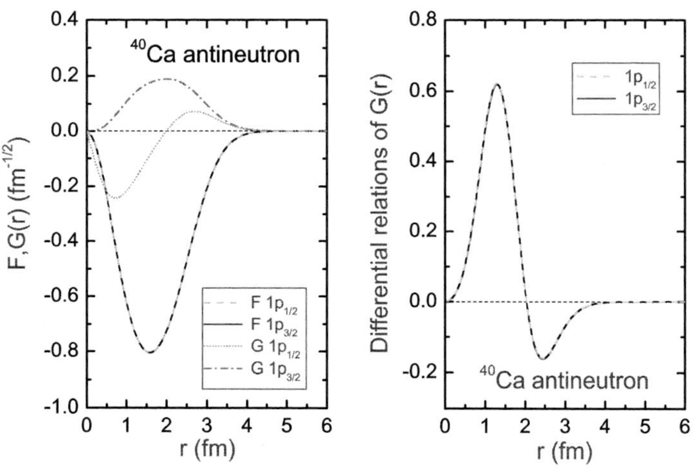

Fig. 8. (Color online) Radial wave functions (the left panel) and the differential relation (28) of the lower components (the right panel) for the anti-neutron spin doublets $1p_{1/2}$ and $1p_{3/2}$ in ^{40}Ca. Taken from Ref. 38.

In Ref. 38, it was examined to what extent the relations (27) and (28) are fulfilled in realistic nuclei. The RMF calculations were performed for ^{40}Ca, ^{90}Zr, ^{124}Sn, and ^{208}Pb with the effective interaction NL3.[20] A good SS is found in both the anti-proton and anti-neutron spectra. The radial wave functions for the anti-neutron $1p$ doublets in ^{40}Ca are shown in the left panel of Fig. 8, where the energies of these two states are 271.91 and 271.55 MeV, respectively. One can see that their upper components $F(r)$ are almost identical to each other due to the good SS, but their lower components $G(r)$ deviate dramatically from each other. In the right panel of Fig. 8, the differential relation of the lower components in Eq. (28) is presented. This differential relation is satisfied remarkably well, which gives further support to the SS in the anti-nucleon spectra in nuclei.

5.2. *Pseudospin symmetry in single-particle resonant states*

The PSS was initially found in the bound states. Does this symmetry also exist in the unbound states of quantum systems? Weakly bound or unbound nuclei with unusual N/Z ratios are open quantum many-body systems in which the continuum plays an important role.[39] In these nuclei, the neutron (or proton) Fermi surface is close to the particle threshold, thus the contribution of the continuum is crucial. Many methods or models developed for the studies of resonances,[40] e.g., the analytical continuation in coupling constant (ACCC) method,[41] the real stabilization method,[42] the complex scaling method,[43] and the coupled channels method,[44–46] have been used to study nuclear single-particle resonant states. Based on some of these methods, the PSS in the single-particle resonant states have been numerically investigated.[47–50] Recently, Lu, Zhao, and Zhou[51] have verified the existence of exactly conserved PSS under the same condition for the PSS in bound states, i.e., $\Sigma(r) = 0$ or $d\Sigma(r)/dr = 0$ given in Eq. (15). Further, they have examined in detail the exact conservation and breaking mechanism of the PSS of single-particle resonant states in the square-well potentials.[52] In this Section, we will sketch some of these researches in the single-particle resonant states.

5.2.1. *Approximate degeneracy for resonant energy*

By combining the ACCC into the Dirac equation, the energy, width, and wave function of the single-particle resonance states can be obtained, and the symmetry hidden in the Dirac equation can be disclosed. The details can be found in Refs. 47, 53.

Assuming a nucleon with mass M moving in an attractive scalar potential $S(\mathbf{r})$ and a repulsive vector potential $V(\mathbf{r})$, the Dirac equation is written as Eq. (2),

$$H = \boldsymbol{\alpha} \cdot \mathbf{p} + \beta[M + S(\mathbf{r})] + V(\mathbf{r})\,. \tag{29}$$

For the spherical nuclei, the Dirac equation is simplified to Eq. (7),

$$\begin{pmatrix} M + \Sigma(r) & -\frac{d}{dr} + \frac{\kappa}{r} \\ \frac{d}{dr} + \frac{\kappa}{r} & -M + \Delta(r) \end{pmatrix} \begin{pmatrix} G(r) \\ F(r) \end{pmatrix} = \epsilon \begin{pmatrix} G(r) \\ F(r) \end{pmatrix}. \tag{30}$$

The bound solutions of this equation can be obtained by usual methods. In order to obtain the resonant states, one of the options is the ACCC method, in which the resonant states are treated as the analytic continuation of bound states. Namely, by increasing the attractive potential as $\Sigma(r) \to \lambda\Sigma(r)$, a resonant state will be lowered and become a bound state if the coupling constant λ is large enough. Near the branch point λ_0, defined by the scattering threshold $k(\lambda_0) = 0$, the wave number $k(\lambda)$ behaves as[40]

$$k(\lambda) \sim \begin{cases} i\sqrt{\lambda - \lambda_0}, & l > 0, \\ i(\lambda - \lambda_0), & l = 0. \end{cases} \tag{31}$$

These properties suggest an analytic continuation of the wave number k in the complex λ plane from the bound-state region into the resonance region by the second kind of Padé approximate[40]

$$k(x) \simeq k^{[L,N]}(x) = i\frac{c_0 + c_1 x + c_2 x^2 + \cdots + c_L x^L}{1 + d_1 x + d_2 x^2 + \cdots + d_N x^N}, \tag{32}$$

where $x \equiv \sqrt{\lambda - \lambda_0}$, and $c_0, c_1, \ldots, c_L, d_1, d_2, \ldots, d_N$ are the coefficients of Padé approximate. These coefficients can be determined by a set of reference points x_i and $k(x_i)$ obtained from the Dirac equation with $\lambda_i > \lambda_0$, $i = 1, 2, \ldots, L + N + 1$. With the complex wave number $k(\lambda = 1) = k_r + ik_i$, the resonant energy E and width Γ can be extracted from the relations $\varepsilon = E - i\frac{\Gamma}{2}$ $(E, \Gamma \in \mathbb{R})$ and $k^2 = \varepsilon^2 - M^2$, i.e.,

$$E = \sqrt{\frac{1}{2}\left[\sqrt{\left(M^2 + k_r^2 - k_i^2\right)^2 + 4k_r^2 k_i^2} + \left(M^2 + k_r^2 - k_i^2\right)^2\right]} - M,$$

$$\Gamma = \sqrt{2\sqrt{\left(M^2 + k_r^2 - k_i^2\right)^2 + 4k_r^2 k_i^2} - 2\left(M^2 + k_r^2 - k_i^2\right)}. \tag{33}$$

In the non-relativistic limit $(k \ll M)$, Eq. (33) reduces to

$$E = \frac{k_r^2 - k_i^2}{2M}, \qquad \Gamma = \frac{2k_r k_i}{M}. \tag{34}$$

In Ref. 47, ^{208}Pb was taken as an example, and all the resonant states in the Woods–Saxon typed scalar and vector potentials were obtained by ACCC. With these resonant states, the energy and width splittings between the pseudospin doublets were extracted. Their evolutions with respect to the surface diffuseness a are shown in Figs. 9 and 10.

From Fig. 9, it is seen that the trend for the change of the pseudospin splittings with a is agreeable for all the pseudospin doublets except for $(3d_{5/2}, 2g_{7/2})$. As a increases, the pseudospin energy splittings decrease, i.e., PSS becomes better; but as a increases further, the pseudospin energy

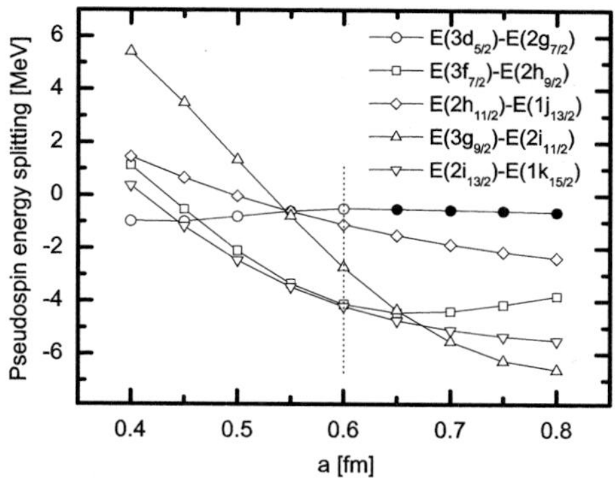

Fig. 9. Pseudospin energy splittings for the neutron resonant states in ^{208}Pb as a function of the potential diffuseness. The filled symbols represent that the corresponding pseudospin doublets have been lowered to the bound states. The vertical line corresponds to the fitted Woods–Saxon parameters for ^{208}Pb. Taken from Ref. 47.

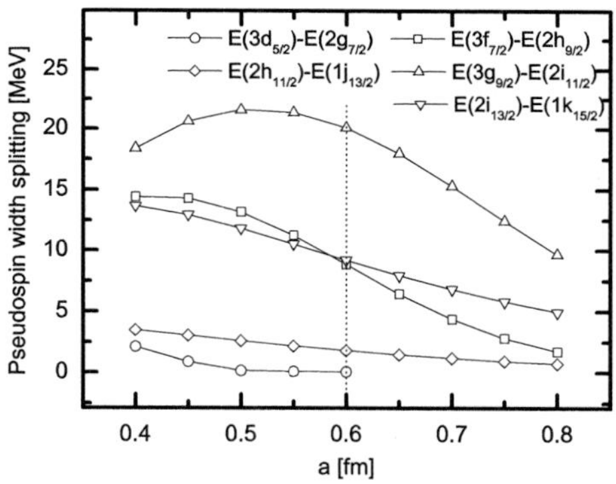

Fig. 10. Same as Fig. 9 but for the width splittings of pseudospin partners in ^{208}Pb. Taken from Ref. 47.

splittings will be inverted, i.e., it changes from $E_{j_<} > E_{j_>}$ to $E_{j_<} < E_{j_>}$. This inversion has been also observed experimentally. However, for the widths, the widths for the $j_<$ states are always larger than those for the $j_>$ states no matter their order in energy, as shown in Fig. 10. This implies that the resonant states with higher orbital angular momentum have longer decay time than their pseudospin partners. The PSS is also correlated with the central depth Σ_0 and the radius R. In other words, the quality of PSS depends on the shape of potential for resonant states as that for bound states.

By combining the ACCC into the CDFT, the isospin dependence of PSS for the resonant states was studied in Ref. 48. The energies and widths of single-particle resonant states in the Sn isotopes were investigated systematically. An isospin dependence of PSS is clearly shown in the resonant states and it is consistent with that observed in the bound states. Although the isospin dependence of the pseudospin splitting in energy is rather complicated, the width splitting decreases almost monotonically with increasing mass number.

5.2.2. *Approximate degeneracy for resonant wave functions*

As there exists the quasi-degeneracy in the resonant energy, are the wave functions also similar to each other for the resonant pseudospin doublets?

In Ref. 50, the ACCC and CDFT were combined to explore the resonant states for not only the resonant energies and widths, but also the wave functions for pseudospin doublets. Taking ^{120}Sn as an example, in Fig. 11 are plotted the real parts of the upper and lower components of the radial wave functions for the proton $1\tilde{g}$ pseudospin doublets. It is seen that the lower components have the same number of radial nodes and agree very well in the region where nuclear potential dominates, except for on the surface.

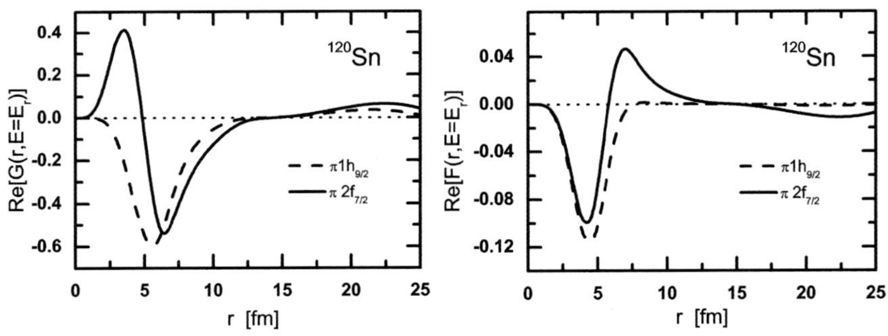

Fig. 11. Real part of the upper component $G(r)$ (the left panel) and the lower component $F(r)$ (the right panel) for the proton resonant state $1h_{9/2}$ (dashed line) compared with the proton resonant state $2f_{7/2}$ (solid line) in ^{120}Sn. Taken from Ref. 50.

5.2.3. *Exactly conserved PSS in resonant states*

The approximate PSS could be found by the ACCC calculations, by examining the resonant energies, widths, and single-particle wave functions. Note that there are no bound states in the PSS limit, so is there exactly conserved PSS in the resonant states? In 2012, a verification of the existence of exact PSS in the resonant states was presented for the first time by Lu, Zhao, and Zhou.[51] Here we briefly summarize this verification.

In the PSS limit, the Schrödinger-like equation for the small component becomes

$$\left\{-\frac{1}{M_-}\frac{d^2}{dr^2}+\left[(-M+\Delta)+\frac{1}{M_-}\frac{\kappa(\kappa-1)}{r^2}\right]\right\}F=\epsilon F\,. \qquad (35)$$

For the bound states, there is a singularity in $1/M_-(r)$, but for the resonant states or continuum, the singularity in $1/M_-(r)$ disappears.

For the continuum in the Fermi sea, i.e., $\epsilon \geq M$, there exist two independent solutions for Eq. (35). The physically acceptable solution is the one that vanishes at the origin. As usual the regular solution $F(r)$ is defined as the one that behaves like $j_{\tilde{l}}(pr)$ function as $r \to 0$,[54]

$$\lim_{r\to 0} F(r)/j_{\tilde{l}}(pr) = 1\,, \qquad p=\sqrt{\epsilon^2-M^2}\,. \qquad (36)$$

At large r the potential for neutrons vanishes and the radial wave functions oscillate. Equation (35) becomes a Ricatti–Bessel equation with angular momentum $\tilde{l}$, and the solutions can be written as a combination of the Ricatti–Hankel functions,

$$F(r)=\frac{i}{2}\left[\mathcal{J}_\kappa^F(p)h_{\tilde{l}}^-(pr)-\mathcal{J}_\kappa^F(p)^* h_{\tilde{l}}^+(pr)\right]\,, \qquad r\to\infty\,, \qquad (37)$$

where $\mathcal{J}_\kappa^F(p)$ is the Jost function for the small component and $h_{\tilde{l}}^{\pm}(pr)$ the Ricatti–Hankel functions.

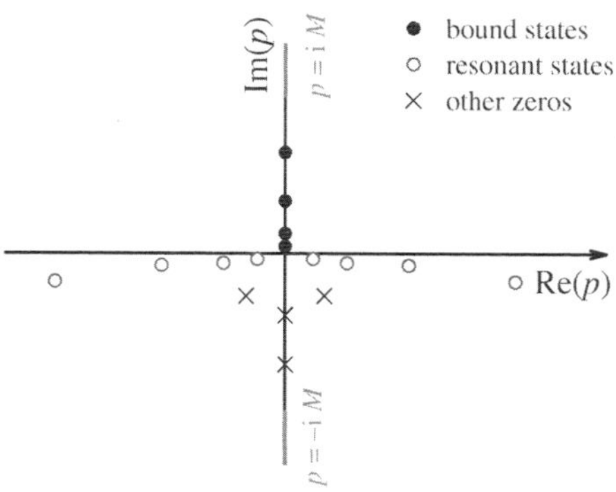

Fig. 12. (Color online) Schematic picture of the zeros of the Jost function $\mathcal{J}_\kappa^F$ on the complex momentum plane. A cut is made on the imaginary axis, from $p = iM$ to infinity and back to $p = -iM$. Taken from Ref. 51.

In nuclei the vector $V(r)$ and scalar $S(r)$ potentials are analytical functions of r. Hence, the Jost function is an analytic function of p, and it can be analytically continued to a large area in the complex p plane.[51] In Fig. 12, the zeros of the Jost function $\mathcal{J}_\kappa^F(p)$ on the complex momentum plane are schematically shown for the first sheet with $\mathrm{Re}(\epsilon) \geq 0$, where the full circles, open circles, and crosses denote the bound states, resonant states, and other zeros, respectively. The resonant energy $E_{\rm res}$ and width $\Gamma_{\rm res}$ are determined by the relation $E = E_{\rm res} - i\Gamma_{\rm res}/2 = \sqrt{p^2 + M^2} - M$. At the PSS limit, for the states with the quantum numbers κ_a and κ_b ($\kappa_a + \kappa_b = 1$), the small components satisfy the same equation because they have the same pseudo-orbital angular momentum $\tilde{l}$.[24] Hence, for a resonant state with the resonant energy and width $(E_{\rm res}, \Gamma_{\rm res})$ and the quantum number κ_a, there must be another state with the quantum number κ_b with the same energy and width. Namely, the PSS in the single-particle resonant states is exactly conserved in nuclei when the attractive scalar and repulsive vector potentials have the same magnitude but opposite sign.

In order to obtain the resonant parameters, the scattering phase shift method can be adopted. One can also verify the PSS in the single-particle resonant states by examining the phase shift. The details can be found in Ref. 51.

To check the conservation and breaking of the PSS in the single-particle resonant states, an illustrated example was introduced in Ref. 51, where the square-well potentials were used to replace realistic potentials. The spherical square-well potentials for $\Sigma(r)$ and $\Delta(r)$ read

$$\Sigma(r) = \begin{cases} C\,, & r < R\,, \\ 0\,, & r \geq R\,, \end{cases} \qquad \Delta(r) = \begin{cases} D\,, & r < R\,, \\ 0\,, & r \geq R\,, \end{cases} \tag{38}$$

where C and D are depths and R is the width. The Jost function $\mathcal{J}_\kappa^F(p)$ is derived as[51]

$$\begin{aligned} \mathcal{J}_\kappa^F(p) = -\frac{p^{\tilde{l}}}{2ik^{\tilde{l}+1}} \Big\{ & j_{\tilde{l}}(kR) p h_{\tilde{l}}^{+\prime}(pR) - k j_{\tilde{l}}^{\prime}(kR) h_{\tilde{l}}^{+}(pR) \\ & - \frac{C}{\epsilon - M - C} \left[k j_{\tilde{l}}^{\prime}(kR) - \frac{\kappa}{R} j_{\tilde{l}}(kR) \right] h_{\tilde{l}}^{+}(pR) \Big\}\,, \end{aligned} \tag{39}$$

with $k = \sqrt{(\epsilon - C - M)(\epsilon - D + M)}$. The PSS in both bound states and resonant states can be then explained explicitly. If $C = 0$, the second term in $\mathcal{J}_\kappa^F(p)$ vanishes and the first term only depends on the pseudo-orbital angular momentum $\tilde{l}$, then the Jost functions with different κ but the same $\tilde{l}$ are identical. The energies and widths of resonant pseudospin partners are exactly the same.

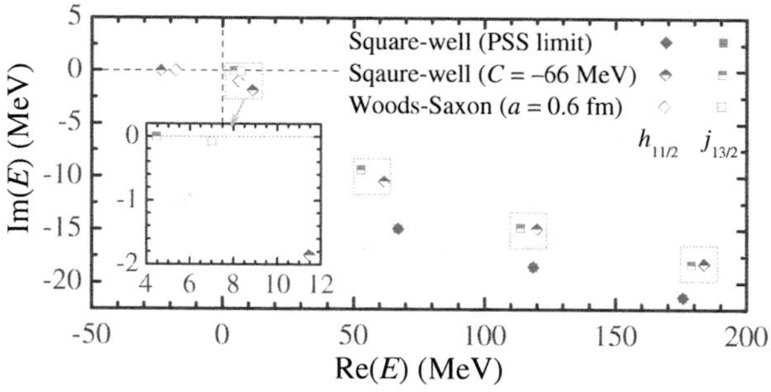

Fig. 13. (Color online) The zeros of the Jost function $\mathcal{J}_\kappa^F$ on the complex energy plane in the square-well potentials (38) with $C = 0$ (solid symbols) and $C = -66$ MeV (half-filled symbols) for pseudospin doublets $h_{11/2}$ (diamond) and $j_{13/2}$ (square). The results with Woods–Saxon-like scalar and vector potentials are also shown as the open symbols. Taken from Ref. 51.

Although the solutions of $\mathcal{J}_\kappa^F(p) = 0$ cannot be derived analytically, the roots can be searched for by the secant method, because the Jost function is analytical near its zeros. In Fig. 13, three pairs of pseudospin resonant partners are shown by the full diamonds and squares. The conservation of the PSS for the single-particle resonant states is clearly seen. When $C = -66$ MeV, there is one bound state only for $h_{11/2}$. Three pairs of pseudospin partners of resonant states are shown by the half-filled diamonds and squares. One finds the breaking of the PSS both in the bound states and in the resonant states. For pseudospin doublets with other values of $\tilde{l}$, similar behaviors are observed concerning the exact conservation and the breaking of the PSS.

5.3. *Pseudospin symmetry with similarity renormalization group*

Although the relativistic characteristics of PSS is recognized by transforming the Dirac equation into a Schrödinger-like form, one encounters inevitably the singularity in calculating the contribution of every component to the pseudospin splitting, and also encounters the coupling between the operator and its eigenenergy in solving the Schrödinger-like equation for the lower component of Dirac spinor. Recent works in Refs. 55–57 have cured these defects by using the SRG. The idea of SRG[58] is to drive the Hamiltonian towards a block-diagonal form via the flow equation and unitary transformations that suppress the off-diagonal matrix elements. The

Dirac Hamiltonian shown in Eq. (2) can be transformed with the SRG into a diagonal form and expanded in a series of $1/M$. It is important that the effective Hamiltonian in the Schrödinger-like equation thus obtained is Hermitian. This makes the perturbation calculations possible.

Fist of all, according to the commutation and anti-commutation relations with respect to the β matrix, the Dirac Hamiltonian in Eq. (2) is separated into the diagonal ε and off-diagonal o parts, $H = \varepsilon + o$, which satisfy

$$[\varepsilon, \beta] = 0 \qquad \text{and} \qquad \{o, \beta\} = 0\,. \tag{40}$$

In order to obtain the equivalent Schrödinger-like equation for nucleons, the main task is to decouple the eigenvalue equations for the upper and lower components of the Dirac spinor. A possible way is to make the off-diagonal part of the Dirac Hamiltonian vanish with a proper unitary transformation. According to the SRG,[58] the Hamiltonian H is transformed by a unitary operator $U(l)$ as

$$H(l) = U(l)HU^{\dagger}(l)\,, \tag{41}$$

with $H(l) = \varepsilon(l) + o(l)$, $H(0) = H$, and a flow parameter l. Then, the so-called Hamiltonian flow equation can be obtained by taking the differential of the above equation,

$$\frac{d}{dl}H(l) = [\eta(l), H(l)]\,, \tag{42}$$

with the anti-Hermitian generator

$$\eta(l) = \frac{dU(l)}{dl}U^{\dagger}(l)\,. \tag{43}$$

As pointed out in Ref. 59, one of the proper choices of $\eta(l)$ for letting the off-diagonal part $o(l) = 0$ as $l \to \infty$ reads

$$\eta(l) = [\beta M, H(l)]\,. \tag{44}$$

Then, the diagonal part of the Dirac Hamiltonian $\varepsilon(l)$ at the $l \to \infty$ limit can be derived analytically in a series of $1/M$,[55]

$$
\begin{aligned}
\varepsilon(\infty) &= M\varepsilon_0(\infty) + \varepsilon_1(\infty) + \frac{\varepsilon_2(\infty)}{M} + \frac{\varepsilon_3(\infty)}{M^2} + \frac{\varepsilon_4(\infty)}{M^3} + \cdots \\
&= \beta M + (\beta S + V) + \frac{1}{2M}\beta(\boldsymbol{\alpha}\cdot\mathbf{p})^2 + \frac{1}{8M^2}\left[\left[\boldsymbol{\alpha}\cdot\mathbf{p}, (\beta S + V)\right], \boldsymbol{\alpha}\cdot\mathbf{p}\right] \\
&\quad + \frac{1}{32M^3}\beta\Big(-4(\boldsymbol{\alpha}\cdot\mathbf{p})^4 + \left\{\boldsymbol{\alpha}\cdot\mathbf{p}, \left[\left[\boldsymbol{\alpha}\cdot\mathbf{p}, (\beta S + V)\right], (\beta S + V)\right]\right\} \\
&\qquad - 2\left[\boldsymbol{\alpha}\cdot\mathbf{p}, (\beta S + V)\right]^2 \Big) + \cdots
\end{aligned}
\tag{45}
$$

In such a way, the eigenvalue equations for the upper and lower components of the Dirac spinor are decoupled. The equivalent Schrödinger-like equations for nucleons with Hermitian effective Hamiltonian can be obtained. The corresponding details can be found in Refs. 55, 56 for the spherical case and Ref. 57 for the axially deformed case.

In particular, for the spherical case, the effective Hamiltonian for the nucleons in the Fermi sea up to the $1/M^3$ order reads[55]

$$
\begin{aligned}
H_F = {} & M + \Sigma + \frac{p_F^2}{2M} - \frac{1}{2M^2}\left(Sp_F^2 - S'\frac{d}{dr}\right) - \frac{\kappa}{r}\frac{\Delta'}{4M^2} + \frac{\Sigma''}{8M^2} \\
& + \frac{S}{2M^3}\left(Sp_F^2 - 2S'\frac{d}{dr}\right) + \frac{\kappa}{r}\frac{S\Delta'}{2M^3} - \frac{(\Sigma')^2 - 2\Sigma'\Delta' + 4S\Sigma''}{16M^3} - \frac{p_F^4}{8M^3},
\end{aligned}
\tag{46}
$$

with the operator $p_F^2 = -d^2/(dr^2) + \kappa(\kappa+1)/r^2$. Correspondingly, the effective Hamiltonian for the nucleons in the Dirac sea reads[55]

$$
\begin{aligned}
H_D = {} & -M + \Delta - \frac{p_D^2}{2M} + \frac{1}{2M^2}\left(Sp_D^2 - S'\frac{d}{dr}\right) + \frac{\kappa}{r}\frac{\Sigma'}{4M^2} + \frac{\Delta''}{8M^2} \\
& - \frac{S}{2M^3}\left(Sp_D^2 - 2S'\frac{d}{dr}\right) - \frac{\kappa}{r}\frac{S\Sigma'}{2M^3} + \frac{(\Delta')^2 - 2\Sigma'\Delta' - 4S\Delta''}{16M^3} + \frac{p_D^4}{8M^3},
\end{aligned}
\tag{47}
$$

with the operator $p_D^2 = -d^2/(dr^2) + \kappa(\kappa-1)/r^2$. For Eqs. (46) and (47), one can see that the operators H_F and H_D are Hermitian and the singularity disappears in every component. In addition, there is no coupling between the energy ϵ and the operator H_F (H_D). Thus, the energy spectra of H_F (H_D) can be calculated straightforwardly.

The energy spectra of H_F (H_D) agree with the results of the original Dirac equations very well, and the energy splittings of pseudospin doublets are also in agreement with the exact relativistic case. To show this,

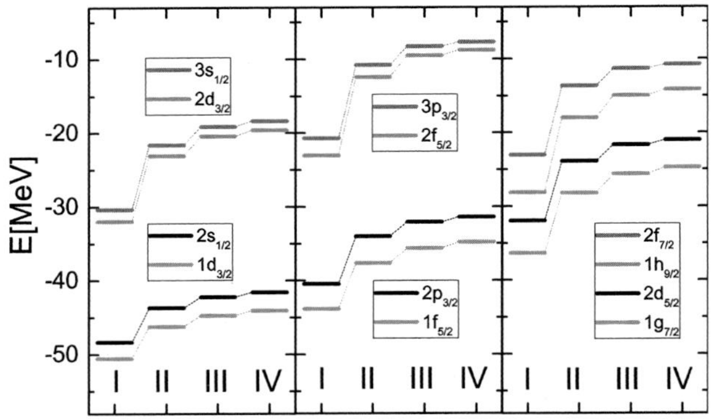

Fig. 14. (Color online) Energy spectra of H_F for six pseudospin doublets. The first column in each panel corresponds to the spectrum of H_F approximated to the non-relativistic limit, while the second and third columns correspond to the spectrum approximated to the $1/M^2$ and $1/M^3$ orders, respectively, and the exact relativistic spectra are displayed in the fourth column. Taken from Ref. 55.

the Woods–Saxon potentials for $\Sigma(r)$ and $\Delta(r)$ are taken as example, i.e., $\Sigma(r) = \Sigma_0 f(a_\Sigma, r_\Sigma, r)$ and $\Delta(r) = \Delta_0 f(a_\Delta, r_\Delta, r)$ with

$$f(a_0, r_0, r) = \frac{1}{1 + \exp\left(\frac{r - r_0}{a_0}\right)}. \tag{48}$$

The corresponding parameters are determined by fitting the energy spectrum from the RMF calculations for ^{208}Pb.[47] The energy spectra of the six pseudospin doublets are shown in Fig. 14, where the first, second, and third columns in each panel correspond to the spectra of H_F approximated to the third-, fourth-, and fifth-order perturbation, respectively. The exact relativistic spectra are displayed in the fourth column. From Fig. 14, it can be seen that the deviations between the non-relativistic limit (the first column) and the exact relativistic case (the fourth column) are substantial, i.e., the relativistic effect is profound in the present system. With the increasing perturbation order, the calculated results get closer to the exact ones. When H_F is approximated to the fifth order, the calculated spectra are considerably in agreement with those from the exact relativistic calculations. In particular, the pseudospin energy splittings are in good agreement with the exact relativistic results. This shows that the operator H_F presents a good description for the Dirac particle and it can be used to analyze the PSS.

In order to explore the breaking mechanism of SS and PSS, one can decompose the H_F into five components as follows,

$$\begin{cases} H_{\rm nr} &= \Sigma(r) + \frac{p^2}{2M}\,, \\ H_{\rm dy} &= -\frac{1}{2M^2}\left(Sp^2 - S'\frac{d}{dr}\right) + \frac{S}{2M^3}\left(Sp^2 - 2S'\frac{d}{dr}\right)\,, \\ H_{\rm sl} &= -\frac{\Delta'}{4M^2}\frac{\kappa}{r} + \frac{S\Delta'}{2M^3}\frac{\kappa}{r}\,, \\ H_{\rm km} &= -\frac{p^4}{8M^3}\,, \\ H_{\rm dw} &= \frac{\Sigma''}{8M^2} - \frac{\Sigma'^2 - 2\Sigma'\Delta' + 4S\Sigma''}{16M^3}\,. \end{cases} \tag{49}$$

Here, $H_{\rm nr}$ corresponds to the operator describing the Dirac particle in the non-relativistic limit; $H_{\rm dy}$ is related to the dynamical effect; the SO interaction is reflected in $H_{\rm sl}$; $H_{\rm km}$ represents the relativistic modification of kinetic energy; and $H_{\rm dw}$ can be regarded as the Darwin term. The contribution of every component to the pseudospin energy splitting is listed in Table 2 for the $1\tilde{f}$ and $1\tilde{g}$ pseudospin doublets. It is found that the contributions of $H_{\rm nr}$ to $\Delta\epsilon$ are large, in accordance with the fact that there is no PSS in the non-relativistic limit. The contributions of $H_{\rm dy}$ to $\Delta\epsilon$ are positive, which means the pseudospin splittings are enhanced by the dynamical terms, and which supports a relativistic origin for this symmetry. The contributions of $H_{\rm sl}$ to $\Delta\epsilon$ are negative, which implies the pseudospin splittings are reduced by the spin-orbit terms. Compared with $H_{\rm nr}$, $H_{\rm dy}$, and $H_{\rm sl}$, the contributions of $H_{\rm km}$ and $H_{\rm dw}$ to $\Delta\epsilon$ are relatively minor. The above discussions clearly show that the quality of PSS originates mainly from the competition between the SO interactions and the dynamical effects.

Table 2. Contributions of operators O_i to the single-particle energies E_k, $\epsilon_i(k) = \langle k| O_i |k\rangle$, where O_i correspond to $H_{\rm nr}$, $H_{\rm dy}$, $H_{\rm sl}$, $H_{\rm km}$, and $H_{\rm dw}$, k represents the single-particle states $2d_{5/2}$, $1g_{7/2}$, $2f_{7/2}$, and $1h_{9/2}$, and $\Delta\epsilon = \epsilon_i(a) - \epsilon_i(b)$. The total sums are listed in the last line.

i	$2d_{5/2}$ $\epsilon_i(a)$	$1g_{7/2}$ $\epsilon_i(b)$	$\Delta\epsilon$	$2f_{7/2}$ $\epsilon_i(a)$	$1h_{9/2}$ $\epsilon_i(b)$	$\Delta\epsilon$
1	−30.872	−35.649	4.777	−21.303	−27.035	5.732
2	7.511	6.746	0.765	8.448	8.003	0.445
3	−0.509	0.721	−1.230	−0.732	1.083	−1.815
4	0.018	0.042	−0.023	0.004	0.042	−0.038
5	2.732	2.349	0.383	3.021	2.717	0.303
6	−0.240	0.410	−0.650	−0.303	0.581	−0.884
7	0.001	0.018	−0.017	−0.006	0.015	−0.022
8	−0.316	−0.257	−0.059	−0.440	−0.389	−0.051
total	−21.675	−25.621	3.946	−11.312	−14.982	3.671

For the deformed nuclei, the components of H_F are represented as

$$\begin{cases} H_{\rm nr} &= \Sigma + \frac{p^2}{2M} , \\ H_{\rm dy} &= -\frac{1}{2M^2}\left(Sp^2 - \nabla S\cdot\nabla\right) + \frac{S}{2M^3}\left(Sp^2 - 2\nabla S\cdot\nabla\right) , \\ H_{\rm sl} &= \frac{1}{4M^2}\left(1-\frac{2S}{M}\right)\vec{\sigma}\cdot\left(\nabla\Delta\times\vec{p}\right) , \\ H_{\rm km} &= -\frac{p^4}{8M^3} , \\ H_{\rm dw} &= \frac{1}{16M^3}\left[2\left(M-2S\right)\nabla^2\Sigma + \left(\nabla\Sigma\right)^2 + 2\nabla\Sigma\cdot\nabla\Delta\right] , \end{cases} \tag{50}$$

which are applicable for any deformed system. As an example, when it is applied to an axially quadrupole-deformed nucleus, the corresponding potentials are adopted as

$$\begin{aligned} S\left(\vec{r}\right) &= S_0\left(r\right) + S_2\left(r\right)P_2\left(\theta\right) , \\ V\left(\vec{r}\right) &= V_0\left(r\right) + V_2\left(r\right)P_2\left(\theta\right) , \end{aligned} \tag{51}$$

where $P_2\left(\theta\right) = \frac{1}{2}\left(3\cos^2\theta - 1\right)$. The radial parts in Eq. (51) take the Woods–Saxon form,

$$\begin{aligned} S_0\left(r\right) &= S_{\rm WS}f(r) , & S_2\left(r\right) &= -\beta_2 S_{\rm WS}k\left(r\right) , \\ V_0\left(r\right) &= V_{\rm WS}f(r) , & V_2\left(r\right) &= -\beta_2 V_{\rm WS}k\left(r\right) , \end{aligned} \tag{52}$$

with $f\left(r\right) = 1/[1+\exp\left(\frac{r-R}{a}\right)]$ and $k\left(r\right) = r\,df(r)/dr$. Here $V_{\rm WS}$ and $S_{\rm WS}$ are chosen as the typical depths of the scalar and vector potentials in the RMF model, i.e., 350 MeV and −405 MeV, respectively, and the potential diffuseness a is fixed as 0.67 fm. The β_2 is the axial deformation parameter of the potential. The radius $R \equiv r_0 A^{1/3}$ with $r_0 = 1.27$ fm. Here, ^{154}Dy is chosen as an example, and the energy spectra of H_F are calculated by the expansion in harmonic oscillator basis.

In Fig. 15, the energy splitting between the (5/2[303], 7/2[303]) spin doublets is shown as a function of β_2. In the prolate side, the spin energy splitting comes almost entirely from the contribution of the spin-orbit term $H_{\rm sl}$. The contributions of the non-relativistic term $H_{\rm nr}$ and the dynamical term $H_{\rm dy}$ to the energy splitting nearly cancel out each other, while those from the relativistic modification of kinetic energy $H_{\rm km}$ and the Darwin term $H_{\rm dw}$ are negligible. Furthermore, the spin energy splitting increases with the increasing deformation. In the oblate side, the spin energy splitting is rather sensitive to β_2, which comes from the combination of the spin-orbit term and the non-relatvistic term.

Compared with the SS, the origin of PSS is more complicated for the deformed nuclei. In Fig. 16, the energy splitting between the

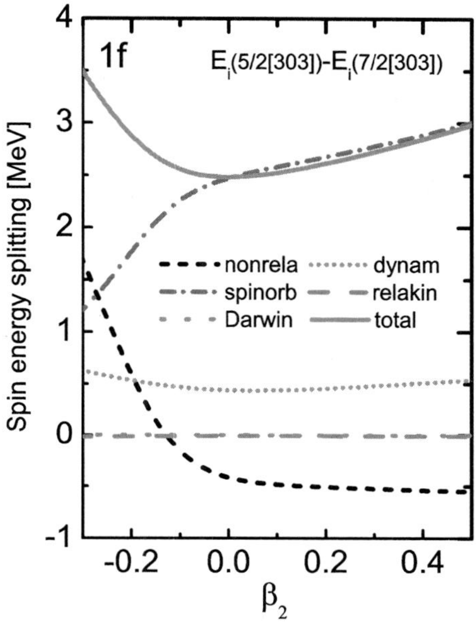

Fig. 15. (Color online) Contributions of each term in H_F to the spin energy splitting and their correlations with the deformation parameter β_2 for the $(5/2[303], 7/2[303])$ spin doublets. The "nonrela, dynam, spinorb, relakin, and Darwin" denote the non-relativistic part, the dynamical term, the spin-orbit term, the relativistic modification of kinetic energy, and the Darwin term, respectively. The total pseudospin energy splitting is marked as "total". Taken from Ref. 57.

$(5/2[402], 7/2[404])$ pseudospin doublets is shown as a function of β_2. The variation of total energy splitting with β_2 is dominated by three parts, i.e., the non-relativistic term $H_{\rm nr}$, the spin-orbit term $H_{\rm sl}$, and the dynamical term $H_{\rm dy}$. The influence from the relativistic modification of kinetic energy $H_{\rm km}$ and the Darwin term $H_{\rm dw}$ are negligible. Over the range of β_2 shown in the figure, the energy splitting from the non-relativistic term is the most remarkable. The relativistic PSS is significantly improved by the spin-orbit interaction and sometimes the dynamical effect. The spin-orbit term always plays a role in favor of the PSS, while the dynamical effect depends on the particular doublets and the deformation.[57] For the doublets near the continuum, they evolve away from the continuum with the increasing deformation β_2, and the contribution of the dynamical term evolves from an improvement to a destruction for the PSS. This explains the reason why the PSS becomes better for the single-particle states closer to the continuum.

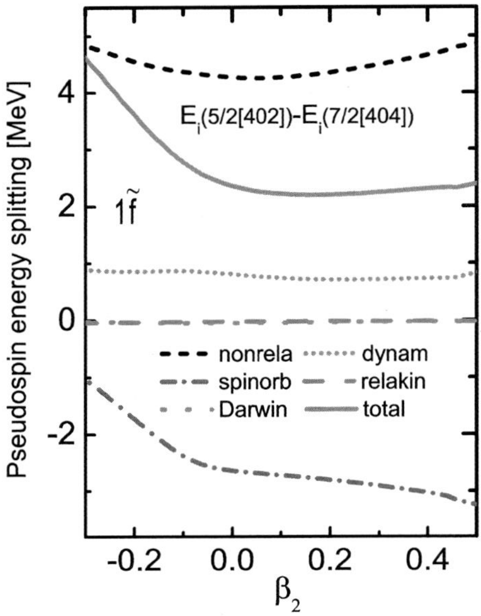

Fig. 16. (Color online) Same as Fig. 15, but for the energy splitting of the (5/2[402], 7/2[404]) pseudospin doublets. Taken from Ref. 57.

5.4. *Pseudospin symmetry in supersymmetric representation*

Works are also in progress for understanding the origin of PSS and its breaking mechanism in a perturbative way. On the one hand, the perturbation theory was used in Ref. 30 to investigate the symmetries of the Dirac Hamiltonian and their breaking in realistic nuclei, which provided a clear way for investigating the perturbative nature of PSS. On the other hand, the supersymmetric (SUSY) quantum mechanics[60] was used to investigate the symmetries of the Dirac Hamiltonian.[61–63] In particular, by employing both exact and broken supersymmetries, the phenomenon that all states with $\tilde{l} > 0$ have their own pseudospin partners except for the so-called intruder states can be interpreted naturally within a unified scheme.

Recent works shown in the previous Section bridged the perturbation calculations and the supersymmetric descriptions by using the SRG. With this method, the Dirac Hamiltonian is transformed into a diagonal form, expanding in a series of $1/M$. The effective Hamiltonian in the Schrödinger-like equation thus obtained is Hermitian, which makes the perturbation calculations possible.

By combining the SRG, SUSY quantum mechanics, and perturbation theory, Liang and coauthors[64,65] pointed out that it is promising to understand the PSS and its breaking mechanism in a fully quantitative way. In this Section, we will introduce the corresponding progress.

5.4.1. *Supersymmetric quantum mechanics*

It has been shown that every second-order differential Hamiltonian can be factorized into a product of two Hermitian conjugate first-order differential operators, i.e.,[66]

$$H_1 = B^+ B^- , \tag{53}$$

with $B^- = [B^+]^\dagger$. Its SUSY partner Hamiltonian can thus be constructed by[60]

$$H_2 = B^- B^+ . \tag{54}$$

The Hermitian operators

$$Q_1 = \begin{pmatrix} 0 & B^+ \\ B^- & 0 \end{pmatrix} , \qquad Q_2 = \begin{pmatrix} 0 & -iB^+ \\ iB^- & 0 \end{pmatrix} , \tag{55}$$

are the so-called supercharges, and the extended SUSY Hamiltonian H_S is the square of these Hermitian supercharges,

$$H_S = Q_1^2 = Q_2^2 = \begin{pmatrix} H_1 & 0 \\ 0 & H_2 \end{pmatrix} . \tag{56}$$

Since the extended Hamiltonian H_S is the square of the supercharges, all eigenvalues $E_S(n)$ of the eigenvalue equation

$$H_S \Psi_S(n) = E_S(n) \Psi_S(n) \tag{57}$$

are non-negative, and the two-component wave functions read

$$\Psi_S(n) = \begin{pmatrix} \psi_1(n) \\ \psi_2(n) \end{pmatrix} , \tag{58}$$

where $\psi_1(n)$ and $\psi_2(n)$ are the eigenfunctions of H_1 and H_2, respectively. For each eigenstate with $E_S(n) > 0$, it is an eigenstate for both H_1 and H_2, and the corresponding eigenfunctions satisfy

$$\psi_2(n) = \frac{B^-}{\sqrt{E_S(n)}} \psi_1(n) , \qquad \psi_1(n) = \frac{B^+}{\sqrt{E_S(n)}} \psi_2(n) , \tag{59}$$

with the normalization factor $1/\sqrt{E_S(n)}$.

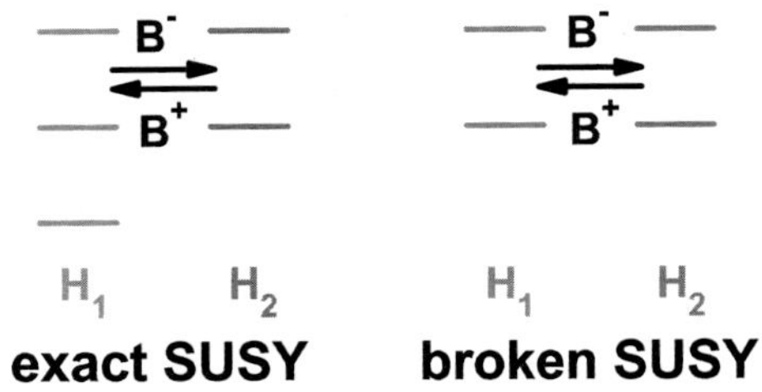

Fig. 17. (Color online) Schematic patterns of the exact and broken supersymmetries. Taken from Ref. 64.

The property of SUSY can be either exact (also called unbroken) or broken.[60] The SUSY is exact when the eigenvalue equation (57) has a zero energy eigenstate $E_S(0) = 0$. In this case, as a usual convention, the Hamiltonian H_1 has an additional eigenstate at zero energy that does not appear in its partner Hamiltonian H_2, because

$$B^-\psi_1(0) = 0\,, \qquad \psi_2(0) = 0\,, \tag{60}$$

i.e., the trivial eigenfunction of H_2 identically equals zero. The SUSY is broken when the eigenvalue equation (57) does not have any zero energy eigenstate. In this case, the partner Hamiltonians H_1 and H_2 have identical spectra. The schematic patterns of both cases are illustrated in Fig. 17.

5.4.2. *Perturbation with SRG and SUSY*

Gathering all pieces presented above, it is promising to understand the PSS and its breaking mechanism in a fully quantitative way by combining the SRG, SUSY quantum mechanics, and perturbation theory.[64,65]

As shown in Eq. (46), the Dirac equation can be transformed into a diagonal form in a series of $1/M$ by using the SRG, and its lowest-order approximation corresponds to a Schrödinger equation. In Ref. 64, taking this lowest-order approximation as an example, the idea for applying the SUSY quantum mechanics to trace the origin of the PSS was illustrated and the PSS breaking mechanism was explored quantitatively by the perturbation theory.

By adopting the spherical symmetry, the radial Schrödinger equation reads

$$H(\kappa)R(r) = ER(r) \tag{61}$$

with the single-particle Hamiltonian

$$H(\kappa) = -\frac{d^2}{2Mdr^2} + \frac{\kappa(\kappa+1)}{2Mr^2} + V(r)\,, \tag{62}$$

and the single-particle wave functions

$$\psi_\alpha(\mathbf{r}) = \frac{R_a(r)}{r} Y^{l_a}_{j_a m_a}(\hat{\mathbf{r}})\,. \tag{63}$$

Here $V(r)$ is the non-relativistic central potential standing for the sum of the scalar and vector potentials $\Sigma(r)$ in Eq. (46).

In order to investigate the origin of PSS and its breaking mechanism, it is crucial to identify the relevant term proportional to $\tilde{l}(\tilde{l}+1) = \kappa(\kappa-1)$, which holds the same values for the pseudospin doublets a and b with $\kappa_a + \kappa_b = 1$. The SUSY quantum mechanics is one of several promising approaches for identifying such $\kappa(\kappa-1)$ structure.

A couple of Hermitian conjugate first-order operators are set as[64]

$$B^+_\kappa = \left[Q_\kappa(r) - \frac{d}{dr}\right]\frac{1}{\sqrt{2M}}\,, \qquad B^-_\kappa = \frac{1}{\sqrt{2M}}\left[Q_\kappa(r) + \frac{d}{dr}\right]\,, \tag{64}$$

and the reduced superpotentials $q_\kappa(r)$ are assumed as

$$q_\kappa(r) = Q_\kappa(r) - \frac{\kappa}{r}\,. \tag{65}$$

In this way, the Hamiltonians H_1 and H_2 can be expressed as

$$\begin{aligned} H_1(\kappa) = B^+_\kappa B^-_\kappa &= \frac{1}{2M}\left[-\frac{d^2}{dr^2} + \frac{\kappa(\kappa+1)}{r^2} + q^2_\kappa + \frac{2\kappa}{r}q_\kappa - q'_\kappa\right]\,, \\ H_2(\kappa) = B^-_\kappa B^+_\kappa &= \frac{1}{2M}\left[-\frac{d^2}{dr^2} + \frac{\kappa(\kappa-1)}{r^2} + q^2_\kappa + \frac{2\kappa}{r}q_\kappa + q'_\kappa\right]\,, \end{aligned} \tag{66}$$

and

$$\begin{aligned} H(\kappa) &= H_1(\kappa) + e(\kappa)\,, \\ \tilde{H}(\kappa) &= H_2(\kappa) + e(\kappa) = -\frac{d^2}{2Mdr^2} + \frac{\kappa(\kappa-1)}{2Mr^2} + \tilde{V}_\kappa(r)\,, \end{aligned} \tag{67}$$

where $e(\kappa)$ is the energy shift, whose details can be found in Ref. 64. Thus, the reduced superpotentials $q_\kappa(r)$ satisfy the first-order differential equation,[64]

$$\frac{1}{2M}\left[q^2_\kappa(r) + \frac{2\kappa}{r}q_\kappa(r) - q'_\kappa(r)\right] + e(\kappa) = V(r)\,. \tag{68}$$

In this Section, we use a tilde to denote the operators, potentials, and wave functions belonging to $\tilde{H}$.

In the following calculations, the mass of a nucleon is taken as $M = 939.0$ MeV, and the central potential $V(r) = V_0/[1+e^{(r-R)/a}]$ is chosen as a Woods–Saxon form with the parameters $V_0 = -63.297$ MeV, $R = 6.278$ fm, and $a = 0.615$ fm, which corresponds to the neutron potential provided in Ref. 67 by taking $N = 82$ and $Z = 50$. This potential is illustrated as the solid line in Fig. 18 below.

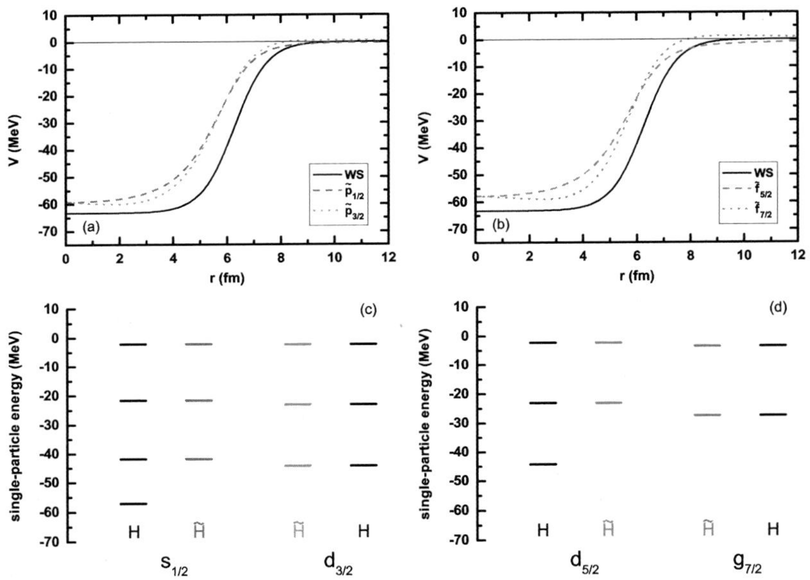

Fig. 18. (Color online) Upper panels: κ-dependent central potentials $\tilde{V}_\kappa(r)$ in $\tilde{H}$ as a function of r for the (a) $\tilde{p}$ and (b) $\tilde{f}$ orbitals, while the Woods–Saxon potential in H is shown for comparison. Lower panels: The corresponding single-particle energies obtained with H and $\tilde{H}$. Taken from Ref. 64.

In Fig. 19, the reduced PSO splittings $(E_{j_<} - E_{j_>})/(2\tilde{l}+1)$ versus their average single-particle energies $E_{\rm av} = (E_{j_<} + E_{j_>})/2$ are plotted. It is seen that the amplitude of the reduced PSO splittings is less than 1 MeV. Moreover, as a general tendency, the splittings become smaller with increasing single-particle energies, which is in agreement with the results shown in Fig. 4 by the self-consistent RMF theory. It is very interesting to investigate the physical mechanism for such energy-dependent behavior. This also helps to figure out whether or not the PSS is an accidental symmetry.[68]

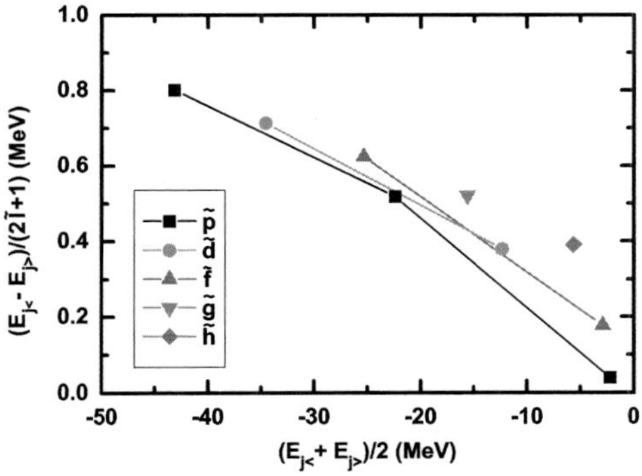

Fig. 19. (Color online) Reduced PSO splittings $(E_{j_<} - E_{j_>})/(2\tilde{l}+1)$ versus their average single-particle energies $(E_{j_<} + E_{j_>})/2$. Taken from Ref. 64.

To obtain the SUSY partner Hamiltonian $\tilde{H}$ in Eq. (67), one solves the first-order differential equation (68) for the reduced superpotentials $q_\kappa(r)$ with the boundary condition $q_\kappa(0) = 0$. Then the κ-dependent central potentials $\tilde{V}_\kappa(r)$ in $\tilde{H}$ can be obtained. In the upper panels of Fig. 18, these central potentials $\tilde{V}_\kappa(r)$ are shown, while the Woods–Saxon potential $V(r)$ in H is also shown for comparison. For all κ, the potentials $\tilde{V}_\kappa(r)$ remain approximately a Woods–Saxon shape, and they become shallower than the original potential $V(r)$. By comparing two upper panels, it is seen that the amplitude of the difference between $\tilde{V}_\kappa(r)$ for a pair of pseudospin doublets increases with the difference of their quantum numbers $|\kappa_a - \kappa_b|$.

After getting the central potentials $\tilde{V}_\kappa(r)$, one is ready to calculate the single-particle energies and wave functions of the SUSY partner Hamiltonians $\tilde{H}(\kappa)$. In the lower panels of Fig. 18, the discrete single-particle energies obtained with $\tilde{H}$ are compared with those obtained with H. It is clear that the eigenstates of Hamiltonians H and $\tilde{H}$ are identical, except for the lowest eigenstates with $\kappa < 0$ in H, which are the so-called intruder states. In other words, the fact that the intruder states have no pseudospin partners can be interpreted as a natural result of the exact SUSY for $\kappa < 0$ and broken SUSY for $\kappa > 0$. By holding the one-to-one mapping relation in the two sets of spectra, the origin of PSS, which is deeply hidden in H, can now be traced by employing its SUSY partner Hamiltonian $\tilde{H}$.

The single-particle radial wave functions $\tilde{R}_{\tilde{n}\tilde{l}j}(r)$ of $\tilde{H}$ for the $2\tilde{p}$ and $1\tilde{f}$ pseudospin doublets are shown in Fig. 20. It is found that the wave functions of pseudospin doublets are almost identical to each other. Therefore,

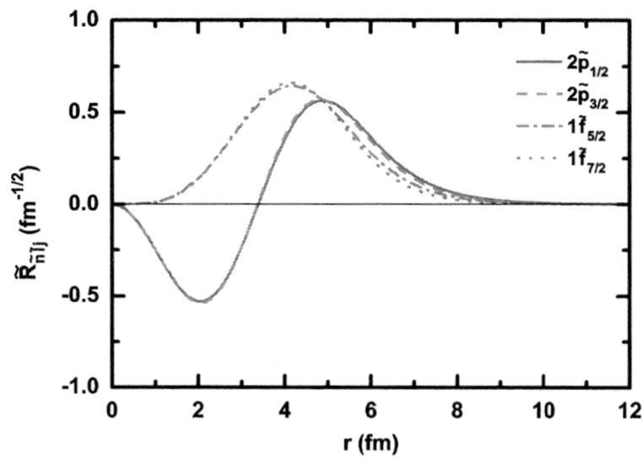

Fig. 20. (Color online) Single-particle wave functions $\tilde{R}_{\tilde{n}\tilde{l}j}(r)$ of $\tilde{H}$ for the $3s_{1/2}$, $2d_{3/2}$, $2d_{5/2}$, and $1g_{7/2}$ states. Taken from Ref. 64.

within this representation, the quasi-degeneracy of pseudospin doublets is closely related to the similarity of their wave functions, and vice versa.[64]

To perform the perturbation calculations, the Hamiltonian $\tilde{H}$ is split as

$$\tilde{H} = \tilde{H}_0^{\rm PSS} + \tilde{W}^{\rm PSS}, \tag{69}$$

where $\tilde{H}_0^{\rm PSS}$ and $\tilde{W}^{\rm PSS}$ are the corresponding PSS-conserving and PSS-breaking terms, respectively. By requiring $\tilde{W}^{\rm PSS}$ proportional to κ, which is similar to the case of the spin-orbit term in the conventional scheme, one has

$$\begin{aligned} \tilde{H}_0^{\rm PSS} &= \frac{1}{2M}\left[-\frac{d^2}{dr^2} + \frac{\kappa(\kappa-1)}{r^2}\right] + \tilde{V}_{\rm PSS}(r)\,, \\ \tilde{W}^{\rm PSS} &= \kappa \tilde{V}_{\rm PSO}(r)\,. \end{aligned} \tag{70}$$

In such a way, the PSS-conserving $\tilde{V}_{\rm PSS}(r)$ and PSS-breaking $\tilde{V}_{\rm PSO}(r)$ potentials can be determined as[64]

$$\tilde{V}_{\rm PSS}(r) = \frac{\kappa_b \tilde{V}_{\kappa_a}(r) - \kappa_a \tilde{V}_{\kappa_b}(r)}{\kappa_a - \kappa_b} \quad \text{and} \quad \tilde{V}_{\rm PSO}(r) = \frac{1}{M}\frac{q'_{\kappa_a}(r) - q'_{\kappa_b}(r)}{\kappa_a - \kappa_b}\,. \tag{71}$$

In Fig. 21, the $\tilde{V}_{\rm PSS}(r)$ and $\tilde{V}_{\rm PSO}(r)$ potentials are illustrated by taking the $\tilde{p}$ and $\tilde{f}$ orbitals as examples. On the one hand, it can be seen that the PSS-conserving potentials remain an approximate Woods–Saxon shape, and they are κ-dependent to a small extent. On the other hand, the PSS-breaking potentials show several special features:[64] (i) These PSS-breaking

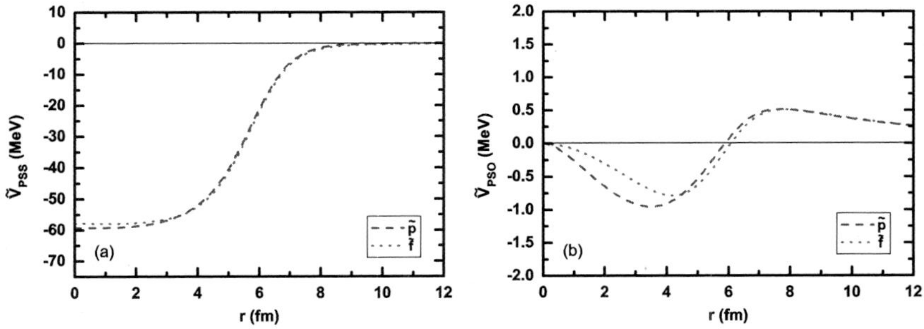

Fig. 21. (Color online) PSS-conserving potentials $\tilde{V}_{\rm PSS}(r)$ (the left panel) and PSS-breaking potentials $\tilde{V}_{\rm PSO}(r)$ (the right panel) for the $\tilde{p}$ and $\tilde{f}$ orbitals. Taken from Ref. 64.

potentials are regular functions of r, in particular, they vanish at $r \to \infty$. (ii) It can be seen that the amplitude of $\tilde{V}_{\rm PSO}$ is around 1 MeV, which directly leads to the reduced PSO splittings $\Delta E_{\rm PSO} \lesssim 1$ MeV as shown in Fig. 19. (iii) More importantly, the PSO potentials $\tilde{V}_{\rm PSO}(r)$ are negative at small radius but positive at large radius with a node at the surface region, which is totally different from the usual SO potentials with a surface-peak shape. The particular shape of the PSO potentials can explain well the variations of the PSO splitting with the single-particle energy.

Finally, the perturbation calculations can be performed based on the pseudospin symmetric Hamiltonian $\tilde{H}_0^{\rm PSS}$ with the perturbation $\tilde{W}^{\rm PSS}$. It is found that the PSO splittings are excellently reproduced by the first-order perturbation calculations.[64]

In such an explicit and quantitative way, the PSO splittings $\Delta E_{\rm PSO}$ can be directly understood by the PSS-breaking term $\tilde{W}^{\rm PSS}$ within the representation of the SUSY partner Hamiltonian $\tilde{H}$. Furthermore, this symmetry-breaking term can be treated as a very small perturbation on the exact PSS limit $\tilde{H}_0^{\rm PSS}$. Therefore, the PSS discussed here is of pertubative nature.[64]

Recently, it has been demonstrated in Ref. 65 that the perturbative nature of PSS remains even when a substantial spin-orbit potential is included. The spin-orbit term shows both indirect and direct effects on the PSS-breaking potentials $\tilde{V}_{\rm PSO}(r)$. The indirect effect due to the changes of the reduced superpotentials $q_\kappa(r)$ is rather small. In contrast, the direct effect, i.e., the spin-orbit potential itself appearing in both H and $\tilde{H}$, reduces the PSO splittings $\Delta E_{\rm PSO}$ systematically.

6. Summary and perspectives

After the independent observation of the near degeneracy between pairs of single-particle states $(n, l, j = l+1/2)$ and $(n-1, l+2, j = l+3/2)$, i.e., the so-called pseudospin symmetry, by Hecht and Adler[5] and by Arima, Harvey, and Shimizu[6] in 1969, it raised a fascinating question whether such near degeneracy is accidental, a degeneracy not explained by an obvious symmetry; or due to symmetry breaking, more descriptively hidden symmetry. The pioneering work by examining the ratio between the strengths of the spin-orbit and orbit-orbit potentials by Bohr, Hamamoto, and Mottelson[19] in 1982 opened the door to explore the hidden symmetry in understanding the origin of PSS.

With the success of the RMF theory,[25] it was quite encouraging to find that such a special ratio between the strengths of the spin-orbit and orbit-orbit interactions can be reproduced by the RMF theory.[69] The revelation of the pseudo-orbital angular momentum as the orbital angular momentum of the lower component of the Dirac spinor by Ginocchio[24] in 1997 proved to be an unexpected success of the RMF theory. It was then followed by lots of exciting discoveries. For example, the vanishing of the derivative for the sum of the scalar and vector potentials, i.e., $d\Sigma(r)/dr = 0$, can lead to the exact PSS,[27] which indicates that the PSS becomes much better for exotic nuclei with a highly diffused potential.[29] While developing the RMF theory in a Dirac Woods–Saxon basis,[36] Zhou, Meng, and Ring[37] discovered the spin symmetry in the anti-nucleon spectra. A rigorous verification of the PSS in the single-particle resonant states is given by Lu, Zhao, and Zhou[51] by examining the Jost functions corresponding to the small components of the radial Dirac wave functions. By using the similarity renormalization group, Guo and coauthors have made a new exploration of the PSS in the spherical[55,56] and axially deformed[57] systems. Combining the SRG, supersymmetric quantum mechanics, and perturbation theory, Liang and coauthors[64,65] have provided a promising way to understand the origin of PSS and its breaking quantitatively.

From the experimental point of view, while introducing some themes in the study of very deformed rotating nuclei, Mottelson[70] preluded the link between the PSS and observed experiments, including nuclear superdeformed configurations,[7] identical bands,[8,9] quantized alignment,[10] pseudospin partner bands,[11,12] and so on. The experimental verification of the PSS can also be done from the single-particle energies. The observed single-particle energies obtained by the pickup or knockout reactions contain the polarization effects, except for the cases that the spectroscopic factors of the single-particle states are close to one.[71] Therefore, one should take into account the particle-vibration coupling and polarization effects, see, e.g., Refs. 72, 73.

Finally, exploring the deep reasons for the approximate PSS observed in realistic nuclei is also a challenging problem. There are already some discussions on the physics behind the near equality of the vector and scalar potentials in the Dirac Hamiltonian[74–76] and more investigations are expected. In particular, it is highly necessary to search for the links with more fundamental models on strong interactions or realistic interactions between nucleons.

Acknowledgment

We would like to express our gratitude to all the collaborators and colleagues who contributed to the investigations presented here, in particular to A. Arima, T. S. Chen, L. S. Geng, J. N. Ginocchio, X. T. He, R. V. Jolos, A. Leviatan, F. Q. Li, W. H. Long, B. N. Lu, H. F. Lü, P. Ring, H. Sagawa, W. Scheid, S. H. Shen, C. Y. Song, K. Sugawara-Tanabe, H. Toki, N. Van Giai, S. Yamaji, S. C. Yang, J. M. Yao, S. Q. Zhang, Y. Zhang, E. G. Zhao, and P. W. Zhao. We acknowledge the fruitful discussions with K. Arita, I. Hamamoto, A. Ikeda, K. Matsuyanagi, T. Nakatsukasa, K. Sato, T. T. Sun, T. Suzuki, and Y. Zhang during the iTHES workshop: Exploration of hidden symmetries in atomic nuclei at RIKEN. This work was supported in part by the Major State 973 Program of China (Grant No. 2013CB834400), the Natural Science Foundation of China (Grants No. 10975008, No. 11175002, No. 11105006, No. 11335002, and No. 11525524), the Research Fund for the Doctoral Program of Higher Education (RFDP) (Grant No. 20110001110087), the Grant-in-Aid for JSPS Fellows under Grant No. 24-02201, the RIKEN iTHES Project, and the RIKEN Foreign Postdoctoral Researcher Program.

References

1. W. M. Elsasser, Sur le principe de pauli dans les noyaux, *J. Phys. Radium.* **4**, 549–556 (1933). doi: 10.1051/jphysrad:01933004010054900.
2. O. Haxel, J. H. D. Jensen, and H. E. Suess, On the "magic numbers" in nuclear structure, *Phys. Rev.* **75**, 1766–1766 (Jun, 1949). doi: 10.1103/PhysRev.75.1766.2.
3. M. G. Mayer, On closed shells in nuclei. II, *Phys. Rev.* **75**, 1969–1970 (June, 1949). doi: 10.1103/PhysRev.75.1969.

4. O. Sorlin and M.-G. Porquet, Nuclear magic numbers: New features far from stability, *Prog. Part. Nucl. Phys.* **61**, 602–673 (Oct., 2008). ISSN 0146-6410. doi: 10.1016/j.ppnp.2008.05.001.
5. K. T. Hecht and A. Adler, Generalized seniority for favored $J \neq 0$ pairs in mixed configurations, *Nucl. Phys. A.* **137**, 129–143 (Nov., 1969). doi: 10.1016/0375-9474(69)90077-3.
6. A. Arima, M. Harvey, and K. Shimizu, Pseudo LS coupling and pseudo SU(3) coupling schemes, *Phys. Lett. B.* **30**, 517–522 (Dec., 1969). doi: 10.1016/0370-2693(69)90443-2.
7. J. Dudek, W. Nazarewicz, Z. Szymanski, and G. A. Leander, Abundance and systematics of nuclear superdeformed states; relation to the pseudospin and pseudo-SU(3) symmetries, *Phys. Rev. Lett.* **59**, 1405–1408 (Sept., 1987). doi: 10.1103/PhysRevLett.59.1405.
8. W. Nazarewicz, P. J. Twin, P. Fallon, and J. D. Garrett, Natural-parity states in superdeformed bands and pseudo SU(3) symmetry at extreme conditions, *Phys. Rev. Lett.* **64**, 1654–1657 (Apr, 1990). doi: 10.1103/PhysRevLett.64.1654.
9. J. Y. Zeng, J. Meng, C. S. Wu, E. G. Zhao, Z. Xing, and X. Q. Chen, Spin determination and quantized alignment in the superdeformed bands in ^{152}Dy, ^{151}Tb, and ^{150}Gd, *Phys. Rev. C.* **44**, R1745–R1748 (Nov., 1991). doi: 10.1103/PhysRevC.44.R1745.
10. F. S. Stephens, M. A. Deleplanque, J. E. Draper, R. M. Diamond, A. O. Macchiavelli, C. W. Beausang, W. Korten, W. H. Kelly, F. Azaiez, J. A. Becker, E. A. Henry, S. W. Yates, M. J. Brinkman, A. Kuhnert, and J. A. Cizewski, Pseudospin symmetry and quantized alignment in nuclei, *Phys. Rev. Lett.* **65**, 301–304 (Jul, 1990). doi: 10.1103/PhysRevLett.65.301.
11. Q. Xu, S. J. Zhu, J. H. Hamilton, A. V. Ramayya, J. K. Hwang, B. Qi, J. Meng, J. Peng, Y. X. Luo, J. O. Rasmussen, I. Y. Lee, S. H. Liu, K. Li, J. G. Wang, H. B. Ding, L. Gu, E. Y. Yeoh, and W. C. Ma, Identification of pseudospin partner bands in ^{108}Tc, *Phys. Rev. C.* **78**, 064301 (Dec., 2008). doi: 10.1103/PhysRevC.78.064301.
12. W. Hua, X. H. Zhou, Y. H. Zhang, Y. Zheng, M. L. Liu, F. Ma, S. Guo, L. Ma, S. T. Wang, N. T. Zhang, Y. D. Fang, X. G. Lei, Y. X. Guo, M. Oshima, Y. Toh, M. Koizumi, Y. Hatsukawa, B. Qi, S. Q. Zhang, J. Meng, and M. Sugawara, Properties of the rotational bands in the transitional nucleus ^{189}Pt, *Phys. Rev. C.* **80**, 034303 (Sept., 2009). doi: 10.1103/PhysRevC.80.034303.
13. W. H. Long, P. Ring, J. Meng, N. Van Giai, and C. A. Bertulani, Nuclear halo structure and pseudospin symmetry, *Phys. Rev. C.* **81**, 031302(R) (Mar., 2010). doi: 10.1103/PhysRevC.81.031302.
14. R. Jolos and V. Voronov, Pseudospin symmetry and structure of nuclei with $Z \geq 100$, *Phys. At. Nucl.* **70**, 812–817 (May, 2007). doi: 10.1134/S106377880705002X.
15. J. J. Li, W. H. Long, J. Margueron, and N. Van Giai, Superheavy magic structures in the relativistic Hartree–Fock–Bogoliubov approach, *Phys. Lett. B.* **732**, 169–173 (2014). ISSN 0370-2693. doi: http://dx.doi.org/10.1016/j.physletb.2014.03.031.

16. J. N. Ginocchio, Relativistic symmetries in nuclei and hadrons, *Phys. Rep.* **414**, 165–261 (Aug., 2005). ISSN 0370-1573. doi: 10.1016/j.physrep.2005.04.003.
17. H. Liang, J. Meng, and S.-G. Zhou, Hidden pseudospin and spin symmetries and their origins in atomic nuclei, *Phys. Rep.* **570**, 1–84 (Mar., 2015). ISSN 0370-1573. doi: 10.1016/j.physrep.2014.12.005. URL `http://www.sciencedirect.com/science/article/pii/S0370157315000502`.
18. S. G. Nilsson, C. F. Tsang, A. Sobiczewski, Z. Szymanski, S. Wycech, C. Gustafson, I.-L. Lamm, P. Möller, and B. Nilsson, On the nuclear structure and stability of heavy and superheavy elements, *Nucl. Phys. A.* **131**, 1–66 (June, 1969). doi: 10.1016/0375-9474(69)90809-4.
19. A. Bohr, I. Hamamoto, and B. R. Mottelson, Pseudospin in rotating nuclear potentials, *Phys. Scr.* **26**, 267–272 (1982). ISSN 1402-4896. doi: 10.1088/0031-8949/26/4/003.
20. G. A. Lalazissis, J. Konig, and P. Ring, New parametrization for the Lagrangian density of relativistic mean field theory, *Phys. Rev. C.* **55**, 540–543 (Jan., 1997). doi: 10.1103/PhysRevC.55.540.
21. J. N. Ginocchio, A. Leviatan, J. Meng, and S.-G. Zhou, Test of pseudospin symmetry in deformed nuclei, *Phys. Rev. C.* **69**, 034303 (Mar., 2004). doi: 10.1103/PhysRevC.69.034303.
22. O. Castaños, M. Moshinsky, and C. Quesne, Transformation to pseudo-SU(3) in heavy deformed nuclei, *Phys. Lett. B.* **277**, 238–242 (Mar., 1992). ISSN 0370-2693. doi: 10.1016/0370-2693(92)90741-L.
23. A. L. Blokhin, C. Bahri, and J. P. Draayer, Origin of pseudospin symmetry, *Phys. Rev. Lett.* **74**, 4149–4152 (May, 1995). doi: 10.1103/PhysRevLett.74.4149.
24. J. N. Ginocchio, Pseudospin as a relativistic symmetry, *Phys. Rev. Lett.* **78**, 436–439 (Jan., 1997). doi: 10.1103/PhysRevLett.78.436.
25. P. Ring, Relativistic mean field theory in finite nuclei, *Prog. Part. Nucl. Phys.* **37**, 193–263 (1996). ISSN 0146-6410. doi: 10.1016/0146-6410(96)00054-3.
26. D. A. Varshalovich, A. N. Moskalev, and V. K. Khersonskii, *Quantum theory of angular momentum.* World Scientific, Singapore (1988).
27. J. Meng, K. Sugawara-Tanabe, S. Yamaji, P. Ring, and A. Arima, Pseudospin symmetry in relativistic mean field theory, *Phys. Rev. C.* **58**, R628–R631 (Aug., 1998). doi: 10.1103/PhysRevC.58.R628.
28. A. Leviatan and J. N. Ginocchio, Consequences of a relativistic pseudospin symmetry for radial nodes and intruder levels in nuclei, *Phys. Lett. B.* **518**, 214–220 (Oct., 2001). doi: 10.1016/S0370-2693(01)01039-5.
29. J. Meng, K. Sugawara-Tanabe, S. Yamaji, and A. Arima, Pseudospin symmetry in Zr and Sn isotopes from the proton drip line to the neutron drip line, *Phys. Rev. C.* **59**, 154–163 (Jan., 1999). doi: 10.1103/PhysRevC.59.154.
30. H. Liang, P. Zhao, Y. Zhang, J. Meng, and N. Van Giai, Perturbative interpretation of relativistic symmetries in nuclei, *Phys. Rev. C.* **83**, 041301(R) (Apr., 2011). doi: 10.1103/PhysRevC.83.041301.
31. W. Long, J. Meng, N. Van Giai, and S.-G. Zhou, New effective interactions in relativistic mean field theory with nonlinear terms and density-dependent

meson-nucleon coupling, *Phys. Rev. C.* **69**, 034319 (Mar., 2004). doi: 10.1103/PhysRevC.69.034319.
32. J. N. Ginocchio and D. G. Madland, Pseudospin symmetry and relativistic single-nucleon wave functions, *Phys. Rev. C.* **57**, 1167–1173 (Mar., 1998). doi: 10.1103/PhysRevC.57.1167.
33. J. N. Ginocchio, Pseudospin symmetry and relativistic mean field eigenfunctions, *Phys. Rev. C.* **66**, 064312 (Dec., 2002). doi: 10.1103/PhysRevC.66.064312.
34. P. W. Zhao, Z. P. Li, J. M. Yao, and J. Meng, New parametrization for the nuclear covariant energy density functional with a point-coupling interaction, *Phys. Rev. C.* **82**, 054319 (Nov., 2010). doi: 10.1103/PhysRevC.82.054319.
35. J. N. Ginocchio, A relativistic symmetry in nuclei, *Phys. Rep.* **315**, 231–240 (July, 1999). doi: 10.1016/S0370-1573(99)00021-6.
36. S.-G. Zhou, J. Meng, and P. Ring, Spherical relativistic Hartree theory in a Woods–Saxon basis, *Phys. Rev. C.* **68**, 034323 (Sept., 2003). doi: 10.1103/PhysRevC.68.034323.
37. S.-G. Zhou, J. Meng, and P. Ring, Spin symmetry in the antinucleon spectrum, *Phys. Rev. Lett.* **91**, 262501 (Dec., 2003). doi: 10.1103/PhysRevLett.91.262501.
38. X. T. He, S. G. Zhou, J. Meng, E. G. Zhao, and W. Scheid, Test of spin symmetry in anti-nucleon spectra, *Eur. Phys. J. A.* **28**, 265–269 (June, 2006). doi: 10.1140/epja/i2006-10066-0.
39. N. Michel, W. Nazarewicz, M. Ploszajczak, and T. Vertse, Shell model in the complex energy plane, *J. Phys. G: Nucl. Part. Phys.* **36**, 013101 (Jan., 2009). ISSN 0954-3899. doi: 10.1088/0954-3899/36/1/013101.
40. V. I. Kukulin, V. M. Krasnopol'sky, and J. Horácek, *Theory of Resonances: Principles and Applications*. Kluwer Academic, Dordrecht (1989).
41. S.-C. Yang, J. Meng, and S.-G. Zhou, Exploration of unbound states by analytical continuation in the coupling constant method within relativistic mean field theory, *Chin. Phys. Lett.* **18**, 196–198 (Feb., 2001). doi: 10.1088/0256-307X/18/2/314.
42. L. Zhang, S.-G. Zhou, J. Meng, and E.-G. Zhao, Real stabilization method for single particle resonances, *Acta Phys. Sinica.* **56**, 3839–3844 (July, 2007). doi: 10.7498/aps.56.3839.
43. J.-Y. Guo, X.-Z. Fang, P. Jiao, J. Wang, and B.-M. Yao, Application of the complex scaling method in relativistic mean-field theory, *Phys. Rev. C.* **82**, 034318 (Sept., 2010). doi: 10.1103/PhysRevC.82.034318.
44. K. Hagino and N. Van Giai, Structure of positive energy states in a deformed mean-field potential, *Nucl. Phys. A.* **735**, 55–76 (Apr., 2004). ISSN 0375-9474. doi: 10.1016/j.nuclphysa.2004.02.002.
45. Z. P. Li, J. Meng, Y. Zhang, S. G. Zhou, and L. N. Savushkin, Single-particle resonances in a deformed Dirac equation, *Phys. Rev. C.* **81**, 034311 (Mar., 2010). doi: 10.1103/PhysRevC.81.034311.
46. Z.-P. Li, Y. Zhang, D. Vretenar, and J. Meng, Single-particle resonances in a deformed relativistic potential, *Sci. China-Phys. Mech. Astron.* **53**, 773–778 (Apr., 2010). ISSN 1674-7348. doi: 10.1007/s11433-010-0161-7.

47. J.-Y. Guo, R.-D. Wang, and X.-Z. Fang, Pseudospin symmetry in the resonant states of nuclei, *Phys. Rev. C.* **72**, 054319 (Nov., 2005). doi: 10.1103/PhysRevC.72.054319.
48. J. Y. Guo and X. Z. Fang, Isospin dependence of pseudospin symmetry in nuclear resonant states, *Phys. Rev. C.* **74**, 024320 (Aug., 2006). doi: 10.1103/PhysRevC.74.024320.
49. Q. Liu, Z.-M. Niu, and J.-Y. Guo, Resonant states and pseudospin symmetry in the Dirac–Morse potential, *Phys. Rev. A.* **87**, 052122 (May, 2013). doi: 10.1103/PhysRevA.87.052122.
50. S.-S. Zhang, B.-H. Sun, and S.-G. Zhou, Exploration of pseudospin symmetry in the resonant states, *Chin. Phys. Lett.* **24**, 1199–1202 (May, 2007). ISSN 0256-307X. doi: 10.1088/0256-307X/24/5/020.
51. B.-N. Lu, E.-G. Zhao, and S.-G. Zhou, Pseudospin symmetry in single particle resonant states, *Phys. Rev. Lett.* **109**, 072501 (Aug., 2012). doi: 10.1103/PhysRevLett.109.072501.
52. B.-N. Lu, E.-G. Zhao, and S.-G. Zhou, Pseudospin symmetry in single-particle resonances in spherical square wells, *Phys. Rev. C.* **88**, 024323 (Aug, 2013). doi: 10.1103/PhysRevC.88.024323.
53. T.-S. Chen, H.-F. Lü, J. Meng, S.-Q. Zhang, and S.-G. Zhou, Pseudospin symmetry in relativistic framework with harmonic oscillator potential and Woods–Saxon potential, *Chin. Phys. Lett.* **20**, 358–361 (Mar., 2003). ISSN 0256-307X. doi: 10.1088/0256-307X/20/3/312.
54. J. R. Taylor, *Scattering Theory: The Quantum Theory on Non-relativistic Collisions.* John-Wiley & Sons, Inc., New York (1972).
55. J.-Y. Guo, Exploration of relativistic symmetry by the similarity renormalization group, *Phys. Rev. C.* **85**, 021302(R) (Feb., 2012). doi: 10.1103/PhysRevC.85.021302.
56. D.-P. Li, S.-W. Chen, and J.-Y. Guo, Further investigation of relativistic symmetry with the similarity renormalization group, *Phys. Rev. C.* **87**, 044311 (Apr., 2013). doi: 10.1103/PhysRevC.87.044311.
57. J.-Y. Guo, S.-W. Chen, Z.-M. Niu, D.-P. Li, and Q. Liu, Probing the symmetries of the Dirac Hamiltonian with axially deformed scalar and vector potentials by similarity renormalization group, *Phys. Rev. Lett.* **112**, 062502 (Feb., 2014). doi: 10.1103/PhysRevLett.112.062502.
58. F. Wegner, Flow-equations for Hamiltonians, *Ann. Phys. (Berlin).* **506**, 77–91 (1994). doi: 10.1002/andp.19945060203.
59. A. B. Bylev and H. J. Pirner, Hamiltonian flow equations for a Dirac particle in an external potential, *Phys. Lett. B.* **428**, 329–333 (June, 1998). doi: 10.1016/S0370-2693(98)00417-1.
60. F. Cooper, A. Khare, and U. Sukhatme, *Supersymmetry in Quantum Mechanics.* World Scientific, Singapore (2001).
61. A. Leviatan, Supersymmetric patterns in the pseudospin, spin, and Coulomb limits of the Dirac equation with scalar and vector potentials, *Phys. Rev. Lett.* **92**, 202501 (May, 2004). doi: 10.1103/PhysRevLett.92.202501.
62. S. Typel, Pseudospin, supersymmetry and the shell structure of atomic nuclei, *Nucl. Phys. A.* **806**, 156–178 (June, 2008). doi: 10.1016/j.nuclphysa.2008.03.018.

63. A. Leviatan, Symmetries and supersymmetries of the Dirac Hamiltonian with axially deformed scalar and vector potentials, *Phys. Rev. Lett.* **103**, 042502 (July, 2009). doi: 10.1103/PhysRevLett.103.042502.
64. H. Liang, S. Shen, P. Zhao, and J. Meng, Pseudospin symmetry in supersymmetric quantum mechanics: Schrödinger equations, *Phys. Rev. C.* **87**, 014334 (Jan., 2013). doi: 10.1103/PhysRevC.87.014334.
65. S. Shen, H. Liang, P. Zhao, S. Zhang, and J. Meng, Pseudospin symmetry in supersymmetric quantum mechanics. II. spin-orbit effects, *Phys. Rev. C.* **88**, 024311 (Aug, 2013). doi: 10.1103/PhysRevC.88.024311.
66. L. Infeld and T. E. Hull, The factorization method, *Rev. Mod. Phys.* **23**, 21–68 (Jan, 1951). doi: 10.1103/RevModPhys.23.21.
67. W. Koepf and P. Ring, The spin-orbit field in superdeformed nuclei: a relativistic investigation, *Z. Phys. A.* **339**, 81–90 (Mar., 1991). doi: 10.1007/BF01282936.
68. S. Marcos, M. López-Quelle, R. Niembro, and L. N. Savushkin, Pseudospin symmetry as an accidental symmetry in the relativistic framework, *Eur. Phys. J. A.* **37**, 251–262 (Aug., 2008). ISSN 1434-6001. doi: 10.1140/epja/i2008-10619-1.
69. C. Bahri, J. P. Draayer, and S. A. Moszkowski, Pseudospin symmetry in nuclear physics, *Phys. Rev. Lett.* **68**, 2133–2136 (Apr., 1992). doi: 10.1103/PhysRevLett.68.2133.
70. B. Mottelson, Some themes in the study of very deformed rotating nuclei, *Nucl. Phys. A.* **522**, 1c–12c (1991). ISSN 0375-9474. doi: 10.1016/0375-9474(91)90048-B.
71. P. Cottle, Nuclear physics: Doubly magic tin, *Nature.* **465**, 430–431 (May, 2010). ISSN 0028-0836. doi: 10.1038/465430a.
72. E. V. Litvinova and A. V. Afanasjev, Dynamics of nuclear single-particle structure in covariant theory of particle-vibration coupling: From light to superheavy nuclei, *Phys. Rev. C.* **84**, 014305 (July, 2011). doi: 10.1103/PhysRevC.84.014305.
73. Y. F. Niu, G. Colò, M. Brenna, P. F. Bortignon, and J. Meng, Gamow-Teller response within Skyrme random-phase approximation plus particle-vibration coupling, *Phys. Rev. C.* **85**, 034314 (Mar., 2012). doi: 10.1103/PhysRevC.85.034314.
74. T. D. Cohen, R. J. Furnstahl, D. K. Griegel, and X. Jin, QCD sum rules and applications to nuclear physics, *Prog. Part. Nucl. Phys.* **35**, 221–298 (1995). ISSN 0146-6410. doi: 10.1016/0146-6410(95)00043-I.
75. R. Furnstahl and B. D. Serot, Large Lorentz scalar and vector potentials in nuclei, *Nucl. Phys. A.* **673**, 298–310 (June, 2000). ISSN 0375-9474. doi: 10.1016/S0375-9474(00)00146-9.
76. H.-W. Ke, Z. Li, J.-L. Chen, Y.-B. Ding, and X.-Q. Li, Symmetry of Dirac equation and corresponding phenomenology, *Int. J. Mod. Phys. A.* **25**, 1123–1134 (Mar., 2010). doi: 10.1142/S0217751X1004783X.

Chapter 7

Structure of hypernuclei in relativistic approaches

Kouichi Hagino

Department of Physics, Tohoku University, Sendai 980-8578, Japan
Research Center for Electron Photon Science, Tohoku University, 1-2-1 Mikamine, Sendai 982-0826, Japan
hagino@nucl.phys.tohoku.ac.jp

Jiangming Yao

School of Physical Science and Technology, Southwest University, 400715 Chongqing, China
Department of Physics, Tohoku University, Sendai 980-8578, Japan
jmyao@nucl.phys.tohoku.ac.jp

We review the relativistic mean-field approaches to hypernuclear physics. This includes Lambda hypernuclei, anti-Lambda hypernuclei, and multi-strangeness hypernuclei. We particularly focus on the properties of both ground state and collective excitations, hyperon binding energies, spin-orbit splittings, magnetic moments, a stabilization of drip-line nuclei, and the hyperon impurity effect on nuclear collectivity. We also discuss briefly the influence of hyperons on neutron stars. We conclude that the relativistic mean-field approaches have achieved a great success in the studies of hypernuclear physics.

Contents

1. Introduction

Hypernuclei consist of neutrons, protons, and one or more hyperons such as Λ, Σ, and Ξ. In Table 1, we summarize properties of those hyperons from the Particle Data Group.[1] The most extensively studied hypernuclei are single-Λ hypernuclei, which contain one Λ particle inside ordinary nuclei. Since the first discovery of Λ hypernucleus by Danysz and Pniewski in 1953[2] (see also Ref. 3), more than thirty Λ hypernuclei ranging from ${}^{3}_{\Lambda}$H up to ${}^{208}_{\Lambda}$Pb have been produced. Several properties of hypernuclei, such as the mass number dependence of Λ binding energy, spin-orbit splittings, and electromagnetic transitions, have been revealed using reaction spectroscopy with (K^-, π^-), (π^+, K^+), and $(e, e'K^+)$ reactions as well as γ-ray spectroscopy.[4] Decay properties of hypernuclei have also been studied.[5,6]

An important motivation to study hypernuclei is to extract information on baryon-baryon interactions including the strangeness degrees of freedom. Such information is crucial in order to understand neutron stars, in which hyperons may emerge in the inner part,[7,8] since the first hyperon to appear may be different depending on the properties of baryon-baryon interaction. Because hyperon-nucleon and hyperon-hyperon scattering experiments are difficult to perform, the structure of hypernuclei has been playing a vital role in order to shed light on baryon-baryon interactions.

Table 1. Properties of nucleons and hyperons from the Particle Data Group.[1]

particle	spin and parity	charge	mass (MeV/c^2)	mean-life (sec.)	quark content
p	$1/2^+$	$+e$	938.27	$> 2.1 \times 10^{29}$	uud
n	$1/2^+$	0	939.57	880	udd
Λ	$1/2^+$	0	1115.68	$2.63(2) \times 10^{-10}$	uds
Σ^+	$1/2^+$	$+e$	1189.37	0.80×10^{-10}	uus
Σ^0	$1/2^+$	0	1192.64	$7.4(7) \times 10^{-20}$	uds
Σ^-	$1/2^+$	$-e$	1197.45	$1.48(1) \times 10^{-10}$	dds
Ξ^0	$1/2^+$	0	1314.86	$2.90(9) \times 10^{-10}$	uss
Ξ^-	$1/2^+$	$-e$	1321.71	$1.64(1) \times 10^{-10}$	dss

From the nuclear structure point of view, an important feature of the Λ particle is that it carries a strangeness degree of freedom and therefore it does not suffer from the Pauli principle from other nucleons and can thus deeply penetrate into the nuclear interior. This may induce a change of *e.g.*, nuclear size,[9,10] the density distribution,[11] deformation properties,[12–21] collective excitations,[17,22–25] the neutron drip-line,[26,27] and fission barrier.[28,29] This property is referred to as the impurity effect of Λ particle, which has also been one of the most important topics in hypernuclear physics. Moreover, some new states called *supersymmetric states*[30–32] that cannot exist in ordinary nuclei can appear in hypernuclei, the study of which greatly broadens our knowledge on nuclear physics.

Theoretically, the structure of hypernuclei has been studied with the cluster model,[6,9,10,32] the shell model,[33–36] the mean-field approaches,[13,14,16,18,26,37–43] and the antisymmetrized molecular dynamics (AMD).[19–21,25] Recently, an *ab initio* method has also been applied to hypernuclei.[44] Among these theoretical methods, the mean-field approach has an advantage in that it can be globally applied to hypernuclei from light to heavy mass regions. The mean-field approach can also be directly applied to neutron stars, in which hyperons play an important role. Both non-relativistic Skyrme-Hartree-Fock[37,38] and relativistic mean-field method[40–42] have been applied to hypernuclear physics.

In this Chapter, we review the relativistic mean-field approach to various aspects of hypernuclei. It was Brockmann and Weise who first applied this approach to hypernuclei.[40] At that time, it had been already observed experimentally that the spin-orbit splittings in hypernuclei are significantly smaller than that in ordinary nuclei.[45] The relativistic approach was suitable for a discussion of spin-orbit splittings in hypernuclei, as the spin-orbit interaction naturally emerges with the relativistic framework. Since this seminal work, many developments have occurred and the relativistic approach has been applied both to the ground state and to excited states of hypernuclei. We summarize those developments in this Chapter. We shall concentrate on Λ-hypernuclei and do not discuss Σ hypernuclei,[46–50] since the theoretical treatment is rather similar to each other.

The Chapter is organized as follows. In Sec. 2, we detail how the relativistic Lagrangian can be extended to hypernuclei. We will discuss both the meson exchange and the point coupling approaches. In Sec. 3, we discuss the ground state properties of hypernuclei. These include the binding energy of Λ particle, the spin-orbit splittings, the hypernucler magnetic moments, an extension of neutron drip-line, and the deformation properties.

In Sec. 4, we discuss how collective excitations are altered in hypernuclei, employing the beyond mean-field approach. In Sec. 5, we focus our attention on anti-Λ hypernuclei. We will discuss the spin symmetry and the tensor coupling effects. In Sec. 6, we will briefly discuss multi-Λ hypernuclei as well as neutron stars. We will then summarize this Chapter in Sec. 7.

2. Relativistic mean-field approaches to hypernuclei

2.1. *Lagrangian density with hyperons*

In this section, we will introduce the general framework of RMF approaches to hypernuclei. The starting point of the RMF approaches for hypernuclei is the following Lagrangian density

$$\mathcal{L} = \mathcal{L}^{\text{free}} + \mathcal{L}^{\text{em}} + \mathcal{L}^{NN} + \mathcal{L}^{NY}, \tag{1}$$

where the free $\mathcal{L}^{\text{free}}$ and electromagnetic $\mathcal{L}^{\text{em}}$ parts have the standard form.

The last two terms $\mathcal{L}^{NN}$ and $\mathcal{L}^{NY}$ in Eq. (1) describe the effective nucleon-nucleon (NN) and nucleon-hyperon (NY) strong interactions and they are parameterized phenomenologically into different forms in different versions of RMF approaches. Generally speaking, these terms can be classified into two types, i.e., the meson-exchange version and the point-coupling version, according to the way how the nucleons and hyperons interact in the hypernuclei.

2.1.1. *Meson-exchange version*

In the meson-exchange version of RMF model for hypernuclei, nucleons and hyperons interact effectively via an exchange of mesons (ϕ_m), including the scalar σ, vector ω_μ, and isovector-vector $\boldsymbol{\rho}_\mu$ mesons, which are employed in order to simulate different characters of nuclear forces. The electromagnetic interaction between protons is described in terms of photon field A_μ. The bare charge e^B is adopted for nucleons ($e^n = 0, e^p = 1$) and Λ hyperon ($e^\Lambda = 0$). The NN interaction term in Eq.(1) has the same form as that for ordinary nuclei (see Chap. 2).

In this model, the Lagrangian density for the NY interaction has the following form

$$\begin{aligned}\mathcal{L}^{NY} = \bar{\psi}^Y \left(-g_{\sigma Y}\sigma - g_{\omega Y}\gamma^\mu \omega_\mu + \frac{f_{\omega YY}}{4m_Y}\sigma^{\mu\nu}\Omega_{\mu\nu} \right) \psi^Y \\ + \mathcal{L}^{\rho Y} + \mathcal{L}^{AY},\end{aligned} \tag{2}$$

where $g_{\sigma Y}$ and $g_{\omega Y}$ are the coupling constants of the hyperon with the scalar and vector meson fields, respectively. The term proportional to $f_{\omega YY}$ with $\sigma^{\mu\nu} = \frac{i}{2}[\gamma^\mu, \gamma^\nu]$ represents the tensor coupling between the hyperon and the ω field. For Λ hyperon ($Y = \Lambda$), since it is isoscalar and charge neutral, the last two terms $\mathcal{L}^{\rho Y}$ and $\mathcal{L}^{AY}$ vanish, which describes the couplings of Λ with the ρ meson and the photon. In the following, we will replace the index Y with Λ since we mainly focus on the studies of Λ hypernuclei in this Chapter.

Notice that the NN and NY interactions in this model have finite range. Therefore, the meson-exchange version of RMF model is also referred to as finite-range RMF model.

With the *no-sea* and *mean-field approximation* at the Hartree level, the contributions of anti(quasi)-particles and quantum fluctuations of meson fields are neglected and the meson fields are thus replaced with their expectation values. Following the standard procedure,[51] one can derive an energy functional corresponding to the Lagrangian density (1). Using the Kohn–Sham ansatz, the density of strongly interacting many-body system can be generated from a non-interacting many-body system. The equations of motion for the nucleons and hyperons are obtained by minimization of the energy functional with respect to the auxiliary single-particle wave function ψ_k^B, the bilinear of which gives the densities and currents in the energy functional. The resultant Dirac equation for nucleons in hypernuclei has the same form as that for atomic nuclei and it has been introduced in the previous Chapters. Therefore, here we present the equation only for the Λ hyperon, which has the following form

$$\left[\alpha \cdot \mathbf{p} + (m_\Lambda + U_S) + U_V - \frac{f_{\omega\Lambda\Lambda}}{2m_\Lambda}\sigma_{\mu\nu}\partial^\mu\omega^\nu\right]\psi_k^\Lambda(\boldsymbol{r}) = \epsilon_k^\Lambda \psi_k^\Lambda(\boldsymbol{r}), \tag{3}$$

with the scalar potential $U_S = g_{\sigma\Lambda}\sigma$ and vector potential $U_V = g_{\omega\Lambda}\gamma_\mu\omega^\mu$.

We note that in the finite-range RMF model, the nucleons and hyperon interact with each other through the exchange of mesons. Therefore, the behavior of nucleons in hypernuclei is modified by the changes in the meson fields due to the presence of hyperons, which provides additional contribution to the sources S_{ϕ_m} for the meson field ϕ_m, fulfilling the Klein–Gordon equation

$$\partial_\mu\partial^\mu\phi_m + U'(\phi_m) = \pm S_{\phi_m}, \tag{4}$$

where the (+) sign is for vector fields and the (−) sign for the scalar field. The source terms S_{ϕ_m} in Eq. (4) are sums of bilinear products of Dirac

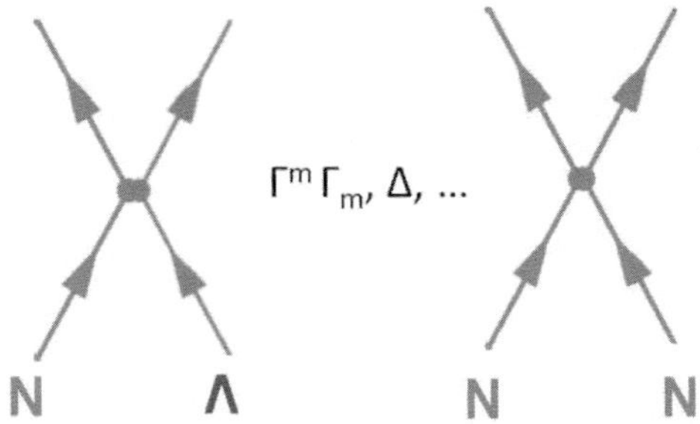

Fig. 1. (Color online) Diagrams for the nucleon-Λ and nucleon-nucleon (NN) interactions in the point-coupling RMF model.

spinors

$$S_{\phi_m} = \begin{cases} \sum_{B,k>0} v_k^2 \bar{\psi}_k^B \psi_k^B, & \phi_m = \sigma, \\ \sum_{B,k>0} v_k^2 \bar{\psi}_k^B \gamma_\mu \psi_k^B - \frac{f_{\omega\Lambda\Lambda}}{2m_\Lambda} j_{T,\mu}^\Lambda, & \phi_m = \omega_\mu, \\ \sum_{B,k>0} v_k^2 \bar{\psi}_k^B \gamma_\mu \boldsymbol{\tau} \psi_k^B, & \phi_m = \boldsymbol{\rho}_\mu, \end{cases} \tag{5}$$

where the sums run over only the positive-energy states ($k > 0$) (i.e., no-sea approximation) and the occupation probability of the single-particle energy level k, i.e., v_k^2, is evaluated within the BCS or the generalized Bogoliubov transformation method. The tensor current $j_{T,\mu}^\Lambda$ is defined as,

$$j_{T,\mu}^\Lambda = \sum_{\Lambda,k} v_k^2 \partial^\nu (\bar{\psi}_k^\Lambda \sigma_{\mu\nu} \psi_k^\Lambda). \tag{6}$$

2.1.2. *Point-coupling version*

In the point-coupling version of RMF model for hypernuclei, nucleons (N) and hyperons (Y) interact effectively via contact couplings (zero-range) with different vertices, see Fig. 1.

The NN interaction term in Eq. (1) has been given in Chap. 2. Similarly to the NN interaction, one can construct the point-coupling terms for the $N\Lambda$ interaction, $\mathcal{L}^{N\Lambda}$. In Ref. 52, the $\mathcal{L}^{N\Lambda}$ has been constructed as follows,

$$\mathcal{L}^{N\Lambda} = \mathcal{L}_{4f}^{N\Lambda} + \mathcal{L}_{der}^{N\Lambda} + \mathcal{L}_{ten}^{N\Lambda}, \tag{7}$$

with

$$\mathcal{L}_{4\text{f}}^{N\Lambda} = -\alpha_S^{(N\Lambda)}(\bar{\psi}^N\psi^N)(\bar{\psi}^\Lambda\psi^\Lambda) - \alpha_V^{(N\Lambda)}(\bar{\psi}^N\gamma_\mu\psi^N)(\bar{\psi}^\Lambda\gamma^\mu\psi^\Lambda), \quad (8\text{a})$$

$$\begin{aligned}\mathcal{L}_{\text{der}}^{N\Lambda} &= -\delta_S^{(N\Lambda)}(\partial_\mu\bar{\psi}^N\psi^N)(\partial^\mu\bar{\psi}^\Lambda\psi^\Lambda)\\ &\quad -\delta_V^{(N\Lambda)}(\partial_\mu\bar{\psi}^N\gamma_\nu\psi^N)(\partial^\mu\bar{\psi}^\Lambda\gamma^\nu\psi^\Lambda),\end{aligned} \quad (8\text{b})$$

$$\mathcal{L}_{\text{ten}}^{N\Lambda} = \alpha_T^{(N\Lambda)}(\bar{\psi}^\Lambda\sigma^{\mu\nu}\psi^\Lambda)(\partial_\nu\bar{\psi}^N\gamma_\mu\psi^N). \quad (8\text{c})$$

Similar to the finite-range RMF model for hypernuclei, the tensor coupling term $\mathcal{L}_{\text{ten}}^{N\Lambda}$, which plays an important role in reproducing the small spin-orbit splittings in the Λ single-particle spectrum,[53,54] has been introduced.

In the point-coupling RMF model, there is no mesonic degree-of-freedom. The nucleons interact with hyperons directly with the zero-range forces. Therefore, the behavior of nucleons is modified by changes in the mean-field potentials due to the presence of hyperons. This picture is equivalent to that presented in the meson-exchange RMF model in the sense that the meson fields play the same role as that of mean-field potentials.

The Dirac equation for nucleons in the hypernuclear system reads

$$[\boldsymbol{\alpha}\cdot\boldsymbol{p} + V_V + V_{TV}\tau_3 + V_C + (m_N + V_S + V_{TS}\tau_3)\beta]\psi_k^N = \epsilon_k^N\psi_k^N, \quad (9)$$

where the potentials are functionals of various densities and currents. We note that the time-reversal invariance is violated in mean-field approximation for odd-mass systems, for which, one has to take into account the contribution of time-odd fields, namely, the currents generated by the unpaired particles. The two-fold degeneracy of time-reversal single-particle states will be lost. The study of hypernuclei in the RMF approach with time-odd fields have been carried out only in Ref. 55 within a time-odd triaxial RMF approach,[56] where the time-odd fields separate the single-particle energies of time-reversal partner states by up to 100 keV. In most calculations, for the sake of simplicity, an *equal-filling approximation*, i.e., the unpaired particle is equally distributed on the time-reversal partner states, is adopted to keep the time-reversal invariance, except in the studies of hypernuclear magnetic moments.[55] We will discuss the effects of time-odd fields on hypernuclear magnetic moments in Section 3.3.

In the following, we adopt the equal-filling approximation if not mentioned explicitly. In this case, the currents vanish and the potentials in (9) are simply functionals of densities,

$$\begin{aligned} V_S &= \delta_S \Delta\rho_S^N + \alpha_S \rho_S^N + \beta_S(\rho_S^N)^2 + \gamma_S(\rho_S^N)^3 \\ &\quad + \delta_S^{(N\Lambda)} \Delta\rho_S^\Lambda + \alpha_S^{(N\Lambda)} \rho_S^\Lambda, \end{aligned} \tag{10a}$$

$$\begin{aligned} V_V &= \delta_V \Delta\rho_V^N + \alpha_V \rho_V^N + \gamma_V(\rho_V^\Lambda)^3 \\ &\quad + \delta_V^{(N\Lambda)} \Delta\rho_V^\Lambda + \alpha_T^{(N\Lambda)} \rho_T^\Lambda + \alpha_V^{(N\Lambda)} \rho_V^\Lambda, \end{aligned} \tag{10b}$$

$$V_{TS} = \delta_{TS} \Delta\rho_{TS}^N + \alpha_{TS} \rho_{TS}^N, \tag{10c}$$

$$V_{TV} = \delta_{TV} \Delta\rho_{TV}^N + \alpha_{TV} \rho_{TV}^N, \tag{10d}$$

$$V_C = eA_0 \frac{1-\tau_3}{2}. \tag{10e}$$

The density $\rho_{m=S,TS,V,TV}^B$ in the different channels and the Λ-tensor density ρ_T^Λ are defined as

$$\rho_m^B = \sum_k \bar{\psi}_k^B \Gamma_m \psi_k^B, \quad \rho_T^\Lambda = \nabla \cdot \left(\sum_k \bar{\psi}_k^\Lambda i\boldsymbol{\alpha} \psi_k^\Lambda \right), \tag{11}$$

where Γ_m is $1, \tau_3, \gamma_0, \gamma_0\tau_3$, respectively.

The Dirac equation for Λ hyperon reads

$$[\boldsymbol{\alpha} \cdot \boldsymbol{p} + U_V + U_T + (U_S + m_\Lambda)\beta]\psi_k^\Lambda = \epsilon_k^\Lambda \psi_k^\Lambda, \tag{12}$$

with the scalar, vector and tensor potentials given by

$$U_S = \delta_S^{(N\Lambda)} \Delta\rho_S^N + \alpha_S^{(N\Lambda)} \rho_S^N, \tag{13a}$$

$$U_V = \delta_V^{(N\Lambda)} \Delta\rho_V^N + \alpha_V^{(N\Lambda)} \rho_V^N, \tag{13b}$$

$$U_T = -i\alpha_T^{(N\Lambda)} \beta\boldsymbol{\alpha} \cdot (\boldsymbol{\nabla}\rho_V^N). \tag{13c}$$

The Dirac equations for nucleons (9) and hyperons (12) are solved iteratively until a convergence is reached with a desired precision.

The RMF energy $E_{\rm RMF}$ for a single-Λ hypernucleus ($^{A+1}_{\ \ \Lambda}Z$ composed of A nucleons and one Λ) is determined as

$$\begin{aligned} E_{\rm RMF} &= \sum_{k=1} v_k^2 \epsilon_k + \epsilon_\Lambda - Am_N - m_\Lambda \\ &\quad - \int d^3\mathbf{r} \left[\frac{1}{2} \sum_m (\alpha_m (\rho_m^N)^2 + \delta_m \rho_m^N \Delta\rho_m^N) + \frac{1}{2} eA^0 \rho_V^{(p)} \right. \\ &\quad \left. + \frac{2}{3}\beta_S(\rho_S^N)^3 + \frac{3}{4}\gamma_S(\rho_S^N)^4 + \frac{3}{4}\gamma_V(\rho_V^N)^4 \right] \\ &\quad - \int d^3\mathbf{r} \left[\sum_{m=S,V} (\alpha_m^{(N\Lambda)} \rho_m^N \rho_m^\Lambda + \delta_m^{(N\Lambda)} \rho_m^N \Delta\rho_m^\Lambda) + \alpha_T^{(N\Lambda)} \rho_V^N \rho_T^\Lambda \right], \end{aligned} \tag{14}$$

where the summation over m in the second line runs over all the channels, S, V, TS, and TV. Alternatively, the energy of single-Λ hypernucleus (14) can also be evaluated as a sum of kinetic energy of nucleons T_N and hyperons T_Λ, NN interaction energy E_{NN}, and $N\Lambda$ interaction energy $E_{N\Lambda}$ terms

$$\begin{aligned}
E_{\text{RMF}} &= T_N + T_\Lambda + E_{NN} + E_{N\Lambda} \\
&= \int d^3\mathbf{r}\left[\sum_{k=1}^{N} v_k^2 \psi_k^\dagger(\boldsymbol{\alpha}\cdot\mathbf{p} + m_N\beta - m_N)\psi_k + \psi_\Lambda^\dagger(\boldsymbol{\alpha}\cdot\mathbf{p} + m_\Lambda\beta - m_\Lambda)\psi_\Lambda\right] \\
&\quad + \int d^3\mathbf{r}\left[\frac{1}{2}\sum_m (\alpha_m(\rho_m^N)^2 + \delta_m\rho_m^N\Delta\rho_m^N) + \frac{1}{2}eA^0\rho_V^{(p)}\right. \\
&\quad \left. + \frac{1}{3}\beta_S(\rho_S^N)^3 + \frac{1}{4}\gamma_S(\rho_S^N)^4 + \frac{1}{4}\gamma_V(\rho_V^N)^4\right] \\
&\quad + \int d^3\mathbf{r}\left[\sum_{m=S,V}(\alpha_m^{(N\Lambda)}\rho_m^N\rho_m^\Lambda + \delta_m^{(N\Lambda)}\rho_m^N\Delta\rho_m^\Lambda) + \alpha_T^{(N\Lambda)}\rho_V^N\rho_T^\Lambda\right]. \qquad (15)
\end{aligned}$$

The center-of-mass correction (CM) energy E_{CM} is calculated by taking the expectation value of the kinetic energy for the center of mass motion with respect to the many-body ground state wave function. For a single-Λ hypernucleus, it is given by

$$E_{\text{CM}} = \frac{\langle \boldsymbol{P}_{\text{CM}}^2 \rangle}{2[Am_N + m_\Lambda]}, \qquad (16)$$

where $\boldsymbol{P}_{\text{CM}}$ is the total momentum of a hypernucleus with A nucleons and one Λ hyperon.

For Λ hypernuclei with open-shell nuclear core, the pairing correlation among nucleons needs to be taken into account. In this case, the pairing energy E_{pair} related to pairing density should also be added to the total energy. Therefore, the total binding energy $B({}^{A+1}_{\Lambda}Z)$ of a single-Λ hypernucleus is finally given by

$$B({}^{A+1}_{\Lambda}Z) = -(E_{\text{RMF}} - E_{\text{CM}} + E_{\text{pair}}). \qquad (17)$$

3. Ground state properties of hypernuclei

3.1. *Systematics of Λ binding energy*

Let us now discuss applications of the relativistic approaches to the ground state properties of hypernuclei. We first discuss the systematics of binding energy of Λ particle in single-Λ hypernuclei.

The Λ binding energy is defined as

$$B_\Lambda = M(^AZ)c^2 + m_\Lambda c^2 - M(^{A+1}_{\ \ \Lambda}Z)c^2, \tag{18}$$

$$= B(^{A+1}_{\ \ \Lambda}Z) - B(^AZ), \tag{19}$$

where $M(^{A+1}_{\ \ \Lambda}Z)c^2$ and $M(^AZ)c^2$ are the mass of a hypernucleus and that of the corresponding core nucleus, respectively, while $m_\Lambda c^2$ is the mass of a Λ particle (see Table 1). $B(^{A+1}_{\ \ \Lambda}Z)$ and $B(^AZ)$ are the binding energy of the hypernucleus and the core nucleus, respectively. The binding energies for light hypernuclei with Λ particle in the 1s state (that is, the lowest single-particle state) were determined by early emulsion experiments.[57] The Λ binding energies for higher l single-particle states, from light to medium-heavy and heavy hypernuclei, such as $^{28}_{\Lambda}$Si, $^{89}_{\Lambda}$Y, $^{139}_{\ \Lambda}$La, and $^{209}_{\ \Lambda}$Pb, have been obtained with the (π^+, K^+) reaction spectroscopy.[4] More recently, the $(e, e'K^+)$ reaction spectroscopy has also been developed,[58,59] which provides a higher energy resolution than the (π^+, K^+) reaction spectroscopy. One of the most important achievements with these reaction spectroscopies is that single-particle Λ states have been clearly identified in hypernuclei from s (l=0) to h (l=5) states.[4]

The mass number dependence of Λ binding energies have been well fitted with a simple Woods–Saxon potential for a Λ particle.[61,62] See the dashed lines in Fig. 2 for a fit with a Woods–Saxon potential in which the depth, the diffuseness, and the radius parameters are $V_0 = -30$ MeV, $a = 0.6$ fm, and $R = 1.1 \cdot A^{1/3}$ fm, respectively.[4,62] If one approximates a Woods–Saxon potential with an infinite square well potential with the depth of $-V_0$ and the range of R, the energy of the 1s state reads

$$\epsilon_{1s} = -V_0 + \frac{\hbar^2}{2m_\Lambda} \cdot \frac{\pi^2}{R^2}. \tag{20}$$

If one assumes that the depth parameter V_0 is independent of the mass number of hypernucleus, A, and the range parameter R is proportional to $A^{1/3}$, this equation implies that the Λ single-particle energies scale as $A^{-2/3}$. This has indeed been observed experimentally, at least for medium-heavy

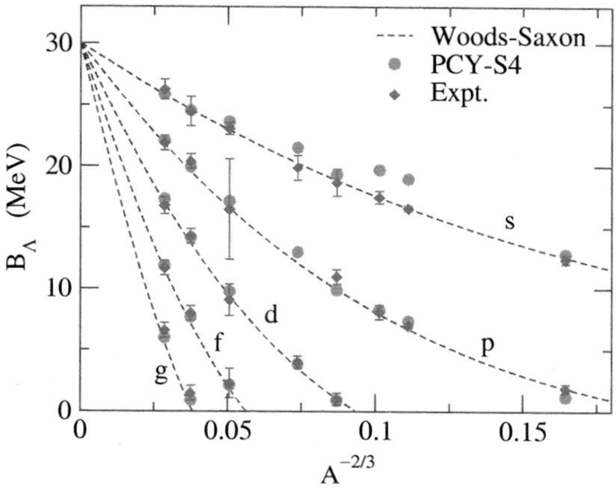

Fig. 2. (Color online) The mass number dependence of Λ binding energies plotted as a function of $A^{-2/3}$. The dashed lines are obtained with a Woods–Saxon potential with a depth parameter of 30 MeV, the diffuseness parameter of 0.6 fm, and the radius parameter of 1.1 fm. The filled circles show the results of spherical mean-field calculations with the relativistic point coupling model with the PCY-S4 parameter set.[52] The filled diamonds show the experimental data taken from Ref. 4 (see also Table I in Ref. 60).

and heavy hypernuclei, as can be seen in Fig. 2. The Λ binding energies are therefore often plotted as a function of $A^{-2/3}$. Moreover, Eq. (20) suggests that the depth parameter V_0 can be obtained by extrapolating the single-particle energies to $A = \infty$. The empirical value of the depth parameter is about $V_0 = 30$ MeV (see Fig. 2). This value is about 2/3 of the depth parameter for nucleons, indicating that the ΛN interaction is weaker than the NN interaction. A similar value of the depth parameter was obtained earlier in Refs. 63–65.

A microscopic understanding of the Λ binding energies can be achieved with the self-consistent mean-field approaches. As a matter of fact, the experimental Λ binding energies have been used as important empirical inputs in determining the value of parameters of an effective $N\Lambda$ interaction for mean-field calculations. This has been done both for the non-relativistic Skyrme–Hartree–Fock apprach[39] and for the relativistic mean-field approach.[42,43,52,66,67] As an example, Fig. 2 shows the results of spherical mean-field calculations with the PCY-S4 parameter set for the relativistic point-coupling Lagrangian.[52] One can see that the overall fit is good except for a few deformed hypernuclei.

3.2. *Spin-orbit splittings*

We next discuss the spin-orbit splittings of hypernuclei. In ordinary nuclei, large spin-orbit splittings are one of the most important quantities in order to understand the shell structures of atomic nuclei. In contrast, in hypernuclei, the (K^-, π^-) reaction spectroscopies have revealed already by the late 1970's that the spin-orbit splittings are much smaller.[45] More recently, the γ-ray spectroscopy has been carried out for the ${}^{13}_{\Lambda}$C hypernucleus, which has provided more quantitative information on the spin-orbit splittings.[68] In the experiment, γ-rays from the excited $1/2^-$ and $3/2^-$ states to the ground state were measured following the ^{13}C$(K^-,\pi^-){}^{13}_{\Lambda}$C reaction.[68] From this experiment, the spin-orbit splitting between the $1p_{1/2}$ and the $1p_{3/2}$ hyperon states in ${}^{13}_{\Lambda}$C was determined to be $152 \pm 54 \pm 36$ keV, which is smaller than the spin-orbit splitting in ordinary nuclei by a factor of 20–30.[4,68]

As we have mentioned in the Introduction, early applications of the relativistic approaches focused on discussions on the spin-orbit splittings. Brockmann and Weise first argued that the particular structure of the 2π and 3π exchange interactions lead to a reduction of $N\Lambda$ coupling constants from the corresponding NN coupling constants by a factor of about 3, considerably quenching the spin-orbit splittings[40] (notice that soon after the publication of Brockmann and Weise, Pirner pointed out that the scaling factor is about 2/3 rather than 1/3 if the quark model is adopted[69]). See also Ref. 41 for a similar argument. Brockmann and Weise also investigated the effect of Fock exchange terms on the spin-orbit splittings.[70] They have shown that the K and K* meson exchange contributions almost cancel each other, and have concluded that the essential features of the spin-orbit splittings are well represented already in the Hartree approximation. Subsequently, Noble pointed out the importance of the tensor coupling (see Eq. (2)),

$$\mathcal{L}_{\rm tensor} = \frac{f_\omega}{2m_\Lambda}\bar{\psi}_\Lambda\, \sigma^{\mu\nu}\psi_\Lambda\, \partial_\nu\omega_\mu, \tag{21}$$

in the spin-orbit splittings.[53] That is, Noble showed that the contribution of the tensor coupling term largely cancels the Thomas precession term of the spin-orbit potential, resulting in the small spin-orbit splittings in Λ hypernuclei. The tensor coupling has further been investigated *e.g.*, in Refs. 43, 47, 48, 52, 71 and its role in the spin-orbit splittings has by now been well established.

3.3. *Hypernuclear magnetic moments*

The magnetic moments of hypernuclei are important physics observables, since they are sensitive to the spin and angular momentum structure of hypernuclei as well as the spin-dependent hyperon-nucleon interactions. In particular, these quantities provide direct information on the properties of hadrons in the nuclear medium.

The first study of hypernuclear magnetic moments was performed for light p-shell hypernuclei within a three-body cluster model.[32] Thereafter, magnetic moments of light hypernuclei were theoretically studied within the shell model, and predicted to be close to the Schmidt lines,[72] which are obtained with the extreme single-particle model neglecting the core polarization due to a particle or a hole added. However, the magnetic moments of hypernuclei may deviate from the Schmidt values, especially if the meson (K and π) exchange currents are considered.[73]

In the RMF description of magnetic moments in ordinary nuclei near double-closed shell, the contribution of the polarized current from core nucleons is compensated by the enhanced Dirac current from the valence charged particle due to the reduction of the nucleon mass.[56] However, this cancelation is not expected in Λ-hypernuclei due to the charge neutrality of the Λ hyperon. Therefore, the polarized proton current induced by the valence hyperon causes the total magnetic moment to deviate from the Schmidt value. Fig. 3 shows the dependence of the deviation, $\Delta\mu$, on the ratio of the vector meson coupling constants, $\alpha_\omega (\equiv g_{\omega\Lambda}/g_{\omega N})$. It is seen that the deviations are proportional to α_ω. These deviations were suggested as an indicator of relativistic effects in nuclei.[74,75] However, it has been eventually realized that the tensor coupling of vector field to the Λ hyperon could significantly change the hypernuclear magnetic moments by renormalizing the electromagnetic current vertex in nuclear medium, and bring the magnetic moment of Λ-hypernucleus with $\ell_\Lambda = 0$ close to the Schmidt value again, although this is not the case for $\ell_\Lambda \neq 0$.[76,77]

Table 2. Magnetic moments of Λ hypernuclei ${}^{17}_{\Lambda}$O and ${}^{41}_{\Lambda}$Ca taken from Ref. 77.

	${}^{17}_{\Lambda}$O			${}^{41}_{\Lambda}$Ca		
μ (n.m.)	$1s_{1/2}$	$1p_{3/2}$	$1p_{1/2}$	$1s_{1/2}$	$1p_{3/2}$	$1p_{1/2}$
Schmidt	−0.613	−0.613	0.204	−0.613	−0.613	0.204
RMF	−0.660	−0.662	0.170	−0.682	−0.718	0.153
RMF+Tensor	−0.611	−0.643	0.186	−0.611	−0.667	0.170

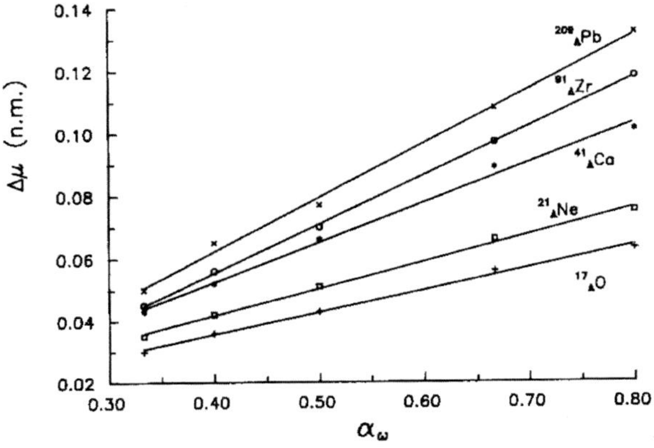

Fig. 3. The dependence of the deviation of a magnetic moment from the Schmidt value, $\Delta\mu$, on the ratio of the vector meson coupling constants, $\alpha_\omega(\equiv g_{\omega\Lambda}/g_{\omega N})$. Magnetic moments (with Λ in the $s_{l/2}$ state) are calculated within the RMF model without the non-linear couplings of mesons for the NN part and several pairs of ratios of coupling constants $(\alpha_\sigma, \alpha_\omega)$ for the $N\Lambda$ interaction. The figure is taken from Ref. 75.

It is shown in Table 2 that there is an evident difference between the magnetic moments by the RMF and the Schmidt formula, which mainly comes from the polarized Dirac magnetic moments. After taking into account the tensor effect on the current, the difference almost disappears for the Λ hyperon in the $1s$ state. For the $1p$ states, this difference is greatly suppressed, but there is still a $5\% - 25\%$ difference for a Λ hyperon in the $1p$ state.

We note that the magnetic moments of Λ hypernuclei from light to heavy mass have been systematically investigated recently based on the perturbation consideration of core polarization effect within the spherical RMF model[78] as well as the self-consistent solution of the time-odd triaxial RMF model.[55] Figure 4 displays the dependence of $\Delta\mu$ on the ratio of meson-hyperon coupling strengths $R_\sigma \equiv g_{\sigma\Lambda}/g_{\sigma N}\omega$ and $R_\omega \equiv g_{\omega\Lambda}/g_{\omega N}\omega$ for $^{17}_{\Lambda}$O hypernucleus. As expected, the time-odd triaxial RMF calculation gives similar results as that of time-odd axial RMF calculation for $^{17}_{\Lambda}$O if the same effective interaction is used. In both calculations, the vector meson tensor coupling is not taken into account. Figure 4 also shows the results by the time-odd triaxial RMF calculation using the parameter set PK1-Y1, PK1-Y2 and PK1-Y3.[79] After taking into account the tensor coupling term, the deviation still increases with the vector coupling strength $g_{\omega\Lambda}$, but with a smaller slope. The self-consistent deformed RMF calculations with time-odd components have confirmed the conclusion from the analytical expression for spherical nuclei. Moreover, Fig. 4 indicates that the deviation $\Delta\mu$ also depends on the scalar coupling strength $g_{\sigma\Lambda}$ through the effective scalar mass M^*.[74]

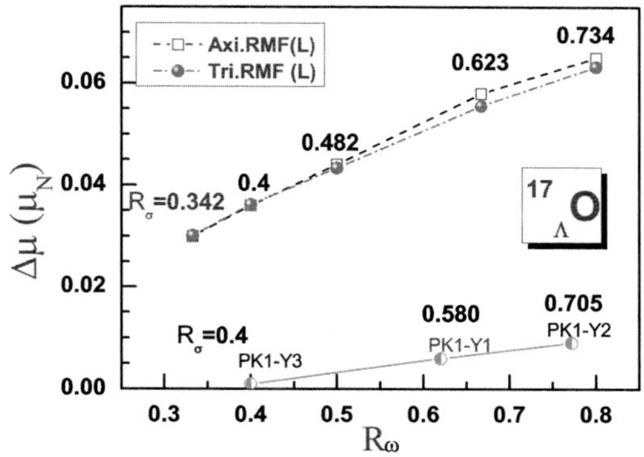

Fig. 4. (Color online) Same as Fig. 3, but only for $^{17}_{\Lambda}$O obtained with the time-odd axial (labeled with open squares, taken from Ref. 75) and with the time-odd triaxial RMF (labeled with bullets) calculations. The results from the time-odd triaxial RMF calculation with the $\omega\Lambda\Lambda$ tensor coupling are also plotted with half-filled circles. Here, $R_\sigma \equiv g_{\sigma\Lambda}/g_{\sigma N}\omega$ and $R_\omega \equiv g_{\omega\Lambda}/g_{\omega N}$. This figure is taken from Ref. 55.

Compared with the magnetic moments, the spin-flip $M1$ transitions between hypernuclear doublet states are relatively easier to measure. A few data points have already been accumulated.[80,81] Therefore, it is an important future work to extend the RMF approach to study the $M1$ transition strengths in Λ hypernuclei.

3.4. *Stabilization of neutron-rich nuclei on the drip line*

Physics of unstable nuclei is one of the most important topics in current nuclear physics.[82–85] Many interesting phenomena have been found in neutron-rich nuclei, such as a halo and skin structure, soft dipole excitations, and changes of magic numbers, all of which originate from the weakly bound character of neutron-rich nuclei. In particular, the halo structure is one of the most important phenomena in neutron-rich nuclei, in which the density distribution is largely extended up to large distances.

An interesting question is how the halo structure of neutron-rich nuclei is influenced by adding a Λ particle. This has been studied in Ref. 86

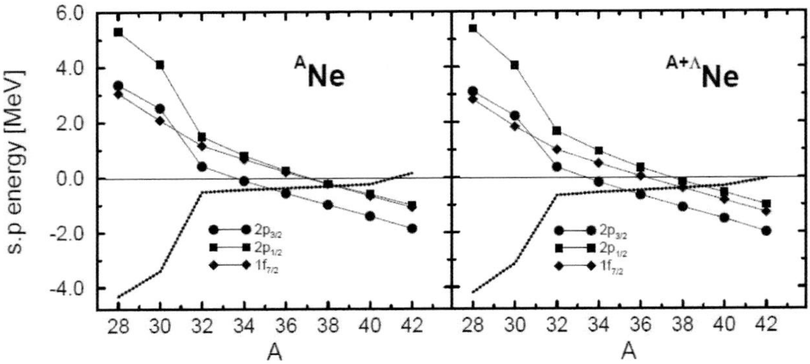

Fig. 5. Single-particle energies for neutron levels in Ne isotopes (the left panel) and Ne hypernuclei (the right panel) obtained with the relativistic Hartree–Bogoliubov method. The thick solid lines denote the Fermi energy. Taken from Ref. 26.

using a three-body model, in which the authors have shown that the Λ particle induces an extra binding and as a consequence the halo structure disappears. In terms of mean-field theory, the extra Λ particle leads to a modification of single-particle potentials for nucleons. Vretenar *et al.* have used the relativistic Hartree–Bogoliubov method and demonstrated that the change in the spin-orbit potential plays an important role in stabilizing neutron-rich nuclei.[26] This is one of the most successful applications of the relativistic approach, since the conclusion is difficult to reach with the non-relativistic approaches, in which the spin-orbit potential is treated phenomenologically. Vretenar *et al.* have also shown that the stabilization of neutron-rich nuclei leads to an extension of the neutron-drip line[26] (see Fig. 5). A similar study has been carried out also by Lü *et al.*[87] (see also Ref. 27 for a study with a non-relativistic approach).

3.5. *Deformation of hypernuclei*

It is well known that many open-shell nuclei are deformed in the ground state. A clear experimental evidence for nuclear deformation is provided by a rotational spectrum, which scales as $E_I \propto I(I+1)$ as a function of the angular momentum I, as well as by strongly enhanced electric quadrupole transition probabilities. Theoretically, a standard method to discuss nuclear deformation is a self-consistent mean-field theory. In working with the intrinsic (body-fixed) frame, in which the rotational symmetry is broken in the mean-field potential, the mean-field theory provides an intuitive and transparent view of the nuclear deformation.

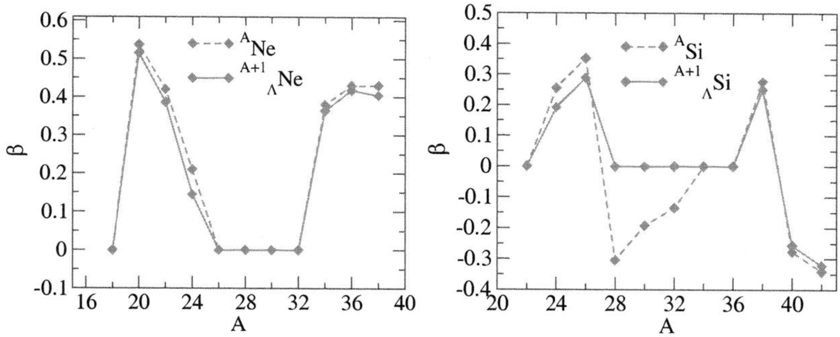

Fig. 6. (Color online) Quadrupole deformation parameter β for Ne (left panel) and Si (right panel) isotopes obtained with the RMF method with the NL3 parameter set. The dashed lines show the deformation parameter for the core nucleus, while the solid lines show the deformation parameter for the corresponding hypernucleus.

The mean-field approach is a suitable tool also to investigate how the nuclear deformation is affected by a Λ particle, because the optimum density distribution is obtained automatically by minimizing the total energy. Žofka was the first person who applied the self-consistent method to deformed hypernuclei.[12] He used non-relativistic Gaussian interactions for nucleon-nucleon (NN) and nucleon-Lambda ($N\Lambda$) interactions, and showed that a Λ particle changes the quadrupole moment at most by 5 % in the *sd*-shell region. This result has been confirmed with more recent Skyrme–Hartree–Fock (SHF) calculations for axially deformed hypernuclei.[13]

An application of the relativistic mean field approach to deformed hypernuclei was first carried out in Ref. 14. In these calculations, axial symmetry for the density distribution was assumed, and a Λ particle was put in the lowest single-particle orbit. The NL3 parameter set of RMF[88] was employed. Figure 6 shows the quadrupole deformation parameter β for the ground state of Ne and Si isotopes obtained with these calculations, as well as that for the ground state of corresponding hypernuclei. The deformation parameter is defined as

$$\beta \equiv \frac{4\pi}{3AR^2} Q_{20}, \tag{22}$$

with the expectation value of mass quadrupole moments,

$$Q_{20} = \sqrt{\frac{5}{16\pi}} \int d^3\mathbf{r}(2z^2 - x^2 - y^2)\rho_{\rm tot}(\mathbf{r}), \tag{23}$$

where the total density $\rho_{\rm tot}(\mathbf{r})$ is contributed from neutrons and protons

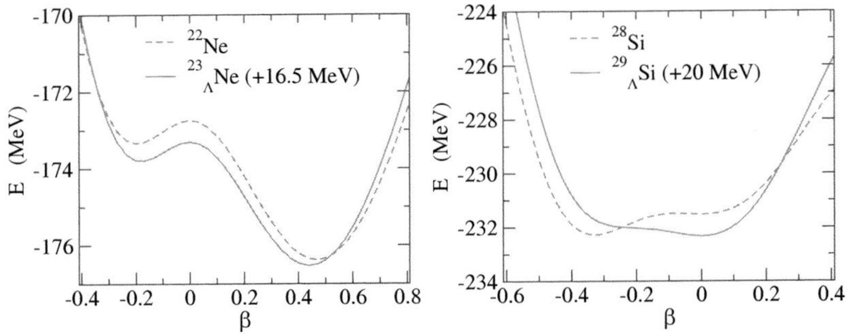

Fig. 7. (Color onlne) (Left panel) The potential energy surface for ^{22}Ne (the dashed line) and $^{23}_{\Lambda}$Ne (the solid line) nuclei obtained with the RMF method. The energy surface for $^{23}_{\Lambda}$Ne is shifted by a constant amount as indicated in the figure. (Right panel) The same as the left panel, but for ^{28}Si and $^{29}_{\Lambda}$Si nuclei.

for ordinary nuclei, and additionally from hyperons for hypernuclei.

One can see that the change in the deformation parameter for most of the nuclei shown in the figure is small, as in the non-relativistic self-consistent calculations.[12,13] However, in the relativistic approach, certain nuclei, that is, ^{28,30,32}Si, show a drastic change of the deformation parameter, from oblate deformation to spherical when a Λ particle is added. It was shown that a similar disappearance of nuclear deformation due to a Λ particle also takes place in ^{12}C, both with RMF[14] and anti-symmetrized molecular dynamics (AMD).[19]

The disppearance of nuclear deformation was not observed in the non-relativistic Skyrme–Hartree–Fock (SHF) calculations.[13] In Ref. 15, it has been shown that the difference comes about because the RMF yields a somewhat stronger polarization effect of the Λ hyperon than that of the SHF approach.

Figure 7 shows the potential energy surfaces for the $^{23}_{\Lambda}$Ne and $^{29}_{\Lambda}$Si nuclei obtained with the RMF method. The energy surfaces for the corresponding core nuclei are also shown. In contrast to the ^{22}Ne nucleus, which has a deep prolate minimum in the energy surface, the energy surface for the ^{28}Si nucleus shows a relatively shallow oblate minimum, with a shoulder at the spherical configuration. Notice that the Λ particle gains the largest binding energy at the spherical configuration as the overlap with the core nucleus is maximum. The energy difference between the oblate and the spherical configurations in ^{28}Si is as small as 0.754 MeV, and could be easily inverted when a Λ particle is added. This explains why the disappearance of deformation occurs in ^{28}Si but not in ^{22}Ne.

Those calculations have been extended in Ref. 18 by including the triaxial degree of freedom (see Ref. 16 for a similar 3D calculation with the non-relativistic approach). The main conclusion remained the same as in Ref. 14, although the results are slightly different in details. A general conclusion is that the potential energy surface tends to be softer both in the β (the axial symmetric deformation) and the γ (the triaxial deformation) directions when a Λ particle is embedded in a nucleus. This would indicate that the shape fluctuation effect, that is, the effect beyond the mean-field approximation, is more important in hypernuclei as compared to ordinary nuclei. We will discuss this in the next section.

More recently, Lu *et al.* applied the RMF method to superdeformation of hypernuclei.[89] They pointed out that there are a few nuclei which show a bubble-like density distribution for the superdeformed configuration and that the Λ binding energy becomes larger at the superdeformed state compared to that at a normal deformation in that situation. This implies that it might be possible to study the density distribution of the superdeformed state of ordinary nuclei by measuring the Λ binding energy.

4. Collective excitations

4.1. *Impurity effect of Lambda hyperon on nuclear collective excitations*

Let us next consider the collective excitations of hypernuclei. To this end, it is important to take into account the effect beyond the mean-field approximation. We first discuss the impurity effect of Λ hyperon on nuclear collective excitation. In Ref. 17, the collective excitations of nuclear core in Λ hypernuclei including rotation and vibration of quadrupole deformed shapes have been described with a five-dimensional collective Bohr Hamiltonian (5DCH), in which the moments of inertia and mass parameters are functions of deformation parameters (β, γ) determined by self-consistent mean-field calculations. The presence of a Λ particle in the atomic nucleus modifies the mean-field potentials for nucleons and thus changes the energies and wave functions of single-particle states. These changes will enter into the mass parameters in the 5DCH, which are usually calculated with the cranking approximation. Subsequently, a similar 5DCH study has also been carried out based on the solutions of a triaxially deformed RMF approach for Λ hypernuclei.[90]

Figure 8 displays the low-spin spectra of the ground-state band for ^{24}Mg

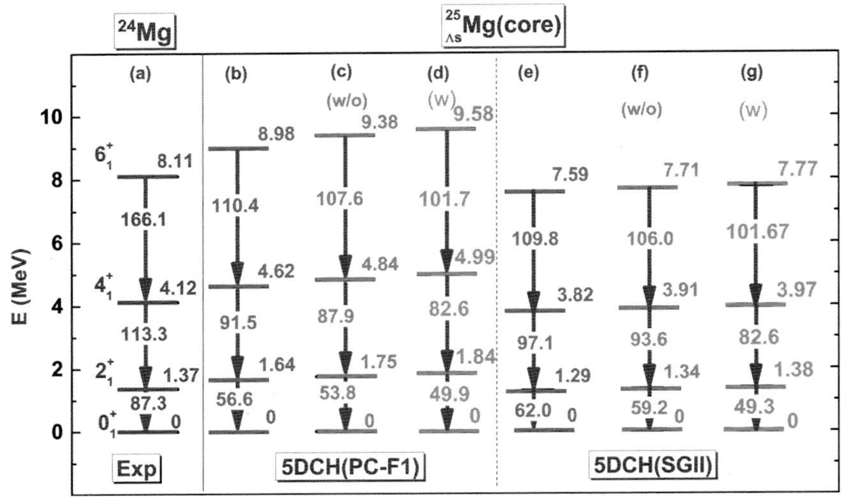

Fig. 8. (Color online) The low-spin spectra of the ground-state rotational band for ^{24}Mg (a,b,e) and for the nuclear core of $^{25}_{\Lambda}$Mg (c, d, f, g). The theoretical results are obtained by the 5DCH calculation with the parameters determined by the triaxial RMF (b, c, d) and by the Skyrme–Hartree–Fock+BCS (e, f, g) calculations. The difference between (c) and (d), and between (f) and (g) is whether the ΛN interaction term is included (w) or not (w/o) in the potential energy for the nuclear core part. The $B(E2)$ values are in units of e^2 fm^4.

and for the nuclear core of $^{25}_{\Lambda}$Mg calculated by the 5DCH based on the relativistic point-coupling EDF with PC-F1 parameter set and the non-relativistic EDF with the SGII parameter set. It is noted that the Λ stretches the spectra of the ground-state band. Comparing the columns (b) and (d), one finds that the Λ increases the excitation energy of 2^+_1 state, $E_x(2^+_1)$, by $\sim$ 12% and reduces the $B(E2 : 2^+_1 \to 0^+_1)$ value by $\sim$ 12%. The reduction effect in the relativistic calculation is much larger than that found in the non-relativistic calculation,[17] as illustrated in the comparison of columns (e) and (g). It is shown that the Λ increases $E_x(2^+_1)$ by $\sim$ 7% and reduces the $B(E2 : 2^+_1 \to 0^+_1)$ by $\sim$ 9%. Considering the fact that the magnitude of impurity effect of Λ hyperon on nuclear collective properties is not much dependent on the underlying EDF used to determine the parameters of the 5DCH,[91] one may ascribe this significant difference in the results by the 5DCH(PC-F1) and 5DCH(SGII) to the relativistic effect, which enhances the impurity effect in atomic nuclei. Moreover, the comparison of columns (c) and (d), as well as (f) and (g) demonstrates the important influence of ΛN interaction on nuclear collective excitations.

4.2. *Microscopic description of low-energy spectra in Lambda hypernuclei*

We next discuss a spectrum of a whole hypernucleus, which is intimately related to the γ-ray spectroscopy, the measurement of which has been done with a germanium detector array (Hyperball) for p-shell hypernuclei.[81] To describe the energy spectrum of a whole Lambda hypernucleus based on a covariant energy density functional, one needs to treat properly the coupling between the unpaired valence hyperon and nuclear core excitations. To this end, one can construct the wave function for single-Λ hypernuclei with an even-even nuclear core as follows,

$$\Psi_{IM}(\boldsymbol{r}_\Lambda, \{\boldsymbol{r}_N\}) = \sum_{j\ell I_c} \mathscr{R}_{j\ell I_c}(r_\Lambda) \mathscr{F}^{IM}_{j\ell I_c}(\hat{\boldsymbol{r}}_\Lambda, \{\boldsymbol{r}_N\}), \tag{24}$$

where

$$\mathscr{F}^{IM}_{j\ell I_c}(\hat{\boldsymbol{r}}_\Lambda, \{\boldsymbol{r}_N\}) = [\mathscr{Y}_{j\ell}(\hat{\boldsymbol{r}}_\Lambda) \otimes \Phi_{I_c}(\{\boldsymbol{r}_N\})]^{(IM)} \tag{25}$$

with $\boldsymbol{r}_\Lambda$ and $\boldsymbol{r}_N$ being the coordinates for the Λ hyperon and the nucleons, respectively. In this equation, I is the total angular momentum and M is its projection onto the z-axis for the whole Λ hypernucleus. $\mathscr{R}_{j\ell I_c}(r_\Lambda)$ and $\mathscr{Y}_{j\ell}(\hat{\boldsymbol{r}}_\Lambda)$ are the four-component radial wave function and the spin-angular wave function for the Λ hyperon, respectively.

The wave function of the nuclear core part, $\Phi_{I_cM_c}(\{\boldsymbol{r}_N\})$, is given as a superposition of particle-number and angular-momentum projected RMF+BCS states, $|\varphi(\beta)\rangle$, that is,

$$|\Phi_{I_cM_c}\rangle = \sum_{\beta} f_{I_cNZ}(\beta) \hat{P}^{I_c}_{M_cK} \hat{P}^N \hat{P}^Z |\varphi(\beta)\rangle, \tag{26}$$

where $\hat{P}^{I_c}_{M_cK}$, $\hat{P}^N$, $\hat{P}^Z$ are the projection operators onto good numbers of angular momentum, neutrons and protons, respectively. The mean-field wave functions $|\varphi(\beta)\rangle$ are a set of Slater determinants of quasi-particle states with different quadrupole deformation β. For simplicity, we consider only the axial deformation for the nuclear core and thus the K quantum number is zero in Eq. (26). The weight factor $f_{I_cNZ}(\beta)$ is determined by solving the Hill–Wheeler–Griffin equation.

Substituting Eq. (24) into the Dirac equation for the whole hypernucleus, $H|\Psi_{IM}\rangle = E_I|\Psi_{IM}\rangle$, where H is the relativistic Hamiltonian derived from the effective Lagrangian density for the whole Λ hypernucleus, one can derive the coupled-channels equations for $\mathscr{R}_{j\ell I_c}(r_\Lambda)$, in which the coupling potentials are given in terms of the transition densities.

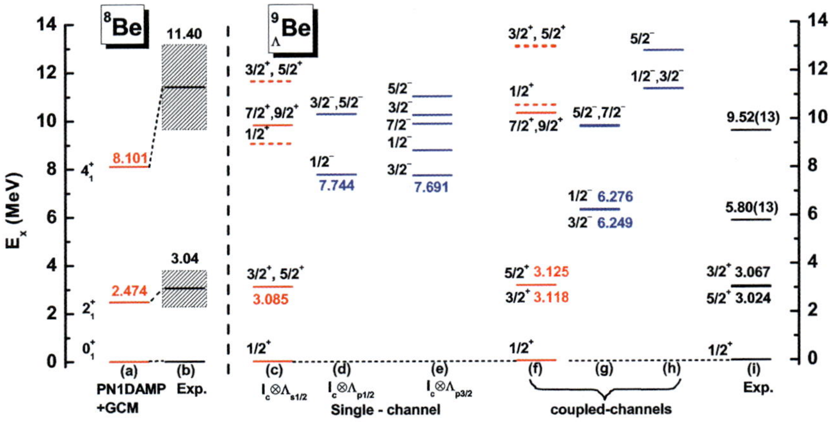

Fig. 9. The low-energy excitation spectra of ^{8}Be (the columns (a) and (b)) and $^9_\Lambda$Be. For ^{8}Be, the GCM+PN1DAMP calculations shown in the column (a) are compared with the experimental data. For $^9_\Lambda$Be, the columns (c), (d), and (e) show the results of the single-channel calculations for the Λ particle in $s_{1/2}$, $p_{1/2}$, and $p_{3/2}$ channels, respectively. The columns (f), (g), and (h) show the results of the coupled-channels equations, which are compared with the experimental data[81] shown in the column (i). This figure is taken from Ref. 24.

Figure 9 displays the low-energy excitation spectra of ^{8}Be and $^9_\Lambda$Be, in comparison with the available data.[24] For the $N\Lambda$ effective interaction term, the vector and scalar types of four-fermion coupling terms in Ref. 52 are adopted. The coupling strengths are fitted to the Λ binding energy. One can see that a good agreement with the data is obtained for both ^{8}Be and $^9_\Lambda$Be. The low-lying states of $^9_\Lambda$Be are categorized into three rotational bands, whose structures are confirmed by the calculated $B(E2)$ relations. It is remarkable that the present calculation reconfirms an interesting prediction of the cluster model that the strong coupling of a hyperon to the collective rotation is realized when the Λ is in the p-orbit.[9] Among these rotational bands, the column (g) corresponds to the so-called genuine hypernuclear states, which are also referred to as the supersymmetric states having the SU3 symmetry $(\lambda\mu) = (50)$ of s^4p^5 shell-model configuration. These states do not have corresponding states in the ordinary nucleus, ^{9}Be, because of the Pauli principle of the valence neutron. Moreover, according to our calculations, the measured state in $^9_\Lambda$Be with excitation energy of 5.80(13) MeV is actually a mixture of two negative-parity states with $J^\pi = 3/2^-$ and $1/2^-$.

In view of the success of this novel method for the low-energy spectrum of Lambda hypernuclei, it would be interesting to extend this method to study hypernuclear magnetic moments and $M1$ transitions in the low-lying states. In particular, the determination of ΛN interaction using the spectroscopic data of hypernuclei based on this method is an interesting future work.

5. Anti-Lambda hypernuclei

5.1. *Polarization effect of anti-Lambda in nucleus*

Antibaryons are interesting particles for nuclear physics. They are building blocks of antimatter that can be produced in the laboratory.[92] In free space, baryons and antibaryons promptly annihilate each other when they come into contact. Attempts have been made to find $\bar{N}N$ bound states with mass close to the threshold.[93] In a nuclear medium, many interesting predictions concerning the antibaryon behavior have been made. In particular, the appearance of antinucleon or antihyperon bound states in nuclei is one of the most popular conjectures.[94–100] It is demonstrated that the presence of a real antibaryon leads to a drastic rearrangement of a target nucleus, resulting in a significant increase of its binding energy and local compression.

In the RMF approach with meson-exchange for anti-Λ hypernuclei, the coupling strengths of $\bar{\Lambda}$ and mesons are related to those of Λ and mesons according to the G-parity transformation. To take into account possible deviations from the G-parity symmetry in a many-body system, one can introduce an overall scaling $0 \leq \xi \leq 1$,[97] namely,

$$g_{\sigma\bar{\Lambda}} = \xi g_{\sigma\Lambda}, \quad g_{\omega\bar{\Lambda}} = -\xi g_{\omega\Lambda}. \tag{27}$$

The polarization effect due to the $\bar{\Lambda}$ on the nuclear density distribution of ^{16}O is illustrated in Fig. 10. It is seen that the presence of $\bar{\Lambda}$ compresses the protons and neutrons into the center of the nucleus with the central nuclear density (the sum of densities for proton and neutron) up to 2–3 times of saturation density. The distribution of $\bar{\Lambda}$ indicates its interaction range, which is between 1–2 fm. It means that only the nucleons close to nuclear center will be strongly influenced by the $\bar{\Lambda}$. The distribution of the net baryon density ($\rho_N - \rho_{\bar{\Lambda}}$) in the ^{4}He+$\bar{\Lambda}$ and ^{16}O+$\bar{\Lambda}$ systems from the RMF model calculation with the NLZ[101] parameter set is shown in Fig. 11. Because the $\bar{\Lambda}$ has -1 baryon number, it depresses the density of baryon

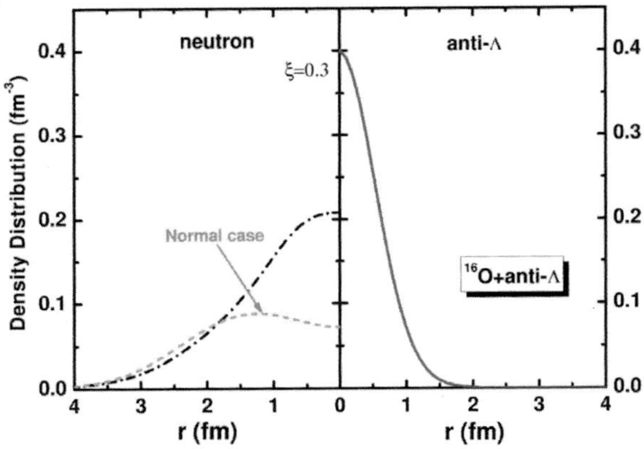

Fig. 10. (Color online) Density distribution for neutrons (left panel) and $\bar{\Lambda}$ (right panel) in $^{16}O+\bar{\Lambda}$ system, where the $\bar{\Lambda}$ occupies the lowest s orbit. The results are given by the RMF calculations with the PK1 force for the effective NN interaction and $g_{\sigma\Lambda}/g_{\sigma N} = g_{\omega\Lambda}/g_{\omega N} = 2/3$ (based on the SU(3) flavor symmetry), $\xi = 0.3$ for the $\bar{\Lambda}N$ interaction. The density distribution for neutron in ^{16}O is plotted for comparison. Taken from Ref. 100.

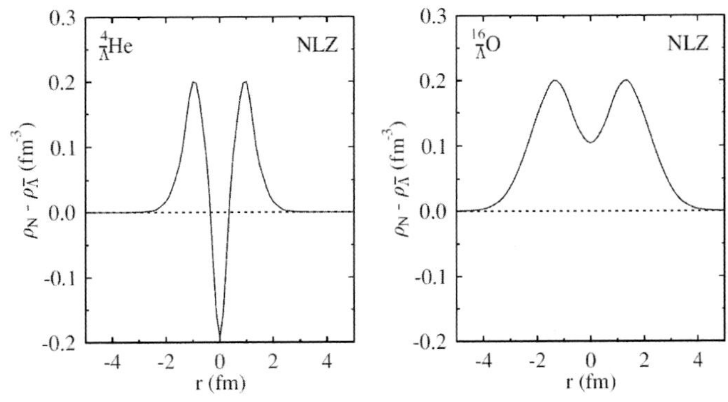

Fig. 11. Distribution of the net baryon density ($\rho_N - \rho_{\bar{\Lambda}}$) in the $^4He+\bar{\Lambda}$ (left) and $^{16}O+\bar{\Lambda}$ (right) systems calculated within the RMF model with the NLZ[101] parameter set. Taken from Ref. 97.

numbers in the center, even to a negative value in $^4He+\bar{\Lambda}$ due to the strong localization of $\bar{\Lambda}$[97] as also shown in Fig. 10.

The nucleons in $1s_{1/2}$ state distribute mainly around the nuclear center and thus feel a deeper potential due to the presence of the $\bar{\Lambda}$, which results in a larger binding energy, as seen in Fig. 12. However, the nucleons at higher energy states are far away from the center and therefore are not much influenced by the $\bar{\Lambda}$, except the changes in spin-orbit splitting. That is, the polarization effect of anti-Lambda on single-particle structure of atomic nuclei varies depending on which state it occupies. For nucleons in 1s1/2 state, the anti-lambda makes it deepest bound when it occupies 1p3/2 state. For the nucleons in 1p3/2 state, on the other hand, the largest polarization effect is found when the anti-Lambda occupies 2s1/2 state. This phenomenon can also be understood from the range of the polarization effect by $\bar{\Lambda}$.

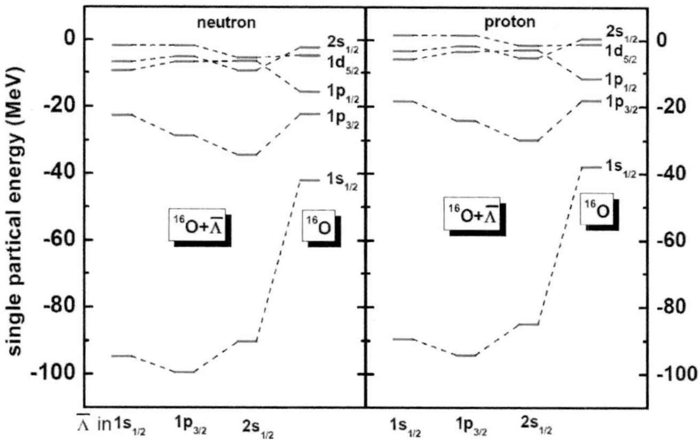

Fig. 12. (Color online) The single-particle energy of neutron and proton in ^{16}O+$\bar{\Lambda}$ system, where the $\bar{\Lambda}$ occupies $1s_{1/2}, 1p_{3/2}$ and $2s_{1/2}$, respectively. The single-particle energy in ^{16}O (last column) is plotted for comparison.

5.2. *Spin symmetry in the spectrum of anti-Lambda hyperon*

The potential and single $\bar{\Lambda}$ spectrum in ^{16}O and those of $\bar{\Lambda}$ in ^{16}O+$\bar{\Lambda}$ system are plotted in Fig. 13. As seen in the left panel of Fig. 13, when the rearrangement effect of $\bar{\Lambda}$ or Λ is not taken into account in ^{16}O, the energies of $\bar{\Lambda}$ spin-orbit doublets are almost identical. The energy differences $\epsilon_{\bar{\Lambda}(nl_{j=l-1/2})} - \epsilon_{\bar{\Lambda}(nl_{j=l+1/2})}$ in the anti-Lambda spectrum are actually around 0.09–0.17 MeV for p states, which are much smaller than that in the Λ spectrum, 2.26 MeV, estimated without the tensor coupling. It has been discussed in ordinary atomic nuclei that there is good pseudo-spin symmetry in single-nucleon spectrum and spin symmetry in the spectrum

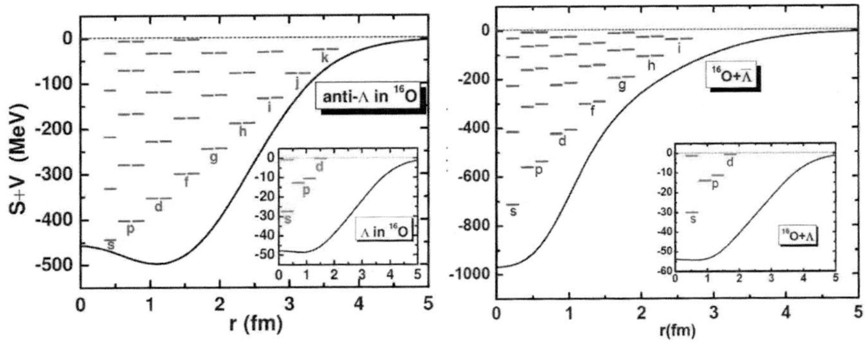

Fig. 13. (Color online) Mean-field potential and single-particle energy spectrum of $\bar{\Lambda}$ in ^{16}O (the left panel) and those of $\bar{\Lambda}$ in ^{16}O+$\bar{\Lambda}$ system (the right panel). The inset gives the potential and spectrum of Λ in ^{16}O and that of Λ in ^{16}O+$\bar{\Lambda}$ system. The polarization effect of $\bar{\Lambda}$ or Λ in ^{16}O is taken into account in the right panel. $\xi = 1$ is used for the $\bar{\Lambda}N$ interaction. Taken from Ref. 102.

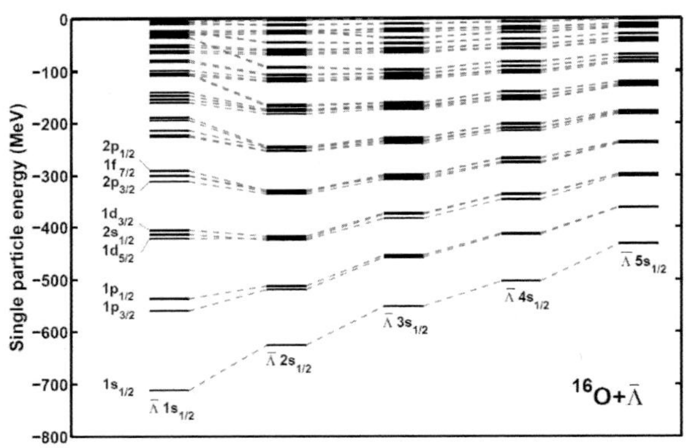

Fig. 14. (Color online) Single-particle energy spectrum of $\bar{\Lambda}$ in ^{16}O+$\bar{\Lambda}$ system, where the $\bar{\Lambda}$ occupies different s orbits, respectively.

of anti-nucleon.[103] However, if the polarization effect of $\bar{\Lambda}$ on the mean-field is taken into account, the spin symmetry will be destroyed as illustrated in the right panel of Fig. 13. With the increase of the energy of the occupied $\bar{\Lambda}$ state, the polarization effect of $\bar{\Lambda}$ decreases, and thus the spin symmetry in the single-particle energy spectrum of $\bar{\Lambda}$ is recovered gradually as exhibited in Fig. 14.

A further study on the spin symmetry in the single-particle energy spectrum of $\bar{\Lambda}$ was made in Ref. 104. It has been found that the tensor coupling enlarges the spin-orbit splittings for $\bar{\Lambda}$ by a factor of 5 but has a negligible effect on the wave functions of $\bar{\Lambda}$. This indicates that the spin symmetry in anti-Λ-nucleus systems is still a good approximation.

6. Multi-strange systems

6.1. *Multi-Λ hypernuclei*

So far, we have focused on the properties of single-Λ hypernuclei. In this section, we will discuss properties of multi-Λ hypernuclei. Among multi-Λ hypernuclei, so far only double-Λ hypernuclei have been experimentally found in an emulsion.[105,106] The mass of double-Λ hypernuclei is often discussed in terms of the $\Lambda - \Lambda$ interaction energy defined as

$$\Delta B_{\Lambda\Lambda}({}^{A}_{\Lambda\Lambda}Z) \equiv B_{\Lambda\Lambda}({}^{A}_{\Lambda\Lambda}Z) - 2B_{\Lambda}({}^{A-1}_{\Lambda}Z), \tag{28}$$

where $B_{\Lambda\Lambda}$ and B_{Λ} are the two-Λ and one-Λ binding energies, respectively. The experimental value of $\Delta B_{\Lambda\Lambda}$ for the ${}^{6}_{\Lambda\Lambda}$He hypernucleus is 0.67±0.17 MeV,[106] which indicates that the $\Lambda\Lambda$ interaction is attractive but weak.

For the meson exchange version of relativistic mean-field theory, it is straightforward to extend it from single-Λ to multi-Λ systems.[107] That is, only the Λ density as a source of the Klein–Gordon equations for mesons is changed and the structure of the Dirac equation remains the same. Since the $\Lambda\Lambda$ interaction is weak, a residual $\Lambda\Lambda$ pairing interaction is usually neglected. This simply leads to a larger impurity effect of Λ particle.[15,23] Fig. 15 shows the root-mean-square radii and the total energy per baryon as a function of the number of Λ particles, n, in ${}^{48}_{n\Lambda}$Ca hypernuclei.[66] Notice that the neutron number is varied for each n so that the total mass number is kept at 48. The energy and the radii show non-monotonic behaviors. The Λ particles first increase the binding energy and the radius decreases as its consequence. When the Λ particle start filling weakly bound single-particle Λ levels, the binding then becomes weaker and the radius increases. This behavior is understandable since there is no Pauli principle between the Λ particles and the nucleons. A similar conclusion has been reached also in Refs. 42, 108 and in a very recent study.[109]

6.2. *Neutron stars*

We next discuss neutron stars, which offer natural laboratories to explore the equation of state (EOS) at low temperature and high densities.[110–113] In particular, the observed masses of neutron stars have provided a strong

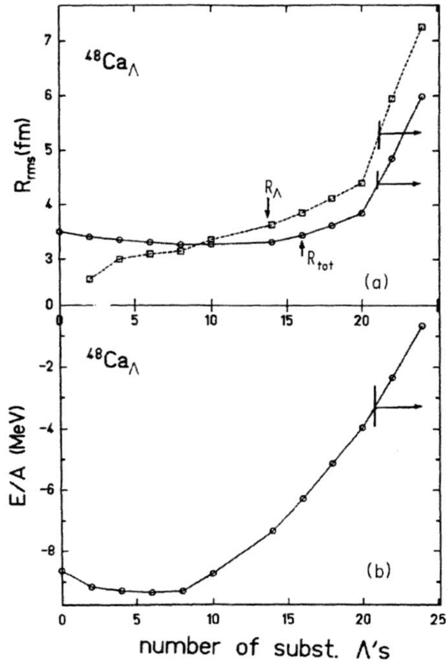

Fig. 15. The root mean square radii (the upper panel) and the binding energies per baryon (the lower panel) as a function of the number n of Λ particles in $^{48}_{n\Lambda}$Ca hypernuclei. The mass number of each hypernucleus is kept at 48 by changing the neutron number according to the number of n. The arrows indicate the particle unbound region. Taken from Ref. 66.

constraint on the EOS. Many neutron stars have been observed to have a mass of around 1.4 $M_\odot$,[111] where $M_\odot$ is the solar mass, and this value has been regarded as a standard value for many years.

In addition to the non-relativistic approaches,[115–120] the relativistic approach has also been applied both to strange matter[66,121] and to neutron star matter with hyperons.[43,114,122–129] One of the most important conclusions of these studies is that hyperons should admix in dense neutron star matter (see Fig. 16), in which the first hyperon to appear is either Λ or Σ^- depending on the properties of single-particle potential for Σ particle. The hyperon admixture can be easily understood using the non-relativistic Thomas–Fermi approximation (see *e.g.,* Refs. 130, 131 for the Thomas–Fermi approximation). In the Thomas–Fermi approximation, the total energy (including the mass energies) for matter with N_n neutrons and N_Λ Λ

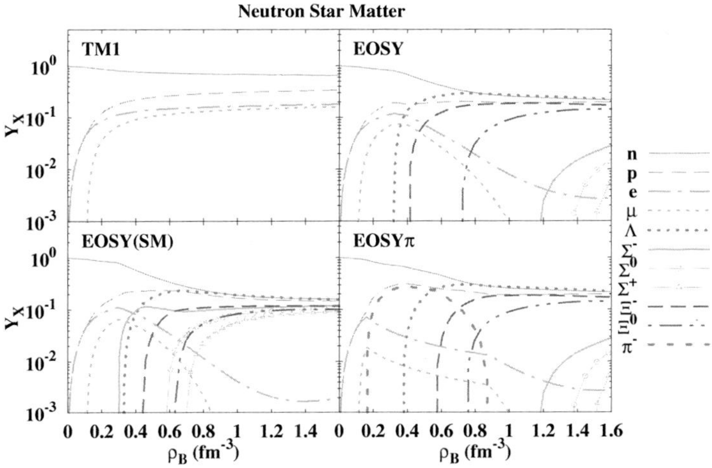

Fig. 16. (Color online) The fraction of baryons and leptons in neutron star matter obtained with the relativistic mean field calculations. The upper left panel (TM1) shows the result without hyperons, while the right upper panel (EOSY) and the left lower panel (EOSY(SM)) show the results with hyperons, with a repulsive and an attractive Σ potential, respectively. The right lower panel (EOSYπ) shows the result with pion contribution. Taken from Ref. 114.

particles is given by

$$E = m_N c^2 \cdot N_n + \frac{3}{5} N_n E_{Fn} + m_\Lambda c^2 \cdot N_\Lambda + \frac{3}{5} N_\Lambda E_{F\Lambda}, \tag{29}$$

where E_{Fn} and $E_{F\Lambda}$ are the Fermi energy for the neutrons and the Λ particles, respectively, and m_N is the nucleon mass. For $N_n = (1-x)N_B$ and $N_\Lambda = xN_B$, the total energy per baryon then reads

$$\frac{E}{N_B} = m_N c^2 + \frac{3}{5} E_{Fn} + \left[(m_\Lambda - m_N)c^2 + \frac{3}{5}(E_{F\Lambda} - E_{Fn}) \right] x, \tag{30}$$

with

$$E_{Fn} = \frac{\hbar^2}{2m_N} \left(3\pi^2 \cdot (1-x)\rho_B \right)^{2/3}, \tag{31}$$

$$E_{F\Lambda} = \frac{\hbar^2}{2m_\Lambda} \left(3\pi \cdot x\rho_B \right)^{2/3}, \tag{32}$$

where ρ_B is the baryon density. This equation indicates that the total energy is minimized with $x = 0$ (and thus $E_{F\Lambda} = 0$) when the neutron Fermi energy E_{Fn} is smaller than $5/3 \cdot (m_\Lambda - m_N)c^2$ while x becomes a finite value

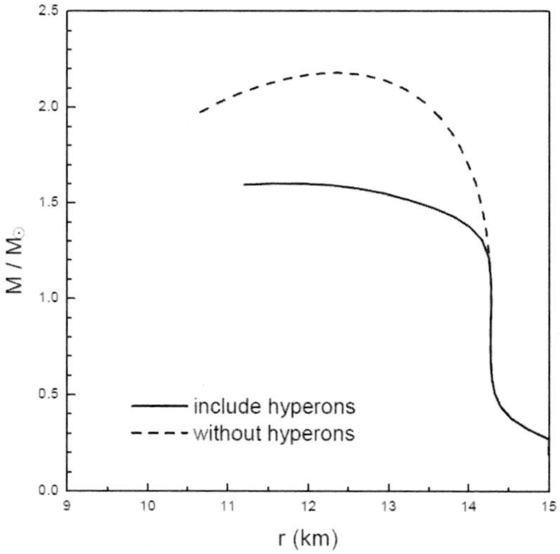

Fig. 17. A mass-radius relation for neutron stars obtained with RMF. Taken from Ref. 126.

after E_{Fn} exceeds $5/3 \cdot (m_\Lambda - m_N)c^2$. This result is a physical consequence of the fact that the increase of neutron energy due to the Pauli principle can be avoided by converting a fraction of neutrons to Λ particles, although the actual value of the critical density for the hyperon admixture depends on details of baryon-baryon interactions.

Another important conclusion on the previous theoretical studies is that the EOS is softened when the admixture of hyperons is taken into account. This results in a smaller maximum mass of neutron stars. Fig. 17 shows a mass-radius relation obtained by solving the Tolman–Oppenheimer–Volkov equation[113] with the EOS of Shen *et al.*,[126,127] which was constructed based on RMF. One can see that the maximum mass of neutron stars is around 2.2 $M_\odot$ when the hyperon admixture is not considered whereas it decreases to 1.6 $M\odot$ when the hyperon admixture is taken into account. Very recently, a large neutron mass of $(1.97\pm\ 0.04)M_\odot$ was observed using the Shapiro delay for the binary millisecond pulsar J1614-2230.[132] Most of EOS for hyperonic neutron star matter is not compatible with such a large neutron star mass, and it has remained a big open problem in present nuclear physics.

More detailed discussions on neutron stars will be given in Chap. 15.

7. Summary

Hypernuclei with one or more hyperons have provided interesting and unique opportunities to study nuclear many-body problems, in addition to the main motivation of hypernuclear physics, that is, to extract the information on nucleon-hyperon and hyperon-hyperon interactions. One of the important subjects along this direction is to clarify the impurity effect of Λ hyperons, that is, to clarify how several properties, such as deformation and collective excitations, of atomic nuclei are influenced by adding a Λ particle. A relativistic mean-field approach has been one of the standard tools for this purpose. An important point is that the spin-orbit splitting is described naturally with the relativistic approach. It has been known experimentally that the spin-orbit splittings for Λ particles are much smaller than that for nucleons, and the relativistic approach has played an important role in hypernuclear physics, especially at the early stage of research in the late 1970's.

The relativistic approach has predicted an interesting impurity effect on nuclear deformation. That is, there are certain nuclei for which the deformation disappears when a Λ particle is added to them. This phenomenon is difficult to see with non-relativistic approaches since the polarization effect of Λ particle is weaker than the relativistic approach.

A drawback of the mean-field approach is that the pure mean-field approximation does not yield a spectrum of hypernuclei due to the broken symmetries (this applies both to the non-relativistic and the relativistic approaches). In order to cure this problem, the so called beyond mean-field approach has been developed for ordinary nuclei (see Chap. 10 in this book). The beyond mean-field approach includes a restoration of the broken symmetries by angular momentum and particle number projections. It also includes the shape fluctuation effect, which is taken into account with the generator coordinate method (GCM) or its approximation, the collective Hamiltonian approach. The beyond mean-field approach has been applied to hypernuclei and the spectra of hypernuclei have now been constructed based on (Covariant) Density Functional Theory, using the microscopic particle-rotor model. A nice feature of hypernuclei is that the problem becomes simpler because of the absence of Pauli principle between a Λ particle and nucleons in a core nucleus.

Another important subject of hypernuclear physics is the neutron star. There is almost no doubt that hyperons appear in dense neutron star matter, since it is a natural consequence of Pauli principle. A problem is that the hyperon admixture softens the equation of state and decreases the maximum neutron star mass. In fact, most EOS for hyperonic neutron star matter are incompatible with the recently observed large neutron star mass ($\sim 2M_\odot$), and it has remained a big challenge in the nuclear physics community to account for the large neutron mass together with the hyperon admixture. Hopefully, new experimental investigations of hypernuclei at the next generation experimental facilities, *e.g.*, the J-PARC facility, and theoretical studies with such new data will resolve this big problem in the near future.

Acknowledgment

We thank T. Koike, Z. P. Li, H. F. Lü, H. Mei, J. Meng, T. Motoba, H. Sagawa, H.-J. Schulze, C. Y. Song, H. Tamura, Y. Tanimura, M. T. Win, W. X. Xue for useful discussions and their contributions to the work reviewed in this chapter. This work was supported in part by the Tohoku University Focused Research Project "Understanding the origins for matters in universe", JSPS KAKENHI Grant Numbers 25105503 and 26400263, the National Natural Science Foundation of China under Grant Nos. 11305134, 11105111, and the Fundamental Research Funds for the Central Universities (XDJK2010B007 and XDJK2013C028).

References

1. J. Beringer, J. F. Arguin, R. M. Barnett, K. Copic, O. Dahl, D. E. Groom, C. J. Lin, J. Lys, H. Murayama, C. G. Wohl, W. M. Yao, P. A. Zyla, C. Amsler, M. Antonelli, D. M. Asner, H. Baer, H. R. Band, T. Basaglia, C. W. Bauer, J. J. Beatty, V. I. Belousov, E. Bergren, G. Bernardi, W. Bertl, S. Bethke, H. Bichsel, O. Biebel, E. Blucher, S. Blusk, G. Brooijmans, O. Buchmueller, R. N. Cahn, M. Carena, A. Ceccucci, D. Chakraborty, M. C. Chen, R. S. Chivukula, G. Cowan, G. D'Ambrosio, T. Damour, D. de Florian, A. de Gouvêa, T. DeGrand, P. de Jong, G. Dissertori, B. Dobrescu, M. Doser, M. Drees, D. A. Edwards, S. Eidelman, J. Erler, V. V. Ezhela, W. Fetscher, B. D. Fields, B. Foster, T. K. Gaisser, L. Garren, H. J. Gerber, G. Gerbier, T. Gherghetta, S. Golwala, M. Goodman, C. Grab, A. V. Gritsan, J. F. Grivaz, M. Grünewald, A. Gurtu, T. Gutsche, H. E. Haber, K. Hagiwara, C. Hagmann, C. Hanhart, S. Hashimoto, K. G. Hayes, M. Heffner, B. Heltsley, J. J. Hernández-Rey, K. Hikasa, A. Höcker, J. Holder, A. Holtkamp, J. Huston, J. D. Jackson, K. F. Johnson, T. Junk, D. Karlen, D. Kirkby, S. R. Klein, E. Klempt, R. V. Kowalewski, F. Krauss, M. Kreps, B. Krusche, Y. V. Kuyanov, Y. Kwon, O. Lahav, J. Laiho, P. Langacker, A. Liddle, Z. Ligeti, T. M. Liss, L. Littenberg, K. S. Lugovsky,

S. B. Lugovsky, T. Mannel, A. V. Manohar, W. J. Marciano, A. D. Martin, A. Masoni, J. Matthews, D. Milstead, R. Miquel, K. Mönig, F. Moortgat, K. Nakamura, M. Narain, P. Nason, S. Navas, M. Neubert, P. Nevski, Y. Nir, K. A. Olive, L. Pape, J. Parsons, C. Patrignani, J. A. Peacock, S. T. Petcov, A. Piepke, A. Pomarol, G. Punzi, A. Quadt, S. Raby, G. Raffelt, B. N. Ratcliff, P. Richardson, S. Roesler, S. Rolli, A. Romaniouk, L. J. Rosenberg, J. L. Rosner, C. T. Sachrajda, Y. Sakai, G. P. Salam, S. Sarkar, F. Sauli, O. Schneider, K. Scholberg, D. Scott, W. G. Seligman, M. H. Shaevitz, S. R. Sharpe, M. Silari, T. Sjöstrand, P. Skands, J. G. Smith, G. F. Smoot, S. Spanier, H. Spieler, A. Stahl, T. Stanev, S. L. Stone, T. Sumiyoshi, M. J. Syphers, F. Takahashi, M. Tanabashi, J. Terning, M. Titov, N. P. Tkachenko, N. A. Törnqvist, D. Tovey, G. Valencia, K. van Bibber, G. Venanzoni, M. G. Vincter, P. Vogel, A. Vogt, W. Walkowiak, C. W. Walter, D. R. Ward, T. Watari, G. Weiglein, E. J. Weinberg, L. R. Wiencke, L. Wolfenstein, J. Womersley, C. L. Woody, R. L. Workman, A. Yamamoto, G. P. Zeller, O. V. Zenin, J. Zhang, R. Y. Zhu, G. Harper, V. S. Lugovsky, and P. Schaffner, Review of particle physics, *Phys. Rev. D.* **86**, 010001 (Jul, 2012). doi: 10.1103/PhysRevD.86.010001. URL `http://link.aps.org/doi/10.1103/PhysRevD.86.010001`.

2. M. Danysz and J. Pniewski, Delayed disintegration of a heavy nuclear fragment. 1., *Phil. Mag.* **44**, 348 (1953).
3. A. Wroblewski, Hypernuclei (and strange particles) - how it all began?, *Act. Phys. Pol. B.* **35**, 1 (2004).
4. O. Hashimoto and H. Tamura, Spectroscopy of hypernuclei, *Prog. Part. Nucl. Phys.* **57**, 564 (2006).
5. W. Alberico and G. Garbarino, Weak decay of Lambda-hypernuclei, *Phys. Rep.* **369**, 1 (2002).
6. H. Bando, T. Motoba, and J. Zofka, Production, structure and decay of hypernuclei, *Int. J. Mod. Phys. A.* **5**, 4021 (1990).
7. I. Vidana, Hyperons and neutron stars, *Nucl. Phys. A.* **914**, 367 (2013).
8. J. Schaffner-Bielich, Hypernuclear physics for neutron stars, *Nucl. Phys. A.* **804**, 309 (2008).
9. T. Motoba, H. Bandō, and K. Ikeda, Light p-shell Λ-hypernuclei by the microscopic three-cluster model, *Prog. Theor. Phys.* **70**, 189 (1983).
10. E. Hiyama, M. Kamimura, K. Miyazaki, and T. Motoba, γ transitions in $A = 7$ hypernuclei and a possible derivation of hypernuclear size, *Phys. Rev. C.* **59**, 2351 (1999).
11. E. Hiyama, M. Kamimura, Y. Yamamoto, T. Motoba, and T. Rijken, S=−1 hypernuclear structure, *Prog. Theor. Phys. Suppl.* **185**, 106 (2010).
12. J. Zofka, Light deformed hypernuclei, *Czech. J. Phys.* **B30**, 95 (1980).
13. X. R. Zhou, H.-J. Schulze, H. Sagawa, C. X. Wu, and E. G. Zhao, Hypernuclei in the deformed Skyrme–Hartree–Fock approach, *Phys. Rev. C.* **76**, 034312 (2007).
14. M. T. Win and K. Hagino, Deformation of Λ-hypernuclei, *Phys. Rev. C.* **78**, 054311 (2008).

15. H.-J. Schulze, M. Win, K. Hagino, and H. Sagawa, Hyperons as a probe of nuclear deformation, *Prog. Theo. Phys.* **123**, 569 (2010).
16. M. Win, K. Hagino, and T. Koike, Shape of lambda hypernuclei in the (β, γ) deformation plane, *Phys. Rev. C.* **83**, 014301 (2011).
17. J. M. Yao, Z. P. Li, K. Hagino, M. T. Win, Y. Zhang, and J. Meng, Impurity effect of Λ hyperon on collective excitations of atomic nuclei, *Nucl. Phys. A.* **868-869**, 12 (2011).
18. B.-N. Lu, E.-G. Zhao, and S.-G. Zhou, Quadrupole deformation (β, γ) of light Λ hypernuclei in a constrained relativistic mean field model: Shape evolution and shape polarization effect of the Λ hyperon, *Phys. Rev. C.* **84**, 014328 (2011).
19. M. Isaka, M. Kimura, A. Dote, and A. Ohnishi, Deformation of hypernuclei studied with antisymmetrized molecular dynamics, *Phys. Rev. C.* **83**, 044323 (2011).
20. M. Isaka, H. Homma, M. Kimura, A. Dote, and A. Ohnishi, Modification of triaxial deformation and change of spectrum in ${}^{25}_{\Lambda}$Mg caused by the Λ hyperon, *Phys. Rev. C.* **85**, 034303 (2012).
21. M. Isaka, K. Fukukawa, M. Kimura, E. Hiyama, H. Sagawa, and Y. Yamamoto, Superdeformed Λ hypernuclei with antisymmetrized molecular dynamics, *Phys. Rev. C.* **89**, 024310 (2014).
22. K. Hagino, J. M. Yao, F. Minato, Z. P. Li, and M. T. Win, Collective excitations of Λ hypernuclei, *Nucl. Phys. A.* **914**, 151 (2013).
23. F. Minato and K. Hagino, Application of random phase approximation to vibrational excitations of double-Λ hypernuclei, *Phys. Rev. C.* **85**, 024316 (2012).
24. H. Mei, K. Hagino, J. M. Yao, and T. Motoba, Microscopic particle-rotor model for the low-lying spectrum of ${}^{9}{}_{\Lambda}$Be hypernuclei, *Phys. Rev. C.* **90**, 064302 (Dec, 2014). doi: 10.1103/PhysRevC.90.064302. URL `http://link.aps.org/doi/10.1103/PhysRevC.90.064302`.
25. M. Isaka, M. Kimura, A. Dote, and A. Ohnishi, Splitting of the p orbit in triaxially deformed ${}^{25}_{\Lambda}$Mg, *Phys. Rev. C.* **87**, 021304(R) (2013).
26. D. Vretenar, W. Pöschl, G. Lalazissis, and P. Ring, Relativistic mean-field description of light Λ hypernuclei with large neutron excess, *Phys. Rev. C.* **57**, R1060 (1998).
27. X.-R. Zhou, A. Polls, H.-J. Schulze, and I. Vidana, Λ hyperons and the neutron drip line, *Phys. Rev. C.* **78**, 054306 (2008).
28. F. Minato, S. Chiba, and K. Hagino, Fission of heavy Λ hypernuclei with the Skyrme–Hartree–Fock approach, *Nucl. Phys. A.* **831**, 150 (2009).
29. F. Minato and S. Chiba, Fission barrier of actinide nuclei with double-Λ particles with the Skyrme–Hartree–Fock method, *Nucl. Phys. A.* **856**, 55 (2011).
30. R. H. Dalitz and A. Gal, Supersymmetric and strangeness analog states in p-shell Λ hypernuclei, *Phys. Rev. Lett.* **36**, 362–365 (Feb, 1976). doi: 10.1103/PhysRevLett.36.362. URL `http://link.aps.org/doi/10.1103/PhysRevLett.36.362`.
31. Z. Y. Zhang, G. L. Li, and Y. W. Yu, Su(3) group classification

for the energy spectrum and production probabilities of ${}^{13}_{\Lambda}$C, *Phys. Lett. B.* **108**(4-5), 261 – 264 (1982). ISSN 0370-2693. doi: http://dx.doi.org/10.1016/0370-2693(82)91188-1. URL `http://www.sciencedirect.com/science/article/pii/0370269382911881`.
32. T. Motoba, H. Bando, K. Ikeda, and T. Yamada, Production, structure and decay of light p-shell Λ-hypernuclei, *Prog. Theo. Phys. Suppl.* **81**, 42 (1985).
33. R. Dalitz and A. Gal, The formation of, and the γ-radiation from, the p-shell hypernuclei, *Ann. Phys. (N.Y.).* **116**, 167 (1978).
34. A. Gal, J. Soper, and R. Dalitz, A shell-model analysis of Λ binding energies for the p-shell hypernuclei. i. basic formulas and matrix elements for ΛN and ΛNN forces, *Ann. Phys. (N.Y.).* **63**, 53 (1971).
35. D. J. Millener, Shell-model interpretation of γ-ray transitions in p-shell hypernuclei, *Nucl. Phys. A.* **804**, 84 (2008).
36. D. J. Millener, Shell-model calculations applied to the γ-ray spectroscopy of light Λ hypernuclei, *Nucl. Phys. A.* **914**, 109 (2013).
37. M. Rayet, Self-consistent calculations of hypernuclear properties with the skyrme interaction, *Ann. of Phys. (N.Y.).* **102**, 226 (1976).
38. M. Rayet, Skyrme parametrization of an effective Λ-nucleon interaction, *Nucl. Phys. A.* **367**, 381 (1981).
39. Y. Yamamoto, H. Bando, and J. Zofka, On the Λ-hypernuclear single particle energies, *Prog. Theo. Phys.* **80**, 757 (1988).
40. R. Brockmann and W. Weise, ΛN-interaction and spin-orbit coupling in hypernuclei, *Phys. Lett.* **69B**, 167 (1977).
41. J. Boguta and S. Bohrmann, Relativistic quantum field theory of Λ hypernuclei, *Phys. Lett.* **102B**, 93 (1981).
42. J. Mares and J. Zofka, On Λ-hyperon(s) in the nuclear medium, *Z. Phys. A.* **333**, 209 (1989).
43. Y. Sugahara and H. Toki, Relativistic mean field theory for Lambda hypernuclei and neutron stars, *Prog. Theo. Phys.* **92**, 803 (1994).
44. R. Wirth, D. Gazda, P. Navratil, A. Calci, J. Langhammer, and R. Roth, Ab initio description of p-shell hypernuclei, *arXiv:1403.3067* (2014).
45. W. Brückner, M. Faessler, T. Ketel, K. Kilian, J. Niewisch, B. Pietrzyk, B. Povh, H. Ritter, M. Uhrmacher, P. Birien, H. Catz, A. Chaumeaux, J. Durand, B. Mayer, J. Thirion, R. Bertini, and O. Bing, Spin-orbit interaction of Lambda particle in nuclei, *Phys. Lett.* **79B**, 157 (1978).
46. R. Brockmann, Spin-orbit coupling in Σ-hypernuclei, *Phy. Lett.* **104B**, 256 (1981).
47. A. Boussy, Meson exchange contributions in a relativistic description of nuclei and Λ- or Σ-hypernuclei, *Phys. Lett.* **99B**, 305 (1981).
48. J.-K. Zhang and X.-J. Qiu, An approach of relativistic mean field theory to hypernuclei, *Phys. Lett.* **152B**, 153 (1985).
49. J. Mares and B. Jennings, Relativistic description of Λ,Σ, and Ξ hypernuclei, *Phys. Rev. C.* **49**, 2472 (1994).
50. N. K. Glendenning, D. Von-Eiff, M. Haft, H. Lenske, and M. Weigel, Relativistic mean-field calculations of Λ and Σ hypernuclei, *Phys. Rev. C.* **48**, 889 (1993).

51. J. Meng, H. Toki, S. Zhou, S. Zhang, W. H. Long, and L. Geng, Relativistic continuum Hartree Bogoliubov theory for ground-state properties of exotic nuclei, *Prog. Part. Nucl. Phys.* **57**(2), 470 – 563 (2006). ISSN 0146-6410. URL `http://www.sciencedirect.com/science/article/pii/S014664100500075X`.
52. Y. Tanimura and K. Hagino, Description of single-Λ hypernuclei with a relativistic point-coupling model, *Phys. Rev. C.* **85**, 014306 (Jan, 2012).
53. J. V. Noble, Nuclear Dirac phenomenology and the Λ-nucleus potential, *Phys. Lett. B.* **89B**, 325 (1980).
54. B. K. Jennings, The dirac equation and Λ-nucleus systematics, *Phys. Lett. B.* **246**, 325–328 (1990). ISSN 0370-2693. URL `http://www.sciencedirect.com/science/article/pii/037026939090607B`.
55. H. Y. Sang, X. S. Wang, H. F. Lü, J. M. Yao, and H. Sagawa, Magnetic moments of Λ hypernuclei within the time-odd triaxial relativistic mean-field approach, *Phys. Rev. C.* **88**, 064304 (Dec, 2013). URL `http://link.aps.org/doi/10.1103/PhysRevC.88.064304`.
56. J. M. Yao, H. Chen, and J. Meng, Time-odd triaxial relativistic mean field approach for nuclear magnetic moments, *Phys. Rev. C.* **74**(2), 024307 (2006).
57. D. Davis, Hypernuclei-the early days, *Nucl. Phys. A.* **547**, 369c (1992).
58. S. N. Nakamura, A. Matsumura, Y. Okayasu, T. Seva, V. M. Rodriguez, P. Baturin, L. Yuan, A. Acha, A. Ahmidouch, D. Androic, A. Asaturyan, R. Asaturyan, O. K. Baker, F. Benmokhtar, P. Bosted, R. Carlini, C. Chen, M. Christy, L. Cole, S. Danagoulian, A. Daniel, V. Dharmawardane, K. Egiyan, M. Elaasar, R. Ent, H. Fenker, Y. Fujii, M. Furic, L. Gan, D. Gaskell, A. Gasparian, E. F. Gibson, T. Gogami, P. Gueye, Y. Han, O. Hashimoto, E. Hiyama, D. Honda, T. Horn, B. Hu, E. V. Hungerford, C. Jayalath, M. Jones, K. Johnston, N. Kalantarians, H. Kanda, M. Kaneta, F. Kato, S. Kato, D. Kawama, C. Keppel, K. J. Lan, W. Luo, D. Mack, K. Maeda, S. Malace, A. Margaryan, G. Marikyan, P. Markowitz, T. Maruta, N. Maruyama, T. Miyoshi, A. Mkrtchyan, H. Mkrtchyan, S. Nagao, T. Navasardyan, G. Niculescu, M.-I. Niculescu, H. Nomura, K. Nonaka, A. Ohtani, M. Oyamada, N. Perez, T. Petkovic, S. Randeniya, J. Reinhold, J. Roche, Y. Sato, E. K. Segbefia, N. Simicevic, G. Smith, Y. Song, M. Sumihama, V. Tadevosyan, T. Takahashi, L. Tang, K. Tsukada, V. Tvaskis, W. Vulcan, S. Wells, S. A. Wood, C. Yan, and S. Zhamkochyan, Observation of the ${}^{7}_{\Lambda}$He hypernucleus by the $(e, e'K^{+})$ reaction, *Phys. Rev. Lett.* **110**, 012502 (2013).
59. T. Gogami. *Spectroscopic research of hypernuclei up to medium-heavy mass region with the $(e, e'K^{+})$ reaction.* PhD thesis, Tohoku University (2014).
60. Q. N. Usmani and A. Bodmer, Λ single particle energies, *Phys. Rev. C.* **60**, 055215 (1999).
61. D. J. Millener, C. Dover, and A. Gal, Λ-nucleus single-particle potentials, *Phys. Rev. C.* **38**, 2700 (1988).
62. T. Motoba, H. Bando, R. Wünsch, and J. Zofka, Hypernuclear production by the (π^{+}, K^{+}) reaction, *Phys. Rev. C.* **28**, 1322 (1988).

63. A. Boussy, Shell-model calculation of strangeness exchange reactions on nuclei, *Nucl. Phys. A.* **290**, 324 (1977).
64. A. Boussy, Strangeness exchange reactions and hypernuclear spectroscopy, *Phys. Lett.* **84B**, 41 (1979).
65. A. Boussy, ΛN residual interaction and $^{16}_{\Lambda}$O hypernucleus, *Phys. Lett.* **91B**, 15 (1980).
66. M. Rufa, J. Schaffner, J. Maruhn, H. Stöcker, W. Greiner, and P.-G. Reinhard, Multi-lambda hypernuclei and the equation of state of hypermatter, *Phys. Rev. C.* **42**, 2469 (1990).
67. C. Y. Song, J. M. Yao, H. F. Lu, and J. Meng, Lambda and anti-lambda hypernuclei in relativistic mean-field theory, *Int. J. Mod. Phys. E.* **19**, 2538 (2010). doi: http://dx.doi.org/10.1142/S0218301310017058.
68. S. Ajimura, H. Hayakawa, T. Kishimoto, H. Kohri, K. Matsuoka, S. Minami, T. Mori, K. Morikubo, E. Saji, A. Sakaguchi, Y. Shimizu, M. Sumihama, R. E. Chrien, M. May, P. Pile, A. Rusek, R. Sutter, P. Eugenio, G. Franklin, P. Khaustov, K. Paschke, B. P. Quinn, R. A. Schumacher, J. Franz, T. Fukuda, H. Noumi, H. Outa, L. Gan, L. Tang, L. Yuan, H. Tamura, J. Nakano, T. Tamagawa, K. Tanida, and R. Sawafta, Observation of spin-orbit splitting in Λ single-particle states, *Phys. Rev. Lett.* **86**, 4255 (May, 2001).
69. H. J. Pirner, Quantum chromodynamics and the spin-orbit splitting in nuclei and Λ- and Σ-hypernuclei, *Phys. Lett.* **85B**, 190 (1979).
70. R. Brockmann and W. Weise, Relativistic single particle motion and spin-orbit coupling in nuclei and hypernuclei, *Nucl. Phys. A.* **1981**, 365 (1981).
71. A. Boussy, A simplified relativistic model for nuclei and hypernuclei, *Nucl. Phys. A.* **381**, 445 (1982).
72. Y. Tanaka, The schmidt diagrams and the configuration mixing effects on hypernuclear magnetic moments, *Phys. Lett. B.* **227**(2), 195–198 (1989). ISSN 0370-2693. doi: http://dx.doi.org/10.1016/S0370-2693(89)80021-8. URL `http://www.sciencedirect.com/science/article/pii/S0370269389800218`.
73. K. Saito, M. Oka, and T. Suzuki, Exchange currents for hypernuclear magnetic moments, *Nucl. Phys. A.* **625**, 95–106 (1997). ISSN 0375-9474. doi: http://dx.doi.org/10.1016/S0375-9474(97)00493-4. URL `http://www.sciencedirect.com/science/article/pii/S0375947497004934`.
74. J. Cohen and R. J. Furnstahl, Hypernuclear currents in a relativistic mean-field theory, *Phys. Rev. C.* **35**, 2231–2235 (Jun, 1987). doi: 10.1103/PhysRevC.35.2231. URL `http://link.aps.org/doi/10.1103/PhysRevC.35.2231`.
75. J. Mares and J. Zofka, Hypernuclear magnetic moments, *Phys. Lett. B.* **249**(2), 181–185 (1990). ISSN 0370-2693. doi: http://dx.doi.org/10.1016/0370-2693(90)91239-8. URL `http://www.sciencedirect.com/science/article/pii/0370269390912398`.
76. A. O. Gattone, M. Chiapparini, and E. D. Izquierdo, Λ-hypernuclei magnetic moments in a relativistic model, *Phys. Rev. C.* **44**, 548–551 (Jul, 1991). doi: 10.1103/PhysRevC.44.548. URL `http://link.aps.org/doi/10.1103/PhysRevC.44.548`.

77. J. M. Yao, H. F. Lu, G. Hillhouse, and J. Meng, Core polarization and tensor coupling effects on magnetic moments of hypernuclei, *Chinese Physics Letters.* 25(5):1629 (2008). URL `http://cpl.iphy.ac.cn/EN/abstract/article_43629.shtml`.
78. X. S. Wang, H. Y. Sang, H. F. Lü, J. M. Yao, and H. Sagawa, Systematic study of hypernuclear magnetic moments under a perturb treatment, *Eur. Phys. J. A.* **49**, 101 (2013). doi: 10.1140/epja/i2013-13101-1.
79. X. S. Wang, H. Y. Sang, J. H. Wang, and H. F. Lü, A new determination of the lambda-nucleon coupling constants in relativistic mean field theory, *Commun. Theor. Phys.* **60**(3), 479 (2013). URL `10.1088/0253-6102/60/4/16`.
80. H. Tamura, K. Tanida, D. Abe, H. Akikawa, K. Araki, H. Bhang, T. Endo, Y. Fujii, T. Fukuda, O. Hashimoto, K. Imai, H. Hotchi, Y. Kakiguchi, J. H. Kim, Y. D. Kim, T. Miyoshi, T. Murakami, T. Nagae, H. Noumi, H. Outa, K. Ozawa, T. Saito, J. Sasao, Y. Sato, S. Satoh, R. I. Sawafta, M. Sekimoto, T. Takahashi, L. Tang, H. H. Xia, S. H. Zhou, and L. H. Zhu, Observation of a spin-flip $m1$ transition in ${}^{7}_{\Lambda}$Li, *Phys. Rev. Lett.* **84**, 5963–5966 (Jun, 2000). doi: 10.1103/PhysRevLett.84.5963. URL `http://link.aps.org/doi/10.1103/PhysRevLett.84.5963`.
81. O. Hashimoto and H. Tamura, Spectroscopy of hypernuclei, *Prog. Part. Nucl. Phys.* **57**, 564 (2006).
82. I. Tanihata, H. Savajols, and R. Kanungo, Recent experimental progress in nuclear halo structure studies, *Prog. in Part. and Nucl. Phys.* **68**, 215 (2013).
83. K. Hagino, I. Tanihata, and H. Sagawa. Exotic nuclei far from the stability line. In eds. E. Henley and S. Ellis, *100 years of subatomic physics*, p. 231. World Scientific, Singapore (2013). arXiv:1208.1583.
84. T. Nakamura and Y. Kondo, Neutron halo and breakup reactions, *Lec. Notes in Physics.* **848**, 67 (2012).
85. K. Ikeda, T. Myo, K. Kato, and H. Toki, Di-neutron clustering and deuteron-like tensor correlation in nuclear structure focusing on ^{11}Li, *Lec. Notes in Physics.* **818**, 165 (2010).
86. E. Hiyama, M. Kamimura, T. Motoba, T. Yamada, and Y. Yamamoto, Three-body model study of $A = 6 - 7$ hypernuclei: halo and skin structures, *Phys. Rev. C.* **53**, 2075 (1996).
87. H. Lü, J. Meng, S. Zhang, and S.-G. Zhou, Neutron halos in hypernuclei, *Euro. Phys. J. A.* **17**, 19 (2003).
88. G. Lalazissis, J. König, and P. Ring, New parametrization for the lagrangian density of relativistic mean field theory, *Phys. Rev. C.* **55**, 540 (1997).
89. B.-N. Lu, E. Hiyama, H. Sagawa, and S.-G. Zhou, Superdeformed Λ hypernuclei within relativisitic mean field models, *Phys. Rev. C.* **89**, 044307 (2014).
90. W. X. Xue, J. M. Yao, K. Hagino, Z. P. Li, H. Mei, and Y. Tanimura, Triaxially deformed relativistic point-coupling model for Λ hypernuclei: a quantitative analysis of hyperon impurity effect on nuclear collective properties, *Phys. Rev. C.* **91**, 024327 (2015).

91. H. Mei, Z. P. Li, J. M. Yao, and K. Hagino, Impurity effect of Λ hyperon on shape-coexistence nucleus ^{44}S in the energy functional based collective hamiltonian, *Int. J. Mod. Phys. E.* **21**, 1250024 (2012).
92. N. Martin and the ALICE Collaboration, (anti-)matter and hyper-matter production at the lhc with alice, *Journal of Physics: Conference Series.* **455** (1), 012007 (2013). URL http://stacks.iop.org/1742-6596/455/i=1/a=012007.
93. J. Z. *et al.* Bai, Observation of a near-threshold enhancement in the $p\bar{p}$ mass spectrum from radiative $j/\psi \to \gamma p\bar{p}$ decays, *Phys. Rev. Lett.* **91**, 022001 (Jul, 2003). doi: 10.1103/PhysRevLett.91.022001. URL http://link.aps.org/doi/10.1103/PhysRevLett.91.022001.
94. N. Auerbach, A. S. Goldhaber, M. B. Johnson, L. Miller, and A. Picklesimer, Probing for dirac vacuum structure in nuclei, *Physics Letters B.* **182**, 221–225 (1986). ISSN 0370-2693. doi: http://dx.doi.org/10.1016/0370-2693(86)90078-X. URL http://www.sciencedirect.com/science/article/pii/037026938690078X.
95. G. Mao, H. Stocker, and W. Greiner, Relativistic Hartree approach including both positive- and negative-energy bound states, *Int. J. Mod. Phys. E.* **8**, 389–416 (1999). doi: 10.1142/S0218301399000288.
96. T. Burvenich, I. N. Mishustin, L. M. Satarov, J. Maruhn, H. Stacker, and W. Greiner, Enhanced binding and cold compression of nuclei due to admixture of antibaryons, *Physics Letters B.* **542**, 261 – 267 (2002). ISSN 0370-2693. doi: http://dx.doi.org/10.1016/S0370-2693(02)02351-1. URL http://www.sciencedirect.com/science/article/pii/S0370269302023511.
97. I. N. Mishustin, L. M. Satarov, T. J. Bürvenich, H. Stöcker, and W. Greiner, Antibaryons bound in nuclei, *Phys. Rev. C.* **71**, 035201 (Mar, 2005). doi: 10.1103/PhysRevC.71.035201. URL http://link.aps.org/doi/10.1103/PhysRevC.71.035201.
98. A. B. Larionov, I. N. Mishustin, L. M. Satarov, and W. Greiner, Dynamical simulation of bound antiproton-nuclear systems and observable signals of cold nuclear compression, *Phys. Rev. C.* **78**, 014604 (Jul, 2008). doi: 10.1103/PhysRevC.78.014604. URL http://link.aps.org/doi/10.1103/PhysRevC.78.014604.
99. C. Xi, L. Ning, Y. Hai-Bo, Q. Xu-Ming, and W. Shi-Shu, Lifetime of antibaryon bound in finite nuclei, *Communications in Theoretical Physics.* **53** (1), 128 (2010). URL http://stacks.iop.org/0253-6102/53/i=1/a=27.
100. C. Y. Song and J. M. Yao, Polarization effect on the spin symmetry for anti-lambda spectrum in $^{16}\mathrm{O} + \bar{\Lambda}$ system, *Chin. Phys. C.* **34**, 1425 (2010). doi: 10.1088/1674-1137/34/9/061.
101. M. Rufa, P.-G. Reinhard, J. A. Maruhn, W. Greiner, and M. R. Strayer, Optimal parametrization for the relativistic mean-field model of the nucleus, *Phys. Rev. C.* **38**, 390–409 (Jul, 1988). doi: 10.1103/PhysRevC.38.390. URL http://link.aps.org/doi/10.1103/PhysRevC.38.390.
102. C. Y. Song, J. M. Yao, and J. Meng, Spin symmetry for anti-lambda spectrum in atomic nucleus, *Chin, Phys. Lett.* **26**, 122102 (2009). doi: 10.1088/0256-307X/26/12/122102.

103. S.-G. Zhou, J. Meng, and P. Ring, Spin symmetry in the antinucleon spectrum, *Phys. Rev. Lett.* **91**(26), 262501 (Dec, 2003).
104. C. Y. Song, J. M. Yao, and J. Meng, Tensor coupling effects on spin symmetry in the anti-lambda spectrum of hypernuclei, *Chin. Phys. Lett.* **28**, 092101 (2011). doi: 10.1088/0256-307X/28/9/092101.
105. H. Takahashi, J. K. Ahn, H. Akikawa, S. Aoki, K. Arai, S. Y. Bahk, K. M. Baik, B. Bassalleck, J. H. Chung, M. S. Chung, D. H. Davis, T. Fukuda, K. Hoshino, A. Ichikawa, M. Ieiri, K. Imai, Y. H. Iwata, Y. S. Iwata, H. Kanda, M. Kaneko, T. Kawai, M. Kawasaki, C. O. Kim, J. Y. Kim, S. J. Kim, S. H. Kim, Y. Kondo, T. Kouketsu, Y. L. Lee, J. W. C. McNabb, M. Mitsuhara, Y. Nagase, C. Nagoshi, K. Nakazawa, H. Noumi, S. Ogawa, H. Okabe, K. Oyama, H. M. Park, I. G. Park, J. Parker, Y. S. Ra, J. T. Rhee, A. Rusek, H. Shibuya, K. S. Sim, P. K. Saha, D. Seki, M. Sekimoto, J. S. Song, T. Takahashi, F. Takeutchi, H. Tanaka, K. Tanida, J. Tojo, H. Torii, S. Torikai, D. N. Tovee, N. Ushida, K. Yamamoto, N. Yasuda, J. T. Yang, C. J. Yoon, C. S. Yoon, M. Yosoi, T. Yoshida, and L. Zhu, Observation of a ${}^{6}_{\Lambda\Lambda}$He double hypernucleus, *Phys. Rev. Lett.* **87**, 212502 (2001).
106. J. K. Ahn, H. Akikawa, S. Aoki, K. Arai, S. Y. Bahk, K. M. Baik, B. Bassalleck, J. H. Chung, M. S. Chung, D. H. Davis, T. Fukuda, K. Hoshino, A. Ichikawa, M. Ieiri, K. Imai, K. Itonaga, Y. H. Iwata, Y. S. Iwata, H. Kanda, M. Kaneko, T. Kawai, M. Kawasaki, C. O. Kim, J. Y. Kim, S. H. Kim, S. J. Kim, Y. Kondo, T. Kouketsu, H. N. Kyaw, Y. L. Lee, J. W. C. McNabb, A. A. Min, M. Mitsuhara, K. Miwa, K. Nakazawa, Y. Nagase, C. Nagoshi, Y. Nakanishi, H. Noumi, S. Ogawa, H. Okabe, K. Oyama, B. D. Park, H. M. Park, I. G. Park, J. Parker, Y. S. Ra, J. T. Rhee, A. Rusek, A. Sawa, H. Shibuya, K. S. Sim, P. K. Saha, D. Seki, M. Sekimoto, J. S. Song, H. Takahashi, T. Takahashi, F. Takeutchi, H. Tanaka, K. Tanida, K. T. Tint, J. Tojo, H. Torii, S. Torikai, D. N. Tovee, T. Tsunemi, M. Ukai, N. Ushida, T. Wint, K. Yamamoto, N. Yasuda, J. T. Yang, C. J. Yoon, C. S. Yoon, M. Yosoi, T. Yoshida, and L. Zhu, Double-Λ hypernuclei observed in a hybrid emulsion experiment, *Phys. Rev. C.* **88**, 014003 (2013).
107. H. Shen, F. Yang, and H. Toki, Double-Λ hypernuclei in the relativistic mean-field theory, *Prog. Theo. Phys.* **115**, 325 (2006).
108. J. Mares and J. Zofka, Multi-strange systems in relativistic mean fields, *Z. Phys. A.* **345**, 47 (1993).
109. M. Ikram, S. K. Singh, A. A. Usmani, and S. K. Patra, A relativistic mean field study of multi-strange system, *arXiv:1406.4612 [nucl-th]* (2014).
110. H. Heiselberg and M. Hjorth-Jensen, Phases of dense matter in neutron stars, *Phys. Rep.* **328**, 237 (2000).
111. L. M. Lattimer and M. Prakash, Neutron star observations: prognosis for equation of state constraints, *Phy. Rep.* **442**, 109 (2007).
112. J. M. Lattimer, The nuclear equation of state and neutron star masses, *Annu. Rev. Nucl. Part. Sci.* **62**, 485 (2012).
113. R. R. Silbar and S. Reddy, Neutron stars for undergraduates, *Am. J. Phys.* **72**, 892 (2004).

114. C. Ishizuka, A. Ohnishi, K. Tsubakihara, K. Sumiyoshi, and S. Yamada, Tables of hyperonic matter equation of state for core-collapse supernovae, *J. of Phys. G.* **35**, 085201 (2008).
115. H.-J. Schulze, M. Baldo, U. Lombardo, J. Cugnon, and A. Lejeune, Hyperonic nuclear matter in Brueckner theory, *Phys. Rev. C.* **57**, 704 (1998).
116. M. Baldo, G. Burgio, and H.-J. Schulze, Onset of hyperon formation in neutron star matter from Brueckner theory, *Phys. Rev. C.* **58**, 3688 (1998).
117. I. Vidana, A. Polls, A. Ramos, M. Hjorth-Jensen, and V. Stoks, Strange nuclear matter with Brueckner–Hartree–Fock theory, *Phys. Rev. C.* **61**, 025802 (2000).
118. H.-J. Schulze, A. Polls, A. Ramos, and I. Vidana, Maximum mass of neutron stars, *Phys. Rev. C.* **058801**, 73 (2006).
119. S. Nishizaki, Y. Yamamoto, and T. Takatsuka, Hyperon-mixed neutron star matter and neutron stars, *Prog. Theo. Phys.* **108**, 703 (2002).
120. T. Takatsuka, S. Nishizaki, Y. Yamamoto, and R. Tamagaki, Occurrence of hyperon superfluidity in neutron star cores, *Prog. Theo. Phys.* **115**, 355 (2006).
121. M. Barranco, R. Lombard, S. Marcos, and S. Moszkowski, Multi-lambda matter in a derivative coupling model, *Phys. Rev. C.* **44**, 178 (1991).
122. N. Glendenning and S. Moszkowski, Reconciliation of neutron-star masses and binding of the Λ in hypernuclei, *Phys. Rev. Lett.* **67**, 2414 (1991).
123. R. Knorren, M. Prakash, and P. Ellis, Strangeness in hadronic stellar matter, *Phys. Rev. C.* **52**, 3470 (1995).
124. J. Schaffner and I. Mishustin, Hyperon-rich matter in neutron stars, *Phys. Rev. C.* **53**, 1416 (1996).
125. W. H. Long, B. Y. Sun, K. Hagino, and H. Sagawa, Hyperon effects in covariant density functional theory and recent astrophysical observations, *Phys. Rev. C.* **85**, 025806 (2012).
126. H. Shen, Complete relativistic equation of state for neutron stars, *Phys. Rev. C.* **65**, 035802 (2002).
127. H. Shen, H. Toki, K. Oyamatsu, and K. Sumiyoshi, Relativistic equation of state for core-collapse supernova simulations, *Astrophys. J. Suppl. Series.* **197:20**, 1 (2011).
128. K. Tsubakihara, H. Maekawa, H. Matsumiya, and A. Ohnishi, Λ hypernuclei and neutron star matter in a chiral SU(3) relativistic mean field model with a logarithmic potential, *Phys. Rev. C.* **81**, 065206 (2010).
129. K. Tsubakihara and A. Ohnishi, Three-body couplings in RMF and its effects on hyperonic star equation of state, *Nucl. Phys. A.* **914**, 438 (2013).
130. J. Walecka, *Introduction to Modern Physics.* World Scientific, Singapore (2008).
131. J. A. Maruhn, P.-G. Reinhard, and E. Suraud, *Simple Models of Many-Fermion Systems.* Springer, Berlin (2010).
132. P. B. Demorest, T. Pennucci, S. Ransom, M. Roberts, and J. Hessels, A two-solar-mass neutron star measured using Shapiro decay, *Nature (London).* **467**, 1081 (2011).

Chapter 8

Rotating nuclei: from ground state to the extremes of spin and deformation

A. V. Afanasjev

Department of Physics and Astronomy, Mississippi State University, MS 39762, USA

Anatoli.Afanasjev@gmail.com

The rotating nuclei represent one of most interesting subjects for theoretical and experimental studies. They open a new dimension of nuclear landscape, namely, spin direction. Contrary to the majority of nuclear systems, their properties sensitively depend on time-odd mean fields and currents in density functional theories. Moreover, they show a considerable interplay of collective and single-particle degrees of freedom. In this chapter, I discuss the basic features of the description of rotating nuclei in one-dimensional cranking approximation of covariant density functional theory. The successes of this approach to the description of rotating nuclei at low spin in pairing regime and at high spin in unpaired regime in wide range of deformations (from normal to hyperdeformation) are illustrated. I also discuss the recent progress and open questions in our understanding of the role of proton-neutron pairing in rotating nuclei at $N \approx Z$, the physics of band termination and other phenomena in rotating nuclei.

Contents

1. Introduction

The development of self-consistent many-body theories aiming at the description of low-energy nuclear phenomena provides the necessary theoretical tools for an exploration of the nuclear chart into known and unknown regions. Theoretical methods (both relativistic[1] and non-relativistic[2]) formulated within the framework of density functional theory (DFT) are the most promising tools for the global investigation of the properties of atomic nuclei. The power of the DFT models is essentially unchallenged in medium and heavy mass nuclei where "ab-initio" type few-body calculations are computationally impossible and the applicability of spherical shell model is restricted to a few regions in the vicinity of doubly shell closures.

One can consider a nuclear chart as a three-dimensional object in which the charge, isospin and spin play the role of the coordinates. The studies of superheavy and neutron-rich nuclei allow us to extend this chart in the directions of the charge and isospin, respectively. On the other hand, the investigations of rotating nuclei explore the nuclear chart in the spin direction. Over the decades, the studies of rotating nuclei have revealed a lot of new interesting nuclear phenomena such as superdeformation,[3] smooth band termination,[4] magnetic rotation,[5] etc. Many of them can be successfully studied in the framework of one-dimensional cranking approximation.[a] The realization of this approximation in the framework of covariant density functional theory (CDFT) and its application to rotating nuclei in different regimes of spin, deformation and pairing and in different regions of the nuclear chart are reviewed in the present chapter. This review mostly focuses on the physical phenomena which were studied from 2005 since pre-2005 studies of rotating nuclei in the CDFT have been overviewed in Ref. 1. However, a few examples of the studies of rotating nuclei from this time period are given in Sec. 8 for completeness.

2. One-dimensional cranking approximation

The description of rotating nuclei requires a transformation of the relativistic Hartree–Bogoliubov (RHB) equations to the rotating frame. The concept of a frame, which rotates around a fixed axis with constant rotational

[a] A review of two- and three-dimensional cranking approximations in the CDFT is presented in Ref. 6 and in Chapter 9 of this book.

frequency Ω, is not strictly compatible with Lorentz invariance because a rotating frame is an accelerated rather than an inertial one. Strictly speaking such a problem should be treated in the framework of general relativity. However, one can also look on this problem differently. Points connected with such a frame move with the velocity $r\Omega$, where r is the distance of such a point from the axis of rotation. In nuclei the relevant distances r are not much larger than the nuclear radius R. Since rotational frequencies in nuclei usually do not exceed $\Omega \sim 1-2$ MeV, the linear velocities at the surface of rotating nuclei are of the order of only a few percents of the velocity of light. Therefore, the effects of general relativity can be neglected and we just have to transform the system to a coordinate frame rotating with constant rotational frequency Ω. Such transformations have been studied in detail within the semi-classical approximation,[7] within the formalism of special relativity[8] and with techniques of general relativity.[9] All these investigations lead to identical results.

The transformation of the RHB equations to the rotating frame leads to the cranked RHB (CRHB) equations.[10,11] Note that in the present manuscript only one-dimensional rotation with rotational frequency Ω_x around x-axis (*one-dimensional cranking approximation*) is considered. The nucleonic part of the RHB equations has the form

$$\begin{pmatrix} \hat{h}_D - \lambda_\tau - \Omega_x \hat{j}_x & \hat{\Delta} \\ -\hat{\Delta}^* & -\hat{h}_D^* + \lambda_\tau + \Omega_x \hat{j}_x^{\,*} \end{pmatrix} \begin{pmatrix} U(\mathbf{r}) \\ V(\mathbf{r}) \end{pmatrix}_k = E_k \begin{pmatrix} U(\mathbf{r}) \\ V(\mathbf{r}) \end{pmatrix}_k \tag{1}$$

with the chemical potentials λ_τ ($\tau = n, p$) for neutrons and protons and the single-particle angular momentum operators $\hat{j}_x$ for fermions with spin $\frac{1}{2}$. The Dirac Hamiltonian $\hat{h}_D$

$$\hat{h}_D = \boldsymbol{\alpha}(-i\boldsymbol{\nabla} - \boldsymbol{V}(\mathbf{r})) \; + \; V_0(\mathbf{r}) \; + \; \beta(m + S(\mathbf{r})) \tag{2}$$

contains the average fields determined by the mesons, i.e. the attractive scalar field $S(\mathbf{r})$

$$S(\mathbf{r}) = g_\sigma \sigma(\mathbf{r}), \tag{3}$$

and the repulsive time-like component of the vector field $V_0(\mathbf{r})$

$$V_0(\mathbf{r}) = g_\omega \omega_0(\mathbf{r}) + g_\rho \tau_3 \rho_0(\mathbf{r}) + e\frac{1-\tau_3}{2} A_0(\mathbf{r}). \tag{4}$$

A magnetic potential $\boldsymbol{V}(\mathbf{r})$

$$\boldsymbol{V}(\mathbf{r}) = g_\omega \boldsymbol{\omega}(\mathbf{r}) + g_\rho \tau_3 \boldsymbol{\rho}(\mathbf{r}) + e\frac{1-\tau_3}{2} \boldsymbol{A}(\mathbf{r}), \tag{5}$$

originates from the space-like components of the vector mesons and behaves in the Dirac equation like a magnetic field. Therefore the effect produced by it is called *nuclear magnetism*.[8]

Note that in these equations, the four-vector components of the vector fields ω^μ, ρ^μ, and A^μ are separated into the time-like (ω_0, ρ_0 and A_0) and space-like [$\boldsymbol{\omega} = (\omega^x, \omega^y, \omega^z)$, $\boldsymbol{\rho} = (\rho^x, \rho^y, \rho^z)$, and $\boldsymbol{A} = (A^x, A^y, A^z)$] components.

The cranked RHB-equations (1) contain three constraints characterized by three Lagrange parameters, the chemical potentials λ_τ ($\tau = n, p$) and the rotational frequency Ω_x. The chemical potentials λ_τ are determined by the average particle numbers for neutrons and protons ($\tau = n, p$)

$$\langle \Phi_\Omega | \hat{N}_n | \Phi_\Omega \rangle = N, \qquad \langle \Phi_\Omega | \hat{N}_p | \Phi_\Omega \rangle = Z. \tag{6}$$

The rotational frequency Ω_x along the x-axis is defined from the condition[12]

$$J_x(\Omega_x) = \langle \Phi_{\Omega_x} \mid \hat{J}_x \mid \Phi_{\Omega_x} \rangle = \sqrt{I(I+1)}, \tag{7}$$

where $J_x(\Omega_x)$ is the expectation value of the total angular momentum $\hat{\boldsymbol{J}}$ of the system along the x-axis and I is the total nuclear spin. The angular momentum is carried essentially by the fermions. As a result, the contributions of the meson fields to $J_x(\Omega_x)$ are neglected. The Coriolis term is given by

$$-\Omega_x \hat{J}_x = -\Omega_x \left(\hat{L}_x + \frac{1}{2}\hat{\Sigma}_x \right). \tag{8}$$

The time-independent inhomogeneous Klein-Gordon equations for the mesonic fields obtained by means of variational principle are given in the CRHB theory by[10,11]

$$\begin{aligned}
\left\{-\Delta - (\Omega_x \hat{L}_x)^2 + m_\sigma^2\right\} \ \sigma(\mathbf{r}) &= -g_\sigma \rho_s(\mathbf{r}) - g_2 \sigma^2(\mathbf{r}) - g_3 \sigma^3(\mathbf{r}), \\
\left\{-\Delta - (\Omega_x \hat{L}_x)^2 + m_\omega^2\right\} \omega_0(\mathbf{r}) &= g_\omega \rho_v^{is}(\mathbf{r}), \\
\left\{-\Delta - [\Omega_x(\hat{L}_x + \hat{S}_x)]^2 + m_\omega^2\right\} \ \boldsymbol{\omega}(\mathbf{r}) &= g_\omega \mathbf{j}^{is}(\mathbf{r}), \\
\left\{-\Delta - (\Omega_x \hat{L}_x)^2 + m_\rho^2\right\} \rho_0(\mathbf{r}) &= g_\rho \rho_v^{iv}(\mathbf{r}), \\
\left\{-\Delta - [\Omega_x(\hat{L}_x + \hat{S}_x)]^2 + m_\rho^2\right\} \ \boldsymbol{\rho}(\mathbf{r}) &= g_\rho \mathbf{j}^{iv}(\mathbf{r}), \\
-\Delta \ A_0(\mathbf{r}) = e\rho_v^p(\mathbf{r}), \qquad -\Delta \ \mathbf{A}(\mathbf{r}) &= e\mathbf{j}^p(\mathbf{r}),
\end{aligned} \tag{9}$$

where the source terms are sums of bilinear products of baryon amplitudes

$$\begin{aligned}
\rho_s(\mathbf{r}) &= \sum_{k>0} [V_k^n(\mathbf{r})]^\dagger \hat{\beta} V_k^n(\mathbf{r}) + [V_k^p(\mathbf{r})]^\dagger \hat{\beta} V_k^p(\mathbf{r}), \\
\rho_v^{is}(\mathbf{r}) &= \sum_{k>0} [V_k^n(\mathbf{r})]^\dagger V_k^n(\mathbf{r}) + [V_k^p(\mathbf{r})]^\dagger V_k^p(\mathbf{r}), \\
\rho_v^{iv}(\mathbf{r}) &= \sum_{k>0} [V_k^n(\mathbf{r})]^\dagger V_k^n(\mathbf{r}) - [V_k^p(\mathbf{r})]^\dagger V_k^p(\mathbf{r}), \\
\mathbf{j}^{is}(\mathbf{r}) &= \sum_{k>0} [V_k^n(\mathbf{r})]^\dagger \hat{\boldsymbol{\alpha}} V_k^n(\mathbf{r}) + [V_k^p(\mathbf{r})]^\dagger \hat{\boldsymbol{\alpha}} V_k^p(\mathbf{r}), \\
\mathbf{j}^{iv}(\mathbf{r}) &= \sum_{k>0} [V_k^n(\mathbf{r})]^\dagger \hat{\boldsymbol{\alpha}} V_k^n(\mathbf{r}) - [V_k^p(\mathbf{r})]^\dagger \hat{\boldsymbol{\alpha}} V_k^p(\mathbf{r}).
\end{aligned} \tag{10}$$

The sums over $k > 0$ run over all quasiparticle states corresponding to positive energy single-particle states (*no-sea approximation*).[13,14] In Eqs. (9,10), the indexes n and p indicate neutron and proton states, respectively, and the indexes *is* and *iv* are used for isoscalar and isovector quantities. $\rho_v^p(\mathbf{r})$, $\mathbf{j}^p(\mathbf{r})$ in Eq. (9) correspond to $\rho_v^{is}(\mathbf{r})$ and $\mathbf{j}^{is}(\mathbf{r})$ defined in Eq. (10), respectively, but with the sums over neutron states neglected. $\hat{\boldsymbol{L}} = -i\mathbf{r} \times \boldsymbol{\nabla}$ is the orbital angular momentum operator of the scalar and the time-like meson fields and $\hat{\boldsymbol{L}} + \hat{\boldsymbol{S}}$ is the total angular momentum of the space-like parts of the vector fields. The operators $\hat{\mathbf{S}}$ are three 3×3 matrices representing the spin-matrices of vector fields (see Ref. 15). Note that only the x-components of these operators are considered in Eq. (10). The Coriolis terms $\Omega_x(\hat{L}_x + \hat{S}_x)$ and $\Omega_x \hat{L}_x$ can be neglected for reasons discussed in Sec. 8.1 of Ref. 1. The Coriolis term for the Coulomb potential $A_0(\mathbf{r})$ and the spatial components of the vector potential $\mathbf{A}(\mathbf{r})$ are neglected in Eqs. (9) since the coupling constant of the electromagnetic interaction is small compared with the coupling constants of the meson fields.

Two terms in the Dirac equation, namely, the Coriolis operator $\hat{J}_x$ and the magnetic potential $\mathbf{V}(\mathbf{r})$ (as well as the currents $\mathbf{j}^{n,p}(\mathbf{r})$ in the Klein–Gordon equations) break time-reversal symmetry.[17] Their presence leads to the appearance of time-odd mean fields. However, one should distinguish time-odd mean fields originating from the Coriolis operator and the magnetic potential. The Coriolis operator is always present in the description of rotating nuclei in the framework of the cranking model. However, the cranked relativistic mean field (CRMF) calculations, with only these time-odd fields accounted for, underestimate the experimental moments of inertia.[17–19] A similar situation also holds in non-relativistic theories.[20,21] The inclusion of the currents $\mathbf{j}^{n,p}(\mathbf{r})$ into the Klein–Gordon equations, which

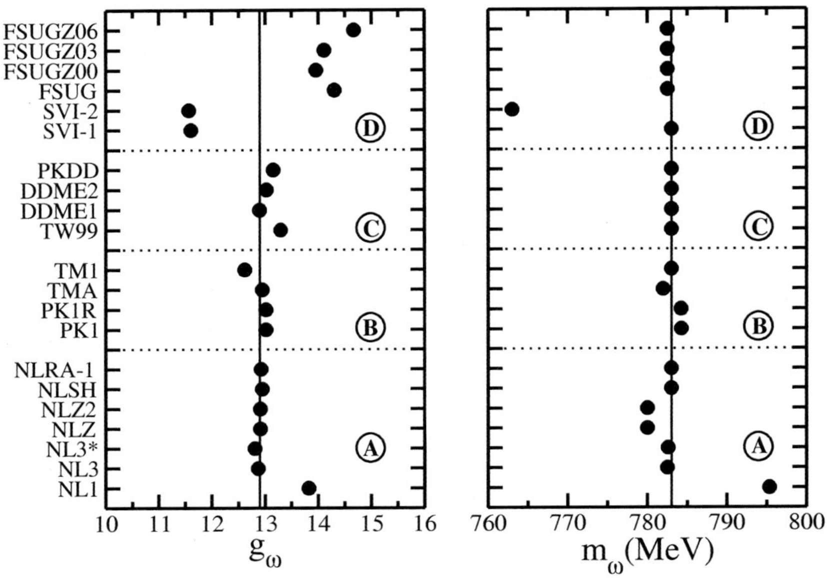

Fig. 1. The m_ω and g_ω parameters of different CEDF's. These CEDF's are combined into four groups dependent on how self- and mixed-couplings are introduced. Group A represents the CEDF's which include non-linear self-couplings only for the σ-meson. Group B contains the CEDF's which include self-couplings for the σ- and ω-mesons (and ρ-mesons in the case of PK1R). Group C represents the CEDF's which include density-dependent meson-nucleon couplings for the σ-, ω-, and ρ-mesons. The other CEDF's are included into group D. For details see Ref. 16 from which this figure is taken.

leads to the space-like components of the vector ω and ρ mesons and thus to the magnetic potential $\mathbf{V}(\mathbf{r})$, considerably improves the description of experimental moments of inertia.

Note that time-odd mean fields related to nuclear magnetism are defined through the Lorentz invariance[1,16,19] and thus they do not require additional coupling constants: the coupling constants of time-even mean fields are used also for time-odd mean fields. This is because the densities ρ and currents $\boldsymbol{j}$ (Eq. (10)) do not transform independently under Lorentz transformation since they form covariant four-vector $j^\mu = \{\rho, \mathbf{j}\}$. This fact explains why the structure of the Klein–Gordon equations for time-like and space-like components of vector mesons are the same (Eq. (10)) and why the same coupling constant stands in front of the densities and currents on the right hand side of these equations.[16]

The currents are isoscalar and isovector in nature for the ω and ρ mesons (Eqs. (9) and (10)), respectively. As a consequence, the contribution of the

ρ-meson to magnetic potential is marginal.[16] Thus, time-odd mean fields in the CDFT framework depend predominantly on the spatial components of the ω meson. Neglecting the contribution of the ρ meson, one can see that only two parameters, namely, the mass m_ω and coupling constant g_ω of the ω meson, define the properties of time-odd mean fields (Eqs. (5) and (9)). Figure 1 clearly indicates that these parameters are well localized in the parameter space for the covariant energy density functionals (CEDF's) in the groups A, B, and C. These are precisely the CEDF's which have been extensively tested on nuclear structure data sensitive to time-even mean fields. For these CEDF's parameter dependence of the impact of time-odd mean fields on the physical observables is quite weak.[16,19,22] This parameter dependence is expected to be larger in group D. However, at present it is not clear whether the CEDF's in this group are reasonable ones since they have not been tested on nuclear structure data.

Time-odd mean fields appear only in nuclear systems with broken time-reversal symmetry. They affect magnetic moments,[23] isoscalar monopole vibrations,[24] electric giant resonances,[25] large amplitude collective dynamics,[26] fusion process,[27] the strengths and energies of Gamow–Teller resonances,[28] the binding energies of odd-mass nuclei[29–31] and the definition of the strength of pairing correlations.[16,31,32] However, as discussed in detail in Sec. 3 they are especially pronounced in rotating nuclei.

The pairing potential (field) $\hat{\Delta}$ in Eq. (1) is given by

$$\hat{\Delta} \equiv \Delta_{ab} = \frac{1}{2}\sum_{cd} V^{pp}_{abcd}\kappa_{cd}, \tag{11}$$

where the indices $a, b, \ldots$ denote quantum numbers which specify the single-particle states with the space coordinates $\mathbf{r}$ as well as the Dirac and isospin indices s and τ. It contains the pairing tensor κ[b]

$$\kappa = V^* U^T \tag{12}$$

and the matrix elements V^{pp}_{abcd} of the effective interaction in the pp-channel. The phenomenological Gogny D1S finite range interaction[33] is used as such effective interaction in all CRHB calculations. It is given by

$$V^{pp}(1,2) = f \sum_{i=1,2} e^{-[(\mathbf{r}_1-\mathbf{r}_2)/\mu_i]^2} \times (W_i + B_i P^\sigma - H_i P^\tau - M_i P^\sigma P^\tau), \tag{13}$$

where μ_i, W_i, B_i, H_i and M_i ($i = 1, 2$) are the parameters of the force and P^σ and P^τ are the exchange operators for the spin and isospin variables,

[b]This quantity is sometimes called as *abnormal density*.

respectively. This interaction is density-independent. Note also that an additional factor f affecting the strength of the Gogny force is introduced in Eq. (13) (see Refs. 31, 34, 35 for the reasons of its introduction).

In the CRHB calculations, the size of the pairing correlations is measured in terms of the pairing energy defined as

$$E_{pairing} = -\frac{1}{2}\mathrm{Tr}(\Delta\kappa). \tag{14}$$

Note that this is not an experimentally accessible quantity.

One should note that the Bogoliubov transformation is not commutable with the nucleon number operator and consequently the resulting wave function does not correspond to a system having a definite number of protons and neutrons. The best way to deal with this problem would be to perform an exact particle number projection before the variation;[36] however, such calculations are expected to be extremely time-consuming for realistic interactions. As a result, an approximate particle number projection by means of the Lipkin–Nogami (LN) method[37–40] is used in the CRHB calculations because of its simplicity. The details of the implementation of this method into the CRHB framework are given in Ref. 11; the CRHB calculations with the LN method included are abbreviated as CRHB+LN ones. The application of the LN method considerably improves agreement with experiment for rotational properties (see Refs. 11, 41, 42).

The total energy of system in the laboratory frame is given as a sum of fermionic (E^{F}) and bosonic (E^{B}) contributions

$$E_{\mathrm{CRHB}} = E^{\mathrm{F}} + E^{\mathrm{B}}. \tag{15}$$

The fermionic energies E^{F} are given by

$$E^{\mathrm{F}} = E_{part} + \Omega_x J + E_{pairing} + E_{cm} \tag{16}$$

where

$$E_{part} = Tr(h_D\rho), \qquad J = Tr(j_x\rho), \tag{17}$$

are the particle energy and the expectation value of the total angular momentum along the rotational axis and E_{cm} is the correction for the spurious center-of-mass motion.

The bosonic energies $E^{\rm B}$ in the laboratory frame are given by

$$
\begin{aligned}
E^{\rm B} = & -\frac{1}{2}\int d\mathbf{r}\,[g_\sigma\,\sigma(\mathbf{r})\rho_s(\mathbf{r}) + \frac{1}{3}g_2\sigma^3(\mathbf{r}) + \frac{1}{2}g_3\sigma^4(\mathbf{r})] \\
& -\frac{1}{2}g_\omega\int d\mathbf{r}\,[\omega_0(\mathbf{r})\rho_v^{is}(\mathbf{r}) - \boldsymbol{\omega}(\mathbf{r})\mathbf{j}^{is}(\mathbf{r})] \\
& -\frac{1}{2}g_\rho\int d\mathbf{r}\,[\rho_0(\mathbf{r})\rho_v^{iv}(\mathbf{r}) - \boldsymbol{\rho}(\mathbf{r})\mathbf{j}^{iv}(\mathbf{r})] \\
& -\frac{1}{2}e\int d\mathbf{r}\,[A_0(\mathbf{r})\rho_v^{p}(\mathbf{r}) + \boldsymbol{A}(\mathbf{r})\mathbf{j}^{p}(\mathbf{r})] \\
& + \Omega_x^2\int d\mathbf{r}\,[\sigma(\mathbf{r})\hat{L}_x^2\sigma(\mathbf{r}) - \omega_0(\mathbf{r})\hat{L}_x^2\omega_0(\mathbf{r}) + \boldsymbol{\omega}(\mathbf{r})(\hat{L}_x+\hat{S}_x)^2\boldsymbol{\omega}(\mathbf{r}) \\
& \qquad -\rho_0(\mathbf{r})\hat{L}_x^2\rho_0(\mathbf{r}) + \boldsymbol{\rho}(\mathbf{r})(\hat{L}_x+\hat{S}_x)^2\boldsymbol{\rho}(\mathbf{r})]. \qquad (18)
\end{aligned}
$$

Cranked relativistic mean field (CRMF) theory is a limiting case of the CRHB theory in which the pairing correlations are neglected; the details of the formalism can be found in Refs. 8, 18, 43. It has been an important step in the development of the CDFT theory to the description of rotating nuclei and, as exemplified in the present paper, still remains a powerful theoretical tool for the study of rotating nuclei in unpaired regime at high spin.

3. Time-odd mean fields in rotating nuclei

Nuclear magnetism, i.e. the time-odd component of the mean fields $\boldsymbol{V}(\mathbf{r})$, appears only in nuclear systems with broken time-reversal symmetry in the intrinsic frame. Rotating nuclei represent a system which is strongly affected by time-odd mean fields. In rotating nuclei, the average field has two sources of time-reversal symmetry breaking, the Coriolis operator $\Omega\hat{J}_x$ and the magnetic part of the vector fields $\boldsymbol{V}(\mathbf{r})$ (Eq. (5)) induced in a self-consistent way by the currents. In this section, the results obtained with and without nuclear magnetism are denoted as NM and WNM, respectively.

The physical observables, most frequently used in the analysis of rotating nuclei, are kinematic ($J^{(1)}$) and dynamic ($J^{(2)}$) moments of inertia which are defined as

$$
J^{(1)}(\Omega_x) = \frac{J}{\Omega_x}, \qquad J^{(2)}(\Omega_x) = \frac{dJ}{d\Omega_x} \qquad (19)
$$

where J is the expectation value of the total angular momentum along the x-axis. In the CRMF theory, this quantity is defined as a sum of the

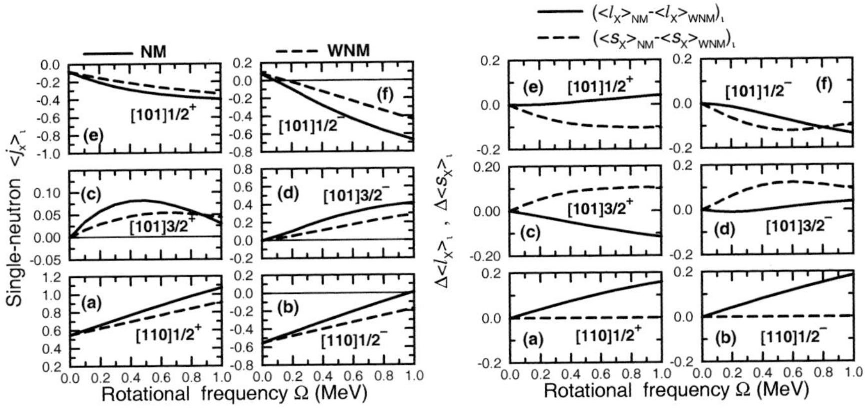

Fig. 2. Left two columns: expectation values $\langle \hat{j}_x \rangle_i$ of the neutron orbitals forming the $N = 1$ shell calculated with and without nuclear magnetism. The orbitals are labeled by the Nilsson quantum numbers $[Nn_z\Lambda]\Omega$ and the sign of the signature $r = \pm i$. Right two columns: the changes of the expectation values of spin and orbital angular momenta caused by nuclear magnetism. The results presented are from the study of doubly magic superdeformed configuration in ^{152}Dy. From Ref. 17.

expectation values of the single-particle angular momentum operators $\hat{j}_x$ of the occupied states

$$J = \sum_i \langle i|\hat{j}_x|i\rangle. \tag{20}$$

Thus, the modifications of the moments of inertia due to NM, which as discussed below are important in rotating nuclei, can be traced back to the changes of the single-particle expectation values $\langle \hat{j}_x \rangle_i = \langle i|\hat{j}_x|i\rangle$ and the corresponding contributions of spin ($\langle \hat{s}_x \rangle_i$) and orbital ($\langle \hat{l}_x \rangle_i$) angular momenta.[17]

On the microscopic level, the contribution to $\langle \hat{j}_x \rangle_i$ due to NM is defined as[17]

$$\Delta\langle j_x \rangle_i = \langle \hat{j}_x \rangle_i^{NM} - \langle \hat{j}_x \rangle_i^{WNM}. \tag{21}$$

Figure 2 shows that the $\Delta\langle j_x \rangle_i$ is positive at the bottom and negative at the top of the N-shell.[17] The absolute value of $\Delta\langle j_x \rangle_i$ correlates with the absolute value of $\langle \hat{j}_x \rangle_i$. Note that the contributions to $\langle \hat{j}_x \rangle_i$ due to NM are small in the middle of the shell. The $\Delta\langle j_x \rangle_i$ contributions can be decomposed into the contributions due to spin ($\Delta\langle s_x \rangle_i$) and orbital ($\Delta\langle l_x \rangle_i$) angular momenta using equations similar to Eq. (21). As shown in right panels of Fig. 2 these contributions have complicated dependences both on the frequency and the structure of the single-particle orbital under study.[17]

Oscillator shells with higher N-values show a similar behavior. At a given rotational frequency the modifications of single-particle $\langle \hat{j}_x \rangle_i$ values induced by NM lead to an increase of the total J and of the moments of inertia (see Figs. 3 and 5 below). This means that NM enhances the angular momentum in rotating nuclei in addition to the Coriolis term.

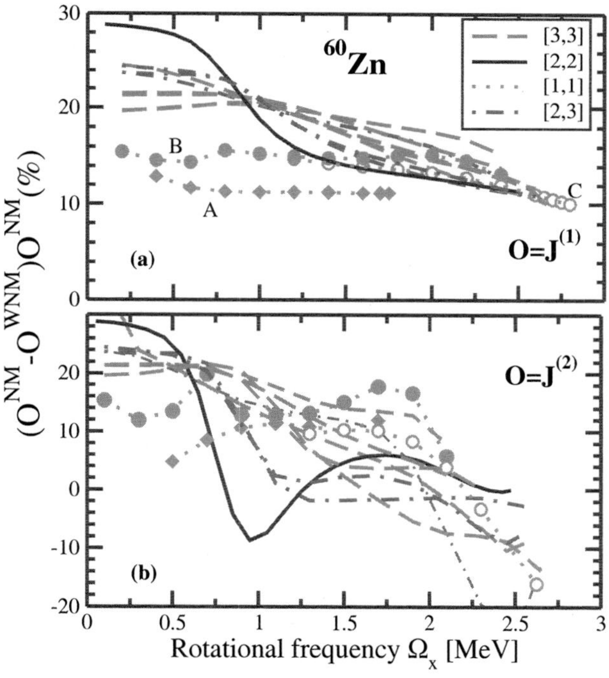

Fig. 3. (Color online) The contributions of NM to the dynamic ($J^{(2)}$) (panel (b)) and kinematic ($J^{(1)}$) (panel (a)) moments of inertia as a function of rotational frequency for highly-deformed and superdeformed configurations of ^{60}Zn obtained in the CRMF calculations. Different line types are used for different groups of configurations. The configurations are labelled by the shorthand notation $[n, p]$, where $n(p)$ is the number of occupied $g_{9/2}$ neutrons (protons). From Ref. 19.

The most important impact of NM on physical observables is revealed in the moments of inertia. To quantify this impact, the contribution $\Delta O^{NM-contr}$ (in percentage) of NM to the physical observable O is

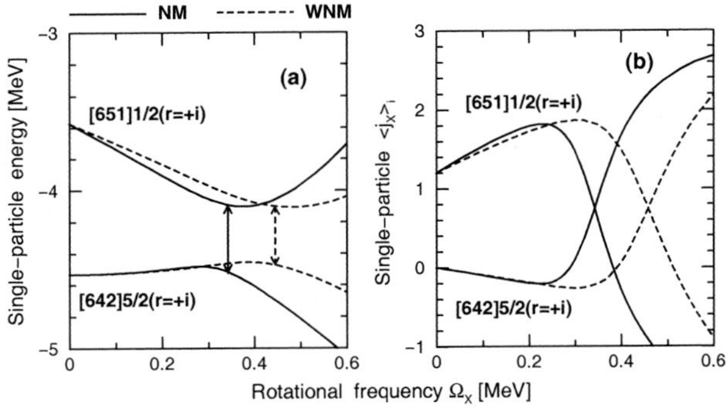

Fig. 4. (a) Proton single-particle energies (Routhians) in the self-consistent rotating potential as a function of rotational frequency Ω_x obtained in the CRMF calculations with and without NM. They are given along the deformation path of the lowest SD configuration in ^{194}Pb. Only interacting $[651]1/2^+$ and $[642]5/2^+$ orbitals are shown, see Fig. 1 in Ref. 11 for full spectra. (b) The expectation values $\langle \hat{j}_x \rangle_i$ of the single-particle angular momentum operator $\hat{j}_x$ of the orbitals shown in panel (a). Solid and dashed arrows are used to indicate the frequencies (as well as the energy gap between the interacting orbitals in panel (a)) at which the band crossings take place in the calculations with and without NM, respectively. From Ref. 19.

defined as

$$\Delta O^{NM-contr} = \frac{O^{NM} - O^{WNM}}{O^{NM}} \times 100\%. \tag{22}$$

The increase of the moments of inertia due to NM has been studied first in a semiclassical approximation in Ref. 44, then in fully self-consistent CRMF calculations on the example of the yrast SD band in ^{152}Dy in Ref. 18 and finally in a systematic way in Refs. 17, 19, 22. NM typically increases the calculated kinematic moments of inertia of normal-deformed rotational bands in the rare-earth region by 10-30%.[19] Considerable fluctuations of the $(J^{(1)}_{NM} - J^{(1)}_{WNM})/J^{(1)}_{NM}$ quantity as a function of proton and neutron numbers seen in some isotonic and isotopic chains at normal deformation are due to the changes in underlying single-particle structure. The $(J^{(1)}_{NM} - J^{(1)}_{WNM})/J^{(1)}_{NM}$ quantity is around 20% in the superdeformed bands of the $A \sim 150$ mass region and around 20–25% in the hyperdeformed bands in the $Z = 40 - 58$ part of nuclear chart at typical frequencies at which these bands are either observed or expected to be observed.

The configuration and frequency dependence of the impact of NM on the moments of inertia is shown in Fig. 3a. With increasing rotational frequency, the average contribution of NM into kinematic moments of inertia decreases and it falls below 15% at $\Omega_x \sim 2.5$ MeV . In addition, the configuration dependence of the $\Delta J^{(1)}_{NM-contr}$ quantities is weaker than

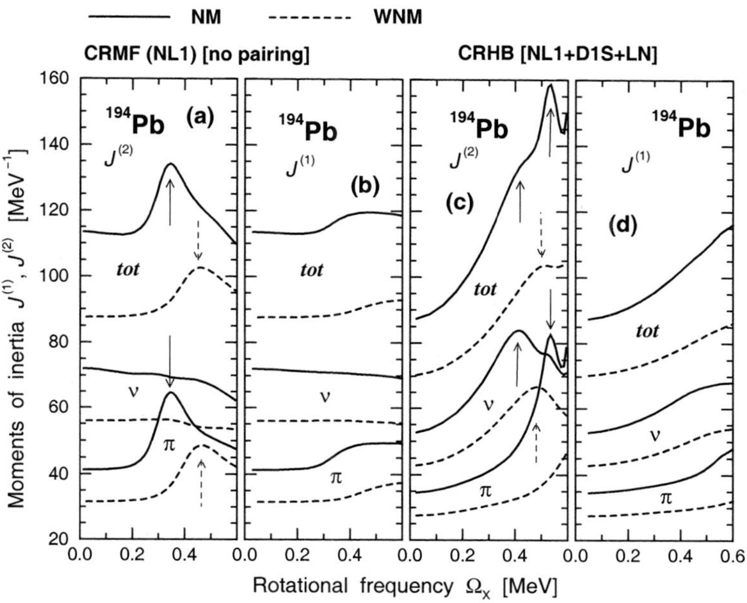

Fig. 5. Kinematic ($J^{(1)}$) and dynamic ($J^{(2)}$) moments of inertia for the lowest SD configuration in ^{194}Pb obtained in the calculations with and without NM. Proton and neutron contributions to these quantities are indicated by π and ν, while total moments are indicated by "*tot*". Panels (a) and (b) show the results obtained in the calculations without pairing, while panels (c) and (d) show the results of the calculations within the CRHB+LN framework. Solid and dashed arrows are used to indicate the frequencies at which the band crossings take place in the calculations with and without NM, respectively. From Ref. 19.

the one at low frequencies. At these frequencies, the majority of occupied single-particle orbitals are either completely aligned or very close to complete alignment. However, NM do not modify the expectation values of the single-particle angular momenta $\langle j_x \rangle_i$ of completely aligned orbitals.[22] As a result, only the remaining orbitals, which are still aligning, contribute to $\Delta J^{(1)}_{NM-contr}$. The combined contribution of these orbitals to $\Delta J^{(1)}_{NM-contr}$ is smaller than the one at lower frequencies because the alignment of these orbitals is not far away from complete. The impact of NM on the dynamic moments of inertia is shown in Fig. 3b and it clearly displays a much more complicated pattern as compared with the impact of NM on the kinematic moments of inertia (see Ref. 19 for details).

The modification of the single-particle alignments and energies in the presence of NM leads to substantial impact on the band crossing properties. This is illustrated in Figs. 4 and 5 for the example of the lowest superdeformed (SD) band in ^{194}Pb.

In the CRMF calculations, the unpaired proton band crossing originates from the interaction between the $\pi[642]5/2^+$ and $\pi[651]1/2^+$ orbitals (Fig. 4a). Since NM increases somewhat the single-particle alignment $\langle \hat{j}_x \rangle_i$ (Fig. 4b) and the slope of the routhian for the $\pi[651]1/2^+$ orbital (Fig. 4a), the band crossing takes place at lower frequency. The shift of crossing frequency due to NM is considerable (120 keV) from 0.465 MeV (WNM) down to 0.345 MeV (NM), Fig. 4a. The calculations also suggest that the strength of the interaction between two interacting orbitals at the band crossing is modified in the presence of NM as seen in the change of the energy distance (gap) between these two orbitals at the crossing frequency (Fig. 4a).

The impact of NM on band crossing features is also seen in the CRHB+LN calculations where the alignment of the pairs of $j_{15/2}$ neutrons and $i_{13/2}$ protons causes the shoulder and peak in total dynamic moment of inertia $J^{(2)}$ (Fig. 5c) (see also Ref. 11). Note that each of these two alignments creates a peak in the dynamic moment of inertia of the corresponding subsystem. NM shifts the paired neutron band crossing to lower frequencies by 70 keV from 0.485 MeV (WNM) to 0.415 MeV (NM). Paired proton band crossing lies in the calculations with NM at $\Omega_x = 0.535$ MeV, while only the beginning of this crossing is seen in the calculations without NM (Fig. 5c).

The origin of this effect is twofold. Similar to the unpaired calculations, the part of it can be traced to the fact that NM increases the expectation values $\langle \hat{j}_x \rangle_i$ of the orbitals located at the bottom of the shell (the discussed orbitals are of this kind).[17] The corresponding larger slope of the quasiparticle routhians causes the shift of the crossing to lower frequencies.

However, an additional contribution comes from the modification of the pairing by NM. There is a difference in the pairing energies calculated with and without NM which increases with rotational frequency, see Fig. 3c in Ref.[19] The pairing in the calculations with NM is weaker. This can be explained by the increase of $\langle \hat{j}_x \rangle_i$ of the orbitals located at the bottom of the shell due to NM (see above). Thus in the presence of NM the gradual breaking of high-j pairs proceeds faster which leads to a faster reduction of pairing with increasing Ω_x as compared with the WNM calculations. This effect is called *an anti-pairing effect induced by NM.*[19]

The CRMF and CRHB+LN examples discussed above clearly show that the modifications of band crossing features (crossing frequencies and the features of the kinematic and dynamic moments of inertia in band crossing region) caused by NM are substantial and depend on the underlying modifications of single-particle properties such as alignments and single-particle (quasi-particle) energies.

The changes of single-particle properties induced by nuclear magnetism affect also other physical observables in rotating nuclei, the discussion of which is not possible due to space limitations. These are

- *modifications of the energy splittings between signature partner orbitals (signature splitting).*[17] These modifications represent an additional source of band crossing frequency changes in odd-even and odd-odd nuclei and excited configurations of even-even ones in the presence of NM (Sec. III in Ref. 19).
- *modifications of the effective alignments* i_{eff}*.*[17] This observable is often used in the analysis of the single-particle structure of superdeformed bands (see Ref. 45)
- *the existence of signature-separated rotational bands.*[19,46,47] They reveal themselves in a considerable energy splitting of the $r_{tot} = +1$ and $r_{tot} = -1$ branches of the configurations which have the same structure in terms of occupation of single-particle states with given Nilsson labels. *This feature is a strong spectroscopic fingerprint of the presence of time-odd mean fields.*
- *Within specific configuration the impact of NM on the binding energies reaches its maximum at the terminating state.*[22] Underlying microscopic mechanism for additional binding due to NM at such states has the same features as those seen in low-spin one- and two-particle configurations of odd and odd-odd nuclei.[16] However, the magnitude of the effects is significantly larger.
- The values of kinematic moment of inertia calculated with NM are typically within 5% of the rigid body value for the moments of inertia at super- and hyperdeformation,[19] but the deviations from the rigid-body value are significantly larger for normal-deformed bands.
- NM has very small effect on the deformation properties of nuclei.[16,17,19]

4. Currents in the intrinsic (rotating) frame

The Coriolis term is present in NM and WNM calculations. This means that the currents (Eq. (10)) are always present in rotating nuclei. However, it is

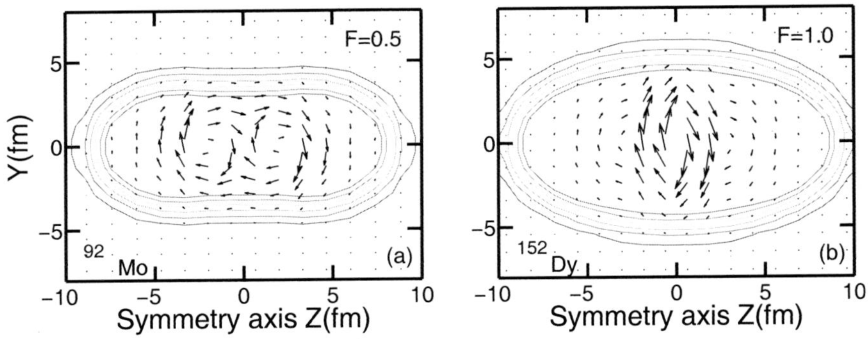

Fig. 6. (Color online) Total neutron current distributions $\mathbf{j}^n(\mathbf{r})$ in the intrinsic frame in the y-z plane for the yrast hyperdeformed (^{92}Mo) and superdeformed (^{152}Dy) configurations at rotational frequencies $\Omega_x \sim 1.0$ MeV and $\Omega_x = 0.5$ MeV, respectively. The currents in panel (b) are plotted at arbitrary units for better visualization. The currents in panel (a) are normalized to the currents in panel (b) by using factor F. The shape and size of the nucleus are indicated by density lines which are plotted in the range 0.01–0.06 fm^{-3} in steps of 0.01 fm^{-3}. Based on Fig. 8 of Ref. 19.

important to distinguish the currents induced by the Coriolis term and the ones which appear due to magnetic potential. The currents, which appear in the WNM calculations, are generated by the Coriolis term. Thus, following Ref. 19 they are called as *Coriolis induced currents.* On the contrary, the currents in the NM calculations are generated by both the Coriolis term and magnetic potential. The difference of the currents in the NM and WNM calculations is attributable to magnetic potential. Thus, the currents $[\mathbf{j}^{n,p}(\mathbf{r})]^{NM} - [\mathbf{j}^{n,p}(\mathbf{r})]^{WNM}$ are called *magnetic potential induced currents.*[19]

The detailed analysis of the currents in the rotating nuclei in the CDFT framework was first performed in Ref. 19.[c] The currents in the rotating frame of reference that is fixed to the body is caused by quantized motion of the fermions. The complicated structure of the currents in the rotating systems of independent fermions visible in Fig. 6 is the consequence of the fact that total current is the sum of the single-particle currents. The single-particle currents show vortices (circulations), the existence of which implies non-vanishing current circulations defined as $\mathbf{C}(\mathbf{r}) = \nabla \times \mathbf{j}(\mathbf{r})$.[54] Note that the strength and localization of vortices depends on the single-particle state (see Fig. 7). As a result, the differences seen in total currents of different nuclei/structures in Fig. 6 can be traced back to the single-particle content of underlying single-particle configuration.

[c]For earlier studies of the currents in rotating nuclei in non-relativistic frameworks see Refs. 48–56.

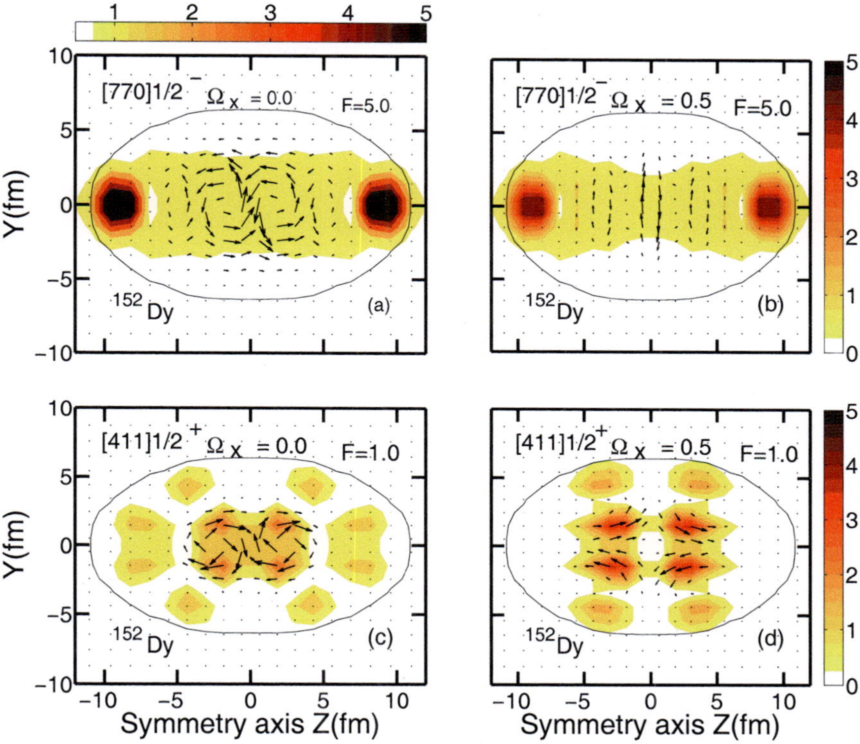

Fig. 7. Current distributions $\mathbf{j}^n(\mathbf{r})$ produced by single neutron in indicated single-particle states of the yrast SD configuration in ^{152}Dy at rotational frequencies Ω_x = 0.0 MeV (left panels) and Ω_x = 0.5 MeV (right panels). The shape and size of the nucleus are indicated by density line which is plotted at $\rho = 0.01$ fm^{-3}. The currents in panels (c) and (d) are plotted at arbitrary units for better visualization. The currents in panels (a) and (b) are normalized to the currents in panels (c) and (d) by using factor F. The currents and densities are shown in the intrinsic frame in the y-z plane at $x = 0.48$ fm. The single-neutron density distributions due to the occupation of the indicated Nilsson state are shown by colormap. Note that slightly different colormap is used in panel (a) for better visualization of densities. Based on Fig. 9 of Ref. 19.

The localization, the strength and the structure of the current vortices created by a particle in a specific single-particle state depend on its nodal structure (see Ref. 54 and Sec. IIIC in Ref. 16). In a slowly rotating anisotropic harmonic oscillator potential Coriolis induced current for a single particle shows a rather simple structure with the centers of the circulations found at the nodes and peaks of the oscillator eigenfunctions;[54] this structure forms a rectangular array somewhat similar to a crystal lattice. The analysis of single-particle vortices in rotating nuclei in the CRMF framework in general confirms these results (see Fig. 7 and Ref. 19). In part, this is due to the fact that magnetic potential induced currents are weaker than Coriolis induced ones.

The typical features of the single-particle currents in the CRMF approach are seen in Fig. 7. The comparison of left and right columns of this figure clearly indicates that for a given single-particle state the increase of rotational frequency (i) does not lead to appreciable modifications of the density distribution but (ii) considerably modifies the strength of the currents and changes the shape of circulations. The latter is due to two factors. First, the Coriolis induced currents become active at $\Omega_x \neq 0.0$ MeV and at $\Omega_x = 0.5$ they are the dominant type of currents. Second, the wave function undergoes considerable modifications with increasing rotational frequency. For example, the wave function (in terms of two largest components) of the $\nu[770]1/2^-$ state changes from 62%[770]1/2+17%[761]1/2 at $\Omega_x = 0.0$ MeV to 39%[770]1/2+28%[761]3/2 and $\Omega_x = 0.5$ MeV.

The total current is the sum of Coriolis induced and magnetic potential induced currents. In the majority of the cases total current is dominated by the Coriolis induced currents; magnetic potential induced currents represent approximately 5-'20% [30%] of total current in the HD and SD [ND] nuclei.[19] The spatial distribution of Coriolis induced and magnetic potential induced currents is similar in the majority of nuclei. However, there are cases in which the spatial distribution of these two types of currents differ substantially.[19] Note that current is weak in the surface area (Fig. 6). This is contrary to the semiclassical description of currents in normal and superfluid rotating nuclei[49] according to which the average intrinsic current flows mainly in the nuclear surface area. This underlines the importance of quantum mechanical treatment of the currents.

It is well known that there are no currents in the intrinsic frame if the rigid non-spherical body rotates uniformly (rigid rotation) (see Sec. 6A-5 in Ref. 57). The moments of inertia of super- and hyperdeformed configurations in unpaired regime come very close to the rigid-body values in the CRMF calculations.[19] However, the intrinsic currents display dramatic deviations from rigid rotation (Fig. 6). This clearly shows that the closeness of the moments of inertia to rigid body value does not necessarily imply that the current distribution should correspond to rigid rotation. On a microscopic level, the building blocks of the total current, namely, the single-particle currents certainly do not have a rigid-flow character; on the contrary, they have the vortex-flow character (see Fig. 7).

5. Ground state rotational bands in normal-deformed even-even nuclei

Despite the considerable amount of experimental data on normal-deformed nuclei and the availability of relevant theoretical frameworks, only recently and only in a single region (actinides) of the nuclear chart has the systematic investigation of rotational properties been performed in the DFT framework. These are the studies of Refs. 42, 58 performed in the CDFT framework. On the contrary, only a few nuclei have been studied at normal deformation in non-relativistic DFT.[59–63] In the current section I will review the major features of the rotational bands in the actinides and assess the accuracy of the description of these bands in the CRHB+LN framework.

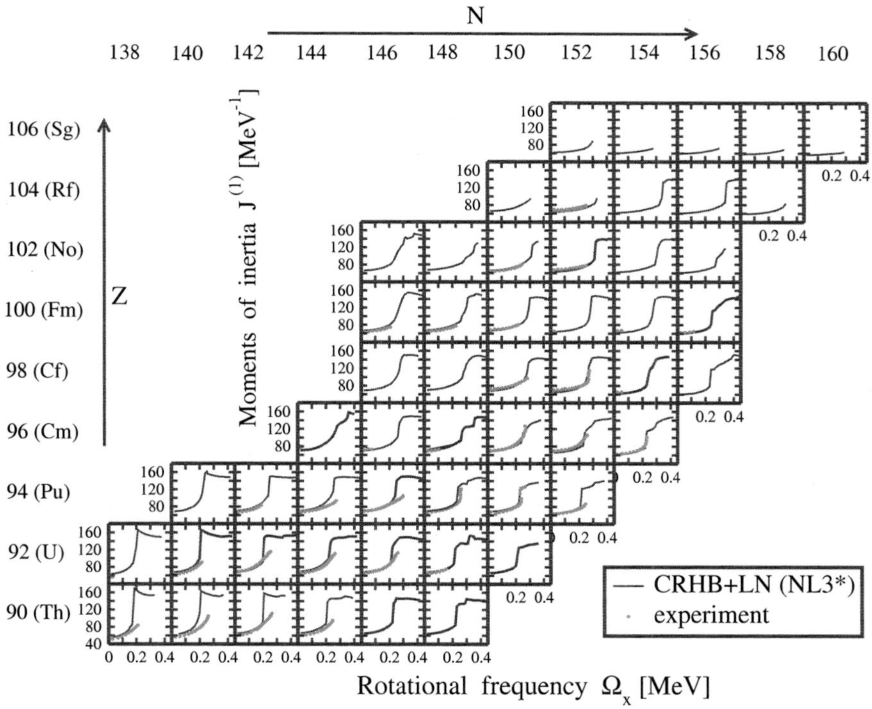

Fig. 8. (Color online) The experimental and calculated moments of inertia $J^{(1)}$ as a function of rotational frequency Ω_x. The calculations are performed with the NL3* CEDF.[64] Calculated results and experimental data are shown by black lines and red dots, respectively. Cyan dots show new experimental data from Ref. 65 which were not included in Ref. 42. From Refs. 42, 58.

Figure 8 shows the results of the first ever (in any DFT framework) systematic investigation of rotational properties of even-even nuclei at normal deformation.[42] The calculations are performed within the CRHB+LN approach.[11] One can see that the gradual increase of the moments of inertia below band crossings are reproduced well. The upbendings observed in a number of rotational bands of the $A \geq 242$ nuclei are also reasonably well described in model calculations. However, the calculations also predict similar upbendings in lighter nuclei, but they have not been seen in experiment. Note that similar problems exist also in the cranking model calculations based on phenomenological potentials.[42] The analysis of Ref. 42 suggests that the stabilization of octupole deformation at high spin, not included in the present CRHB+LN calculations, could be responsible for this discrepancy between theory and experiment.

Either sharp or more gradual increases of the kinematic moments of inertia are observed at $\Omega_x \sim 0.2$ MeV in light nuclei (Fig. 8); this crossing frequency gradually increases with mass and reaches $\Omega_x \sim 0.3$ MeV in heavy nuclei. They are due to the alignments of the neutron $j_{15/2}$ and proton $i_{13/2}$ orbitals which in many cases take place at similar rotational frequencies. The situation is more complicated than in the rare-earth region in which the $h_{11/2}$ protons align substantially later than the $i_{13/2}$ neutrons. For example, the routhian diagrams (Fig. 9) show simultaneous alignment of the proton $i_{13/2}$ pair ($\pi[633]7/2$ Nilsson orbit) and neutron $j_{15/2}$ pair ($\nu[734]9/2$ orbit) at $\Omega_x \approx 0.32$ MeV in ^{254}No. In the calculations with NL3 CEDF, the total angular momenta gain at the band crossing is $\approx 17\hbar$, with proton and neutron contributions of $\approx 7\hbar$ and $\approx 10\hbar$, respectively.[31] The alignment of these orbitals leads to a decrease of the mass hexadecapole moment Q_0, to a sign change of the mass hexadecapole moment Q_{40} and to an appreciable increase of the absolute value of the γ-deformation (Figs. 10a,b and c); this type of behavior is typical for the bands in this mass region. In addition, the pairing energies decrease with increasing rotational frequency due to the Coriolis anti-pairing effect and the pairing becomes very weak above the band crossing region (Fig. 10d).

Figure 11 illustrates the accuracy of the description of band crossing features in the Pu and Cm nuclei. Here I will discuss in detail only the band crossings in the Pu isotopes. One can see that apart from ^{250}Cm rather good description of experimental data has been obtained in the CRHB+LN calculations.

The CRHB+LN calculations predict sharp upbend $\Omega_x \sim 0.25$ MeV in all three even-even Pu isotopes.[42] The upbending is complete in ^{244}Pu and

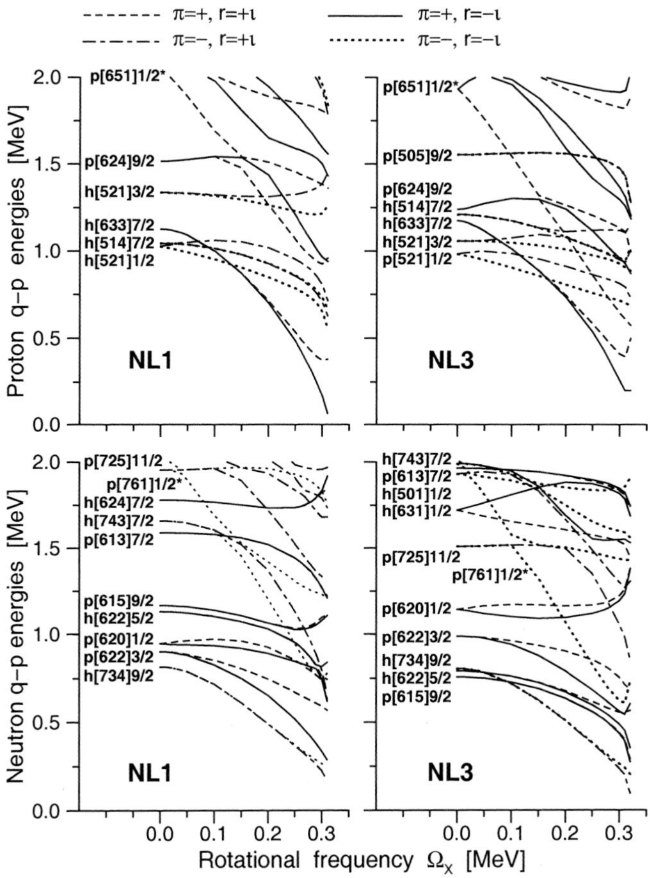

Fig. 9. Proton (top panels) and neutron (bottom panels) quasiparticle energies corresponding to the lowest configuration in ^{254}No. The CRHB+LN calculations have been performed with the NL1 (left panels) and NL3 (right panels) parametrizations. The letters "p" and "h" before the Nilsson labels are used to indicate whether a given routhian is of particle or hole type. From Ref. 31.

the CRHB+LN(NL3*) calculations rather well describe it (Fig. 11b); the sharp alignment of the proton $i_{13/2}$ orbitals is a source of this backbending and the neutron $j_{15/2}$ alignment proceeds gradually over extended frequency range. On the contrary, sharp alignments of the proton and neutron pairs take place at the same frequency in the CRHB+LN(NL1) calculations (Fig. 11b) and they somewhat overestimate the kinematic moment of inertia above the band crossing. The same situation with the alignments of the proton $i_{13/2}$ and neutron $j_{15/2}$ pairs exists also in the CRHB+LN(NL1) and

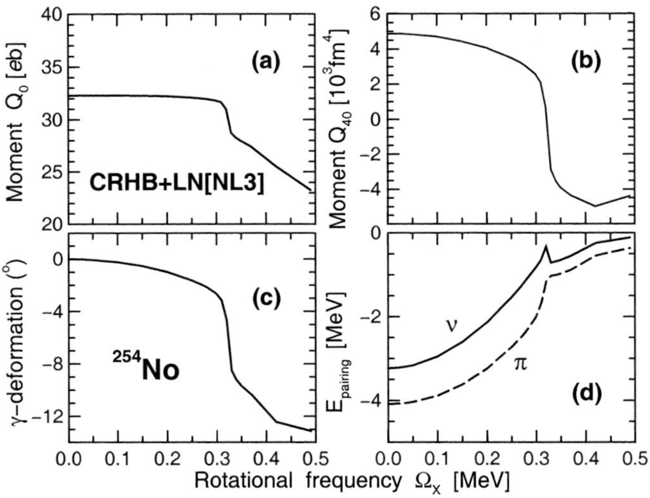

Fig. 10. Calculated mass quadrupole (Q_0) and hexadecapole (Q_{40}) moments, the γ-deformation and proton and neutron pairing energies ($E_{pairing}$) of the ground state rotational band in ^{254}No as a function of rotational frequency Ω_x. The results are obtained in the CRHB+LN calculations with the NL3 CEDF. From Ref. 31.

CRHB+LN(NL3*) calculations for ^{242}Pu. They accurately reproduce the evolution of kinematic moments of inertia with frequency and the frequency of the paired band crossing (Figs. 8 and 11a). However, since upbending is not complete in experiment it is impossible to judge whether the simultaneous sharp alignments of proton and neutron pairs really take place in nature. The kinematic moment of inertia of the ground state rotational band in ^{246}Pu shows a rapid increase at the highest observed frequencies similar to the one seen before upbendings in ^{242,244}Pu. However, the ^{246}Pu data does not reveal an upbend yet. The upbend in ^{242,244}Pu is predicted to be 0.01-0.02 MeV earlier in the CRHB+LN(NL3*) calculations as compared with experiment. A similar situation is expected in ^{246}Pu. Considering this and the fact that the last observed point in ^{246}Pu is tentative, one can conclude that there are no significant discrepancies with experimental data. Even better agreement with this new data is seen in the case of the CRHB+LN(NL1) calculations.

Similar CRHB+LN analysis of the paired band crossings in ^{248,250}Cf and ^{246,248,250}Cf (see also Figs. 11d, e and f) has been presented in Ref. 58. In this reference, the CRHB+LN results for the Cf and Cm nuclei were compared with the ones obtained in the cranked shell model with

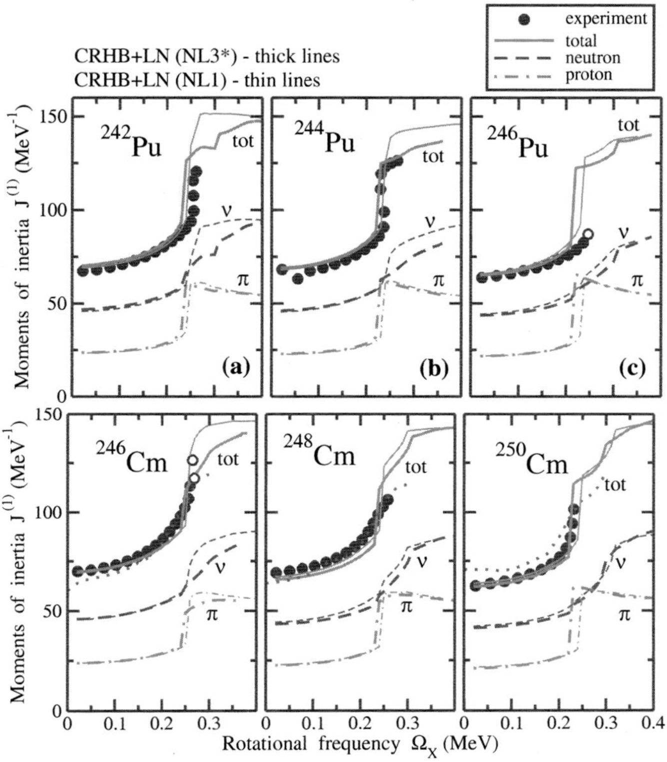

Fig. 11. (Color online) The experimental and calculated kinematic moments of inertia $J^{(1)}$ of ground state rotational bands in indicated nuclei as a function of rotational frequency Ω_x. Proton and neutron contributions to the kinematic moment of inerta are presented. Open circles are used for tentative experimental points. Total kinematic moments of inertia obtained in the CSM+PNC approach[66] are shown by blue solid lines. Based on Ref. 58.

the pairing correlations treated by a particle-number conserving method (further CSM+PNC).[66] There are three important differences between the CRHB+LN and CSM+PNC approaches. First, the parameters of the Nilsson potential were carefully adjusted in the CSM+PNC approach to the experimental energies of deformed one-quasiparticle states of actinides in Ref. 66. Second, in the CSM+PNC approach the deformations are chosen to be close to experimental values and they do not change with rotational frequency. These two types of observables are defined fully self-consistently in the CRHB+LN approach without any fit to experimental data, and, in addition, the deformations (mass moments) change with rotational frequency (Fig. 10d). Third, the pairing strength has been fitted to experiment in both approaches.[42,66] However, the pairing strength is different in even-even and odd-mass nuclei in the CSM+PNC approach;[66] this is a well known deficiency of the cranked shell model (see, for example, Ref. 67). In contrast, the same pairing strength is used in even-even and odd-mass nuclei in the CRHB+LN approach; it leads to a consistent and accurate description of odd-even mass staggerings (the $\Delta^{(3)}$ indicators) and the moments of inertia in even-even and odd-mass actinides (see Ref. 42 and Sec. 6).

Thus, the CRHB+LN approach provides much more consistent description of rotational properties in paired regime as compared with the CSM+PNC approach; with no adjustment of single-particle energies and deformations to experiment it obtains a description of experimental rotational properties of actinides that is comparable in accuracy.[58] In addition to these studies in actinides, the CRHB+LN approach has successfully been applied to the investigations of the ground state rotational bands in the $A \sim 70$ $N \sim Z$ region[68] and of the evolutions of the moments of inertia with particle numbers at low frequency ($\Omega_x = 0.02$ MeV) in the rare-earth region.[41]

6. Rotational bands of odd-mass nuclei in paired regime

In the DFT framework, the description of rotational bands in odd-mass nuclei is more technically difficult than the one in even-even nuclei because of the reasons discussed below. The systematic study of such bands has so far been performed only in CDFT.[42]

First, the effects of blocking due to odd particle have to be included in a fully self-consistent way. This is done in the CRHB+LN computer code according to Refs. 36, 71, 72. In addition, it requires the identification of blocked orbital at all frequencies of interest which is a non-trivial problem. In the CRHB+LN code[31,42] the blocked orbital can be specified by different fingerprints such as

- dominant main oscillator quantum number N of the wave function,
- the dominant Ω quantum number (Ω is the projection of the angular momentum on the symmetry axis) of the wave function,
- the particle or hole nature of the blocked orbital,
- the position of the state within specific parity/signature/dominant N/dominant Ω block,

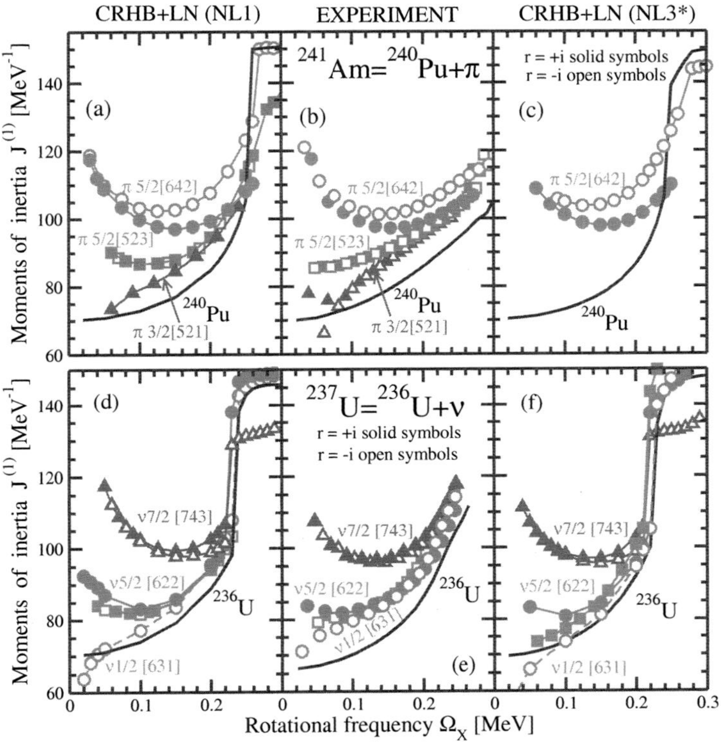

Fig. 12. (Color online) (top panels) Calculated and experimental kinematic moments of inertia $J^{(1)}$ of the indicated one-quasiproton configurations in the ^{241}Am nucleus and ground state rotational band in reference even-even ^{240}Pu nucleus. Experimental data are shown in the middle panel, while the results of the CRHB+LN calculations with the NL1 and NL3* CEDF's in the left and right panels, respectively. The same symbols/lines are used for the same theoretical and experimental configurations. The symbols are used only for the configurations in odd-mass nucleus; the ground state rotational band in reference even-even nucleus is shown by solid black line. The label with the following structure "Odd nucleus = reference even+even nucleus + proton(π)/neutron(ν)" is used in order to indicate the reference even-even nucleus and the type of the particle (proton or neutron) active in odd-mass nucleus. (bottom panels) The same as in top panels but for one-quasineutron configurations in ^{237}U and ground state band in ^{236}U. The experimental data are from Refs. 69, 70. Based on Ref. 42.

or their combination. For a given configuration, possible combinations of the blocked orbital fingerprints were defined from the analysis of calculated quasiparticle spectra in neighboring even-even nuclei and the occupation probabilities of the single-particle orbitals of interest in these nuclei.

Second, variational calculations with blocked orbital(s) are numerically less stable than the ones for the ground state bands in even-even nuclei

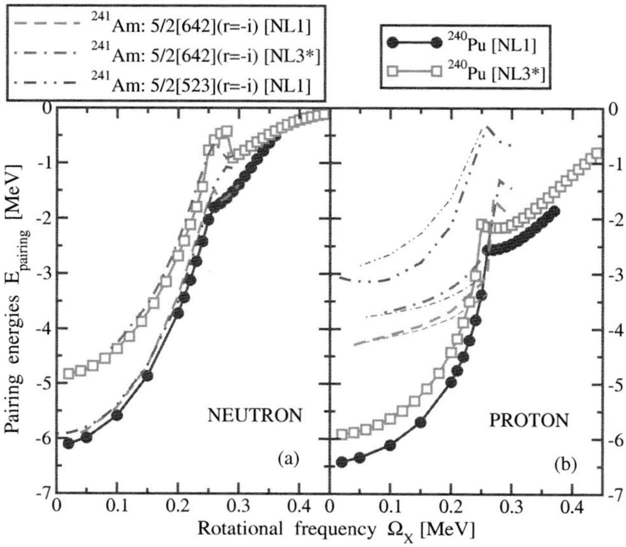

Fig. 13. (Color online) Calculated proton and neutron pairing energies in ground state rotational band of ^{240}Pu and one-quasiproton rotational bands of ^{241}Am. Thick and thin lines are used for the $(r = -i)$ and $(r = +i)$ branches of one-quasiparticle configurations, respectively. Note that neutron pairing almost does not depend on the signature of blocked proton orbital. As a result, only the $(r = -i)$ branches are shown in panel (a). From Ref. 42.

because at each iteration of the variational procedure blocked orbital(s) has(have) to be properly identified. The convergence problems are the consequence of the fact that closely lying orbitals within a given parity/signature block interact and exchange a character; the strength of the interaction is an important factor here. Another ingredient affecting the convergence is the relative energies of interacting orbitals. Different CEDF's are characterized by different single-particle spectra.[73] As a result, the convergence problems for specific blocked solution can show up in one functional but will not affect the solution in another one. A detailed discussion of the convergence problems in the calculations of rotational bands in odd-mass nuclei and the ways to overcome them is presented in Sec. V of Ref. 42.

Figure 12 displays a representative example of the RHB+LN calculations for one-quasiparticle bands in ^{237}Np and ^{241}Am which come from systematical studies of Ref. 42. Additional results for ^{253}No and ^{255}Lr can be found in Refs. 74, 75.[d] In ^{241}Am, the rotational bands based on

[d]To my knowledge, only in a single case of ^{74}Rb^{76} have two-quasiparticle rotational bands been calculated in the DFT framework (namely, in the CRHB+LN approach).

the Nilsson orbitals $\pi 5/2[642]$ (from the $i_{13/2}$ spherical subshell), $\pi 5/2[523]$ (from the $h_{9/2}$ subshell) and $\pi 3/2[521]$ (from the $f_{7/2}$ subshell) have been observed; their kinematic moments of inertia $J^{(1)}$ are distinctly different at low frequencies. Theoretical calculations (Figs. 12a, c) describe well the absolute values of the kinematic moments of inertia of different configurations, their evolution with rotational frequency and signature splitting. They also indicate that the results of the RHB+LN calculations for a specific configuration only weakly depend on CEDF. Above-mentioned features are also clearly seen in ^{237}Np.

Figures 12a and c also show that the convergence depends on CEDF. Indeed, it was not possible to get convergent solutions for the $\pi 5/2[523]$ and $\pi 3/2[521]$ configurations of ^{241}Am in the calculations with the NL3* CEDF.

The increase of the kinematic moment of inertia in the bands of ^{241}Am as compared with the one of the ground state band in ^{240}Pu (Fig. 12) is caused by the blocking effect which results in a decreased proton pairing (see Fig. 13). This figure also shows that the blocking of proton orbitals almost does not affect the pairing in neutron subsystem.

The systematic studies of Ref. 42 allowed us to conclude that rotational properties of one-quasiparticle configurations substantially depend on the structure of blocked orbital. As a result, the rotational properties reflected through the following fingerprints:

- the presence or absence of signature splitting,
- the relative properties of different configurations with respect to each other and/or with respect to the ground state band in reference even-even nucleus,
- the absolute values of the kinematic moments of inertia (especially at low rotational frequencies) and their evolution with rotational frequency,

provide useful tools for quasiparticle configuration assignments. Such configuration assignments are important, for example, for on-going experimental investigations of odd-mass light superheavy nuclei at the edge of the region where spectroscopic studies are still feasible (the nuclei with masses $A \sim 255$ and proton number $Z \geq 102$).[42,77] Ref. 42 clearly showed that with few exceptions these features of rotational bands are well described in the RHB+LN calculations. The presence or absence of signature separation and its magnitude is the most reliable fingerprint which is reproduced in model calculations with good accuracy. The moments of inertia and their evolution with frequency are generally well described in model calculations. As a consequence, the relative properties of different configurations with respect of each other and/or with respect to the ground state band in reference even-even nucleus provide a reasonably reliable fingerprint of configuration.

However, it is necessary to recognize that the configuration assignment based on rotational properties has to be complemented by other independent methods and has to rely on sufficient experimental data.[42] This is because such a method of configuration assignment does not always lead to a unique candidate configuration due to theoretical inaccuracies in the description of the moments of inertia.

The ability to calculate odd-mass nuclei fully self-consistently with allowance of nuclear magnetism and breaking of Kramer's degeneracy has also allowed us to address the question of consistency of the definition of pairing strength in CDFT. The strengths of pairing defined by means of the kinematic moments of inertia $J^{(1)}$ and three-point $\Delta^{(3)}$ indicators

$$\Delta_{\nu}^{(3)}(N) = \frac{(-1)^N}{2}\left[B(N-1) + B(N+1) - 2B(N)\right], \qquad (23)$$

defined from odd-even staggering of binding energies $B(N)$ (similar expression holds also for the proton $\Delta_{\pi}^{(3)}$ indicator), strongly correlate.[42] This is a known result in non-self-consistent models based on phenomenological Woods–Saxon or Nilsson potentials. However, this is a non-trivial result in the DFT framework since time-odd mean fields (absent in phenomenological potentials) strongly affect the moments of inertia[19] and have an impact on three-point $\Delta^{(3)}$ indicators.[16] The definitions of pairing strength via these two observables are complimentary. This is because (i) it is difficult to disentangle proton and neutron contributions to pairing when considering the moments of inertia and (ii) the $\Delta^{(3)}$ indicators are affected by particle-vibration coupling and depend on correct reproduction of the ground states in odd-mass nuclei.[42]

7. Hyperdeformation at high spin

The search for hyperdeformation (HD) at high spin still remains in the focus of attention of the nuclear structure community. Although the attempts to observe discrete HD bands at high spin have not been successful so far, there is hope that the next generation of detectors such as GRETA and AGATA will allow us to observe such bands in the future.

Already, some hints of the presence of hyperdeformation at high spin are available. For example, Hyper-Long-HyperDeformed (HLHD) experiment at the EUROBALL-IV γ-detector array revealed some features expected for HD nuclei.[78–80] Although no discrete HD rotational bands have been identified, rotational patterns in the form of ridge-structures in three-dimensional (3D) rotational mapped spectra have been identified with dynamic moments of inertia $J^{(2)}$ ranging from 71 to 111 MeV^{-1} in 12 different nuclei selected by charged particle- and/or γ-gating. The four nuclei, ^{118}Te, ^{124}Cs, ^{125}Cs and ^{124}Xe, found with moment of inertia $J^{(2)} \sim 110$ MeV^{-1} are most likely hyperdeformed[e] while the remaining nuclei with smaller values of $J^{(2)}$ are expected to be superdeformed. The width in energy of the observed ridges indicates that there are $\approx 6-10$ transitions in the HD cascades, and a fluctuation analysis shows that the number of bands in the ridges exceeds 10. The HD ridges are observed in the frequency range of about 650 to 800 keV, and their dynamic moments of inertia have typical uncertainty of 10% (e.g. 111 ± 11 MeV^{-1} in ^{124}Xe).

The experimental data show unusual features never seen before in the studies of the SD bands. For example, the addition of one neutron on going from ^{124}Cs to ^{125}Cs decreases the experimental $J^{(2)}$ value by $\sim 10\%$ (from 111 MeV^{-1} down to 100 MeV^{-1}, see Fig. 14). It is impossible to find an explanation for such a big impact of the single particle on the properties of nuclei: previous studies in the SD minima in different parts of the nuclear chart showed that the addition or removal of particle affects dynamic moments of inertia less drastically.[45]

The comparative analysis of the CRMF and CRHB+LN results in ^{124}Xe in Ref. 82 reveals that the pairing is reasonably small in the lowest HD configurations and it becomes even smaller in excited configurations due to the blocking effect. The dominant effects in the quenching of pairing correlations are the Coriolis antipairing effect, the quenching due to shell gaps,[83] and the blocking effect.[36] Thus, the pairing has been neglected in systematic studies of HD in Ref. 82 in which unpaired CRMF calculations have been performed.

The results of the CRMF calculations with the NL1 CEDF are confronted with experimental data in Fig. 14. The calculated $J^{(2)}$ moments of inertia somewhat underestimate experimental data. The kinematic moments of inertia of the lowest HD solutions are either nearly constant or very gradually increase with rotational frequency. On the contrary, both

[e]In comparison, the HD ridges in ^{152}Dy are characterized by $J^{(2)} \sim 130$ MeV^{-1}.[81]

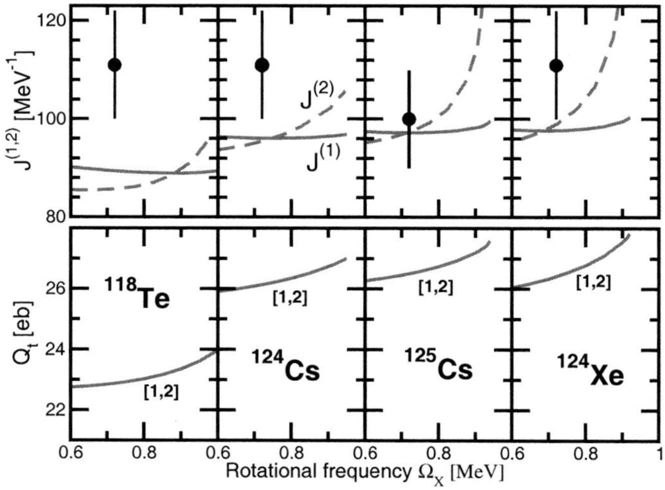

Fig. 14. (Color online) Calculated kinematic and dynamic moments of inertia (top panels) and transition quadrupole moments (bottom panels) as a function of rotational frequency for the lowest HD solutions in ^{118}Te, ^{124,125}Cs and ^{124}Xe. The structure of calculated configurations is indicated in the bottom panels. Experimental data for dynamic moments of inertia of ridge structures are shown in the top panels. From Ref. 82.

the dynamic moments of inertia and the transition quadrupole moment Q_t more rapidly increase with rotational frequency over the calculated frequency range. They are in complete contrast to the features of the SD bands in unpaired regime, in which the Q_t, $J^{(1)}$ and $J^{(2)}$ values (apart from the unpaired band crossing regions) decrease with increasing rotational frequency (see Refs. 1, 43, 84, 85 and references therein). As discussed in detail in Ref. 82 the microscopic origin of these features lies in a centrifugal stretching of the HD shapes with increasing rotational frequency. Systematic analysis of the yrast/near-yrast HD configurations in the part of the nuclear chart studied in Ref. 82 shows that the centrifugal stretching is a general feature of the HD bands.

In order to better understand the general features of HD at high spin I concentrate on the ^{124}Xe nucleus. The results of the CRMF calculations for some HD configurations in ^{124}Xe are displayed in Fig. 15. The calculated configurations are labeled by $[p, n_1 n_2]$, where p, n_1 and n_2 are the number of proton $N = 7$ and neutron $N = 7$ and $N = 8$ hyperintruder orbitals occupied, respectively. For most of the HD configurations, neutron $N = 8$ orbitals are not occupied, so the label n_2 is omitted in the labeling of such

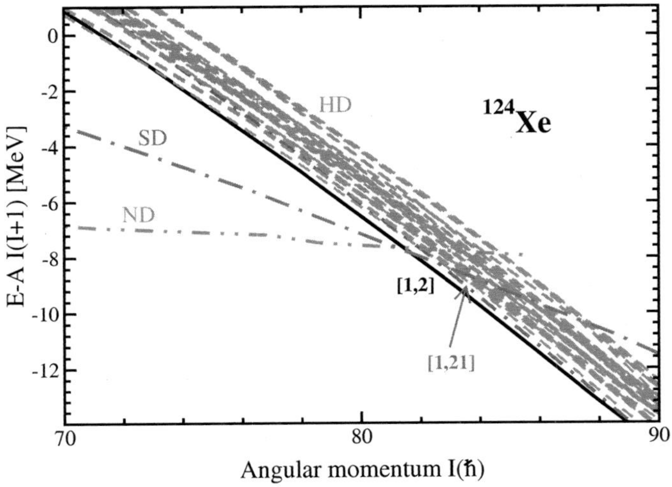

Fig. 15. (Color online) Energies of the calculated configurations relative to a smooth liquid drop reference $AI(I+1)$, with the inertia parameter $A = 0.01$. The ND and SD yrast lines are shown by dotted and dot-dot-dashed lines, respectively. Solid and dot-dashed lines are used for the [1,2] and [1,21] HD configurations, respectively. Dashed lines represent excited HD configurations. From Ref. 82.

configurations. The HD minimum becomes lowest in energy at spin $82\hbar$, and the [1,2] configuration is the yrast HD configuration in the spin range of interest. The excited HD configurations displayed in Fig. 15 are built from this configuration by exciting either one proton or one neutron or simultaneously one proton and one neutron. The total number of excited HD configurations shown is 35. It is interesting to mention that the configuration involving the lowest $N = 8$ neutron orbital (the [1,21] configuration in Fig. 15) is calculated at low excitation energy.

The calculations reveal a high density of the HD configurations which would be even higher if the additional calculations for the excited configurations were performed starting from the low-lying excited HD configurations, such as the [1,21] configuration. This high density is due to two facts: relatively small $Z = 54$ and $N = 70$ HD shell gaps in the frequency range of interest (see Fig. 10 in Ref. 82) and the softness of the potential energy surfaces in the HD minimum.

With a few exceptions mentioned below these two factors are active at hyperdeformation in absolute majority of even-even nuclei in the $Z = 40 - 60$ part of nuclear chart studied in Ref. 82. Note that different proton and neutron HD gaps will be active in different regions.[82]

As a consequence, the density of the HD bands in the spin range where they are yrast or close to yrast is high in the majority of the cases. For such densities, the feeding intensity of an individual HD band will most likely drop below the observational limit of modern experimental facilities so that it will be difficult to observe discrete HD bands. This is because total feeding intensity will be redistributed among many low-lying bands, thus drastically reducing the intensity with which each individual band is populated. On the other hand, the high density of the HD bands may favor the observation of the rotational patterns in the form of ridge-structures in three-dimensional rotational mapped spectra as it has been seen in the HLHD experiment.[80] The study of these patterns as a function of proton and neutron numbers, which seems to be possible with existing facilities, will provide valuable information about hyperdeformation at high spin.

Only in a few nuclei is the density of the HD rotational bands low in the CRMF calculations.[82] These are, for example, ^{96}Cd and ^{107}Cd.[86,87] The latter one is the best candidate for a search of discrete HD bands. An alternative candidate is the doubly magic extremely superdeformed band in ^{111}I,[82] the deformation of which is only slightly lower than that of the HD bands, and which may be observed with existing experimental facilities. In all these cases the low density of the HD bands is due to large shell gaps in the single-particle spectra at HD.

Another interesting question is whether necking degree of freedom is important in the HD bands. Fig. 16 shows some indications of the necking and the clusterization of the density into two fragments in the [1,2] configuration of ^{124}Xe, but this effect is not very pronounced in this nucleus. The systematics of the self-consistent proton density distributions in the HD states has been studied in Ref. 82. In some nuclei (such as ^{124}Te, ^{130}Xe, ^{132}Ba) the necking degree of freedom plays an important role, while others (for example, ^{100}Mo and ^{136}Ce) show no necking. The neck is typically less pronounced in the HD states of the lighter nuclei because of their smaller deformation. In addition, the shell structure also plays a role in a formation of neck. The necking degree of freedom becomes even more important in extremely deformed structures which according to the language of Ref. 88 can be described as megadeformed. Fig. 16 shows an example of density distribution for the megadeformed state in ^{102}Pd, which becomes yrast at $I \sim 85\hbar$ in the CRMF calculations. The neck is more pronounced in the proton subsystem than in the neutron one both in the HD and megadeformed structures due to the Coulomb repulsion of the segments. These results indicate that, in general, the necking degree of freedom is important

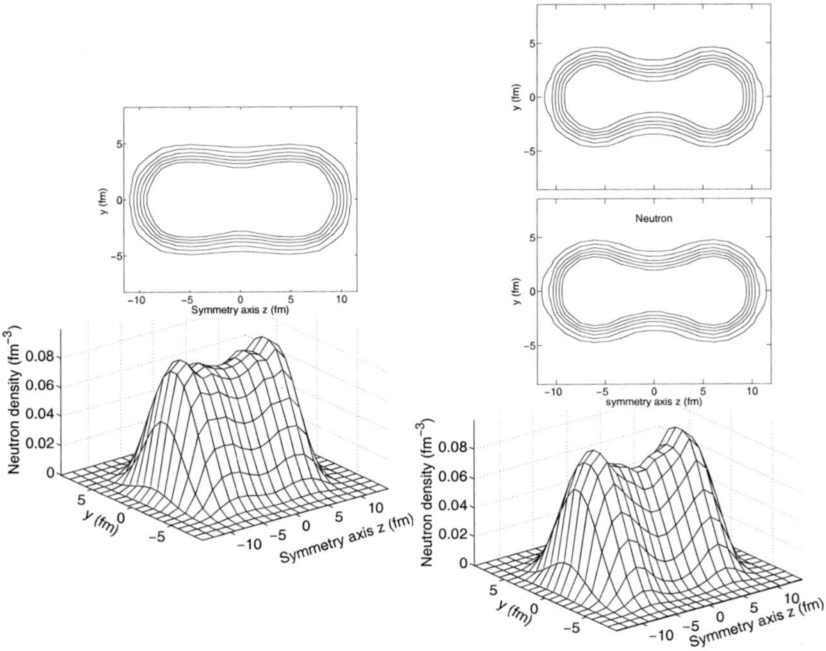

Fig. 16. (Color online) (left panels) The self-consistent neutron density $\rho_n(y,z)$ as a function of y- and z- coordinates for the [1,2] configuration in ^{124}Xe at rotational frequency $\Omega_x = 0.75$ MeV. Top and bottom panels show 2- and 3-dimensional plots of the density distribution, respectively. In the top panel, the densities are shown in steps of 0.01 fm^{-3} starting from $\rho_n(y,z) = 0.01$ fm^{-3}. (right panels) The same as left panels, but for yrast megadeformed state in ^{102}Pd at rotational frequency $\Omega_x = 0.95$ MeV. Two top panels show 2-dimensional plots of the proton and neutron density distribution. From Ref. 82.

in the HD states and that it should be treated within the self-consistent approach which, in particular, allows different necking for the proton and neutron subsystems.

The spins I_{cr}^{HD} at which HD bands become yrast (or shortly "crossing spins") represent an important constraint for observation of such bands. There are several reasons for that. First, the sensitivity of modern γ-ray detectors defines the highest spins which can be studied in experiment. For example, the GAMMASPHERE allows to study discrete rotational bands only up to $\approx 65\hbar$ in medium mass nuclei.[f] The observation of higher spin states will most likely require a next generation of γ-ray tracking detectors

[f]The DFT calculations suggest that the triaxial SD band 1 in ^{158}Er is observed up to spins in excess of $70\hbar$.[89] However, this result requires further experimental confirmation.

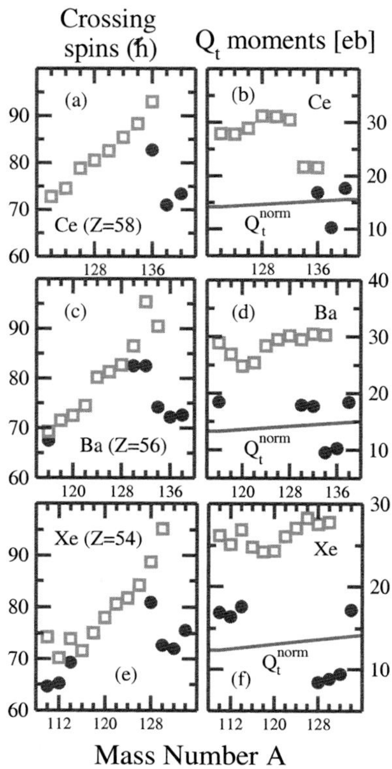

Fig. 17. (Color online) The crossing spins (left panels) at which the SD (solid circles) and HD (open squares) configurations become yrast and their transition quadrupole moments Q_t (right panels) for the Ce, Ba and Xe isotopes. The values for the SD configurations are shown only in the cases when they become yrast at lower spins than the HD configurations. The normalized transition quadrupole moments Q_t^{norm} corresponding to the deformation of the yrast SD band in ^{152}Dy are shown by the solid line. Based on Fig. 4 from Ref. 82.

such as GRETA or AGATA. Second, as suggested by the studies of the Jacobi shape transition in Ref. 90, the coexistence of the SD and HD minima at the feeding spins may have an impact on the survival of the HD minima because of the decay from the HD to SD configurations. If this mechanism is active, then only the nuclei in which the HD minimum is lower in energy than the SD one at the feeding spin and/or the nuclei characterized by the large barrier between the HD and SD minima will be reasonable candidates for a search of the HD bands.

Figure 17 shows that the HD configurations become yrast at lower spin than the SD ones only in a specific mass range which depends on the isotope chain.[82] It also illustrates that the crossing spins I_{cr}^{HD}, at which the HD configurations become yrast, are lower for proton-rich nuclei. This is a feature which is seen in the majority of studied isotope chains; by going from the β-stability valley towards the proton-drip line one can lower I_{cr}^{HD} by approximately $10\hbar$. In addition, the calculated transition quadrupole moments of these configurations at spin values close to the crossing spins are shown. The calculated HD configurations are near-prolate. Additional results for the Zr, Mo, Ru, Pd, Cd, Sn, Te isotope chains are presented in Ref. 82.

In addition, the single-particle properties and their role in future configuration assignments at HD have been investigated in Ref. 82. It was concluded that the individual properties of the single-particle orbitals are not lost at HD. In the future, they will allow the assignment of the configurations to the HD bands using the relative properties of different bands. Such methods of configuration assignment were originally developed for superdeformation. In contrast to the case of SD, the analysis of Ref. 82 has showed that only simultaneous application (by comparing experimental and theoretical $(i_{eff}, \Delta Q_t)$ values) of the methods based on effective alignments i_{eff}[91] and relative transition quadrupole moments ΔQ_t[92] will lead to a reliable configuration assignment for the HD bands. Moreover, additional information on the structure of the HD bands will be obtained from the band crossing features; the cases of strong interaction of the bands in unpaired regime at HD will be more common as compared with the situation at SD.

8. Other phenomena

It is impossible to cover in detail other results obtained for rotational bands within the cranked versions of the CDFT because of space limitations. However, I feel that at least it is necessary to give some short overview of systematic calculations in different mass regions and provide some references on the physical phenomena which have not been discussed in detail in this review. Note that transition quadrupole moments are typically described within the error bars of experimental measurements.[10,11,43,68,93] Thus, we do not consider them below.

8.1. *Superdeformation in paired and unpaired regimes*

Results of systematic CRHB+LN calculations with the NL1 CEDF[10,11] for yrast superdeformed bands in even-even nuclei in the $A \sim 190$ region are shown in Fig. 18. Without any new adjustable parameter a very successful description of rotational features of experimental bands in this region is obtained in the calculations; only in ^{198}Pb the calculations fail to reproduce the kinematic and the dynamic moments of inertia above $\Omega_x \sim 0.25$ MeV (Fig. 18). The increase of the moments of inertia in this mass region can be understood as emerging predominantly from a combination of three effects: the gradual alignment of a pair of $j_{15/2}$ neutrons, the alignment of a pair of $i_{13/2}$ protons at a somewhat higher frequency, and decreasing pairing correlations with increasing rotational frequency. Above $\Omega_x \geq 0.4$ MeV, the $\mathcal{J}^{(2)}$ values determined by the alignment in the neutron subsystem decrease but this process is compensated by the increase of $\mathcal{J}^{(2)}$ due to the continuing alignment of the $i_{13/2}$ proton pair. Thus the shape of the peak (or the plateau) in the total value of $\mathcal{J}^{(2)}$ at these frequencies is determined

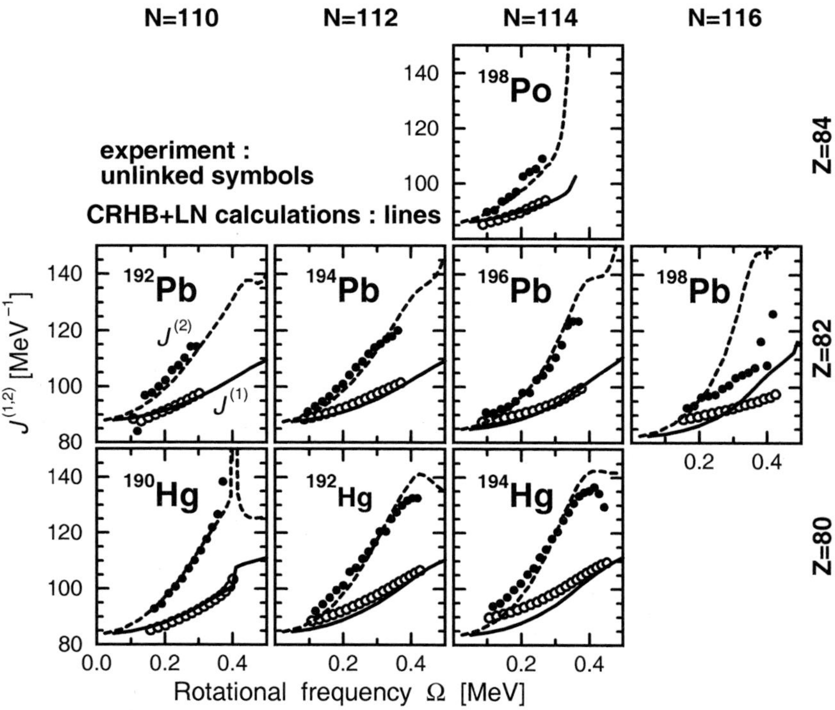

Fig. 18. Kinematic ($\mathcal{J}^{(1)}$) and dynamic ($\mathcal{J}^{(2)}$) moments of inertia of the yrast superdeformed bands in the even-even nuclei of the $A \sim 190$ mass region. Experimental $\mathcal{J}^{(1)}$ and $\mathcal{J}^{(2)}$ values are shown by open and solid circles. Solid and dashed lines are used for the $\mathcal{J}^{(1)}$ and $\mathcal{J}^{(2)}$ values obtained in the CRHB+LN calculations. From Ref. 11.

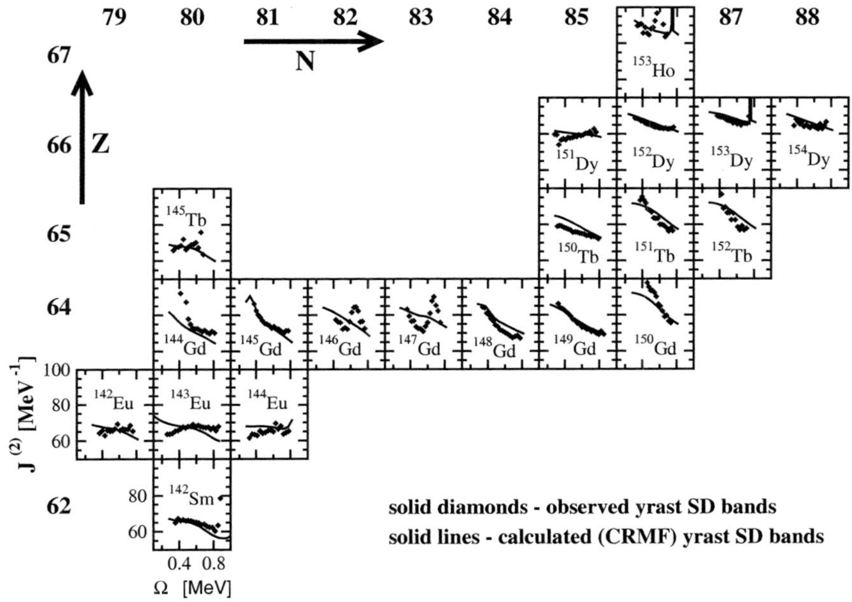

Fig. 19. Dynamic moments of inertia of observed yrast SD bands (solid diamonds) in the $A \sim$ 140-150 mass region of superdeformation compared with the calculations in cranked RMF theory (solid lines). From Ref. 1.

by a delicate balance between alignments in the proton and neutron subsystems which depends on the deformation, the rotational frequency and the position of the Fermi energy. For example, the alignment of a pair of $j_{15/2}$ neutrons at $\Omega_x \sim 0.4$ MeV is visible experimentally in the isotope ^{192}Hg and even more clearly in ^{194}Hg.

In addition to actinides, the CRHB+LN approach has been used for the yrast SD band in ^{60}Zn in Ref. 47.

A systematic investigation of properties of superdeformed bands in unpaired regime has been performed in the CRMF framework in the $A \sim 60$[85] and 150[43,45] mass regions of superdeformation. It was shown that CRMF theory reproduces in general the experimentally observed features well. For example, a summary of these studies for the dynamic moments of inertia $\mathcal{J}^{(2)}$ in the $A \sim 150$ region is shown in Fig. 19 and the left corner of Fig. 20 shows the results for the $A \sim 60$ region. I will discuss here the results for $A \sim 150$ mass region. At high rotational frequencies, where pairing is of minor importance, the dynamic moments of inertia are rather well reproduced in the CRMF calculations. Exceptions are the yrast superdeformed

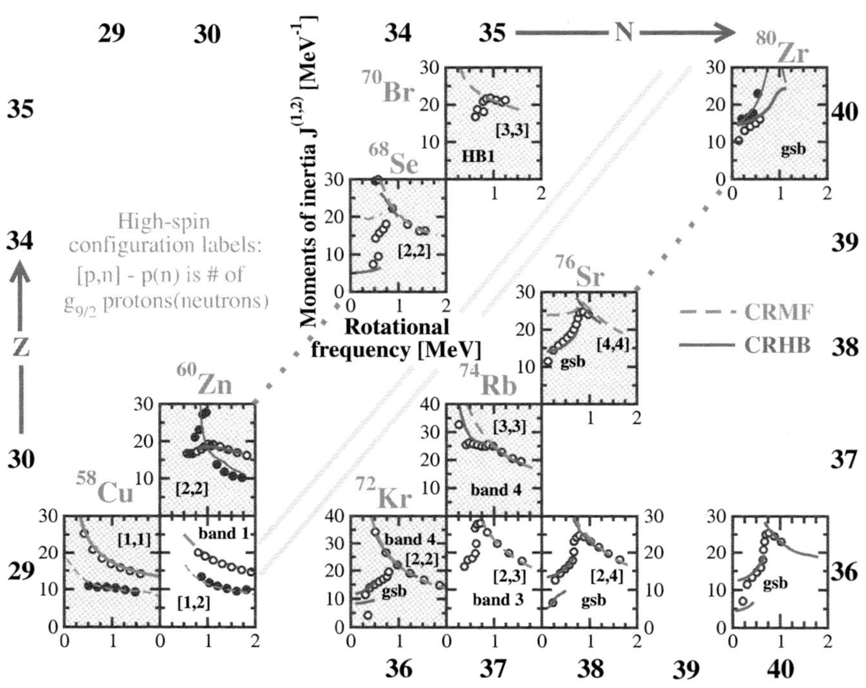

Fig. 20. (Color online) The kinematic moments of inertia $J^{(1)}$ of rotational structures in the $N \approx Z$ nuclei compared with the results of the CRMF and CRHB+LN calculations. The shaded background is used for $N = Z$ nuclei. The vertical scale of the panels for ^{72}Kr and ^{74}Rb is different from the one of the other panels. Note that in a few cases the results for dynamic moments of inertia $J^{(2)}$ are shown. In these cases, thick and thin lines are used for calculated kinematic and dynamic moments of inertia, respectively. Experimental kinematic and dynamic moments of inertia are shown by open and solid circles, respectively. The results of the CRHB+LN calculations at low spin are shown both for prolate and oblate minima in a few cases; in a given nucleus calculated $J^{(1)}$ in oblate minimum is lower than the one in prolate minimum.

bands in ^{146,147}Gd and in ^{153}Ho undergoing unpaired band crossings. For example, in ^{146,147} Gd the peak in $\mathcal{J}^{(2)}$ at $\Omega_x \approx 0.6$ MeV is not reproduced. According to the standard interpretation,[45,91] it originates from the crossing of two specific orbitals with the Nilsson quantum numbers $\nu[651]1/2$ and $\nu[642]5/2$ and signature $r = +i$. The relative position of this pair is not reproduced properly in several RMF parameterizations.[43,45] Even at low spins, the results of the CRMF calculations are close to experiment in most of the cases. The experiment of Ref. 94 has linked the yrast superdeformed band in ^{152}Dy to the low-spin level scheme which allowed the authors to extract experimental kinematic moment of inertia. Its calculated value is by

7–5% higher than in experiment at low spins and the difference decreases with increasing spin (Ref. 1). This suggests some persistence of pairing correlations especially at low spins. Note that the systematics of Fig. 19 has recently been extended by the CRMF studies of excited SD bands in ^{154}Dy.[95]

8.2. *Neutron-proton pairing*

The physics of isoscalar and isovector neutron-proton pairing has been and still is an active topic of nuclear structure studies.[93,96] At present, the existence of isovector *np*-pairing is well established. The isovector *np*-pairing is absolutely necessary in order to restore the isospin symmetry of the total wave function. Its strength is well defined by the isospin symmetry. On the other hand, the observed consequences of the $t = 0$ *np*-pairing still remain illusive. The systematic analysis of the rotational response of $N \approx Z$ nuclei performed in the CRHB+LN and CRMF frameworks (see Refs. 68, 93 and references quoted therein) agrees with the picture which does not involve isoscalar *np*-pairing. According to it (isovector mean-field theory[97]), at low spin, an isoscalar *np*-pair field is absent while a strong isovector pair field exists, which includes a large *np* component, whose strength is determined by isospin conservation. Like in nuclei away from the $N = Z$ line, this isovector pair field is destroyed by rotation. In this high-spin regime, calculations without pairing describe the data accurately (see Fig. 20 and Refs. 68, 93), provided that the shape changes and band termination are taken into account.

8.3. *Band termination*

Another interesting phenomenon is band termination, and, especially smooth band termination. Smooth terminating band shows a continuous and smooth transition within a specific configuration from a collective rotation to a non-collective terminating (single-particle) state.[4] The band terminates in the terminating state which shows full alignment of all angular momentum vectors of the valence particles and holes along the axis of rotation. In the CDFT framework, the band termination phenomenon has been studied only in a few cases. One of them is ground state band in ^{20}Ne terminating at $I = 8^+$; this is a classical case of band termination. The termination of this band and the impact of time-odd mean fields on its rotational properties and terminating state have been studied in Ref. 22. In addition, the impact of these fields on the binding energies of terminating

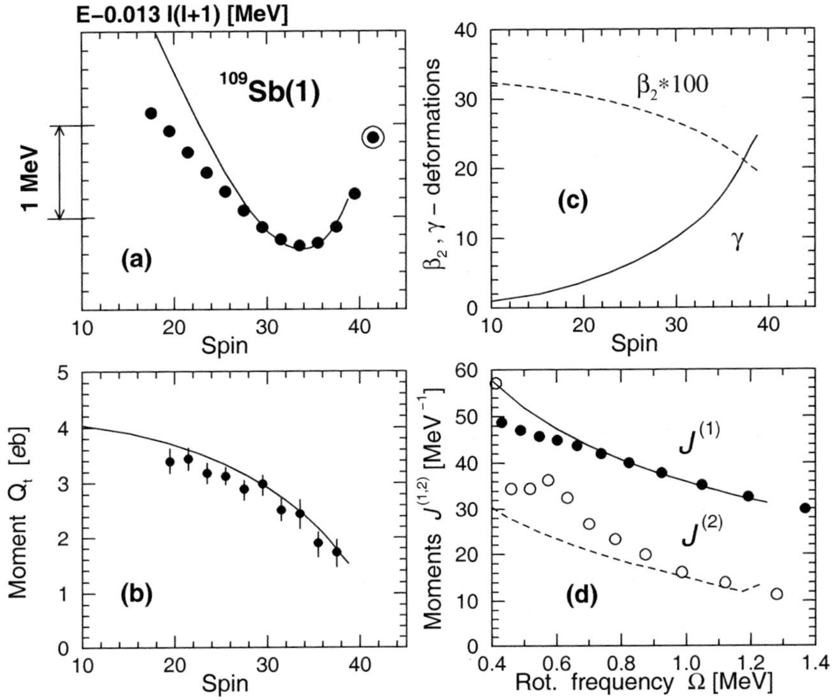

Fig. 21. Smooth terminating band 1 in ^{109}Sb. Circles indicate experimental data, while the results of CRMF calculations are shown by lines. Panel (a) shows the excitation energies relative to a rigid rotor reference $E_{RLD}(I) = 0.013I(I+1)$ MeV. The terminating state is indicated by a large open circle. The results of the calculations are normalized to experiment at the position of the minima in the $(E - E_{RLD})$ curve. Panel (b) compares measured and calculated transition quadrupole moments Q_t. Calculated β_2 and γ deformations are shown in panel (c). Panel (d) shows the calculated and experimental kinematic and dynamic moments of inertia. From Ref. 1.

states in the $A \sim 44$ region has been studied.[22] Figure 21 presents the only case of the study of smooth band termination in the CRMF framework; this is smooth terminating band 1 in ^{109}Sb. One can see that experimental observables shown in Fig. 21 are well described in the CRMF calculations at $\Omega_x \geq 0.8$ MeV (above $I \geq 25\ \hbar$) where pairing is expected to play a minor role. The discrepancies between experiment and theory seen at lower spins are probably due to neglect of the pairing correlations in the CRMF calculations. One should note that without special techniques (as developed in cranked Nilsson–Strutinsky approach[4]) it is impossible to trace this configuration in the CRMF calculations up to its terminating state.

It is necessary to recognize that not all rotational bands will terminate in non-collective terminating state at the maximum spin I_{max}.[4] So far the examples of non-termination of rotational bands at the maximum spin have been experimentally observed and studied in the CRMF framework only in ^{74}Kr and ^{75}Rb.[98,99]

8.4. *Other results*

Other phenomena in rotating nuclei or the features of rotational bands have been studied in the CRMF/CRHB frameworks during last decade. They will only be briefly mentioned here.

The additivity principle for quadrupole moments and relative alignments has been studied on the example of highly-deformed and superdeformed bands in the $A \sim 130$ mass region in Ref. 100. This principle of the extreme shell model stipulates that an average value of a one-body operator be equal to the sum of the core contribution and effective contributions of valence (particle or hole) nucleons. It is only valid in an unpaired regime typical of high angular momenta since the pairing smears out the individuality of each single-particle orbital. The additivity principle for angular momentum alignments does not work as precisely as it does for quadrupole moments.

Triaxial superdeformed (TSD) rotational bands at ultra-high spin in the Er region have been studied on the example of ^{158}Er in Ref. 89. Good description of the deformation and rotational properties of the TSD bands observed in this nucleus has been achieved in the CRMF calculations. Based on the results of covariant and non-relativistic DFT calculations it was suggested that the TSD band 1 in this nucleus is observed up to spins in excess of $70\hbar$[89] which is the highest spin reported so far. The enhanced deformation and SD bands in the Hf isotopes have been investigated in Ref. 101. Contrary to previous claims of triaxiality of the SD bands in ^{175}Hf, the bands in this nucleus are interpreted in the CRMF framework as near-prolate. However, it was concluded that some SD bands in less neutron-rich Hf isotopes may be triaxial.

Theoretical uncertainties in the description of the physical observables of rotational bands (deformations, moments of inertia and their evolution with spins, band crossing frequencies and alignment gains at band crossings etc) and their dependence on CEDF have been analyzed in Ref. 102.

9. Conclusions

The systematic studies of different types of rotational bands (normal-deformed, smooth terminating, superdeformed and hyperdeformed) have been performed in one-dimensional cranking approximation of the CDFT in different regions of the nuclear chart from ground states to the extremes of deformation and spin. Contrary to the studies of rotating nuclei in non-relativistic DFT models, the CDFT studies have been performed systematically covering large amounts of experimental data in the region of study. The CRMF and CRHB+LN calculations succesfully describe the experimental situation and in many cases they outperform non-relativistic models. The rotating nuclei offer a unique laboratory for testing of the channels of the CDFT which are not accessible by other physical observables. This is because their properties sensitively depend on nuclear magnetism (time-odd mean fields) and underlying single-particle structure. The part of the success of the CDFT in the description of rotating nuclei is definitely attributable to the fact that time-odd mean fields are uniquely defined via Lorentz covariance.

Acknowledgment

This material is based upon work supported by the U.S. Department of Energy, Office of Science, Office of Nuclear Physics under Award Number DE-SC0013037.

References

1. D. Vretenar, A. V. Afanasjev, G. A. Lalazissis, and P. Ring, Relativistic Hartree–Bogoliubov theory: Static and dynamic aspects of exotic nuclear structure, *Phys. Rep.* **409**, 101 (2005).
2. M. Bender, P.-H. Heenen, and P.-G. Reinhard, Self-consistent mean-field models for nuclear structure, *Rev. Mod. Phys.* **75**, 121 (2003).
3. C. Baktash, B. Haas, and W. Nazarewicz, Identical bands in deformed and superdeformed nuclei, *Annu. Rev. Nucl. Part. Sci.* **45**, 485 (1995).
4. A. V. Afanasjev, D. B. Fossan, G. J. Lane, and I. Ragnarsson, Termination of rotational bands: Disappearance of quantum many-body collectivity, *Phys. Rep.* **322**, 1 (1999).
5. S. Frauendorf, Spontaneous symmetry breaking in rotating nuclei, *Rev. Mod. Phys.* **73**, 463–514 (2001). doi: 10.1103/RevModPhys.73.463. URL `http://link.aps.org/doi/10.1103/RevModPhys.73.463`.
6. J. Meng, J. Peng, S.-Q. Zhang, and P.-W. Zhao, Progress on tilted axis cranking covariant density functional theory for nuclear magnetic and

antimagnetic rotation, *Front. Phys.* **8**, 55–79 (2013). ISSN 2095-0462. doi: 10.1007/s11467-013-0287-y.

7. J. König, diploma thesis, Technical University of Munich (unpublished) (1996).
8. W. Koepf and P. Ring, A relativistic description of rotating nuclei: the yrast line of ^{20}Ne, *Nucl. Phys. A.* **493**, 61 (1989).
9. H. Madokoro and M. Matsuzaki, General relativistic mean field theory for rotating nuclei, *Phys. Rev. C.* **56**, R2934–R2937 (1997). URL `http://link.aps.org/doi/10.1103/PhysRevC.56.R2934`.
10. A. V. Afanasjev, J. König, and P. Ring, Cranked relativistic Hartree–Bogoliubov theory: Superdeformed bands in the $A \sim 190$ region, *Phys. Rev. C.* **60**, 051303 (1999). URL `http://link.aps.org/doi/10.1103/PhysRevC.60.051303`.
11. A. V. Afanasjev, P. Ring, and J. König, Cranked relativistic Hartree–Bogoliubov theory: formalism and application to the superdeformed bands in the $A \sim 190$ region., *Nucl. Phys.* **A676**, 196 (2000).
12. D. R. Inglis, Nuclear moments of inertia due to nucleon motion in a rotating well, *Phys. Rev.* **103**, 1786–1795 (1956). URL `http://link.aps.org/doi/10.1103/PhysRev.103.1786`.
13. B. D. Serot and J. D. Walecka, The relativistic nuclear many body problem, *Adv. Nucl. Phys.* **16**, 1 (1986).
14. P.-G. Reinhard, M. Rufa, J. Maruhn, W. Greiner, and J. Friedrich, Nuclear ground state properties in a relativistic meson field model, *Z. Phys. A.* **323**, 13 (1986).
15. A. R. Edmonds, *Angular Momentum in Quantum Mechanics.* Princeton University Press (1957).
16. A. V. Afanasjev and H. Abusara, Time-odd mean fields in covariant density functional theory: nonrotating systems, *Phys. Rev. C.* **81**, 014309 (2010).
17. A. V. Afanasjev and P. Ring, Time-odd mean fields in the rotating frame: Microscopic nature of nuclear magnetism, *Phys. Rev. C.* **62**, 031302(R) (2000).
18. J. König and P. Ring, Identical bands in superdeformed nuclei: A relativistic description, *Phys. Rev. Lett.* **71**, 3079–3082 (1993) URL `http://link.aps.org/doi/10.1103/PhysRevLett.71.3079`.
19. A. V. Afanasjev and H. Abusara, Time-odd mean fields in covariant density functional theory: Rotating systems, *Phys. Rev. C.* **82**, 034329 (2010).
20. J. Dobaczewski and J. Dudek, Time-odd components in the mean field of rotating superdeformed nuclei, *Phys. Rev. C.* **52**, 1827 (1995).
21. M. Yamagami and K. Matsuyanagi, High-spin yrast structure of ^{32}S suggested by symmetry-unrestricted, cranked Hartree–Fock calculations, *Nucl. Phys. A.* **672**, 123–140 (2000). ISSN 0375-9474. URL `http://www.sciencedirect.com/science/article/pii/S0375947499003917`.
22. A. V. Afanasjev, Band terminations in density functional theory, *Phys. Rev. C.* **78**, 054303 (2008).
23. U. Hofmann and P. Ring, A new method to calculate magnetic moments of relativistic mean field theories, *Phys. Lett. B.* **214**, 307 (1988).

24. Y. M. Engel, D. M. Brink, K. Geoke, S. J. Krieger, and D. Vauterin, Time-dependent Hartree–Fock theory with Skyrme's interaction, *Nucl. Phys. A.* **249**, 215 (1975).
25. V. O. Nesterenko, W. Kleinig, J. Kvasil, P. Vesely, and P.-G. Reinhard, TDDFT with Skyrme forces: effect of time-odd densities on electric giant resonances., *Int. J. Mod. Phys. E.* **17**, 89 (2008).
26. N. Hinohara, T. Nakatsukasa, M. Matsuo, , and K. Matsuyanagi, Effects of time-odd components in mean field on large amplitude collective dynamics, *Prog. Theor. Phys.* **115**, 567 (2006).
27. A. S. Umar and V. E. Oberacker, Three-dimensional unrestricted time-dependent Hartree–Fock fusion calculations using the full Skyrme interaction, *Phys. Rev. C.* **73**, 054607 (May, 2006). URL `http://link.aps.org/doi/10.1103/PhysRevC.73.054607`.
28. M. Bender, J. Dobaczewski, J. Engel, and W. Nazarewicz, Gamow–Teller strength and the spin-isospin coupling constants of the Skyrme energy functional, *Phys. Rev. C.* **65**, 054322 (2002).
29. W. Satula, in *Nuclear Structure 98*, edited by C. Baktash, AIP conf. proc. no. 481 (AIP, new york, 1999). p. 114 (1999).
30. T. Duguet, P. Bonche, P.-H. Heenen, and J. Meyer, Pairing correlations. i. description of odd nuclei in mean-field theories, *Phys. Rev. C.* **65**, 014310 (Dec, 2001). URL `http://link.aps.org/doi/10.1103/PhysRevC.65.014310`.
31. A. V. Afanasjev, T. L. Khoo, S. Frauendorf, G. A. Lalazissis, and I. Ahmad, Cranked relativistic Hartree–Bogoliubov theory: Probing the gateway to superheavy nuclei, *Phys. Rev. C.* **67**, 024309 (2003).
32. K. Rutz, M. Bender, P.-G. Reinhard, and J. A. Maruhn, *Phys. Lett. B.* **468**, 1 (1999).
33. J. F. Berger, M. Girod, and D. Gogny, Microscopic analysis of collective dynamics in low energy fission, *Nucl. Phys.* **A428**, 23c (1984).
34. L. J. Wang, B. Y. Sun, J. M. Dong, and W. H. Long, Odd-even staggering of the nuclear binding energy described by covariant density functional theory with calculations for spherical nuclei, *Phys. Rev. C.* **87**, 054331 (2013).
35. S. E. Agbemava, A. V. Afanasjev, D. Ray, and P. Ring, Global performance of covariant energy density functionals: Ground state observables of even-even nuclei and the estimate of theoretical uncertainties, *Phys. Rev. C.* **89**, 054320 (2014).
36. P. Ring and P. Schuck, *The nuclear many-body problem.* Springer-Verlag, New York (1980).
37. H. J. Lipkin, Collective motion in many-particle systems: Part 1. the violation of conservation laws, *Ann. Phys.* **9**, 272 (1960).
38. Y. Nogami, Improved superconductivity approximation for the pairing interaction in nuclei, *Phys. Rev.* **134**, 313 (1964).
39. Y. Nogami and I. Zucker, A note on the pairing interaction in nuclei, *Nucl. Phys.* **60**, 203 (1964).
40. H. C. Pradhan, Y. Nogami, and J. Law, Study of approximations in the nuclear pairing-force problem, *Nucl. Phys. A.* **201**, 357 (1973).

41. A. V. Afanasjev, J. König, P. Ring, L. M. Robledo, and J. L. Egido, Moments of inertia of nuclei in the rare earth region: A relativistic versus nonrelativistic investigation, *Phys. Rev. C.* **62**, 054306 (2000).
42. A. V. Afanasjev and O. Abdurazakov, Pairing and rotational properties of actinides and superheavy nuclei in covariant density functional theory, *Phys. Rev. C.* **88**, 014320 (2013).
43. A. V. Afanasjev, J. König, and P. Ring, Superdeformed rotational bands in the $A \sim 140-150$ mass region: A cranked relativistic mean field description, *Nucl. Phys. A.* **608**, 107 (1996).
44. W. Koepf and P. Ring, A relativistic theory of superdeformations in rapidly rotating nuclei, *Nucl. Phys. A.* **511**(2), 279–300 (1990). ISSN 0375-9474. URL http://www.sciencedirect.com/science/article/pii/037594749090160N.
45. A. V. Afanasjev, G. Lalazissis, and P. Ring, Relativistic mean field theory in rotating frame: Single-particle properties at superdeformation, *Nucl. Phys. A.* **634** (1998).
46. H. Molique, J. Dobaczewski, and J. Dudek, Superdeformed bands in ^{32}S and neighboring nuclei predicted within the Hartree–Fock method, *Phys. Rev. C.* **61**, 044304 (2000). URL http://link.aps.org/doi/10.1103/PhysRevC.61.044304.
47. A. V. Afanasjev, P. Ring, and I. Ragnarsson, Mean field studies of high-spin properties in the $A \sim 30$ and 60 regions of superdeformation, *Proc. Int. Workshop PINGST2000 "Selected topics on $N = Z$ nuclei", 2000, Lund, Sweden, Eds. D. Rudolph and M. Hellström.* p. 183 (2000).
48. K.-K. Kan and J. J. Griffin, Single-particle Schrödinger fluid. I. formulation, *Phys. Rev. C.* **15**, 1126 (1977). URL http://link.aps.org/doi/10.1103/PhysRevC.15.1126.
49. M. Durand, P. Schuck, and J. Kunz, Semiclassical description of currents in normal and superfluid rotating nuclei, *Nucl. Phys. A.* **439**, 263 (1985). ISSN 0375-9474. URL http://www.sciencedirect.com/science/article/pii/0375947485904312.
50. I. Mikhailov, P. Quentin, and D. Samsoen, Nuclear intrinsic vorticity and its coupling to global rotations, *Nucl. Phys. A.* **627**(2), 259 (1997). ISSN 0375-9474. URL http://www.sciencedirect.com/science/article/pii/S0375947497004016.
51. H. Laftchiev, D. Samsœn, P. Quentin, and I. N. Mikhailov, Equivalence of pairing correlations and intrinsic vortical currents in rotating nuclei, *Phys. Rev. C.* **67**, 014307 (2003). URL http://link.aps.org/doi/10.1103/PhysRevC.67.014307.
52. M. Radomski, Nuclear rotational current and velocity fields, the cranking model, and transverse electroexcitation of the first excited state of ^{12}C, *Phys. Rev. C.* **14**, 1704 (1976). URL http://link.aps.org/doi/10.1103/PhysRevC.14.1704.
53. P. Gulshani and D. J. Rowe, Quantum mechanics in rotating frames. I. The impossibility of rigid flow, *Can. J. Phys.* **56**, 468 (1978).
54. P. Gulshani and D. J. Rowe, Quantum mechanics in rotating frames. II. The

lattice structure of current circulations for a rotating single-particle fluid, *Can. J. Phys.* **56**, 480 (1978).
55. J. Kunz and U. Mosel, Flow patterns for collective rotations in heavy nuclei, *Nucl. Phys. A.* **323**, 271 (1979). ISSN 0375-9474. URL http://www.sciencedirect.com/science/article/pii/0375947479901106.
56. J. Fleckner, J. Kunz, U. Mosel, and E. Wuest, Current distributions for rotating nuclei, *Nucl. Phys. A.* **339**(2), 227 (1980). ISSN 0375-9474. URL http://www.sciencedirect.com/science/article/pii/0375947480900895.
57. A. Bohr and B. R. Mottelson, *NUCLEAR STRUCTURE Volume II: Nuclear Deformation.* W. A. Benjamin, Inc. (1975).
58. A. V. Afanasjev, Microscopic description of rotation: From ground states to the extremes of ultra-high spin, *Phys. Scr.* **89**, 054001 (2014).
59. J. Egido and L. Robledo, High spin properties of the Er and Yb isotopes with the Gogny force, *Nucl. Phys. A.* **570**, 69–76 (1994). ISSN 0375-9474. URL http://www.sciencedirect.com/science/article/pii/0375947494902704.
60. M. Anguiano, J. Egido, and L. Robledo, Coulomb exchange and pairing contributions in nuclear Hartree–Fock–Bogoliubov calculations with the Gogny force, *Nucl. Phys. A.* **683**, 227–254 (2001). ISSN 0375-9474. URL http://www.sciencedirect.com/science/article/pii/S0375947400004450.
61. M. Anguiano, J. Egido, and L. Robledo, Particle number projection with effective forces, *Nucl. Phys. A.* **696**, 467–493 (2001). ISSN 0375-9474. URL http://www.sciencedirect.com/science/article/pii/S0375947401012192.
62. T. Duguet, P. Bonche, and P.-H. Heenen, Rotational properties of ^{252,253,254}No: influence of pairing correlations, *Nucl. Phys. A.* **679**, 427–440 (2001). ISSN 0375-9474. URL http://www.sciencedirect.com/science/article/pii/S0375947400003705.
63. M. Bender, P. Bonche, T. Duguet, and P.-H. Heenen, Skyrme mean-field study of rotational bands in transfermium isotopes, *Nucl. Phys. A.* **723**, 354–364 (2003). ISSN 0375-9474. URL http://www.sciencedirect.com/science/article/pii/S0375947403010819.
64. G. A. Lalazissis, S. Karatzikos, R. Fossion, D. P. Arteaga, A. V. Afanasjev, and P. Ring, The effective force NL3 revisited, *Phys. Lett.* **B671**, 36 (2009).
65. S. Hota, *PhD thesis, University of Massachussets, Lowell* (2012).
66. Z.-H. Zhang, X.-T. He, J.-Y. Zeng, E.-G. Zhao, and S.-G. Zhou, Systematic investigation of the rotational bands in nuclei with $Z \sim 100$ using a particle-number conserving method based on a cranked shell model, *Phys. Rev. C.* **85**, 014324 (2012).
67. S. K. Tandel, P. Chowdhury, S. Lakshmi, U. S. Tandel, I. Ahmad, M. P. Carpenter, S. Gros, R. V. F. Janssens, T. L. Khoo, F. G. Kondev, J. P. Greene, D. J. Hartley, T. Lauritsen, C. J. Lister, D. Peterson, A. Robinson, D. Seweryniak, and S. Zhu, Rotational bands in odd-A Cm and Cf isotopes: Exploring the highest neutron orbitals, *Phys. Rev. C.* **82** .

68. A. V. Afanasjev and S. Frauendorf, Description of rotating $N = Z$ nuclei in terms of isovector pairing, *Phys. Rev. C.* **71**, 064318 (2005).
69. X. Wang, R. V. F. Janssens, M. P. Carpenter, S. Zhu, I. Wiedenhöver, U. Garg, S. Frauendorf, T. Nakatsukasa, I. Ahmad, A. Bernstein, E. Diffenderfer, S. J. Freeman, J. P. Greene, T. L. Khoo, F. G. Kondev, A. Larabee, T. Lauritsen, C. J. Lister, B. Meredith, D. Seweryniak, C. Teal, and P. Wilson, Structure of ^{240}Pu: Evidence for octupole phonon condensation?, *Phys. Rev. Lett.* **102**, 122501 (2009). URL `http://link.aps.org/doi/10.1103/PhysRevLett.102.122501`.
70. K. Abu Saleem, R. V. F. Janssens, M. P. Carpenter, F. G. Kondev, I. Wiedenhöver, I. Ahmad, J. Caggiano, P. Chowdhury, J. A. Cizewski, D. Cline, M. Devlin, N. Fotiades, J. P. Greene, G. Hackman, A. Heinz, T. L. Khoo, T. Lauritsen, C. J. Lister, A. O. Macchiavelli, E. H. Seabury, D. Seweryniak, A. Sonzogni, and C. Y. Wu, Alignments in the odd-proton actinides ^{237}Np and ^{241}Am, *Phys. Rev. C.* **70**, 024310 (2004). URL `http://link.aps.org/doi/10.1103/PhysRevC.70.024310`.
71. P. Ring, R. Beck, and H. J. Mang, On the application of the Hartree–Fock–Bogolyubov-equations to a microscopic theory of nuclear rotations, *Z. Phys.* **231**, 10 (1970).
72. J. L. Egido, H. J. Mang, and P. Ring, Selfconsistent treatment of excited rotational bands in deformed nuclei, *Nucl. Phys. A.* **334**, 1 (1980).
73. A. V. Afanasjev and S. Shawaqfeh, Deformed one-quasiparticle states in covariant density functional theory, *Phys. Lett. B.* **706**, 177 (2011).
74. R.-D. Herzberg, S. Moon, S. Eeckhaudt, P. T. Greenlees, P. A. Butler, T. Page, A. V. Afanasjev, N. Amzal, J. E. Bastin, F. Becker, M. Bender, B. Bruyneel, J. F. C. Cocks, I. G. Darby, O. Dorvaux, K. Eskola, J. Gerl, T. Grahn, C. Gray-Jones, N. J. Hammond, K. Hauschild, P.-H. Heenen, K. Helariutta, A. Herzberg, F. Hessberger, M. Houry, A. Hürstel, R. D. Humphreys, G. D. Jones, P. M. Jones, R. Julin, S. Juutinen, H. Kankanpää, H. Kettunen, T. L. Khoo, W. Korten, P. Kuusiniemi, Y. LeCoz, M. Leino, A.-P. Leppanen, C. J. Lister, R. Lucas, M. Muikku, P. Nieminen, M. Nyman, R. D. Page, T. Page, J. Pakarinen, A. Pritchard, P. Rahkila, P. Reiter, M. Sandzelius, J. Saren, C. Schlegel, C. Scholey, C. Theisen, W. H. Trzaska, J. Uusitalo, A. Wiens, and H. J. Wollersheim, Structure of rotational bands in ^{253}No, *Eur. Phys. J.* **A42**, 333–337 (2009).
75. H. B. Jeppesen, R. M. Clark, K. E. Gregorich, A. V. Afanasjev, M. N. Ali, J. M. Allmond, C. W. Beausang, M. Cromaz, M. A. Deleplanque, I. Dragojevic, J. Dvorak, P. A. Ellison, P. Fallon, M. A. Garcia, J. M. Gates, S. Gros, I. Y. Lee, A. O. Macchiavelli, S. L. Nelson, H. Nitsche, L. Stavsetra, F. S. Stephens, and M. Wiedeking, High-K multi-quasiparticle states and rotational bands in $_{103}$Lr^{255}, *Phys. Rev. C.* **80**, 034324 (2009).
76. C. D. O'Leary, C. E. Svensson, S. G. Frauendorf, A. V. Afanasjev, D. E. Appelbe, R. A. E. Austin, G. C. Ball, J. A. Cameron, R. M. Clark, M. Cromaz, P. Fallon, D. F. Hodgson, N. S. Kelsall, A. O. Macchiavelli, I. Ragnarsson, D. Sarantites, J. C. Waddington, and R. Wadsworth, Evi-

dence for isovector neutron-proton pairing from high-spin states in $N = Z$ ^{74}Rb, *Phys. Rev. C.* **67**, 021301(R) (2003).
77. R. D. Herzberg and P. T. Greenlees, In-beam and decay spectroscopy of transfermium nuclei, *Prog. Part. Nucl. Phys.* **61**, 674 (2008).
78. B. M. Nyakó, F. Papp, J. Gal, J. Molnár, J. Timár, A. Algora, Z. Dombrádi, G. Kalinka, L. Zolnai, K. Juhász, A. K. Singh, H. Hüebel, A. Al-Khatib, P. Bringel, A. Büerger, A. Neusser, G. Schöenwasser, B. Herskind, G. B. Hagemann, C. R. Hansen, G. Sletten, J. N. Scheurer, F. Hannachi, M. Kmiecik, A. Maj, J. Styczeń, K. Zuber, K. Hauschild, A. Korichi, A. Lopez-Martens, J. Roccaz, S. Siem, P. Bednarczyk, T. Byrski, D. Curien, O. Dorvaux, G. Ducheńe, B. Gall, F. Khalfallah, I. Piqueras, J. Robin, S. B. Patel, A. O. Evans, G. Rainovski, A. Airoldi, G. Benzoni, A. Bracco, F. Camera, B. Million, P. Mason, A. Paleni, R. Sacchi, O. Wieland, G. L. Rana, R. Moro, C. M. Petrache, D. Petrache, G. D. Angelis, P. Fallon, I.-Y. Lee, J. C. Lisle, B. Cederwall, K. Lagergren, R. M. Lieder, E. Podsvirova, W. Gast, H. Jäeger, N. Redon, and A. Göergen, Search for hyperdeformation in light Xe nuclei, *Acta Phys. Pol.* **B36**, 1033 (2005).
79. H. Hübel, Search for hyperdeformation, *Acta Phys. Pol.* **B36**, 1015 (2005).
80. B. Herskind, G. B. Hagemann, G. Sletten, T. Døssing, C. R. Hansen, N. Schunck, S. Ødegård, H. Hübel, P. Bringel, A. Bürger, A. Neusser, A. K. Singh, A. Al-Khatib, S. B. Patel, A. Bracco, S. Leoni, F. Camera, G. Benzoni, P. Mason, A. Paleni, B. Million, O. Wieland, P. Bednarczyk, F. Azaiez, T. Byrski, D. Curien, O. Dakov, G. Duchene, F. Khalfallah, B. Gall, L. Piqeras, J. Robin, J. Dudek, N. Rowley, B. M. Nyakó, A. Algora, Z. Dombradi, J. Gal, G. K. andD. Sohler, J. Molnár, J. Timár, L. Zolnai, K. Juhász, N. Redon, F. Hannachi, J. N. Scheurer, J. N. Wilson, A. Lopez-Martens, A. Korichi, K. Hauschild, J. Roccaz, S. Siem, P. Fallon, I. Y. Lee, A. Görgen, A. Maj, M. Kmiecik, M. Brekiesz, J. Styczen, K. Zuber, J. C. Lisle, B. Cederwall, K. Lagergren, A. O. Evans, G. Rainovski, G. D. Angelis, G. L. Rana, R. Moro, W. Gast, R. M. Lieder, E. Podsvirova, H. Jäger, C. M. Petrache, and D. Petrache, Charged particle feeding of hyperdeformed nuclei in the $A = 118$–126 region, *Phys. Scr.* **T125**, 108 (2006).
81. G. Viesti, M. Lunardon, D. Bazzacco, R. Burch, D. Fabris, S. Lunardi, N. H. Medina, G. Nebbia, C. Rossi-Alvarez, G. de Angelis, M. De Poli, E. Fioretto, G. Prete, J. Rico, P. Spolaore, G. Vedovato, A. Brondi, G. La Rana, R. Moro, and E. Vardaci, Population of hyperdeformed structures in ^{152}Dy from proton-gamma coincidence experiments, *Phys. Rev. C.* **51**, 2385–2393 (1995). URL `http://link.aps.org/doi/10.1103/PhysRevC.51.2385`.
82. A. V. Afanasjev and H. Abusara, Hyperdeformation in the cranked relativistic mean field theory: The $Z = 40$–58 region of the nuclear chart, *Phys. Rev. C.* **78**, 014315 (2008). URL `http://link.aps.org/doi/10.1103/PhysRevC.78.014315`.
83. Y. R. Shimizu, J. D. Garrett, R. A. Broglia, M. Gallardo, and E. Vigezzi, Pairing fluctuations in rapidly rotating nuclei, *Rev. Mod. Phys.* **61**, 131 (1989).

84. T. Bengtsson, I. Ragnarsson, and S. Åberg, The role of high-N orbits in superdeformed states, *Phys. Lett. B.* **208**, 39 (1988).
85. A. V. Afanasjev, I. Ragnarsson, and P. Ring, Comparative study of superdeformed and highly deformed bands in the $A \sim 60$ mass region, *Phys. Rev. C.* **59**, 3166 (1999).
86. A. V. Afanasjev and S. Frauendorf, Superdeformation and hyperdeformation in the ^{108}Cd nucleus, *Phys. Rev. C.* **72**, 031301 (2005). URL `http://link.aps.org/doi/10.1103/PhysRevC.72.031301`.
87. H. Abusara and A. V. Afanasjev, Hyperdeformation in the Cd isotopes: A microscopic analysis, *Phys. Rev. C.* **79**, 024317 (2009). URL `http://link.aps.org/doi/10.1103/PhysRevC.79.024317`.
88. J. Dudek, K. Pomorski, N. Schunck, and N. Dubray, Hyperdeformed and megadeformed nuclei: Lessons from the slow progress and emerging new strategies, *Eur. Phys. J.* **A20**, 15 (2004).
89. A. V. Afanasjev, Y. Shi, and W. Nazarewicz, Description of ^{158}Er at ultrahigh spin in nuclear density functional theory, *Phys. Rev. C.* **86**, 031304(R) (2012).
90. N. Schunck, J. Dudek, and B. Herskind, Nuclear hyperdeformation and the Jacobi shape transition, *Phys. Rev. C.* **75**, 054304 (2007). URL `http://link.aps.org/doi/10.1103/PhysRevC.75.054304`.
91. I. Ragnarsson, Orbital and spin assignment of superdeformed bands in the Dy/Gd region — identical bands, *Nucl. Phys. A.* **557**, c167 (1993).
92. W. Satuła, J. Dobaczewski, J. Dudek, and W. Nazarewicz, Additivity of quadrupole moments in superdeformed bands: Single-particle motion at extreme conditions, *Phys. Rev. Lett.* **77**, 5182–5185 (1996). URL `http://link.aps.org/doi/10.1103/PhysRevLett.77.5182`.
93. A. V. Afanasjev, Isoscalar and isovector neutron-proton pairing, *chapter 11 in the book "50 Years of Nuclear BCS", (World Scientific Publishing Co, Singapore, 2013), p. 138, see also nuclear theory arkhive arXiv:1205.2134.* (2013).
94. T. Lauritsen, M. P. Carpenter, T. Døssing, P. Fallon, B. Herskind, R. V. F. Janssens, D. G. Jenkins, T. L. Khoo, F. G. Kondev, A. Lopez-Martens, A. O. Macchiavelli, D. Ward, K. S. Abu Saleem, I. Ahmad, R. Clark, M. Cromaz, J. P. Greene, F. Hannachi, A. M. Heinz, A. Korichi, G. Lane, C. J. Lister, P. Reiter, D. Seweryniak, S. Siem, R. C. Vondrasek, and I. Wiedenhöver, Direct decay from the superdeformed band to the yrast line in $^{152}_{66}Dy_{86}$, *Phys. Rev. Lett.* **88**, 042501 (2002). URL `http://link.aps.org/doi/10.1103/PhysRevLett.88.042501`.
95. Q. A. Ijaz, W. C. Ma, H. Abusara, A. V. Afanasjev, Y. B. Xu, R. B. Yadav, Y. C. Zhang, M. P. Carpenter, R. V. F. Janssens, T. L. Khoo, T. Lauritsen, and D. T. Nisius, Excited superdeformed bands in ^{154}Dy and cranked relativistic mean field interpretation, *Phys. Rev. C.* **80**, 034322 (2009). URL `http://link.aps.org/doi/10.1103/PhysRevC.80.034322`.
96. S. Frauendorf and A. Macchiavelli, Proton-neutron pairing, *Prog. Part. Nucl. Phys.* **78**, 24 (2014).

97. S. G. Frauendorf and J. A. Sheikh, Cranked shell model and isospin symmetry near $N = Z$, *Nucl. Phys. A.* **645**, 509–535 (1999).
98. J. J. Valiente-Dobón, T. Steinhardt, C. E. Svensson, A. V. Afanasjev, I. Ragnarsson, C. Andreoiu, R. A. E. Austin, M. P. Carpenter, D. Dashdorj, G. de Angelis, F. Dönau, J. Eberth, E. Farnea, S. J. Freeman, A. Gadea, P. E. Garrett, A. Görgen, G. F. Grinyer, B. Hyland, D. Jenkins, F. Johnston-Theasby, P. Joshi, A. Jungclaus, K. P. Lieb, A. O. Macchiavelli, E. F. Moore, G. Mukherjee, D. R. Napoli, A. A. Phillips, C. Plettner, W. Reviol, D. Sarantites, H. Schnare, M. A. Schumaker, R. Schwengner, D. Seweryniak, M. B. Smith, I. Stefanescu, O. Thelen, and R. Wadsworth, Evidence for non-termination of rotational bands in ^{74}Kr, *Phys. Rev. Lett.* **95**, 232501 (Nov, 2005). URL `http://link.aps.org/doi/10.1103/PhysRevLett.95.232501`.
99. P. J. Davies, A. V. Afanasjev, R. Wadsworth, C. Andreoiu, R. A. E. Austin, M. P. Carpenter, D. Dashdorj, P. Finlay, S. J. Freeman, P. E. Garrett, A. Görgen, J. Greene, G. F. Grinyer, B. Hyland, D. G. Jenkins, F. L. Johnston-Theasby, P. Joshi, A. O. Macchiavelli, F. Moore, G. Mukherjee, A. A. Phillips, W. Reviol, D. Sarantites, M. A. Schumaker, D. Seweryniak, M. B. Smith, C. E. Svensson, J. J. Valiente-Dobon, and D. Ward, Evidence of nontermination of collective rotation near the maximum angular momentum in ^{75}Rb, *Phys. Rev. C.* **82**, 061303 (2010). URL `http://link.aps.org/doi/10.1103/PhysRevC.82.061303`.
100. M. Matev, A. V. Afanasjev, J. Dobaczewski, G. A. Lalazissis, and W. Nazarewicz, Additivity of effective quadrupole moments and angular momentum alignments in $A \sim 130$ nuclei, *Phys. Rev. C.* **76**, 034304 (2007).
101. Y. C. Zhang, W. C. Ma, A. V. Afanasjev, G. B. Hagemann, J. Begnaud, M. P. Carpenter, P. Chowdhury, D. M. Cullen, M. K. Djongolov, D. J. Hartley, R. V. F. Janssens, T. L. Khoo, F. G. Kondev, T. Lauritsen, E. F. Moore, E. Ngijoi-Yogo, S. Ødegård, L. L. Riedinger, S. V. Rigby, D. G. Roux, D. T. Scholes, R. B. Yadav, J.-Y. Zhang, and S. Zhu, Nuclear shapes of highly deformed bands in ^{171,172}Hf and neighboring Hf isotopes, *Phys. Rev. C.* **76**, 064321 (2007). URL `http://link.aps.org/doi/10.1103/PhysRevC.76.064321`.
102. A. V. Afanasjev, Addressing spectroscopic quality of covariant density functional theory, *J. Phys. G.* **42**, 034002 (2015).

Chapter 9

Novel rotational excitations

Jie Meng
School of Physics, Peking University, Beijing 100871, P.R.China
mengj@pku.edu.cn

Shuangquan Zhang
School of Physics, Peking University, Beijing 100871, P.R.China
sqzhang@pku.edu.cn

Pengwei Zhao
School of Physics, Peking University, Beijing 100871, P.R.China
pwzhao@pku.edu.cn

Novel rotational excitations like magnetic and antimagnetic rotations observed in the near spherical nuclei and chiral rotation in the triaxial nuclei have attracted a lot of attention. Unlike the conventional rotation, here the rotational axis does not necessarily coincide with any principle axis of nuclear density distribution, and thus a tilted axis cranking is mandatory to describe them self-consistently in the framework of covariant density functional theory. In this chapter, we introduce the related formalism with the point-coupling covariant density functional and discuss its applications for the magnetic and antimagnetic rotation phenomena. The configuration-fixed covariant density functional theory developed to search for nuclear chiral configurations and favorable triaxial deformation parameters is included as well, which leads to the discoveries of the multiple chiral doublets (MχD) in ^{133}Ce and ^{103}Rh.

Contents

1. Introduction

In Chapter 8, the rotational excitations of well deformed or superdeformed nuclei are discussed in terms of one-dimensional cranking approximation based on the framework of CDFT. In this chapter, we will focus on the novel rotational excitations observed in the near spherical nuclei and the triaxial nuclei.

It is well known that for a spherical system no rotational excitation is allowed by the principle of quantum mechanics i.e., the statistical explanation of wave function. However in the early 1990s, the rotational-like sequences of strongly enhanced magnetic dipole (M1) transitions ($\Delta I = 1$) were systematically observed in several light-mass Pb isotopes,[1–5] which are known to be spherical or near-spherical. A problem arises whether these observations violate the principle of quantum mechanics or not.

These rotational-like sequences observed in near-spherical nuclei reflect the complexity of nuclear system. They are based on a configuration of relatively few high-j proton particles and high-j neutron holes close to a nearly spherical core.[6] Configurations with maximally aligned single-particle angular momentum have an oblate density distribution for particles and a prolate density distribution for holes with respect to the rotating axis. At the band head, the neutron holes have angular momenta $\mathbf{J}_n$ aligned in the direction of the z axis producing in this way a prolate density distribution. The proton particles have an oblate density distribution. To maximize the overlap of the orbits in this configuration, at the band head, the angular momenta $\boldsymbol{J}_p$ of the proton particles point in a direction almost orthogonal to the neutrons, i.e., in the direction of the x axis. This coupling results in a total angular momentum $\mathbf{J} = \mathbf{J}_p + \mathbf{J}_n$ tilted from the principal axes and pointing in the xz plane. Within the cranking model, the rotational axis is therefore tilted (TAC) pointing in the xz plane.

For a triaxial nucleus, in which the three principal axes of spatial distribution are different in length, no rotational excitations around any of three principal axes are excluded by the principle of quantum mechanics, i.e. the statistical explanation of wave function. Although the rotation along the principal axis with the largest moment of inertia (MOI) is energy favored, the angular momentum contributions along the other axes induced by the core excitations and/or valence nucleon(s) could also play an important role. Thus, it is possible for triaxial nuclei to produce a collective rotation around an axis arbitary to the spatial distribution. In the language of cranking mean field, this corresponds to the three-dimensional TAC solution.

The semiclassical mean field approximation for tilted axis rotation can be traced back to the 1980s.[7,8] Kerman and Onishi[7] pointed out the possibility of uniform rotation about a non-principal axis. Frisk and Bengtsson[8] demonstrated the existence of such solutions for realistic nuclei. However, the physical interpretation of these TAC solutions was missing and whether they are self-consistent with respect to the shape degrees of freedom was open. In 1993, Frauendorf found the first self-consistent TAC solutions in the framework of the pairing plus quadrupole model and interpreted them in terms of $\Delta I = 1$ rotational bands, which marks the origin of the fully-fledged TAC approach.[9] By comparing with the quantal particle rotor model, Fruendorf and Meng examined the qualities of TAC approximation and discussed its physical interpretations in Ref. 10.

Historically, the rotational bands with strongly enhanced M1 transitions has been observed in a number of nearly spherical light-mass Pb nuclei and attracted lots of attention (for reviews see Refs. 11–13). In the framework of the pairing plus quadrupole model, the TAC approach[9,13,14] was used to describe these dipole bands and reproduced the experimental reduced $B(M1)$ values in ^{198,199}Pb.[15] In order to distinguish this kind of rotation from the usual collective rotation in well-deformed nuclei (called electric rotation), the name "magnetic rotation" (MR) was introduced in Ref. 16, which alludes to the fact that the magnetic moment is the order parameter inducing a violation of rotational symmetry and thus causing rotational-like structures in the spectrum.[17] This forms an analogy to a ferromagnet where the total magnetic moment, the sum of the atomic dipole moments, is the order parameter.

Similar to ferromagnetism and antiferromagnetism in condensed matter physics, the existence of the antimagnetic rotation (AMR) in nuclear physics is also an interesting issue. For an antiferromagnet, one-half of the atomic dipole moments are aligned on one sublattice and the other half are aligned in the opposite direction on the other sublattice. In this way, the net magnetic moment in an antiferromagnet is canceled out. However, it is still an ordered state since the isotropy of such a state is also broken like a ferromagnet. In analogy with an antiferromagnet, antimagnetic rotation[13,18] is predicted to occur in some specific nearly spherical nuclei, in which the subsystems of valence protons (neutrons) are aligned back to back in opposite directions and nearly perpendicular to the orientation of the total spin of the valence neutrons (protons).

The MR and AMR[11–13] discussed above are based on the assumption that the nucleus involved is not triaxially deformed. If the nucleus is triaxially deformed, novel rotational excitations such as wobbling motion[19] and nuclear chirality[20,21] may occur.

The first attempt to develop the tilted axis cranking covariant density functional theory (TAC-CDFT) for describing these novel rotational excitations can be seen in Ref. 22. Based on the TAC-CDFT, the novel excitations spreading over different nuclear mass regions can be consistently described within a unified covariant density functional, and it is possible to construct classical vector diagrams showing the angular momentum composition from the calculated results. Furthermore, it is of particular importance that the CDFT includes nuclear magnetism,[23] and provides a consistent description of currents and time-odd fields, which plays an important role in the nuclear rotations.

The TAC-CDFT has been realized based on either the meson-exchange density functional[6,22] or the point-coupling one[24] and was extensively applied to investigate the magnetic rotation as well as the anti-magnetic rotation. A comprehensive review of the TAC-CDFT and its applications to MR and AMR has been given in Ref. 25.

In section 2, the formalism of TAC-CDFT will be firstly introduced. Its applications to magnetic rotation and anti-magnetic rotation will be presented in sections 3 and 4 respectively. Section 5 is devoted to discussing the novel chiral rotation in triaxial nucleus and a summary is given in section 6.

2. Tilted axis cranking covariant density functional

The initial version of the TAC approach was developed based on the potentials of the Nilsson type, which are combined with a pairing plus quadrupole model (PQTAC)[9] or the shell correction method (SCTAC)[14] for finding the deformation.

Symmetries play a central role in the interpretation of the tilted axis cranking solutions.[13,19] There are symmetries maintained in the laboratory reference frame but broken in the nuclear intrinsic one. This corresponds to the spontaneous symmetry breaking in the intrinsic frame, and the restoration of breaking symmetries in laboratory frame would manifest excitation spectra corresponding to the particular symmetry. The spontaneous symmetry breaking in rotating nuclei by considering the orientation of the angular momentum $\boldsymbol{J}$ in the intrinsic frame has been discussed in Ref. 20 for principal axis rotation and TAC with planar and aplanar rotation.

Cranking model is based on the classical treatment of the angular momentum and the assumption of uniform rotation, which have the consequence that the angular momentum is not conserved. The connection with the quantal spectra is made by means of semiclassical expressions for the energy and transition. Hence, it is important to investigate how well these approximations are in describing the experimental observables. By comparing with the particle rotor model which treats the quantal angular momentum dynamics properly, the above questions have been discussed in Ref. 10. It was found that the TAC approach quantitatively accounts both for the energies and the intra-band transition rates of the lowest bands generated by one or two quasi particles coupled to an axial rotor.

2.1. *Tilted axis cranking CDFT*

The tilted axis cranking CDFT, in a covariant and self-consistent way, describes the nuclear rotations where the rotational axis does not coincide with any principal axis of nuclear density distribution. This approach has been realized based on either the meson-exchange interaction[6,22] or the point-coupling interaction.[24] In the following, taking the two-dimensional cranking as an example, the TAC-CDFT formalism based on the point-coupling interaction will be presented.[24] A parallel formalism based on the meson-exchange interaction can be found in Ref. 6.

Assuming that the nucleus uniformly rotates around an axis in the xz plane, firstly it is necessary to transform the effective Lagrangian into a frame rotating with a constant rotational frequency,

$$\boldsymbol{\Omega} = (\Omega_x, 0, \Omega_z) = (\Omega\cos\theta_\Omega, 0, \Omega\sin\theta_\Omega), \tag{1}$$

where $\theta_\Omega := \sphericalangle(\boldsymbol{\Omega}, \boldsymbol{e}_x)$ is the tilted angle between the cranking axis and the x axis. As a rotating frame is an accelerated non-inertial one, one needs the techniques of general relativity. It has been shown that, from the transformed rotating Lagrangian, the equation of motion for the nucleons can be derived equivalently by either starting from a special relativistic transformation,[23] or more generally adopting the tetrad formalism in the framework of general relativistic theory.[26] One can thus find[23,26]

$$[\boldsymbol{\alpha}\cdot(-i\boldsymbol{\nabla} - \boldsymbol{V}) + \beta(m+S) + V - \boldsymbol{\Omega}\cdot\hat{\boldsymbol{J}}]\psi_k = \epsilon_k\psi_k, \tag{2}$$

where $\hat{\boldsymbol{J}} = \hat{\boldsymbol{L}} + \frac{1}{2}\hat{\boldsymbol{\Sigma}}$ is the total angular momentum of the nucleon spinors, and ϵ_k represents the single-particle Routhians for nucleons. The relativistic scalar and vector fields $S(\boldsymbol{r})$ and $V^\mu(\boldsymbol{r})$ read

$$S(\boldsymbol{r}) = \alpha_S\rho_S + \beta_S\rho_S^2 + \gamma_S\rho_S^3 + \delta_S\triangle\rho_S, \tag{3a}$$

$$\begin{aligned} V(\boldsymbol{r}) &= \alpha_V\rho_V + \gamma_V\rho_V^3 + \delta_V\triangle\rho_V \\ &\quad + \tau_3\alpha_{TV}\rho_{TV} + \tau_3\delta_{TV}\triangle\rho_{TV} + eA^0, \end{aligned} \tag{3b}$$

$$\begin{aligned} \boldsymbol{V}(\boldsymbol{r}) &= \alpha_V\boldsymbol{j}_V + \gamma_V(\boldsymbol{j}_V)^3 + \delta_V\triangle\boldsymbol{j}_V \\ &\quad + \tau_3\alpha_{TV}\boldsymbol{j}_{TV} + \tau_3\delta_{TV}\triangle\boldsymbol{j}_{TV} + e\boldsymbol{A}. \end{aligned} \tag{3c}$$

As usual, it is assumed that the nucleon single-particle states do not mix isospin, i.e., the single-particle states are eigenstates of τ_3. Therefore only the third component of isovector potentials survives. The Coulomb field $A^0(\boldsymbol{r})$ is determined by Poisson's equation

$$-\triangle A^0(\boldsymbol{r}) = e\rho_c. \tag{4}$$

The spatial components of the electromagnetic vector potential $\boldsymbol{A}(\boldsymbol{r})$ are neglected since their contributions are extremely small.[23,27]

Since the Coriolis term $\boldsymbol{\Omega}\cdot\hat{\boldsymbol{J}}$ in the Dirac equation (2) breaks time reversal symmetry in the intrinsic frame, the nucleon currents are induced and as a consequence the spatial components of the vector potential $\boldsymbol{V}(\boldsymbol{r})$

emerge. The densities and currents in Eqs. (3) have the form

$$\rho_S(\boldsymbol{r}) = \sum_{i=1}^{A} \bar{\psi}_i(\boldsymbol{r})\psi_i(\boldsymbol{r}), \tag{5a}$$

$$\rho_V(\boldsymbol{r}) = \sum_{i=1}^{A} \psi_i^\dagger(\boldsymbol{r})\psi_i(\boldsymbol{r}), \tag{5b}$$

$$\boldsymbol{j}_V(\boldsymbol{r}) = \sum_{i=1}^{A} \psi_i^\dagger(\boldsymbol{r})\boldsymbol{\alpha}\psi_i(\boldsymbol{r}), \tag{5c}$$

$$\rho_{TV}(\boldsymbol{r}) = \sum_{i=1}^{A} \psi_i^\dagger(\boldsymbol{r})\tau_3\psi_i(\boldsymbol{r}), \tag{5d}$$

$$\boldsymbol{j}_{TV}(\boldsymbol{r}) = \sum_{i=1}^{A} \psi_i^\dagger(\boldsymbol{r})\boldsymbol{\alpha}\tau_3\psi_i(\boldsymbol{r}), \tag{5e}$$

$$\rho_c(\boldsymbol{r}) = \sum_{i=1}^{A} \psi_i^\dagger(\boldsymbol{r})\frac{1-\tau_3}{2}\psi_i(\boldsymbol{r}). \tag{5f}$$

Here, the "no-sea" approximation is adopted, i.e., the sums run over only the particle states in the Fermi sea and the contribution of the negative-energy states are neglected.

By solving the equation of motion iteratively, one finally obtains the total energy in the laboratory frame

$$E_{\text{tot}} = E_{\text{kin}} + E_{\text{int}} + E_{\text{cou}} + E_{\text{c.m.}}, \tag{6}$$

which is composed of a kinetic part

$$E_{\text{kin}} = \int d^3\boldsymbol{r} \sum_{i=1}^{A} \psi_i^\dagger[\boldsymbol{\alpha}\cdot\boldsymbol{p} + \beta m]\psi_i, \tag{7}$$

an interaction part

$$\begin{aligned} E_{\text{int}} = \int d^3\boldsymbol{r} \Big\{ &\frac{1}{2}\alpha_S\rho_S^2 + \frac{1}{3}\beta_S\rho_S^3 + \frac{1}{4}\gamma_S\rho_S^4 + \frac{1}{2}\delta_S\rho_S\Delta\rho_S \\ &+\frac{1}{2}\alpha_V(\rho_V^2 - \boldsymbol{j}\cdot\boldsymbol{j}) + \frac{1}{2}\alpha_{TV}(\rho_{TV}^2 - \boldsymbol{j}_{TV}\cdot\boldsymbol{j}_{TV}) \\ &+\frac{1}{4}\gamma_V(\rho_V^2 - \boldsymbol{j}\cdot\boldsymbol{j})^2 + \frac{1}{2}\delta_V(\rho_V\Delta\rho_V - \boldsymbol{j}\Delta\boldsymbol{j}) \\ &+\frac{1}{2}\delta_{TV}(\rho_{TV}\Delta\rho_{TV} - \boldsymbol{j}_{TV}\Delta\boldsymbol{j}_{TV})\Big\}, \end{aligned} \tag{8}$$

an electromagnetic part

$$E_{\text{cou}} = \int d^3\boldsymbol{r}\frac{1}{2}eA_0\rho_c, \tag{9}$$

and the center-of-mass (c.m.) correction energy $E_{\text{c.m.}}$ accounting for the treatment of center-of-mass motion.

One can express the total energy in the laboratory frame in an equivalent way with the summation of the single-particle Routhians for nucleons,

$$\begin{aligned} E_{\text{tot}} = & \sum_{k=1}^{A} \epsilon_k \\ & - \int d^3\boldsymbol{r} \left\{ \frac{1}{2}\alpha_S\rho_S^2 + \frac{2}{3}\beta_S\rho_S^3 + \frac{3}{4}\gamma_S\rho_S^4 + \frac{1}{2}\delta_S\rho_S\Delta\rho_S \right. \\ & + \frac{1}{2}\alpha_V(\rho_V^2 - \boldsymbol{j}\cdot\boldsymbol{j}) + \frac{1}{2}\alpha_{TV}(\rho_{TV}^2 - \boldsymbol{j}_{TV}\cdot\boldsymbol{j}_{TV}) \\ & + \frac{3}{4}\gamma_V(\rho_V^2 - \boldsymbol{j}\cdot\boldsymbol{j})^2 + \frac{1}{2}\delta_V(\rho_V\Delta\rho_V - \boldsymbol{j}\Delta\boldsymbol{j}) \\ & \left. + \frac{1}{2}\delta_{TV}(\rho_{TV}\Delta\rho_{TV} - \boldsymbol{j}_{TV}\Delta\boldsymbol{j}_{TV}) + \frac{1}{2}eA_0\rho_c \right\} \\ & + \sum_{k=1}^{A}\langle k|\boldsymbol{\Omega}\cdot\hat{\boldsymbol{J}}|k\rangle + E_{c.m.} \,. \end{aligned} \tag{10}$$

For each rotational frequency Ω, the expectation values of the angular momentum components $\boldsymbol{J} = (J_x, J_y, J_z)$ in the intrinsic frame are given by

$$J_x = \langle \hat{J}_x \rangle = \sum_{i=1}^{A} j_x^{(i)}, \tag{11a}$$

$$J_y = 0, \tag{11b}$$

$$J_z = \langle \hat{J}_z \rangle = \sum_{i=1}^{A} j_z^{(i)}, \tag{11c}$$

and by means of the semiclassical cranking condition

$$J = \sqrt{\langle \hat{J}_x \rangle^2 + \langle \hat{J}_z \rangle^2} \equiv \sqrt{I(I+1)}, \tag{12}$$

one can relate the rotational frequency Ω to the angular momentum quantum number I in the rotational band.

The orientation of the angular momentum vector $\boldsymbol{J}$ is represented by the angle $\theta_J := \sphericalangle(\boldsymbol{J}, \boldsymbol{e}_x)$ between the angular momentum vector $\boldsymbol{J}$ and the x axis. In a fully self-consistent calculation, the orientation θ_J of the angular momentum $\boldsymbol{J}$ should be identical to the orientation θ_Ω of the angular velocity $\boldsymbol{\Omega}$.[7] Thus for the tilted axis cranking calculations, the tilted angle θ_Ω for a given Ω is determined self-consistently either by minimizing the total Routhian

$$E(\Omega, \theta_\Omega) = \hat{H} - \cos\theta_\Omega \Omega \hat{J}_x - \sin\theta_\Omega \Omega \hat{J}_z, \tag{13}$$

with respect to the angle θ_Ω or by requiring that $\boldsymbol{\Omega}$ is parallel to $\boldsymbol{J}$. If one verifies that both the conditions are fullfilled, it means that the self-consistency has been fully achieved by the tilted axis cranking calculations.

The quadrupole moments Q_{20} and Q_{22} are calculated by

$$Q_{20} = \sqrt{\frac{5}{16\pi}} \langle 3z^2 - r^2 \rangle, \tag{14a}$$

$$Q_{22} = \sqrt{\frac{15}{32\pi}} \langle x^2 - y^2 \rangle, \tag{14b}$$

and the deformation parameters β and γ can thus be extracted from

$$\beta = \sqrt{a_{20}^2 + 2a_{22}^2}, \tag{15a}$$

$$\gamma = \arctan\left[\sqrt{2}\frac{a_{22}}{a_{20}}\right], \tag{15b}$$

by using the relations

$$Q_{20} = \frac{3A}{4\pi} R_0^2 a_{20}, \tag{16a}$$

$$Q_{22} = \frac{3A}{4\pi} R_0^2 a_{22}, \tag{16b}$$

with $R_0 = 1.2A^{1/3}$ fm. Note that the sign convention in Ref. 28 is adopted for the definition of γ here.

The nuclear magnetic moment, in units of the nuclear magneton, is given by

$$\boldsymbol{\mu} = \sum_{i=1}^{A} \int d^3r \left[\frac{mc^2}{\hbar c} q\psi_i^\dagger(\boldsymbol{r})\boldsymbol{r} \times \boldsymbol{\alpha}\psi_i(\boldsymbol{r}) + \kappa\psi_i^\dagger(\boldsymbol{r})\beta\boldsymbol{\Sigma}\psi_i(\boldsymbol{r})\right], \tag{17}$$

where the charge q ($q_p = 1$ for protons and $q_n = 0$ for neutrons) is in units of e, m the nucleon mass, and κ is the free anomalous gyromagnetic ratio of the nucleon ($\kappa_p = 1.793$ and $\kappa_n = -1.913$).

From the quadrupole moments and the magnetic moment, the $B(M1)$ and $B(E2)$ transition probabilities can be derived in semiclassical approximation

$$B(M1) = \frac{3}{8\pi}\mu_\perp^2 = \frac{3}{8\pi}(\mu_x \sin\theta_J - \mu_z \cos\theta_J)^2, \tag{18a}$$

$$B(E2) = \frac{3}{8}\left[Q_{20}^p \cos^2\theta_J + \sqrt{\frac{2}{3}}Q_{22}^p(1+\sin^2\theta_J)\right]^2, \tag{18b}$$

where Q_{20}^p and Q_{22}^p corresponds to the quadrupole moments of protons.

2.2. *Numerical techniques*

Orientation constraint. In the usual PAC programs (one-dimensional cranking), the principal axes of the densities and fields are implemented to be along the x, y, and z axis. For the TAC code (two-dimensional cranking), it allows for arbitrary rotations of the density distributions around the intrinsic y axis. The freedom of rotations around the y axis can lead (in particular for $\Omega = 0$ and for small Ω values) to instabilities during the iterative solution because the solutions with different orientations in the xz plane are degenerate. Therefore, the x, y, and z axes are enforced to be identical with the principal axes of the density distribution by introducing a quadratic constraint[28] for the expectation value of the quadrupole moment

$$\langle Q_{2-1}\rangle = -\sqrt{\frac{15}{8\pi}}\,\langle xz\rangle = 0, \tag{19}$$

i.e., by minimizing

$$\langle H'\rangle = \langle H\rangle + \frac{1}{2}C\left(\langle Q_{2-1}\rangle - a_{2-1}\right)^2, \tag{20}$$

with $a_{2-1} = 0$, and C being a spring constant, which, if properly chosen, has no influence on the final result.

Expansion in harmonic oscillator basis. For the solution of the relativistic tilted axis cranking equations, the Dirac spinors are normally expanded in terms of three-dimensional harmonic oscillator wave functions in Cartesian coordinates,

$$\varphi_{n_x n_y n_z}(\boldsymbol{r}) = \langle \boldsymbol{r}|n_x, n_y, n_z\rangle = \varphi_{n_x}(x)\varphi_{n_y}(y)\varphi_{n_z}(z). \tag{21}$$

The normalized oscillator functions $\varphi_{n_k}(x_k)$ in k-direction ($x_k = x, y, z$) are given by

$$\varphi_{n_k}(x_k) = \frac{N_{n_k}}{\sqrt{b_k}}H_{n_k}\left(\frac{x_k}{b_k}\right)\exp\left[-\frac{1}{2}\left(\frac{x_k}{b_k}\right)^2\right], \tag{22}$$

where $N_n = (\sqrt{\pi}2^n n!)^{-1/2}$ is a normalization factor and

$$H_n(\xi) = (-1)^n e^{\xi^2} \frac{d^n}{d\xi^n} e^{-\xi^2} \tag{23}$$

are the Hermite polynomials.[29]

One can take the following basis states

$$\varphi_\alpha(\mathbf{r}, s) = \langle \mathbf{r}, s|\alpha\rangle = i^{n_y} \varphi_{n_x}(x)\varphi_{n_y}(y)\varphi_{n_z}(z)\frac{1}{\sqrt{2}} \begin{pmatrix} 1 \\ (-1)^{n_x+1} \end{pmatrix}, \tag{24}$$

and

$$\varphi_{\overline{\alpha}}(\mathbf{r}, s) = \langle \mathbf{r}, s|\overline{\alpha}\rangle = (-i)^{n_y} \varphi_{n_x}(x)\varphi_{n_y}(y)\varphi_{n_z}(z)\frac{1}{\sqrt{2}} \begin{pmatrix} (-1)^{n_x+1} \\ -1 \end{pmatrix}, \tag{25}$$

which correspond to the eigenfunctions of the simplex operation with the positive ($+i$) and negative ($-i$) eigenvalues, respectively. The phase factor i^{n_y} has been added in order to have real matrix elements for the Dirac equation.[30]

The Dirac spinor for the nucleon has the form

$$\psi_i(\mathbf{r}) = \begin{pmatrix} f_i(\mathbf{r}, s) \\ ig_i(\mathbf{r}, s) \end{pmatrix} \chi_i(t), \tag{26}$$

where $\chi_i(t)$ is the isospin part. In the tilted axis cranking calculations, the simplex symmetry is violated, and therefore the large and small components of the wave function in Eq. (26) have to be written as linear combinations of the sets (24) and (25) with different simplex:

$$\begin{aligned} f_i(\mathbf{r}, s) &= \sum_\alpha f_{\alpha i}|\alpha\rangle + \sum_{\bar{\alpha}} f_{\bar{\alpha} i}|\bar{\alpha}\rangle, \\ g_i(\mathbf{r}, s) &= \sum_{\tilde{\alpha}} g_{\tilde{\alpha} i}|\tilde{\alpha}\rangle + \sum_{\bar{\tilde{\alpha}}} g_{\bar{\tilde{\alpha}} i}|\bar{\tilde{\alpha}}\rangle. \end{aligned} \tag{27}$$

Since the large and small components in the Dirac equation have different parity, the sums in the expansions for the large and the small components have to run over oscillator quantum numbers with even $N = n_x + n_y + n_z$ or odd N respectively. This is indicated in Eq. (27) by the indices α and $\tilde{\alpha}$.

On this basis, the solution of Dirac equation (2) is obtained by the matrix diagonalization

$$\mathcal{H} \begin{pmatrix} f_{\alpha i} \\ f_{\overline{\alpha} i} \\ g_{\tilde{\alpha} i} \\ g_{\overline{\tilde{\alpha}} i} \end{pmatrix} = \varepsilon_i \begin{pmatrix} f_{\alpha i} \\ f_{\overline{\alpha} i} \\ g_{\tilde{\alpha} i} \\ g_{\overline{\tilde{\alpha}} i} \end{pmatrix}, \tag{28}$$

where the Hamiltonian matrix $\mathcal{H}$ has the form

$$\begin{pmatrix} \langle\alpha|M^*+V-\boldsymbol{\Omega}\hat{\boldsymbol{J}}|\alpha'\rangle & \langle\alpha|M^*+V-\boldsymbol{\Omega}\hat{\boldsymbol{J}}|\overline{\alpha}'\rangle & \langle\alpha|\boldsymbol{\sigma}(\boldsymbol{\nabla}-i\boldsymbol{V})|\tilde{\alpha}'\rangle & \langle\alpha|\boldsymbol{\sigma}(\boldsymbol{\nabla}-i\boldsymbol{V})\Big|\overline{\tilde{\alpha}}'\Big\rangle \\ \langle\overline{\alpha}|M^*+V-\boldsymbol{\Omega}\hat{\boldsymbol{J}}|\alpha'\rangle & \langle\overline{\alpha}|M^*+V-\boldsymbol{\Omega}\hat{\boldsymbol{J}}|\overline{\alpha}'\rangle & \langle\overline{\alpha}|\boldsymbol{\sigma}(\boldsymbol{\nabla}-i\boldsymbol{V})\,|\tilde{\alpha}'\rangle & \langle\overline{\alpha}|\boldsymbol{\sigma}(\boldsymbol{\nabla}-i\boldsymbol{V})\Big|\overline{\tilde{\alpha}}'\Big\rangle \\ \langle\tilde{\alpha}|-\boldsymbol{\sigma}(\boldsymbol{\nabla}-i\boldsymbol{V})|\alpha'\rangle & \langle\tilde{\alpha}|-\boldsymbol{\sigma}(\boldsymbol{\nabla}-i\boldsymbol{V})|\overline{\alpha}'\rangle & \langle\tilde{\alpha}|-M^*+V-\boldsymbol{\Omega}\hat{\boldsymbol{J}}|\tilde{\alpha}'\rangle & \langle\tilde{\alpha}|-M^*+V-\boldsymbol{\Omega}\hat{\boldsymbol{J}}\Big|\overline{\tilde{\alpha}}'\Big\rangle \\ \langle\overline{\tilde{\alpha}}|-\boldsymbol{\sigma}(\boldsymbol{\nabla}-i\boldsymbol{V})|\alpha'\rangle & \langle\overline{\tilde{\alpha}}|-\boldsymbol{\sigma}(\boldsymbol{\nabla}-i\boldsymbol{V})|\overline{\alpha}'\rangle & \langle\overline{\tilde{\alpha}}|-M^*+V-\boldsymbol{\Omega}\hat{\boldsymbol{J}}|\tilde{\alpha}'\rangle & \langle\overline{\tilde{\alpha}}|-M^*+V-\boldsymbol{\Omega}\hat{\boldsymbol{J}}\Big|\overline{\tilde{\alpha}}'\Big\rangle \end{pmatrix}. \tag{29}$$

Note that the Coriolis term $\boldsymbol{\Omega}\hat{\boldsymbol{J}}$ breaks the invariance with respect to time reversal and with respect to rotations around the x axis as well as around the z axis. Therefore, only the invariance of space reflection $\mathcal{P}$ and the combination of time reversal and reflection in y direction $\mathcal{P}_y\mathcal{T}$ are valid.

For the evaluation of the Coulomb field, due to its long range character, an expansion in harmonic oscillator states is very difficult and therefore the standard Green function method[31] is used for the calculation of the Coulomb field in each step of the iteration.

Configuration constraint. Normally the rotational bands are built on specific proton and neutron configurations. The orbits to be blocked are usually given in the spherical basis, i.e., by the spherical quantum numbers $|nljm\rangle$. The equations of motion are solved by expanding the Dirac spinors in terms of the three-dimensional harmonic oscillator functions in the Cartesian basis Eq. (27) labeled by the quantum number $|n_x, n_y, n_z, \pm i\rangle$. In order to identify which orbits have to be blocked, one needs to transform the wave functions from the Cartesian basis with the quantum number $|n_x n_y n_z \pm i\rangle$ to a spherical basis with the quantum numbers $|nljm\rangle$ using the techniques given in Refs. 32, 33. Consequently, one can block the levels which have the maximal overlap with the required $|nljm\rangle$ orbits. These techniques are considerably simplified if we work in an isotropic Cartesian basis with identical basis parameters $b_x = b_y = b_z = b$ in Eq. (22).

To describe a rotational band, one should keep the corresponding configuration fixed for a set of increasing values of rotational frequency Ω. Because of the considerable K-mixing in TAC solutions and the high level density in the 3-dimensional calculations, it may occur that the configuration is changing with the iteration and also the different rotational frequency. To constrain the specific configuration we are interested in, one can adopt the following prescription: starting from the Dirac level $|\psi_i(\Omega_n)\rangle$ blocked for $\Omega = \Omega_n$, one can block for $\Omega = \Omega_{n+1}$ the level $|\psi_j(\Omega_{n+1})\rangle$ which maximizes the overlap $\langle\psi_i(\Omega_n)|\psi_j(\Omega_{n+1})\rangle$, i.e.,

$$\langle\psi_i(\Omega)|\psi_j(\Omega + \delta\Omega)\rangle = 1 + \mathcal{O}(\delta\Omega). \tag{30}$$

For infinitesimal step sizes, this condition corresponds to the so-called parallel transport.[34,35]

3. Magnetic rotation

3.1. *Basic concepts*

In the early 1990s, the rotational-like sequences of strongly enhanced magnetic dipole (M1) transitions were surprisingly observed in several light-mass Pb isotopes, which are known to be spherical or near-spherical. Sequentially, this new type of rotational bands which have strong M1 and very weak E2 transitions attracted great attentions and has been discovered in a number of nearly spherical nuclei (for reviews see Refs. 11–13, 25). Besides the outstanding magnetic dipole transitions, the intriguing feature here is that the orientation of the rotor is not specified by the deformation of the overall density but rather by the current distribution induced by specific nucleons moving in high-j orbits.

The explanation of such bands was given in terms of the shears mechanism.[9] In this interpretation, the magnetic dipole vector, which arises from proton particles (holes) and neutron holes (particles) in high-j orbits, rotates around the total angular momentum vector. See Fig. 1 for a schematic illustration. At the bandhead, the proton and neutron angular momenta are almost perpendicular to each other. Along the bands, energy and angular momentum are increased by an alignment of the proton and neutron angular momenta along the total angular momentum. Consequently, the orientation of the total angular momentum in the intrinsic frame does not change so much and regular rotational bands are formed in spite of the fact that the density distribution of the nucleus is almost spherical or only weakly deformed.

Let us further consider the magnetic dipole vector of the system illustrated in Fig. 1, where the total magnetic dipole vector is mainly contributed by the proton particles (holes) and neutron holes (particles). Since the magnetic moments of a proton and a neutron are different in sign, near the bandhead, it, to a large extent, cancels the longitudinal component of the magnetic dipole moment vector $\mu_{\parallel}$ while producing a large transverse component of the magnetic dipole moment vector $\mu_{\perp}$ which leads to the enhanced M1 transitions. Along the bands, the alignment of the proton and neutron angular momenta along the total angular momentum decreases the angles between them, and gradually reduces the transverse component $\mu_{\perp}$ which leads to the decrease of the strong M1 transitions in the bands.

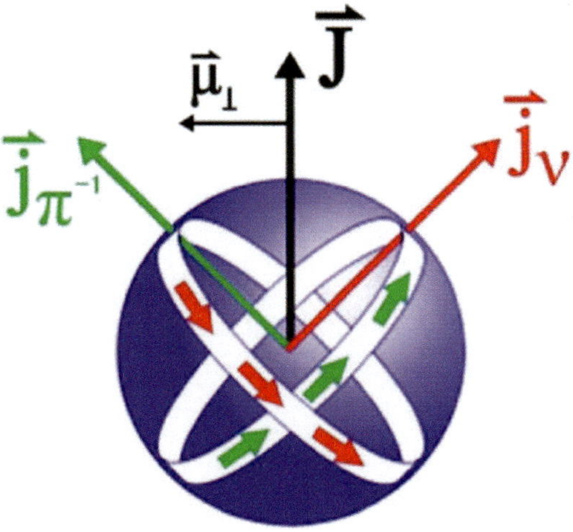

Fig. 1. A schematic illustration for the spin-coupling scheme of magnetic rotation. For a near-spherical nuclei, the coupling of the proton-hole j_π^{-1} and neutron-particle j_ν, each in high-j orbital, gives the total angular momentum J. As a result, a large transverse component of the magnetic dipole moment vector, $\mu_\perp$, rotates around the total angular momentum vector, and creates the enhanced M1 transitions. Taken from Ref. 25.

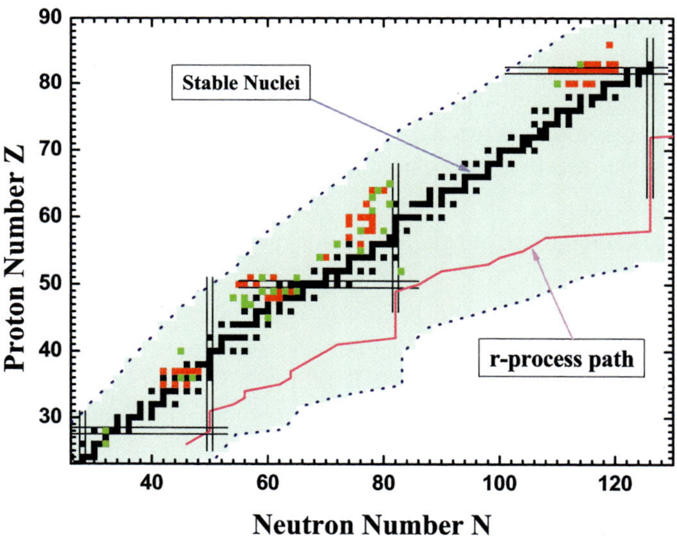

Fig. 2. The nuclides with magnetic rotation observed in the nuclear chart. The red squares represent 56 nuclides with 120 magnetic dipole bands observed before 2000 and compiled in Ref. 45. The green squares represent the corresponding data after 2000. Taken from Ref. 25.

The experimental indicators for magnetic rotation have been summarized at several places in the literature,[12,13,25] such as:
1) a $\Delta I = 1$ sequence of strong magnetic dipole transitions, corresponding to a reduced transition probability $B(M1) \sim$ a few μ_N^2, which decreases with increasing spin;
2) weak or absent quadrupole transitions, corresponding to a deformation parameter $|\beta| \lesssim 0.15$, which combined with strong M1 transitions results in large $B(M1)/B(E2)$ ratios, $\gtrsim 20\ \mu_N^2/(eb)^2$;
3) a smooth variation in the γ transition energy with angular momentum;
4) a substantial dynamic moment of inertia, corresponding to the large ratio of the $\mathcal{J}^{(2)}/B(E2) \gtrsim 100$ MeV$^{-1}(eb)^{-2}$, compared with the values in well-deformed [$\sim$10 MeV$^{-1}(eb)^{-2}$] or superdeformed [$\sim$5 MeV$^{-1}(eb)^{-2}$] rotational bands.

It is difficult to define how much contribution from the collective rotation can be expected in the MR bands. Obviously either the alignment of the angular momenta for the valent nucleons or the collective rotation due to deformation will cost energy. For a given angular momentum, the nuclear system will try to minimize its energy via a competition between the kinetic energy due to the collective rotation and the potential energy due to the closing of the blades of the shears blades. Weaker quadrupole deformation will lead to purer MR band and larger quadrupole deformation will reduce the $B(M1)/B(E2)$ ratios. Therefore it might be possible to observe the competition and transition between the electric and the magnetic rotations. The experimental information for the lifetime measurement in ^{106}Sn has suggested an extremely large ratio $\mathcal{J}^{(2)}/B(E2) > 1000$ MeV$^{-1}(eb)^{-2}$, which may provide an example of almost pure magnetic rotation in a spherical nucleus.[36]

Following the observation of many long cascades of magnetic dipole γ-ray transitions in the neutron deficient Pb nuclei in the early 1990s,[37–43] several attempts[1–5] have been made to measure the lifetimes for the states in MR bands. In 1997, the high-accuracy lifetime measurements for four M1 bands in ^{198,199}Pb performed with GAMMASPHERE provided a clear evidence for shears mechanism.[15] Additional evidence for the shears mechanism was provided by measuring the g-factor of a dipole band in ^{193}Pb.[44] It is demonstrated that at the bandhead the longitudinal component of the magnetic moment $\mu_{\parallel}$ has the value expected for an opening angle of 90° of the two blades composed of the suggested particles and holes.

From then on, more and more MR bands have been observed not only in the mass region of $A \sim 190$ but also in $A \sim 80$, $A \sim 110$, and $A \sim 140$

regions. A compilation in these four mass regions up to December 2006 including 178 magnetic dipole bands in 76 nuclides is given in Ref. 45 and additional data can be found in Refs. 46–57. Recent observations in ^{58}Fe[58] and ^{60}Ni[59] have extended the observed MR mass region to $A = 60$. In total, more than 195 magnetic dipole bands spread over 85 nuclides have been observed,[25] which have been summarized in the nuclear chart in Fig. 2.

3.2. *A representative example:* ^{198}*Pb*

The observation of the $\Delta I = 1$ rotational-like structures in neutron deficient Pb nuclei in the early 1990s opened a new era for magnetic rotation.[11–13] Long cascades of M1 transitions were first observed in Pb nuclei in the early 1990s.[41–43,60,61] With improved detector techniques and lots of efforts the lifetime measurements based on the Doppler-shift attenuation method (DSAM) for four M1-bands in the nuclei ^{198}Pb and ^{199}Pb provided clear evidence for magnetic rotation.[15] Subsequently, another experiment using the recoil distance method (RDM) in ^{198}Pb together with the results of the DSAM experiment provided further support to the shears mechanism.[62] Many works along this line for almost 10 years have been devoted to the magnetic rotation in Pb isotopes which has become a classic example and has received wide attention from theoretical and experimental aspects. Naturally, it is a good test ground for various theories aiming at the description of the MR.

In the framework of the pairing plus quadrupole model, the tilted axis cranking approach[9,13,14] has been used to describe the magnetic rotation from the very beginning and reproduce the experimental reduced $B(M1)$ values of magnetic dipole bands in ^{198,199}Pb very well.[15] In addition, the shell model[63] and the many-particles-plus-rotor model[64] have also been applied to investigate magnetic rotation in the Pb region.

As the magnetic rotation in Pb isotopes is a classic example, in the following the band 1 in ^{198}Pb[60] will be chosen as an example to demonstrate the relativistic self-consistent description of the MR characteristics. The calculations have been carried out with the covariant point-coupling density functional PC-PK1[65] in Ref. 66. The detailed and numerical techniques can be found there.

Configuration and single particle Routhian. In order to describe a specific rotational band in a nucleus, one should apply the technique of configuration constraint to keep the corresponding configuration fixed for a set of rotational frequency Ω. In most cases, it is necessary to perform the self-consistent TAC calcualtions for a number of possible configurations and figure out the most likely configuration by comparing the theoretical results with the experimental charateristics, such as total Routhian, energy spectra, alignments, moments of inertia, and eletromagnetic transitions.

In band 1 of ^{198}Pb,[60] a backbending phenomenon has been observed caused by the alignment of a pair of $i_{13/2}$ neutrons. Before the backbending, the configuration AE11 is assigned, while after the backbending, the configuration becomes ABCE11.[45,67] Note that a short hand notation is used for the configurations:[61] A, B, C and D denote $\nu i_{13/2}$ holes with positive parity and E denotes a neutron hole with negative parity (mainly of $f_{5/2}$ and $p_{3/2}$ origin); the proton configuration $\pi[s_{1/2}^{-2}h_{9/2}i_{13/2}]11^-$ is abbreviated by its spin number 11. Thus, "AE" denotes the neutron configuration $\nu[i_{13/2}^{-1}(fp)^{-1}]$ before the backbending and "ABCE" denotes $\nu[i_{13/2}^{-3}(fp)^{-1}]$ after the backbending.

With the constraint of the two configurations AE11 and ABCE11 in ^{198}Pb, the single particle Routhians for the neutrons are self-consistently obtained by the TAC-CDFT calculations with the interaction PC-PK1, which are respectively shown as a function of the rotational frequency Ω in Fig. 3. It is shown that due to the self-consistency of mean field solutions the single particle Routhians for configurations AE11 and ABCE11 are slightly different from each other.

Total Routhian and energy spectra. For ^{198}Pb, the calculated total Routhian and energy spectra with configurations AE11 and ABCE11 are shown in comparison with the data of band 1[67] in Fig. 4. It can be seen from Fig. 4a that the TAC-CDFT calculations reproduce well the experimental total Routhian for the band before the band crossing. However, one can see that the band crossing would appear at too low frequency due to the neglect of pairing. In order to achieve the correct band crossing frequency, a renormalization of the total Routhians ($\sim$ 3 MeV) after band crossing is taken in Fig. 4a and then the experimental total Routhian could be also reproduced after the band crossing. In the PQTAC results,[15] which include pairing, the proper band crossing frequency can be gotten.

For the energy spectra shown in Fig. 4b, it can be seen that for certain regions of angular momenta, the calculated energy values for I = 19-21$\hbar$ are missing. This behavior is common and has been discussed in Ref. 6, which is due to the level crossing connected with the backbending phenomenon, and no converged solutions could be found for these angular momentum values. It can be seen that the TAC-CDFT calculations reproduce well the experimental energies but underestimate the particle-hole excitation

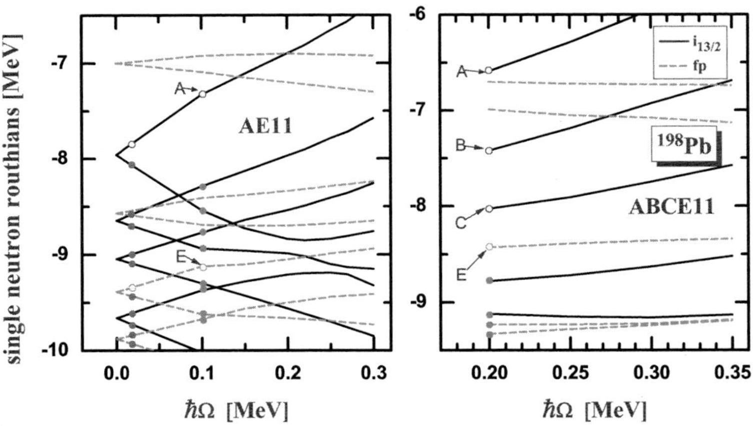

Fig. 3. (Color online) Single particle Routhians for the neutrons in ^{198}Pb as a function of the rotational frequency based on the configurations AE11 and ABCE11. The positive parity levels belonging to the $\nu i_{13/2}$ orbit are given by full black curves and the neutron levels with negative parity (pf) are indicated by dashed red curves. The blue dots indicate the occupied levels at $\Omega = 0$ and the green dots indicate the occupied levels at the band heads with the configuration AE11 (left panel) and ABCE11 (right panel). Taken from Ref. 66.

energies at the band head of the configuration ABCE11 in ^{198}Pb. Like the Routhians shown in Fig. 4a, this underestimation is due to neglect of the pairing correlations. By choosing different references for the configurations involved, the underestimation is compensated in Fig. 4b.

The experimental rotational frequency $\Omega_{\rm exp}$ is extracted from the energy spectra by the relation

$$\hbar\Omega_{\rm exp} \approx \frac{dE}{dI} = \frac{1}{2}[E_\gamma(I+1 \to I) + E_\gamma(I \to I-1)]. \tag{31}$$

In Fig. 5, the calculated total angular momenta of the band 1 in ^{198}Pb as functions of the rotational frequency are shown in comparison with the experimental data[67] and the PQTAC results.[68] It is found that both the TAC-CDFT and the PQTAC results agree well with the experimental data. This shows that the TAC calculations can reproduce the relative changes of the moment of inertia within the different bands rather well. The TAC calculations support the hypothesis that the backbending arises through an excitation of a neutron-hole pair in the $i_{13/2}$ shell, i.e. by the transitions in the configurations AE11→ABCE11 in band 1 of ^{198}Pb. In detail, before the backbending the spin values found in the TAC-CDFT and PQTAC

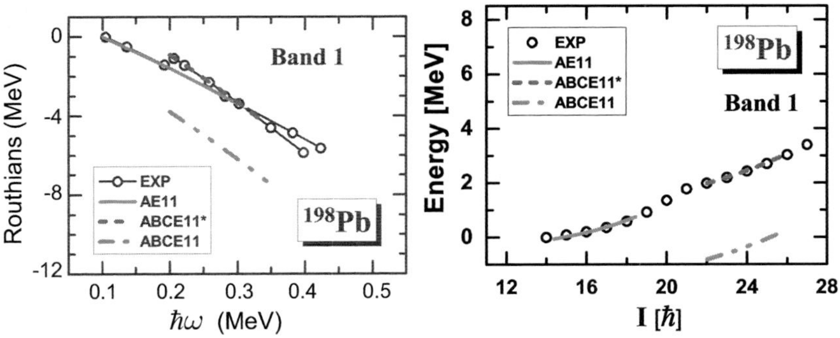

Fig. 4. (Color online) Total Routhian (left panel) and energy spectra (right panel) in the TAC-CDFT calculations compared with the data[67] for band 1 in ^{198}Pb. For the total Routhian, the value at $\hbar\Omega = 0.1$ MeV is taken as the reference and the Routhian for configuration ABCE11* is renormalized to the values at $\hbar\Omega = 0.2$ MeV, respectively. For the energy spectra, the energies at $I = 15\hbar$ are taken as references and the energies for configuration ABCE11* in ^{198}Pb are renormalized to the energies at $I = 22\hbar$. Taken from Ref. 66.

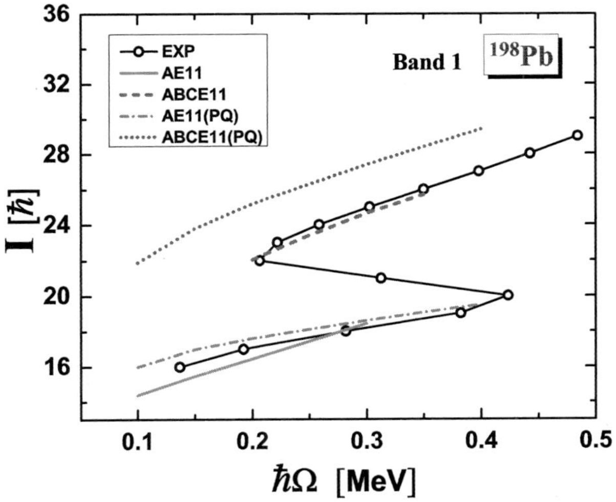

Fig. 5. (Color online) Angular momenta as functions of the rotational frequency in the TAC-CDFT calculations compared with the data[67] and the PQTAC results[68] for band 1 in ^{198}Pb. The configurations with "(PQ)" denote the corresponding results of PQTAC calculations. Taken from Ref. 66.

models differ from experimental values by up to $2\hbar$. After the backbending, the PQTAC result for the band 1 in ^{198}Pb is nearly $3\hbar$ larger than the experimental values and the TAC-CDFT results. Comparing with the

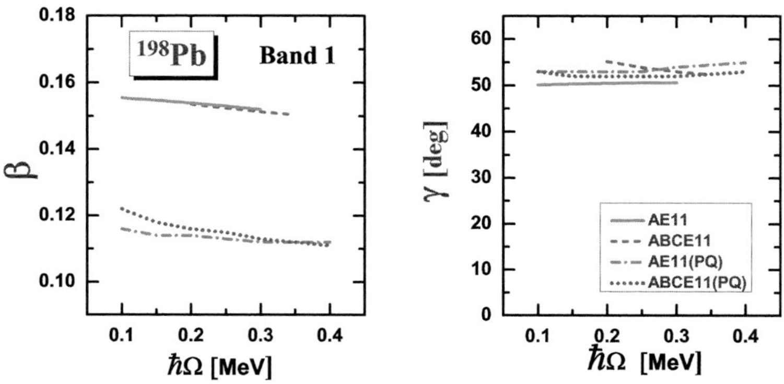

Fig. 6. (Color online) Deformation parameters β (left panels) and γ (right panels) as functions of the rotational frequency in the TAC-CDFT calculations compared with the PQTAC results[68] for band 1 in ^{198}Pb. Taken from Ref. 66.

experimental values in Fig. 5, the appearance of backbending is clearly seen for this band.

Shape evolution. The advantage of the TAC-CDFT calculations is that the nuclear shape and deformation can be obtained self-consistently and automatically as outputs. The deformation parameters β and γ as well as their evolutions for band 1 in ^{198}Pb obtained in the TAC-CDFT calculations are compared with the PQTAC results[68] in Fig. 6. In the TAC-CDFT calculations, the quadrupole deformations are around $\beta = 0.15$ and remain almost constant. The PQTAC calculations produce the same tendency with slightly smaller deformations around $\beta = 0.11$. Meanwhile, the deformation γ varies between 47° and 59° which means small triaxiality close to oblate axial symmetry in the TAC-CDFT calculations. This is consistent with the PQTAC results of Ref. 68.

3.3. *Shears mechanism*

One of the fundamental and important characteristics in magnetic rotation is the shears mechanism. In Fig. 7, the proton and neutron angular momentum vectors $\boldsymbol{J}_\pi$ and $\boldsymbol{J}_\nu$ as well as the total angular momentum vectors $\boldsymbol{J}_{\rm tot} = \boldsymbol{J}_\pi + \boldsymbol{J}_\nu$ at both the minimum and the maximum rotational frequencies in TAC-CDFT calculations for the band 1 in ^{198}Pb are shown. The proton and neutron angular momenta $\boldsymbol{J}_\pi$ and $\boldsymbol{J}_\nu$ are defined as

$$\boldsymbol{J}_\pi = \langle \hat{\boldsymbol{J}}_\pi \rangle = \sum_{p=1}^{Z} \langle p|\hat{J}|p\rangle, \qquad \boldsymbol{J}_\nu = \langle \hat{\boldsymbol{J}}_\nu \rangle = \sum_{n=1}^{N} \langle n|\hat{J}|n\rangle, \tag{32}$$

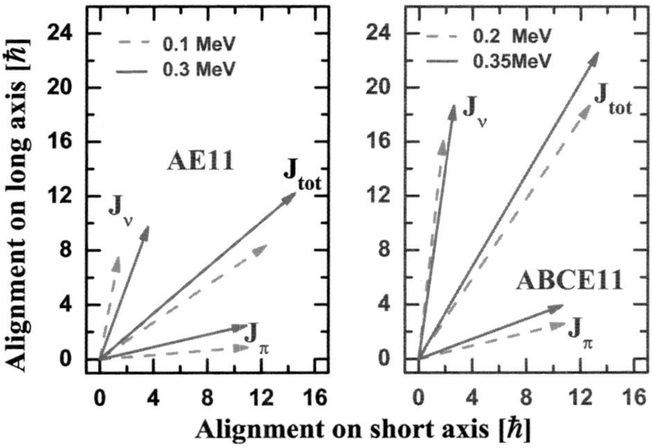

Fig. 7. (Color online) Composition of the total angular momentum at both the minimum and the maximum rotational frequencies in TAC-CDFT calculations for band 1 in ^{198}Pb. Left (right) panel is result for the rotation before (after) backbending. Taken from Ref. 66.

where the sum runs over all the proton (or neutron) levels occupied in the cranking wave function in the intrinsic system.

For the magnetic dipole bands in ^{198}Pb, the contributions to the angular momenta come mainly from the high-j orbits, i.e., the $i_{13/2}$ neutron(s) as well as $h_{9/2}$ and $i_{13/2}$ protons. At the band head, the proton particles excited across the closed $Z = 82$ shell gap into the $h_{9/2}$ and $i_{13/2}$ orbitals contribute to the proton angular momentum along the short axis, and the neutron hole(s) at the upper end of the $i_{13/2}$ shell contribute to the neutron angular momentum along the long axis. By comparing the left panel (before backbending) with the right one (after backbending) in Fig. 7, one finds that after the backbending the neutron angular momentum vectors are considerably larger, because they contain the contributions of an aligned pair of $i_{13/2}$ neutron holes. Therefore, the proton and neutron angular momentum vectors form the two blades of the shears. With the frequency increasing, the two blades move toward each other and the closing of shears increases the angular momentum, while the direction of the total angular momentum stays nearly unchanged. In such a way, the shears mechanism is clearly observed.

3.4. *Electric and magnetic transitions*

A typical characteristic of magnetic rotation is the strongly enhanced M1 transition probabilities which decrease with the spin. In Fig. 8, the calcu-

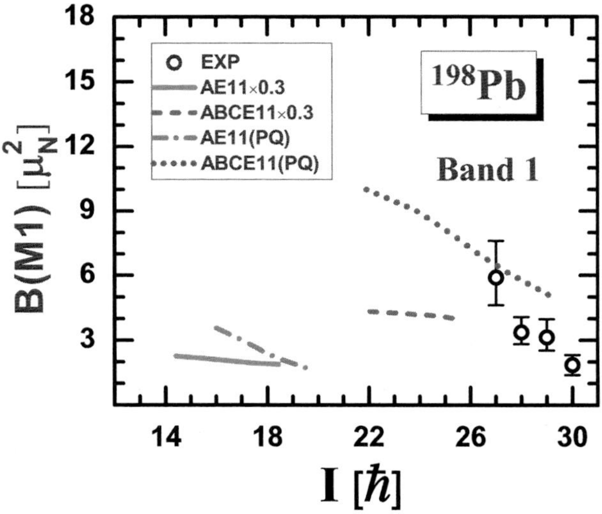

Fig. 8. (Color online) $B(M1)$ values as functions of the total angular momentum in the TAC-CDFT calculations compared with the data and the PQTAC results[68] for band 1 in ^{198}Pb. Circles and squares denote experimental data from DSAM[15] and RDM,[62] respectively. Taken from Ref. 66.

lated $B(M1)$ values as functions of the total angular momentum for the band 1 in ^{198}Pb are shown in comparison with the data[15,62] and the PQ-TAC results.[68] The TAC-CDFT calculations reproduce the decrease of the observed $B(M1)$ values with increasing spin. However, as observed already in earlier investigation,[22,24] the absolute values show discrepancies. As shown in Fig. 8, one has to attenuate the results by a factor 0.3 in order to reproduce the absolute $B(M1)$ values. The same factor has been used in Refs. 22, 24. So far the origin of this attenuation factor is not understood in detail. As discussed in Ref. 24, there are however several reasons: (a) Pairing correlations strongly affect the deformation and the levels in the neighborhood of the Fermi surface. This causes a strong reduction for the $B(M1)$ values with major contributions from the valence particles or holes. (b) The coupling to complex configurations such as particle vibrational coupling (Arima–Horie effect[69,70]) leads in all cases to a quenching of the $B(M1)$ values for neutron configurations.[71,72] (c) Meson exchange currents and higher corrections also cause a reduction of the effective g-factors for the neutrons.[73–75] However, it is not the absolute $B(M1)$ value which characterize the shear bands, but rather the behavior of these values with increasing angular momentum. On the other side, the absolute

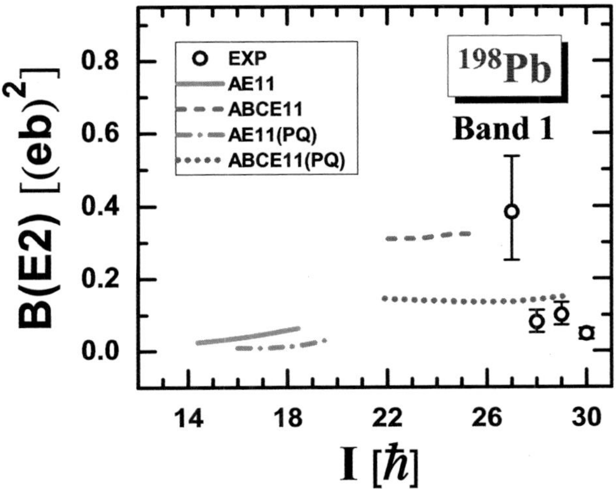

Fig. 9. (Color online) $B(E2)$ values as functions of the total angular momentum in the TAC-CDFT calculations compared with the data from DSAM[15] and the PQTAC results[68] for band 1. Taken from Ref. 66.

values of PQTAC results agree with the observed $B(M1)$ data and the attenuated TAC-CDFT results. However, they show a sharper decreasing trend as compared with the TAC-CDFT calculations. The agreement between the calculated and experimental $B(M1)$ values and their trend shows a convincing confirmation of the shears mechanism.

In contrast to the enhanced M1 transitions, the E2 transitions are weak for magnetic rotational bands. In Fig. 9, the calculated $B(E2)$ values as functions of the total angular momentum are shown and compared with the DSAM-data of Ref. 15 and the PQTAC results of Ref. 68 for band 1 in ^{198}Pb. The $B(E2)$ values in the TAC-CDFT calculations are in reasonable agreement with the data and show a roughly constant trend. This is consistent with the nearly constant quadrupole deformation in each configuration calculated. Compared to the PQTAC results, the TAC-CDFT calculations predict larger $B(E2)$ values, in accordance with the larger deformations shown in Fig. 5.

3.5. *Magnetic rotation in other mass regions*

To date, the magnetic dipole bands observed have been summarized in the nuclear chart in Fig. 2.

$A\sim 140$ mass region. For $A \sim 140$ mass region, $\pi h_{11/2}$ particles combined with $\nu h_{11/2}$ holes satisfy the high-j configurations for magnetic rotation. In fact, magnetic rotation in $A \sim 140$ mass region has been identified in Tb, Gd, Eu, Sm, Pm, Nd, Pr, Ce, La, Ba, Cs, Xe, Te and Dy isotopes (see Ref. 45 and the references therein).

In Ref. 6, TAC-CDFT with the meson-exchange interaction PK1[76] has been applied to one of magnetic dipole bands in ^{142}Gd based on the configuration $\pi h^2_{11/2} \otimes \nu h^{-2}_{11/2}$. The shears mechanism and the spectrum of this band was previously investigated by the tilted axis cranking Skyrme Hartree–Fock method in Ref. 77. It was shown that the spectrum calculated with TAC-CDFT achieves better agreement with the data as compared with the SHF results and the characteristic shears mechanism of magnetic rotation was provided.[6]

$A\sim 110$ mass region. For $A \sim 110$ mass region, $\pi h_{11/2}$ particles combined with $\nu g_{9/2}$ holes satisfy the high-j configurations for magnetic rotation. In fact, the magnetic rotation in $A \sim 110$ mass region has been identified in Cd, In, Sn, Sb, and Te isotopes (see Ref. 45 and the references therein). Using the recently developed TAC-CDFT based on point-coupling interactions, the magnetic rotation bands in ^{113,114}In in $A \sim 110$ mass region are well reproduced.[54,57]

$A\sim 80$ mass region. For $A \sim 80$ mass region, $\pi g_{9/2}$ particles combined with $\nu g_{9/2}$ holes satisfy the high-j configurations for magnetic rotation. In fact, the magnetic rotation in $A \sim 80$ mass region has been identified in Rb, Kr and Br isotopes (see Ref. 45 and the references therein).

As the first relativistic investigation of the magnetic rotation, three-dimensional cranking CDFT has been developed and applied for ^{84}Rb.[22] The proton configuration is fixed to be $\pi(pf)^7(1g_{9/2})^2$ with respect to the $Z = 28$ magic number and $\nu(1g_{9/2})^{-3}$ with respect to the $N = 50$ magic number is adopted for the neutron configuration. The signals of the shears mechanism, such as the nearly constant tilt angle and the smooth decreases of the shears angle and of the $B(M1)/B(E2)$ ratio, are well reproduced. Note that because of the numerical complexity, so far, the three-dimensional cranking CDFT has been applied only for the magnetic rotation in ^{84}Rb.[22]

$A\sim 60$ mass region. The recent observations in ^{58}Fe and ^{60}Ni are identified as the lightest mass region to exhibit magnetic rotation phenomenon and have extended the observed MR mass region to $A \sim 60$ mass region.[58,59]

In Ref. 24, TAC-CDFT calculation with PC-PK1[65] has been performed for four magnetic dipole bands, denoted as M-1, M-2, M-3, and M-4, re-

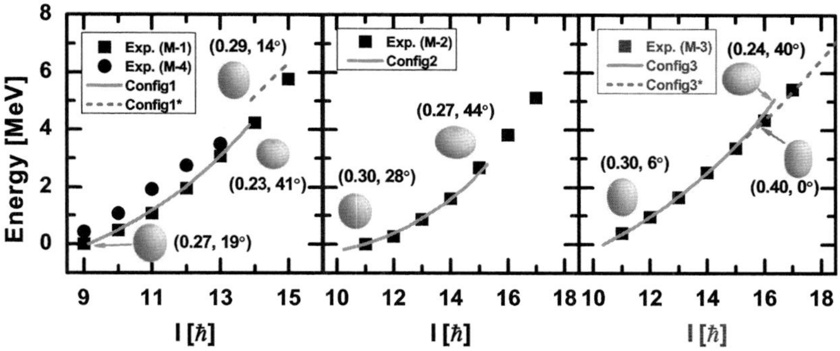

Fig. 10. (Color online) Energy spectra obtained from the TAC-CDFT calculations in comparison with the available data for bands M-1 and M-4 (left panel), M-2 (middle panel), as well as M-3 (right panel). The energies at $I = 9\hbar$, $I = 11\hbar$, and $I = 15\hbar$ are taken as references in the left, middle and right panels, respectively. The evolution of the nuclear shape (β, γ) for bands M-1, M-2, and M-3 are also illustrated with the schematic pictures. Taken from Ref. 24.

ported in ^{60}Ni.[59] The bands M-1 and M-4 are suggested[59] to be built from the same type of configurations, i.e., $\pi[(1f_{7/2})^{-1}(fp)^1] \otimes \nu[(1g_{9/2})^1(fp)^3]$. For the bands M-2 and M-3, TAC-CDFT calculation clearly indicates that they are respectively built from the configuration $\pi[(1f_{7/2})^{-1}(1g_{9/2})^1] \otimes \nu[(1g_{9/2})^1(fp)^3]$ and $\pi[(1f_{7/2})^{-1}(fp)^1] \otimes \nu[(1g_{9/2})^2(fp)^2]$. For simplicity, the above configurations are referred as Config1, Config2, and Config3, respectively.

The calculated energy spectra are shown in comparison with the available data for the bands M-1 and M-4 (left panel), M-2 (middle panel), and M-3 (right panel) for ^{60}Ni in Fig. 10. In general, the experimental energies of the bands M-1, M-2, and M-3 are reproduced very well by these TAC-CDFT calculations. However, the assigned configuration for each of these bands could not be followed in the calculations up to the highest spin observed, i.e., convergent results could be obtained only up to $\sim 14\hbar$ for Config1, $\sim 15\hbar$ for Config2, and $\sim 16\hbar$ for Config3. These are connected with the configuration change and shape evolution.[24] With the increase of the rotational frequency, the configurations, $\pi[(1f_{7/2})^{-1}(fp)^1] \otimes \nu[(1g_{9/2})^1(fp)^4(1f_{7/2})^{-1}]$ (Config1*) and $\pi[(1f_{7/2})^{-2}(fp)^2] \otimes \nu[(1g_{9/2})^2(fp)^3(1f_{7/2})^{-1}]$ (Config3*) will strongly compete with Config1 and Config3, respectively. In other words, one observes a neutron pair broken in the $f_{7/2}$ shell at $I = 15\hbar$ in band M-1, and the excitation of a unpaired proton from the $f_{7/2}$ shell to the fp orbital and a neutron pair broken in the $f_{7/2}$ shell at $I = 16\hbar$ in the band M-3.

The shape evolutions of bands M-1, M-2, and M-3 are also shown in Fig. 10. It is interesting to note that the nucleus changes its shape from prolate-like to oblate-like with the frequency in Config1, Config2, and Config3, and comes back to a prolate-like deformation with the configuration changing from Config1 to Config1*, and Config3 to Config3*. In particular, the nucleus with Config3* has a relatively large deformation ($\beta \sim 0.4$) with axial symmetry. The mechanism of producing higher spin states with Config3* is electric rotation rather than magnetic rotation. Therefore, a transition from magnetic rotation to electric rotation is observed in Config3*.[24]

Recently, the high spin structure in ^{58}Fe has been investigated by heavy-ion induced fusion-evaporation reactions at Gammasphere.[58] The magnetic rotational bands observed have been interpreted with the TAC-CDFT, which concludes that ^{58}Fe is the lightest nucleus exhibiting magnetic rotation.

***A* ~ 20 mass region.** In order to explore the MR in the lightest mass region, the TAC-CDFT calculation has been performed for ^{22}F with the configuration $\pi d_{5/2} \otimes \nu d^{-1}_{5/2}$ in Ref. 78. The possible existence of magnetic rotation is suggested for ^{22}F via investigating the spectra, the relation between the rotational frequency and the angular momentum, the electromagnetic transition probabilities $B(M1)$ and $B(E2)$, and the shears mechanism characteristic of magnetic rotation.

4. Anti-magnetic rotation

4.1. *Basic concepts*

In analogy with an antiferromagnet, antimagnetic rotation[13,18] is predicted to occur in some specific nearly spherical nuclei, in which the subsystems of valence protons (neutrons) are aligned back to back in opposite directions and nearly perpendicular to the orientation of the total spin of the valence neutrons (protons). Such arrangement of the proton and neutron angular momenta also breaks the rotational symmetry in these nearly spherical nuclei and causes excitations with rotational character on top of this bandhead. Along this band, energy and angular momentum are increased by simultaneous closing of the two proton (neutron) blades toward the neutron (proton) angular momentum vector. Consequently, a new kind of rotational bands in nearly spherical nuclei is formed showing some analogy with an antiferromagnet. A schematic illustration for AMR is shown in Fig. 11.

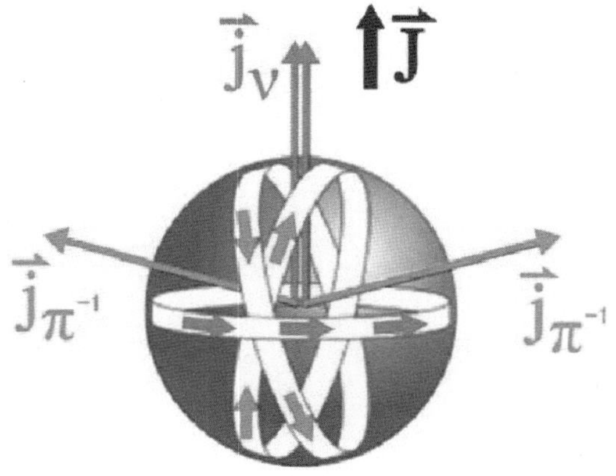

Fig. 11. (Color online) A schematic illustration for the spin-coupling scheme of antimagnetic rotation. Instead of the shears mechanism for magnetic rotation, the two shearslike configurations here have the result that the magnetic moments are antialigned and cancel, and therefore the $B(M1)$ values vanish. Taken from Ref. 25.

The experimental indicators for antimagnetic rotation can be summarized as:[25]
1) a $\Delta I = 2$ sequence with E2 transitions only, as the cancellation of the magnetic moments leads to the absence of the M1 transition;
2) weak E2 transitions corresponding to the small deformation parameter $|\beta| < 0.15$, and the $B(E2)$ values decrease with spin;
3) a smooth variation in the γ transition energy with angular momentum;
4) a substantial dynamic moment of inertia, similar to the MR band, corresponding to the large ratio of the $\mathcal{J}^{(2)}/B(E2) \gtrsim 100$ MeV$^{-1}(eb)^{-2}$.

AMR is expected to be observed in the same regions as MR in the nuclear chart.[13] However, it differs from magnetic rotation in two aspects. Firstly, there are no M1 transitions in the AMR band since the transverse magnetic moments of the two subsystems are antialigned and canceled out. Secondly, as the antimagnetic rotor is symmetric with respect to a rotation by π about the rotating axis, the energy levels in AMR band differ in spin by $2\hbar$ and are connected by weak E2 transitions reflecting the nearly spherical core. Moreover, the AMR phenomenon is characterized by a decrease of the $B(E2)$ values with spin, which has been confirmed by lifetime measurements.[79] To date, AMR has attracted lots of attention and has been observed in Cd isotopes including ^{105}Cd,[80] ^{106}Cd,[79] ^{108}Cd,[81,82] ^{109}Cd,[83] ^{110}Cd,[84] and ^{112}Cd.[85] The other candidates include ^{100}Pd,[86] and ^{144}Dy.[87]

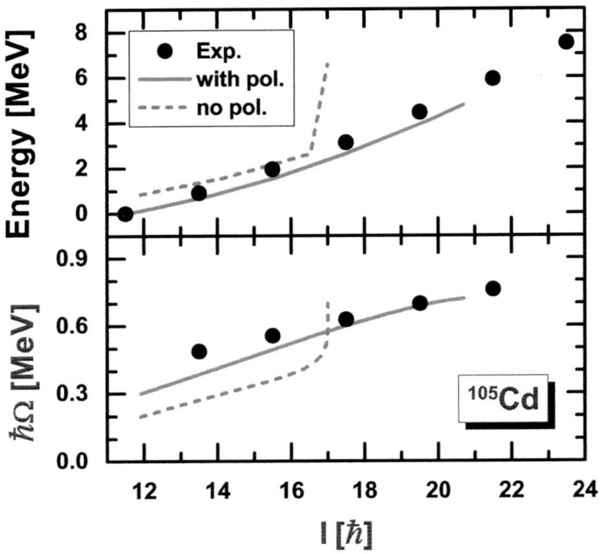

Fig. 12. (Color online) Energy (upper panel) and rotational frequency (lower panel) as functions of the total angular momentum. The fully self-consistent solution (solid lines) and that neglecting polarization (dashed lines) are compared with the data[80] (solid dots). The energy at $I = 23/2\hbar$ is taken as reference in the upper panel. Taken from Ref. 88.

In order to apply the CDFT for antimagnetic rotation phenomenon, the newly developed TAC-CDFT based on point-coupling interactions are used to investigate antimagnetic rotation (AMR) in ^{105}Cd in a fully self-consistent and microscopic way in Refs. 88, 89.

4.2. *A representative example:* ^{105}*Cd*

For the TAC-CDFT calculation[88,89] based on point-coupling interactions PC-PK1[65] for AMR band in ^{105}Cd, the odd neutron occupies the lowest level in the $h_{11/2}$ shell and the remaining nucleons are treated self-consistently by filling the orbitals according to their energy from the bottom of the well. This automatically leads to the configuration for AMR.

In Fig. 12, the TAC-CDFT calculated energy and the rotational frequency (solid lines)[88] are compared with data.[80] In the upper panel it can be clearly seen that, apart from the bandhead, the experimental energies are reproduced excellently by the self-consistent calculations. In the lower panel it is found that the calculated total angular momenta agree well with the data and increase almost linearly with increasing frequency. This indi-

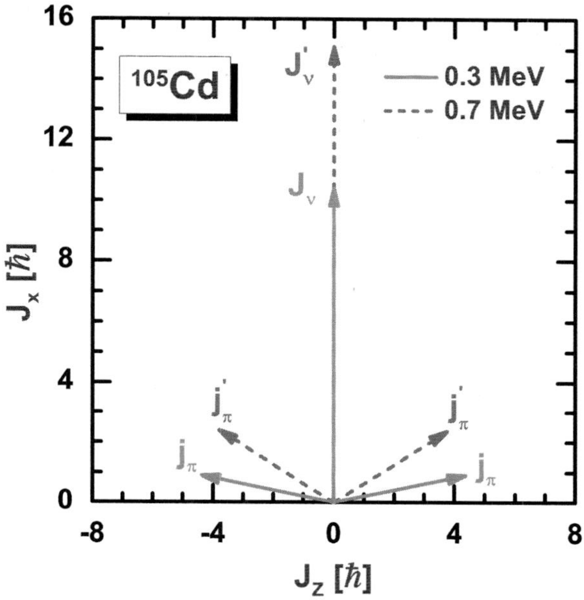

Fig. 13. (Color online) Angular momentum vectors of neutrons $\boldsymbol{J}_\nu$ and the two $g_{9/2}$ proton holes $\boldsymbol{j}_\pi$ at both the bandhead ($\hbar\Omega = 0.3$ MeV) and the maximum rotational frequency. Taken from Ref. 88.

cates that the moment of inertia is nearly constant and well reproduced by the TAC-CDFT calculations.

4.3. *Two shearslike mechanism*

In order to demonstrate the two shearslike mechanism in ^{105}Cd, both at the bandhead and at the maximum rotational frequency are shown in Fig. 13, the angular momentum vectors of the two $g_{9/2}$ proton-holes $\boldsymbol{j}_\pi$ and of the neutrons $\boldsymbol{J}_\nu = \sum_n \boldsymbol{j}_\nu^{(n)}$, where n runs over all the occupied neutron levels. At the bandhead, the two proton angular momentum vectors $\boldsymbol{j}_\pi$ are pointing opposite to each other and are nearly perpendicular to the vector $\boldsymbol{J}_\nu$. They form the blades of the two shears. With increasing Ω the gradual alignment of the vectors $\boldsymbol{j}_\pi$ of the two $g_{9/2}$ proton holes toward the vector $\boldsymbol{J}_\nu$ generates angular momentum while the direction of the total angular momentum stays unchanged. This leads to the closing of the two shears. The two shearslike mechanism can thus be clearly seen, and it is consistent with the previous works.[13,79]

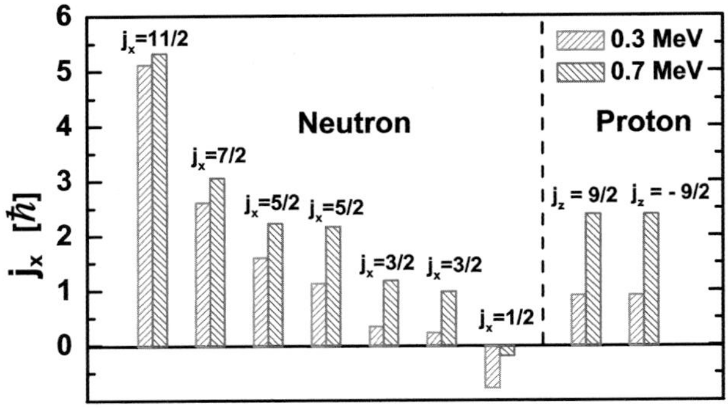

Fig. 14. (Color online) Alignment of the valence neutrons (left side) and proton holes (right side) at both the bandhead ($\hbar\Omega = 0.3$ MeV) and the maximum rotational frequency. Taken from Ref. 88.

In a microscopic calculation, there is no inert core and all the energy and angular momentum comes from the particles. The contributions of the valence neutrons and proton holes to the angular momentum J_x at both the bandhead and the maximum rotational frequency are shown in Fig. 14. It is found that the contributions come mainly from high-j orbitals, i.e., from $g_{9/2}$ proton holes as well as from $h_{11/2}$ and $g_{7/2}$ neutrons. In order to provide a simple picture which can be compared with the core angular momentum given in Ref. 13, one can estimate the "corelike" angular momentum by excluding the contributions of three valence neutrons, shown in the left three columns in Fig. 14, from the total neutron angular momentum. It is found that the "core" contributes about $3\hbar$ when the frequency Ω increases from the bandhead to the maximum value.

For the protons, only the two holes in the $g_{9/2}$ shell contribute. As shown in Fig. 13, they cancel each other in the z direction giving non-negligible contributions to the angular momentum along the x axis even at the bandhead. With growing frequency, the proton angular momentum in the x direction increases because of the alignment of the two proton hole blades. For the neutrons, on the other hand, we have only contributions above the $N = 50$ shell. One neutron sits in the $h_{11/2}$ orbit and the other six are, because of considerable mixing, distributed over the $g_{7/2}$ and $d_{5/2}$ orbitals. As Ω grows, the contributions of the aligned orbitals with $j_x = 11/2$ and $9/2$ do not change much and the increase in angular momentum is generated mostly by the alignment of orbitals with low j_x values. This microscopic calculation shows that the phenomenological interpreta-

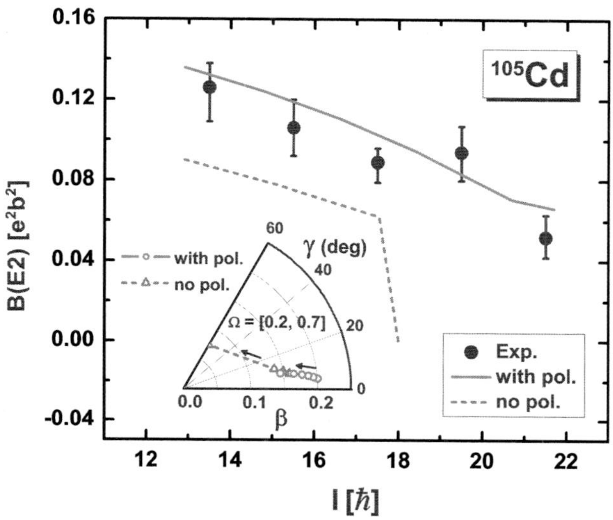

Fig. 15. (Color online) $B(E2)$ values as a function of the angular momentum. Solutions with (solid line) and without (dashed line) polarization are compared with data[80] (solid dots). Inset: Deformations β and γ driven by the increasing rotational frequency whose direction is indicated by arrows. Taken from Ref. 88.

tion given in Ref. 80 is only partially justified: it does have two proton holes in the $g_{9/2}$ and one neutron particle in the $h_{11/2}$ orbit, but, due to the mixing of orbits with lower j values the other neutrons are distributed over several subshells above the $N = 50$ core and the increasing angular momentum results from the alignment of the proton holes and the mixing within the neutron orbitals. Because of this strong mixing between the neutrons, a core needed for the phenomenological model in Ref. 12 cannot really be defined.

4.4. *Electric transition probability*

AMR is characterized by weak E2 transitions decreasing with increasing spin. In Fig. 15, the calculated $B(E2)$ values (solid lines) are compared with the available data.[80] It is found that the resulting $B(E2)$ values are very small ($< 0.14\, e^2b^2$) and in very good agreement with the data. Furthermore, the fact that the $B(E2)$ values decrease with spin is in agreement with the interpretation by the two shearslike mechanism.

The decrease of the $B(E2)$ values can be understood by the changes in the nuclear deformation. As shown in the inset of Fig. 15, with increasing frequency, the nucleus undergoes a rapid decrease of β deformation from 0.2 to 0.14 at a small and near-constant triaxiality ($\gamma \leq 9°$). As usual, it is found that the deformation of the charge distribution, responsible for the $B(E2)$ values, changes in a similar manner. Therefore, one can conclude that the alignment of the proton and neutron angular momenta, i.e., the two shearslike mechanisms, is accompanied by a transition from prolate towards nearly spherical shape.

4.5. *Core polarization*

In order to investigate the importance of the polarization effects induced by the two proton holes, which is taken into account fully in the TAC-CDFT calculation, an additional calculation without polarization has been carried out in Ref. 88. For this purpose, at each frequency the ^{107}Sn core is firstly calculated in Ref. 88, where the two proton holes are filled. This results in a filled nearly spherical $g_{9/2}$ shell. In the next step, the self-consistency (dashed lines in Figs. 12 and 15) is neglected and the band in ^{105}Cd is calculated, in which two protons are removed from the $g_{9/2}$ shell using, at each frequency, the corresponding nearly spherical potentials S and V^{μ} obtained in the calculations of the ^{107}Sn core. As shown in the upper panel of Fig. 12, the energy is reduced only slightly by polarization in the lower part of the spectrum. At the same time the relation between angular velocity and angular momentum is considerably changed in the lower part of Fig. 12. Without polarization, a much smaller frequency Ω is necessary to reach the same angular momentum as with polarization. In addition, without polarization, there is a maximal angular momentum of roughly $17\hbar$. Higher values cannot be reached even at rather high frequencies.

This behavior can be well understood from the evolution of the deformation parameters shown in the inset of Fig. 15. Without polarization we use at each frequency the potentials of the core-nucleus ^{107}Sn, where the deformation is relatively small. Angular momentum can only be produced by alignment of neutron particles along the rotational axis leading at $\hbar\Omega = 0.5$ MeV to an oblate shape with a rotation around the symmetry axis. Removing two protons would lead, if polarization is taken into account, to a larger prolate configuration with lower energy and hindering alignment. Therefore, in the lower part of the band, it is easier to produce angular momentum without polarization, where the deformation is small. On the other side the oblate deformation keeps the high-j proton holes in the $j_x = \pm 1/2$ orbitals of the $g_{9/2}$ shell pairwise occupied and hinders their alignment. Above $\hbar\Omega = 0.5$ MeV we reach the maximum angular momentum of the neutron configuration. With polarization, because of the prolate deformation, we have much more mixing and can therefore reach larger angular momentum by aligning the protons. This can also be seen in Fig. 15 that the E2 transitions do survive with polarization when $\hbar\Omega \geq 0.5$ MeV. Therefore, it is of importance to emphasize that polarization effects play a very important role in the self-consistent microscopic description of AMR bands.

5. Chirality in triaxial nuclei

5.1. *Basic concepts*

The term "chiral" was first used by Lord Kelvin in 1893:[90] "I call any geometrical figure, or group of points, 'chiral', and say that it has chirality if its image in a plane mirror, ideally realized, cannot be brought to coincide with itself". Static chiral symmetries exist commonly in nature, including the macroscopic spirals of snail shells and the microscopic handedness of certain molecules, and now spontaneous chiral-symmetry breaking has become a phenomenon of general interest in many branches of physics, such as particle physics, optics, condensed matter physics, and nuclear physics.

In nuclear physics, Frauendorf and Meng originally suggested the existence of spontaneous chiral-symmetry breaking in rotating triaxial odd-odd nuclei in 1997.[20] A schematic illustration for nuclear chirality is shown in Fig. 16. If the two odd nucleons have a particle- and hole-like nature respectively, they align their angular momentum vectors along the short and long axes of the nucleus, while the collective rotation vector aligns along the intermediate axis, which could have the largest moment of inertia. A left- and right-handed system will be generated in the intrinsic frame depending on which side of the short-long plane the rotation vector projects from. These two systems are transformed into each other by the chiral operator which combines time reversal and spatial rotation of 180°, $\chi = \mathcal{TR}(\pi)$. In the laboratory frame, the restoration of spontaneous chiral-symmetry breaking manifests itself as a pair of $\Delta I = 1$ bands, called *chiral doublet bands*, with the same parity and near degenerate in energy due to the effect of quantum tunneling.

Experimentally, evidences of chiral doublet bands have been reported in the $A \approx 80$, 100, 130 and 190 mass regions.[91–108] However, the lifetime measurements which are essential to extract the absolute electromagnetic

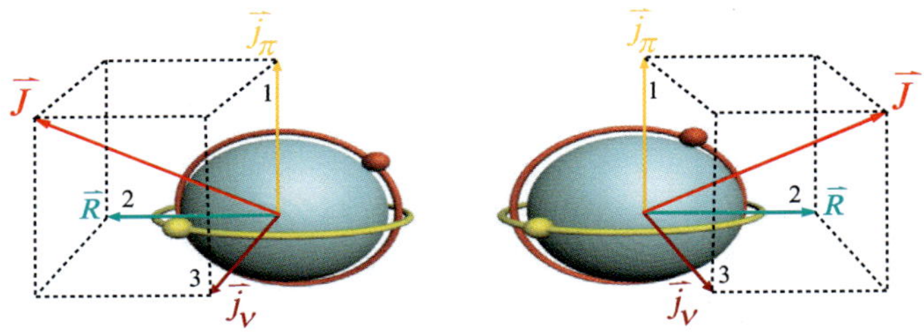

Fig. 16. Left- and right-handed chiral systems for a triaxial odd-odd nucleus in the intrinsic frame. The vector $\boldsymbol{J}$ represents the total angular momentum of the nucleus, and $\boldsymbol{j}_{\pi}$, $\boldsymbol{j}_{\nu}$, $\boldsymbol{R}$ respectively the angular momenta of the odd-proton, odd-neutron and collective core. Here axis-2 indicates the intermediate axis, while axis-1 and axis-3 expand the short-long plane of the triaxial nucleus. Taken from Ref. 21.

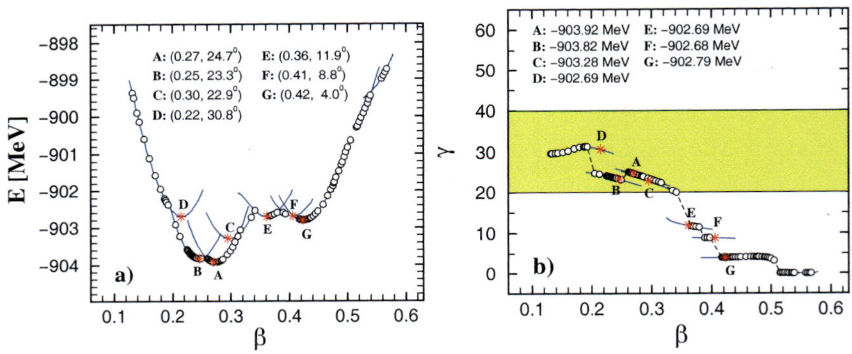

Fig. 17. The energy surfaces (a) and the deformations γ (b) as functions of deformation β in adiabatic (open circles) and configuration-fixed (solid lines) constrained triaxial RMF calculation with PK1 for ^{106}Rh. The minima in the energy surfaces are represented as stars and labelled respectively as A, B, C, D, E, F and G. Their corresponding deformations β and γ together with their energies are respectively given in (a) and (b). Taken from Ref. 126.

transition probabilities are still rare for the candidate chiral bands. Examples in $A \sim 130$ mass region include ^{134}Pr,[109] ^{128}Cs,[100] ^{135}Nd,[110] ^{126}Cs.[111] In $A \sim 100$ mass region, the lifetime measurements have been performed for ^{103,104}Rh,[112] ^{102}Rh[113] and ^{106}Ag.[114,115]

Theoretically, the chiral doublets bands have been successfully described by a variety of approaches including the triaxial particle rotor model (PRM),[20,116–118] the interacting boson fermion-fermion model (IBFFM),[109,119,120] the tilted axis cranking (TAC) model with shell correction (SCTAC),[121] Skyrme–Hartree–Fock,[122,123] and the random phase approximation.[110,124] Recently, a new collective Hamiltonian to describe the chiral rotation and vibration was proposed and applied to the system with one $h_{11/2}$ proton particle and one $h_{11/2}$ neutron hole coupled to a triaxial rotor.[125]

Based on the covariant density functional theory, in order to describe the chiral rotation self-consistently, one should introduce the three-dimensional cranking techniques. The cranked RMF theory in the context of three-dimensional rotation has been developed.[26] However, due to numerical complexity, it was restricted to two-dimensional studies only, i.e., the magnetic rotation.[22,26]

On the other hand, for the study of nuclear chirality, it is quite helpful and constructive to search for nuclear configurations and triaxial deformation parameters suitable for chirality. For this purpose, the adiabatic and configuration-fixed constrained triaxial CDFT approaches were first used to investigate the triaxial shape coexistence and possible chiral doublet bands in ^{106}Rh.[126] It was demonstrated that it is possible to have multiple pairs of chiral doublet bands in a single nucleus, and the acronym MχD was introduced for this phenomenon.[126] This prediction will remain with the time-odd fields included,[127] and also holds true for other rhodium isotopes, ^{104,106,108,110}Rh.[128] Subsequently, with the available experimental information, candidates of MχD bands were suggested in ^{105}Rh[129] and ^{107}Ag.[130]

In 2013, the first experimental evidence for the MχD was reported in ^{133}Ce.[131] Calculations based on a combination of the constrained triaxial CDFT and the triaxial particle rotor model (PRM)[116–118] were performed to investigate the nature of the observed pairs of doublet bands in ^{133}Ce. It was found both the negative parity bands 5 and 6 with $\pi(1h_{11/2})^2 \otimes \nu(1h_{11/2})^{-1}$ configuration and the positive parity bands 2 and 3 with $\pi[(1g_{7/2})^{-1}(1h_{11/2})^1] \otimes \nu(1h_{11/2})^{-1}$ configuration can be interpreted as a pair of chiral doublet bands. The parity of one pair of chiral bands distinguishes it with the other. Thus two distinct chiral doublet bands were observed for the first time in ^{133}Ce, and were interpreted in the context of the MχD phenomenon. This observation further confirms the manifestation of triaxial shape coexistence in this nucleus.

It is also very interesting to study the robustness of chiral geometry against the increase of the intrinsic excitation energy, i.e., whether the chiral geometry is sustained or not in the higher excited bands with suitable chiral configuration. In fact, the MχD predicted based on the same configuration[132–134] provides a good probe to verify such robustness. Very recently, the first experimental evidence for such a new type of MχD have been reported in ^{103}Rh.[135]

5.2. *Multiple chiral bands (MχD)*

The phenomenon of MχD, i.e., more than one pair of chiral doublet bands exist in a single nucleus, was first predicted based on triaxial CDFT in ^{106}Rh,[126] where each pair of chiral doublet bands differ from each other in their triaxial deformations and multiparticle configurations. This phenomenon represents an important confirmation of triaxial shape coexistence in nuclei.

In general, the triaxial CDFT calculation leads to only some local extrema. In order to get the ground state for the triaxial deformed nucleus, constrained calculations are necessary, and in principle, such calculations should be carried out in the two-dimensional β and γ plane. However, the two-dimensional constrained calculations are time-consuming; for simplicity in Ref. 126 the alternative $\hat{Q}_{20}^2 + \hat{Q}_{22}^2$, i.e., β^2 constrained calculations were carried out to search for the ground state for a triaxially deformed nucleus.

The energy surface and the deformation γ as functions of deformation β in adiabatic constrained triaxial RMF calculation with PK1[76] for ^{106}Rh are presented as open circles in Figs. 17(a) and (b) respectively. Some irregularities in the energy surface are seen. Furthermore some local minima are too obscure to be recognized and it is technically difficult to understand their corresponding single particle configurations. To solve these difficulties, the technique of configuration-fixed constraint[34,35,136] was introduced to the triaxial CDFT calculations in Ref. 126. Here the configuration-fixed constrained calculation requires that the single particle orbits occupied are fixed during the constrained calculation with different β, i.e.,

$$|\langle\psi_j(\beta+\delta\beta)|\psi_i(\beta)\rangle| \approx 1. \tag{33}$$

Starting from the points in the energy surface given by the adiabatic constrained calculations, the energy surfaces and the deformations γ in configuration-fixed calculations with PK1 for ^{106}Rh are given as solid lines in Figs. 17(a) and (b) respectively. For each fixed configuration, the constrained calculation gives a continuous and smooth curve for the energy surface and the deformation γ as a function of deformation β. The irregularities in the adiabatic energy surface disappear. The minima in the configuration-fixed energy surfaces become obvious, which are respectively represented by stars and labelled as A, B, C, D, E, F and G. Their

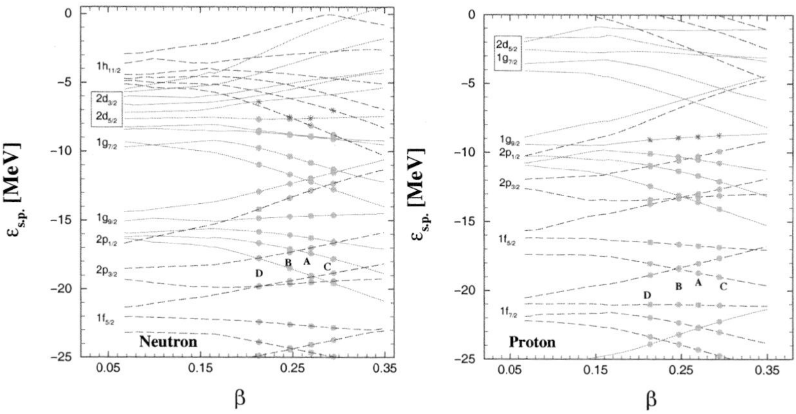

Fig. 18. (Color online) The neutron and proton single particle levels obtained in configuration-fixed constrained triaxial RMF calculations with PK1 for ^{106}Rh as functions of the deformation β. Positive (negative) parity states are marked by solid (dashed) lines. The corresponding quantum numbers for spherical case are labelled at the left side of the levels. The occupations corresponding to the minima in Fig. 17 are represented by filled circles (two particles) and stars (one particle). Taken from Ref. 126.

corresponding deformations β and γ together with their binding energies are respectively given in Figs. 17(a) and (b). One can find that, for each fixed configuration, the deformation γ is approximately a constant (as in Fig 17(b)), indicating that the triaxial deformation is relatively stable for given configuration.[126] Here the nucleus ^{106}Rh provides a good example of the triaxial shape coexistence, as the energies of these minima are within 1.3 MeV of each other but correspond to different deformations β and γ. In addition, the states A, B, C, and D have deformation parameters β and γ suitable for chirality.[20,116]

The neutron and proton single particle levels as functions of deformation β, obtained in the configuration-fixed constrained calculations for the ground state, are presented in Fig. 18. The difference in single particle levels obtained by choosing other minima is negligible. As the energy curves become very stiff for small deformation in Fig. 17, the configuration-fixed constrained calculations for the ground state of ^{106}Rh cannot be performed for deformation $\beta < 0.06$. Therefore the occupation of some low-j orbits is indistinguishable, e.g., the last occupied neutron orbit for ground state A may come from $2d_{5/2}$ or $2d_{3/2}$, as marked in the figures.

Table 1 summarizes the binding energies, deformations β and γ as well as the corresponding configurations obtained for the minima A, B, C, and D

Table 1. The binding energies, deformations β and γ and the corresponding configurations for the minima A, B, C, and D in ^{106}Rh obtained in the configuration-fixed constrained triaxial CDFT calculations with PK1. Taken from Ref. 126.

State	E (MeV)	β	γ	Configurations
A	903.9150	0.270	24.7°	$\pi(1g_{9/2})^{-3} \otimes \nu(1h_{11/2})^2[(2d_{5/2})^1$ or $(2d_{3/2})^1]$
B	903.8196	0.246	23.3°	$\pi(1g_{9/2})^{-3} \otimes \nu(1h_{11/2})^1[(2d_{5/2})^2$ or $(2d_{3/2})^2]$
C	903.2790	0.295	22.9°	$\pi(1g_{9/2})^{-3} \otimes \nu(1h_{11/2})^3$
D	902.6960	0.215	30.8°	$\pi(1g_{9/2})^{-3} \otimes \nu[(2d_{5/2})^3$ or $(2d_{3/2})^3]$

in ^{106}Rh. Except the state D, in which there is no high-j neutron valence particle, the high-j proton and neutron configuration was found for the ground state (state A) as $\pi(1g_{9/2})^{-3} \otimes \nu(1h_{11/2})^2$, state B as $\pi(1g_{9/2})^{-3} \otimes \nu(1h_{11/2})^1$, and state C as $\pi(1g_{9/2})^{-3} \otimes \nu(1h_{11/2})^3$, respectively. These high-j proton holes and high-j neutron particles configurations, together with their triaxial deformations, favor the construction of the chiral doublet bands and led to the prediction of the new phenomenon MχD in Ref. 126.

The prediction of the triaxial shape coexistence and possible chiral doublet bands in ^{106}Rh remains even with the inclusion of the time-odd fields,[127] and also holds true for other rhodium isotopes, ^{104,106,108,110}Rh.[128] Subsequently, the MχD candidates were suggested in ^{105}Rh[129] and ^{107}Ag.[130]

5.3. *MχD with distinct configurations: ^{133}Ce*

After the prediction of MχD,[126] efforts have been made to search for the corresponding experimental evidences. Using the ATLAS facility at the Argonne National Laboratory, high-spin states in ^{133}Ce were populated following the ^{116}Cd(^{22}Ne, 5n) reaction. A more complete level scheme was obtained, of which two distinct sets of doublet bands (respectively labeled as bands 5 and 6 and bands 2 and 3) were identified.[131]

Calculations based on a combination of the constrained triaxial CDFT[126–129] and the triaxial PRM[116–118] were performed to investigate the observed pairs of doublet bands in ^{133}Ce. In Fig. 19, the energy spectra, $S(I)$ parameters, and $B(M1)/B(E2)$ calculated by PRM are in comparison with the experimental values for the negative parity bands 5 and 6 as well as the positive parity bands 2 and 3. For bands 5 and 6, they are based on a $\pi(1h_{11/2})^2 \otimes \nu(1h_{11/2})^{-1}$ configuration. The theoretical results show an impressive agreement with the data. The energy separation between the two bands is about 400 keV at $I = 29/2\hbar$, and remains relatively constant over an extended spin range. The energy variation between

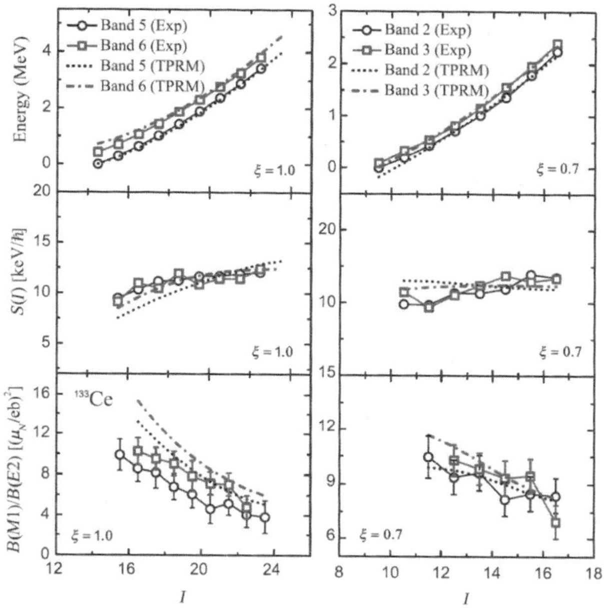

Fig. 19. (Color online) Experimental excitation energies, $S(I)$ parameters, and $B(M1)/B(E2)$ ratios for the negative-parity chiral doublet (left panels) and positive-parity chiral doublet (right panels) in ^{133}Ce. Also shown are results of PRM calculations with the indicated attenuation factors ξ (see text). Taken from Ref. 131.

bands 5 and 6 similarly suggests a form of chiral vibration, a tunneling between the left- and right-handed configurations, such that bands 5 and 6 are associated with the zero- and one-phonon state, respectively. Furthermore, the calculated staggering parameter $S(I)$ varies smoothly with spin. This is expected since the Coriolis interaction is substantially reduced for a three-dimensional coupling of angular momentum vectors in a chiral geometry. The calculated electromagnetic transition probability ratio in the PRM has no obvious odd-even staggering of the $B(M1)/B(E2)$ values, although a small effect is apparent experimentally. This further supports the interpretation in terms of a chiral vibration between bands 5 and 6.[137]

The positive parity bands 2 and 3 are associated with the configuration $\pi[(1g_{7/2})^{-1}(1h_{11/2})^1] \otimes \nu(1h_{11/2})^{-1}$. When comparing the experimental energies with the PRM calculations, a Coriolis attenuation factor, $\xi = 0.7$, was employed considering the fact that the configuration contains a low-j orbital $\pi(1g_{7/2})$ which has large admixtures with other low-j orbitals. The energy separation between bands 2 and 3 was found to be nearly constant at $\sim$ 100 keV, which, combined with the spin-independent $S(I)$ parameter, leads to the interpretation of these bands being chiral partners as well. The similar behavior of $B(M1)/B(E2)$ ratios is also seen. Similar to bands 5 and 6, there is no perceptible staggering in the calculated $B(M1)/B(E2)$ ratios, but such an effect is apparent experimentally. The $B(M1)/B(E2)$ ratio depends sensitively on the details of the transition from the vibrational to the tunneling regime,[137] which may account for the deviations of the PRM calculation from experiment.

5.4. *MχD with identical configuration: ^{103}Rh*

The phenomenon of MχD may exist with the same particle-hole configuration, i.e., not only the yrast and yrare bands but also highly excited doublets bands might be chiral partners.[20,132–134]

Medium- and high-spin states of ^{103}Rh were populated using the ^{96}Zr(^{11}B,4n) reaction at a beam energy of 40 MeV.[135] The beam, provided by the 88-inch Cyclotron of the Lawrence Berkeley National Laboratory, impinged upon an enriched 500 μg/cm^2 thick self-supporting Zr foil. The emitted γ-rays were detected by the Gammasphere spectrometer. A more complete level scheme of ^{103}Rh, including three positive parity and five negative parity bands, was constructed using the observed coincidence relations and relative intensities of the gamma transitions and based on the formerly reported states.[106,138]

In order to understand the observed band structure in ^{103}Rh, the adiabatic and configuration-fixed constrained CDFT calculations[126] were performed to search for the possible configurations and deformations. Subsequently, the configurations and deformations were further confirmed by tilted axis cranking CDFT (TAC-CDFT) calculations[24,25,88,89] by examining the energy spectra, Routhians, spin-frequency relations, deformations, and alignments. Finally, with the obtained configurations and deformations, quantum PRM[116–118] calculations were performed to study the energy spectra and *B(M1)/B(E2)* ratios for both the positive- and the negative-parity bands.

The obtained energy spectra by PRM are shown in comparison with data in Fig. 20. For bands 1 and 2, the PRM results excellently reproduce the data. These two bands are separated by $\sim$500 keV at $I = 29/2\,\hbar$. They approach each other with increasing spin and the separation finally goes to $\sim$360 keV at $I = 39/2$. The $B(M1)/B(E2)$ values of bands 1 and 2 are similar. The observation that the experimental $B(M1)/B(E2)$ values for

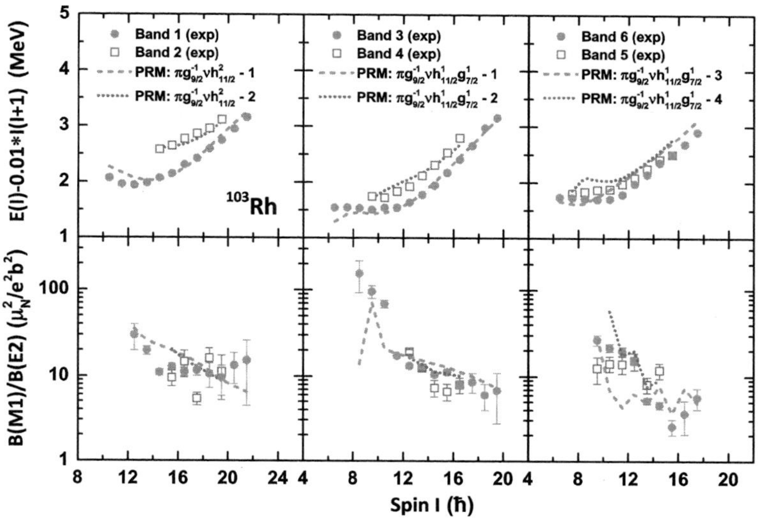

Fig. 20. (Color online) Experimental excitation energies and *B(M1)/B(E2)* ratios for the positive-parity chiral bands 1–2 (left panels) and negative-parity multiple chiral bands 3–6 (middle and right panels) in ^{103}Rh together with the results of triaxial particle rotor model. The number following the configuration label of the theoretical curve corresponds to the energy ordering of the calculated band with the given configuration. Taken from Ref. 135.

bands 1 and 2 do not fall off as quickly with spin as the theoretical values comes from the frozen rotor assumption adopted in PRM. As discussed in Ref. 139, the chiral bands with positive parity change from chiral vibration to nearly static chirality at spin $I = 37/2$ and back to another type of chiral vibration at higher spins. Such a conclusion is still held here for the positive-parity doublet.

Special attention was paid to the four negative-parity bands 3–6, where the configuration $\pi(1g_{9/2})^{-1}\otimes \nu(1h_{11/2})^1(1g_{7/2})^1$ is adopted to describe the experimental data.[135] The four calculated bands form two chiral doublets bands, of which the first ones fit the experimental bands 3 and 4, while the second ones reasonably reproduce the trend of bands 6 and 5. The calculated energies for bands 5 and 6 are higher than the experimental values of about 200 keV, which might be because the complex correlations not fully taken into account in the PRM calculations with single-j shell Hamiltonian. The corresponding calculated electromagnetic transition probabilities, shown in Fig. 20, are also able to reproduce the data reasonably. The weak odd-even $B(M1)/B(E2)$ staggering for bands 3 and 4 is consistent with the case of chiral vibration as discussed in Ref. 137. For bands 5 and 6, the $B(M1)/B(E2)$ values show a staggering at $I = 15.5\ \hbar$, which is also reproduced by the PRM.

In contrast with the multiple chiral doublets predicted in Ref. 126 and experimentally reported in ^{133}Ce,[131] the observed MχD in the negative-parity bands of ^{103}Rh is built from the same configuration. Observation of MχD with the same configuration shows that the chiral geometry in nuclei can be robust against the increase of the intrinsic excitation energy.

6. Summary and prespective

In the past decades, the rotational-like sequences in near-spherical or weakly deformed nuclei have attracted significant attention. This phenomenon, known as magnetic rotation, has been extensively explored experimentally and theoretically. With its many successes in describing nuclear phenomena in stable as well as in exotic nuclei, the CDFT has been generalized to the cranking CDFT, the tilted axis cranking CDFT and three-dimensional cranking CDFT and applied for electric and magnetic rotations all over the nuclear chart. In particular, the newly developed TAC-CDFT based on point-coupling interactions includes significant improvements and reduces computation time, which makes it possible to perform systematic investigation. A number of MR and AMR bands have been described successfully with the TAC-CDFT, including the most well-known examples, the shears bands in nuclei ^{198,199}Pb.

One should note that the pairing correlations have been neglected in the TAC-CDFT calculations. In fact, the pairing correlations are significantly suppressed for MR and AMR bands since the high j oribitals near the Fermi surface are blocked. However, there might be some special MR and AMR bands in which the influence of the pairing correlations on the rotation exists. Therefore, it is very important to investigate the pairing effects on the MR and AMR in the framework of the tilted axis cranking covariant density functional theory. Due to the time-reversal symmetry broken, the simple BCS method is not valid and one should resort to the Bogoliubov method. The most recent progress along this line can be found in Ref. 140.

Since the particle number conservation is violated in the Bogoliubov transformation, the particle number projection is usually necessary. The so-called Particle Number Conservation (PNC) method[141,142] can treat the pairing correlation and blocking effect exactly and has been implemented successfully in the cranking Nilsson model.[142–144] The same method, but named as the Shell-model-like Approach (SLAP), has been applied to treat the pairing correlations in the framework of CDFT.[145] The SLAP has the advantage that the paring correlation is treated exactly and thus the particle number conservation is not violated in the calculations. Therefore, it would be very interesting to implement SLAP to the present tilted axis cranking covariant density functional theory to investigate the pairing effects in the tilted axis cranking calculations.

The aplanar rotation of the triaxial odd-odd nuclei leads to the spontaneous chiral-symmetry breaking. The restoration of spontaneous chiral-symmetry breaking manifests itself as chiral doublet bands. The adiabatic and configuration-fixed constrained triaxial CDFT approaches were developed and used to investigate the triaxial shape coexistence and possible chiral doublet bands in ^{106}Rh.[126] It is predicted that multiple pairs of chiral doublet bands exist in a single nucleus, i.e., the MχD phenomenon.[126] The first experimental evidence of MχD was reported in ^{133}Ce[131] with distinct configurations in 2013. This observation further confirms the phenomenon of triaxial shape coexistence. Later, a new type of MχD was first reported in ^{103}Rh with identical configuration,[135] which indicates the robustness of chiral geometry against the increase of the intrinsic excitation energy.

In order to describe the novel chiral rotation in nuclei self-consistently, the three-dimensional cranking density functional theories are necessary. Such three-dimensional TAC investigations have been carried out with the phenomenological mean field combining a spherical Woods–Saxon potential with a deformed Nilsson potential[121] and Hartree–Fock self-consistent field with Skyrme interaction.[122] Similar work has been also performed within the framework of CDFT with the meson-exchange interactions earlier.[22] However, due to the numerical complexities, the calculations of three-dimensional TAC-CDFT have not yet been performed for realistic chiral nucleus. It would be very interesting to extend the tilted axis cranking covariant density functional theory to three-dimensional cranking case with the point-coupling interaction, and efforts should be made to simplify the numerical complexities to investigate the chiral rotation.

It should be further noted that the full description of chiral doublet bands requires a theory going beyond the mean field approximation. With the mean field approximation, the three-dimensional TAC calculations cannot describe the quantum tunneling between the left-handed and right-handed states in the body-fixed frame and would not yield the chiral partners. However, combining with the collective Hamiltonian proposed recently[125] might provide a powerful tool to describe the chiral rotation and vibration microscopically and self-consistently.

Acknowledgment

We would like to thank all the friends and collaborators, who contributed to the investigations presented in this chapter, in particular to Q. B. Chen, S. Frauendorf, H. Z. Liang, H. Madokoro, M. Matsuzaki, J. Peng, P. Ring, S. Yamaji, and L. F. Yu. This work was supported in part by the Major State 973 Program of China (Grant No. 2013CB834400), the Natural Science Foundation of China (Grants Nos. 10975007, 10975008, 11175002, 11105005, 11335002, 11375015, 11461141002), the Research Fund for the Doctoral Program of Higher Education under Grant No. 20110001110087, and the China Postdoctoral Science Foundation Grant No. 2012M520101.

References

1. T. F. Wang, E. A. Henry, J. A. Becker, A. Kuhnert, M. A. Stoyer, S. W. Yates, M. J. Brinkman, J. A. Cizewski, A. O. Macchiavelli, F. S. Stephens, M. A. Deleplanque, R. M. Diamond, J. E. Draper, F. A. Azaiez, W. H. Kelly, W. Korten, E. Rubel, and Y. A. Akovali, First lifetime measurement of dipole collective bands in neutron-deficient lead nuclei, *Phys. Rev. Lett.* **69**, 1737–1740 (Sep, 1992). doi: 10.1103/PhysRevLett.69.1737. URL `http://link.aps.org/doi/10.1103/PhysRevLett.69.1737`.
2. J. R. Hughes, J. A. Becker, M. J. Brinkman, E. A. Henry, R. W. Hoff, M. A. Stoyer, T. F. Wang, B. Cederwall, M. A. Deleplanque, R. M. Diamond, P. Fallon, I. Y. Lee, J. R. B. Oliveira, F. S. Stephens, J. A. Cizewski, L. A. Bernstein, J. E. Draper, C. Duyar, E. Rubel, W. H. Kelly, and D. Vo, Lifetime measurements in the regular $\Delta\ I = 1$ oblate band in ^{197}Pb, *Phys. Rev. C.* **48**, R2135–R2139 (Nov, 1993). doi: 10.1103/PhysRevC.48.R2135. URL `http://link.aps.org/doi/10.1103/PhysRevC.48.R2135`.
3. R. M. Clark, R. Wadsworth, H. R. Andrews, C. W. Beausang, M. Bergstrom, S. Clarke, E. Dragulescu, T. Drake, P. J. Dagnall, A. Galindo-Uribarri, G. Hackman, K. Hauschild, I. M. Hibbert, V. P. Janzen, P. M. Jones, R. W. MacLeod, S. M. Mullins, E. S. Paul, D. C. Radford, A. Semple, J. F. Sharpey-Schafer, J. Simpson, D. Ward, and G. Zwartz, Recoil distance lifetime measurements of states in the oblate dipole bands of ^{197,198}pb, *Phys. Rev. C.* **50**, 84–92 (Jul, 1994). doi: 10.1103/PhysRevC.50.84. URL `http://link.aps.org/doi/10.1103/PhysRevC.50.84`.
4. M. Neffgen, G. Baldsiefen, S. Frauendorf, H. Grawe, J. Heese, H. Hübel, H. Kluge, A. Korichi, W. Korten, K. Maier, D. Mehta, J. Meng, N. Nenoff, M. Piiparinen, M. Schönhofer, R. Schubart, U. van Severen, N. Singh, G. Sletten, B. T. Rao, and P. Willsau, Lifetimes of shears bands in

^{199}Pb, *Nuclear Physics A.* **595**(4), 499–512 (1995). ISSN 0375-9474. doi: 10.1016/0375-9474(95)00375-4. URL `http://www.sciencedirect.com/science/article/pii/0375947495003754`.

5. E. F. Moore, M. P. Carpenter, Y. Liang, R. V. F. Janssens, I. Ahmad, I. G. Bearden, P. J. Daly, M. W. Drigert, B. Fornal, U. Garg, Z. W. Grabowski, H. L. Harrington, R. G. Henry, T. L. Khoo, T. Lauritsen, R. H. Mayer, D. Nisius, W. Reviol, and M. Sferrazza, Collectivity of dipole bands in ^{196}pb, *Phys. Rev. C.* **51**, 115–124 (Jan, 1995). doi: 10.1103/PhysRevC.51.115. URL `http://link.aps.org/doi/10.1103/PhysRevC.51.115`.
6. J. Peng, J. Meng, P. Ring, and S. Q. Zhang, Covariant density functional theory for magnetic rotation, *Phys. Rev. C.* **78**(2), 024313 (Aug, 2008). doi: 10.1103/PhysRevC.78.024313.
7. A. K. Kerman and O. Naoki, Nuclear rotations studied by the time-dependent variational method, *Nucl. Phys.* **A361**(1), 179–191 (1981). doi: DOI:10.1016/0375-9474(81)90475-9.
8. H. Frisk and R. Bengtsson, Spin orientation in rotating nuclei, *Phys. Lett.* **B196**(1), 14–18 (1987). doi: 10.1016/0370-2693(87)91667-4.
9. S. Frauendorf, Tilted cranking, *Nucl. Phys. A.* **557**, 259c–276c (may, 1993). doi: 10.1016/0375-9474(93)90546-A.
10. S. Frauendorf and J. Meng, Interpretation and quality of the tilted axis cranking approximation, *Z. Phys. A.* **356**(3), 263–279 (Dec., 1996). doi: 10.1007/s002180050179.
11. H. Hübel, Magnetic rotation in nuclei, *Prog. Part. Nucl. Phys.* **54**(1), 1–69 (2005). doi: 10.1016/j.ppnp.2004.06.002.
12. R. M. Clark and A. O. Macchiavelli, The shears mechanism in nuclei, *Annu. Rev. Nucl. Part. Sci.* **50**, 1 (2000). doi: 10.1146/annurev.nucl.50.1.1.
13. S. Frauendorf, Spontaneous symmetry breaking in rotating nuclei, *Rev. Mod. Phys.* **73**(2), 463 (June, 2001). doi: 10.1103/RevModPhys.73.463.
14. S. Frauendorf, Description of multi-quasiparticle bands by the tilted axis cranking model, *Nucl. Phys. A.* **677**(1-4), 115–170 (2000). doi: 10.1016/S0375-9474(00)00308-0.
15. R. M. Clark et al., Evidence for "magnetic rotation" in nuclei: Lifetimes of states in the M1 bands of ^{198,199}Pb, *Phys. Rev. Lett.* **78**(10), 1868–1871 (Mar, 1997). doi: 10.1103/PhysRevLett.78.1868.
16. S. Frauendorf, J. Meng, and J. Reif. In ed. M. A. Deleplanque, *Proceedings of the Conference on Physics From Large γ-Ray Detector Arrays*, vol. II of Report LBL35687, p. 52, Univ. of California, Berkeley (1994).
17. S. Frauendorf, Tilted rotation of weakly deformed and triaxial nuclei, *Z. Phys. A.* **358**, 163 (1997).
18. S. Frauendorf. In eds. M. A. Deleplanque, I. Y. Lee, and A. O. Macchiavelli, *Proceedings of the Workshop on Gammasphere Physics, Berkeley, 1995*, p. 272, World Scientific, Singapore (1996).
19. A. Bohr and B. R. Mottelson, *Nuclear Structure.* vol. II, Benjamin, New York (1975).
20. S. Frauendorf and J. Meng, Tilted rotation of triaxial nuclei, *Nucl. Phys. A.* **617**(2), 131–147 (May, 1997). doi: 10.1016/S0375-9474(97)00004-3.

21. J. Meng and S. Q. Zhang, Open problems in understanding the nuclear chirality, *Journal of Physics G: Nuclear and Particle Physics.* **37**(6), 064025 (2010). URL `http://stacks.iop.org/0954-3899/37/i=6/a=064025`.
22. H. Madokoro, J. Meng, M. Matsuzaki, and S. Yamaji, Relativistic mean field description for the shears band mechanism in ^{84}Rb, *Phys. Rev. C.* **62** (6), 061301 (Nov, 2000). doi: 10.1103/PhysRevC.62.061301.
23. W. Koepf and P. Ring, A relativistic description of rotating nuclei: The yrast line of ^{20}Ne, *Nucl. Phys. A.* **493**(1), 61–82 (1989). doi: 10.1016/0375-9474(89)90532-0.
24. P. W. Zhao, S. Q. Zhang, J. Peng, H. Z. Liang, P. Ring, and J. Meng, Novel structure for magnetic rotation bands in ^{60}Ni, *Phys. Lett. B.* **699**(3), 181–186 (2011). doi: 10.1016/j.physletb.2011.03.068.
25. J. Meng, J. Peng, S. Q. Zhang, and P. W. Zhao, Progress on tilted axis cranking covariant density functional theory for nuclear magnetic and antimagnetic rotation, *Frontiers of Physics.* **8**, 55–79 (2013). ISSN 2095-0462. doi: 10.1007/s11467-013-0287-y. URL `http://dx.doi.org/10.1007/s11467-013-0287-y`.
26. H. Madokoro and M. Matsuzaki, General relativistic mean field theory for rotating nuclei, *Phys. Rev. C.* **56**, R2934–R2937 (Dec, 1997). doi: 10.1103/PhysRevC.56.R2934.
27. W. Koepf and P. Ring, A relativistic theory of superdeformations in rapidly rotating nuclei, *Nucl. Phys. A.* **511**(2), 279–300 (1990). doi: 10.1016/0375-9474(90)90160-N.
28. P. Ring and P. Schuck, *The nuclear many-body problem.* Springer-Verlag, New York (1980).
29. M. Abramowitz and I. A. Stegun, *Handbook of Mathematical Functions.* New York: Dover (1965).
30. M. Girod and B. Grammaticos, Triaxial Hartree–Fock–Bogolyubov calculations with D1 effective interaction, *Phys. Rev. C.* **27**, 2317–2339 (May, 1983). doi: 10.1103/PhysRevC.27.2317. URL `http://link.aps.org/doi/10.1103/PhysRevC.27.2317`.
31. D. Vautherin, Hartree–Fock calculations with Skyrme's interaction. II. axially deformed nuclei, *Phys. Rev. C.* **7**, 296–316 (Jan, 1973). doi: 10.1103/PhysRevC.7.296. URL `http://link.aps.org/doi/10.1103/PhysRevC.7.296`.
32. R. Chasman and S. Wahlborn, Transformation scheme for harmonic-oscillator wave functions, *Nuclear Physics A.* **90**(2), 401–406 (1967). ISSN 0375-9474. doi: http://dx.doi.org/10.1016/0375-9474(67)90242-4. URL `http://www.sciencedirect.com/science/article/pii/0375947467902424`.
33. J. D. Talman, Some properties of three-dimensional harmonic oscillator wave functions, *Nuclear Physics A.* **141**(2), 273–288 (1970). ISSN 0375-9474. doi: http://dx.doi.org/10.1016/0375-9474(70)90847-X. URL `http://www.sciencedirect.com/science/article/pii/037594747090847X`.
34. B. Simon, Holonomy, the quantum adiabatic theorem, and Berry's phase, *Phys. Rev. Lett.* **51**, 2167–2170 (Dec, 1983). doi: 10.1103/PhysRevLett.51.2167. URL `http://link.aps.org/doi/10.1103/PhysRevLett.51.2167`.

35. T. Bengtsson, A method to remove virtual interactions with applications to nuclear spectroscopy, *Nuclear Physics A*. **496**(1), 56–92 (1989). ISSN 0375-9474. doi: http://dx.doi.org/10.1016/0375-9474(89)90216-9. URL http://www.sciencedirect.com/science/article/pii/0375947489902169.
36. D. G. Jenkins, R. Wadsworth, J. A. Cameron, R. M. Clark, D. B. Fossan, I. M. Hibbert, V. P. Janzen, R. Krücken, G. J. Lane, I. Y. Lee, A. O. Macchiavelli, C. M. Parry, J. M. Sears, J. F. Smith, and S. Frauendorf, Confirmation of the shears mechanism in near-spherical tin nuclei, *Phys. Rev. Lett.* **83**, 500–503 (Jul, 1999). doi: 10.1103/PhysRevLett.83.500. URL http://link.aps.org/doi/10.1103/PhysRevLett.83.500.
37. H. Hübel, *Contribution to Weatherill Symposium, Philadelphia, (unpublished)* (1991).
38. W. Andreitscheff and D. Elenkov, eds., in Proceedings X. International School on Nuclear Physics and Nuclear Energy. Varna (1991).
39. B. Fant, R. Tanner, P. Butler, A. James, G. Jones, R. Poynter, C. White, K. Ying, D. Love, J. Simpson, and K. Connell, Search for shape coexistence in ^{194}Pb, *J. Phys. G*. **17**(3), 319–340 (MAR, 1991). doi: 10.1088/0954-3899/17/3/013.
40. H. Hübel, G. Baldsiefen, D. Mehta, B. T. Rao, U. Birkental, G. Fröhlingsdorf, M. Neffgen, N. Nenoff, S. Pancholi, N. Singh, W. Schmitz, K. Theine, P. Willsau, H. Grawe, J. Heese, H. Kluge, K. Maier, M. Schramm, R. Schubart, and H. Maier, Collective structures in ^{199}Pb and ^{200}Pb, *Progress in Particle and Nuclear Physics*. **28**(0), 427–428 (1992). ISSN 0146-6410. doi: 10.1016/0146-6410(92)90043-2. URL http://www.sciencedirect.com/science/article/pii/0146641092900432.
41. R. M. Clark et al., First observation of a collective dipole rotational band in the $A \sim 200$ mass region, *Phys. Lett. B*. **275**(3-4), 247–251 (1992). doi: 10.1016/0370-2693(92)91585-W.
42. G. Baldsiefen et al., Oblate collective bands in ^{199}Pb and ^{200}Pb, *Phys. Lett. B*. **275**(3-4), 252–258 (1992). doi: 10.1016/0370-2693(92)91586-X.
43. A. Kuhnert et al., Oblate collectivity in ^{197}Pb, *Phys. Rev. C*. **46**(1), 133–143 (Jul, 1992). doi: 10.1103/PhysRevC.46.133.
44. S. Chmel, F. Brandolini, R. V. Ribas, G. Baldsiefen, A. Görgen, M. De Poli, P. Pavan, and H. Hübel, g factor of a magnetic-rotational band head in ^{193}Pb, *Phys. Rev. Lett.* **79**, 2002–2005 (Sep, 1997). doi: 10.1103/PhysRevLett.79.2002. URL http://link.aps.org/doi/10.1103/PhysRevLett.79.2002.
45. Amita, A. K. Jain, and B. Singh, Table of magnetic dipole rotational bands, *At. Data Nucl. Data Tables*. **74**(2), 283–331 (2000). doi: 10.1006/adnd.2000.0831. Revised edition at [http://www.nndc.bnl.gov/publications/preprints/mag-dip-rot-bands.pdf].
46. P. Agarwal, S. Kumar, S. Singh, R. K. Sinha, A. Dhal, S. Muralithar, R. P. Singh, N. Madhavan, R. Kumar, R. K. Bhowmik, S. S. Malik, S. C. Pancholi, L. Chaturvedi, H. C. Jain, and A. K. Jain, Bandcrossing of magnetic rotation bands in ^{137}Pr, *Phys. Rev. C*. **76**, 024321 (Aug, 2007). doi: 10.1103/PhysRevC.76.024321. URL http://link.aps.org/doi/10.1103/PhysRevC.76.024321.

47. D. Yuan, Y. Zheng, D. Zhou, Y. Zuo, P. Fan, M. Liu, X. Wu, L. Zhu, G. Li, G. Xu, Q. Fan, X. Zhang, and S. Zhu, g-factors of magnetic–rotational states in ^{85}Zr, *Hyperfine Interactions.* **180**, 49–54 (2007). ISSN 0304-3843. doi: 10.1007/s10751-008-9684-y. URL `http://dx.doi.org/10.1007/s10751-008-9684-y`.
48. T. Bhattacharjee, S. Chanda, S. Bhattacharyya, S. Basu, R. Bhowmik, J. Das, U. D. Pramanik, S. Ghugre, N. Madhavan, A. Mukherjee, G. Mukherjee, S. Muralithar, and R. Singh, Band structures in near spherical ^{138}Ce, *Nuclear Physics A.* **825**(12), 16–38 (2009). ISSN 0375-9474. doi: 10.1016/j.nuclphysa.2009.04.007. URL `http://www.sciencedirect.com/science/article/pii/S0375947409002188`.
49. R. Schwengner, G. Rainovski, H. Schnare, A. Wagner, S. Frauendorf, F. Dönau, A. Jungclaus, M. Hausmann, O. Yordanov, K. P. Lieb, D. R. Napoli, G. de Angelis, M. Axiotis, N. Marginean, F. Brandolini, and C. R. Alvarez, Magnetic dipole sequences in ^{83}Rb, *Phys. Rev. C.* **80**, 044305 (Oct, 2009). doi: 10.1103/PhysRevC.80.044305. URL `http://link.aps.org/doi/10.1103/PhysRevC.80.044305`.
50. A. Y. Deo, R. Palit, Z. Naik, S. Sihotra, S. Kumar, P. K. Joshi, I. Mazumdar, R. Chakrabarti, R. Kshetri, D. Mehta, and H. C. Jain, Structure of dipole bands in ^{106}In, *Phys. Rev. C.* **79**, 067304 (Jun, 2009). doi: 10.1103/PhysRevC.79.067304. URL `http://link.aps.org/doi/10.1103/PhysRevC.79.067304`.
51. D.-Q. Yuan, Y.-N. Zheng, Y. Zuo, P. Fan, D.-M. Zhou, X.-G. Wu, L.-H. Zhu, G.-S. Li, G.-J. Xu, Q.-W. Fan, X.-Z. Zhang, and S.-Y. Zhu, The g-factors and magnetic rotation in ^{82}Rb, *Chinese Physics B.* **19**(6), 062701 (2010). URL `http://stacks.iop.org/1674-1056/19/i=6/a=062701`.
52. C. Y. He, X. Q. Li, L. H. Zhu, X. G. Wu, B. Qi, Y. Liu, B. Pan, G. S. Li, L. H. Li, Z. M. Wang, Z. Y. Li, S. Y. Wang, Q. Xu, J. G. Wang, H. B. Ding, and J. Zhai, Magnetic rotation in ^{112}In, *Phys. Rev. C.* **83**, 024309 (Feb, 2011). doi: 10.1103/PhysRevC.83.024309. URL `http://link.aps.org/doi/10.1103/PhysRevC.83.024309`.
53. T. Trivedi, R. Palit, J. Sethi, S. Saha, S. Kumar, Z. Naik, V. V. Parkar, B. S. Naidu, A. Y. Deo, A. Raghav, P. K. Joshi, H. C. Jain, S. Sihotra, D. Mehta, A. K. Jain, D. Choudhury, D. Negi, S. Roy, S. Chattopadhyay, A. K. Singh, P. Singh, D. C. Biswas, R. K. Bhowmik, S. Muralithar, R. P. Singh, R. Kumar, and K. Rani, Small quadrupole deformation for the dipole bands in ^{112}In, *Phys. Rev. C.* **85**, 014327 (Jan, 2012). doi: 10.1103/PhysRevC.85.014327. URL `http://link.aps.org/doi/10.1103/PhysRevC.85.014327`.
54. C. Li, J. Li, X. Wu, X. Li, Y. Zheng, C. He, G. Li, S. Yao, B. Yu, X. Cao, S. Hu, J. Wang, C. Xu, and Y. Cheng, Magnetic rotation in doubly odd nucleus ^{114}In, *Nuclear Physics A.* **892**(0), 34–42 (2012). ISSN 0375-9474. doi: 10.1016/j.nuclphysa.2012.07.013. URL `http://www.sciencedirect.com/science/article/pii/S0375947412002199`.
55. N. T. Zhang, Y. H. Zhang, X. H. Zhou, M. L. Liu, Y. Zheng, J. G. Wang, Y. D. Fang, B. Ding, W. J. Huang, Y. X. Guo, X. G. Lei, L. Chen, S. T. Wang, X. G. Wu, and Y. Zheng, In-beam γ spectroscopy

of the odd-Z nucleus ^{139}Pm, *Phys. Rev. C.* **84**, 057302 (Nov, 2011). doi: 10.1103/PhysRevC.84.057302. URL `http://link.aps.org/doi/10.1103/PhysRevC.84.057302`.

56. S. Y. Wang, D. P. Sun, B. Qi, Z. Q. Chen, X. B. Hu, G. Wang, C. Liu, C. J. Xu, L. Liu, P. Zhang, Z. Q. Li, M. Z. Guo, Y. Y. Cai, Y. Q. Li, S. Y. Liu, S. M. Wyngaardt, L. H. Zhu, X. G. Wu, C. Y. He, Y. Zheng, and G. S. Li, Collective and noncollective states in ^{116}Sb, *Phys. Rev. C.* **86**, 064302 (Dec, 2012). doi: 10.1103/PhysRevC.86.064302. URL `http://link.aps.org/doi/10.1103/PhysRevC.86.064302`.
57. K. Ma, J. Lu, D. Yang, H. Wang, Y. Liu, J. Li, L. Zhu, X. Wu, Y. Zheng, and C. He, Level structures of ^{113}In, *The European Physical Journal A.* **48**, 82 (2012). ISSN 1434-6001. doi: 10.1140/epja/i2012-12082-9. URL `http://dx.doi.org/10.1140/epja/i2012-12082-9`.
58. D. Steppenbeck, R. V. F. Janssens, S. J. Freeman, M. P. Carpenter, P. Chowdhury, A. N. Deacon, M. Honma, H. Jin, T. Lauritsen, C. J. Lister, J. Meng, J. Peng, D. Seweryniak, J. F. Smith, Y. Sun, S. L. Tabor, B. J. Varley, Y.-C. Yang, S. Q. Zhang, P. W. Zhao, and S. Zhu, Magnetic rotation and quasicollective structures in ^{58}Fe: Influence of the $\nu g_{9/2}$ orbital, *Phys. Rev. C.* **85**, 044316 (Apr, 2012). doi: 10.1103/PhysRevC.85.044316. URL `http://link.aps.org/doi/10.1103/PhysRevC.85.044316`.
59. D. A. Torres et al., Deformations and magnetic rotations in the ^{60}Ni nucleus, *Phys. Rev. C.* **78**(5), 054318 (Nov, 2008). doi: 10.1103/PhysRevC.78.054318.
60. R. Clark, R. Wadsworth, E. Paul, C. Beausang, I. Ali, A. Astier, D. Cullen, P. Dagnall, P. Fallon, M. Joyce, M. Meyer, N. Redon, P. Regan, J. Sharpey-Schafer, W. Nazarewicz, and R. Wyss, Collective oblate dipole rotational bands in ^{198}Pb, *Nuclear Physics A.* **562**(1), 121–156 (1993). ISSN 0375-9474. doi: http://dx.doi.org/10.1016/0375-9474(93)90036-W. URL `http://www.sciencedirect.com/science/article/pii/037594749390036W`.
61. G. Baldsiefen, H. Hubel, W. Korten, D. Mehta, N. Nenoff, B. Rao, P. Willsau, H. Grawe, J. Heese, H. Kluge, K. Maier, R. Schubart, S. Frauendorf, and H. Maier, Shears bands in ^{199}Pb and ^{200}Pb, *Nuclear Physics A.* **574**(3), 521–558 (1994). ISSN 0375-9474. doi: http://dx.doi.org/10.1016/0375-9474(94)90243-7. URL `http://www.sciencedirect.com/science/article/pii/0375947494902437`.
62. R. Krücken, R. M. Clark, A. Dewald, M. A. Deleplanque, R. M. Diamond, P. Fallon, K. Hauschild, I. Y. Lee, A. O. Macchiavelli, R. Peusquens, G. J. Schmid, F. S. Stephens, K. Vetter, and P. von Brentano, Lifetimes near the bandhead of a shears band in ^{198}Pb, *Phys. Rev. C.* **58**, R1876–R1879 (Oct, 1998). doi: 10.1103/PhysRevC.58.R1876. URL `http://link.aps.org/doi/10.1103/PhysRevC.58.R1876`.
63. S. Frauendorf, J. Reif, and G. Winter, Shell-model study of shears bands in light Pb nuclei, *Nuclear Physics A.* **601**(1), 41–55 (1996). ISSN 0375-9474. doi: http://dx.doi.org/10.1016/0375-9474(96)00012-7. URL `http://www.sciencedirect.com/science/article/pii/0375947496000127`.
64. B. G. Carlsson and I. Ragnarsson, Many-particles-plus-rotor description of

magnetic bands at high spin, *Phys. Rev. C.* **74**, 044310 (Oct, 2006). doi: 10.1103/PhysRevC.74.044310. URL `http://link.aps.org/doi/10.1103/PhysRevC.74.044310`.

65. P. W. Zhao, Z. P. Li, J. M. Yao, and J. Meng, New parametrization for the nuclear covariant energy density functional with a point-coupling interaction, *Phys. Rev. C.* **82**(5), 054319 (Nov, 2010). doi: 10.1103/PhysRevC.82.054319.
66. L. F. Yu, P. W. Zhao, S. Q. Zhang, P. Ring, and J. Meng, Magnetic rotations in ^{198}Pb and ^{199}Pb within covariant density functional theory, *Phys. Rev. C.* **85**, 024318 (Feb, 2012). doi: 10.1103/PhysRevC.85.024318.
67. A. Görgen, N. Nenoff, H. Hübel, G. Baldsiefen, J. Becker, A. Byrne, S. Chmel, R. Clark, M. Deleplanque, R. Diamond, P. Fallon, K. Hauschild, I. Hibbert, W. Korten, R. Krücken, I. Lee, A. Macchiavelli, E. Paul, U. van Severen, F. Stephens, K. Vetter, R. Wadsworth, A. Wilson, and J. Wilson, Magnetic rotation in ^{197}Pb and ^{198}Pb, *Nuclear Physics A.* **683** (1–4), 108–144 (2001). ISSN 0375-9474. doi: http://dx.doi.org/10.1016/S0375-9474(00)00470-X. URL `http://www.sciencedirect.coM/science/article/pii/S037594740000470X`.
68. S. Chmel, S. Frauendorf, and H. Hübel, Deformation parameters and transition probabilities for shears bands in Pb isotopes, *Phys. Rev. C.* **75**, 044309 (Apr, 2007). doi: 10.1103/PhysRevC.75.044309. URL `http://link.aps.org/doi/10.1103/PhysRevC.75.044309`.
69. A. Arima and H. Horie, Configuration mixing and magnetic moments of nuclei, *Prog. Theor. Phys.* **11**, 509 (1954).
70. A. Arima, A short history of nuclear magnetic moments and GT transitions, *Science China Physics, Mechanics & Astronomy.* **54**(2), 188 (2011).
71. R. Bauer, J. Speth, V. Klemt, P. Ring, E. Werner, and T. Yamazaki, Magnetic multipole moments and transition probabilities of single-particle states around ^{208}Pb, *Nuclear Physics A.* **209**(3), 535–556 (1973). ISSN 0375-9474. doi: http://dx.doi.org/10.1016/0375-9474(73)90845-2. URL `http://www.sciencedirect.com/science/article/pii/0375947473908452`.
72. M. Matsuzaki, Y. R. Shimizu, and K. Matsuyanagi, Quasiparticle-vibration couplings in rotating triaxial odd-A nuclei, *Progress of Theoretical Physics.* **79**(4), 836–862 (1988). doi: 10.1143/PTP.79.836. URL `http://ptp.oxfordjournals.org/content/79/4/836.abstract`.
73. I. Towner, Quenching of spin matrix elements in nuclei, *Physics Reports.* **155**(5), 263–377 (1987). ISSN 0370-1573. doi: http://dx.doi.org/10.1016/0370-1573(87)90138-4. URL `http://www.sciencedirect.com/science/article/pii/0370157387901384`.
74. J. Li, J. Meng, P. Ring, J. Yao, and A. Arima, Relativistic description of second-order correction to nuclear magnetic moments with point-coupling residual interaction, *Science China Physics, Mechanics & Astronomy.* 54(2): 204 (2011). doi: 10.1007/s11433-010-4215-7. URL `http://phys.scichina.com:8083/sciGe/EN/abstract/article_501875.shtml`.
75. J. Li, J. M. Yao, J. Meng, and A. Arima, One-pion exchange current corrections for nuclear magnetic moments in relativistic mean field

theory, *Progress of Theoretical Physics*. **125**(6), 1185–1192 (2011). doi: 10.1143/PTP.125.1185. URL `http://ptp.oxfordjournals.org/content/125/6/1185.abstract`.

76. W. Long, J. Meng, N. Van Giai, and S.-G. Zhou, New effective interactions in relativistic mean field theory with nonlinear terms and density-dependent meson-nucleon coupling, *Phys. Rev. C*. **69**(3), 034319 (Mar, 2004). doi: 10.1103/PhysRevC.69.034319.
77. P. Olbratowski, J. Dobaczewski, J. Dudek, T. Rzaca-Urban, Z. Marcinkowska, and R. M. Lieder, The shears mechanism in ^{142}Gd in the Skyrme–Hartree–Fock method with the tilted-axis cranking, *ACTA PHYSICA POLONICA B*. **33**(1), 389–396 (JAN, 2002). 27th Mazurian-Lakes-School-of-Physics Meeting, KRZYZE, POLAND, SEP 02-09, 2001.
78. J. Peng, J.-M. Yao, S.-Q. Zhang, and J. Meng, Exotic magnetic rotation in ^{22}F, *Chinese Physics Letters*. **27**(12), 122101 (2010). URL `http://stacks.iop.org/0256-307X/27/i=12/a=122101`.
79. A. J. Simons et al., Evidence for a new type of shears mechanism in ^{106}Cd, *Phys. Rev. Lett.* **91**(16), 162501 (Oct, 2003). doi: 10.1103/PhysRevLett.91.162501.
80. D. Choudhury et al., Evidence of antimagnetic rotation in odd-A ^{105}Cd, *Phys. Rev. C*. **82**(6), 061308 (Dec, 2010). doi: 10.1103/PhysRevC.82.061308.
81. A. J. Simons et al., Investigation of antimagnetic rotation in light cadmium nuclei: ^{106,108}Cd, *Phys. Rev. C*. **72**(2), 024318 (Aug, 2005). doi: 10.1103/PhysRevC.72.024318.
82. P. Datta et al., Observation of antimagnetic rotation in ^{108}Cd, *Phys. Rev. C*. **71**(4), 041305 (Apr, 2005). doi: 10.1103/PhysRevC.71.041305.
83. C. J. Chiara et al., Shears mechanism in ^{109}Cd, *Phys. Rev. C*. **61**(3), 034318 (Feb, 2000). doi: 10.1103/PhysRevC.61.034318.
84. S. Roy et al., Systematics of antimagnetic rotation in even–even Cd isotopes, *Phys. Lett. B*. **694**(4-5), 322–326 (2011). doi: 10.1016/j.physletb.2010.10.018.
85. X. W. Li, J. Li, J. B. Lu, K. Y. Ma, Y. H. Wu, L. H. Zhu, C. Y. He, X. Q. Li, Y. Zheng, G. S. Li, X. G. Wu, Y. J. Ma, and Y. Z. Liu, Candidate antimagnetic rotational band in ^{112}In, *Phys. Rev. C*. **86**, 057305 (Nov, 2012). doi: 10.1103/PhysRevC.86.057305. URL `http://link.aps.org/doi/10.1103/PhysRevC.86.057305`.
86. S. Zhu et al., Investigation of antimagnetic rotation in ^{100}Pd, *Phys. Rev. C*. **64**(4), 041302 (Sep, 2001). doi: 10.1103/PhysRevC.64.041302.
87. M. Sugawara et al., Possible magnetic and antimagnetic rotations in ^{144}Dy, *Phys. Rev. C*. **79**(6), 064321 (Jun, 2009). doi: 10.1103/PhysRevC.79.064321.
88. P. W. Zhao, J. Peng, H. Z. Liang, P. Ring, and J. Meng, Antimagnetic rotation band in nuclei: A microscopic description, *Phys. Rev. Lett.* **107**, 122501 (Sep, 2011). doi: 10.1103/PhysRevLett.107.122501.
89. P. W. Zhao, J. Peng, H. Z. Liang, P. Ring, and J. Meng, Covariant density functional theory for antimagnetic rotation, *Phys. Rev. C*. **85**, 054310 (May, 2012). doi: 10.1103/PhysRevC.85.054310.

90. W. T. Kelvin, *The Molecular Tactics of a Crystal.* Clarendon Press (1894).
91. K. Starosta, T. Koike, C. J. Chiara, D. B. Fossan, D. R. LaFosse, A. A. Hecht, C. W. Beausang, M. A. Caprio, J. R. Cooper, R. Krücken, J. R. Novak, N. V. Zamfir, K. E. Zyromski, D. J. Hartley, D. L. Balabanski, J.-y. Zhang, S. Frauendorf, and V. I. Dimitrov, Chiral doublet structures in odd-odd $N = 75$ isotones: Chiral vibrations, *Phys. Rev. Lett.* **86**(6), 971–974 (Feb, 2001). doi: 10.1103/PhysRevLett.86.971.
92. A. Hecht, C. Beausang, K. Zyromski, D. Balabanski, C. Barton, M. Caprio, R. Casten, J. Cooper, D. Hartley, R. Krucken, D. Meyer, H. Newman, J. Novak, E. Paul, N. Pietralla, A. Wolf, N. Zamfir, J. Zhang, and F. Donau, Evidence for chiral symmetry breaking in ^{136}Pm and ^{138}Eu, *Phys. Rev. C.* **63**(5) (May, 2001). ISSN 0556-2813. doi: {10.1103/PhysRevC.63.051302}.
93. D. J. Hartley, L. L. Riedinger, M. A. Riley, D. L. Balabanski, F. G. Kondev, R. W. Laird, J. Pfohl, D. E. Archer, T. B. Brown, R. M. Clark, M. Devlin, P. Fallon, I. M. Hibbert, D. T. Joss, D. R. LaFosse, P. J. Nolan, N. J. O'Brien, E. S. Paul, D. G. Sarantites, R. K. Sheline, S. L. Shepherd, J. Simpson, R. Wadsworth, J.-y. Zhang, P. B. Semmes, and F. Dönau, Detailed spectroscopy of the chiral-twin candidate bands in ^{136}Pm, *Phys. Rev. C.* **64**(3), 031304 (Aug, 2001). doi: 10.1103/PhysRevC.64.031304.
94. T. Koike, K. Starosta, C. J. Chiara, D. B. Fossan, and D. R. LaFosse, Observation of chiral doublet bands in odd-odd $N = 73$ isotones, *Phys. Rev. C.* **63**(6), 061304 (May, 2001). doi: 10.1103/PhysRevC.63.061304.
95. R. Bark, A. Baxter, A. Byrne, G. Dracoulis, T. Kibdi, T. McGoram, and S. Mullins, Candidate chiral band in ^{134}La, *Nuclear Physics A.* **691** (3?), 577–598 (2001). ISSN 0375-9474. doi: http://dx.doi.org/10.1016/S0375-9474(01)00592-9. URL `http://www.sciencedirect.com/science/article/pii/S0375947401005929`.
96. K. Starosta, C. J. Chiara, D. B. Fossan, T. Koike, T. T. S. Kuo, D. R. LaFosse, S. G. Rohoziński, C. Droste, T. Morek, and J. Srebrny, Role of chirality in angular momentum coupling for $A \sim 130$ odd-odd triaxial nuclei: ^{132}La, *Phys. Rev. C.* **65**(4), 044328 (Apr, 2002). doi: 10.1103/PhysRevC.65.044328.
97. T. Koike, K. Starosta, C. J. Chiara, D. B. Fossan, and D. R. LaFosse, Systematic search of $\pi h_{11/2} \bigotimes \nu h_{11/2}$ chiral doublet bands and role of triaxiality in odd-odd $Z = 55$ isotopes: ^{128,130,132,134}Cs, *Phys. Rev. C.* **67**(4), 044319 (Apr, 2003). doi: 10.1103/PhysRevC.67.044319.
98. G. Rainovski, E. S. Paul, H. J. Chantler, P. J. Nolan, D. G. Jenkins, R. Wadsworth, P. Raddon, A. Simons, D. B. Fossan, T. Koike, K. Starosta, C. Vaman, E. Farnea, A. Gadea, T. Kröll, R. Isocrate, G. d. Angelis, D. Curien, and V. I. Dimitrov, Candidate chiral twin bands in the odd-odd nucleus ^{132}Cs: Exploring the limits of chirality in the mass $A \approx 130$ region, *Phys. Rev. C.* **68**(2), 024318 (Aug, 2003). doi: 10.1103/PhysRevC.68.024318.
99. S. Zhu, U. Garg, B. K. Nayak, S. S. Ghugre, N. S. Pattabiraman, D. B. Fossan, T. Koike, K. Starosta, C. Vaman, R. V. F. Janssens, R. S. Chakrawarthy, M. Whitehead, A. O. Macchiavelli, and S. Frauendorf, A

composite chiral pair of rotational bands in the odd-A nucleus ^{135}Nd, *Phys. Rev. Lett.* **91**(13), 132501 (Sep, 2003). doi: 10.1103/PhysRevLett.91.132501.

100. E. Grodner, J. Srebrny, A. A. Pasternak, I. Zalewska, T. Morek, C. Droste, J. Mierzejewski, M. Kowalczyk, J. Kownacki, M. Kisieliński, S. G. Rohoziński, T. Koike, K. Starosta, A. Kordyasz, P. J. Napiorkowski, M. Woli ńska Cichocka, E. Ruchowska, W. Płóciennik, and J. Perkowski, ^{128}Cs as the best example revealing chiral symmetry breaking, *Phys. Rev. Lett.* **97**(17), 172501 (Oct, 2006). doi: 10.1103/PhysRevLett.97.172501.
101. C. Vaman, D. B. Fossan, T. Koike, K. Starosta, I. Y. Lee, and A. O. Macchiavelli, Chiral degeneracy in triaxial ^{104}Rh, *Phys. Rev. Lett.* **92**(3), 032501 (Jan, 2004). doi: 10.1103/PhysRevLett.92.032501.
102. P. Joshi, D. Jenkins, P. Raddon, A. Simons, R. Wadsworth, A. Wilkinson, D. Fossan, T. Koike, K. Starosta, C. Vaman, J. Timr, Z. Dombrdi, A. Krasznahorkay, J. Molnr, D. Sohler, L. Zolnai, A. Algora, E. Paul, G. Rainovski, A. Gizon, J. Gizon, P. Bednarczyk, D. Curien, G. Duchne, and J. Scheurer, Stability of chiral geometry in the odd–odd Rh isotopes: spectroscopy of ^{106}Rh, *Physics Letters B*. **595**(1–4), 135–142 (2004). ISSN 0370-2693. doi: http://dx.doi.org/10.1016/j.physletb.2004.05.066. URL `http://www.sciencedirect.com/science/article/pii/S0370269304008810`.
103. P. Joshi, A. Wilkinson, T. Koike, D. Fossan, S. Finnigan, E. Paul, P. Raddon, G. Rainovski, K. Starosta, A. Simons, C. Vaman, and R. Wadsworth, First evidence for chirality in Tc isotopes: Spectroscopy of ^{100}Tc, *Eur. Phys. J. A*. **24**(1), 23–29 (2005). doi: 10.1140/epja/i2004-10120-y.
104. J. A. Alcántara-Núñez, J. R. B. Oliveira, E. W. Cybulska, N. H. Medina, M. N. Rao, R. V. Ribas, M. A. Rizzutto, W. A. Seale, F. Falla-Sotelo, K. T. Wiedemann, V. I. Dimitrov, and S. Frauendorf, Magnetic dipole and electric quadrupole rotational structures and chirality in ^{105}Rh, *Phys. Rev. C*. **69** (2), 024317 (Feb, 2004). doi: 10.1103/PhysRevC.69.024317.
105. J. Timár, P. Joshi, K. Starosta, V. Dimitrov, D. Fossan, J. Molnár, D. Sohler, R. Wadsworth, A. Algora, P. Bednarczyk, D. Curien, Z. Dombrádi, G. Duchene, A. Gizon, J. Gizon, D. Jenkins, T. Koike, A. Krasznahorkay, E. Paul, P. Raddon, G. Rainovski, J. Scheurer, A. Simons, C. Vaman, A. Wilkinson, L. Zolnai, and S. Frauendorf, Experimental evidence for chirality in the odd-A ^{105}Rh, *Physics Letters B*. **598**(34), 178–187 (2004). ISSN 0370-2693. doi: http://dx.doi.org/10.1016/j.physletb.2004.07.050. URL `http://www.sciencedirect.com/science/article/pii/S0370269304011402`.
106. J. Timár, C. Vaman, K. Starosta, D. B. Fossan, T. Koike, D. Sohler, I. Y. Lee, and A. O. Macchiavelli, Role of the core in degeneracy of chiral candidate band doubling: ^{103}Rh, *Phys. Rev. C*. **73**(1), 011301 (Jan, 2006). doi: 10.1103/PhysRevC.73.011301.
107. D. L. Balabanski, M. Danchev, D. J. Hartley, L. L. Riedinger, O. Zeidan, J.-y. Zhang, C. J. Barton, C. W. Beausang, M. A. Caprio, R. F. Casten, J. R. Cooper, A. A. Hecht, R. Krücken, J. R. Novak, N. V. Zamfir, and K. E. Zyromski, Possible assignment of chiral twin bands in ^{188}Ir, *Phys. Rev. C*. **70**(4), 044305 (Oct, 2004). doi: 10.1103/PhysRevC.70.044305.

108. S. Wang, B. Qi, L. Liu, S. Zhang, H. Hua, X. Li, Y. Chen, L. Zhu, J. Meng, S. Wyngaardt, P. Papka, T. Ibrahim, R. Bark, P. Datta, E. Lawrie, J. Lawrie, S. Majola, P. Masiteng, S. Mullins, J. Gl, G. Kalinka, J. Molnr, B. Nyak, J. Timr, K. Juhsz, and R. Schwengner, The first candidate for chiral nuclei in the mass region: ^{80}Br, *Physics Letters B*. **703**(1), 40–45 (2011). ISSN 0370-2693. doi: http://dx.doi.org/10.1016/j.physletb.2011.07.055. URL http://www.sciencedirect.com/science/article/pii/S0370269311008598.

109. D. Tonev, G. de Angelis, P. Petkov, A. Dewald, S. Brant, S. Frauendorf, D. L. Balabanski, P. Pejovic, D. Bazzacco, P. Bednarczyk, F. Camera, A. Fitzler, A. Gadea, S. Lenzi, S. Lunardi, N. Marginean, O. Möller, D. R. Napoli, A. Paleni, C. M. Petrache, G. Prete, K. O. Zell, Y. H. Zhang, J.-y. Zhang, Q. Zhong, and D. Curien, Transition probabilities in ^{134}Pr: A test for chirality in nuclear systems, *Phys. Rev. Lett.* **96**(5), 052501 (Feb, 2006). doi: 10.1103/PhysRevLett.96.052501.

110. S. Mukhopadhyay, D. Almehed, U. Garg, S. Frauendorf, T. Li, P. V. M. Rao, X. Wang, S. S. Ghugre, M. P. Carpenter, S. Gros, A. Hecht, R. V. F. Janssens, F. G. Kondev, T. Lauritsen, D. Seweryniak, and S. Zhu, From chiral vibration to static chirality in ^{135}Nd, *Phys. Rev. Lett.* **99**(17), 172501 (Oct, 2007). doi: 10.1103/PhysRevLett.99.172501.

111. E. Grodner, I. Sankowska, T. Morek, S. Rohoziski, C. Droste, J. Srebrny, A. Pasternak, M. Kisieliski, M. Kowalczyk, J. Kownacki, J. Mierzejewski, A. Krl, and K. Wrzosek, Partner bands of ^{126}Cs first observation of chiral electromagnetic selection rules, *Physics Letters B*. **703**(1), 46–50 (2011). ISSN 0370-2693. doi: http://dx.doi.org/10.1016/j.physletb.2011.07.062. URL http://www.sciencedirect.com/science/article/pii/S0370269311008665.

112. T. Suzuki, G. Rainovski, T. Koike, T. Ahn, M. P. Carpenter, A. Costin, M. Danchev, A. Dewald, R. V. F. Janssens, P. Joshi, C. J. Lister, O. Möller, N. Pietralla, T. Shinozuka, J. Timár, R. Wadsworth, C. Vaman, and S. Zhu, Lifetime measurement of candidate chiral doublet bands in the ^{103,104}Rh isotopes with the recoil-distance doppler-shift method in inverse kinematics, *Phys. Rev. C*. **78**, 031302 (Sep, 2008). doi: 10.1103/PhysRevC.78.031302. URL http://link.aps.org/doi/10.1103/PhysRevC.78.031302.

113. D. Tonev, S. Yavahchova, M. N. Goutev, G. de Angelis, P. Petkov, K. Bhowmik, R. P. Singh, R. S. Muralithar, N. Madhavan, R. Kumar, M. Kumar Raju, J. Kaur, G. Mohanto, A. Singh, N. Kaur, R. Garg, A. Shukla, K. Marinov, Ts. and S. Brant, Candidates for twin chiral bands in ^{103}Rh, *Phys. Rev. Lett.* **112**, 052501 (Feb, 2014). doi: 10.1103/PhysRevLett.112.052501. URL http://link.aps.org/doi/10.1103/PhysRevLett.112.052501.

114. O. Lieder, E. M. Lieder, R. A. Bark, R. B. Chen, Q. Q. Zhang, S. J. Meng, A. Lawrie, E. J. Lawrie, J. P. Bvumbi, S. Y. Kheswa, N. S. Ntshangase, S. E. Madiba, T. L. Masiteng, P. M. Mullins, S. S. Murray, P. Papka, G. Roux, D. O. Shirinda, H. Zhang, Z. W. Zhao, P. P. Li, Z. J. Peng, B. Qi, Y. Wang, S. G. Xiao, Z. and C. Xu, Resolution of

chiral conundrum in ^{106}Ag: Doppler-shift lifetime investigation, *Phys. Rev. Lett.* **112**, 202502 (May, 2014). doi: 10.1103/PhysRevLett.112.202502. URL `http://link.aps.org/doi/10.1103/PhysRevLett.112.202502`.
115. N. Rather, P. Datta, S. Chattopadhyay, S. Rajbanshi, A. Goswami, H. Bhat, G. A. Sheikh, J. S. Roy, R. Palit, S. Pal, S. Saha, J. Sethi, S. Biswas, P. Singh, and C. Jain, H. Exploring the origin of nearly degenerate doublet bands in ^{106}Ag, *Phys. Rev. Lett.* **112**, 202503 (May, 2014). doi: 10.1103/PhysRevLett.112.202503. URL `http://link.aps.org/doi/10.1103/PhysRevLett.112.202503`.
116. J. Peng, J. Meng, and S. Q. Zhang, Description of chiral doublets in $A \sim 130$ nuclei and the possible chiral doublets in $A \sim 100$ nuclei, *Phys. Rev. C.* **68** (4), 044324 (Oct, 2003). doi: 10.1103/PhysRevC.68.044324.
117. S. Q. Zhang, B. Qi, S. Y. Wang, and J. Meng, Chiral bands for a quasi-proton and quasi-neutron coupled with a triaxial rotor, *Phys. Rev. C.* **75** (4), 044307 (Apr, 2007). doi: 10.1103/PhysRevC.75.044307.
118. B. Qi, S. Zhang, J. Meng, S. Wang, and S. Frauendorf, Chirality in odd-A nucleus ^{135}Nd in particle rotor model, *Physics Letters B.* **675**(2), 175–180 (2009). ISSN 0370-2693. doi: http://dx.doi.org/10.1016/j.physletb.2009.02.061. URL `http://www.sciencedirect.com/science/article/pii/S0370269309003979`.
119. D. Tonev, G. d. Angelis, S. Brant, S. Frauendorf, P. Petkov, A. Dewald, F. Dönau, D. L. Balabanski, Q. Zhong, P. Pejovic, D. Bazzacco, P. Bednarczyk, F. Camera, D. Curien, F. D. Vedova, A. Fitzler, A. Gadea, G. L. Bianco, S. Lenzi, S. Lunardi, N. Marginean, O. Möller, D. R. Napoli, R. Orlandi, E. Sahin, A. Saltarelli, J. V. Dobon, K. O. Zell, J.-y. Zhang, and Y. H. Zhang, Question of dynamic chirality in nuclei: The case of ^{134}Pr, *Phys. Rev. C.* **76**(4), 044313 (Oct, 2007). doi: 10.1103/PhysRevC.76.044313.
120. S. Brant, D. Tonev, G. De Angelis, and A. Ventura, Dynamic chirality in the interacting boson fermion-fermion model, *Phys. Rev. C.* **78**(3), 034301 (Sep, 2008). doi: 10.1103/PhysRevC.78.034301.
121. V. I. Dimitrov, S. Frauendorf, and F. Dönau, Chirality of nuclear rotation, *Phys. Rev. Lett.* **84**(25), 5732–5735 (Jun, 2000). doi: 10.1103/PhysRevLett.84.5732.
122. P. Olbratowski, J. Dobaczewski, J. Dudek, and W. Płóciennik, Critical frequency in nuclear chiral rotation, *Phys. Rev. Lett.* **93**, 052501 (Jul, 2004). doi: 10.1103/PhysRevLett.93.052501. URL `http://link.aps.org/doi/10.1103/PhysRevLett.93.052501`.
123. P. Olbratowski, J. Dobaczewski, and J. Dudek, Search for the Skyrme–Hartree–Fock solutions for chiral rotation in $N = 75$ isotones, *Phys. Rev. C.* **73**, 054308 (May, 2006). doi: 10.1103/PhysRevC.73.054308. URL `http://link.aps.org/doi/10.1103/PhysRevC.73.054308`.
124. D. Almehed, F. Dönau, and S. Frauendorf, Chiral vibrations in the $A = 135$ region, *Phys. Rev. C.* **83**(5), 054308 (May, 2011). doi: 10.1103/PhysRevC.83.054308.
125. Q. B. Chen, S. Q. Zhang, P. W. Zhao, R. V. Jolos, and J. Meng, Collective hamiltonian for chiral modes, *Phys. Rev. C.* **87**, 024314 (Feb, 2013). doi:

10.1103/PhysRevC.87.024314. URL http://link.aps.org/doi/10.1103/PhysRevC.87.024314.

126. J. Meng, J. Peng, S. Q. Zhang, and S.-G. Zhou, Possible existence of multiple chiral doublets in ^{106}Rh, *Phys. Rev. C.* **73**(3), 037303 (Mar, 2006). doi: 10.1103/PhysRevC.73.037303.
127. J. M. Yao, B. Qi, S. Q. Zhang, J. Peng, S. Y. Wang, and J. Meng, Candidate multiple chiral doublets nucleus ^{106}Rh in a triaxial relativistic mean-field approach with time-odd fields, *Phys. Rev. C.* **79**, 067302 (Jun, 2009). doi: 10.1103/PhysRevC.79.067302. URL http://link.aps.org/doi/10.1103/PhysRevC.79.067302.
128. J. Peng, H. Sagawa, S. Q. Zhang, J. M. Yao, Y. Zhang, and J. Meng, Search for multiple chiral doublets in rhodium isotopes, *Phys. Rev. C.* **77** (2), 024309 (Feb, 2008). doi: 10.1103/PhysRevC.77.024309.
129. J. Li, S. Q. Zhang, and J. Meng, Multiple chiral doublet candidate nucleus ^{105}Rh in a relativistic mean-field approach, *Phys. Rev. C.* **83**, 037301 (Mar, 2011). doi: 10.1103/PhysRevC.83.037301. URL http://link.aps.org/doi/10.1103/PhysRevC.83.037301.
130. B. Qi, H. Jia, N. B. Zhang, C. Liu, and S. Y. Wang, Possible multiple chiral doublet bands in ^{107}Ag, *Phys. Rev. C.* **88**, 027302 (Aug, 2013). doi: 10.1103/PhysRevC.88.027302. URL http://link.aps.org/doi/10.1103/PhysRevC.88.027302.
131. A. D. Ayangeakaa, U. Garg, M. D. Anthony, S. Frauendorf, J. T. Matta, B. K. Nayak, D. Patel, Q. B. Chen, S. Q. Zhang, P. W. Zhao, B. Qi, J. Meng, R. V. F. Janssens, M. P. Carpenter, C. J. Chiara, F. G. Kondev, T. Lauritsen, D. Seweryniak, S. Zhu, S. S. Ghugre, and R. Palit, Evidence for multiple chiral doublet bands in ^{133}Ce, *Phys. Rev. Lett.* **110**, 172504 (Apr, 2013). doi: 10.1103/PhysRevLett.110.172504. URL http://link.aps.org/doi/10.1103/PhysRevLett.110.172504.
132. C. Droste, S. Rohoziski, K. Starosta, L. Prchniak, and E. Grodner, Chiral bands in odd-odd nuclei with rigid or soft cores, *The European Physical Journal A.* **42**(1), 79–89 (2009). ISSN 1434-6001. doi: 10.1140/epja/i2009-10860-0. URL http://dx.doi.org/10.1140/epja/i2009-10860-0.
133. Q. B. Chen, J. M. Yao, S. Q. Zhang, and B. Qi, Chiral geometry of higher excited bands in triaxial nuclei with particle-hole configuration, *Phys. Rev. C.* **82**, 067302 (Dec, 2010). doi: 10.1103/PhysRevC.82.067302. URL http://link.aps.org/doi/10.1103/PhysRevC.82.067302.
134. I. Hamamoto, Possible presence and properties of multi-chiral-pair bands in odd-odd nuclei with the same intrinsic configuration, *Phys. Rev. C.* **88**, 024327 (Aug, 2013). doi: 10.1103/PhysRevC.88.024327. URL http://link.aps.org/doi/10.1103/PhysRevC.88.024327.
135. I. Kuti, Q. B. Chen, J. Timár, D. Sohler, S. Q. Zhang, Z. H. Zhang, P. W. Zhao, J. Meng, K. Starosta, T. Koike, E. S. Paul, D. B. Fossan, and C. Vaman, Multiple chiral doublet bands of identical configuration in ^{103}Rh, *Phys. Rev. Lett.* **113**, 032501 (Jul, 2014). doi: 10.1103/PhysRevLett.113.032501. URL http://link.aps.org/doi/10.1103/PhysRevLett.113.032501.
136. L. Guo, F. Sakata, and E.-G. Zhao, Characteristic feature of self-

consistent mean-field in level crossing region, *Nuclear Physics A*. **740** (12), 59–76 (2004). ISSN 0375-9474. doi: http://dx.doi.org/10.1016/j.nuclphysa.2004.04.117. URL `http://www.sciencedirect.com/science/article/pii/S0375947404006955`.

137. B. Qi, S. Q. Zhang, S. Y. Wang, J. M. Yao, and J. Meng, Examining $B(M1)$ staggering as a fingerprint for chiral doublet bands, *Phys. Rev. C*. **79**, 041302 (Apr, 2009). doi: 10.1103/PhysRevC.79.041302. URL `http://link.aps.org/doi/10.1103/PhysRevC.79.041302`.
138. H. Dejbakhsh, R. P. Schmitt, and G. Mouchaty, Collective and single particle structure in ^{103}Rh, *Phys. Rev. C*. **37**, 621–636 (Feb, 1988). doi: 10.1103/PhysRevC.37.621. URL `http://link.aps.org/doi/10.1103/PhysRevC.37.621`.
139. B. Qi, S. Q. Zhang, S. Y. Wang, J. Meng, and T. Koike, Chirality in odd-A Rh isotopes within the triaxial particle rotor model, *Phys. Rev. C*. **83**, 034303 (Mar, 2011). doi: 10.1103/PhysRevC.83.034303. URL `http://link.aps.org/doi/10.1103/PhysRevC.83.034303`.
140. P. W. Zhao, S. Q. Zhang, and J. Meng, Impact of pairing anti-viscosity on the orientation of the nuclear spin, *Phys. Rev. C* . **92**, 034319 (2015).
141. J. Zeng, T. Cheng, L. Cheng, and C. Wu, Pairing reduction due to the blocking effect, *Nuclear Physics A*. **411**(1), 49–64 (1983). ISSN 0375-9474. doi: http://dx.doi.org/10.1016/0375-9474(83)90507-9. URL `http://www.sciencedirect.com/science/article/pii/0375947483905079`.
142. J. Y. Zeng, T. H. Jin, and Z. J. Zhao, Reduction of nuclear moment of inertia due to pairing interaction, *Phys. Rev. C*. **50**, 1388–1397 (Sep, 1994). doi: 10.1103/PhysRevC.50.1388. URL `http://link.aps.org/doi/10.1103/PhysRevC.50.1388`.
143. X. Wu, Z. H. Zhang, J. Y. Zeng, and Y. A. Lei, Nuclear pairing reduction due to rotation and blocking, *Phys. Rev. C*. **83**, 034323 (Mar, 2011). doi: 10.1103/PhysRevC.83.034323. URL `http://link.aps.org/doi/10.1103/PhysRevC.83.034323`.
144. Z.-H. Zhang, J.-Y. Zeng, E.-G. Zhao, and S.-G. Zhou, Particle-number conserving analysis of rotational bands in ^{247,249}Cm and ^{249}Cf, *Phys. Rev. C*. **83**, 011304 (Jan, 2011). doi: 10.1103/PhysRevC.83.011304. URL `http://link.aps.org/doi/10.1103/PhysRevC.83.011304`.
145. J. Meng, J. Y. Guo, L. Liu, and S. Q. Zhang, Shell-model-like approach (SLAP) for the nuclear properties in relativistic mean field theory, *Frontiers of Physics in China*. **1**(1), 38–46 (2006). ISSN 1673-3487. doi: 10.1007/s11467-005-0013-5. URL `http://dx.doi.org/10.1007/s11467-005-0013-5`.

Chapter 10

Small amplitude motion

Nils Paar

Physics Department, Faculty of Science, University of Zagreb
Zagreb, HR-10002, Croatia
npaar@phy.hr

Yifei Niu

Institute of Fluid Physics, China Academy of Engineering Physics
Mianyang, 621900, China
nyfster@gmail.com

In this chapter, we introduce the collective small amplitude motion in atomic nucleus, i.e., the vibrational modes which are short time responses of the system to external perturbations. Extensive studies over past several decades have provided valuable information on the structure of the nucleus and the forces of cohesion that are responsible for nuclear binding. This chapter introduces in detail the microscopic theory for the description of the small amplitude motion, the random phase approximation model and its realization within covariant density functional theory. Recent and typical applications of such a model in giant resonances, pygmy strengths and spin-isospin resonances are illustrated, among which, the exotic modes in nuclei far from β-stability line and the inclusion of temperature degree of freedom for astrophysical applications are emphasized.

Contents

1. Introduction

Collective small amplitude motion in atomic nuclei is associated with short time response of the system to external perturbation. A typical example of small amplitude vibration is giant resonance, a collective mode at energy above the particle separation threshold, involving coherent transitions of many particles in the nucleus.[1,2] A giant resonance usually corresponds to a high-frequency, damped, nearly harmonic density or shape vibration around the equilibrium of the nuclear system. The vibration amplitude is small, only a few percent of the nuclear radius.[1] The first evidence of giant resonances in nuclei was discovered by Bothe and Gentner in 1937.[3] The observation of giant resonance phenomenon has been confirmed in 1947 by Baldwin and Klaiber with photons from a betatron. The measurement was based on bremsstrahlung scattering on uranium target.[4] In the model established by Goldhaber and Teller,[5] it was suggested that pronounced peak in the measured cross section was due to a resonance-like behavior of protons moving against the neutrons. Two-fluid model by Steinwedel and Jensen,[6] where the variation of the neutron and proton densities was described using hydrodynamical equations, indicated a similar mode to Goldhaber and Teller. This mode, corresponding to isovector giant dipole resonance, has been extensively studied with photon beams.[7] A classical example of giant dipole resonance in heavy nuclei is demonstrated in the photo-neutron cross section for ^{208}Pb, resulting in a resonance that fits well a Lorentzian with a mean energy $E = 13.5$ MeV and a width $\Gamma = 4$ MeV.[8] Another type of giant resonances has also been discovered in the early studies, which was identified as isoscalar giant quadrupole resonance.[9–11]

Over the past decades, giant resonances have been a topic of intensive research both experimentally and theoretically.[1,7] Many different types of giant resonances have been observed and studied, which are classified according to their multipolarity, spin, and isospin. The giant resonances of various multipolarities have common features of excitation. They occur systematically for all nuclei across the mass table, and their excitation energy and width vary smoothly as a function of the nuclear mass number A. Their transition strength exhausts a large fraction of the corresponding sum rule.

Giant resonances are interesting not only as unique modes of nuclear excitation, but also due to their relevance in many applications, e.g., probing the properties of effective nuclear interaction, constraining the nuclear symmetry energy and neutron skin thickness in finite nuclei, modeling astrophysical weak interaction processes, etc. The properties of giant resonances also provide important constraints on the properties of neutron stars, in particular the density and pressure that determine the phase transition at the inner edge separating the liquid core and the solid crust of a neutron star.[12]

Over the past years, another very active field of research related to small amplitude oscillations developed, involving multipole response in nuclei away from the valley of stability. In particular, in neutron-rich nuclei the modification of the effective nuclear potential leads to the formation of nuclei with very diffuse neutron densities and the occurrence of the neutron skin and halo structures. These phenomena will also affect the multipole response of unstable nuclei.[13,14] In particular, the monopole, dipole and quadrupole transition strengths become modified, and new modes of excitation might arise in nuclei at the frontiers of nuclear binding.

Collective excitations of a quantum many-body system, i.e., the nucleus, can be described as the response of the nucleus to a weak external field, such that a linear approximation can be applied. Following this spirit, random phase approximation (RPA) is introduced as a microscopic theory framework based on the mean-field concept, to describe the microscopic structure of giant resonances. Microscopic description of ground-states and properties of excited states in nuclei characterized by the closeness of the Fermi surface to the particle continuum, necessitates a unified description of mean-field and pairing correlations, as for example in the framework of the Hartree–Fock–Bogoliubov (HFB) theory. Collective low-lying excited states in weakly bound nuclei are best described by the quasiparticle random phase approximation (QRPA) based on the HFB framework.

In Sec. 2 we introduce the general formalism of the RPA, and its extension to the relativistic QRPA (RQRPA) in the canonical basis is given in Sec. 3. Sections 4 and 5 include several applications of RQRPA approach, including giant resonances and pygmy strengths in neutron-rich nuclei. Further extensions of the theory framework, including charge exchange proton-neutron (pn) RQRPA and finite temperature RPA (FTRPA) with applications are described in Secs. 6 and 7, respectively. Section 8 contains the concluding remarks and perspectives.

2. Random phase approximation

2.1. *Basic concepts*

In the mean-field approximation, the Hartree–Fock (HF) ground state is formed by filling up the mean-field potential with A nucleons from bottom up to a certain Fermi level. Due to the existence of the residual interaction,

$$V_R = \frac{1}{2}\sum_{kk'll'} V_{kk'll'} a_k^\dagger a_{k'}^\dagger a_{l'} a_l - V_{\mathrm{HF}}, \tag{1}$$

the real ground state should also include the 1 particle-1 hole (1p-1h), 2 particle-2 hole (2p-2h), ..., n particle-n hole (np-nh) excitations based on the HF ground state, see Fig. 1. The HF ground state $|\mathrm{HF}\rangle$ and 1p-1h, ..., np-nh excitations form a complete orthogonal basis (i.e., full shell model space) which can be used to expand the true many-nucleon wave functions of the ground state $|0\rangle$ or the excited state $|\nu\rangle$,[15,16]

$$\begin{aligned}
|0\rangle &= C_0^0|\mathrm{HF}\rangle + \sum_{mi} C_{mi}^0 a_m^\dagger a_i|\mathrm{HF}\rangle + \frac{1}{4}\sum_{mnij} C_{mn,ij}^0 a_m^\dagger a_n^\dagger a_i a_j|\mathrm{HF}\rangle + \cdots,\\
|\nu\rangle &= C_0^\nu|\mathrm{HF}\rangle + \sum_{mi} C_{mi}^\nu a_m^\dagger a_i|\mathrm{HF}\rangle + \frac{1}{4}\sum_{mnij} C_{mn,ij}^\nu a_m^\dagger a_n^\dagger a_i a_j|\mathrm{HF}\rangle + \cdots,
\end{aligned} \tag{2}$$

where m, n (i, j) denote the state above (below) Fermi level.

The exact solution of the Schroedinger equation can be obtained by a diagonalization of H in the full shell model space or equivalently by a variation of the above expansion coefficients, which is actually the no-core shell model. However, with the increase of nucleon number A, the dimension of the matrix increases exponentially, and this task becomes impossible to solve. Let us suppose it is reasonable to take the expansion

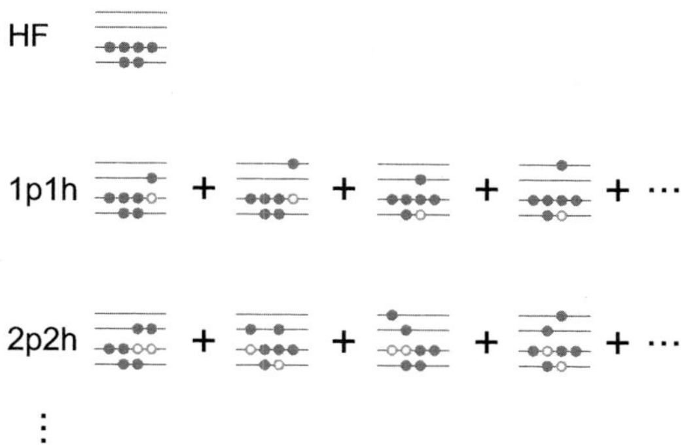

Fig. 1. (Color online) The schematic picture for HF ground state, 1p-1h excitations, and 2p-2h excitations.

of the excited state $|\nu\rangle$ only up to 1p-1h excitations, which is

$$|\nu\rangle = C_0^\nu|\text{HF}\rangle + \sum_{mi} C_{mi}^\nu a_m^\dagger a_i|\text{HF}\rangle. \tag{3}$$

Since $\langle\text{HF}|Ha_m^\dagger a_i|\text{HF}\rangle = 0$, we see that the ansatz (3) implies that the ground state remains a Slater determinant, whereas it is sufficient for the excited state to retain only[16]

$$|\nu\rangle = \sum_{mi} C_{mi}^\nu a_m^\dagger a_i|\text{HF}\rangle = Q_\nu^\dagger|\text{HF}\rangle. \tag{4}$$

$Q_\nu^\dagger = \sum_{mi} C_{mi}^\nu a_m^\dagger a_i$ is the operator that creates an excited state $|\nu\rangle$. In this case the correlations are only taken into account for the excited states, whereas the ground state is not changed. This is the basic idea for the Tamm–Dancoff approximation (TDA). For a better approximation, the correlations should also be built into the ground state. If we think of a ground state containing 2p-2h correlations indicated by Eq. (2), we can not only create a particle-hole (ph) pair but also destroy one. The most straightforward way is to generalize the operator $Q_\nu^\dagger$ to

$$Q_\nu^\dagger = \sum_{mi} X_{mi}^\nu a_m^\dagger a_i - \sum_{mi} Y_{mi}^\nu a_i^\dagger a_m, \tag{5}$$

where the minus sign has been chosen for convenience. The excited RPA state is $|\nu\rangle = Q_\nu^\dagger|\text{RPA}\rangle$, and the RPA ground state is defined as $Q_\nu|\text{RPA}\rangle = 0$.

2.2. *RPA equations*

In this section, we derive the RPA equations by the linear response to the external field. This method is general in the sense that it could also be applied to theories with density dependent forces.[16]

As in the previous subsection, we use i, j to label the hole states (occupied states), and m, n for particle state (unoccupied states).

Considering a time-dependent external field $\mathcal{W}(t)$, which depends on time in a harmonic way,

$$\mathcal{W}(t) = W(\boldsymbol{r})e^{-i\omega t} + W^{\dagger}(\boldsymbol{r})e^{i\omega t}, \tag{6}$$

the system Hamiltonian H becomes

$$H \to H + \mathcal{W}(t). \tag{7}$$

It also leads to the changes of single particle wave function ψ_i and the HF Hamiltonian h_0,

$$\psi_i \to \varphi_i = \psi_i + \sum_m \beta_{mi}(t)\psi_m, \tag{8a}$$

$$h_0[\psi] \to h_0[\varphi] + \mathcal{W}. \tag{8b}$$

Thus, the time-dependent Hartree–Fock (TDHF) equation becomes

$$i\frac{\partial}{\partial t}\varphi_i = (h_0[\varphi] + \mathcal{W} - \epsilon_i)\varphi_i. \tag{9}$$

In the small amplitude limit, only the linear response to the external field is taken into account, so just the linear terms of β_{mi} are kept.

Firstly, write the time dependent part of expansion coefficients β in the same form of $\mathcal{W}(t)$,

$$\beta_{mi}(t) = X_{mi}e^{-i\omega t} + Y^*_{mi}e^{i\omega t}. \tag{10}$$

Then, using $\langle\psi_m|$ to act on the TDHF equation (9), one can obtain that

$$\text{lhs} = \langle\psi_m|i\frac{\partial}{\partial t}|\varphi_i\rangle = \omega(X_{mi}e^{-i\omega t} - Y^*_{mi}e^{i\omega t}), \tag{11}$$

and

$$\text{rhs} = \langle\psi_m|h_0[\varphi] + \mathcal{W} - \epsilon_i|\psi_i + \sum_n \beta_{ni}(t)\psi_n\rangle. \tag{12}$$

In the following, one can calculate the 6 terms one by one on the right hand side. The first term is

$$\langle\psi_m|h_0[\varphi]|\psi_i\rangle = \langle\psi_m|h_0[\psi] + \delta h_0|\psi_i\rangle. \tag{13}$$

Its zero-order term $\langle\psi_m|h_0[\psi]|\psi_i\rangle$ vanishes due to the orthogonality of the single-particle wave functions, and its first-order term reads

$$\begin{aligned}&\langle\psi_m|\delta h_0|\psi_i\rangle\\&\quad=\sum_{nj}\left(\beta^*_{nj}\langle mn|V|ij-ji\rangle+\beta_{nj}\langle mj|V|in-ni\rangle\right).\end{aligned}\tag{14}$$

The other five terms are much simpler. The non-vanishing terms are

$$\langle\psi_m|\mathcal{W}|\psi_i\rangle=\langle\psi_m|We^{-i\omega t}+W^\dagger e^{i\omega t}|\psi_i\rangle,\tag{15a}$$

$$\langle\psi_m|h_0[\varphi]|\sum_n\beta_{ni}\psi_n\rangle=\epsilon_m\beta_{mi},\tag{15b}$$

$$\langle\psi_m|-\epsilon_i|\sum_n\beta_{ni}\psi_n\rangle=-\epsilon_i\beta_{mi}.\tag{15c}$$

Combining all the non-vanishing terms together, and separating the coefficients of $e^{i\omega t}$ and $e^{-i\omega t}$, one has

$$\begin{aligned}-\langle m|W|i\rangle&=[(\epsilon_m-\epsilon_i)-\omega]X_{mi}\\&\quad+\sum_{nj}[\langle mj|V|in-ni\rangle X_{nj}+\langle mn|V|ij-ji\rangle Y_{nj}],\\-\langle m|W^\dagger|i\rangle^*&=-[-(\epsilon_m-\epsilon_i)-\omega]Y_{mi}\\&\quad+\sum_{nj}[\langle mn|V|ij-ji\rangle X_{nj}+\langle mj|V|in-ni\rangle Y_{nj}].\end{aligned}$$

When the external field approaches zero, the above equation becomes the RPA equation

$$\begin{pmatrix}\mathcal{A}_{mi,nj}&\mathcal{B}_{mi,nj}\\-\mathcal{B}_{mi,nj}&-\mathcal{A}_{mi,nj}\end{pmatrix}\begin{pmatrix}X_{nj}\\Y_{nj}\end{pmatrix}=\Omega\begin{pmatrix}X_{mi}\\Y_{mi}\end{pmatrix},\tag{17}$$

where the matrix elements for particle-hole residual interaction are

$$\mathcal{A}_{mi,nj}=(\epsilon_m-\epsilon_i)\delta_{mi,nj}+\langle mj|V|in-ni\rangle,\tag{18a}$$

$$\mathcal{B}_{mi,nj}=\langle mn|V|ij-ji\rangle.\tag{18b}$$

In the matrix element, both the direct term and exchange term are included. If the Hartree approximation is adopted for the ground state calculation, the exchange term in the particle-hole residual interaction here should also be neglected. The matrix $\mathcal{A}$ is Hermitian and the matrix $\mathcal{B}$ is symmetric.

The matrix $\begin{pmatrix}\mathcal{A}&\mathcal{B}\\\mathcal{B}^*&\mathcal{A}^*\end{pmatrix}$ is Hermitian, and it leads to the orthogonality relation

$$\delta_{\nu\nu'}=\sum_{mi}(X^{\nu*}_{mi}X^{\nu'}_{mi}-Y^{\nu*}_{mi}Y^{\nu'}_{mi}).\tag{19}$$

The above relation could also be obtained from

$$\langle \nu|\nu' \rangle = \delta_{\nu\nu'} = \langle \text{RPA}|[Q_\nu, Q^\dagger_{\nu'}]|\text{RPA}\rangle \simeq \langle \text{HF}|[Q_\nu, Q^\dagger_{\nu'}]|\text{HF}\rangle. \quad (20)$$

For $\nu = \nu'$, this gives us the possibility of normalizing the vector (X^ν, Y^ν). For the solutions of RPA, to each eigenvector (X^ν, Y^ν) with eigenvalue Ω_ν, we have another eigenvector (Y^ν, X^ν) with eigenvalue $-\Omega_\nu$. Both eigenvectors have the same absolute norm in the sense of (19) but with different sign. We usually choose eigenvectors giving positive norms as the physical solutions.

The X^ν_{mi} and Y^ν_{mi} are called the forward- and backward-going p-h amplitudes, and they have a direct physical meaning. Their absolute squares give the probability of finding the states $a^\dagger_m a_i|0\rangle$ and $a^\dagger_i a_m|0\rangle$ in the excited state $|\nu\rangle$, that is,

$$\begin{aligned} X^\nu_{mi} &= \langle \nu|a^\dagger_m a_i|0\rangle, \\ Y^\nu_{mi} &= \langle \nu|a^\dagger_i a_m|0\rangle. \end{aligned} \quad (21)$$

2.3. *RPA equations in angular momentum coupled form*

For the spherical case, the single-particle states as the basis vectors of the RPA equation are given by $|jm\rangle$ according to the operators $\hat{j}^2$ and $\hat{j}_z$. Making use of this, the angular integrals in the RPA matrix elements $\mathcal{A}$ and $\mathcal{B}$ can be calculated independently and analytically by the angular momentum coupling. Therefore in this subsection, the RPA equations will be written in the angular momentum coupled form.

We define the angular momentum coupled form of X and Y as following,

$$\begin{aligned} X^{JM}_{mi} &= \sum_{m_m m_i} \langle 0|\langle j_m m_m j_i - m_i|JM\rangle(-)^{j_i - m_i} a^\dagger_{j_m m_m} a_{j_i m_i}|\nu, JM\rangle \\ &= \sum_{m_m m_i} \langle j_m m_m j_i - m_i|JM\rangle(-)^{j_i - m_i} X_{j_m m_m j_i m_i}, \end{aligned} \quad (22a)$$

$$\begin{aligned} Y^{JM}_{mi} &= (-)^{J-M} \sum_{m_m m_i} \langle 0|\langle j_m m_m j_i - m_i|J - M\rangle(-)^{j_i - m_i} a^\dagger_{j_i m_i} a_{j_m m_m}|\nu, JM\rangle \\ &= (-)^{J-M} \sum_{m_m m_i} \langle j_m m_m j_i - m_i|J - M\rangle(-)^{j_i - m_i} Y_{j_m m_m j_i m_i}, \end{aligned} \quad (22b)$$

and the angular momentum coupled form of p-h matrix elements,

$$\mathcal{A}^J_{minj} = \sum_{m_m m_i} \sum_{m_n m_j} \langle j_m m_m j_i - m_i | JM\rangle (-)^{j_i - m_i}$$
$$\langle j_n m_n j_j - m_j | JM\rangle (-)^{j_j - m_j} \mathcal{A}_{j_m m_m j_i m_i j_n m_n j_j m_j}, \tag{23a}$$
$$\mathcal{B}^J_{minj} = \sum_{m_m m_i} \sum_{m_n m_j} \langle j_m m_m j_i - m_i | JM\rangle (-)^{j_i - m_i}$$
$$(-)^{J-M} \langle j_n m_n j_j - m_j | J - M\rangle (-)^{j_j - m_j} \mathcal{B}_{j_m m_m j_i m_i j_n m_n j_j m_j}. \tag{23b}$$

The respective RPA equations in the angular momentum coupled form are given by,

$$\begin{pmatrix} \mathcal{A}^J_{minj} & \mathcal{B}^J_{minj} \\ -\mathcal{B}^{J*}_{minj} & -\mathcal{A}^{*J}_{minj} \end{pmatrix} \begin{pmatrix} X^{JM}_{nj} \\ Y^{JM}_{nj} \end{pmatrix} = \Omega \begin{pmatrix} X^{JM}_{mi} \\ Y^{JM}_{mi} \end{pmatrix}. \tag{24}$$

After simplification, the RPA matrix elements become

$$\mathcal{A}^J_{minj} = (\epsilon_m - \epsilon_i)\delta_{j_m j_n}\delta_{j_i j_j} + V^J_{mjin}, \tag{25}$$
$$\mathcal{B}^J_{minj} = (-)^{j_j - j_n + J} V^J_{mnij}, \tag{26}$$

where the angular momentum coupled matrix element of two-body residual interaction is

$$V^J_{mjin} = \sum_{m_m m_i} \sum_{m_n m_j} \langle j_m m_m j_i - m_i | JM\rangle (-)^{j_i - m_i}$$
$$\langle j_n m_n j_j - m_j | JM\rangle (-)^{j_j - m_j} V_{j_m m_m j_j m_j j_i m_i j_n m_n}. \tag{27}$$

2.4. *Transition probabilities*

For a one-body operator $\hat{F}$ with specific quantum numbers JM, it reads

$$\hat{F}_{JM} = \sum_{nj} \langle n | F_{JM} | j\rangle a^\dagger_n a_j. \tag{28}$$

The transition matrix element involving the operator $\hat{F}_{JM}$ is

$$\begin{aligned}\langle \nu_{JM} | \hat{F}_{JM} | 0\rangle &= \langle 0 | [Q_\nu, \hat{F}_{JM}] | 0\rangle \simeq \langle \mathrm{HF} | [Q_\nu, \hat{F}_{JM}] | \mathrm{HF}\rangle \\ &= \left\langle \mathrm{HF} \left| \left[\sum_{mi} X^\nu_{mi} a^\dagger_i a_m - Y^\nu_{mi} a^\dagger_m a_i, \sum_{nj} \langle n | F_{JM} | j\rangle a^\dagger_n a_j \right] \right| \mathrm{HF} \right\rangle \\ &= \sum_{mi} \left(X^\nu_{mi} \langle m | F_{JM} | i\rangle - Y^\nu_{mi} \langle i | F_{JM} | m\rangle \right), \end{aligned} \tag{29}$$

where we used the commutator relation

$$\langle \mathrm{HF} | [a^\dagger_i a_m, a^\dagger_n a_j] | \mathrm{HF}\rangle = \delta_{mn}\delta_{ij}. \tag{30}$$

Making use of the angular momentum coupled X, Y in Eq. (22), and the Wigner–Eckart theorem, we obtain

$$\begin{aligned}\langle \nu_{JM}|\hat{F}_{JM}|0\rangle &= \hat{J}^{-1}\sum_{mi}\left(X^{\nu}_{mi}\langle m||F_{JM}||i\rangle + (-)^{j_i-j_m+J}Y^{\nu}_{mi}\langle i||F_{JM}||m\rangle\right)\\ &= \hat{J}^{-1}\sum_{mi}\left(X^{\nu}_{mi} + (-)^{J}Y^{\nu}_{mi}\right)\langle m||F_{JM}||i\rangle,\end{aligned} \tag{31}$$

where $\hat{J} = \sqrt{2J+1}$. The second step is true only for electric transitions.

The reduced transition probabilities caused by the operator $\hat{F}_J$ from the ground state $|0\rangle$ to excited state $|\nu_J\rangle$ are given by

$$B_\nu = \sum_M \left|\langle \nu_{JM}|\hat{F}_{JM}|0\rangle\right|^2 = \hat{J}^2\left|\langle \nu_{JM}|\hat{F}_{JM}|0\rangle\right|^2. \tag{32}$$

In order to obtain transition strength as a smooth function of the excitation energy, one usually calculates the Lorentzian-averaged strength distribution

$$R(E) = \sum_\nu B_\nu \frac{\Gamma/2\pi}{(E-\Omega_\nu)^2 + \Gamma^2/4} \tag{33}$$

with the width Γ, which is introduced artificially to mimic the spreading and escaping width of giant resonances.

2.5. *Transition operators and sum rules*

The kth-order energy weighted sum rule for the operator $\hat{F}$ is defined as

$$S_k \equiv \sum_\nu \Omega_\nu^k |\langle \nu|\hat{F}|0\rangle|^2, \tag{34}$$

where $|\nu\rangle$ represents a complete set of eigenstates of the system's Hamiltonian $\hat{H}$ with eigenenergy E_ν and the corresponding excitation energy $\Omega_\nu = E_\nu - E_0$. Making use of the relation of completeness $\sum_\nu |\nu\rangle\langle\nu| = 1$, one can obtain

$$S_k = \langle 0|\hat{F}^\dagger(\hat{H} - E_0)^k\hat{F}|0\rangle. \tag{35}$$

The most important sum rule is the energy weighted sum rule S_1. It can be written as a double commutator

$$S_1 = \sum_\nu \Omega_\nu|\langle\nu|\hat{F}|0\rangle|^2 = \frac{1}{2}\langle 0|[\hat{F},[H,\hat{F}]]|0\rangle. \tag{36}$$

The left-hand side is evaluated with RPA wave functions and energies while the right-hand side is calculated using the HF ground state wave function.

According to the third component of isospin T_0, the collective modes can be classified as non-charge-exchange channel and charge-exchange channel.

In the non-charge-exchange channel, the nucleus is excited from the ground state to a higher energy state of the same nucleus, i.e., the proton number Z and neutron number N are not changed. According to their multipolarity L, spin S and isospin T quantum numbers, the non-charge-exchange excitation modes can be further classified. The $\Delta S = 0$, $\Delta T = 0$ modes are electric, isoscalar vibrations in which the protons and neutrons oscillate in phase according to a multipole pattern defined by $\Delta L = 0, 2, 3, \ldots$. The $\Delta L = 1$ vibration is missing because to the first order it corresponds to a translational motion of the nucleus and it is not an intrinsic excitation. However, there is a higher-order $\Delta L = 1$ vibration, for which we will give its transition operator later. The $\Delta S = 0$, $\Delta T = 1$ modes are electric, isovector vibrations in which the protons and neutrons oscillate out of phase against each other according to a multipole pattern defined by $\Delta L = 0, 1, 2, 3, \ldots$. The $\Delta S = 1$, $\Delta T = 0$ modes are magnetic, isoscalar vibrations in which nucleons with spin up vibrate against nucleons with spin down, in a multipole pattern given by ΔL. The $\Delta S = 1$, $\Delta T = 1$ modes are magnetic, isovector vibrations in which protons with spin up (down) vibrate against neutrons with spin down (up), in a multipole pattern given by ΔL.[1]

In the following, we give the transition operators for electric transitions. The operators for isoscalar and isovector multipole transitions with multipolarity $L = J$ are

$$\hat{F}_{JM}^{T=0} = \sum_{i=1}^{A} r_i^L Y_{LM}, \tag{37}$$

$$\hat{F}_{JM}^{T=1} = \sum_{i=1}^{A} \tau_3^{(i)} r_i^L Y_{LM}. \tag{38}$$

The corresponding sum rule of isoscalar transitions is[1,16]

$$m_1^{T=0} = \frac{\hbar^2 A}{8\pi M} L(2L+1)^2 \langle r^{2L-2} \rangle. \tag{39}$$

However, there are two exceptions to the above operators. One is the operator for isoscalar monopole giant resonance (ISGMR), where the second-order term in the expansion of the spherical Bessel function is used for the transition operator,

$$\hat{F}_{\rm ISGMR} = \sum_{i=1}^{A} r_i^2, \tag{40}$$

and its corresponding sum rule is

$$m_1 = \frac{2\hbar^2 A}{M}\langle r^2\rangle. \tag{41}$$

The other one is the operator for giant dipole resonance (GDR). In order to remove the spurious state of translational motion, the operators are written as[17]

$$\hat{F}_{1M}^{T=0} = \sum_{i=1}^{A}\left(r_i^3 - \frac{5}{3}\langle r^2\rangle r_i\right)Y_{1M}, \tag{42}$$

$$\hat{F}_{1M}^{T=1} = \frac{N}{A}\sum_{i=1}^{Z} r_i Y_{1M} - \frac{Z}{A}\sum_{i=1}^{N} r_i Y_{1M}. \tag{43}$$

The respective sum rules are

$$m_1^{T=0} = \frac{3\hbar^2 A}{8\pi M}\left[11\langle r^4\rangle - \frac{25}{3}\langle r^2\rangle^2\right], \tag{44}$$

$$m_1^{T=1} = \frac{9\hbar^2}{8\pi M}\frac{NZ}{A}(1+\kappa). \tag{45}$$

The latter one corresponds to the Thomas–Reiche–Kuhn (TRK) sum rule. κ is introduced because of the presence of charge-dependent terms in the nucleon-nucleon interaction.[1] For $A \geq 90$ nuclei the experimental value of $(1+\kappa)$ deduced from experiment varies from about 1 for $A \approx 100$ to 1.3 ± 0.2 for heavy nuclei such as the actinide nuclei with an average value of 1.20 ± 0.1.[1,7] Theoretical estimates for κ due to the exchange part of the nucleon-nucleon interaction give $\kappa \approx 0.3 - 0.5$ while the velocity dependent part has been estimated to give 0.2 (See Ref. 1 and references therein). Measured energy weighted strength for the isovector GDR amounts to $\approx$ 60–120% TRK sum rule, depending on the nucleus under consideration, and the range of excitation energies taken into account.[18] In the case of isovector GDR in ^{208}Pb, the analysis based on three truncated Lorentzian shapes resulted in 122% of the TRK sum rule.[19]

In the charge-exchange channel, the third component of the isospin T_0 is changed by ± 1. The $T_0 \pm 1$ transitions are strongly excited in the $\Delta T = 1$ nucleon charge-exchange reactions (p,n) for T_- channel and (n,p) for T_+ channel. In the following, some basic operators for charge-exchange transitions are given.

The operator for the isobaric analog resonance (IAS) is

$$\hat{F}^{\pm}_{\text{IAS}} = \sum_{i=1}^{A} \tau_{\pm}(i), \tag{46}$$

and its corresponding non-energy weighted sum rule is

$$m_0^- - m_0^+ = \sum_{\nu} B_\nu^- - \sum_{\nu} B_\nu^+ = N - Z. \tag{47}$$

The operator for the Gamow–Teller resonance (GTR) is

$$\hat{F}^{\pm}_{\text{GTR}} = \sum_{i=1}^{A} \boldsymbol{\sigma}(i)\tau_{\pm}(i), \tag{48}$$

and its corresponding non-energy weighted sum rule, known as Ikeda sum rule,[20] reads

$$m_0^- - m_0^+ = 3(N - Z). \tag{49}$$

The operator for the spin dipole resonance (SDR) is

$$\hat{F}^{\pm}_{\text{SDR}} = \sum_{i=1}^{A} [r_i Y_1(i) \otimes \boldsymbol{\sigma}(i)]_{J=0,1,2}\tau_{\pm}(i), \tag{50}$$

and its corresponding non-energy weighted sum rule is

$$m_0^- - m_0^+ = \frac{9}{4\pi}\left(N\langle r^2\rangle_n - Z\langle r^2\rangle_p\right). \tag{51}$$

3. Relativistic quasiparticle random phase approximation

When considering open shell nuclei, description of multipole response necessitates inclusion of pairing correlations, both in modeling the nuclear ground state properties and excitations. In the former case, the relativistic Hartree–Bogoliubov (RHB) model includes both the relativistic mean field and pairing correlations within a unified framework.[21] In order to describe the excitations, the relativistic RPA (RRPA) is extended to relativistic quasiparticle random phase approximation (RQRPA), with configuration space composed of two-quasiparticle pairs.

In the RHB model, the ground state of a nucleus $|\Phi\rangle$ is represented as the vacuum with respect to independent quasiparticle operators

$$\alpha_k^+ = \sum_l U_{lk} c_l^+ + V_{lk} c_l, \tag{52}$$

where U_{lk}, V_{lk} are the Hartree–Bogoliubov coefficients. They determine the Hermitian single-particle density matrix

$$\hat{\rho} = V^* V^T, \tag{53}$$

and the antisymmetric pairing tensor

$$\hat{\kappa} = V^* U^T. \tag{54}$$

The generalized density matrix $\mathcal{R}$ reads,

$$\mathcal{R} = \begin{pmatrix} \rho & \kappa \\ -\kappa^* & 1-\rho^* \end{pmatrix}, \tag{55}$$

which obeys the equation of motion

$$i\partial_t \mathcal{R} = [\mathcal{H}(\mathcal{R}), \mathcal{R}]. \tag{56}$$

The generalized Hamiltonian $\mathcal{H}$ is a functional derivative of the energy with respect to the generalized density

$$\mathcal{H} = \frac{\delta E}{\delta \mathcal{R}} = \begin{pmatrix} \hat{h}_D - m - \lambda & \hat{\Delta} \\ -\hat{\Delta}^* & -\hat{h}_D + m + \lambda \end{pmatrix}. \tag{57}$$

It contains two average potentials: the single-particle mean field $\hat{h}_D = \frac{\delta E}{\delta \hat{\rho}}$, and the pairing field $\hat{\Delta} = \frac{\delta E}{\delta \hat{\kappa}}$.

The RQRPA represents a limit of small amplitude oscillations of the time-dependent RHB model. The generalized density matrix $\mathcal{R}$ and the meson and photon fields $\phi_m = \sigma, \omega^\mu, \boldsymbol{\rho}^\mu, A^\mu$ have been considered as independent variables related only by the equations of motion. One can use the Klein–Gordon equations to eliminate the meson degrees of freedom, but this is only possible in the small amplitude limit. Thus the generalized Hamiltonian $\mathcal{H}$ can be expressed as a functional of the generalized density $\mathcal{R}$ only.[22] In the linear approximation the generalized density matrix is expanded

$$\mathcal{R} = \mathcal{R}_0 + \delta\mathcal{R}(t), \tag{58}$$

where $\mathcal{R}_0$ is the stationary ground-state generalized density. Since $\mathcal{R}(t)$ is a projector at all times, in linear order

$$\mathcal{R}_0 \delta\mathcal{R} + \delta\mathcal{R}\mathcal{R}_0 = \delta\mathcal{R}. \tag{59}$$

In the quasiparticle basis the matrices $\mathcal{R}_0$ and $\mathcal{H}_0 = \mathcal{H}(\mathcal{R}_0)$ are diagonal

$$\mathcal{R}_0 = \begin{pmatrix} 0 & 0 \\ 0 & 1 \end{pmatrix} \quad \text{and} \quad \mathcal{H}_0 = \begin{pmatrix} E_n & 0 \\ 0 & -E_n \end{pmatrix}. \tag{60}$$

From Eq. (59) it follows that the matrix $\delta\mathcal{R}$ has the form

$$\delta\mathcal{R} = \begin{pmatrix} 0 & \delta R \\ -\delta R^* & 0 \end{pmatrix}. \tag{61}$$

The linearized equation of motion (56) reduces to

$$i\partial_t \mathcal{R} = [\mathcal{H}_0, \delta\mathcal{R}] + \left[\frac{\delta\mathcal{H}}{\delta\mathcal{R}}\delta\mathcal{R}, \mathcal{R}_0\right]. \tag{62}$$

Assuming an oscillatory solution

$$\delta\mathcal{R}(t) = \sum_\nu \delta\mathcal{R}^{(\nu)} e^{i\omega_\nu t} + h.c., \tag{63}$$

the RQRPA equation is obtained

$$\begin{pmatrix} A & B \\ -B^* & -A^* \end{pmatrix}\begin{pmatrix} X^\nu \\ Y^\nu \end{pmatrix} = \omega_\nu \begin{pmatrix} X^\nu \\ Y^\nu \end{pmatrix}. \tag{64}$$

For $k < k'$, $l < l'$ the RQRPA matrix elements read

$$A_{kk'll'} = (E_k + E_{k'})\delta_{kl}\delta_{k'l'} + \frac{\delta^2 E}{\delta R^*_{kk'}\delta R_{ll'}}, \tag{65}$$

$$B_{kk'll'} = \frac{\delta^2 E}{\delta R^*_{kk'}\delta R^*_{ll'}}. \tag{66}$$

If the two-body Hamiltonian is density independent, the matrices A and B have the simple forms[16]

$$\begin{aligned} A_{kk',ll'} &= \langle\Phi| [\alpha_{k'}\alpha_k, [\hat{H}, \alpha_l^+\alpha_{l'}^+]] |\Phi\rangle , \\ B_{kk',ll'} &= -\langle\Phi| [\alpha_{k'}\alpha_k, [\hat{H}, \alpha_{l'}\alpha_l]] |\Phi\rangle . \end{aligned} \tag{67}$$

Using the representation of the Hamiltonian in the quasiparticle basis

$$\begin{aligned} \hat{H} = E_0 &+ \sum_{kk'} H^{11}_{kk'}\alpha_k^+\alpha_{k'} + \frac{1}{4}\sum_{kk'll'} H^{22}_{kk'll'}\alpha_k^+\alpha_{k'}^+\alpha_{l'}\alpha_l \\ &+ \sum_{kk'll'} \left(H^{40}_{kk'll'}\alpha_k^+\alpha_{k'}^+\alpha_{l'}^+\alpha_l^+ + h.c.\right) \\ &+ \sum_{kk'll'} \left(H^{31}_{kk'll'}\alpha_k^+\alpha_{k'}^+\alpha_{l'}^+\alpha_l + h.c.\right), \end{aligned} \tag{68}$$

we find

$$A_{kk'll'} = H^{11}_{kl}\delta_{k'l'} - H^{11}_{k'l}\delta_{kl'} - H^{11}_{kl'}\delta_{k'l} + H^{11}_{k'l'}\delta_{kl} + H^{22}_{kk'll'}$$
$$B_{kk'll'} = 4H^{40}_{kk'll'}. \tag{69}$$

In the quasiparticle representation the matrix H^{11} is diagonal, i.e. $H^{11}_{kl} = E_k\delta_{kl}$. The matrices H^{22} and H^{40} are rather complicated expressions containing the two-body *ph* and particle-particle (*pp*) matrix elements and the coefficients U and V (for details see Ref. 16). In the more general case of a density dependent Hamiltonian the same expressions can be used, but one has to take into account the rearrangement terms originating from the variation of the interaction with respect to the density $\hat{\rho}$, which will be discussed later in detail.

The full RQRPA equations require the evaluation of the matrix elements $H^{22}_{kk'll'}$ and $H^{40}_{kk'll'}$ in the basis of the Hartree–Bogoliubov spinors $U_k(\boldsymbol{r})$ and $V_k(\boldsymbol{r})$. It is considerably simpler to solve these equations in the canonical basis, in which the RHB wave functions can be expressed in the form of BCS-like wave functions. The canonical basis is specified by the requirement that it diagonalizes the single-nucleon density matrix. The transformation to the canonical basis determines the energies and occupation probabilities of single-nucleon states that correspond to the self-consistent solution for the ground state of a nucleus. Since it diagonalizes the density matrix, the canonical basis is always localized. In this case one needs only the matrix elements $V^{ph}_{\kappa\lambda'\kappa'\lambda}$ of the residual *ph* interaction, and the matrix elements $V^{pp}_{\kappa\kappa'\lambda\lambda'}$ of the pairing *pp* interaction, as well as certain combinations of the occupation factors u_κ, v_κ. In the following we use the indices κ, λ, κ' and λ' to denote states in the canonical basis. We emphasize that the solution of the RQRPA equations in the canonical basis does not represent an approximation. We obtain a full solution and the results do not depend on this special choice of the basis.

Taking into account the rotational invariance of the nuclear system, the quasiparticle pairs can be coupled to good angular momentum and the matrix equations of the RQRPA read

$$\begin{pmatrix} A^J & B^J \\ B^{*J} & A^{*J} \end{pmatrix} \begin{pmatrix} X^{\nu,JM} \\ Y^{\nu,JM} \end{pmatrix} = \omega_\nu \begin{pmatrix} 1 & 0 \\ 0 & -1 \end{pmatrix} \begin{pmatrix} X^{\nu,JM} \\ Y^{\nu,JM} \end{pmatrix}. \tag{70}$$

For each RQRPA energy ω_ν, X^ν and Y^ν denote the corresponding forward- and backward-going two-quasiparticle amplitudes, respectively. The coupled RQRPA matrices in the canonical basis read

$$\begin{aligned} A^J_{\kappa\kappa'\lambda\lambda'} &= H^{11(J)}_{\kappa\lambda}\delta_{\kappa'\lambda'} - H^{11(J)}_{\kappa'\lambda}\delta_{\kappa\lambda'} - H^{11(J)}_{\kappa\lambda'}\delta_{\kappa'\lambda} + H^{11(J)}_{\kappa'\lambda'}\delta_{\kappa\lambda} \\ &+\frac{1}{2}(\xi^+_{\kappa\kappa'}\xi^+_{\lambda\lambda'} + \xi^-_{\kappa\kappa'}\xi^-_{\lambda\lambda'})V^{ppJ}_{\kappa\kappa'\lambda\lambda'} \\ &+\zeta_{\kappa\kappa'\lambda\lambda'}V^{phJ}_{\kappa\lambda'\kappa'\lambda}, \qquad (71) \\ B^J_{\kappa\kappa'\lambda\lambda'} &= \frac{1}{2}(\xi^+_{\kappa\kappa'}\xi^+_{\lambda\lambda'} - \xi^-_{\kappa\kappa'}\xi^-_{\lambda\lambda'})V^{ppJ}_{\kappa\kappa'\lambda\lambda'} \\ &+\zeta_{\kappa\kappa'\lambda\lambda'}(-1)^{j_\lambda - j_{\lambda'}+J}V^{phJ}_{\kappa\lambda\kappa'\lambda'}\,. \qquad (72) \end{aligned}$$

H^{11} denotes the one-quasiparticle terms

$$H^{11}_{\kappa\lambda} = (u_\kappa u_\lambda - v_\kappa v_\lambda)h_{\kappa\lambda} - (u_\kappa v_\lambda + v_\kappa u_\lambda)\Delta_{\kappa\lambda}\,, \qquad (73)$$

i.e. the canonical RHB basis does not diagonalize the Dirac single-nucleon mean-field Hamiltonian $\hat{h}_D$ and the pairing field $\hat{\Delta}$. The occupation probabilities v_k^2 of the canonical states are eigenvalues of the density matrix. V^{ph} and V^{pp} are the *ph* and *pp* residual interactions, respectively. Their matrix elements are multiplied by the pairing factors $\xi^{\pm}$ and ζ, defined by the occupation amplitudes of the canonical states. The pairing factors are

$$\zeta_{\kappa\kappa'\lambda\lambda'} = \begin{cases} \eta^+_{\kappa\kappa'}\eta^+_{\lambda\lambda'} & \text{for } \sigma,\ \omega^0\ ,\rho^0,\ A^0;\ \text{if J is even} \\ & \text{for } \boldsymbol{\omega},\ \boldsymbol{\rho},\ \boldsymbol{A};\ \text{if J is odd} \\ \eta^-_{\kappa\kappa'}\eta^-_{\lambda\lambda'} & \text{for } \sigma,\ \omega^0,\ \rho^0,\ A^0;\ \text{if J is odd} \\ & \text{for } \boldsymbol{\omega},\ \boldsymbol{\rho},\ \boldsymbol{A};\ \text{if J is even} \end{cases} \qquad (74)$$

with the η-coefficients defined by

$$\eta^{\pm}_{kk'} = u_k v_{k'} \pm v_k u_{k'}\,, \qquad (75)$$

and

$$\xi^{\pm}_{kk'} = u_k u_{k'} \mp v_k v_{k'}\,. \qquad (76)$$

σ, ω^0, ρ^0, and A^0 denote the time-like components, and $\boldsymbol{\omega}$, $\boldsymbol{\rho}$, $\boldsymbol{A}$ the spatial components of the meson and photon fields, respectively. The relativistic *ph* interaction V^{ph} is derived from the density dependent relativistic nuclear energy density functional,[23] i.e., the same functional used in the mean-field Dirac single-nucleon Hamiltonian $\hat{h}_D$. More details about the derivation of V^{ph} for the functional with density dependent meson-nucleon vertex functions are given in Ref. 23. In addition to the direct contribution of meson exchange interactions, the explicit density dependence of the meson-nucleon couplings introduces a number of rearrangement terms in the residual two-body interaction. These rearrangement terms are essential for fully consistent RRPA calculations. Only when their contribution is included in the matrix elements of the residual interaction, does it become possible to reproduce reasonably well the excitation energies of giant multipole resonances. Without rearrangement terms, one finds discrepancies of the order of several MeV between the experimental excitation energies and the RRPA peak energies, calculated with relativistic effective interactions that are adjusted to ground state properties of spherical nuclei.

The RQRPA configuration space includes the Dirac sea of negative energy states. In addition to the configurations built from two-quasiparticle states of positive energy, the RQRPA configuration space must also contain pair-configurations formed from the fully or partially occupied states of positive energy and the empty negative-energy states from the Dirac sea. The inclusion of configurations built from occupied positive-energy states and empty negative-energy states is essential for current conservation and the decoupling of spurious states.[24] The RRPA calculation also showed that the fully consistent inclusion of the Dirac sea of negative energy states in the RRPA configuration space is essential for a quantitative comparison with the experimental excitation energies of giant resonances.[25,26]

It should be emphasized that the RQRPA established in this way is fully consistent: the same interactions, both in the particle-hole and particle-particle channels, are used in the RHB equation that determines the canonical quasiparticle basis, and in the RQRPA equation. In both channels the same strength parameters of the interactions are used in the RHB and RQRPA calculations. No additional adjustment of the parameters is needed in RQRPA calculations. This essential feature ensures that RQRPA amplitudes do not contain spurious components associated with the mixing of the nucleon number in the RHB ground state (for 0^+ excitations), or with the center-of-mass translational motion (for 1^- excitations).

Multipole transition strength is governed by the multipole operators $\hat{F}_{JM}$ introduced in Sec. 2.5. The respective transition probability is given by

$$B(J,\omega_\nu) = \left| \sum_{\kappa\kappa'} \left\{ X^{\nu,J0}_{\kappa\kappa'} \langle\kappa||\hat{F}_J||\kappa'\rangle + (-1)^{j_\kappa - j_{\kappa'} + J}\, Y^{\nu,J0}_{\kappa\kappa'} \langle\kappa'||\hat{F}_J||\kappa\rangle \right\} \right.$$
$$\left. (u_\kappa v_{\kappa'} + (-1)^J v_\kappa u_{\kappa'}) \right|^2 . \tag{77}$$

The discrete strength function is folded by a Lorentzian from Eq.(33) using an arbitrary width, for example, Γ=1 MeV.

Another quantity of interest is transition density that describes variations of the nuclear density for a specific excitation mode.[27] For the state

$|J,\nu\rangle$, the RQRPA transition density reads

$$\delta\rho_J^\nu(r) = \sum_{\kappa\kappa'} \Big\{ \langle\kappa||Y_J||\kappa'\rangle f_\kappa(r) f_{\kappa'}(r) + \langle\hat{\kappa}||Y_J||\hat{\kappa}'\rangle g_\kappa(r) g_{\kappa'}(r) \Big\}$$
$$\cdot \Big(X_{\kappa\kappa'}^{\nu,J0} + (-1)^J Y_{\kappa\kappa'}^{\nu,J0} \Big) (u_\kappa v_{\kappa'} + (-1)^J v_\kappa u_{\kappa'}), \tag{78}$$

where κ and $\hat{\kappa}$ denote the quantum numbers of the large and small components of the Dirac spinors, respectively. $f_\kappa(r)$ and $g_\kappa(r)$ are the corresponding large and small radial components.

4. Giant resonances

4.1. *Giant monopole resonances*

One of the basic collective modes of excitation in nuclei is giant monopole resonance (GMR). The isoscalar GMR (ISGMR) corresponds to the compressional breathing mode, where the nuclear density oscillates around the ground state density, while the spherical shape is preserved. Compressional modes in nuclei can be used to deduce the value of the nuclear matter compression modulus K_{nm} from a comparison of experimental excitation energies with those predicted by microscopic nuclear effective interactions.[28,29] Inelastic α-scattering experiments have been employed in high precision studies of the systematics of ISGMR.

Theoretical studies of nuclear compressional modes include various approaches, e.g., the fluid dynamics approach,[30] the Hartree–Fock + RPA with Skyrme interactions,[13,31–33] the RPA based on separable Hamiltonians,[34] linear response within a stochastic one-body transport theory,[35] the relativistic transport approach,[36] and the self-consistent relativistic RPA.[37–40] As emphasized in Refs. 33, 41, fully self-consistent implementation of the RPA is essential to determine the value of K_{nm} from the ISGMR. Both non-relativistic and relativistic frameworks have been developed and employed to constrain the value of K_{nm}. The value of the incompressibility coefficient of symmetric nuclear matter at $K_{\mathrm{nm}} = 240 \pm 10$ MeV is obtained (see Ref. 42 and references therein).

In Fig. 2 we display the isoscalar monopole strength distributions for ^{208}Pb. The relativistic RPA calculation has been performed with the density dependent DD-ME2 effective interaction.[43] The calculation predicts pronounced collective peak in the strength distribution, located at 13.82 MeV, in reasonable agreement with the experimental data 13.9±0.3 MeV[29] and 14.18±0.11 MeV.[44] The respective nuclear matter compression modulus for the DD-ME2 effective interaction amounts $K_{nm} = 251$ MeV. Fig. 2 also shows the transition densities for the main ISGMR peak. The transition densities are shown separately for protons and neutrons, as well as the overall transition density. In the case of isoscalar monopole mode, protons and neutrons oscillate in phase. As expected for the ISGMR resonance, densities in the interior of the nucleus increase while those at the surface decrease (or the other way around), and a characteristic node at the nuclear surface appears. The transition strength distribution and transition density establish the ISGMR as highly collective compression mode in ^{208}Pb.

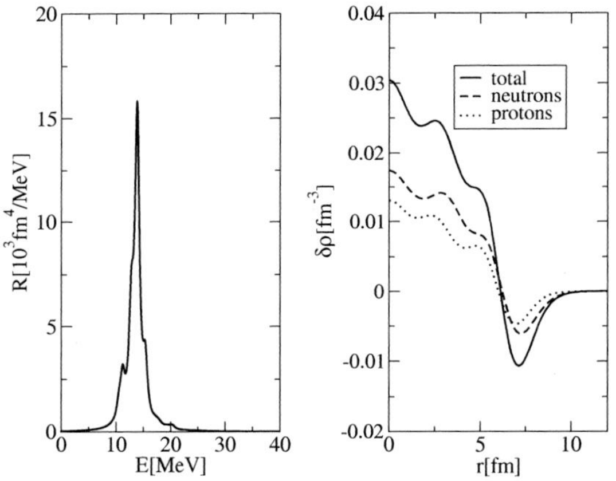

Fig. 2. The ISGMR transition strength distribution for ^{208}Pb, calculated with the DD-ME2 effective interaction (left panel). Right panel includes the transition densities (total, neutron, proton) for the state at 13.82 MeV.

4.2. *Giant dipole resonances*

Isoscalar and isovector giant dipole resonances represent another case of collective excitations in nuclei. In addition to the ISGMR from Sec. 4.1, compressional isoscalar giant dipole resonance (ISGDR) also provides information to constrain the range of allowed values of K_{nm}.[45,46] However, the isoscalar E1 strength distributions show a characteristic bimodal structure with two broad components. Theoretical analysis have shown that only the high-energy component represents compressional vibrations,[25,47] whereas the broad structure in the low-energy region corresponds to a vortical nuclear flow associated with the toroidal dipole moment.[48–50] A strong mixing between compressional and vorticity vibrations in the isoscalar E1 states can be expected up to the highest excitation energies in the region $\approx 3\hbar\omega$.[49,50] Nevertheless, models which use effective interactions with K_{nm} adjusted to ISGMR excitation energies in heavy nuclei, also reproduce the structure of the high-energy portion of the ISGDR data.[31,51,52]

In Fig. 3 the isoscalar dipole transition strength is shown for ^{208}Pb, obtained using the RRPA with DD-ME2 parameterization. The transition strength is basically concentrated in two broad structures: one in the region 10 MeV $\leq E_x \leq$ 20 MeV, and the high-energy component above 20 MeV and extending up to $\approx$30 MeV excitation energy. Only the high-energy portion of the calculated E1 strength is sensitive to the nuclear matter compression modulus of the effective interaction, i.e., it corresponds to dipole compression mode.[53] On the other side, the main low-energy state is of different nature, dominated by vortex collective motion.[53] In the right panel of Fig. 3 the transition densities are shown for the most collective ISGDR state at 26.10 MeV. The transition densities display a characteristic radial dependence for the compression mode.

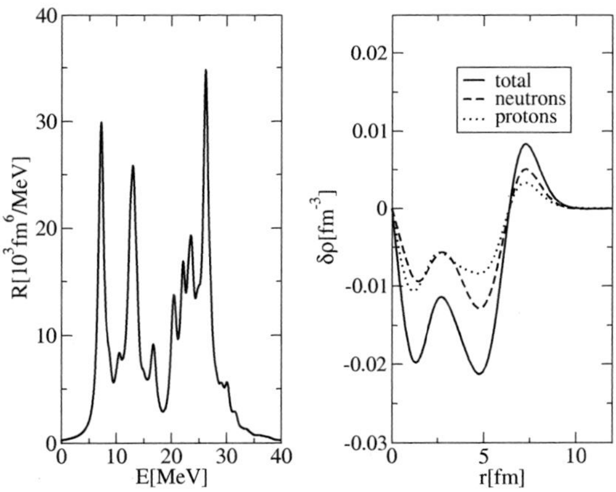

Fig. 3. The ISGDR transition strength distribution for ^{208}Pb, calculated with the DD-ME2 effective interaction (left panel). Right panel includes the transition densities (total, neutron, proton) for the main ISGDR state at 26.10 MeV.

The isovector giant dipole resonance (IVGDR) represents a basic example of nuclear collective motion that has been extensively studied over the past decades (see Sec. 1). It corresponds to a vibration mode where protons coherently oscillate against neutrons. The IVGDR provides an important source of information on the isovector sector of the effective nuclear interactions as well as the symmetry energy. The isovector dipole transitions also play an important role in astrophysical scenarios governing the nucleosynthesis through (n,γ) and (γ,n) reactions.[54] In Fig. 4 the calculated isovector dipole transition strength is shown for ^{208}Pb, using the same framework as in the isoscalar case. It is characterized by a pronounced peak at 13.45 MeV corresponding to the IVGDR. In the right panel in Fig. 4 the transition densities for the state at 13.45 MeV demonstrate the nature of this collective mode. Neutron and proton transition densities have the opposite phases to a large extent, thus indicating an out-of-phase motion of neutrons against protons.

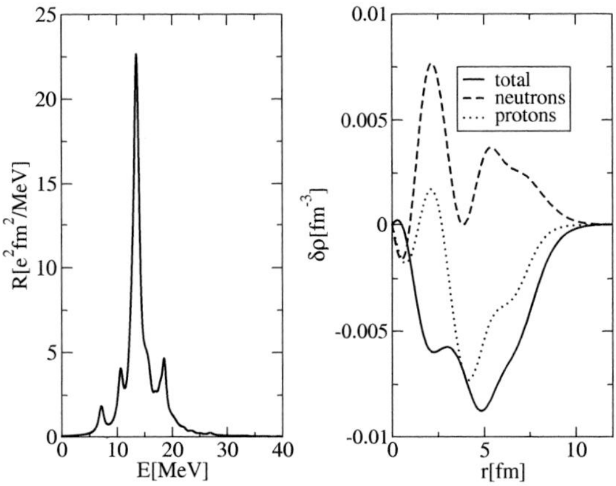

Fig. 4. The IVGDR transition strength distribution for ^{208}Pb, calculated with the DD-ME2 effective interaction (left panel). The transition densities for the main IVGDR peak at 13.45 MeV are shown in the right panel.

4.3. *Giant quadrupole resonances*

The isoscalar giant quadrupole resonance (ISGQR) represents a collective mode that was experimentally discovered decades ago in inelastic electron and proton scattering experiments.[55,56] In fact, the isoscalar transition strength distribution is characterized by two pronounced structures (i) low-lying 2^+ states and (ii) the ISGQR. As pointed out in Ref. 57, low-lying quadrupole excitations depend on the number of particles outside closed shells. On the other side, the ISGQR represents a collective mode with the excitation energy smoothly varying with the mass number of the nucleus. In the isovector channel, the quadrupole response also includes collective excitations, i.e. isovector giant quadrupole resonance (IVGQR). This resonance was rather difficult to measure, and only recently the accuracy in experiment considerably improved.[58] Recently, it has been shown that data on the IVGQR can be used together with the experimental information on the ISGQR to estimate the density slope of the symmetry energy close to saturation.[59]

In the following we illustrate the properties of quadrupole response in ^{208}Pb. In Fig 5 the isoscalar and isovector quadrupole transition strengths are shown (left panel), with the corresponding transition densities for the main ISGQR and IVGQR states (right panel). As shown in the figure, the isoscalar transition strength is composed of the low-lying 2^+ state at $\approx$5 MeV and a collective ISGQR peak at 12.76 MeV. The isovector response is more distributed, it is composed of three separated structures. The first one is also the low-energy 2^+, while the second pronounced state appears at the same energy as the ISGQR. Rather broad IVGQR structure is located at higher energies, and peaked at 24.04 MeV, in close agreement with the experimental data, 23.0$\pm$0.13(stat.)$\pm$0.18(sys.)[58] As pointed out in Ref. 59, the two lower structures arise because the RPA states contain some isospin mixture and could be excited both by isoscalar and isovector probes.

A more detailed insight into the structure of the quadrupole resonances is provided by the transition densities associated with the main ISGQR and IVGQR states. In Fig. 5 (b) the neutron and proton transition densities are shown for the ISGQR state at 12.76 MeV, as well as the overall isoscalar transition density. One can observe a typical response of isoscalar resonance where protons and neutrons oscillate in phase. In Fig. 5 (d) the respective transition densities are shown for the IVGQR state at 24.04 MeV. As expected for the isovector mode, the protons and neutrons mainly oscillate against each other, except in the bulk of nucleus where some isoscalar component is also present.

5. Pygmy dipole strength in neutron-rich nuclei

In addition to giant resonances, in specific nuclei other modes of excitation may also appear. Of particular interest are excitations in the regions of the nuclide map away from the valley of β stability, that also play an important role in the synthesis of chemical elements in stellar environment. In halo nuclei, radial wave functions of weakly-bound or unbound neutron states are extended far beyond the nuclear surface, resulting in diffuse neutron density distributions and unique structure properties.[60] Medium-heavy and heavy neutron-rich isotopes are characterized by the appearance

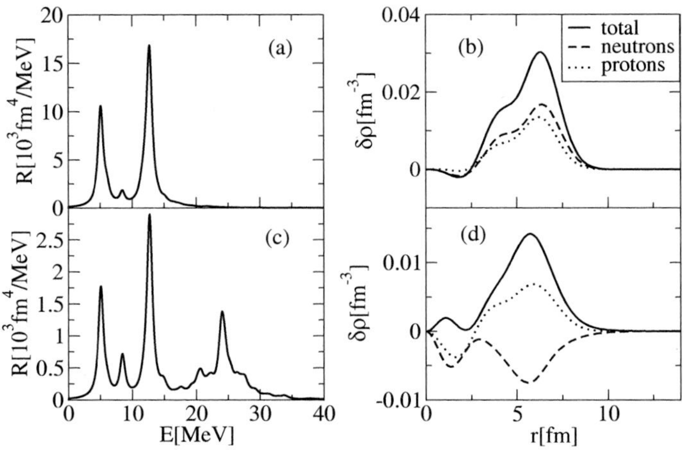

Fig. 5. The isoscalar (a) and isovector (c) quadrupole transition strength distributions for ^{208}Pb using the DD-ME2 effective interaction. The respective transitions densities are shown for the isoscalar state at 12.76 MeV (b) and isovector state at 24.04 MeV (d). The total transition densities are isoscalar and isovector transition densities for panel (b) and (d) respectively.

of a neutron skin, i.e. a layer of excess neutrons on the nuclear surface.[61,62] The question whether the excess neutrons in the skin can be excited to perform oscillations against the rest of the nucleus, or whether they only contribute to the non-collective threshold strength, has attracted considerable interest in recent years. In the former case one expects that, because the outer neutron orbitals are weakly bound, the resulting dipole mode will be rather soft, i.e. its excitation energy will be far below the giant resonance region. From the theoretical point of view, such a soft dipole mode also provides a unique test of the isospin-dependent components of effective nuclear interactions, which are particularly pronounced in nuclei with a large proton-neutron asymmetry. Besides being intrinsically interesting as an exotic mode of excitation, the occurrence of low-lying dipole strength plays an important role in predictions of neutron capture rates in the r-process nucleosynthesis, and consequently in the calculated elemental abundance distribution. Namely, although its transition strength is small compared to the total dipole strength, the low-lying collective dipole state located close to the neutron threshold can significantly enhance the radiative neutron capture cross section for neutron-rich nuclei, as shown in large-scale QRPA calculations.[54]

The properties of low-lying strength in neutron-rich nuclei and its relation to the soft mode known as pygmy dipole resonance (PDR) have been explored over the past years in a variety of theoretical (see Ref. 63–65 and references therein) and experimental studies.[66] Early studies of the PDR were based on rather simple hydrodynamical models which involve classical oscillations of the nucleon fluids. These include the three-fluid (protons, neutrons in the same orbitals as protons, and excess neutrons) hydrodynamical model,[67] the Steinwedel–Jensen[68] and Goldhaber–Teller models.[69] The low-lying mode has been qualitatively described as a collective oscillation of the neutron enriched surface layer against the core nucleons. It was also suggested, however, that the PDR could arise in nuclear systems with only moderate neutron excess, for instance in Ca isotopes.[70] More recently, microscopic theoretical approaches have been developed and employed in description of low-lying dipole strength (e.g., Refs. 63–65). These include the density functional theory, RPA, shell model, phonon coupling model, relativistic quasiparticle time blocking approximation, etc.

On the experimental side, extensive studies of low-lying electric dipole excitations have been performed in recent years, for more details see Ref. 66. Radioactive nuclear beams provide new opportunities for studies of low-lying dipole excitations in heavier nuclei with large proton-neutron asymmetry.[66,71] For example, by using the Coulomb dissociation of secondary Sn beams produced by in-flight fission of a primary ^{238}U beam, the dipole strength distribution above the one-neutron separation energy was measured in the unstable ^{130}Sn and the doubly-magic ^{132}Sn.[72] In addition to the giant dipole resonance, evidence was reported for a PDR structure at excitation energy around 10 MeV both in ^{130}Sn and ^{132}Sn, exhausting a few percent of the E1 energy-weighted sum rule.

Similar to the results obtained with non-relativistic models, the relativistic QRPA dipole response of neutron-rich nuclei is characterized by the fragmentation of the strength distribution and its spreading into the low-energy region. Fully consistent R(Q)RPA calculations have shown that with the increase of the number of neutrons along an isotopic chain, a relatively strong E1 peak appears below 10 MeV. The dynamics of this peak is very different from that of the isovector giant dipole resonance (IVGDR).[22] This is illustrated for ^{132}Sn in Fig. 6, where the RRPA strength distribution is shown for the isovector dipole transition operator, using the DD-ME2 effective interaction.[43] In the inserted panels the neutron and proton ground-state density distributions are shown, as well as the neutron and proton transition densities for the low-lying state at 7.8 MeV, and for the IVGDR at 15.3 MeV. For the main peak at 15.3 MeV the transition

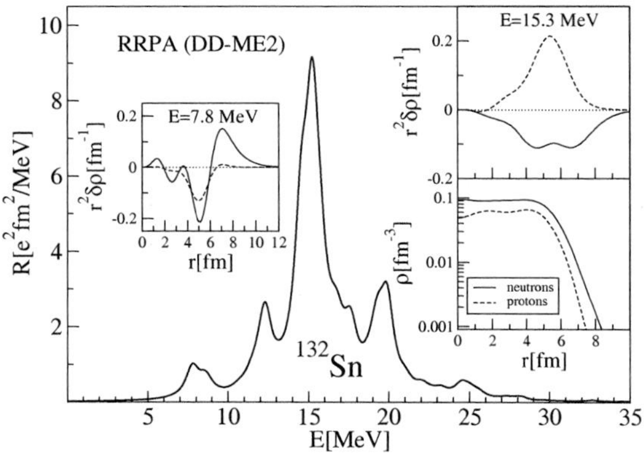

Fig. 6. The calculated isovector dipole transition strength distribution for ^{132}Sn (DD-ME2). The insertions include the ground-state proton and neutron density profiles, and the proton and neutron transition densities for the peaks at 7.8 MeV and 15.3 MeV excitation energy. Taken from Ref. 63.

densities display a radial dependence which is characteristic for the IVGDR mode: the proton and neutron densities oscillate with opposite phases. The dynamics of the state at 7.8 MeV is completely different: the proton and neutron transition densities are in phase in the bulk of the nucleus, whereas only neutron excitations contribute to the transition density in the surface region. The structure of the pygmy dipole strength (PDS) is rather complex, i.e. there are partial cancelations of contributions from various *ph* configurations that reduce the amount of the isovector low-energy transition strength.[73] The corresponding isoscalar dipole transition strength for the same PDS energy is collective, i.e. several neutron *ph* configurations coherently contribute to the transition strength.[73]

The theoretical analysis of low-lying excitations and the comparison with data, can be used as a sensitive tool to constrain the isovector channel of effective nuclear interactions. In the lower panel of Fig. 7 the calculated B(E1) strength in the region below 10 MeV is compared with data. The theoretical values have been obtained in the consistent RHB+RQRPA model with the non-linear meson-exchange effective interaction NL3,[76] and with the density dependent interaction DD-ME2.[43] One can observe that the NL3 interaction, which is known to overestimate the size of the neutron skin not only in exotic neutron-rich nuclei but also in ^{208}Pb, predicts too much low-lying B(E1) strength. On the other hand, an interaction like DD-ME2, that is adjusted to the empirical differences between the radii of

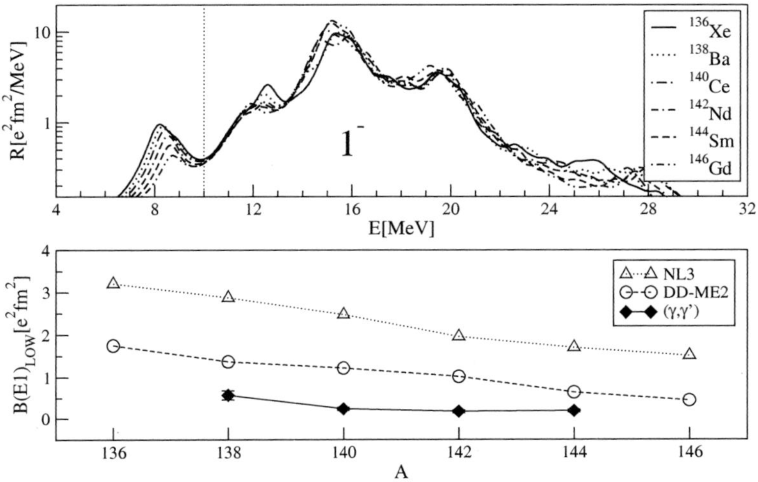

Fig. 7. The isovector dipole strength distributions in N=82 isotones calculated in the RHB+RQRPA model, with the DD-ME2 effective interaction (upper panel). The integrated B(E1) transition strength below 10 MeV, for the NL3 and DD-ME2 interactions, is compared with the data from (γ, γ') scattering[74,75] (lower panel). Taken from Ref. 63.

neutron and proton density distributions, significantly improves the agreement of the calculated low-energy dipole strength with data. The remaining difference might be caused by the coupling with more complex phonon configurations,[77] not taken into account in the RHB+RQRPA, or in the missing E1 strength in (γ, γ') scattering which may be considerable when dealing with end-point energies close to the neutron separation threshold.[78]

6. Proton-neutron QRPA and spin-isospin resonances in charge-exchange reactions

Spin-isospin resonances represent another class of collective nuclear transitions, involving oscillations in isospin and spin space.[79] Spin-isospin resonances are interesting not only as unique modes of excitation, but they also play an important role in various applications in astrophysics and particle physics. Spin-isospin resonances provide valuable information to tune the spin-isospin channel of the effective nucleon-nucleon interactions in nuclear medium and to study the nuclear equation of state. The spin-dipole transition strength is a good indicator of the neutron skin, since the model-independent non-energy-weighted sum rule of the spin-dipole excitations is directly related to the neutron skin thickness. The neutron skin provides important information about the constraints on the symmetry energy. Besides, spin-isospin excitations also play an important role in the weak-interaction processes, e.g., electron capture, β decay and neutrino-nucleus reactions, which are of great interest for nuclear astrophysics.[80,81] In the core-collapse supernovae, the electron capture rates govern the evolution of the system and consequently, Gamow–Teller (GT) transition matrix elements must be accurately known.[80–84] The β-decay half-lives set the time scale of the rapid neutron capture process (r-process), and hence influence the production of heavy elements in the universe.[85–87] Last but not least, a very accurate knowledge of spin-isospin matrix elements is also instrumental to extract the properties of the neutrinos from the measured half-life of double-β decay.[88,89] Furthermore, the isospin corrections can be extracted from the $0^+ \to 0^+$ superallowed Fermi transitions, which play a key role in determining the V_{ud} matrix element in Cabibbo–Kobayashi–Maskawa (CKM) matrix.[90,91]

6.1. *Proton-neutron QRPA equations*

We consider transitions between 0^+ ground state of a spherical even-even parent nucleus and the J^π excited state of the corresponding odd-odd daughter nucleus. These transitions are induced by the charge-exchange operator $\hat{F}_{JM}$. Taking into account the rotational invariance of the nuclear system, the quasiparticle pairs can be coupled to a good angular momentum and the matrix equations of the proton-neutron(PN)-QRPA read

$$\begin{pmatrix} A^J_{pnp'n'} & B^J_{pnp'n'} \\ -B^{J*}_{pnp'n'} & -A^{J*}_{pnp'n'} \end{pmatrix} \begin{pmatrix} X^J_{p'n'} \\ Y^J_{p'n'} \end{pmatrix} = \Omega_\nu \begin{pmatrix} X^J_{pn} \\ Y^J_{pn} \end{pmatrix}. \tag{79}$$

The matrix elements $\mathcal{A}^J$ and $\mathcal{B}^J$ are defined in the canonical basis

$$\begin{aligned} \mathcal{A}^J_{pnp'n'} &= H^{11(J)}_{pp'}\delta_{nn'} + H^{11(J)}_{nn'}\delta_{pp'} + V^{phJ}_{pn'np'}(u_p v_n u_{p'} v_{n'} + v_p u_n v_{p'} u_{n'}) \\ &\quad + V^{ppJ}_{pnp'n'}(u_p u_n u_{p'} u_{n'} + v_p v_n v_{p'} v_{n'}), \end{aligned} \tag{80}$$

$$\begin{aligned} \mathcal{B}^J_{pnp'n'} &= V^{phJ}_{pn'np'}(u_p v_n v_{p'} u_{n'} + v_p u_n u_{p'} v_{n'}) \\ &\quad - V^{ppJ}_{pnp'n'}(u_p u_n v_{p'} v_{n'} + v_p v_n u_{p'} u_{n'}). \end{aligned} \tag{81}$$

Here p, p', and n, n' denote proton and neutron quasiparticle canonical states, respectively. V^{ph} is the proton-neutron particle-hole residual interaction, and V^{pp} is the corresponding particle-particle interaction. The canonical basis diagonalizes the density matrix and the occupation amplitudes $v_{p,n}$ are the corresponding eigenvalues. The canonical basis, however, does not diagonalize the Dirac single-nucleon mean-field Hamiltonian $\hat{h}_D$ and the pairing field $\hat{\Delta}$, and therefore the off-diagonal matrix elements $H^{11}_{nn'}$ and $H^{11}_{pp'}$ appear in Eq. (79):

$$H^{11}_{\kappa\kappa'} = (u_\kappa u_{\kappa'} - v_\kappa v_{\kappa'})h_{\kappa\kappa'} - (u_\kappa v_{\kappa'} + v_\kappa u_{\kappa'})\Delta_{\kappa\kappa'} \, . \tag{82}$$

The total strength for the transition between the ground state of the even-even (N,Z) nucleus and the excited state of the odd-odd (N+1,Z-1) or (N-1,Z+1) nucleus, induced by the operator $\hat{F}_{JM}$, reads

$$B^{\pm}_\nu = |\sum_{pn} [u_p v_n X^J_{pn} + v_p u_n Y^J_{pn}]\langle p||\hat{F}_J||n\rangle|^2. \tag{83}$$

In the framework of covariant density functional, the spin-isospin dependent interaction terms are generated by the ρ- and π-meson exchange.[92] Because of parity conservation, the one-pion direct contribution vanishes in the mean-field calculation of a nuclear ground state. Its inclusion is important, however, in calculations of excitations that involve spin and isospin degrees of freedom. The particle-hole residual interaction in the PN-RQRPA is derived from the Lagrangian density

$$\mathcal{L}^{int}_{\pi+\rho} = -g_\rho \bar{\psi}\gamma^\mu \boldsymbol{\rho}_\mu \boldsymbol{\tau}\psi - \frac{f_\pi}{m_\pi}\bar{\psi}\gamma_5\gamma^\mu \partial_\mu \boldsymbol{\pi}\boldsymbol{\tau}\psi \, . \tag{84}$$

Vectors in isospin space are denoted by arrows, and boldface symbols will indicate vectors in ordinary three-dimensional space.

The coupling between the ρ-meson and the nucleon is assumed to be a vertex function of the vector density $\rho_v = \sqrt{j_\mu j^\mu}$, with $j_\mu = \bar{\psi}\gamma_\mu\bar{\psi}$. As emphasized in Sec. 3 the explicit density dependence of the meson-nucleon couplings introduces additional rearrangement terms in the residual two-body interaction of the RRPA, and their contributions are essential for a quantitative description of excited states. However, since the rearrangement terms include the corresponding isoscalar ground-state densities, one concludes that they are absent in the charge exchange channel. The residual two-body interaction reads

$$V_\rho(\boldsymbol{r}_1, \boldsymbol{r}_2) = \boldsymbol{\tau}_1 \cdot \boldsymbol{\tau}_2 (\beta\gamma^\mu)_1 (\beta\gamma_\mu)_2 g_\rho(\rho_v(\boldsymbol{r}_1)) g_\rho(\rho_v(\boldsymbol{r}_2)) D_\rho(\boldsymbol{r}_1, \boldsymbol{r}_2), \tag{85}$$

$$V_\pi(\boldsymbol{r}_1, \boldsymbol{r}_2) = -(\frac{f_\pi}{m_\pi})^2 \boldsymbol{\tau}_1 \cdot \boldsymbol{\tau}_2 (\boldsymbol{\Sigma}_1 \cdot \boldsymbol{\nabla}_1)(\boldsymbol{\Sigma}_2 \cdot \boldsymbol{\nabla}_2) D_\pi(\boldsymbol{r}_1, \boldsymbol{r}_2). \tag{86}$$

$D_{\rho(\pi)}$ denotes the meson propagator

$$D_{\rho(\pi)} = \frac{1}{4\pi}\frac{e^{-m_{\rho(\pi)}|\boldsymbol{r}_1 - \boldsymbol{r}_2|}}{|\boldsymbol{r}_1 - \boldsymbol{r}_2|} \, , \tag{87}$$

and

$$\boldsymbol{\Sigma} = \begin{pmatrix} \boldsymbol{\sigma} & 0 \\ 0 & \boldsymbol{\sigma} \end{pmatrix} . \tag{88}$$

For the ρ-meson coupling the functional form of the density-dependent effective interaction is used

$$g_\rho(\rho_v) = g_\rho(\rho_{sat}) exp[-a_\rho(x-1)] , \tag{89}$$

where $x = \rho_v/\rho_{sat}$, and ρ_{sat} denotes the saturation vector nucleon density in symmetric nuclear matter. For the pseudovector pion-nucleon coupling standard values are employed,

$$m_\pi = 138.0 \text{ MeV} \qquad \frac{f_\pi^2}{4\pi} = 0.08 . \tag{90}$$

The derivative type of the pion-nucleon coupling necessitates the inclusion of the zero-range Landau–Migdal term, which accounts for the contact part of the nucleon-nucleon interaction

$$V_{\delta\pi}(\boldsymbol{r}_1, \boldsymbol{r}_2) = g' \left(\frac{f_\pi}{m_\pi}\right)^2 \boldsymbol{\tau}_1 \cdot \boldsymbol{\tau}_2 \boldsymbol{\Sigma}_1 \cdot \boldsymbol{\Sigma}_2 \delta(\boldsymbol{r}_1 - \boldsymbol{r}_2), \tag{91}$$

with the parameter $g' = 0.52$ adjusted to reproduce the experimental data on GTR excitation energy in ^{208}Pb for the DD-ME2 effective interaction.

For the particle-particle interaction V^{pp}, it not only includes the usual $T = 1$ channel, which has the same form as the pairing interaction for the ground state, for example, the Gogny pairing, but also includes the $T = 0$ channel, for which we could adopt a similar form as the Gogny pairing force,

$$V_{12} = -V_0 \sum_{j=1}^{2} g_j \, \mathrm{e}^{-r_{12}^2/\mu_j^2} \, \hat{\Pi}_{S=1,T=0} \quad , \tag{92}$$

where $\hat{\Pi}_{S=1,T=0}$ projects onto states with $S = 1$ and $T = 0$. Usually, one takes the ranges μ_1=1.2 fm and μ_2=0.7 fm of the two Gaussians from the Gogny interaction, together with the relative strengths $g_1 = 1$ and $g_2 = -2$ so that the force is repulsive at small distances.[92,93] The only remaining free parameter is V_0, the overall strength.

6.2. *Isobaric analog resonance*

The isobaric analog resonance (IAR) represents the simplest charge-exchange excitation mode, induced by the isospin-flip transition operator $\tau_\pm$. This mode has been detected already fifty years ago in experiments on low-energy proton elastic scattering from heavy nuclei. The observed resonances were at energies consistent with the interpretation that they were isobaric analog states in the compound nucleus. One of the important characteristics of the IAR is its narrow width. This is because it has the same isospin as the parent state, while the neighboring states have the isospin of the ground state of the daughter nucleus, i.e., they differ in isospin by one unit. This means that they will couple only weakly with the IAR.

If the system's Hamiltonian commutes with the operator $\tau_{\pm}$, then the energy of IAR state is zero, and the excitation strength exhausts 100% of the sum rule. The system's Hamiltonian will commute with the operator $\tau_{\pm}$ if the Coulomb field is turned off. In this case although the single-particle Hamiltonian does not commute with $\tau_{\pm}$, the broken symmetry could be restored through the self-consistent RPA calculation. In this way one can check the accuracy and self-consistency of the RPA calculation. In Fig. 8, the isobaric analog state (IAS) transition strength calculated by the RRPA with parameterization DD-ME2 is shown. After the Coulomb field is turned off, the IAS excited state appears at $E = 0$ MeV and exhausts 100% of the sum rule $N - Z = 8$.

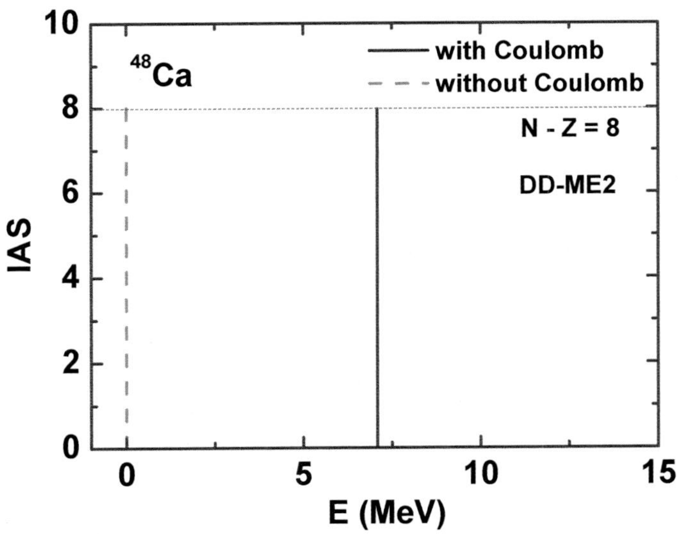

Fig. 8. (Color online) The IAS transition strength calculated by RRPA with parameterization DD-ME2. The solid line and dashed line are the results obtained with Coulomb and without Coulomb interaction. Taken from Ref. 94.

In Fig. 9 we display the IAS strength distributions for ^{48}Ca, ^{90}Zr and ^{208}Pb, calculated by the RQRPA with parameterization DD-ME2. The strength distributions are dominated by a single IAR peak, which corresponds to a coherent superposition of $\pi p - \nu h$ (or proton-neutron 2qp) excitations. The calculated excitation energies (evaluated with respect to the ground state of the parent nucleus) are compared with the corresponding experimental values (thick arrows) from (p, n) scattering data for ^{48}Ca,[95] ^{90}Zr,[96,97] and ^{208}Pb.[98] The agreement between the PN-RQRPA and experimental data is indeed very good.

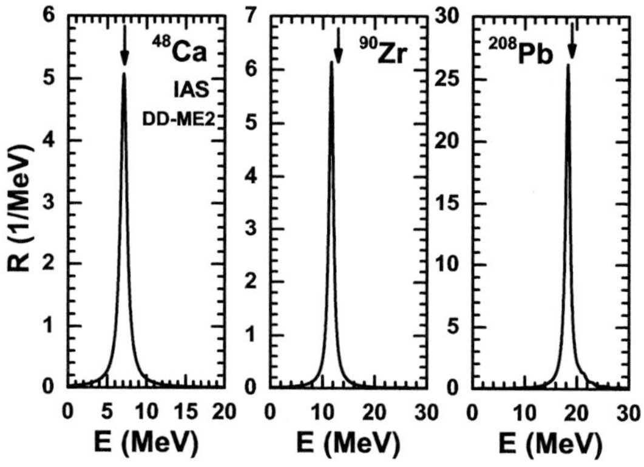

Fig. 9. PN-RQRPA $J^\pi = 0^+$ strength distributions. The excitations of the isobaric analog resonances are compared with experimental data (thick arrows) for ^{48}Ca,[95] ^{90}Zr,[96,97] and ^{208}Pb.[98]

6.3. *Gamow–Teller resonance*

The Gamow–Teller resonance represents a coherent superposition of high-lying $J^\pi = 1^+$ proton-particle neutron-hole configurations associated with charge-exchange excitations of neutrons from orbitals with $j = l + \frac{1}{2}$ into proton orbitals with $j = l - \frac{1}{2}$. This collective mode was predicted already in 1963,[20] but it was only in 1975 that the first experimental indications of the GT resonance were observed in (p, n) charge-exchange reactions at intermediate energies.[99] The detailed knowledge of GT strength distributions is essential for the understanding of nuclear processes in stellar nucleosynthesis. In particular, the low-lying GT strength is directly related to β-decay rates, as well as to the electron-capture process leading to the stellar collapse and supernovae explosion. In many nuclei all over the periodic table the GT strength distribution, when measured in the excitation energy region where the most pronounced GT peaks occur, is quenched by more than 20% when compared to a model independent sum rule. Two physically different mechanisms have been suggested as a possible explanation of the quenching of the total GTR strength: (i) nuclear configuration mixing – the high-lying $2p-2h$ states mix with the $1p-1h$ GT states and shift the GT strength to the energies far beyond the giant resonance region.[100–102] Besides, tensor force is found to be able to couple the GT states with the spin-quadrupole 1^+ states in the high-energy region, which can shift about 10% of the GT strength to high energy region.[103] (ii) the high-energy Δ-isobar nucleon-hole configurations ($\Delta - h$) couple to the GT mode and remove part of the strength from the low-lying excitation spectrum.[104,105] However, (p, n) scattering experiments have shown that only a small fraction of the GT quenching originates from $\Delta - h$ transitions.[97]

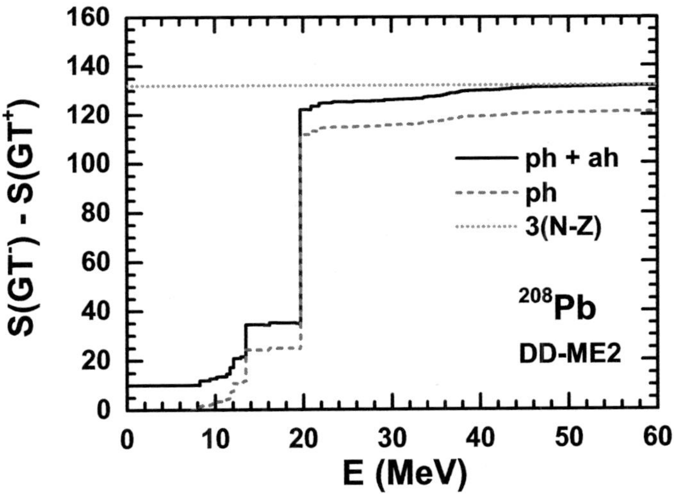

Fig. 10. (Color online) The running sum of the GT strength for ^{208}Pb. The dashed line corresponds to a PN-RRPA calculation with only positive-energy *ph* configurations. For the calculation denoted by the solid line the RRPA space contains configurations formed from occupied states in the Fermi sea and empty negative-energy states in the Dirac sea. The total sum of the GT strength is compared to the model independent Ikeda sum rule (dotted line).

In the relativistic calculations, the configuration of particle-hole excitations includes not only the occupied states and the unoccupied states both in Fermi sea, but also occupied states in the Fermi sea and empty negative-energy states in the Dirac sea. The latter configurations play an important role for satisfying the sum rule. In Fig. 10, the running sum of GT strength is shown for ^{208}Pb. The horizontal dotted line corresponds to the value $3(N - Z) = 132$ of the Ikeda sum rule. The solid and dashed lines corresponds to the values of the GTR running sum with and without the inclusion of the configurations of occupied states in the Fermi sea and negative-energy states in the Dirac sea, respectively. For the second case, even extending the sum up to 60 MeV, the total sum amounts only to ≈122, that is 8% less than the Ikeda sum rule. The total sum rule $3(N - Z)$ is exhausted by the calculated GT strength only when the relativistic RPA/QRPA space contains *ph* excitations formed from ground-state configurations of the fully or partially occupied states of positive energy, and the empty negative-energy states from the Dirac sea (solid line in Fig. 10).

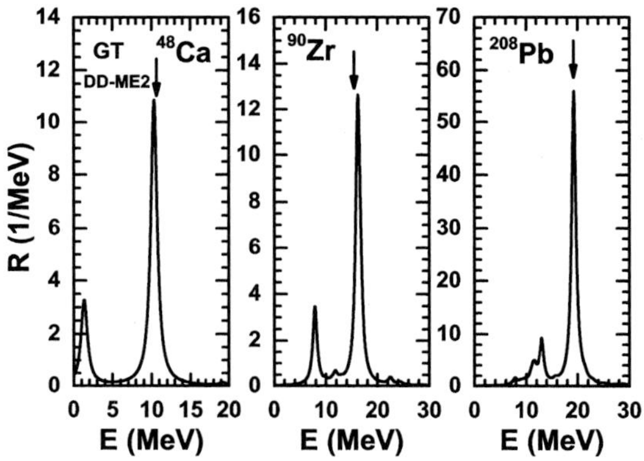

Fig. 11. Gamow–Teller strength distributions for ^{48}Ca, ^{90}Zr and ^{208}Pb. PN-RQRPA results are shown in comparison with experimental data (thick arrows) for the GTR excitation energies in ^{48}Ca,[95] ^{90}Zr,[96,97] and ^{208}Pb.[98,106,107]

In Fig. 11 the GT transition strength distributions calculated using the PN-RQRPA are shown in comparison with the experimental data (thick arrows) for GTR excitation energies in ^{48}Ca,[95] ^{90}Zr,[96,97] and ^{208}Pb.[98,106,107] Although the residual interaction has been adjusted to reproduce the GTR excitation energy in ^{208}Pb, we notice a very good agreement with experiment also for ^{48}Ca and ^{90}Zr.

7. Random phase approximation at finite temperature

The first developments of random phase approximation at finite temperature were motivated by determining the evolution of the giant dipole resonances formed in collisions at increasing excitation energy. Giant resonances in nuclei at finite temperature have been explored in a variety of theoretical frameworks already in the 1980s and 1990s, including RPA with schematic interactions,[108–110] linear response theory,[111,112] extended RPA with the inclusion of collision terms,[113] the self-consistent HF plus RPA model with Skyrme effective interactions,[114] the RPA with phonon coupling,[115–117] etc. More recently, excitations in nuclei at finite temperature attracted new interest due to their relevance and applications in astrophysics. As mentioned in Sec. 5 and 6, nuclear collective excitations play an important role in astrophysics, especially for the dynamics of core-collapse supernovae.[118,119] At the end of its life, a massive star exhausts its nuclear fuel and, therefore, the core can only be stabilized by the electron degeneracy pressure as long as its mass does not exceed the corresponding Chandrasekhar mass $M_{\rm Ch}$. When this mass limit is exceeded, the core cannot attain a stable configuration and it collapses. During the precollapse phase, electron capture reduces the number of electrons available for the pressure support, whereas β-decay acts in the opposite direction. At the same time, the neutrinos produced by electron capture freely escape from the star for values of the matter density $\lesssim 10^{11}$ g cm^{-3}, removing energy and entropy from the core.[80,81,119] During the core-collapse supernovae, the temperature can change from 300–800 keV to 1–2 MeV. To describe the nuclear collective vibrations in stellar environment, the calculation should in principle include the effects of temperature self-consistently. Thus it is necessary to have the RPA model at finite temperature.

Experimentally it was first observed that the measured γ spectrum accompanying the spontaneous fission of ^{252}Cf[120] deviates significantly from the characteristic statistical spectrum at energies between 10 and 15 MeV. In addition to the expected exponential decay probability, there is an enhancement in the IVGDR excitation-energy region. The same phenomenon of increased γ-decay in the IVGDR energy region was confirmed in 1981 in the ^{40}Ar-induced fusion reactions on ^{82}Se, ^{110}Pd and ^{124}Sn targets.[121] The considerable enhancement of the number of photons in the IVGDR energy region is attributed to the decay of isovector giant dipole resonances built on an excited state. Usually, the heavy-ion fusion reaction[121] and inelastic scattering[122–124] experiments are performed for the study of GDR in hot nuclei. In the small-angle light-ion inelastic scattering,[123,124] the angular momentum transferred to the target nucleus is very low compared to heavy-ion fusion reactions, thus it can be used to investigate the influence of temperature on the evolution of GDR in hot nuclei independent of angular momentum effects. The angular-momentum and excitation-energy effects on the GDR width was also disentangled experimentally.[125,126] Reviews on the experimental studies and data of GDR in hot nuclei can be found in Refs. 127–130.

7.1. *Finite temperature RPA equations*

In order to describe nuclear excitations in stellar environment, theoretical framework based on the RPA should be extended to include the finite temperature effects. In the following, the RPA equations at finite temperature are derived,[110,131] using similar approach as in the case of RPA at zero temperature.

We study the influence of a weak time-dependent external field $\mathcal{W}(t)$ with the following form to the nuclear system,

$$\mathcal{W}(t) = \mathcal{W}e^{-i\omega t} + \mathcal{W}^{\dagger}e^{i\omega t}. \tag{93}$$

Suppose $\mathcal{W}$ is a one-body operator, and it can be written as

$$\mathcal{W}(t) = \sum_{kl} \mathcal{W}_{kl} a_k^{\dagger} a_l. \tag{94}$$

The single-particle density matrix of the system is defined as

$$\rho_{kl}(t) = \langle \Phi(t) | a_l^{\dagger} a_k | \Phi(t) \rangle, \tag{95}$$

where $\Phi(t)$ is the wavefunction of the system. Suppose at any time $\rho(t)$ satisfies $\rho^2 = \rho$, then its equation of motion will be

$$i\hbar\dot{\rho} = [h[\rho] + \mathcal{W}(t), \rho], \tag{96}$$

which is the time-dependent Hartree–Fock equation. $h[\rho]$ is the single-particle Hamiltonian. Since the external field is weak, only the linear response of the system will be considered. So the time-dependent density can be written as

$$\rho(t) = \rho^{(0)} + \delta\rho(t). \tag{97}$$

$\rho^{(0)}$ is the static density, and satisfies the static Hartree–Fock equation,

$$[h[\rho^{(0)}], \rho^{(0)}] = 0. \tag{98}$$

$\delta\rho(t)$ has the linear form,

$$\delta\rho = \rho^{(1)}e^{-i\omega t} + \rho^{(1)\dagger}e^{i\omega t}. \tag{99}$$

On the basis of solutions of static HF equation, $\rho^{(0)}$ and $h[\rho^{(0)}]$ will have the diagonal form,

$$\rho_{kl}^{(0)} = \delta_{kl}\rho_k^{(0)} = f_k, \tag{100}$$

and

$$(h_0)_{kl} = (h[\rho^{(0)}])_{kl} = \delta_{kl}\epsilon_k. \tag{101}$$

f_k is the occupation probability of the single-particle state, which has the Fermi–Dirac form

$$f_k = \left[1 + \exp\left(\frac{\epsilon_a - \mu}{T}\right)\right]^{-1}, \tag{102}$$

where ϵ_a is the single-particle energy; μ is the chemical potential; T is the temperature given in MeV units.

Using Eq. (99), if only linear terms are kept, Eq. (96) becomes

$$i\hbar\delta\dot{\rho} = [h_0, \delta\rho] + [\frac{\delta h}{\delta\rho}\delta\rho, \rho^{(0)}] + [\mathcal{W}, \rho^{(0)}], \tag{103}$$

where

$$\frac{\delta h}{\delta\rho}\delta\rho = \sum_{jn}\left(\frac{\partial h}{\partial\rho_{nj}}|_{\rho=\rho^{(0)}}\delta\rho_{nj} + \frac{\partial h}{\partial\rho_{jn}}|_{\rho=\rho^{(0)}}\delta\rho_{jn}\right). \tag{104}$$

Here we always suppose $f_{m,n,l} < f_{i,j,k}$. Writing Eq. (103) in a matrix form on the basis of $\langle m|$ and $|i\rangle$, one obtains

$$\left\{\begin{pmatrix} A & B \\ B^* & A^* \end{pmatrix} - \hbar\omega\begin{pmatrix} 1 & 0 \\ 0 & -1 \end{pmatrix}\right\}\begin{pmatrix} \rho_{nj}^{(1)} \\ \rho_{jn}^{(1)} \end{pmatrix} = -(f_i - f_m)\begin{pmatrix} \mathcal{W}_{mi} \\ \mathcal{W}_{im}^* \end{pmatrix}, \tag{105}$$

where the matrix elements A and B are

$$A_{minj} = (\epsilon_m - \epsilon_i)\delta_{mn}\delta_{ij} + \frac{\partial h_{mi}}{\partial\rho_{nj}}(f_i - f_m), \tag{106}$$

$$B_{minj} = \frac{\partial h_{mi}}{\partial\rho_{jn}}(f_i - f_m). \tag{107}$$

The residual two-body interaction is defined as

$$V_{mnij} = \frac{\partial h_{mi}}{\partial\rho_{jn}} = \frac{\partial^2 E}{\partial\rho_{im}\partial\rho_{jn}}. \tag{108}$$

Since the creation operator $Q_\nu^\dagger$ for RPA state $|\nu\rangle$ is

$$Q_\nu^\dagger = \sum_{mi} X_{mi}^\nu a_m^\dagger a_i - \sum_{mi} Y_{mi}^\nu a_i^\dagger a_m, \tag{109}$$

one should notice that for the finite-temperature case,

$$\rho_{mi}^{(1)} = \langle 0|a_i^\dagger a_m|\nu\rangle \simeq \langle \mathrm{HF}|[a_i^\dagger a_m, Q_\nu^\dagger]|\mathrm{HF}\rangle = (f_i - f_m)X_{mi}, \tag{110}$$

$$\rho_{im}^{(1)} = \langle 0|a_m^\dagger a_i|\nu\rangle \simeq \langle \mathrm{HF}|[a_m^\dagger a_i, Q_\nu^\dagger]|\mathrm{HF}\rangle = (f_i - f_m)Y_{mi}. \tag{111}$$

Here $|0\rangle$ is the RPA ground state, and $|\mathrm{HF}\rangle$ is the HF ground state.

Following this relation, Eq. (105) will have the following form by transforming the transition density $\rho^{(1)}$ to the forth-going and back-going amplitudes X and Y when the external field approaches zero,

$$\begin{pmatrix} \mathcal{A} & \mathcal{B} \\ -\mathcal{B}^* & -\mathcal{A}^* \end{pmatrix} = \Omega \begin{pmatrix} X_{nj} \\ Y_{nj} \end{pmatrix}, \tag{112}$$

where matrix elements $\mathcal{A}$ and $\mathcal{B}$ are

$$\mathcal{A}_{minj} = (\epsilon_m - \epsilon_i)\delta_{mn}\delta_{ij} + \frac{\partial h_{mi}}{\partial \rho_{nj}}(f_j - f_n), \tag{113}$$

$$\mathcal{B}_{minj} = \frac{\partial h_{mi}}{\partial \rho_{jn}}(f_j - f_n). \tag{114}$$

The configuration space includes all the pairs with $f_m < f_i$. Different from the zero-temperature case, the matrix $\begin{pmatrix} \mathcal{A} & \mathcal{B} \\ \mathcal{B}^* & \mathcal{A}^* \end{pmatrix}$ is no longer Hermitian, which leads to the change of orthogonalization relation of the amplitudes X and Y,

$$\sum_{mi} \left(X_{mi}^{\nu *} X_{mi}^{\nu'} - Y_{mi}^{\nu *} Y_{mi}^{\nu'} \right) (f_i - f_m) = \delta_{\nu\nu'}. \tag{115}$$

The transition strength from the ground state $|0\rangle$ to the RPA excited state $|\nu\rangle$ B_ν for the transition operator $\hat{F}_{JM}$ is

$$B_\nu = \left| \sum_{mi} \left(X_{mi}^\nu + (-)^J Y_{mi}^\nu \right) \langle m||F_J||i\rangle (f_i - f_m) \right|^2 . \tag{116}$$

7.2. *Electric dipole excitations in hot nuclei*

In this section, we will give some illustrative calculations of the electric isovetor dipole excitations in hot nuclei, studied using a self-consistent finite-temperature RPA based on relativistic energy density functionals. The self-consistent finite temperature relativistic RPA (FTRRPA) is formulated in the single-nucleon basis of the relativistic mean-field (RMF) model at finite temperature (FTRMF). In nuclei the phase transition from a superfluid to a normal state occurs at temperatures $T \approx 0.5 - 1$ MeV,[132–134] whereas for temperatures above $T \approx 4$ MeV contributions from states in the continuum become large, and additional subtraction schemes have to be implemented to remove the contributions of the external nucleon gas.[135] In the following study we consider nuclear temperatures in the range $T = 1-2$ MeV, for which the FTRMF plus FTRRPA should provide a reasonable description of the evolution of dipole response with temperature.

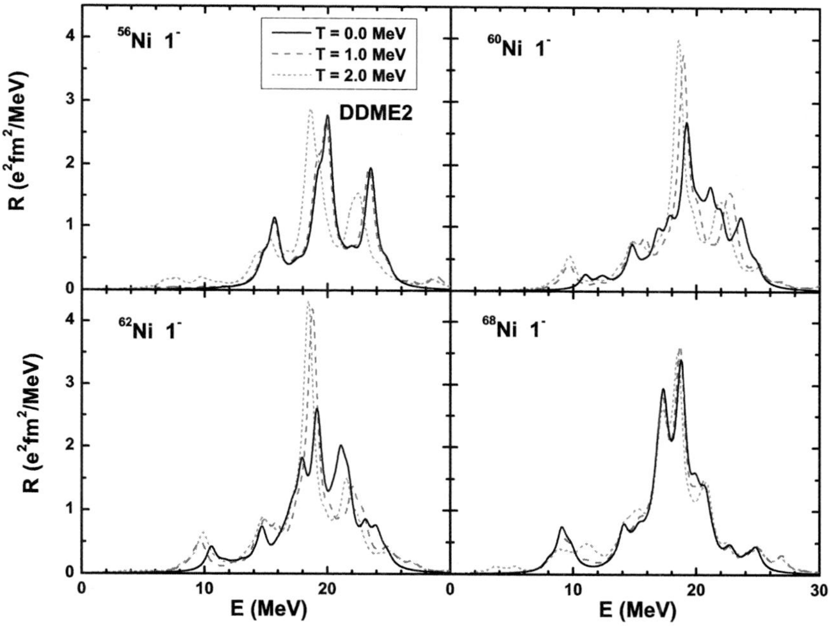

Fig. 12. (Color online) Isovector dipole transition strength distributions for Ni isotopes, calculated with the RQRPA at $T = 0$ MeV and FTRRPA at $T = 1, 2$ MeV with parameterization DD-ME2. Taken from Ref. 136.

Figure 12 illustrates the evolution with temperature of the isovector dipole response in Ni isotopes. At $T = 0$ MeV the strength distributions are calculated using the RQRPA,[22] whereas the FTRRPA is employed at finite temperatures $T = 1$ and 2 MeV. The DD-ME2 effective interaction is used in the particle-hole channel and, at zero temperature, pairing correlations are described by the pairing part of the finite-range Gogny interaction.[137] At zero temperature the main peak of the IVGDR is located at ≈ 18 MeV. For ^{68}Ni, in particular, additional low-energy PDR structure is predicted below 10 MeV at zero temperature, and this result was confirmed in the experimental study reported in Ref. 138. The IVGDR is somewhat modified at finite temperature. At $T = 2$ MeV the main peak is lowered by ≈ 2 MeV in ^{56}Ni, whereas for ^{62}Ni finite temperature reduces the fragmentation of the IVGDR strength in the interval ≈17–20 MeV, and enhances the collectivity of the main resonance peak.

A very interesting finite temperature effect is predicted for the dipole strength distributions of ^{60,62}Ni, for which virtually no excitations at low energy below 10 MeV are present at zero temperature. Already at $T = 1$ MeV additional dipole transition strength develops below 10 MeV. For ^{60}Ni, in particular, at $T = 2$ MeV the main low-energy peak is located at 9.71 MeV, and exhausts 1.54% of the energy-weighted sum of strength.

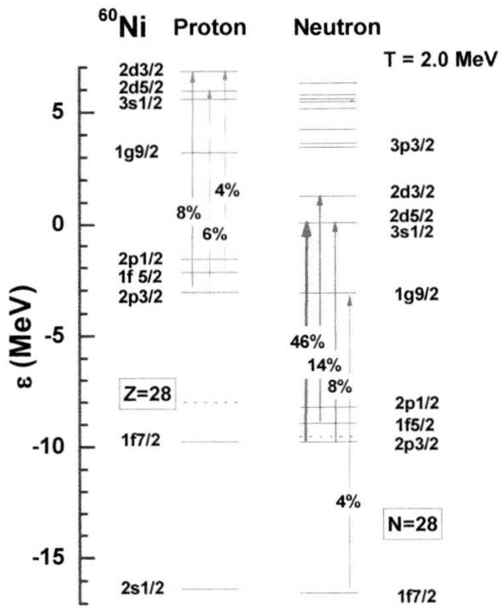

Fig. 13. (Color online) Proton and neutron single particle spectra for ^{60}Ni at temperature $T = 2$ MeV. The dotted lines denote the Fermi surfaces. The arrows represent transitions with dominant contributions to the dipole state at $E = 9.71$ MeV, and the number on each arrow denotes the contribution of respective configuration to the total sum of FTRRPA amplitude. Taken from Ref. 139.

Figure 13 shows the proton and neutron single particle spectra for ^{60}Ni at temperature $T = 2$ MeV. Fermi surfaces are denoted by dotted lines. Four neutron and three proton single-particle transitions give the main contributions to the dipole state at $E = 9.71$ MeV (represented by the arrows in Fig. 13). These transitions appear due to thermal population of single-particle levels around the Fermi surface. The number on each arrow in Fig. 13 shows the contribution of the respective configuration to the total sum of FTRRPA amplitude: $\sum_{mi}(X_{mi}^2 - Y_{mi}^2)(f_i - f_m)$, illustrating the relative importance of the corresponding configuration. The figure includes only configurations with contribution > 2%. Rather rich RPA structure shows that low-energy state at $E = 9.71$ MeV accumulates a small degree of collectivity due to thermal population of both neutron and proton single-particle levels around the Fermi surface.

In the case of ^{62}Ni, two pronounced low-energy peaks are calculated at $T = 2$ MeV: the states at 9.78 MeV and 10.03 MeV, exhausting 1.2% and 1.5% of the energy-weighted sum of strength, respectively. It is interesting to note that both states are dominated by proton transitions. In ^{68}Ni, where the PDR structure is present already at $T = 0$, the increase in temperature leads to fragmentation of the PDR and its spreading to even lower energy. These examples nicely illustrate the variety of effects one can expect for the dipole response of hot nuclei, e.g. the appearance of novel modes of low-energy excitations dominated by proton transitions, and the thermal spreading of PDR structures in neutron-rich nuclei.

7.3. *Gamow–Teller transitions and electron capture in stellar environment*

For the process of electron capture on a nucleus,

$$e^- +^A_Z X_N \rightarrow^A_{Z-1} X^*_{N+1} + \nu_e, \tag{117}$$

the cross section is derived from Fermi's golden rule,

$$\frac{d\sigma}{d\Omega} = \frac{1}{(2\pi)^2} V^2 E_\nu^2 \frac{1}{2} \sum_{\text{lepton spins}} \frac{1}{2J_i + 1} \sum_{M_i M_f} |\langle f|\hat{H}_W|i\rangle|^2, \tag{118}$$

where V is the quantization volume, and E_ν is the energy of the outgoing electron neutrino. The weak-interaction Hamiltonian $\hat{H}_W$ of semileptonic processes is written in the current-current form,[140]

$$\hat{H}_W = -\frac{G}{\sqrt{2}} \int d\boldsymbol{x} \mathcal{J}_\mu(\boldsymbol{x}) j_\mu(\boldsymbol{x}), \tag{119}$$

where $j_\mu(\boldsymbol{x})$ and $\mathcal{J}_\mu(\boldsymbol{x})$ are the weak leptonic and hadronic current density operators, respectively. The matrix elements of leptonic part are evaluated using the standard electroweak model, and contain both vector and axial-vector components.[80] The hadronic current is obtained by using arguments of Lorentz covariance and isospin invariance of the strong interaction. The expression for the electron-capture cross sections (see Refs. 141, 140 for more details) reads

$$\frac{d\sigma_{\rm ec}}{d\Omega} = \frac{G_F^2 \cos^2\theta_c}{2\pi}\frac{F(Z,E_e)}{(2J_i+1)}$$
$$\times\Bigg\{\sum_{J\geq 1}\mathcal{W}(E_e,E_\nu)\Big\{(1-(\hat{\boldsymbol{\nu}}\cdot\hat{\boldsymbol{q}})(\boldsymbol{\beta}\cdot\hat{\boldsymbol{q}}))\Big[|\langle J_f||\hat{\mathcal{T}}_J^{\rm mag}||J_i\rangle|^2$$
$$+|\langle J_f||\hat{\mathcal{T}}_J^{\rm el}||J_i\rangle|^2\Big] - 2\hat{\boldsymbol{q}}\cdot(\hat{\boldsymbol{\nu}}-\boldsymbol{\beta}){\rm Re}\langle J_f||\hat{\mathcal{T}}_J^{\rm mag}||J_i\rangle\langle J_f||\hat{\mathcal{T}}_J^{\rm el}||J_i\rangle^*\Big\}$$
$$+\sum_{J\geq 0}\mathcal{W}(E_e,E_\nu)\Big\{(1-\hat{\boldsymbol{\nu}}\cdot\boldsymbol{\beta}+2(\hat{\boldsymbol{\nu}}\cdot\hat{\boldsymbol{q}})(\boldsymbol{\beta}\cdot\hat{\boldsymbol{q}}))|\langle J_f||\hat{\mathcal{L}}_J||J_i\rangle|^2$$
$$+(1+\hat{\boldsymbol{\nu}}\cdot\boldsymbol{\beta})|\langle J_f||\hat{\mathcal{M}}_J||J_i\rangle|^2$$
$$-2\hat{\boldsymbol{q}}\cdot(\hat{\boldsymbol{\nu}}+\boldsymbol{\beta}){\rm Re}\langle J_f||\hat{\mathcal{L}}_J||J_i\rangle\langle J_f||\hat{\mathcal{M}}_J||J_i\rangle^*\Big\}\Bigg\}, \tag{120}$$

where the momentum transfer $\boldsymbol{q} = \boldsymbol{\nu} - \boldsymbol{k}$ is defined as the difference between neutrino and electron momenta; $\hat{\boldsymbol{q}}$ and $\hat{\boldsymbol{\nu}}$ are the corresponding unit vectors, and $\boldsymbol{\beta} = \boldsymbol{k}/E_e$. The energies of the incoming electron and outgoing neutrino are denoted by E_e and E_ν, respectively. The Fermi function $F(Z, E_e)$ corrects the cross section for the distortion of the electron wave function by the Coulomb field of the nucleus.[142] The explicit energy dependence of the cross section is given by the term

$$\mathcal{W}(E_e,E_\nu) = \frac{E_\nu^2}{(1+E_e/M_T(1-\hat{\boldsymbol{\nu}}\cdot\boldsymbol{\beta}))}, \tag{121}$$

where the phase-space factor $(1 + E_e/M_T(1 - \hat{\boldsymbol{\nu}}\cdot\boldsymbol{\beta}))^{-1}$ accounts for the nuclear recoil, and M_T is the mass of the target nucleus. The nuclear transition matrix elements between the initial state $|J_i\rangle$ and final state $|J_f\rangle$, correspond to the charge $\hat{\mathcal{M}}_J$, longitudinal $\hat{\mathcal{L}}_J$, transverse electric $\hat{\mathcal{T}}_J^{EL}$, and transverse magnetic $\hat{\mathcal{T}}_J^{MAG}$ multipole operators.[140,141] The initial and final nuclear states in the hadronic matrix elements are characterized by angular momentum and parity J^π. The final nuclear states are the spin-isospin excitations of J^π at finite temperature of stellar environment.

In the electron-capture process, the excitation energy of the daughter nucleus ${}^{A}_{Z-1}X_{N+1}$ is obtained by the sum of the RPA energy $E_{\rm RPA}$ given with respect to the ground state of the parent nucleus and the binding energy difference between daughter and parent nucleus. Thus the energy of the outgoing neutrino is determined by the conservation relation:

$$E_\nu = E_e - E_{\rm RPA} - \Delta_{np}, \tag{122}$$

where E_e is the energy of incoming electron, and $\Delta_{np} = 1.294$ MeV is the mass difference between the neutron and the proton. The axial-vector coupling constant $g_A = -1.0$ is quenched for all the multipole excitations with respect to its free-nucleon value $g_A = -1.26$.

The electron-capture rate is expressed in terms of the cross section Eq. (120) and the distribution of electrons $f(E_e, \mu_e, T)$ at a given temperature,

$$\lambda_{\rm ec} = \frac{1}{\pi^2\hbar^3}\int_{E_e^0}^{\infty} p_e E_e \sigma_{ec}(E_e) f(E_e, \mu_e, T) dE_e. \tag{123}$$

$E_e^0 = max(|Q_{if}|, m_e c^2)$ is the minimum electron energy that allows for the capture process, that is, the threshold energy for electrons, where $Q_{if} = -E_{\rm RPA} - \Delta_{np}$. $p_e = (E_e^2 - m_e^2 c^4)^{1/2}$ is the electron momentum. Under stellar conditions that correspond to the core collapse of a supernova, the electron distribution is described by the Fermi–Dirac expression[143]

$$f(E_e, \mu_e, T) = \frac{1}{\exp(\frac{E_e - \mu_e}{kT}) + 1}. \tag{124}$$

T is the temperature, and the chemical potential μ_e is determined from the baryon density ρ by inverting the relation

$$\rho Y_e = \frac{1}{\pi^2 N_A}\left(\frac{m_e c}{\hbar}\right)^3 \int_0^{\infty} (f_e - f_{e^+}) p^2 dp, \tag{125}$$

where Y_e is the ratio of the number of electrons to the number of baryons, N_A is Avogadro's number, and f_{e^+} denotes the positron distribution function similar to Eq. (124), but with $\mu_{e^+} = -\mu_e$. We assume that the phase space is not blocked by neutrinos.

In the following, a few examples are given to illustrate the Gamow–Teller (GT) excitations and electron-capture cross sections in hot nuclei. The single nucleon basis and the thermal occupation factors of the initial nuclear state are determined in the finite-temperature RMF model, and charge-exchange transitions to the excited states are computed using the charge-exchange FTRRPA.

In Fig. 14 we display the GT$^+$ strength distributions for ^{54,56}Fe at $T = 0, 1$, and 2 MeV, as functions of excitation energy with respect to the ground state of the parent nucleus. At zero temperature both the RQRPA and RRPA results are shown, whereas the finite-temperature transition spectra are calculated using only the FTRRPA. One might notice that the RQRPA calculation is in fair agreement with the experimental centroid energies. Compared to the LSSM, the RQRPA excitation energies are ≈1 MeV

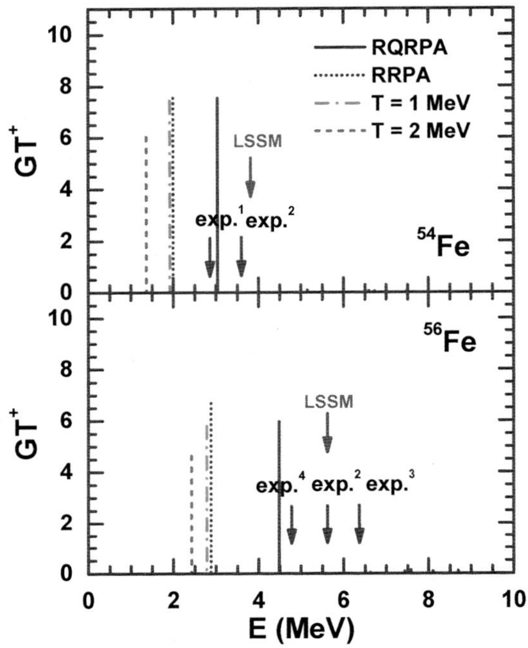

Fig. 14. (Color online) The GT^+ strength distributions for $^{54,56}Fe$ as functions of the excitation energy with respect to the ground state of the parent nucleus, calculated with the proton-neutron RQRPA at zero temperature, and the FTRRPA at $T = 0, 1$, and 2 MeV, for the DD-ME2 relativistic density functional. For comparison, the centroid energies of the LSSM calculation[144] at $T = 0$ MeV and the experimental centroid energies from Refs. 145–148 are indicated by arrows.

lower for both nuclei. By comparing the RRPA and RQRPA, we notice that pairing correlations shift the GT^+ transition to higher energy by $\sim 1 - 1.5$ MeV, because additional energy is needed to break a proton pair. When the temperature is increased to 1 MeV, the transition energy is lowered by ~ 1.1 MeV for ^{54}Fe, and 1.6 MeV for ^{56}Fe. This decrease in energy is mainly caused by the pairing collapse. With a further increase in temperature to 2 MeV, the GT^+ transition energy decreases by $\sim$0.5 MeV in both nuclei. This continuous decrease has its origin in the softening of the repulsive residual interaction because of the occupation factors that appear in the FTRRPA matrix elements. To demonstrate this in a quantitative way, we consider the example of ^{56}Fe, and analyze the unperturbed energies $E_{\rm unper}$, that is, the transition energy without residual interaction, and the energy shift caused by the residual interaction. For ^{56}Fe the principal contribution to the GT^+ comes from the transition from the proton orbital $\pi 1f_{7/2}$

to the neutron orbital $\nu 1f_{5/2}$. In the QRPA the unperturbed energy approximately equals the sum of two quasiparticle energies, and the chemical potential difference of neutrons and protons, resulting in $E_{\text{unper}} \simeq 3.6$ MeV. The energy shift induced by the repulsive residual interaction is 0.9 MeV. If pairing correlations are not included, that is in RPA, the unperturbed energy corresponds to the difference between the single-particle energies of the two orbitals, and this is 1.8 MeV at zero temperature, and 1.7 MeV at $T = 2$ MeV. Therefore the residual interaction shifts the energy by 1.1 MeV at zero temperature, and by 0.7 MeV at $T = 2$ MeV. Obviously the partial occupation factors (the smearing of the Fermi surface), induced either by pairing correlations or by temperature effects, will lead to the weakening of the residual interaction. The temperature effect appears to be more pronounced because the Fermi surface is more diffuse at $T = 2$ MeV. In addition to the excitation energy, the transition strength could also be reduced by the smearing of the Fermi surface through the appearance of occupation factors. Therefore, the transition strength becomes weaker with increasing temperature or with the inclusion of pairing correlations.

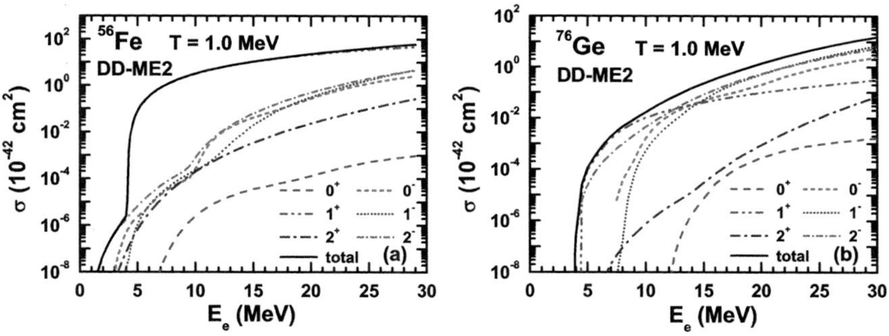

Fig. 15. Electron-capture cross sections for the ^{56}Fe and ^{76}Ge target nuclei at $T = 1$ MeV, calculated with the FTRRPA using the DD-ME2 effective interaction. In addition to the total cross section which includes multipole transitions $J^\pi = 0^\pm$, $1^\pm$, and $2^\pm$, contributions from the individual channels are shown in the plot as functions of the incident electron energy E_e. Taken from Ref. 83.

In Fig. 15 the cross sections for electron capture on ^{56}Fe and ^{76}Ge at $T = 1$ MeV are plotted as functions of the incident electron energy E_e. The cross sections are calculated using the expression of Eq. (120), and the FTRRPA with the DD-ME2 relativistic density functional[43] is used to evaluate the transition matrix elements. In addition to the total cross sections which include multipole transitions $J^{\pi} = 0^{\pm}$, $1^{\pm}$, and $2^{\pm}$, contributions from the individual channels are shown in the plot, as functions of the incident electron energy E_e. For ^{56}Fe the total cross section is completely dominated by the 1^+ channel (GT$^+$) all the way up to $E_e = 30$ MeV, with contributions from other channels being orders of magnitude smaller. In the case of the neutron-rich nucleus ^{76}Ge, on the other hand, forbidden transitions play a more prominent role, already starting from $E_e > 12$ MeV. Their contribution to the total cross section further increases with the electron energy E_e. Obviously in systematic calculations of electron-capture rates on heavier, more neutron-rich nuclei, contributions from forbidden transitions should also be included in addition to the GT$^+$ channel.

8. Concluding remarks and perspectives

As shown in previous examples, small amplitude collective motion in atomic nucleus is represented by appearance of characteristic resonance-like structures in excitation transition spectra. In addition to giant resonances, other modes of excitation may appear, including low-lying transition strength. While the properties of giant resonances are rather well established, of particular interest are studies of exotic modes of excitation in nuclei away from the valley of β-stability, where transitions of nucleons from weakly bound orbitals play an important role. In particular, the focus of current research is on discerning the nature of low-energy pygmy strength of various multipoles and the quest for novel modes of excitation in exotic nuclei.[149] In addition to extreme isospin values, finite temperature, as well as deformation, may open perspectives for new features in modes of nuclear excitation.

Since nuclei at finite temperature and away from the valley of stability participate in astrophysically relevant reactions, an accurate description of the detailed transition strength distribution and new knowledge on exotic nuclear modes of excitation may result in important consequences in understanding the processes in stellar environment. Since models at the RPA level often cannot account for the escape and spreading widths of excitations, reliable microscopic description necessitates a fully self-consistent theory that is able to include the effects of the particle continuum and couplings to complex configurations. These could be formulated by using modern energy density functionals constrained by the strong interaction physics of QCD and/or by the phenomenological input. Joint theoretical and experimental studies, including research with radioactive ion beams, are needed to provide the insight into dynamical properties of nuclei away from the valley of stability, involving the interplay of isospin asymmetry, deformation and finite temperature.

References

1. M. N. Harakeh and A. V. D. Woude, *Giant Resonances — Fundamental High-Frequency Modes of Nuclear Excitation.* Oxford Science Publications (2001).
2. P. F. Bortignon, A. Bracco, and R. A. Broglia, *Giant Resonances: Nuclear Structure at Finite Temperature.* Harwood Academic Publishers (1998).
3. W. Bothe and W. Gentner, *Z. Phys.* **71**, 236 (1937).
4. G. C. Baldwin and G. S. Klaiber, Photo-fission in heavy elements, *Phys. Rev.* **71**, 3 (1947).
5. M. Goldhaber and E. Teller, On nuclear dipole vibrations, *Phys. Rev.* **74**, 1046 (1948).
6. W. Greiner and J. A. Maruhn, *Nuclear Models.* Spinger-Verlag Berlin Heidelberg (1996).
7. B. L. Berman and S. C. Fultz, Measurements of the giant dipole resonance with monoenergetic photons, *Rev. Mod. Phys.* **47**, 713 (1975).
8. P. Gleissl, M. Brack, J. Meyer, and P. Quentin, A density variational approach to nuclear giant resonances at zero and finite temperature, *Ann. Phys.* **197**, 205 (1990).
9. M. B. Lewis and F. E. Bertrand, Evidence from inelastic proton scattering for a giant quadrupole vibration in spherical nuclei, *Nucl. Phys. A.* **196**, 337 (1972).
10. M. Nagao and Y. Torizuka, Electroexcitation of giant resonances in ^{208}Pb, *Phys. Rev. Lett.* **30**, 1068 (1973).
11. D. R. Bes, R. A. Broglia, and B. S. Nilsson, Microscopic description of isoscalar and isovector giant quadrupole resonances, *Phys. Rep.* **16**, 1 (1975).
12. N. Paar, C. C. Moustakidis, T. Marketin, D. Vretenar, and L. G. A., Neutron star structure and collective excitations of finite nuclei, *nucl-th/arXiv.* p. 1403.7574 (2014).
13. I. Hamamoto, H. Sagawa, and X. Z. Zhang, Giant monopole resonances in nuclei near stable and drip lines, *Phys. Rev. C.* **56**, 3121 (1997).
14. I. Hamamoto, H. Sagawa, and X. Z. Zhang, Structure of giant quadrupole resonances in neutron drip line nuclei, *Phys. Rev. C.* **55**, 2361 (1997).
15. D. J. Thouless, Stability conditions and nuclear rotations in the Hartree–Fock theory, *Nucl. Phys.* **21**, 225 (1960).
16. P. Ring and P. Schuck, *The Nuclear Many-Body Problem.* Springer-Varlag New York Inc. (1980).
17. N. Van Giai and H. Sagawa, Monopole and dipole compression modes in nuclei, *Nucl. Phys. A.* **371**, 1 (1981).
18. F. E. Bertrand, Giant multipole resonances perspectives after ten years, *Nucl. Phys. A.* **354**, 129 (1981).
19. K. Boretzky, J. Stroth, E. Wajda, T. Aumann, T. Blaich, J. Cub, T. Elze,

H. Emling, W. Henning, R. Holzmann, H. Klingler, R. Kulessa, J. Kratz, D. Lambrecht, Y. Leifels, E. Lubkiewicz, K. Stelzer, W. Walus, M. Zinser, and E. Zude, Two-phonon giant dipole resonance in ^{208}Pb, *Phys. Lett. B.* **384**, 30 (1996).
20. K. Ikeda, S. Fujii, and J. I. Fujita, The (p,n) reactions and beta decays, *Phys. Lett.* **3**, 271 (1963).
21. D. Vretenar, A. V. Afanasjev, G. A. Lalazissis, and P. Ring, Relativistic Hartree–Bogoliubov theory: static and dynamic aspects of exotic nuclear structure, *Phys. Rep.* **409**, 101 (2005).
22. N. Paar, P. Ring, T. Nikšić, and D. Vretenar, Quasiparticle random phase approximation based on the relativistic Hartree–Bogoliubov model, *Phys. Rev. C.* **67**, 034312 (2003).
23. T. Nikšić, D. Vretenar, and P. Ring, Relativistic random-phase approximation with density-dependent meson-nucleon couplings, *Phys. Rev. C.* **66**, 064302 (2002).
24. J. F. Dawson and R. J. Furnstahl, Relativistic spectral random-phase approximation in finite nuclei, *Phys. Rev. C.* **42**, 2009 (1990).
25. D. Vretenar, A. Wandelt, and P. Ring, Isoscalar dipole mode in relativistic random phase approximation, *Phys. Lett. B.* **487**, 334 (2000).
26. P. Ring, Z. Ma, N. Van Giai, D. Vretenar, A. Wandelt, and L. Cao, The time-dependent relativistic mean-field theory and the random phase approximation, *Nucl. Phys. A.* **694**, 249 (2001).
27. F. E. Serr, T. S. Dumitrescu, T. Suzuki, and C. H. Dasso, Microscopic description of current distributions for collective excitations in spherical nuclei, *Nucl. Phys. A.* **404**, 359 (1983).
28. J. P. Blaizot, Nuclear compressibilities, *Phys. Rep.* **64**, 171 (1980).
29. S. Shlomo and D. H. Youngblood, Nuclear matter compressibility from isoscalar giant monopole resonance, *Phys. Rev. C.* **47**, 529 (1993).
30. V. M. Kolomietz and S. Shlomo, Isoscalar compression modes within fluid dynamic approach, *Phys. Rev. C.* **61**, 064302 (2000).
31. S. Shlomo and A. I. Sanzhur, Isoscalar giant dipole resonance and nuclear matter incompressibility coefficient, *Phys. Rev. C.* **65**, 044310 (2002).
32. B. K. Agrawal, S. Shlomo, and V. Kim Au, Nuclear matter incompressibility coefficient in relativistic and non-relativistic microscopic models, *Phys. Rev. C.* **68**, 031304 (2003).
33. G. Colò, N. Van Giai, J. Meyer, K. Bennaceur, and P. Bonche, Microscopic determination of the nuclear incompressibility within the non-relativistic framework, *Phys. Rev. C.* **70**, 024307 (2004).
34. J. Kvasil, N. L. Iudice, C. Stoyanov, and P. Alexa, Compressional and toroidal dipole modes in nuclei, *J. Phys. G: Nucl. Part. Phys.* **29**, 753 (2003).
35. D. Lacroix, S. Ayik, and P. Chomaz, Collective response of nuclei: Comparison between experiments and extended mean-field calculations, *Phys. Rev. C.* **63**, 064305 (2001).
36. S. Yildirim, T. Gaitanos, M. D. Toro, and V. Greco, Relativistic transport approach to collective nuclear dynamics, *Phys. Rev. C.* **72**, 064317 (2005).

37. Z. Ma, N. Van Giai, A. Wandelt, D. Vretenar, and P. Ring, Isoscalar compression modes in relativistic random phase approximation, *Nucl. Phys. A.* **686**, 173 (2001).
38. J. Piekarewicz, Self-consistent description of nuclear compressional modes, *Phys. Rev. C.* **64**, 024307 (2001).
39. J. Piekarewicz, Correlating the giant-monopole resonance to the nuclear-matter incompressibility, *Phys. Rev. C.* **66**, 034305 (2002).
40. D. Vretenar, T. Nikšić, and P. Ring, A microscopic estimate of the nuclear matter compressibility and symmetry energy in relativistic mean-field models, *Phys. Rev. C.* **68**, 024310 (2003).
41. G. Colò and N. Van Giai, Theoretical understanding of the nuclear incompressibility: where do we stand?, *Nucl. Phys. A.* **731**, 15 (2004).
42. J. Piekarewicz, Do we understand the incompressibility of neutron-rich matter?, *J. Phys. G: Nucl. Part. Phys.* **37**, 064038 (2010).
43. G. A. Lalazissis, T. Nikšić, D. Vretenar, and P. Ring, New relativistic mean-field interaction with density-dependent meson-nucleon couplings, *Phys. Rev. C.* **71**, 024312 (2005).
44. D. H. Youngblood, H. L. Clark, and Y.-W. Lui, Incompressibility of nuclear matter from the giant monopole resonance, *Phys. Rev. Lett.* **82**, 691 (1999).
45. H. L. Clark, Y.-W. Lui, and D. H. Youngblood, Isoscalar giant dipole resonance in ^{90}Zr, ^{116}Sn, and ^{208}Pb, *Phys. Rev. C.* **63**, 031301 (2001).
46. D. H. Youngblood, Y.-W. Lui, B. John, Y. Tokimoto, H. L. Clark, and X. Chen, Compression mode resonances in ^{90}Zr, *Phys. Rev. C.* **69**, 054312 (2004).
47. G. Colò, N. Van Giai, P. Bortignon, and M. Quaglia, On dipole compression modes in nuclei, *Phys. Lett. B.* **485**, 362 (2000).
48. S. I. Bastrukov, S. Misicu, and V. I. Sushkov, Dipole torus mode in nuclear fluid-dynamics, *Nucl. Phys. A.* **562**, 191 (1993).
49. D. Vretenar, N. Paar, P. Ring, and T. Nikšić, Toroidal dipole resonances in the relativistic random phase approximation, *Phys. Rev. C.* **65**, 021301 (2002).
50. S. Misicu, Interplay of compressional and vortical nuclear currents in overtones of the isoscalar giant dipole resonance, *Phys. Rev. C.* **73**, 024301 (2006).
51. M. Uchida, H. Sakaguchi, M. Itoh, M. Yosoi, T. Kawabata, H. Takeda, Y. Yasuda, T. Murakami, T. Ishikawa, T. Taki, N. Tsukahara, S. Terashima, U. Garg, M. Hedden, B. Kharraja, M. Koss, B. Nayak, S. Zhu, M. Fujiwara, H. Fujimura, K. Hara, E. Obayashi, H. Yoshida, H. Akimune, M. Harakeh, and M. Volkerts, Isoscalar giant dipole resonance in ^{208}Pb via inelastic α scattering at 400MeV and nuclear incompressibility, *Phys. Lett. B.* **557**, 12 (2003).
52. M. Uchida, H. Sakaguchi, M. Itoh, M. Yosoi, T. Kawabata, Y. Yasuda, H. Takeda, T. Murakami, S. Terashima, S. Kishi, U. Garg, P. Boutachkov, M. Hedden, B. Kharraja, M. Koss, B. K. Nayak, S. Zhu, M. Fujiwara, H. Fujimura, H. P. Yoshida, K. Hara, H. Akimune, and M. N. Harakeh,

Systematics of the bimodal isoscalar giant dipole resonance, *Phys. Rev. C.* **69**, 051301 (2004).
53. N. Paar, D. Vretenar, T. Nikšić, and P. Ring, Relativistic quasiparticle random-phase approximation description of isoscalar compression modes in open-shell nuclei in the A $\approx$ 60 mass region, *Phys. Rev. C.* **74**, 037303 (2006).
54. S. Goriely and E. Khan, Large-scale QRPA calculation of E1-strength and its impact on the neutron capture cross section, *Nucl. Phys. A.* **706**, 217 (2002).
55. R. Pitthan and T. Walcher, Inelastic electron scattering in the giant resonance region of La, Ce and Pr, *Phys. Lett. B.* **36**, 563 (1971).
56. S. Fukuda and Y. Torizuka, Giant multipole resonances in ^{90}Zr observed by inelastic electron scattering, *Phys. Rev. Lett.* **29**, 1109 (1972).
57. N. Tsoneva and H. Lenske, Pygmy quadrupole resonance in skin nuclei, *Phys. Lett. B.* **695**, 174 (2011).
58. S. S. Henshaw, M. W. Ahmed, G. Feldman, A. M. Nathan, and H. R. Weller, New method for precise determination of the isovector giant quadrupole resonances in nuclei, *Phys. Rev. Lett.* **107**, 222501 (2011).
59. X. Roca-Maza, M. Brenna, B. K. Agrawal, P. F. Bortignon, G. Colò, L.-G. Cao, N. Paar, and D. Vretenar, Giant quadrupole resonances in ^{208}Pb, the nuclear symmetry energy, and the neutron skin thickness, *Phys. Rev. C.* **87**, 034301 (2013).
60. K. Riisager, Nuclear halo states, *Rev. Mod. Phys.* **66**, 1105 (1994).
61. N. Fukunishi, T. Otsuka, and I. Tanihata, Neutron-skin and proton-skin formations in exotic nuclei far from stability, *Phys. Rev. C.* **48**, 1648 (1993).
62. S. Mizutori, J. Dobaczewski, G. A. Lalazissis, W. Nazarewicz, and P.-G. Reinhard, Nuclear skins and halos in the mean-field theory, *Phys. Rev. C.* **61**, 044326 (2000).
63. N. Paar, D. Vretenar, E. Khan, and G. Colò, Exotic modes of excitation in atomic nuclei far from stability, *Rep. Prog. Phys.* **70**, 691 (2007).
64. E. Litvinova, H. P. Loens, K. Langanke, G. Martínez-Pinedo, T. Rauscher, P. Ring, F.-K. Thielemann, and V. Tselyaev, Low-lying dipole response in the relativistic quasiparticle time blocking approximation and its influence on neutron capture cross sections, *Nucl. Phys. A.* **823**, 26 (2009).
65. X. Roca-Maza, G. Pozzi, M. Brenna, K. Mizuyama, and G. Colò, Low-lying dipole response: Isospin character and collectivity in ^{68}Ni, ^{132}Sn, and ^{208}Pb, *Phys. Rev. C.* **85**, 024601 (2012).
66. D. Savran, T. Aumann, and A. Zilges, Experimental studies of the pygmy dipole resonance, *Prog. Part. Nucl. Phys.* **70**, 210 (2013).
67. R. Mohan, M. Danos, and L. C. Biedenharn, Three-fluid hydrodynamical model of nuclei, *Phys. Rev. C.* **3**, 1740 (1971).
68. Y. Suzuki, K. Ikeda, and H. Sato, New type of dipole vibration in nuclei, *Prog. Theor. Phys.* **83**, 180 (1990).
69. P. Van Isacker, M. A. Nagarajan, and D. D. Warner, Effect of the neutron skin on collective states of nuclei, *Phys. Rev. C.* **45**, R13 (1992).

70. J. Chambers, E. Zaremba, J. P. Adams, and B. Castel, Pygmy dipole resonances in the calcium isotopes, *Phys. Rev. C.* **50**, R2671 (1994).
71. T. Aumann, Reactions with fast radioactive beams of neutron-rich nuclei, *Eur. Phys. J. A.* **26**, 441 (2005).
72. P. Adrich, A. Klimkiewicz, M. Fallot, K. Boretzky, T. Aumann, D. Cortina-Gil, U. D. Pramanik, T. W. Elze, H. Emling, H. Geissel, M. Hellström, K. L. Jones, J. V. Kratz, R. Kulessa, Y. Leifels, C. Nociforo, R. Palit, H. Simon, G. Surówka, K. Sümmerer, and W. Waluś, Evidence for pygmy and giant dipole resonances in ^{130}Sn and ^{132}Sn, *Phys. Rev. Lett.* **95**(13), 132501 (2005).
73. D. Vretenar, Y. F. Niu, N. Paar, and J. Meng, Low-energy isovector and isoscalar dipole response in neutron-rich nuclei, *Phys. Rev. C.* **85**, 044317 (2012).
74. A. Zilges, S. Volz, M. Babilon, T. Hartmann, P. Mohr, and K. Vogt, Concentration of electric dipole strength below the neutron separation energy in N=82 nuclei, *Phys. Lett. B.* **542**, 43 (2002).
75. A. Zilges, M. Babilon, T. Hartmann, D. Savran, and S. Volz, Collective excitations close to the particle threshold, *Prog. Part. Nucl. Phys.* **55**, 408 (2005).
76. G. A. Lalazissis, J. König, and P. Ring, New parametrization for the lagrangian density of relativistic mean field theory, *Phys. Rev. C.* **55**, 540 (1997).
77. H. Tsoneva, N.and Lenske and C. Stoyanov, Probing the nuclear neutron skin by low-energy dipole modes, *Phys. Lett. B.* **586**, 213 (2004).
78. K. Govaert, F. Bauwens, J. Bryssinck, D. De Frenne, E. Jacobs, W. Mondelaers, L. Govor, and V. Y. Ponomarev, Dipole excitations to bound states in ^{116}Sn and ^{124}Sn, *Phys. Rev. C.* **57**, 2229 (1998).
79. F. Osterfeld, Nuclear spin and isospin excitations, *Rev. Mod. Phys.* **64**, 491 (1992).
80. K. Langanke and G. Martínez-Pinedo, Nuclear weak-interaction process in stars, *Rev. Mod. Phys.* **75**, 819 (2003).
81. H.-T. Janka, K. Langanke, A. Marek, G. Martínez-Pinedo, and B. Müller, Theory of core-collapse supernovae, *Phys. Rep.* **442**, 38 (2007).
82. N. Paar, G. Colò, E. Khan, and D. Vretenar, Calculation of stellar electron-capture cross sections on nuclei based on microscopic Skyrme functionals, *Phys. Rev. C.* **80**, 055801 (2009).
83. Y. F. Niu, N. Paar, D. Vretenar, and J. Meng, Stellar electron-capture rates calculated with the finite-temperature relativistic random-phase approximation, *Phys. Rev. C.* **83**, 045807 (2011).
84. A. F. Fantina, E. Khan, G. Colò, N. Paar, and D. Vretenar, Stellar electron-capture rates on nuclei based on a microscopic Skyrme functional, *Phys. Rev. C.* **86**, 035805 (2012).
85. E. M. Burbidge, G. R. Burbidge, W. A. Fowler, and F. Hoyle, Synthesis of the elements in stars, *Rev. Mod. Phys.* **29**, 547 (1957).
86. Y.-Z. Qian and G. Wasserburg, Where, oh where has the r-process gone?, *Phys. Rep.* **442**, 237 (2007).

87. Z. Niu, Y. Niu, H. Liang, W. Long, T. Niki, D. Vretenar, and J. Meng, β-decay half-lives of neutron-rich nuclei and matter flow in the r-process, *Phys. Lett. B.* **723**, 172 (2013).
88. F. T. Avignone, S. R. Elliott, and J. Engel, Double beta decay, majorana neutrinos, and neutrino mass, *Rev. Mod. Phys.* **80**, 481 (2008).
89. J. D. Vergados, H. Ejiri, and F. imkovic, Theory of neutrinoless double-beta decay, *Rep. Prog. Phys.* **75**, 106301 (2012).
90. J. C. Hardy and I. S. Towner, Superallowed $0^+ \to 0^+$ nuclear beta decays: A new survey with precision tests of the conserved vector current hypothesis and the standard model, *Phys. Rev. C.* **79**(5), 055502 (2009).
91. H. Liang, N. Van Giai, and J. Meng, Isospin corrections for superallowed Fermi beta decay in self-consistent relativistic random-phase approximation approaches, *Phys. Rev. C.* **79**, 064316 (2009).
92. N. Paar, T. Nikšić, D. Vretenar, and P. Ring, Quasiparticle random phase approximation based on the relativistic Hartree–Bogoliubov model. ii. nuclear spin and isospin excitations, *Phys. Rev. C.* **69**, 054303 (2004).
93. J. Engel, M. Bender, J. Dobaczewski, W. Nazarewicz, and R. Surman, β decay rates of r-process waiting-point nuclei in a self-consistent approach, *Phys. Rev. C.* **60**, 014302 (1999).
94. J. Meng, J. Y. Guo, Z. P. Li, H. Z. Liang, W. H. Long, Y. F. NIU, Z. M. Niu, J. M. Yao, P. W. Zhang, Y. and. Zhao, and S. G. Zhou, Covariant density functional theory in nuclear physics, *Progress in Physics.* **31**, 199 (2011).
95. B. D. Anderson, T. Chittrakarn, A. R. Baldwin, C. Lebo, R. Madey, P. C. Tandy, J. W. Watson, B. A. Brown, and C. C. Foster, Particle-hole strength excited in the ^{48}Ca (p,n) ^{48}Sc reaction at 134 and 160 MeV: Gamow–Teller strength, *Phys. Rev. C.* **31**, 1161 (1985).
96. D. E. Bainum, J. Rapaport, C. D. Goodman, D. J. Horen, C. C. Foster, M. B. Greenfield, and C. A. Goulding, Observation of giant particle-hole resonances in ^{90}Zr(p, n)^{90}Nb, *Phys. Rev. Lett.* **44**, 1751 (1980).
97. T. Wakasa, H. Sakai, H. Okamura, et al., Gamow–Teller strength of ^{90}Nb in the continuum studied via multipole decomposition analysis of the $^{90}\mathrm{Zr}(p,n)$ reaction at 295 MeV, *Phys. Rev. C.* **55**, 2909 (1997).
98. H. Akimune, I. Daito, Y. Fujita, M. Fujiwara, M. B. Greenfield, M. N. Harakeh, T. Inomata, J. Jänecke, K. Katori, S. Nakayama, H. Sakai, Y. Sakemi, M. Tanaka, and M. Yosoi, Direct proton decay from the Gamow–Teller resonance in ^{208}Bi, *Phys. Rev. C.* **52**, 604 (1995).
99. R. R. Doering, A. Galonsky, D. M. Patterson, and G. F. Bertsch, Observation of giant Gamow–Teller strength in (p,n) reactions, *Phys. Rev. Lett.* **35**, 1691 (1975).
100. K. Shimizu, M. Ichimura, and A. Arima, Magnetic moments and GT-type β-decay matrix elements in nuclei with a LS doubly closed shell plus or minus one nucleon, *Nucl. Phys. A.* **226**, 282 (1974).
101. G. F. Bertsch and I. Hamamoto, Gamow–Teller strength at high excitations, *Phys. Rev. C.* **26**, 1323 (1982).

102. S. Drożdż, V. Klemt, J. Speth, and J. Wambach, Giant Gamow–Teller resonances in nuclei described with realistic two-body interactions, *Phys. Lett. B.* **166**, 18 (1986).
103. C. L. Bai, H. Sagawa, H. Q. Zhang, X. Z. Zhang, G. Colò, and F. R. Xu, Effect of tensor correlations on Gamow–Teller states in ^{90}Zr and ^{208}Pb, *Phys. Lett. B.* **675**, 28 (2009).
104. M. Ericson, A. Figureau, and C. Thvenet, Pionic field and renormalization of the axial coupling constant in nuclei, *Phys. Lett. B.* **45**, 19 (1973).
105. W. Knupfer, M. Dillig, and A. Richter, Quenching of the magnetic multipole strength distribution and of the anomalous magnetic moment in complex nuclei and mesonic renormalization of the nuclear spin current, *Phys. Lett. B.* **95**, 349 (1980).
106. D. J. Horen, C. D. Goodman, C. C. Foster, C. A. Goulding, M. B. Greenfield, J. Rapaport, D. E. Bainum, E. Sugarbaker, T. G. Masterson, F. Petrovich, and W. G. Love, Search for isobaric analogues of m1 states and giant spinflip resonances in the ^{208}Pb (p, n) reaction, *Phys. Lett. B.* **95**, 27 (1980).
107. A. Krasznahorkay, H. Akimune, M. Fujiwara, M. N. Harakeh, J. Jänecke, V. A. Rodin, M. H. Urin, and M. Yosoi, Distribution of the Gamow–Teller strength in ^{90}Nb and ^{208}Bi, *Phys. Rev. C.* **64**, 067302 (2001).
108. O. Civitarese, R. A. Broglia, and C. H. Dasso, On the temperature dependence of the nuclear response, *Ann. Phys.* **156**, 142 (1984).
109. D. Vautherin and N. Vinh Mau, Temperature dependence of collective states in the random-phase approximation, *Nucl. Phys. A.* **422**, 140 (1984).
110. W. Besold, P. G. Reinhard, and C. Toepffer, Giant resonances on excited states, *Nucl. Phys. A.* **431**, 1 (1984).
111. M. E. Faber, J. L. Egido, and P. Ring, The giant dipole resonance in hot rotating nuclei, *Phys. Lett. B.* **127**, 5 (1983).
112. P. Ring, L. M. Robledo, J. L. Egido, and M. Faber, Microscopic theory of the isovector dipole resonance at high angular momenta, *Nucl. Phys. A.* **419**, 261 (1984).
113. D. Lacroix, P. Chomaz, and S. Ayik, Finite temperature nuclear response in the extended random phase approximation, *Phys. Rev. C.* **58**, 2154 (1998).
114. H. Sagawa and G. F. Bertsch, Self-consistent calculations of finite temperature nuclear response function, *Phys. Lett. B.* **146**, 138 (1984).
115. P. F. Bortignon, R. A. Broglia, G. F. Bertsch, and J. Pacheco, Damping of nuclear excitations at finite temperature, *Nucl. Phys. A.* **460**, 149 (1986).
116. N. Dinh Dang, On the width of giant dipole resonance in hot nuclei, *Phys. Rep.* **264**, 123 (1996).
117. N. Dinh Dang and F. Sakata, Thermal damping width of the giant dipole resonance in hot nuclei, *Phys. Rev. C.* **55**, 2872 (1997).
118. H. A. Bethe, G. E. Brown, J. Applegate, and J. M. Lattimer, Equation of state in the gravitational collapse of stars, *Nucl. Phys. A.* **324**, 487 (1979).
119. H. A. Bethe, Supernova mechanisms, *Rev. Mod. Phys.* **62**, 801 (1990).
120. F. S. Dietrich, J. C. Browne, W. J. O'Connell, and M. J. Kay, Spectrum of γ rays in the 8- to 20-MeV range from ^{252}Cf spontaneous fission, *Phys. Rev. C.* **10**, 795 (1974).

121. J. O. Newton, B. Herskind, R. M. Diamond, E. L. Dines, J. E. Draper, K. H. Lindenberger, C. Schück, S. Shih, and F. S. Stephens, Observation of giant dipole resonances built on states of high energy and spin, *Phys. Rev. Lett.* **46**, 1383 (1981).
122. J. E. Draper, J. O. Newton, L. G. Sobotka, H. Lindenberger, G. J. Wozniak, L. G. Moretto, F. S. Stephens, R. M. Diamond, and R. J. McDonald, Dependence of the giant dipole strength function on excitation energy, *Phys. Rev. Lett.* **49**, 434 (1982).
123. E. Ramakrishnan, T. Baumann, A. Azhari, R. A. Kryger, R. Pfaff, M. Thoennessen, S. Yokoyama, J. R. Beene, M. L. Halbert, P. E. Mueller, D. W. Stracener, R. L. Varner, R. J. Charity, J. F. Dempsey, D. G. Sarantites, and L. G. Sobotka, Giant dipole resonance built on highly excited states of ^{120}Sn nuclei populated by inelastic α scattering, *Phys. Rev. Lett.* **76**, 2025 (1996).
124. T. Baumann, E. Ramakrishnan, A. Azhari, J. R. Beene, R. J. Charity, J. F. Dempsey, M. L. Halbert, P. F. Hua, R. A. Kryger, P. E. Mueller, R. Pfaff, D. G. Sarantites, L. G. Sobotka, D. W. Stracener, M. Thoennessen, G. V. Buren, R. L. Varner, and S. Yokoyama, Evolution of the giant dipole resonance in excited ^{120}Sn and ^{208}Pb nuclei populated by inelastic alpha scattering, *Nucl. Phys. A.* **635**, 428 (1998).
125. A. Bracco, F. Camera, M. Mattiuzzi, B. Million, M. Pignanelli, J. J. Gaardhøje, A. Maj, T. Ramsøy, T. Tveter, and Z. Źelazny, Increase in width of the giant dipole resonance in hot nuclei: Shape change or collisional damping?, *Phys. Rev. Lett.* **74**, 3748 (1995).
126. M. Mattiuzzi, A. Bracco, F. Camera, W. Ormand, J. Gaardhje, A. Maj, B. Million, M. Pignanelli, and T. Tveter, Angular momentum dependence of the GDR width in Sn nuclei at fixed excitation energy, *Nucl. Phys. A.* **612**, 262 (1997).
127. K. Snover, Giant resonances in excited nuclei, *Annu. Rev. Nucl. Part. Sci.* **36**, 545 (1986).
128. J. J. Gaardhoje, Nuclear structure at high excitation energy studied with giant resonances, *Ann. Rev. Nucl. Part. Sci.* **42**, 483 (1992).
129. D. Santonocito and Y. Blumenfeld, Evolution of the giant dipole resonance properties with excitation energy, *Eur. Phys. J. A.* **30**, 183 (2006).
130. A. Schiller and M. Thoennessen, Compilation of giant electic dipole resonances built on excited states, *At. Data Nucl. Data Tables.* **93**, 549 (2007).
131. H. M. Sommermann, Microscopic description of giant resonances in highly excited nuclei, *Ann. Phys.* **151**, 163 (1983).
132. A. L. Goodman, Finite-temperature HFB theory, *Nucl. Phys. A.* **352**, 30 (1981).
133. E. Khan, N. Van Giai, and M. Grasso, Collective motions in hot exotic nuclei: the finite temperature continuum-QRPA, *Nucl. Phys. A.* **731**, 311 (2004).
134. Y. F. Niu, Z. M. Niu, N. Paar, D. Vretenar, G. H. Wang, J. S. Bai, and J. Meng, Pairing transitions in finite-temperature relativistic Hartree–Bogoliubov theory, *Phys. Rev. C.* **88**, 034308 (2013).

135. S. Bonche, P.and Levii and D. Vautherin, Properties of highly excited nuclei, *Nucl. Phys. A.* **427**, 278 (1984).
136. Y. Niu, N. Paar, D. Vretenar, and J. Meng, Low-energy monopole and dipole response in nuclei at finite temperature, *Phys. Lett. B.* **681**, 315 (2009).
137. J. F. Berger, M. Girod, and D. Gogny, Time-dependent quantum collective dynamics applied to nuclear fission, *Comput. Phys. Commun.* **63**, 365 (1991).
138. O. Wieland, A. Bracco, F. Camera, G. Benzoni, N. Blasi, S. Brambilla, F. C. L. Crespi, S. Leoni, B. Million, R. Nicolini, A. Maj, P. Bednarczyk, J. Grebosz, M. Kmiecik, W. Meczynski, J. Styczen, T. Aumann, A. Banu, T. Beck, F. Becker, L. Caceres, P. Doornenbal, H. Emling, J. Gerl, H. Geissel, M. Gorska, O. Kavatsyuk, M. Kavatsyuk, I. Kojouharov, N. Kurz, R. Lozeva, N. Saito, T. Saito, H. Schaffner, H. J. Wollersheim, J. Jolie, P. Reiter, N. Warr, G. deAngelis, A. Gadea, D. Napoli, S. Lenzi, S. Lunardi, D. Balabanski, G. LoBianco, C. Petrache, A. Saltarelli, M. Castoldi, A. Zucchiatti, J. Walker, and A. Bürger, Search for the pygmy dipole resonance in ^{68}Ni at 600 MeV/nucleon, *Phys. Rev. Lett.* **102**, 092502 (2009).
139. Y. F. Niu, N. Paar, D. Vretenar, and J. Meng, Finite temperature effects on monopole and dipole excitations, *J. Phys.: Conf. Ser.* **312**, 042017 (2011).
140. J. D. Walecka, *Muon Physics.* vol. II, Academic, New York (1975).
141. J. S. O'Connell, T. W. Donnelly, and J. D. Walecka, Semileptonic weak interactions with ^{12}C, *Phys. Rev. C.* **6**, 719 (1972).
142. E. Kolbe, K. Langanke, G. Martínez-Pinedo, and P. Vogel, Neutrinonucleus reactions and nuclear structure, *J. Phys. G.* **29**, 2569 (2003).
143. A. Juodagalvis, K. Langanke, W. R. Hix, G. Martinez-Pinedo, and J. M. Sampaio, Improved estimate of electron capture rates on nuclei during stellar core collapse, *Nucl. Phys. A.* **848**, 454 (2010).
144. E. Caurier, K. Langanke, G. Martínez-Pinedo, and F. Nowacki, Shell-model calculations of stellar weak interaction rates. I. Gamow–Teller distributions and spectra of nuclei in the mass range $A = 45$–65, *Nucl. Phys. A.* **653**, 439 (1999).
145. M. C. Vetterli, O. Häusser, R. Abegg, W. P. Alford, A. Celler, D. Frekers, R. Helmer, R. Henderson, K. H. Hicks, K. P. Jackson, R. G. Jeppesen, C. A. Miller, K. Raywood, and S. Yen, Gamow–Teller strength deduced from charge exchange reactions on ^{54}Fe at 300 MeV, *Phys. Rev. C.* **40**, 559 (1989).
146. T. Rönnqvist, H. Condé, N. Olsson, E. Ramström, R. Zorro, J. Blomgren, A. Håkansson, A. Ringbom, G. Tibell, O. Jonsson, L. Nilsson, P. U. Renberg, S. Y. van der Werf, W. Unkelbach, and F. P. Brady, The ^{54,56}Fe (n, p) ^{54,56}Mn reactions at $E_n = 97$ MeV, *Nucl. Phys. A.* **563**, 225 (1993).
147. S. El-Kateb, K. P. Jackson, W. P. Alford, R. Abegg, R. E. Azuma, B. A. Brown, A. Celler, D. Frekers, O. Häusser, R. Helmer, R. S. Henderson, K. H. Hicks, R. Jeppesen, J. D. King, G. G. Shute, B. M. Spicer, A. Trudel, K. Raywood, M. Vetterli, and S. Yen, Spin-isospin strength distributions for fp shell nuclei: Results for the ^{55}Mn(n,p), ^{56}Fe(n,p), and ^{58}Ni(n,p) reactions at 198 MeV, *Phys. Rev. C.* **49**, 3128 (1994).

148. D. Frekers, Weak interaction processes in supernovae: New probes using charge exchange reaction at intermediate energies, *Nucl. Phys. A.* **752**, 580 (2005).
149. N. Paar, The quest for novel modes of excitation in exotic nuclei, *J. Phys. G: Nucl. Part. Phys.* **37**, 064014 (2010).

Chapter 11

Nuclear shell structure and response with quasiparticle-vibration coupling

Elena Litvinova

Department of Physics, Western Michigan University, Kalamazoo, MI 49008-5252, USA

National Superconducting Cyclotron Laboratory, Michigan State University, East Lansing, MI 48824-1321, USA

elena.litvinova@wmich.edu

Peter Ring

Physics Department, Technical University Munich, 85748 Garching, Germany

School of Physics, Peking University, Beijing 100871, P.R.China

ring@ph.tum.de

Extensions of the covariant density functional theory by quasiparticle-vibration coupling (QVC) are discussed. The formalism for one-body and two-body propagators in the nuclear medium allows calculations of single-particle energies and spectroscopic factors as well as the response to various types of excitations. In both cases QVC leads to a fragmentation of states, in agreement with experimental observations. Peculiarities of various $2p2h$ coupling schemes in the nuclear response function are discussed. The theory of the spin-isospin response includes both QVC and pion exchange and provides a framework for calculations of beta-decay, electron capture and charge-exchange reaction characteristics. The presented approaches are illustrated by realistic calculations for medium-mass and heavy nuclei.

Contents

1. Introduction

Medium-mass and heavy nuclei represent quantum-mechanical Fermi-systems where single-particle and vibrational degrees of freedom are strongly coupled. Collective vibrations lead to shape oscillations of the mean nuclear potential and, therefore, modify the single-particle motion. To take this effect into account, already in Ref. 1 a general concept for the quasiparticle-vibration (phonon) coupling (QVC) part of the single-nucleon self-energy has been proposed. This concept has had various implementations over the years within the Quasiparticle-Phonon Model (QPM),[2–4] Nuclear Field Theory[5–11] and others.[12–21] These studies, however, either are not self-consistent or do not include pairing correlations of the superfluid type. In particular, they use a large number of adjustable parameters in each step of the calculation, first for the single-particle potential, then for the calculation of the vibrational states and, finally, for the quasiparticle-phonon coupling vertices.

Recently, a set of self-consistent approaches to QVC in the relativistic framework has become available,[22–28] where Refs. 24, 26–28 include also superfluid pairing on an equal footing with QVC. The single-particle potentials (self-energies) as well as the vibrations are derived in a fully self-consistent way from well-known universal relativistic energy density functionals discussed in the earlier chapters of this book. No additional parameters are necessary. The particle-phonon coupling vertices are derived from the same density functional. It has been shown that QVC improves considerably the description of the single-particle states around the Fermi surface and explains the strong fragmentation of deep hole states, giant resonances and soft modes.

In this chapter, we review the latter approaches, which implement both superfluid and vibrational correlations in the relativistic framework, and discuss their applications to the description of single-quasiparticle spectra and the response in medium-mass and heavy nuclei. In Section 2 we start with time-dependent density functional theory and discuss ways to connect the QVC methods to static density functional theory, such that we can use the powerful and universal energy density functionals without any further parameters. Section 3 introduces the QVC model for the energy dependence of the relativistic self-energy, which forms the basis of all these methods going beyond mean field. In particular, we discuss in this section also calculations of the fragmentation of the single-particle and quasiparticle spectra. Section 4 is devoted to response theory derived from the energy dependent self energy in the small amplitude limit and applications for the width of giant resonances, for two-phonon configurations and for damping and quenching mechanisms of spin-isocpin resonances. Finally, Section 5 contains a summary and a discussion of perspectives for the future.

2. Time dependent density functional theory

As we saw in the Chapters 1, 2, and 3 density functional theory originally introduced by Kohn and Sham[29,30] in the sixties is based on the mean-field concept and is strictly valid only for a description of the ground states. Over the years it has been extended also for excited states in the framework of time-dependent density functional theory (TDDFT).[31] Runge and Gross derived a theorem[32] which is an extension of the Kohn–Sham method. They consider a many-body problem in a time-dependent external single potential $f_{\text{ext}}(\boldsymbol{r}, t)$ and start from the exact time-dependent Schrödinger equation

$$i\partial_t|\Psi(t)\rangle = (\hat{H} + f_{\text{ext}}(t))|\Psi(t)\rangle \tag{1}$$

with the exact solution $|\Psi(t)\rangle$ for a given initial condition $|\Psi(0)\rangle$. They showed that, for given initial conditions, there is a one-to-one correspondence between the exact time-dependent local single-particle density

$$\rho(\boldsymbol{r}, t) = \langle\Psi(t)|\sum_i^N \delta(\boldsymbol{r} - \boldsymbol{r}_i)|\Psi(t)\rangle \tag{2}$$

and the external potential $f_{\text{ext}}(\boldsymbol{r}, t)$. As in the static Kohn–Sham theory, there exists a fictitious system of non-interacting particles with the single particle wave functions $\varphi_i(\boldsymbol{r}, t)$ $(i = 1 \ldots N)$ that satisfy time-dependent Kohn–Sham equations of the form

$$i\partial_t\varphi_i(\boldsymbol{r}, t) = \left[-\nabla^2/2m + v_{\text{KS}}[\rho](\boldsymbol{r}, t)\right] \varphi_i(\boldsymbol{r}, t) \tag{3}$$

for a well defined time-dependent single-particle potential $v_{KS}[\rho](\boldsymbol{r},t)$ and the density derived from these single-particle wave functions

$$\rho(\boldsymbol{r},t) = \sum_i^N |\varphi_i(\boldsymbol{r},t)|^2 \,, \tag{4}$$

is the exact local density of the interacting many-body system at each time t. The time-dependent Kohn–Sham potential $v_{\rm KS}[\rho](\boldsymbol{r},t)$ is, of course, a function of $\boldsymbol{r}$ and t, but it is in addition a unique functional of the time-dependent density $\rho(\boldsymbol{r},t)$. As in the static case the exchange correlation potential $v_{\rm xc}$ is defined by the condition:

$$v_{\rm KS}[\rho](\boldsymbol{r},t) = v_{\rm H}[\rho](\boldsymbol{r},t) + v_{\rm xc}[\rho](\boldsymbol{r},t) + f_{\rm ext}(\boldsymbol{r},t), \tag{5}$$

where $v_{\rm H}$ is the corresponding Hartree potential. These equations look rather similar to the static Kohn–Sham equations in Chapter 1. The situation is, however, much more complicated than the static case. Since the potential

$$v[\rho](\boldsymbol{r},t) = v_{\rm H}[\rho](\boldsymbol{r},t) + v_{\rm xc}[\rho](\boldsymbol{r},t), \tag{6}$$

which is usually called *self-energy* $\Sigma(\boldsymbol{r},t)$, is a functional of the time-dependent density $\rho(\boldsymbol{r},t)$, it contains the entire history of the system and its derivation would require the full solution of the exact Schrödinger equation (1) for arbitrary systems with the density $\rho(\boldsymbol{r},t)$. In the static case we can obtain $v_{\rm s}[\rho](\boldsymbol{r})$ as the functional derivative of the static energy density functional $E[\rho_s]$, but, as long as we do not want to make further approximations, there is no such relation known in the time-dependent case.

There is, however, a case where the situation can be considerably simplified and this case applies also to nuclear physics. These are problems where the external potential $f_{\rm ext}(\boldsymbol{r},t)$ is rather weak, as for instance in the case of an external electromagnetic field. In this case linear response theory is applicable. Under this condition we do not need to know the functional $v[\rho](\boldsymbol{r},t)$ for arbitrarily changing densities, rather we need it only in the vicinity of the initial density, e.g. of the static ground state density $\rho_{\rm s}(\boldsymbol{r})$ of the system, and we can write

$$\rho(\boldsymbol{r},t) = \rho_{\rm s}(\boldsymbol{r}) + \delta\rho(\boldsymbol{r},t). \tag{7}$$

In this case we obtain to linear order in $f_{\rm ext}$

$$v[\rho_{\rm s}+\delta\rho](\boldsymbol{r},t) = v[\rho_{\rm s}](\boldsymbol{r}) + \int dt' \int d^3r'\, V[\rho_{\rm s}](\boldsymbol{r},t,\boldsymbol{r}',t')\delta\rho(\boldsymbol{r},t'). \tag{8}$$

Here V is an effective two-body interaction and it is obtained as a functional derivative of the mean field (6) with respect to the density

$$V(\boldsymbol{r},\boldsymbol{r}',t-t') = \left.\frac{\delta v(\boldsymbol{r},t)}{\delta\rho(\boldsymbol{r}',t')}\right|_{\rho=\rho_s}, \tag{9}$$

at the value of the static density. It is a functional of the static density ρ_s and for a time-independent Hamiltonian it depends only on the time difference $t-t'$. In Coulombic systems it is the sum of the Coulomb potential and the so-called exchange correlation kernel.[31] Introducing, as usual, the response function $R(\boldsymbol{r},t,\boldsymbol{r}',t')$

$$\delta\rho(\boldsymbol{r},t) = \int d^3r' \int dt' R(\boldsymbol{r},\boldsymbol{r}',t-t') f_{\text{ext}}(\boldsymbol{r}',t'), \tag{10}$$

we find after a Fourier transformation in time the well-known linear response equation[33]

$$R(\boldsymbol{r},\boldsymbol{r}',\omega)=R_0(\boldsymbol{r},\boldsymbol{r}',\omega)+\int d^3r_1\int d^3r_2 R_0(\boldsymbol{r},\boldsymbol{r}_1,\omega)V(\boldsymbol{r}_1,\boldsymbol{r}_2,\omega)R(\boldsymbol{r}_2,\boldsymbol{r}',\omega). \tag{11}$$

Here R_0 is the response function without interaction, i.e. for the response of a system of independent particles obeying the static form of Eq. (3). Of course, all these quantities are functionals of the exact ground state density $\rho_s(\boldsymbol{r})$.

Under the condition that the external field f_{ext} is weak, these equations are exact, but, of course, we do not know the functional $v[\rho](\boldsymbol{r},t)$ in Eq. (6) nor its functional derivative at $\rho = \rho_s$. As usual, in density functional theory the Runge–Gross theorem gives us no hint on how to derive these quantities. It also does not make any statements on the relation between this functional and the exact static density functional $E[\rho_s]$ of Hohenberg and Kohn[34] for the ground state. We therefore need some model for this single particle potential $v[\rho](\boldsymbol{r},t)$ in the Kohn–Sham equation. In any case this potential is a functional of $\rho(\boldsymbol{r},t)$ and therefore it depends on the history of the system, i.e. it has a memory.

There is a very common approximation, the so-called *adiabatic approximation.* Here one neglects the memory and assumes that the density $\rho(\boldsymbol{r},t)$ changes only very slowly with time, such that at each time t the potential is given by the potential of static Kohn–Sham theory $v_s(\boldsymbol{r}) \equiv v[\rho_s]_{\text{KS}}(\boldsymbol{r})$ corresponding to this density:

$$v[\rho](\boldsymbol{r},t) \approx v_s[\rho_s](\boldsymbol{r},t)|_{\rho_s=\rho(t)}. \tag{12}$$

In this approximation $v[\rho]$ is no longer a functional depending on a function $\rho(\boldsymbol{r},t)$ with four variables, but rather it is a functional depending on the function $\rho_s(\boldsymbol{r})$ with three variables. The density $\rho(\boldsymbol{r},t)$ changes slowly with time and at a given time t we use the function $\rho_s(\boldsymbol{r}) = \rho(\boldsymbol{r},t)$ depending only on three variables at that time. Here the time is just a parameter and, therefore, the potential $v[\rho](\boldsymbol{r},t)$ depends on time. Since, in the static case, the potential $v_\mathrm{s}[\rho_\mathrm{s}]$ is just the derivative of the energy density functional $E[\rho_\mathrm{s}]$ with respect to the static density, we obtain from Eq. (9), in the adiabatic approximation, an effective interaction

$$V_\mathrm{ad}(\boldsymbol{r},\boldsymbol{r}',t-t') = \left.\frac{\delta E[\rho_\mathrm{s}]}{\delta\rho_\mathrm{s}(\boldsymbol{r})\delta\rho_\mathrm{s}(\boldsymbol{r}')}\right|_{\rho_\mathrm{s}=\rho(t)} \delta(t-t'), \tag{13}$$

which is local in time. In Fourier space this interaction does not depend on energy. It is just the second derivative of the static energy functional with respect to the static density. This approximation is well known. It corresponds to the small amplitude limit of the time-dependent mean field equations,[33] i.e. to the random phase approximation (RPA), or in superfluid systems to quasiparticle RPA (QRPA). It is extensively used in nuclear physics. In the relativistic case we have relativistic RPA (RRPA)[35] and relativistic QRPA (RQRPA)[36] with many applications in the literature.[37–39]

Here we discuss methods going beyond this time-dependent mean field approximation. They start with a single-particle potential having a memory or, therefore, an energy-dependent self-energy $\Sigma(\omega)$. In the language of Green's functions they solve to some approximation the Dyson equation,[40] which is non-local in time and space. Of course it is, at present, impossible to derive this non-locality from first principles; rather we introduce a model where we decompose the full self-energy, also called the *mass operator*, into two parts:

$$\Sigma(\omega) = \tilde{\Sigma} + \Sigma^{(e)}(\omega). \tag{14}$$

The first part is the static self energy $\tilde{\Sigma}$ derived from static density functional theory. It contains the well-known self-consistent scalar, vector, and pairing potentials of the nuclear ground state. The second part $\Sigma^{(e)}(\omega)$ contains all the rest, in particular, the energy dependence. It will be approximated by the usual diagrams of particle- or quasiparticle vibrational coupling (QVC). They are calculated with the phonon wave functions obtained by self-consistent RPA or QRPA. We thus have a model for an energy-dependent self-energy, which depends only on a static covariant energy density functional and is free of any additional parameters. In the following sections we discuss the details, in particular, how to solve the corresponding Dyson equation and the consequences following from the energy dependence of the self-energy.

3. Quasiparticle-vibration coupling effects in nuclear shell structure

In the approaches based on the density functional concept, single-particle properties, such as energies and spectroscopic amplitudes, are the key ingredients for a description of nuclear masses, decay properties and response to various external fields. In turn, the latter quantities are an essential part of the nuclear physics input for astrophysical applications like r-process nucleosynthesis studies[41] which require accurate information about many nuclei including exotic ones. It has been found recently that shell structure in nuclei with extreme N/Z ratios deviates from the usual picture and magic numbers are shown to change as functions of N and Z[42] at the proton and neutron drip lines. These and the whole series of other theoretical and experimental results (see, for instance, a review of Ref. 43) have shown that habitual concepts in our understanding of nuclear shell structure need to be deeply reconsidered.

The shell structure of superheavy nuclei is another challenge for microscopic models: to define the location of spherical shell gaps in this area of the nuclear chart it is necessary to determine the regions of stability of these nuclei. Up to now, there is no consensus about the spherical shell closures above the proton number $Z = 82$ and the neutron $N = 126$: predictions such as $Z = 114$, $Z = 120$ or $Z = 126$ for the proton and $N = 172$ or $N = 184$ for the neutron magic numbers can be found in the literature.[44–46] The $Z = 120$ and $N = 172$ shell closures predicted by the relativistic and some Skyrme mean-field models are found to be related to a central depression of the nuclear density distribution.[44,47] The $Z = 120$ element represents a challenge for future experimental synthesis since it is located at the limits of accessibility with available cold fusion reactions. Therefore, accurate estimations of its characteristics are needed from the theoretical side.

The predictions made by the mean field models, however, ignore correlations which can play a significant role in the superheavy mass region where the expected spherical shell gaps are considerably smaller (2–3 MeV) than in lighter nuclei and pairing correlations of the superfluid type may not collapse at the shell closures. As a consequence, it has been found in

Refs. 23 and 24 that superheavy nuclei are very soft objects: they possess very rich spectra of low-lying collective vibrations (phonons). Therefore, correlations due to the quasiparticle-vibration coupling are then the next important mechanism having considerable influence on the shell structure.[24] Another important consequence of the QVC is that the leading-order vibrational correction to the total energy is negative, i.e. leads to a stronger binding,[33] thus such nuclei can be, in principle, not only shell-stabilized, but also vibration-stabilized.[24]

3.1. *Temporal and spatial non-localities in the single-nucleon mass operator*

Single-particle degrees of freedom in nuclei are characterized by the single-(quasi)particle energies and the occupation probabilities (or spectroscopic amplitudes) which can be determined in one-nucleon transfer or knockout reactions. In microscopic many-body models these quantities enter the well-known Lehmann expansion of the one-body Green's function on the background of the N-body system over the eigenstates of the N$\pm$1-body systems:[48]

$$G(\xi,\xi';\varepsilon) = \sum_n \frac{(\Psi(\xi))_{0n}(\Psi^\dagger(\xi'))_{n0}}{\varepsilon - (E_n^{(N+1)} - E_0^{(N)}) + i\delta} + \sum_m \frac{(\Psi^\dagger(\xi'))_{0m}(\Psi(\xi))_{m0}}{\varepsilon + (E_m^{(N-1)} - E_0^{(N)}) - i\delta}, \tag{15}$$

$$\begin{aligned}(\Psi^\dagger(\xi))_{n0} &= \langle\Phi_n^{(N+1)}|\Psi^\dagger(\xi)|\Phi_0^{(N)}\rangle,\\ (\Psi(\xi))_{m0} &= \langle\Phi_m^{(N-1)}|\Psi(\xi)|\Phi_0^{(N)}\rangle,\end{aligned} \tag{16}$$

where $\delta \to +0$, $\Phi_0^{(N)}$, $\Phi_n^{(N)}$ are the many-body wave functions of the ground and the excited state n of the N-body system, $E_0^{(N)}$, $E_n^{(N)}$ are its ground state and excited state energies, and the variable ξ includes the full set of the single-particle variables in an arbitrary representation. The numerators of Eq. (15) give the spectroscopic amplitudes of the states n. In the mean-field approximations, such as (relativistic) Hartree or Hartree–Fock (RHF), these amplitudes take the values 0 (1) for the empty (occupied) states. For the mean-field approximations including pairing correlations of the superfluid type, such as Hartree–Bogoliubov (RHB) or Hartree–Fock–Bogoliubov (RHFB) these amplitudes take fractional values between 0 and 1. The many-body wave functions are Slater determinants of the single-(quasi)particle wave functions. In the representation of these functions, the one-body Green's function of the mean-field approximation has the following form:

$$\tilde{G}^{\eta_1\eta_2}_{k_1k_2}(\varepsilon) = \delta_{k_1k_2}\delta_{\eta_1\eta_2}\tilde{G}^{\eta_1}_{k_1}(\varepsilon), \qquad \tilde{G}^{\eta_1}_{k_1}(\varepsilon) = \frac{1}{\varepsilon - \eta_1 E_{k_1} + i\eta_1\delta}, \qquad \delta \to +0. \tag{17}$$

Here and in the subsequent applications the space of the Dirac spinors $|\psi^{(\eta)}_k\rangle$ diagonalizing the RHB Hamiltonian:

$$\mathcal{H}_{RHB}|\psi^{(\eta)}_k\rangle = \eta E_k|\psi^{(\eta)}_k\rangle \tag{18}$$

is taken as the working basis. In this case $\xi = \{k, \eta\}$, where k is the full set of the single-particle quantum numbers in the spherical relativistic mean field (RMF) and $\eta = \pm 1$ denotes the upper and lower components in Bogoliubov's quasiparticle space. Comparing the Eqs. (17) and (15), one can see that the energies E_k correspond to the energy differences in the denominators of Eq. (15). The Green's function (17) and the RHB Hamiltonian are connected by the following operator equation:

$$\big(\varepsilon - \mathcal{H}_{RHB}\big)\tilde{G}(\varepsilon) = 1. \tag{19}$$

The Hamiltonian $\mathcal{H}_{RHB}$ can be written as follows:

$$\mathcal{H}_{RHB} = \begin{pmatrix} h^{\mathcal{D}} - m - \lambda & \Delta \\ -\Delta^* & -h^{\mathcal{D}*} + m + \lambda \end{pmatrix}, \tag{20}$$

where λ is the chemical potential counted from the continuum limit, m is the mass of a nucleon and Δ is the pairing gap. In this Chapter we will consider only systems with time reversal invariance where the currents vanish. In this case, the single nucleon Dirac Hamiltonian reads:

$$h^{\mathcal{D}} = \boldsymbol{\alpha}\mathbf{p} + \beta(m + \tilde{\Sigma}). \tag{21}$$

The term $\tilde{\Sigma}$ denotes the RMF self-energy

$$\tilde{\Sigma}(\boldsymbol{r}) = \sum_m \Gamma_m \phi_m(\boldsymbol{r}) \tag{22}$$

and contains, in general, the summation over all meson fields ϕ_m with the coupling vertices Γ_m calculated in the static approximation. A time-dependent dynamical part containing correlations beyond the mean-field can be included as an additional potential whose Fourier transform $\Sigma^{(e)}(\varepsilon)$ is a function of the external energy variable. In the presence of this potential, the operator equation for the one-body Green's function reads:

$$\big(\varepsilon - \mathcal{H}_{RHB} - \Sigma^{(e)}(\varepsilon)\big)G(\varepsilon) = 1. \tag{23}$$

In the representation of the Dirac spinors $|\psi_k^{(\eta)}\rangle$ all the entities of the Eq. (23) are 2×2 supermatrices in Bogoliubov's quasiparticle space:[26]

$$\sum_{\eta=\pm 1}\sum_{k}\Big((\varepsilon - \eta_1 E_{k_1})\delta_{\eta_1\eta}\delta_{k_1k} - \Sigma^{(e)\eta_1\eta}_{k_1k}(\varepsilon)\Big)G^{\eta\eta_2}_{kk_2}(\varepsilon) = \delta_{\eta_1\eta_2}\delta_{k_1k_2}, \quad (24)$$

which is indicated explicitly by the indices η, η_i. In the present work, the functional form of the self-energy $\Sigma^{(e)}$ is determined by the quasiparticle-vibration coupling which gives the leading polarization correction[1,49] to the unperturbed motion of a quasiparticle in the mean field. The QVC accounts for long-range correlations between nucleons due to their exchange by quanta of collective excitations (phonons, or vibrations). Here we follow the version of the relativistic QVC model which was developed in Ref. 24 as a generalization of Ref. 22 to superfluid systems. In the basis $|\psi_k^{(\eta)}\rangle$ which we will refer to as Dirac–Hartree–BCS (DHBCS) basis the matrix elements of $\Sigma^{(e)}$ are given by:

$$\Sigma^{(e)\eta_1\eta_2}_{k_1k_2}(\varepsilon) = \sum_{\eta=\pm 1}\sum_{\eta_\mu=\pm 1}\sum_{k,\mu}\frac{\delta_{\eta_\mu,\eta}\gamma^{\eta_\mu;\eta_1\eta}_{\mu;k_1k}\gamma^{\eta_\mu;\eta_2\eta *}_{\mu;k_2k}}{\varepsilon - \eta E_k - \eta_\mu(\Omega_\mu - i\delta)}, \qquad \delta \to +0. \quad (25)$$

The index k formally runs over all single-quasiparticle states including antiparticle states with negative energies. In the doubled quasiparticle space we can no longer distinguish occupied and unoccupied states considering that all the orbits are partially occupied. In practical calculations, it is assumed that there are no pairing correlations in the Dirac sea[50] and the orbits with negative energies are treated in the no-sea approximation, although, as it has been shown in calculations for nuclei with closed shells in Ref. 22, the numerical contribution of the diagrams with intermediate states k with negative energies is very small due to the large energy denominators in the corresponding terms of the self-energy (25).[22] The index μ in Eq. (25) labels the set of phonons taken into account, Ω_μ are their frequencies, and $\eta_\mu = \pm 1$ labels their forward and backward going components. The vertices $\gamma^{\eta_\mu;\eta_1\eta_2}_{\mu;k_1k_2}$ determine the coupling of the quasiparticles to the collective vibrational state (phonon) μ and are defined in the following subsection.

3.2. *Phonon spectrum and the Relativistic Quasiparticle Random Phase Approximation (RQRPA)*

Spectra of the vibrational states entering the Eq. (25), namely the frequencies and the coupling vertices, are defined within the framework of a linear response theory. The phonon frequencies Ω_μ are the poles and the vertices are, up to a multiplier, the residues of the nuclear response function, or the two-body Green's function (GF). Both one- and two-body GF's can be derived from the same generating functional by successive differentiation with respect to an auxiliary source field.[51] This concept was generalized in Ref. 52 for superfluid Fermi-systems by introducing an extra dimension into the time-space variable. Following Ref. 52, in the present work this degree of freedom is associated with the index η.

The linear response theory calculates the nuclear response function by solving the Bethe–Salpether equation (BSE) with a given effective interaction. In the approaches based on the covariant density functional theory (CDFT), the effective interaction is the second functional derivative of the covariant density functional or, equivalently, the first functional derivative of the relativistic mean-field self-energy $\tilde{\Sigma}$ with respect to the relativistic generalized density matrix $\mathcal{R}$:

$$\tilde{V}^{\eta_1\eta_4,\eta_2\eta_3}_{k_1k_4,k_2k_3} = \frac{\delta\tilde{\Sigma}^{\eta_4\eta_3}_{k_4k_3}}{\delta\mathcal{R}^{\eta_2\eta_1}_{k_2k_1}}, \tag{26}$$

where $\mathcal{R}^{\eta_2\eta_1}_{k_2k_1}$ is the CDFT ground state density.

The linear response theory defines small-amplitude oscillations of the density around its CDFT value; the amplitudes of these oscillations are the transition densities $\mathcal{R}_\mu$:[33]

$$\mathcal{R}(t) = \mathcal{R}_0 + \sum_\mu(\mathcal{R}_\mu e^{i\Omega_\mu t} + \text{h.c.}). \tag{27}$$

The simplest approach (26) assumes that the effective interaction does not include the QVC explicitly and, thus, the Bethe–Salpeter equation is reduced to the relativistic Quasiparticle Random Phase Approximation (RQRPA):

$$\mathcal{R}^{\eta}_{\mu;k_1k_2} = \tilde{R}^{(0)\eta}_{k_1k_2}(\Omega_\mu)\sum_{k_3k_4}\sum_{\eta'}\tilde{V}^{\eta\eta'}_{k_1k_4,k_2k_3}\mathcal{R}^{\eta'}_{\mu;k_3k_4} \tag{28}$$

where

$$\mathcal{R}^{\eta}_{\mu;k_1k_2} = \mathcal{R}^{\eta,-\eta}_{\mu;k_1k_2}, \qquad \tilde{R}^{(0)\eta}_{k_1k_2}(\omega) = \tilde{R}^{(0)\eta,-\eta}_{k_1k_2}(\omega),$$

$$\tilde{V}^{\eta\eta'}_{k_1k_4,k_2k_3} = \tilde{V}^{\eta,-\eta',-\eta,\eta'}_{k_1k_4,k_2k_3}, \tag{29}$$

which means that we cut out particle-hole channel components of the tensors in the quasiparticle space. The quantity $\tilde{R}^{(0)}$ is the uncorrelated two-quasiparticle propagator, or the mean-field response function, which is a convolution of two single-quasiparticle mean-field Green's functions (17):

$$\tilde{R}^{(0)\eta}_{k_1k_2}(\omega) = \frac{1}{\eta\omega - E_{k_1} - E_{k_2}}. \tag{30}$$

The forward and backward components of the phonon vertices entering Eq. (25) are defined as:

$$\gamma^{\eta_\mu;\eta_1\eta_2}_{\mu;k_1k_2} = \delta_{\eta_\mu,+1}\gamma^{\eta_1\eta_2}_{\mu;k_1k_2} + \delta_{\eta_\mu,-1}\gamma^{\eta_2\eta_1 *}_{\mu;k_2k_1} \tag{31}$$

and calculated from the corresponding transition densities $\mathcal{R}_\mu$ and the static effective interaction:

$$\gamma^{\eta_1\eta_2}_{\mu;k_1k_2} = \sum_{k_3k_4}\sum_{\eta_3\eta_4} \tilde{V}^{\eta_1\eta_4,\eta_2\eta_3}_{k_1k_4,k_2k_3}\mathcal{R}^{\eta_3\eta_4}_{\mu;k_3k_4}. \tag{32}$$

Ideally, in a complete theory, the transition densities and, consequently, the phonon vertices should be calculated taking into account also the additional energy-dependent residual interaction $V^{(e)}$ [see Eq. (42) below] in a self-consistent iteration procedure. However, in the conventional QVC model this is not done. This approach assumes that the QVC corrections to the phonon vertices are minor as compared to the corrections to the one-body GF's (propagators). As we will see from the numerical results, this is a very reasonable approximation. The linear response theory including QVC will be discussed in Sec. 4.1.

3.3. *Splitting of the single-particle states*

In a spherical system, the only non-vanishing off-diagonal matrix elements of the self-energy (25) are those which connect states with different main oscillator quantum numbers. In turn, these components are rather small, so that the use of a diagonal approximation is well justified. In this case, the Green's function is also diagonal and in the $\{k, \eta\}$ representation it can be written in the following simple form:[53]

$$G^{\eta}_{k}(\varepsilon) = \sum_{\nu} \frac{\tilde{S}^{\eta(\nu)}_{k}}{\varepsilon - \eta E^{(\nu)}_{k}}, \tag{33}$$

which is nothing but the Lehmann expansion (15) for the diagonal components of the one-body GF. The poles $E^{(\nu)}_k$ of this propagator are determined by the equation:

$$\varepsilon - \eta E_k - \Sigma^{(e)\eta\eta}_{kk}(\varepsilon) = 0. \tag{34}$$

Substituting the ansatz (33) to Eq. (23) and taking the derivative with respect to the variable ε, one can obtain the expression for the spectroscopic factors:

$$\tilde{S}_k^{\eta(\nu)} = \Big(1 - \frac{d\Sigma_{kk}^{(e)\eta\eta}(\varepsilon)}{d\varepsilon}\,|_{\varepsilon=E_k^{(\nu)}}\Big)^{-1}, \tag{35}$$

and $\tilde{S}_k^{+(\nu)} = \tilde{S}_k^{-(\nu)} = \tilde{S}_k^{(\nu)}$. Thus, each mean-field state with quantum numbers k and energy E_k is split over a manifold of levels $\{\nu\}$ with the same quantum numbers and energies $\lambda \pm E_k^{(\nu)}$. These levels are distributed with the spectroscopic factors $S_k^{(\nu)} = v_k^2 \tilde{S}_k^{(\nu)}$ (for the states below the Fermi energy (FE) λ) and $S_k^{(\nu)} = (1 - v_k^2)\tilde{S}_k^{(\nu)}$ (for the states above the FE), where v^2 are the Bogoliubov's occupation probabilities. It is easy to show that for the self-energy of the form (25) the following relations hold:

$$\sum_\nu \tilde{S}_k^{(\nu)} = 1, \qquad E_k = \sum_\nu E_k^{(\nu)} \tilde{S}_k^{(\nu)}, \tag{36}$$

so that the mean particle number and the mean energy of the fragments are conserved, when the self-energy part $\Sigma^{(e)}$ is added. The present approach has two further simplifications: first, the Pauli principle is violated in the self-energy (25) due to the fact that the phonons are 'made' of quasiparticles. It has been shown in Ref. 54 that the Pauli principle can be taken into account in the QPM implementing the exact commutators of the quasiparticle and the phonon operators. The corresponding corrections are found to be sizeable for the energies of the collective states, but for the single-particle structure they are very small.[55] Second, multiphonon configurations can, in principle, cause further spreading of the single-particle strength. Since the one-phonon self-energy (25) gives the leading-order contribution for spherical nuclei, the study of multiphonon configurations is left beyond the scope of this Chapter although it will be performed elsewhere.

3.4. *Shell structure of open-shell nuclei: interplay of superfluidity and QVC*

Before applying the model to unknown nuclei, benchmarking calculations have been done for nuclei which had been investigated experimentally. Tin isotopes represent a very good reference case as their single-quasiparticle energies and spectroscopic factors in the vicinity of the Fermi energy (FE) are known relatively well. The neutron and proton single-quasiparticle levels in ^{120}Sn calculated within the QVC model are shown in Figure 1. In

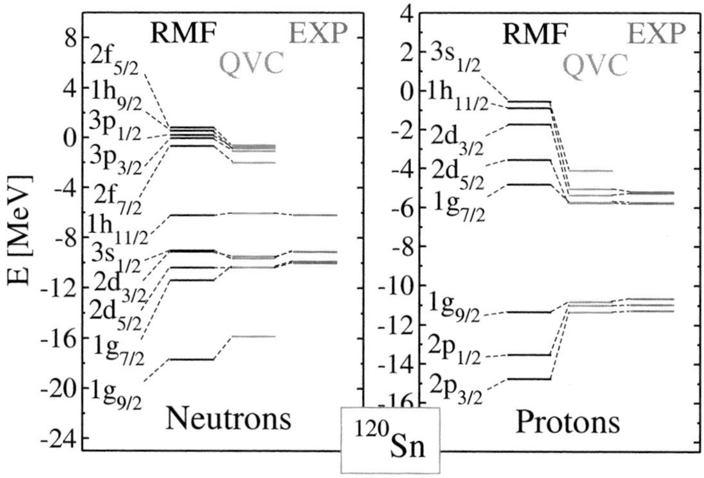

Fig. 1. (Color online) Single-quasiparticle spectrum of ^{120}Sn: RMF (left column), QVC (center) and experimental data (right). In the 'QVC' and 'EXP' cases only the dominant levels are shown. See text for explanations.

each panel, the left columns display the mean-field energies of the Bogoliubov's quasiparticles, $\tilde{E}_k = \lambda + \eta E_k$, the columns in the middle represent the dominant levels $\tilde{E}_k^{(\nu)} = \lambda + \eta E_k^{(\nu)}$ (corresponding to the largest spectroscopic strength $S_k^{(\nu)}$ for each k) obtained within the QVC model and on the right the experimentally observed dominant levels are shown. In the theoretical values $\eta = \pm 1$ are chosen for the states k whose RMF energies in the absence of the pairing field are above/below the Fermi energy, which is equal to the chemical potential λ. In the proton subsystem, where pairing correlations are absent, the quasiparticle energies take the usual non-superfluid RMF values. The calculations are based on the non-linear meson coupling model NL3*[56] for the CDFT.

In the applications to the doubly-magic nuclei,[22,23] as a rule, the QVC correlations push the dominant levels towards the FE. However, for the states, which are very close to the FE, the QVC shift effect on the dominant levels is rather weak. In the open-shell neutron subsystem of ^{120}Sn, where the FE is in the middle of the shell, there are several states of this kind. Because of the relatively high level density inside the valence shell, the leading terms in the sum of Eq. (25) may compensate each other. As a result of the interplay of the two types of correlations, QVC and pairing, the neutron states remain relatively close to their original mean-field positions, as one can see in the left panel of Figure 1. In contrast, in the closed-shell proton subsystem the level density at the FE is smaller, thus, there are weaker compensations and the shifts are considerably larger. For all dominant levels in ^{120}Sn nuclei one can find a very good agreement of the presented QVC results with the data. The obtained spectroscopic factors reproduce the available data also very well, as shown in Table 1. The

Table 1. Spectroscopic factors of the dominant neutron states in ^{120}Sn: Calculations within the QVC model compared to experimental data. The data are taken from Ref. 13.

Neutron state k	S_k^{max} QVC	S_k^{max} Exp.
$2d_{5/2}$	0.32	0.43
$1g_{7/2}$	0.40	0.60
$2d_{3/2}$	0.53	0.45
$3s_{1/2}$	0.43	0.32
$1h_{11/2}$	0.58	0.49
$2f_{7/2}$	0.31	0.35
$3p_{3/2}$	0.58	0.54

success of the response theory built on the nucleonic self-energy of Eq. (25)[26] and discussed in the subsequent sections can be traced back to the results of the Ref. 24 showing that for a proper description of nuclear shell structure and dynamics both types of correlations — pairing and quasiparticle-vibration coupling — should be taken into account self-consistently on an equal footing. At the same time, comparison to the previous calculations for doubly-magic nuclei[22,23] shows that disregarding pairing vibrations in the phonon basis can be the reason why the effect obtained there is not sufficient to reproduce well the observed single-particle level density.

3.5. *Shell structure toward the limits of nuclear mass*

The ^{292,296,300,304}120 nuclei have spherical minima of the potential energy surfaces in both Skyrme and CEDF calculations,[57] therefore, it is justified to keep spherical symmetry for their mean-field potentials. The non-linear meson-exchange interaction NL3*,[56] which is the slightly improved NL3[58] one, is known to give a very good agreement with data for various low-energy phenomena not only in medium-mass nuclei, but also in $A \simeq 250$ mass region.[59] Therefore, it is justified to use this parameter set for the superheavy systems as well as for the tin isotopes. The Bardeen–Cooper–Schrieffer model for pairing correlations as well as the HFB model lead

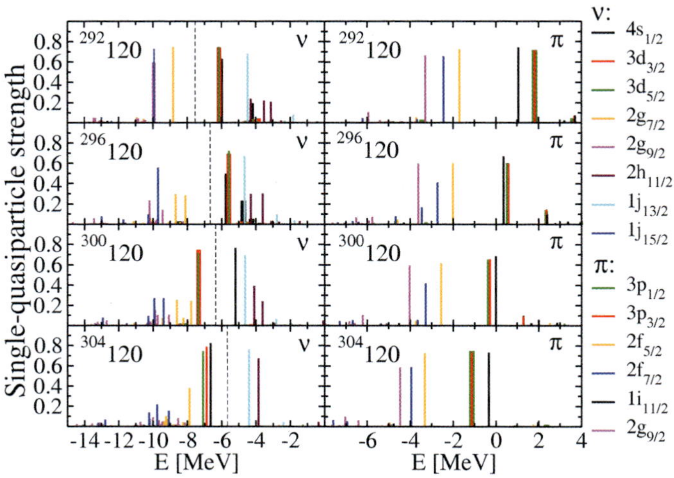

Fig. 2. Single-quasiparticle strength distribution for the orbits around the Fermi surfaces in the neutron (left) and proton (right) subsystems of the Z=120 isotopes calculated in the relativistic QVC model. The dashed lines indicate the chemical potentials of the neutron subsystems.

to a collapsing solution in the Z=120 proton subsystem, so that pairing correlations can be restored only by approximate particle number projection methods.[60] In the neutron subsystems, however, no pairing collapse is found for Z = 120 nuclei, thus, in the present work pairing has been included for the neutrons. In the superheavy mass region the shell gaps are considerably smaller than those between the previous shells. They amount to about 2–3 MeV and, therefore, are compatible with non-vanishing neutron pairing which, in turn, slightly increases the gaps. The phonon spectra calculated with the RQRPA in the chain of Z=120 isotopes show that these nuclei are very soft: many (∼100 in total) rather collective phonons with $J^\pi = 2^+, 3^-, 4^+, 5^-, 6^+$ are found below 15 MeV and included into the self-energy (25).

Selected results on the single-quasiparticle strength distributions in the neutron and the proton subsystems of the Z = 120 isotopic chain are displayed in Figure 2. The distributions for the orbits closest to the neutron and the proton FE's are given and denoted by different colors. The definitions for the single-quasiparticle energies are the same as for Fig. 1. Thus, one can see the evolution of the single-quasiparticle strength distributions with an increase of the neutron number from N = 172 to N = 184. As in the neutron subsystems both pairing and QVC mechanisms are included, and their very delicate interplay is found: pairing correlations tend to increase the shell gap while the QVC alone tends to decrease it and at the same time causes the fragmentation of the states in the middle of the shell. As a result, in the presence of both mechanisms the shell gap in the neutron subsystem remains almost steady while the newly occupied levels jump down over the gap when the neutrons are added. The shell gap in the proton subsystems of the considered nuclei diminishes only slightly when the neutron number increases, so that the proton number $Z = 120$ remains a rather stable shell closure while the detailed structure of the proton levels shows some rearrangements induced by the neutron addition.

4. Response theory with QVC and its applications

Spectra of nuclear excitations are very important for the understanding of the nuclear structure. Apart from particle-hole or few-quasiparticle excitations there are also rotational and vibrational states involving coherent motion of many nucleons. In spherical nuclei collective vibrations like giant resonances dominate in nuclear spectra.[61] They are characterized by high values of electromagnetic transition probabilities and show up in spectra of various nuclei over the entire nuclear chart.[33] The random phase approximation, first proposed in Ref. 62 to describe collective excitations in degenerate electron gas, is widely used for various kinds of correlated Fermi systems including atomic nuclei. The Quasiparticle RPA for superfluid systems has been constructed in a complete analogy to the normal case.[63–65] The effective field equations of the Theory of Finite Fermi Systems[48] developed as an extension of Landau's theory for Fermi liquid are, in fact, the QRPA equations.

The derivation of the relativistic QRPA (RQRPA) equations is a straightforward generalization of the relativistic RPA (RRPA)[35] formulated in the doubled space of Bogoliubov quasiparticles. Both RRPA and RQRPA equations are obtained as a small-amplitude limit of the time-dependent RMF model. In Ref. 36 the RQRPA equations are formulated and solved in the canonical basis of the RHB model.

The common deficiency of (R)QRPA is its inability to describe widths of giant resonances and fine structure of their low-energy regions, which was realized long ago. In medium mass and heavy nuclei, the QVC is found to be the leading mechanism of the further fragmentation of the (R)QRPA states, in addition to the Landau damping. Over the years, various implementations of the QVC for the nuclear response theory have been performed,[4,5,7–11,17–21] but only recently a fully self-consistent response theory in the covariant framework has become available.[25–28] In this Section, a brief review of this theory and some of the key applications are presented.

4.1. *Beyond the Relativistic QRPA: Bethe–Salpeter equation with QVC effects*

A response of a superfluid nucleus to a weak external field is conventionally described by the Bethe–Salpeter equation.[66] The method to derive the BSE for superfluid non-relativistic systems from a generating functional can be found, e.g., in Ref. 52, where the generalized Green's function formalism was introduced. Applying the same technique to the relativistic case, one obtains a similar ansatz for the BSE.[26] For methodological purposes, it is convenient to start in the time representation: let us, therefore, include the time variable and the variable η, which distinguishes components in the doubled quasiparticle space, into the single-quasiparticle indices, so that $1 = \{k_1, \eta_1, t_1\}$. In this notation the BSE for the response function R reads:

$$R(14,23) = G(1,3)G(4,2) - i\sum_{5678} G(1,5)G(6,2)V(58,67)R(74,83), \quad (37)$$

where the summation over the number indices 1, 2, ... implies integration over the respective time variables. The function G is the exact single-quasiparticle Green's function, and V is the amplitude of the effective interaction irreducible in the *ph*-channel. This amplitude is determined as a variational derivative of the full self-energy Σ with respect to the exact single-quasiparticle Green's function:

$$V(14,23) = i\frac{\delta\Sigma(4,3)}{\delta G(2,1)}. \quad (38)$$

Here we introduce the uncorrelated response function $R^0(14,23) = G(1,3)G(4,2)$ and formulate the Bethe–Salpeter equation (37) in shorthand notation, omitting the number indices:

$$R = R^0 - iR^0VR. \quad (39)$$

For the sake of simplicity, we will use this shorthand notation in the following discussion. Since the total self-energy can be decomposed into two parts $\Sigma = \tilde{\Sigma} + \Sigma^{(e)}$, the effective interaction V in Eq. (37) is a sum of the static RMF interaction $\tilde{V}$ and the energy-dependent term $V^{(e)}$:

$$V = \tilde{V} + V^{(e)}, \quad (40)$$

where (with $t_{12} = t_1 - t_2$)

$$\tilde{V}(14,23) = \tilde{V}^{\eta_1\eta_4,\eta_2\eta_3}_{k_1k_4,k_2k_3}\delta(t_{31})\delta(t_{21})\delta(t_{34})\,, \tag{41}$$

$$V^{(e)}(14,23) = i\frac{\delta\Sigma^{(e)}(4,3)}{\delta G(2,1)}\,, \tag{42}$$

and $\tilde{V}^{\eta_1\eta_4,\eta_2\eta_3}_{k_1k_4,k_2k_3}$ is determined by Eq. (26). In the DHBCS basis the Fourier transform of the amplitude $V^{(e)}$ has the form:

$$V^{(e)\eta_1\eta_4,\eta_2\eta_3}_{k_1k_4,k_2k_3}(\omega,\varepsilon,\varepsilon') = \sum_{\mu,\eta_\mu}\frac{\eta_\mu\gamma^{\eta_\mu;\eta_3\eta_1}_{\mu;k_3k_1}\gamma^{\eta_\mu;\eta_4\eta_2*}_{\mu;k_4k_2}}{\varepsilon-\varepsilon'+\eta_\mu(\Omega_\mu - i\delta)}\,, \qquad \delta\to +0\,. \tag{43}$$

In order to make the BSE (39) more convenient for the further analysis, we eliminate the exact Green's function G and rewrite it in terms of the mean field Green's function $\tilde{G}$ which is diagonal in the DHBCS basis. In time representation we have the following ansatz for $\tilde{G}$:

$$\tilde{G}(1,2) = -i\eta_1\delta_{k_1k_2}\delta_{\eta_1\eta_2}\theta(\eta_1\tau)e^{-i\eta_1E_{k_1}\tau}, \qquad \tau = t_1 - t_2. \tag{44}$$

Using the connection between the mean field GF $\tilde{G}$ and the exact GF G in the Nambu form

$$\tilde{G}^{-1}(1,2) = G^{-1}(1,2) + \Sigma^{(e)}(1,2), \tag{45}$$

one can eliminate the unknown exact GF G from the Eq. (39) and rewrite it as follows:

$$R = \tilde{R}^0 - i\tilde{R}^0WR, \qquad W = \tilde{V} + W^{(e)}, \tag{46}$$

with the mean-field response $\tilde{R}^0(14,23) = \tilde{G}(1,3)\tilde{G}(4,2)$ and W as a new interaction, where

$$\begin{aligned} W^{(e)}(14,23) &= V^{(e)}(14,23) + i\Sigma^{(e)}(1,3)\tilde{G}^{-1}(4,2) + \\ &+ i\tilde{G}^{-1}(1,3)\Sigma^{(e)}(4,2) - i\Sigma^{(e)}(1,3)\Sigma^{(e)}(4,2). \end{aligned} \tag{47}$$

Thus, we have obtained the BSE in terms of the mean-field propagator, containing the well-known mean-field Green's functions $\tilde{G}$, and a rather complicated effective interaction W of Eqs. (46), (47), which is also expressed through the mean-field Green's functions and static effective interaction.

The diagrammatic representation of the Eq. (46) with the interaction W is given in Figure 3. The black circle in the second term on the right hand side corresponds to the static effective interaction denoted by $\tilde{V}$ and, in the absence of the third term containing the phonon coupling effects via the amplitude $W^{(e)}$, we would have the QRPA equation. The diagrammatic

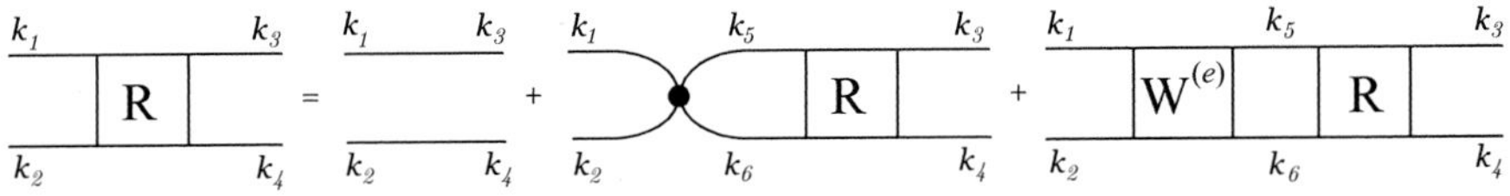

Fig. 3. The Bethe–Salpeter equation for the response function R of the many-body system in diagrammatic representation. The solid lines denote single-quasiparticle propagators. The integral part is divided into two terms, the small black circle represents the static effective interaction $\tilde{V}$ and the energy-dependent block $W^{(e)}(\omega)$ contains the dynamic contributions induced by the coupling to phonons.

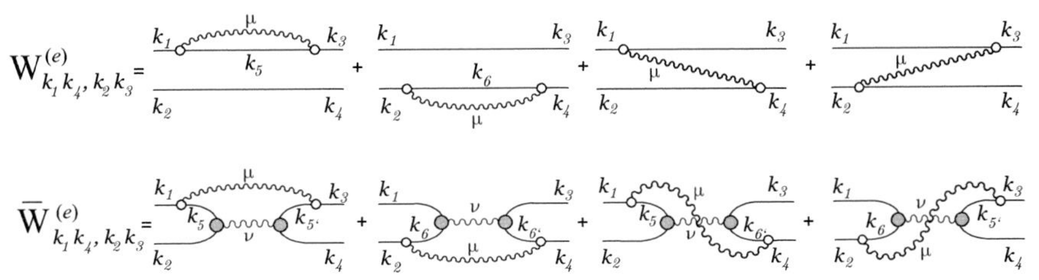

Fig. 4. The correspondence between the 2q⊗phonon amplitude $W^{(e)}$ of the conventional phonon coupling model and the two-phonon amplitude $\bar{W}^{(e)}$ of the two-phonon model in a diagrammatic representation. Solid lines and Latin indices denote the single-quasiparticle nucleonic propagators, wavy curves with Greek indices show the phonon propagators, empty circles represent phonon vertices, and grey circles with the two attached nucleonic lines denote the RQRPA transition densities (see text).

representation of the 2q⊗phonon amplitude $W^{(e)}$ is shown in the upper line of Figure 4. Notice, that the last term is omitted because it brings a second-order effect with respect to the energy-dependent self-energy $\Sigma^{(e)}$. The lower line of this figure represents the two-phonon approach and will be discussed in the next subsection.

In outward appearance these diagrammatic equations written, as in Ref. 26, for systems with pairing correlations have the same form as that for the normal (non-superfluid) system. The formal similarity of the equations for the normal and superfluid systems is achieved by the use of the representation of the basis functions $|\psi_k^{(\eta)}\rangle$ satisfying Eq. (18). This basis is a counterpart of the particle-hole basis of the conventional RPA in which the (Q)RPA equations have the simplest and most compact form. In the representation of the functions $|\psi_k^{(\eta)}\rangle$ the generalized superfluid mean-field Green's function $\tilde{G}$ (often called Gor'kov–Green's function) has a diagonal form and describes the propagation of the quasiparticle with the fixed energy. In this diagonal representation the directions of the fermion lines of

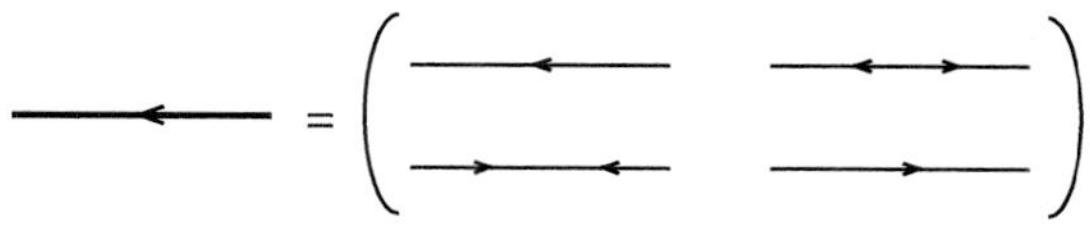

Fig. 5. The 4-component Green's function in the diagrammatic representation.

the diagrams (directions not shown in Figure 3) denote the positive- or the negative-frequency components of the functions $\tilde{G}$. It should be noted that the so-called backward-going diagrams corresponding to the ground-state correlations in the RQRPA are not shown explicitly in Figure 3 though they are included in Eq. (48).

If we come back to the coordinate representation, we get the non-diagonal Green's function $\tilde{G}$ for the quasiparticle which has no definite energy. In the diagrammatic language this Green's function can be represented by the 2×2 block matrix shown in Figure 5. The matrix elements of this matrix are the normal and anomalous Green's functions of the conventional Gor'kov theory. In this representation it is clearly seen that the difference between the normal and the superfluid systems is that in the latter case all the quantities acquire additional components in the doubled quasiparticle space. Obviously, all these components are incorporated in the representation of the functions $|\psi_k^{(\eta)}\rangle$ in the implicit form. On the other hand, the use of the basis $\{|\psi_k^{(\eta)}\rangle\}$ allows us to reduce by a factor of 2 the dimension of the system of the equations for the response function. This property of the diagonal representation of the superfluid mean-field Green's functions has been utilized in Refs. 20, 26, 52, although not discussed in detail.

Then, we apply the quasiparticle time blocking approximation (QTBA) to Eq. (46) employing the time projection operator in the integral part of this equation.[52] The time projection leads, after some algebra and the transformation to the energy domain, to an algebraic equation for the response function. For the *ph*-components of the response function it has the form:

$$R^{\eta\eta'}_{k_1k_4,k_2k_3}(\omega) = \tilde{R}^{(0)\eta}_{k_1k_2}(\omega)\delta_{k_1k_3}\delta_{k_2k_4}\delta_{\eta\eta'} + \tilde{R}^{(0)\eta}_{k_1k_2}(\omega)\sum_{k_5k_6\eta''}\tilde{W}^{\eta\eta''}_{k_1k_6,k_2k_5}(\omega)R^{\eta''\eta'}_{k_5k_4,k_6k_3}(\omega), \quad (48)$$

where we denote *ph*-components as $R^{\eta\eta'}_{k_1k_4,k_2k_3}(\omega) = R^{\eta,-\eta',-\eta,\eta'}_{k_1k_4,k_2k_3}(\omega)$, and

$$\tilde{W}^{\eta\eta'}_{k_1k_4,k_2k_3}(\omega) = \tilde{V}^{\eta\eta'}_{k_1k_4,k_2k_3} + \Big(\Phi^{\eta}_{k_1k_4,k_2k_3}(\omega) - \Phi^{\eta}_{k_1k_4,k_2k_3}(0)\Big)\delta_{\eta\eta'}, \quad (49)$$

where we introduce the tilde sign to distinguish $\tilde{W}$ in the time blocking approximation from the "exact" W of Eqs. (46), (47). In Eq. (49) $\Phi(\omega)$ is the dynamical part of the effective interaction, which is responsible for the QVC effects in the BSE and has the following $\eta = \pm 1$ components:

$$\Phi^{\eta}_{k_1k_4,k_2k_3}(\omega)$$
$$= \sum_{\mu\xi} \delta_{\eta\xi} \Big[\delta_{k_1k_3} \sum_{k_6} \frac{\gamma^{\eta;-\xi}_{\mu;k_6k_2} \gamma^{\eta;-\xi*}_{\mu;k_6k_4}}{\eta\omega - E_{k_1} - E_{k_6} - \Omega_\mu} + \delta_{k_2k_4} \sum_{k_5} \frac{\gamma^{\eta;\xi}_{\mu;k_1k_5} \gamma^{\eta;\xi*}_{\mu;k_3k_5}}{\eta\omega - E_{k_5} - E_{k_2} - \Omega_\mu}$$
$$- \Big(\frac{\gamma^{\eta;\xi}_{\mu;k_1k_3} \gamma^{\eta;-\xi*}_{\mu;k_2k_4}}{\eta\omega - E_{k_3} - E_{k_2} - \Omega_\mu} + \frac{\gamma^{\eta;\xi*}_{\mu;k_3k_1} \gamma^{\eta;-\xi}_{\mu;k_4k_2}}{\eta\omega - E_{k_1} - E_{k_4} - \Omega_\mu} \Big) \Big], \tag{50}$$

where we denote $\gamma^{\eta;\xi}_{\mu;k_1k_2} = \gamma^{\eta;\xi\xi}_{\mu;k_1k_2}$. Indices k_i in Eq. (50) formally run over the whole DHBCS space, but in applications we usually consider that the amplitude $\Phi^{\eta}_{k_1k_4,k_2k_3}(\omega)$ describes phonon coupling only within some energy window (typically 30–50 MeV) around the Fermi surface. This implies, in particular, that this amplitude contains no antiparticle-quasiparticle configurations. Another simplification consists of cutting out only the components without ground state correlations induced by phonon coupling in Eq. (50). Thus, it includes the leading (resonant) contribution of the phonon coupling and neglects the so-called associated terms contained in the QTBA.[52] They can be, however, included straightforwardly. The ground state correlations of the QRPA type are fully taken into account due to the presence of the $\tilde{V}^{\eta\eta'}_{k_1k_4,k_2k_3}$ terms with $\eta' \neq \eta$ in the static interaction in the Eq. (49). By construction, the propagator $R(\omega)$ in Eq. (48) contains only configurations which are not more complicated than 2q$\otimes$phonon, however, they can be taken into account by an iterative procedure.

The subtraction term $\Phi(0)$ in Eq. (49) is introduced to eliminate double counting effects. Since the parameters of the density functional and, as a consequence, the effective interaction $\tilde{V}$ are adjusted to experimental ground state properties at the energy $\omega = 0$, the part of the QVC interaction, which is already contained in $\tilde{V}$ and given approximately by $\Phi(0)$, should be subtracted to avoid double counting of the QVC.[52]

Eventually, to describe the observed spectrum of an excited nucleus in a weak external field P as, for instance, an electromagnetic field, one needs to calculate the strength function:

$$S(E) = -\frac{1}{2\pi} \lim_{\Delta\to+0} Im \sum_{k_1k_2k_3k_4} \sum_{\eta\eta'} P^{\eta*}_{k_1k_2} R^{\eta\eta'}_{k_1k_4,k_2k_3}(E + i\Delta) P^{\eta'}_{k_3k_4}. \tag{51}$$

The imaginary part Δ of the energy variable has the meaning of an additional artificial width for each excitation and emulates effectively contributions from configurations which are not taken into account explicitly in our approach.

4.2. *Relativistic two-phonon model*

In order to go a step further, notice, that in the time blocking approximation the energy-dependent resonant part of the two-quasiparticle amplitude $\Phi(\omega)$ can be factorized in a special way.[52] Namely, the two-quasiparticle intermediate propagator, appearing as the two uncorrelated quasiparticle lines between the phonon emission and absorption vertices in the upper part of the Figure 4, can be extracted. Thus, in the relativistic QTBA the amplitude $\Phi(\omega)$ takes the following form:

$$\Phi^{\eta}_{k_1k_4,k_2k_3}(\omega) = \sum_{k_5k_6,\mu} \zeta^{\mu\eta}_{k_1k_2;k_5k_6}\, \tilde{R}^{(0)\eta}_{k_5k_6}(\omega - \eta\,\Omega_\mu)\, \zeta^{\mu\eta *}_{k_3k_4;k_5k_6}\,, \tag{52}$$

where $\tilde{R}^{(0)\eta}_{k_5k_6}(\omega - \eta\,\Omega_\mu)$ are the matrix elements of the two-quasiparticle propagator in the mean field with the frequency shifted forward or backward by the phonon energy Ω_μ,

$$\begin{aligned}\zeta^{\mu(+)}_{k_1k_2;k_5k_6} &= \delta_{k_1k_5}\,\gamma^{(-)}_{\mu;k_6k_2} - \gamma^{(+)}_{\mu;k_1k_5}\,\delta_{k_6k_2}\,,\\ \zeta^{\mu(-)}_{k_1k_2;k_5k_6} &= \delta_{k_5k_1}\,\gamma^{(+)*}_{\mu;k_2k_6} - \gamma^{(-)*}_{\mu;k_5k_1}\,\delta_{k_2k_6}\,,\end{aligned} \tag{53}$$

so that

$$\zeta^{\mu(-)}_{k_1k_2;k_5k_6} = -\zeta^{\mu(+)*}_{k_2k_1;k_6k_5}\,. \tag{54}$$

In Eq. (53) and below we use a shorthand notation for the phonon amplitudes:

$$\gamma^{\eta}_{\mu;k_1k_2} = \gamma^{\eta_1\eta_2}_{\mu;k_1k_2}\,\delta_{\eta\eta_1}\,\delta_{\eta\eta_2}\,. \qquad \eta = (\pm) \tag{55}$$

In the graphic expression of the amplitude (52) in the upper line of Figure 4 the uncorrelated propagator $\tilde{R}^{(0)\eta}_{k_1k_2}$ is represented by the two straight nucleonic lines between the circles denoting emission and absorption of the phonon by a single quasiparticle with amplitudes $\gamma^{\eta_1\eta_2}_{\mu;k_1k_2}$. The approach to the amplitude $\Phi(\omega)$ expressed by the Eq. (52) represents a version of first-order perturbation theory compared to RQRPA and the amplitude $\Phi(\omega)$ of Eq. (49) is the first-order correction to the effective interaction $\tilde{V}$, because the dimensionless matrix elements of the phonon vertices are such that $\gamma^{\eta_1\eta_2}_{\mu;k_1k_2}/\Omega_\mu \ll 1$ in most of the physical cases. The phonon-coupling term Φ generates fragmentation of nuclear excited states. For giant resonances this fragmentation is the source of the spreading width and in the low-energy region below the neutron threshold this term is responsible almost solely for the appearing strength. In the relativistic framework, this was confirmed and extensively studied[26,67] and verified by comparison to experimental data.[68–70] However, a comparison with high-resolution experiments on the dipole strength below the neutron threshold has revealed that although the total strength and some gross features of the strength are reproduced well, the fine features are sensitive to the truncation of the configuration space by 2q⊗phonon configurations and further extensions of the method should be developed. The first possible extension of this model uses the idea proposed in Ref. 52. It is based on the factorization of Eq. (52): the uncorrelated propagator $\tilde{R}^{(0)\eta}$ in Eq. (52) is replaced by the positive- ($\eta = +1$) or the negative- ($\eta = -1$) frequency part of a correlated one which, in first order approximation, is the antisymmetrized RQRPA propagator $R^{(RQRPA)\eta}$. Instead of the amplitude Φ of Eq. (52), we have the new amplitude $\bar{\Phi}$:

$$\bar{\Phi}^{\eta}_{k_1k_4,k_2k_3}(\omega) = \frac{1}{2} \sum_{k_5k_6,k_{5'}k_{6'}\mu} \zeta^{\mu\eta}_{k_1k_2;k_5k_6} \times R^{(RQRPA)\eta}_{k_5k_{6'},k_6k_{5'}}(\omega - \eta\,\Omega_\mu)\,\zeta^{\mu\eta*}_{k_3k_4;k_{5'}k_{6'}}, \tag{56}$$

containing the intermediate propagator $R^{(RQRPA)\eta}_{k_1k_4,k_2k_3}$ which is non-diagonal in the two-quasiparticle basis. By this substitution, we introduce RQRPA correlations into the intermediate two-quasiparticle propagators, i.e., in diagrammatic language, we perform the operation shown in Figure 6. In the original RQTBA neither ladder nor bubble diagrams were summed up in the intermediate two-quasiparticle (2q) states entering 2q⊗phonon configurations, therefore, the RQTBA amplitude of Eq. (52) contains the unperturbed mean-field propagator $\tilde{R}^{(0)}$. In other words, these 2q states are treated within the independent quasiparticle model in RQTBA. As it can be seen from Figure 4, in the new version of the model the two-phonon configurations are included instead of the 2q⊗phonon ones, but not in addition to them. This ensures that there is no double counting.

The factor 1/2 in Eq. (56) appears due to the antisymmetrization, as was explained in Ref. 52. In particular, this implies that the following relations are fulfilled

$$R^{(RQRPA)\eta}_{k_1k_4,k_2k_3}(\omega) = -R^{(RQRPA)\eta}_{k_2k_4,k_1k_3}(\omega) = -R^{(RQRPA)\eta}_{k_1k_3,k_2k_4}(\omega), \tag{57}$$

which are not valid for the uncorrelated propagator $\tilde{R}^{(0)\eta}$. Thus, two-

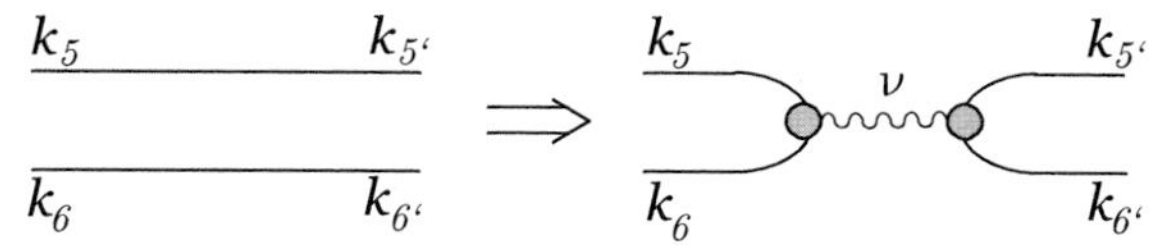

Fig. 6. Replacement of the uncorrelated two-nucleon propagator by the correlated one. Grey circles with the two attached nucleonic lines and wavy curve denote the RQRPA transition densities and the phonon propagator, respectively.

phonon configurations appear in the amplitude $\Phi(\omega)$ as it is clear from the diagrammatic representation of the new amplitude $\bar{\Phi}$ shown in the bottom line of Figure 4. The analytic expression of this amplitude reads:

$$\bar{\Phi}^{\eta}_{k_1k_4,k_2k_3}(\omega) = \frac{1}{2}\sum_{\mu,\nu} \frac{\bar{\zeta}^{\eta}_{\mu\nu;k_1k_2}\,\bar{\zeta}^{\eta*}_{\mu\nu;k_3k_4}}{\eta\,\omega - \Omega_\mu - \Omega_\nu}\,, \tag{58}$$

where

$$\begin{aligned}\bar{\zeta}^{(+)}_{\mu\nu;k_1k_2} &= \sum_{k_6} \mathcal{R}^{(+)}_{\nu;k_1k_6}\,\gamma^{(-)}_{\mu;k_6k_2} - \sum_{k_5} \gamma^{(+)}_{\mu;k_1k_5}\,\mathcal{R}^{(+)}_{\nu;k_5k_2}\\ \bar{\zeta}^{(-)}_{\mu\nu;k_1k_2} &= \bar{\zeta}^{(+)*}_{\mu\nu;k_2k_1}\,,\end{aligned} \tag{59}$$

and $\mathcal{R}^{\eta}_{\nu;k_1k_2}$ are the matrix elements of the RQRPA transition densities defined in Eq. (28). They correspond to grey circles together with two nucleonic lines in Figure 4. One can show that in the limit of vanishing static interaction $\tilde{V}$ between the two intermediate quasiparticles Eq. (58) transforms to the antisymmetrized Eq. (52) of the original (R)QTBA.

As in conventional (R)QTBA, the elimination of double counting effects in the phonon coupling is performed by the subtraction of the static contribution of the amplitude $\bar{\Phi}$ from the residual interaction in Eq. (49), since the parameters of the CEDF have been adjusted to experimental data for ground states and, therefore, include already essential phonon contributions to the ground state. Thus, the BSE in the two-phonon model has the same form as Eq. (48), but it contains the amplitude $\bar{\Phi}$ instead of Φ.

In addition to the elimination of double counting effects, the subtraction of the phonon coupling amplitude at zero frequency $\Phi(0)$ in Eq. (49) acquires another important role for the excitations which have an isoscalar dipole component, for example, the electromagnetic or isoscalar dipole response which is considered as an application of our method in Section 4.3. For such modes of excitation, special care should be undertaken to exclude an admixture of the spurious state, appearing because of broken translation symmetry, to the physical dipole states. In our approach, on the RQRPA level the elimination of the 1^- spurious state is achieved by the use of sufficiently large 2q configuration space within a fully self-consistent approach, similar to Ref. 36. In the RQTBA and RQTBA-2 based on the self-consistent RQRPA the zero energy of the spurious state is ensured by the subtraction of the static contributions of the complex configurations (see Eqs. (11), (22), and (25)): at $\omega = 0$ the phonon coupling amplitude $W(\omega)$ of Eq. (49) vanishes and the BSE (48) reduces to the RQRPA equation, the solutions of which, theoretically, do not contain admixtures of the spurious state as this state is at zero energy. In the numerical implementation, due to numerical inaccuracies, this state appears at a finite energy below 1 MeV already in RQRPA, but due to the subtraction procedure, in the extended theories such as RQTBA and RQTBA-2 the accuracy of elimination of the spurious state is preserved. A detailed description of the subtraction procedure which, in addition, guarantees stability of solutions of the extended RPA theories, is presented in Ref. 71.

The calculations are performed in the following four steps: (i) The RHB equation (18) is solved to determine the single-quasiparticle energies and wave functions. These wave functions serve as a working basis for the subsequent calculations. (ii) The phonon frequencies, their coupling vertices γ^η and the transition densities $\mathcal{R}^\eta$ are calculated within the self-consistent RQRPA using the static residual interaction $\tilde{V}$. (iii) The BSE for the correlated propagator $R^{(e)}(\omega)$

$$R^{(e)\eta}_{k_1k_4,k_2k_3}(\omega) = \tilde{R}^{(0)\eta}_{k_1k_2}(\omega)\delta_{k_1k_3}\delta_{k_2k_4} + \tilde{R}^{(0)\eta}_{k_1k_2}(\omega)\sum_{k_5k_6}\big[\bar{\Phi}^{\eta}_{k_1k_6,k_2k_5}(\omega) - \bar{\Phi}^{\eta}_{k_1k_6,k_2k_5}(0)\big]R^{(e)\eta}_{k_5k_4,k_6k_3}(\omega), \tag{60}$$

is solved in the Dirac–Hartree–BCS basis; (iv) the BSE for the full response function $R(\omega)$

$$R^{\eta\eta'}_{k_1k_4,k_2k_3}(\omega) = R^{(e)\eta}_{k_1k_4,k_2k_3}(\omega)\delta^{\eta\eta'} + \sum_{k_5k_6k_7k_8} R^{(e)\eta}_{k_1k_6,k_2k_5}(\omega)\sum_{\eta''}\tilde{V}^{\eta\eta''}_{k_5k_8,k_6k_7}R^{\eta''\eta'}_{k_7k_4,k_8k_3}(\omega) \tag{61}$$

where

$$\begin{aligned} R^{\eta\eta'}_{k_1k_4,k_2k_3}(\omega) &= R^{\eta,-\eta',-\eta,\eta'}_{k_1k_4,k_2k_3}(\omega),\\ \tilde{V}^{\eta\eta'}_{k_1k_4,k_2k_3} &= \tilde{V}^{\eta,-\eta',-\eta,\eta'}_{k_1k_4,k_2k_3}, \end{aligned} \tag{62}$$

is solved in the momentum-channel representation which is especially convenient because of the structure of the one-boson exchange interaction. Further details on the relativistic two-phonon model are given in Refs. 27, 28.

4.3. *Damping of giant resonances and soft modes in neutral channel*

The isovector giant dipole resonance (GDR) is the most prominent nuclear high-frequency oscillation mode, which appears as a broad peak in spectra of nuclear excitations. The formation of the width of the giant dipole resonance is one of the most complex phenomena in nuclear structure physics. It involves numerous many-body mechanisms and its theoretical description requires a response theory beyond QRPA. Formally, to describe the GDR, one has to calculate the strength function of Eq. (51) as a response to an electromagnetic dipole operator, which in the long wavelength limit reads:

$$P_{1M}^{EM} = \frac{N}{A}\sum_{p=1}^{Z} r_p Y_{1M}(\Omega_p) - \frac{Z}{A}\sum_{n=1}^{N} r_n Y_{1M}(\Omega_n). \tag{63}$$

The cross section of the total dipole photoabsorption is given by:

$$\sigma_{E1} = \frac{16\pi^3 e^2}{9\hbar c} E S(E). \tag{64}$$

Figure 7 shows the cross sections of the total dipole photoabsorption in four medium-mass spherical nuclei obtained within the RQRPA (black dashed curves) and RQTBA (red solid curves), compared to neutron data (blue error bars) from Ref. 72. The details of these calculations are described in Ref. 26. One can clearly see that the QVC included within the RQTBA provides a sizable fragmentation of the GDR. The QVC mechanism of the GDR width formation has been known for decades, see Refs. 4, 5, 19 and references therein. However, the RQTBA is the first fully self-consistent approach which, in contrast to the previously developed ones, accurately reproduces the Lorentzian-like GDR distribution observed in experiments.

In Ref. 68, 69 the RQTBA has been applied to analyse the observed low-lying dipole spectra which are associated with the so-called pygmy dipole resonance (PDR) in ^{124}Sn nucleus. Nuclear response to the two various experimental probes, $(\alpha, \alpha'\gamma)$ and (γ, γ'), has been compared in order to determine the one which is more selective to excite the neutron skin oscillation. The RQTBA calculations were done assuming the response to the electromagnetic and the isoscalar dipole operators which are given by the

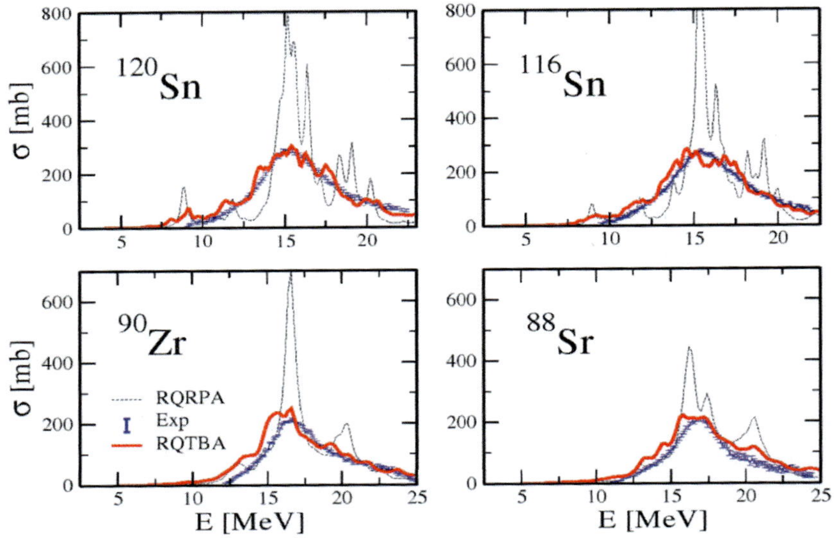

Fig. 7. Total dipole photoabsorption cross section in stable medium-mass nuclei, see text for explanation.

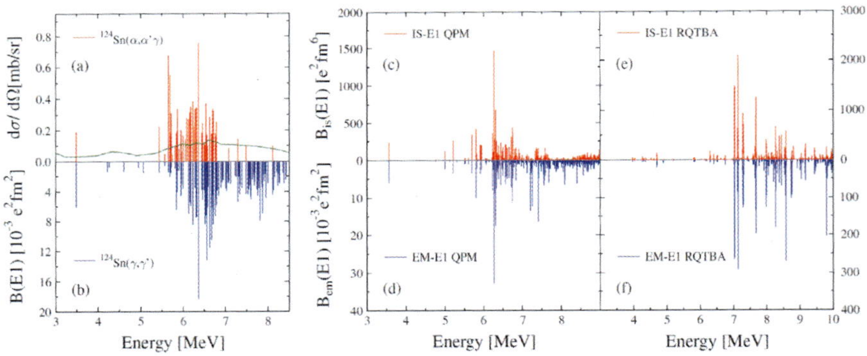

Fig. 8. (a) Singles cross section for the excitation of the $J^\pi = 1^-$ states in ^{124}Sn obtained in the $(\alpha, \alpha'\gamma)$ coincidence experiment.[68,69] The solid green line shows the energy-dependent experimental sensitivity limit. (b) $B(E1) \uparrow$ strength distribution measured with the (γ, γ') reaction. The middle column shows the QPM transition probabilities in ^{124}Sn for the isoscalar (c) and electromagnetic (d) dipole operators. The RQTBA strength functions in ^{124}Sn for the isoscalar and electromagnetic dipole operators are shown in (e) and (f), respectively.

Eq. (63) and the following expression, respectively:

$$P_{1M}^{IS} = \sum_{i=1}^{A} (r_i^3 - \eta r_i) Y_{1M}(\Omega_i) \ . \tag{65}$$

The factor $\eta = \frac{5}{3}\langle r^2 \rangle$ in Eq. (65) (see Ref. 73) and the effective charges in Eq. (63) are used to remove the spurious center-of-mass motion. The electromagnetic dipole operator of Eq. (63) corresponds to the (γ, γ') case and the isoscalar dipole operator of Eq. (65) is related to the case of the $(\alpha, \alpha'\gamma)$ scattering at small angles.

The results are presented in Figure 8. In general, the RQTBA calculation shows a good agreement with the experimental electromagnetic strength which is strongly fragmented into two pronounced peaks at about 7 MeV and 7.5 MeV. Furthermore, the calculated isoscalar strength is suppressed with increasing energy which is also in qualitative agreement with experiment, although the RQTBA strength is shifted by about 600–700 keV towards higher energies. However, the suppression of isoscalar dipole strength at higher energies is clearly visible. In order to investigate the structural difference of lower-lying and higher-lying states, the transition densities are considered. The two RQTBA states at 7.133 MeV and 8.580 MeV have similar $B_{em}(E1)$ values but differ in the $B_{is}(E1)$ case by a factor of four. The corresponding transition densities are depicted in Figure 9.

For the 7.133 MeV state, the neutron and proton contributions are in phase with a dominant neutron contribution at the nuclear surface as expected for a typical state of the PDR. This is compatible with macroscopic interpretation of the PDR as a neutron-skin oscillation against a proton-neutron core. On the other hand, a relative enhancement of the proton contribution in the nuclear medium and a reduction of the neutron component at the surface is visible for the state at 8.580 MeV. Furthermore, the neutron and proton distributions are slightly out of phase in this case, which is expected for a transitional region on the tail of the GDR. This analysis shows that the energetically higher-lying $J^\pi = 1^-$ states contain a contribution from the GDR and are not strongly excited by the α particles. In this way, the observed splitting can be explained by a structural difference of the discussed states while the lower-lying group of states represent the more isoscalar neutron-skin oscillation which is interpreted as the PDR. Thus, the comparison of (γ, γ') and α-scattering experiments supplemented by the RQTBA analysis of the transition densities gave the first justification

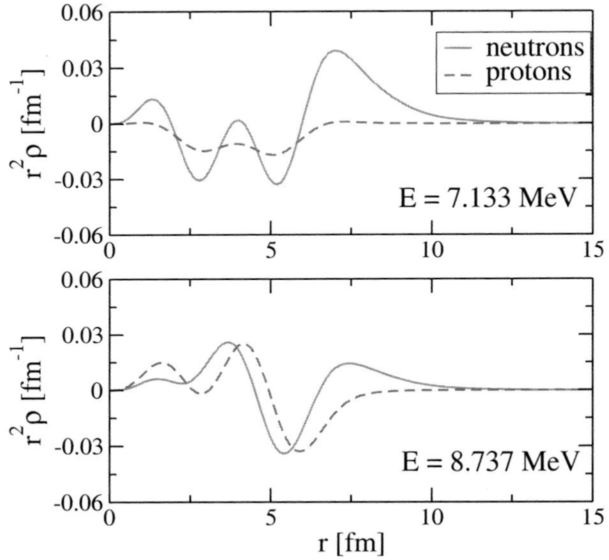

Fig. 9. (Color online) Transition densities for two dipole states in ^{124}Sn calculated in RQTBA at (upper) 7.133 MeV and (lower) 8.580 MeV.[69]

for distinguishing between isoscalar and isovector low-lying dipole states in atomic nuclei.[68,69]

Figure 10 illustrates the effect of two-phonon correlations on spectra of nuclear excitations. It displays the dipole strength functions for ^{68}Ni calculated within the conventional RQTBA and the two-phonon RQTBA-2. The resulting strength functions are compared with the RQRPA strength function because both of them originate from the RQRPA by a similar fragmentation mechanism. Overall, one can see that the GDR in RQTBA-2 is less fragmented than in RQTBA although the central peak remains at the same position. The total energy-weighted sum rule (EWSR) below 30 MeV is found almost unchanged as compared to RQRPA ($\sim$142%), while RQTBA exhausts only 134%, because the stronger fragmentation pushes some strength to higher energy region. The weaker fragmentation in the RQTBA-2 is expected because of geometrical factors in front of each term of $\bar{\Phi}$ amplitude in its expression in the coupled form.[28] These factors are products of two 6j-symbols and, therefore, introduce some smallness as compared to the amplitude Φ of RQTBA taken also in the coupled form.[26] The results shown in Figure 10 confirm this expectation.

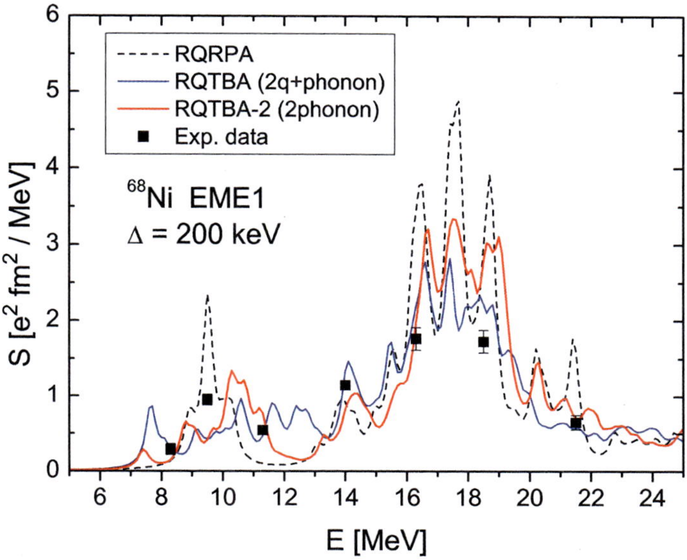

Fig. 10. Electromagnetic dipole strength distribution for ^{68}Ni calculated within the RQRPA (dashed curve), RQTBA (blue solid curve) and RQTBA-2 (red solid curve). A finite smearing parameter $\Delta = 200$ keV has been used in the calculations. The black squares show the experimental data from Ref. 74.

From Eq. (58) one can see that the amplitude $\bar{\Phi}(\omega)$ consists of the pole terms with the poles at the energies which are sums of the two phonon energies. This means that the poles of the response function R originate from these two-phonon poles although they do not coincide but are rather shifted by the static residual interaction $\tilde{V}$ with respect to these sums. Because of this shift, it is not possible to identify pairs of phonons which form particular states in the spectrum of the strength function $S(\omega)$. However, it is clear that the lowest state in the spectrum should be associated with a pair of phonons having the lowest frequencies and compatible spins and parities. Thus, the lowest state in dipole spectrum is formed by 3_1^- and 2_1^+ phonons (lower index 1 indicates the lowest vibrational modes). Though an observation of the whole $2_1^+ \otimes 3_1^-$ quintuplet is a difficult experimental task, the dipole member of this quintuplet is well-known in vibrational nuclei since it appears with relatively high transition probability.

Figure 11 shows the results obtained for the energies and reduced transition probabilities B(E1)↑ of the lowest 1^- state in the chain of tin isotopes ^{112,116,120,124}Sn within the relativistic two-phonon model. The obtained

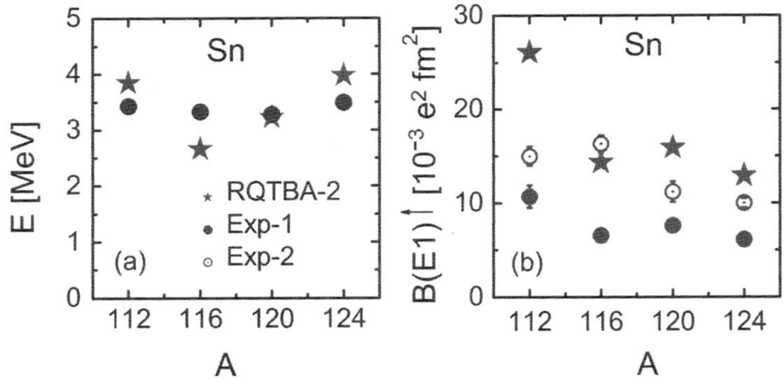

Fig. 11. (Color online) Energies and B(E1)↑ values of the first 1^- states in the chain of tin isotopes ^{112,116,120,124}Sn obtained within the relativistic two-phonon model (stars), compared to data from Refs. 75,76 (filled circles), and[77–79] (open circles).

results are compared to the two sets of experimental data on the photon scattering: Refs. 75, 76 (these data are shown by filled circles) and Refs. 77–79 (shown by open circles). The latter measurements with larger end point energies for the electron bremsstrahlung result in larger $B(E1)\uparrow$ values and agree better with our calculations.

In Ref. 28 the results of calculations within the RQTBA-2 are compared also to those obtained in Quasiparticle-Phonon Model of Ref. 4. The QPM shows a similar level of agreement to data, although they are obtained within a semi-phenomenological scheme with parameters of multipole-multipole forces adjusted to particular states in nuclear spectra. While the RQTBA-2 results for the 1_1^- states in ^{112,116,120,124}Sn are in a better agreement to the data obtained with larger end point energies, the QPM results rather support another data set. For ^{112,124}Sn, however, the QPM transition probabilities are too small.

As mentioned above, in RQTBA-2 the position of the first 1^- state is basically defined by the sum of the energies of the lowest 2^+ and 3^- phonons with some relatively small negative correction introduced by the static residual interaction $\tilde{V}$. Similar conclusions follow from the analysis of the experimentally observed energies of the lowest 2^+, 3^- and 1^- states, see a discussion in Ref. 76. In particular, in ^{120}Sn the energies of the 2_1^+ and 3_1^- phonons calculated within the RQRPA are obtained at 1.48 MeV and 1.90 MeV, respectively, and the 1_1^- state appears at 3.23 MeV in the two-phonon approach. Thus, the quality of description of the first 1^- two-phonon state is mainly determined by the quality of description provided by the RQRPA for the 2_1^+ and 3_1^- phonons, namely, their energies and coupling vertices. In the cases of vibrational nuclei, these quantities are reasonably well described by the RQRPA, however, the description could be further improved by inclusion of correction beyond RQRPA. A relatively small anharmonicity allows an identification of the first experimentally observed 1^- state as a member of the $2_1^+ \otimes 3_1^-$ quintuplet. Theoretically, we have shown that this state appears solely due to the inclusion of the two-phonon correlations and does not appear in the spectra calculated within the conventional RQTBA, although 2q$\otimes$phonon "prototypes" of this two-phonon state are present in the RQTBA model space at higher energies. One can see that for the lowest 1^- state in the considered chain of even-even tin isotopes the obtained agreement of the RQTBA-2 results with the available data is very good in spite of the fact that this tiny structure at about 3 MeV originates by the splitting-out from the very strong RQRPA pygmy mode located at the neutron threshold, due to the two-phonon correlations included consistently without any adjustment procedures. The physical content of the two-phonon RQTBA reminds the two-phonon version of the quasiparticle-phonon model,[4] however, one-to-one correspondence has not been established. Also, the obtained differences between the RQTBA and RQTBA-2 results may occur because of their limitations in terms of the configuration space. Both 2q$\otimes$phonon and phonon$\otimes$phonon configurations are limited by only four quasiparticles and, perhaps, on the higher level of the configuration complexity involving six and more quasiparticles the differences between the coupling schemes will be less pronounced. This is supposed to be clarified in the future studies.

4.4. *Spin-isospin response: damping effects and quenching*

The spin-isospin response associated with finite spin and isospin transfer is one of the most important properties of nuclei. This type of response provides information about a variety of weak interaction processes such as ordinary and double beta-decay, electron capture, neutrino capture and scattering on nuclei and in stars. Lately the models which are commonly used to describe nuclear spin-isospin response, such as Quasiparticle Random Phase Approximation and Shell Model, have advanced considerably. However, a self-consistent model which can simultaneously reproduce data on the overall strength distribution up to high excitation energy, quenching and on the fine structure of the low-lying strength is still a challenge.

In this section, we describe the spin-isospin version of the relativistic time blocking approximation (RTBA) as a model which can pretend to account for all these effects. The spin-isospin RTBA (or proton-neutron RTBA, pn-RTBA) was developed and applied for the first time to the spin-dipole resonance (SDR) in Ref. 80 and, recently, to the Gamow–Teller resonance (GTR) in Ref. 81.

The RTBA for the spin-isospin channel is constructed in a very similar way as the R(Q)TBA for the neutral channel described above. The only limitation so far is that the spin-isospin RTBA does not include pairing correlations, but such a generalization is straightforward and well underway. The calculations presented below are performed in the following three steps: (i) a relativistic mean field (RMF) solution is obtained by minimization of the covariant density functional (here we use the NL3 parametrization[58]), (ii) phonon spectrum and coupling vertices for the phonons with $J^\pi = 2^+, 3^-, 4^+, 5^-, 6^+$ are obtained by the self-consistent relativistic RPA (RRPA) solutions[35] and (iii) the Bethe–Salpeter equation is solved for the proton-neutron response function:

$$R(\omega) = \tilde{R}^0(\omega) + \tilde{R}^0(\omega)W(\omega)R(\omega), \tag{66}$$

where $\tilde{R}^0(\omega)$ is the propagator of two uncorrelated proton and neutron quasiparticles in the static mean field and the second integral part contains the in-medium nucleon-nucleon interaction $W(\omega)$. The active channels of the two-body interaction $W(\omega)$ allowing for spin-flip and isospin-flip include the following static terms and of the terms depending on the frequency ω:

$$W(\omega) = V_\rho + V_\pi + V_{\delta\pi} + \Phi(\omega) - \Phi(0). \tag{67}$$

V_ρ and V_π are the finite-range ρ-meson and the π-meson exchange interactions, respectively. They are derived from the covariant energy density functional and read:[82]

$$\begin{aligned} V_\rho(1,2) &= g_\rho^2 \boldsymbol{\tau}_1\boldsymbol{\tau}_2(\beta\gamma^\mu)_1(\beta\gamma_\mu)_2 D_\rho(\mathbf{r}_1,\mathbf{r}_2) \\ V_\pi(1,2) &= -\left(\frac{f_\pi}{m_\pi}\right)^2 \boldsymbol{\tau}_1\boldsymbol{\tau}_2(\boldsymbol{\Sigma}_1\nabla_1)(\boldsymbol{\Sigma}_2\nabla_2)D_\pi(\mathbf{r}_1,\mathbf{r}_2), \end{aligned} \tag{68}$$

where g_ρ and f_π are the coupling strengths, D_ρ and D_π are the meson propagators and $\boldsymbol{\Sigma}$ is the generalized Pauli matrix.[82] The Landau–Migdal term $V_{\delta\pi}$ is the contact part of the nucleon-nucleon interaction responsible for the short-range repulsion:

$$V_{\delta\pi}(1,2) = g'\left(\frac{f_\pi}{m_\pi}\right)^2 \boldsymbol{\tau}_1\boldsymbol{\tau}_2\boldsymbol{\Sigma}_1\boldsymbol{\Sigma}_2\delta(\mathbf{r}_1 - \mathbf{r}_2), \tag{69}$$

where the parameter $g' = 0.6$ is adjusted to reproduce experimental data on the excitation energies of the Gamow–Teller resonance in ^{208}Pb and kept fixed in the calculations for other nuclei, relying on the results obtained in Ref. 82 within the relativistic QRPA. The amplitude $\Phi(\omega)$ describes the coupling of the nucleons to vibrations generated by the coherent nucleonic oscillations. In the time blocking approximation it has the following operator form:

$$\Phi(\omega) = \sum_{\mu,\eta} g_{\mu}^{(\eta)\dagger} \tilde{R}^{0(\eta)}(\omega - \eta\omega_{\mu}) g_{\mu}^{(\eta)}, \tag{70}$$

where the index μ numerates vibrational modes (phonons) with frequencies ω_{μ} and generalized particle-vibration coupling (PVC) amplitude matrices $g_{\mu}^{(\eta)}$, and the index $\eta = \pm 1$ denotes forward and backward components, in full analogy with the neutral-channel case.[26] The energy-dependent effective interaction of Eq. (70) has the same form as the interaction of Eq. (50), but the pairs of the channel indices $k_1 k_3$ and $k_2 k_4$ have different isospins. As before, this amplitude is responsible for the spreading mechanism caused by the coupling between the ph and ph⊗phonon configurations. The phonon space is truncated by the angular momenta of the phonons at $J^{\pi} = 6^{+}$ and by their frequencies at 15 MeV. The ph⊗phonon configurations are included up to 30 MeV of the excitation energy. The truncation is justified by the subtraction of the term $\Phi(0)$ in Eq. (67). This subtraction removes double counting of the PVC effects from the residual interaction, guarantees the stability of the solutions for the response function and provides faster convergence of the renormalized PVC amplitude $\Phi(\omega) - \Phi(0)$ with respect to the phonon angular momenta and frequencies.

The strength function $S(\omega)$ has the form of Eq. (51) giving the spectral distribution of the nuclear response for a particular external field P. Below, we consider an external field expressed by the Gamow–Teller lowering operator:

$$P_{GT_-} = \sum_{i=1}^{A} \tau_{-}^{(i)} \mathbf{\Sigma}_i \tag{71}$$

and appearing in charge-exchange reactions of the proton-neutron type and various weak interaction processes mentioned above.

In Figure 12, we show the results for the GTR in ^{208}Pb obtained within the RRPA and RTBA, compared to non-relativistic QRPA of Refs. 83, 84 and to data of Ref. 85. The non-relativistic QRPA results are folded by the Lorentz distribution with 1 MeV width which is close to the energy

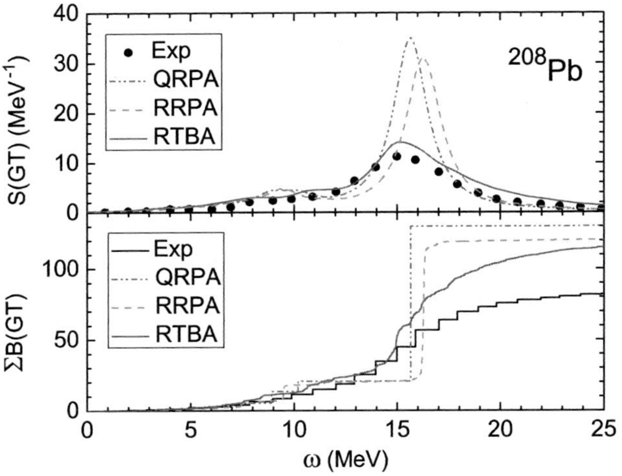

Fig. 12. (Color online) The theoretical and experimental Gamow-Teller strength distributions in ^{208}Pb (upper panel) and their cumulative sums (lower panel).

resolution of the experiment. The parameter of the G-matrix interaction $g_{ph} = 1.15$ is adjusted to reproduce the GTR centroid. The QRPA model space, including pn-configurations up to 45 MeV, accommodates the exact Ikeda sum rule while 3% of the total $B(GT_-)$ is beyond the considered 25 MeV energy interval and the total $B(GT_+)$ is equal to 0.21. Without introducing quenching factors in front of the calculated strength function the experimentally observed total strength[85] is by factor 0.62 smaller than that obtained in the QRPA.

The GTR within the relativistic approaches RRPA and RTBA has been calculated using the same smearing parameter $\Delta = 1$ MeV. The RRPA calculations, neglecting the last two terms of Eq. (67), produce a strength distribution which is very similar to the non-relativistic QRPA calculations with the major peak at 16.5 MeV and a low-energy peak structure around 10 MeV. The exact Ikeda sum rule is accommodated within the model space of pn-configurations between $-$ 1800 MeV and 100 MeV, so that 8% of the $B(GT_-)$ is at large negative energies because of the transitions to the Dirac sea.[82] While both QRPA and RRPA do not account for spreading effects, within the RTBA the GTR acquires the spreading width because of the coupling between the ph and ph⊗phonon configurations, so that the additional 5% of the sum rule goes above the considered energy region, while the total $B(GT_+)$ is equal to 0.34. Comparison to data shows that the spreading effects which are taken into account in the RTBA are reproduced very well. The overall strength is visibly quenched compared to the RRPA one, as seen in Figure 12, however, the cumulative sum shows that the experimental integral strength is still smaller than the RTBA one. The data from Ref. 85 are shown without experimental error bars which are relatively small ($\pm$ 5%). The quenching of the GTR strength below 25 MeV, obtained in the RTBA, is only about a half of the observed quenching, with respect to Ikeda sum rule. A study of Ref. 86 has revealed that the rest of the observed quenching comes from the finite momentum transfer which is not included in our present calculations.

In the context of the quenching problem the size of the basis spanning the ph$\otimes$phonon configurations plays an important role. In the numerical calculations the energy window containing these configurations is that in which the amplitude of Eq. (50) has non-zero matrix elements. The truncation of this energy window is, in some cases, a technical necessity and varies depending on the available computing resources. In the case of ^{208}Pb it turns out that almost the whole effect of coupling to the ph$\otimes$phonon configurations appears within 30 MeV of the excitation energy. An extension of this limit to 60 MeV does not influence the results. The particle-hole proton-neutron configurations are included, however, up to 100 MeV, so that about 10 units of the Ikeda sum rule are absorbed by the GT strength between 25 and 100 MeV. Further increase of the ph-basis size would spread the GT strength wider, although very slowly, so that it is not clear how to justify the cutoff rigorously.

Overall, the extension of the ph$\otimes$phonon window involves larger matrices, but nevertheless, in the present formulation this window is considerably larger than that, for instance, in the shell-model. Moreover, it is possible to investigate the convergence of the results with respect to this window. As discussed above, in the case of ^{208}Pb the value of 30 MeV can be taken as the cutoff energy for the ph$\otimes$phonon configurations.

As the next example, we have chosen a neutron deficient N=Z nucleus ^{28}Si. Being considerably lighter than ^{208}Pb, ^{28}Si has much lower level density, so that one can expect a higher saturation energy for the ph$\otimes$phonon configurations. Indeed, the pn-RTBA calculations confirm this expectation, which is illustrated in Figure 13. In these calculations the ph$\otimes$phonon configurations are accommodated within 90 MeV energy interval, which is close to that where 1p1h pn-RRPA configurations are included (100 MeV). One can see from the left panel of Figure 13 that a part of the GT$_-$ strength appears at large negative energies, which is a typical feature for relativistic approaches using no-sea approximation. In the

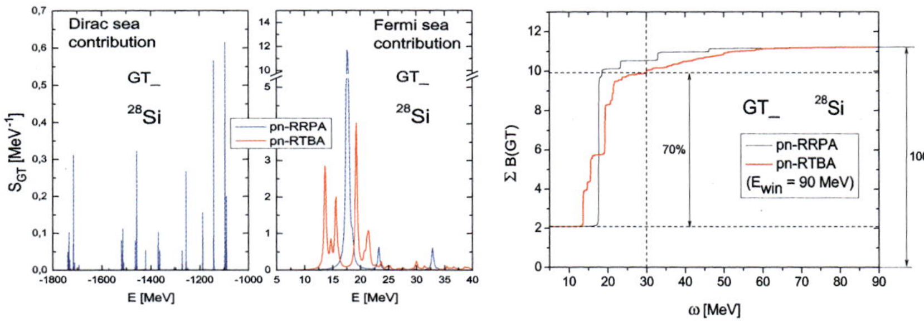

Fig. 13. Gamow–Teller strength distribution for ^{28}Si. Left panel: negative-energy part of the strength originated from the proton-antineutron transitions to the Dirac sea. Middle: positive-energy part of the GTR strength in pn-RRPA and in pn-RTBA. Right: Running sums for the pn-RRPA and pn-RTBA strength distributions.

nucleon sector, the cumulative GT_- sum in the pn-RTBA meets that obtained within the pn-RRPA at E_c ~60 MeV, which is sufficiently smaller than the pn-RTBA window. Thus, the GT_- strength between 0 and 30 MeV (the energy region which is usually studied experimentally) amounts to 70% of the total GT_- strength. This characterizes the ability of the pn-RTBA to describe the quenching, which is observed experimentally in restricted energy intervals, by the damping through 1p1h⊗phonon configurations. The GT_+ strength distribution (not shown) in this nucleus looks very similar although not exactly the same as the GT_-. The unique feature of N=Z nuclei is that the Ikeda sum rule

$$S(GT_-) - S(GT_+) = 3(N - Z) \tag{72}$$

is exactly zero which allows a stringent test of the approach and of its numerical implementation. Although the sum rule gives absolutely no information about the GT^+ and GT^- distributions separately, it is fulfilled in the present calculations for the GTR in ^{28}Si with ~0.2% accuracy, with respect to the total GT^+ or GT^- strength.

An extended discussion of the spin-isospin response calculated within pn-RTBA is presented in Ref. 81. The pn-RTBA results are compared in detail to QRPA with G-matrix effective interaction and to the large-basis shell model. All three models are applied to GTR in ^{132}Sn nucleus allowing, for the first time, a comprehensive comparative study.

Such a comparison turns out to be very constructive in defining strong and weak points of the theory and to determine future directions. It is demonstrated, by the choice of the appropriate physics cases, that very different theoretical models can constrain each other. The RTBA can, to a certain extent, provide information that is missing in QRPA and, in addition, can provide part of the quenching factors that are needed for QRPA and SM. In cases where the RTBA model space includes a sufficiently large amount of configurations and finite momentum transfer is taken into account,[86] RTBA has the potential to describe the overall GTR quenching fully microscopically, except the contribution from the delta-isobar which is found to be small.[87,88] In general, a comparison between the RTBA and SM calculations shows that higher-order correlations should be included into the RTBA that could improve its performance in the low-energy region. In turn, RTBA is an advancement in partly resolving the quenching problem, and, in addition, the relativistic mean field extended by the quasiparticle-vibration coupling[23,24] can provide the SM with the single-particle energies for nuclei where these energies are not available from data.

Based on the presented results, further advancements of the pn-RTBA, first of all, implementation of superfluid-type pairing and higher-order correlations, are anticipated. Experimental data on the spin-isospin strength distributions in broad energy intervals are expected from future measurements of various exotic nuclei at the rare isotope beam facilities. Such data will provide decisive arguments to constrain many-body coupling schemes of the R(Q)TBA as well as the underlying nuclear effective interactions.

5. Summary and perspectives

In this Chapter, we have reviewed possible extensions of the CDFT, their recent developments and applications to various nuclear structure properties. These extensions include spatial and temporal non-localities into the nucleonic self-energy and effective interaction and, thereby, allow accounting for dynamical effects, or correlations, beyond the mean-field and RQRPA approximations. The set of presented approaches is based on, although not confined by the ideas of the nuclear field theory for systems with well-developed collective degrees of freedom. In this work, the nuclear field theory is (i) formulated in the covariant framework and (ii) advanced beyond the previously existing formulation to the inclusion of more complicated correlations which are responsible for fine and exotic features of spectra detectable by modern experimental techniques.

The theory allows for delicate interplay of various correlations while the Lorentz invariance and full self-consistency put stringent restrictions on the number of parameters in the underlying density functional without reducing, but rather improving the quality of agreement with experimental data. Based on the same parametrization of the underlying covariant functional, a wide range of nuclear structure phenomena in medium-mass and heavy nuclei, including exotic ones, is described very successfully.

The covariant nuclear field theory discussed in this Chapter is far from completeness, and the present formulation allows various further developments to include higher-order effects of coupling between single-particle and emergent collective degrees of freedom, coupling to the continuum and the formation of bubbles, skins and other exotic geometries on equal footing. Thus, the approach represents a successful strategy toward a universal and high-precision solution of the nuclear many-body problem. The applications support in part present and future research topics at radioactive beam facilities and have a strong potential to provide a consistent input for astrophysical simulations.

Acknowledgment

This work was supported by US-NSF grants PHY-1404343 and PHY-1204486, by the DFG Cluster of Excellence "Origin and Structure of the Universe"(www.universecluster.de), and by the Overseas Distinguished Professor Project from Ministry of Education (MS2010BJDX001).

References

1. A. Bohr and B. R. Mottelson, *Nuclear Structure Volume II: Nuclear Deformation.* W. A. Benjamin, Inc., Reading, Mass. (1975).
2. V. G. Soloviev, C. Stoyanov, and V. V. Voronov, Nuclear properties in the lead region within the quasiparticle-phonon nuclear model, *Nucl. Phys. A.* **399**(1), 141–162 (1983). ISSN 0375-9474. doi: 10.1016/0375-9474(83)90599-7. URL `http://www.sciencedirect.com/science/article/pii/0375947483905997`.
3. S. Galès, C. Stoyanov, and A. Vdovin, Damping of high-lying single-particle modes in heavy nuclei, *Phys. Rep.* **166**(3), 125–193 (1988). ISSN 0370-1573. doi: http://dx.doi.org/10.1016/0370-1573(88)90066-X. URL `http://www.sciencedirect.com/science/article/pii/037015738890066X`.
4. V. G. Soloviev, *Theory of Atomic Nuclei: Quasiparticles and Phonons.* Institute of Physics, Bristol and Phyladelphia, USA (1992).
5. P. F. Bortignon, R. A. Broglia, D. R. Bes, and R. J. Liotta, Nuclear field theory, *Phys. Rep.* **30C**(4), 305–360 (1977). ISSN 0370-1573. doi: 10.1016/0370-1573(77)90018-7. URL `http://www.sciencedirect.com/science/article/pii/0370157377900187`.
6. G. F. Bertsch, P. F. Bortignon, and R. A. Broglia, Damping of nuclear excitations, *Rev. Mod. Phys.* **55**, 287–314 (Jan, 1983). doi: 10.1103/RevModPhys.55.287. URL `http://link.aps.org/doi/10.1103/RevModPhys.55.287`.

7. C. Mahaux, P. F. Bortignon, R. A. Broglia, and C. H. Dasso, Dynamics of the shell model, *Phys. Rep.* **120**(1), 1–274 (1985). ISSN 0370-1573. doi: http://dx.doi.org/10.1016/0370-1573(85)90100-0. URL `http://www.sciencedirect.com/science/article/pii/0370157385901000`.
8. P. F. Bortignon and R. A. Broglia, Role of the nuclear surface in a unified description of the damping of single-particle states and giant resonances, *Nucl. Phys. A.* **371**(3), 405–429 (1981). ISSN 0375-9474. doi: 10.1016/0375-9474(81)90055-5. URL `http://www.sciencedirect.com/science/article/pii/0375947481900555`.
9. G. Coló, P. F. Bortignon, N. Van Giai, A. Bracco, and R. Broglia, Damping properties of the breathing mode in ^{208}Pb, *Phys. Lett. B.* **276**, 279–284 (1992). ISSN 0370-2693. doi: 10.1016/0370-2693(92)90318-X. URL `http://www.sciencedirect.com/science/article/pii/037026939290318X`.
10. G. Coló and P. Bortignon, QRPA plus phonon coupling model and the photoabsorption cross section for ^{18,20,22}O, *Nucl. Phys. A.* **696**(3–4), 427–441 (2001). ISSN 0375-9474. doi: http://dx.doi.org/10.1016/S0375-9474(01)01217-9. URL `http://www.sciencedirect.com/science/article/pii/S0375947401012179`.
11. D. Sarchi, P. F. Bortignon, and G. Coló, Dipole states in stable and unstable nuclei, *Phys. Lett. B.* **601**, 27 (2004). ISSN 0370-2693. doi: 10.1016/j.physletb.2004.09.019. URL `http://www.sciencedirect.com/science/article/pii/S0370269304013127`.
12. V. Bernard and N. V. Giai, Effects of collective modes on the single-particle states and the effective mass in ^{208}Pb, *Nucl. Phys. A.* **348**(1), 75–92 (1980). ISSN 0375-9474. doi: 10.1016/0375-9474(80)90546-1. URL `http://www.sciencedirect.com/science/article/pii/0375947480905461`.
13. A. Avdeenkov and S. Kamerdzhiev, The role of ground state correlations in the single-particle strength of odd nuclei with pairing, *Phys. Lett. B.* **459**(4), 423–430 (1999). ISSN 0370-2693. doi: 10.1016/S0370-2693(99)00719-4. URL `http://www.sciencedirect.com/science/article/pii/S0370269399007194`.
14. F. Barranco, R. A. Broglia, G. Gori, E. Vigezzi, P. F. Bortignon, and J. Terasaki, Surface vibrations and the pairing interaction in nuclei, *Phys. Rev. Lett.* **83**, 2147–2150 (Sep, 1999). doi: 10.1103/PhysRevLett.83.2147. URL `http://link.aps.org/doi/10.1103/PhysRevLett.83.2147`.
15. C. Barbieri, Role of long-range correlations in the quenching of spectroscopic factors, *Phys. Rev. Lett.* **103**, 202502 (Nov, 2009). doi: 10.1103/PhysRevLett.103.202502. URL `http://link.aps.org/doi/10.1103/PhysRevLett.103.202502`.
16. G. Colò, H. Sagawa, and P. F. Bortignon, Effect of particle-vibration coupling on single-particle states: A consistent study within the Skyrme framework, *Phys. Rev. C.* **82**, 064307 (Dec, 2010). doi: 10.1103/PhysRevC.82.064307. URL `http://link.aps.org/doi/10.1103/PhysRevC.82.064307`.
17. V. I. Tselyaev, *Sov. J. Nucl. Phys.* **50**, 780 (1989). Yad. Fiz. 50, 1252 (1989).

18. S. P. Kamerdzhiev, G. Y. Tertychny, and V. I. Tselyaev, The method of time-ordered graph decoupling and its application to the description of giant resonances in magic nuclei, *Phys. Part. Nucl.* **28**, 134–158 (1997). doi: http://dx.doi.org/10.1134/1.953052. URL `http://scitation.aip.org/content/aip/journal/ppn/28/2/10.1134/1.953052`.
19. S. P. Kamerdzhiev, J. Speth, and G. Y. Tertychny, Extended theory of finite Fermi systems: collective vibrations in closed shell nuclei, *Phys. Rep.* **393**(1), 1–86 (2004). ISSN 0370-1573. doi: 10.1016/j.physrep.2003.11.001. URL `http://www.sciencedirect.com/science/article/pii/S037015730300437X`.
20. E. Litvinova and V. I. Tselyaev, Quasiparticle time blocking approximation in coordinate space as a model for the damping of the giant dipole resonance, *Phys. Rev. C.* **75**, 054318 (2007). doi: 10.1103/PhysRevC.75.054318. URL `http://link.aps.org/doi/10.1103/PhysRevC.75.054318`.
21. S. Drozdz, S. Nishizaki, J. Speth, and J. Wambach, The nuclear response within extended RPA theories, *Phys. Rep.* **197**(1), 1–65 (1990). ISSN 0370-1573. doi: http://dx.doi.org/10.1016/0370-1573(90)90084-F. URL `http://www.sciencedirect.com/science/article/pii/037015739090084F`.
22. E. Litvinova and P. Ring, Covariant theory of particle-vibrational coupling and its effect on the single-particle spectrum., *Phys. Rev. C.* **73**, 044328 (2006). doi: 10.1103/PhysRevC.73.044328.
23. E. Litvinova and A. V. Afanasjev, Dynamics of nuclear single-particle structure in covariant theory of particle-vibration coupling: From light to superheavy nuclei, *Phys. Rev. C.* **84**, 014305 (Jul, 2011). doi: 10.1103/PhysRevC.84.014305. URL `http://link.aps.org/doi/10.1103/PhysRevC.84.014305`.
24. E. Litvinova, Quasiparticle-vibration coupling in a relativistic framework: Shell structure of z=120 isotopes, *Phys. Rev. C.* **85**, 021303 (Feb, 2012). doi: 10.1103/PhysRevC.85.021303. URL `http://link.aps.org/doi/10.1103/PhysRevC.85.021303`.
25. E. Litvinova, P. Ring, and V. I. Tselyaev, Particle-vibration coupling within covariant density function theory., *Phys. Rev. C.* **75**, 064308 (2007). doi: 10.1103/PhysRevC.75.064308.
26. E. Litvinova, P. Ring, and V. I. Tselyaev, Relativistic quasiparticle time blocking approximation: dipole response of open-shell nuclei, *Phys. Rev. C.* **78**, 014312 (2008). doi: 10.1103/PhysRevC.78.014312.
27. E. Litvinova, P. Ring, and V. I. Tselyaev, Mode coupling and the pygmy dipole resonance in a relativistic two-phonon model, *Phys. Rev. Lett.* **105**, 022502 (Jul, 2010). doi: 10.1103/PhysRevLett.105.022502. URL `http://link.aps.org/doi/10.1103/PhysRevLett.105.022502`.
28. E. Litvinova, P. Ring, and V. I. Tselyaev, Relativistic two-phonon model for low-energy nuclear response, *Phys. Rev. C.* **88**, 044320 (Oct, 2013). doi: 10.1103/PhysRevC.88.044320. URL `http://link.aps.org/doi/10.1103/PhysRevC.88.044320`.
29. W. Kohn and L. J. Sham, Self-consistent equations including exchange and correlation effects, *Phys. Rev.* **140**, A1133–A1138 (Nov, 1965).

doi: 10.1103/PhysRev.140.A1133. URL http://link.aps.org/doi/10.1103/PhysRev.140.A1133.
30. W. Kohn, Nobel lecture: Electronic structure of matter–wave functions and density functionals, *Rev. Mod. Phys.* **71**, 1253–1266 (Oct, 1999). doi: 10.1103/RevModPhys.71.1253. URL http://link.aps.org/doi/10.1103/RevModPhys.71.1253.
31. M. A. L. Marques, C. A. Ullrich, F. Nogueira, A. Rubio, K. Burke, and E. K. U. G. (Eds.), *Time-Dependent Density Functional Theory*. Springer, Heidelberg (2006).
32. E. Runge and E. K. U. Gross, Density-functional theory for time-dependent systems, *Phys. Rev. Lett.* **52**, 997–1000 (Mar, 1984). doi: 10.1103/PhysRevLett.52.997. URL http://link.aps.org/doi/10.1103/PhysRevLett.52.997.
33. P. Ring and P. Schuck, *The Nuclear Many-Body Problem*. Springer-Verlag, Berlin (1980).
34. P. Hohenberg and W. Kohn, Inhomogeneous electron gas, *Phys. Rev.* **136**, B864–B871 (Nov, 1964). doi: 10.1103/PhysRev.136.B864. URL http://link.aps.org/doi/10.1103/PhysRev.136.B864.
35. P. Ring, Z.-Y. Ma, N. Van Giai, D. Vretenar, A. Wandelt, and L.-G. Cao, The time-dependent relativistic mean field theory and the random phase approximation., *Nucl. Phys. A*. **694**, 249 (2001).
36. N. Paar, P. Ring, T. Nikšić, and D. Vretenar, Quasiparticle random phase approximation based on the relativistic Hartree–Bogoliubov model, *Phys. Rev. C*. **67**, 034312 (Mar, 2003). doi: 10.1103/PhysRevC.67.034312. URL http://link.aps.org/doi/10.1103/PhysRevC.67.034312.
37. D. Vretenar, A. V. Afanasjev, G. A. Lalazissis, and P. Ring, Relativistic Hartree–Bogoliubov theory: static and dynamic aspects of exotic nuclear structure, *Phys. Rep.* **409**(3-4), 101–259 (Apr., 2005). doi: 10.1016/j.physrep.2004.10.001. URL http://dx.doi.org/10.1016/j.physrep.2004.10.001.
38. D. Pena Arteaga and P. Ring, Relativistic random-phase approximation in axial symmetry, *Phys. Rev. C*. **77**, 034317 (Mar, 2008). doi: 10.1103/PhysRevC.77.034317. URL http://link.aps.org/doi/10.1103/PhysRevC.77.034317.
39. J. Daoutidis and P. Ring, Continuum random-phase approximation for relativistic point coupling models, *Phys. Rev. C*. **80**, 024309 (Aug, 2009). doi: 10.1103/PhysRevC.80.024309. URL http://link.aps.org/doi/10.1103/PhysRevC.80.024309.
40. C. Itzykson and J.-B. Zuber, *Quantum Field Theory*. McGraw-Hill, New York (1980).
41. H. Grawe, K. Langanke, and G. Martinez-Pinedo, Nuclear structure and astrophysics, *Rep. Progr. Phys.* **70**, 1525 (2007). doi: 10.1088/0034-4885/70/9/R02.
42. T. Otsuka, T. Suzuki, M. Honma, Y. Utsuno, N. Tsunoda, K. Tsukiyama, and M. Hjorth-Jensen, Novel features of nuclear forces and shell evolution in exotic nuclei, *Phys. Rev. Lett.* **104**, 012501 (Jan, 2010). doi: 10.

1103/PhysRevLett.104.012501. URL `http://link.aps.org/doi/10.1103/PhysRevLett.104.012501`.

43. O. Sorlin and M.-G. Porquet, Nuclear magic numbers: New features far from stability, *Prog. Part. Nucl. Phys.* **61**(2), 602–673 (Oct., 2008). ISSN 0146-6410. doi: 10.1016/j.ppnp.2008.05.001. URL `http://dx.doi.org/10.1016/j.ppnp.2008.05.001`.
44. M. Bender, K. Rutz, P.-G. Reinhard, J. A. Maruhn, and W. Greiner, Shell structure of superheavy nuclei in self-consistent mean-field models, *Phys. Rev. C.* **60**, 034304 (Aug, 1999). doi: 10.1103/PhysRevC.60.034304. URL `http://link.aps.org/doi/10.1103/PhysRevC.60.034304`.
45. A. Sobiczewski and K. Pomorski, Description of structure and properties of superheavy nuclei, *Prog. Part. Nucl. Phys.* **58**, 292–349 (2007). doi: 10.1016/j.ppnp.2006.05.001.
46. W. Zhang, J. Meng, S. Zhang, L. Geng, and H. Toki, Magic numbers for superheavy nuclei in relativistic continuum Hartree–Bogoliubov theory, *Nucl. Phys. A.* **753**(1–2), 106–135 (2005). ISSN 0375-9474. doi: http://dx.doi.org/10.1016/j.nuclphysa.2005.02.086. URL `http://www.sciencedirect.com/science/article/pii/S0375947405002423`.
47. A. V. Afanasjev and S. Frauendorf, Central depression in nuclear density and its consequences for the shell structure of superheavy nuclei, *Phys. Rev. C.* **71**, 024308 (Feb, 2005). doi: 10.1103/PhysRevC.71.024308. URL `http://link.aps.org/doi/10.1103/PhysRevC.71.024308`.
48. A. B. Migdal, *Theory of Finite Fermi Systems: Applications to Atomic Nuclei.* Wilson Interscience, New York (1967).
49. A. A. Abrikosov, L. P. Gor'kov, and I. E. Dzyaloshincki, *Methods of Quantum Field Theory in Statistical Physics.* Prentice Hall, Engelwood Cliffs, NJ (1963).
50. M. Serra and P. Ring, Relativistic Hartree–Bogoliubov theory for finite nuclei, *Phys. Rev. C.* **65**(6), 064324–9 (June, 2002). doi: 10.1103/PhysRevC.65.064324. URL `http://link.aps.org/abstract/PRC/v65/e064324`.
51. J. Speth, E. Werner, and W. Wild, Theory of finite Fermi systems and application to the lead region, *Phys. Rep.* **33**(3), 127–208 (1977). ISSN 0370-1573. doi: http://dx.doi.org/10.1016/0370-1573(77)90042-4. URL `http://www.sciencedirect.com/science/article/pii/0370157377900424`.
52. V. I. Tselyaev, Quasiparticle time blocking approximation within the framework of generalized greens function formalism, *Phys. Rev. C.* **75**, 024306 (Feb, 2007). doi: 10.1103/PhysRevC.75.024306. URL `http://link.aps.org/doi/10.1103/PhysRevC.75.024306`.
53. V. I. Tselyaev, private communication; unpublished.
54. R. V. Jolos, J. L. Molina, and V. G. Soloviev, Effect of the Pauli principle on the excited states of doubly-even deformed nuclei, *Zeitschrift fr Physik A Atoms and Nuclei.* **295**(2), 147–152 (1980). ISSN 0939-7922. doi: 10.1007/BF01413072. URL `http://dx.doi.org/10.1007/BF01413072`.
55. S. Mishev and V. V. Voronov, Effects of ground state correlations on the structure of odd-mass spherical nuclei, *Phys. Rev. C.* **78**, 024310 (Aug,

2008). doi: 10.1103/PhysRevC.78.024310. URL `http://link.aps.org/doi/10.1103/PhysRevC.78.024310`.

56. G. A. Lalazissis, S. Karatzikos, R. Fossion, D. Peña Arteaga, A. V. Afanasjev, and P. Ring, The effective force NL3 revisited., *Phys. Lett. B.* **671**, 36 (2009). doi: 10.1016/j.physletb.2008.11.070.
57. T. Bürvenich, M. Bender, J. A. Maruhn, and P.-G. Reinhard, Systematics of fission barriers in superheavy elements, *Phys. Rev. C.* **69**, 014307 (Jan, 2004). doi: 10.1103/PhysRevC.69.014307. URL `http://link.aps.org/doi/10.1103/PhysRevC.69.014307`.
58. G. A. Lalazissis, J. König, and P. Ring, New parameterization for the lagrangian density of relativistic mean field theory., *Phys. Rev. C.* **55**(1), 540–543 (Jan., 1997). doi: 10.1103/PhysRevC.55.540. URL `http://link.aps.org/abstract/PRC/v55/p540`.
59. H. Abusara, A. V. Afanasjev, and P. Ring, Fission barriers in actinides in covariant density functional theory: The role of triaxiality, *Phys. Rev. C.* **82**, 044303 (2010). doi: 10.1103/PhysRevC.82.044303. URL `http://link.aps.org/doi/10.1103/PhysRevC.82.044303`.
60. A. V. Afanasjev, Superheavy nuclei: Relativistic mean field outlook, *Phys. Scr. T.* **125**, 62 (2006). doi: 10.1088/0031-8949/2006/T125/001.
61. M. N. Harakeh and A. van der Woude, *Giant Resonances: Fundamental High-Frequency Modes of Nuclear Excitation*. Oxford University Press, Oxford (2001).
62. D. Bohm and D. Pines, A collective description of electron interactions: III. Coulomb interactions in a degenerate electron gas, *Phys. Rev.* **92**, 609–625 (Nov, 1953). doi: 10.1103/PhysRev.92.609. URL `http://link.aps.org/doi/10.1103/PhysRev.92.609`.
63. N. N. Bogoliubov, *Sov. Phys. Usp.* **2**, 236 (1959).
64. M. Baranger, Extension of the shell model for heavy spherical nuclei, *Phys. Rev.* **120**, 957–968 (Nov, 1960). doi: 10.1103/PhysRev.120.957. URL `http://link.aps.org/doi/10.1103/PhysRev.120.957`.
65. V. G. Soloviev, *Theory of Complex Nuclei*. Nauka, Moskow (1971).
66. E. Salpeter and H. A. Bethe, A relativistic equation for bound-state problems, *Phys. Rev.* **84**, 1232–1242 (Dec, 1951). doi: 10.1103/PhysRev.84.1232. URL `http://link.aps.org/doi/10.1103/PhysRev.84.1232`.
67. E. Litvinova, P. Ring, V. I. Tselyaev, and K. Langanke, Relativistic quasiparticle time blocking approximation. ii. pygmy dipole resonance in neutron-rich nuclei, *Phys. Rev. C.* **79**, 054312 (May, 2009). doi: 10.1103/PhysRevC.79.054312. URL `http://link.aps.org/doi/10.1103/PhysRevC.79.054312`.
68. J. Endres, E. Litvinova, D. Savran, P. A. Butler, M. N. Harakeh, S. Harissopulos, R.-D. Herzberg, R. Krücken, A. Lagoyannis, N. Pietralla, V. Y. Ponomarev, L. Popescu, P. Ring, M. Scheck, K. Sonnabend, V. I. Stoica, H. J. Wötche, and A. Zilges, Isospin character of the pygmy dipole resonance in ^{124}Sn, *Phys. Rev. Lett.* **105**, 215503 (2010). doi: 10.1103/PhysRevLett.105.212503. URL `http://www.sciencedirect.com/science/article/pii/S0375947404007122`.
69. J. Endres, D. Savran, P. A. Butler, M. N. Harakeh, S. Harissopulos, R.-D.

Herzberg, R. Krücken, A. Lagoyannis, E. Litvinova, N. Pietralla, V. Ponomarev, L. Popescu, P. Ring, M. Scheck, F. Schlüter, K. Sonnabend, V. I. Stoica, H. J. Wörtche, and A. Zilges, Structure of the pygmy dipole resonance in ^{124}Sn, *Phys. Rev. C.* **85**, 064331 (Jun, 2012). doi: 10.1103/PhysRevC.85.064331. URL `http://link.aps.org/doi/10.1103/PhysRevC.85.064331`.
70. R. Massarczyk, R. Schwengner, F. Dönau, E. Litvinova, G. Rusev, R. Beyer, R. Hannaske, A. R. Junghans, M. Kempe, J. H. Kelley, T. Kögler, K. Kosev, E. Kwan, M. Marta, A. Matic, C. Nair, R. Raut, K. D. Schilling, G. Schramm, D. Stach, A. P. Tonchev, W. Tornow, E. Trompler, A. Wagner, and D. Yakorev, Electromagnetic dipole strength of ^{136}Ba below the neutron separation energy, *Phys. Rev. C.* **86**, 014319 (Jul, 2012). doi: 10.1103/PhysRevC.86.014319. URL `http://link.aps.org/doi/10.1103/PhysRevC.86.014319`.
71. V. I. Tselyaev, Subtraction method and stability condition in extended random-phase approximation theories, *Phys. Rev. C.* **88**, 054301 (Nov, 2013). doi: 10.1103/PhysRevC.88.054301. URL `http://link.aps.org/doi/10.1103/PhysRevC.88.054301`.
72. Experimental Nuclear Reaction Data (EXFOR), http://www-nds.iaea.org/exfor/exfor.htm.
73. N. Van Giai and H. Sagawa, Monopole and dipole compression modes in nuclei, *Nucl. Phys. A.* **371**(1), 1–18 (1981). doi: 10.1016/0375-9474(81)90741-7. URL `http://www.sciencedirect.com/science/article/pii/0375947481907417`.
74. D. M. Rossi, P. Adrich, F. Aksouh, H. Alvarez-Pol, T. Aumann, J. Benlliure, M. Böhmer, K. Boretzky, E. Casarejos, M. Chartier, A. Chatillon, D. Cortina-Gil, U. Datta Pramanik, H. Emling, O. Ershova, B. Fernandez-Dominguez, H. Geissel, M. Gorska, M. Heil, H. T. Johansson, A. Junghans, A. Kelic-Heil, O. Kiselev, A. Klimkiewicz, J. V. Kratz, R. Krücken, N. Kurz, M. Labiche, T. Le Bleis, R. Lemmon, Y. A. Litvinov, K. Mahata, P. Maierbeck, A. Movsesyan, T. Nilsson, C. Nociforo, R. Palit, S. Paschalis, R. Plag, R. Reifarth, D. Savran, H. Scheit, H. Simon, K. Sümmerer, A. Wagner, W. Waluś, H. Weick, and M. Winkler, Measurement of the dipole polarizability of the unstable neutron-rich nucleus ^{68}Ni, *Phys. Rev. Lett.* **111**, 242503 (Dec, 2013). doi: 10.1103/PhysRevLett.111.242503. URL `http://link.aps.org/doi/10.1103/PhysRevLett.111.242503`.
75. I. Pysmenetska, S. Walter, J. Enders, H. von Garrel, O. Karg, U. K. C. K. P. von Neumann-Cosel, H. H. Pitz, V. Y. Ponomarev, M. Scheck, and F. S. S. Volz, Two-phonon 1- state in ^{112}Sn observed in resonant photon scattering, *Phys. Rev. C.* **73**, 017302 (Jan, 2006). doi: 10.1103/PhysRevC.73.017302. URL `http://link.aps.org/doi/10.1103/PhysRevC.73.017302`.
76. J. Bryssinck, L. Govor, D. Belic, F. Bauwens, O. Beck, P. von Brentano, D. D. Frenne, T. Eckert, C. Fransen, K. Govaert, R.-D. Herzberg, E. Jacobs, U. Kneissl, H. Maser, A. Nord, N. Pietralla, H. H. Pitz, V. Y. Ponomarev, and V. Werner, Uniform properties of $J^{\pi} = 1^{-}$ two-phonon states in the semimagic even-even tin isotopes ^{116,118,120,122,124}Sn, *Phys. Rev. C.* **59**, 1930–1934 (Apr, 1999). doi: 10.1103/PhysRevC.59.1930. URL `http://link.aps.org/doi/10.1103/PhysRevC.59.1930`.

77. K. Govaert, F. Bauwens, J. Bryssinck, D. De Frenne, E. Jacobs, W. Mondelaers, L. Govor, and V. Y. Ponomarev, Dipole excitations to bound states in ^{116}Sn and ^{124}Sn, *Phys. Rev. C.* **57**, 2229–2249 (May, 1998). doi: 10.1103/PhysRevC.57.2229. URL `http://link.aps.org/doi/10.1103/PhysRevC.57.2229`.
78. B. Özel. PhD thesis, Cukurova University (2008); private communication (2008).
79. B. Özel, J. Enders, P. von Neumann-Cosel, I. Poltoratska, A. Richter, D. Savran, S. Volz, and A. Zilges, Systematics of the pygmy dipole resonance in stable tin isotopes from resonant photon scattering, *Nucl. Phys. A.* **788**, 385c–388c (2007). doi: 10.1016/j.nuclphysa.2007.01.069. URL `http://www.sciencedirect.com/science/article/pii/S0375947407001297`.
80. T. Marketin, E. Litvinova, D. Vretenar, and P. Ring, Fragmentation of spin-dipole strength in ^{90}Zr and ^{208}Pb, *Phys. Lett. B.* **706**, 477–481 (2012). doi: 10.1016/j.physletb.2011.11.050. URL `http://www.sciencedirect.com/science/article/pii/S0370269311014298`.
81. E. Litvinova, B. A. Brown, D.-L. Fang, T. Marketin, and R. G. T. Zegers, Benchmarking nuclear models for Gamow–Teller response, *Phys. Lett. B.* **730**, 307–313 (2014). doi: http://dx.doi.org/10.1016/j.physletb.2014.02.001. URL `http://www.sciencedirect.com/science/article/pii/S0370269314000926`.
82. N. Paar, T. Nikšić, D. Vretenar, and P. Ring, Quasi-particle random phase approximation based on the relativistic Hartree–Bogoliubov model ii. nuclear spin- and isospin-excitations, *Phys. Rev. C.* **69**, 054303 (May, 2004). doi: 10.1103/PhysRevC.69.054303. URL `http://link.aps.org/doi/10.1103/PhysRevC.69.054303`.
83. J. Suhonen, T. Taigel, and A. Faessler, pnQRPA calculation of the β^+/EC quenching for several neutron-deficient nuclei in mass regions $A = 94$–110 and $A = 146$–156 star, *Nucl. Phys. A.* **486**(1), 91–117 (1988). doi: http://dx.doi.org/10.1016/0375-9474(88)90041-3. URL `http://www.sciencedirect.com/science/article/pii/0375947488900413`.
84. M. S. Yousef, V. Rodin, A. Faessler, and F. Simkovic, Two-neutrino double beta decay of deformed nuclei within the quasiparticle random-phase approximation with a realistic interaction, *Phys. Rev. C.* **79**, 014314 (Jan, 2009). doi: 10.1103/PhysRevC.79.014314. URL `http://link.aps.org/doi/10.1103/PhysRevC.79.014314`.
85. T. Wakasa, M. Okamoto, M. Dozono, K. Hatanaka, M. Ichimura, S. Kuroita, Y. Maeda, H. Miyasako, T. Noro, T. Saito, Y. Sakemi, T. Yabe, and K. Yako, Complete sets of polarization transfer observables for the ^{208}Pb(p,n) reaction at 296 MeV and Gamow–Teller and spin-dipole strengths for ^{208}Pb, *Phys. Rev. C.* **85**, 064606 (Jun, 2012). doi: 10.1103/PhysRevC.85.064606. URL `http://link.aps.org/doi/10.1103/PhysRevC.85.064606`.
86. T. Marketin, G. Martínez-Pinedo, N. Paar, and D. Vretenar, Role of momentum transfer in the quenching of Gamow–Teller strength, *Phys. Rev. C.* **85**, 054313 (May, 2012). doi: 10.1103/PhysRevC.85.054313. URL `http://link.aps.org/doi/10.1103/PhysRevC.85.054313`.

87. A. Arima, K. Shimizu, W. Bentz, and H. Hyuga, Nuclear magnetic properties and Gamov–Teller transitions and gt resonances, *Adv. Nucl. Phys.* **18**, 1–105 (1987).
88. I. S. Towner, Quenching of spin matrix elements in nuclei, *Phys. Rep.* **155**(5), 263–377 (1987). ISSN 0370-1573. doi: http://dx.doi.org/10.1016/0370-1573(87)90138-4. URL `http://www.sciencedirect.com/science/article/pii/0370157387901384`.

Chapter 12

Beyond the relativistic mean-field approximation — collective correlations

Zhipan Li

School of Physical Science and Technology, Southwest University, 400715 Chongqing, China

zpliphy@swu.edu.cn

Tamara Nikšić

Physics Department, Faculty of Science, University of Zagreb, 10000 Zagreb, Croatia

tniksic@phy.hr

Dario Vretenar

Physics Department, Faculty of Science, University of Zagreb, 10000 Zagreb, Croatia

vretenar@phy.hr

Jiangming Yao

School of Physical Science and Technology, Southwest University, 400715 Chongqing, China

Department of Physics, Tohoku University, Sendai 980-8578, Japan

jmyao@nucl.phys.tohoku.ac.jp

Semi-empirical relativistic energy density functionals (EDFs) or effective interactions implicitly comprise short-range correlations related to the repulsive core of the inter-nucleon interaction, and long-range correlations mediated by nuclear resonance modes. To model spectroscopic properties of finite nuclei, the self-consistent mean-field method must be extended to include collective correlations that arise from restoration of broken symmetries and fluctuations in collective coordinates. These correlations are sensitive to shell effects, vary with particle number, and cannot be included in a universal EDF. We review and compare recent advances in "beyond mean-field" methods based on relativistic EDFs: the angular-momentum and particle-number projected triaxial generator coordinate method, the five-dimensional quadrupole collective Hamiltonian and the axial quadrupole-octupole collective Hamiltonian models. Illustrative applications include low-energy collective excitation spectra and electromagnetic transition rates of nuclei characterised by quadrupole and/or octupole deformations: ^{24}Mg, ^{76}Kr, ^{240}Pu and ^{224}Ra, in comparison with available data.

Contents

1. Introduction

To describe complex excitation patterns and transition rates associated with the evolution of shell structures, from stable nuclei to regions of exotic short-lived systems far from β-stability, nuclear structure methods must be developed that are based on a universal microscopic framework. The self-consistent relativistic mean-field (RMF) approach has been applied with considerable success to the analysis of a variety of nuclear structure properties, in nuclei along the valley of β-stability and also in exotic nuclei far from stability and close to the drip lines.[1–4] Various RMF-based effective interactions (density functionals) have been used to compute ground-state properties in the self-consistent Hartree–(Fock)–Bogoliubov framework, and to calculate excitation energies of giant resonances in the (quasiparticle) random-phase approximation (Q)RPA for small-amplitude collective motion.

For a microscopic self-consistent description of collective excitation spectra and electromagnetic transition rates, however, the challenge for relativistic energy density functionals (REDFs) is the systematic treatment of collective correlations related to restoration of broken symmetries and fluctuations in collective coordinates. In practical implementations the EDF framework is realized on two specific levels. The basic implementation is in terms of self-consistent mean-field (SCMF) models, in which an EDF is constructed as a functional of one-body nucleon density matrices that correspond to a single product state — Slater determinant of single-particle or single-quasiparticle states. The SCMF level is characterized by symmetry breaking — translational, rotational, particle number — and the equilibrium solution only provides an approximate description of bulk ground-state properties. To calculate excitation spectra and transition rates it is necessary to include correlations that arise from symmetry restoration and fluctuations around the mean-field equilibrium solution. On the second level on which broken symmetries are restored and configuration mixing of symmetry-breaking product states is taken into account, the many-body energy takes the form of a functional of all transition density matrices that can be constructed from the chosen set of product states. This set is chosen to restore symmetries or/and to perform a mixing of configurations that correspond to specific collective modes using, for instance, the (quasiparticle) random-phase approximation (QRPA), the Generator Coordinate Method (GCM), or the collective Hamiltonian. Symmetry restoration and configuration mixing models based on non-relativistic density functionals or effective interactions (Skyrme, Gogny) have been used for quite some time now, but only in recent years have this type of structure models been developed using relativistic density functionals.[5–10]

In this chapter we review recent advances in the extension of the self-consistent relativistic mean-field framework to include the treatment of collective correlations. In section 2 we introduce the theoretical framework and consider a particular class of REDFs that will be used in illustrative calculations discussed in the following sections. The treatment of collective correlations is reviewed and illustrated with a number of examples in section 3 (symmetry restoration and configuration mixing calculations, both for axial and triaxial quadrupole shapes) and in section 4 (quadrupole collective Hamiltonian in five dimensions and quadrupole-octupole collective Hamiltonian for axial shapes). Section 5 summarizes the review.

2. Relativistic mean-field models with point-coupling effective interactions

At relatively low energies characteristic for nuclear binding and low-lying collective excitations, the meson exchange (σ, ω, ρ, ...) picture is just a convenient representation of the effective nuclear interaction. The exchange of heavy mesons is associated with short-distance dynamics that cannot be resolved at these energies, and therefore in each channel (scalar-isoscalar, vector-isoscalar, scalar-isovector, and vector-isovector) meson exchange can be replaced by the corresponding local four-point (contact) interactions between nucleons. The self-consistent relativistic mean-field framework can be formulated in terms of point-coupling effective nucleon interactions. Medium effects can be taken into account by the inclusion of higher-order interaction terms, or can be encoded in the density dependence of strength parameters of the interaction in the isoscalar and isovector channels. In this section we outline the basic features of the relativistic mean-field point-coupling (RMF-PC) model. More details can be found in Refs. 11–13 and references cited therein.

The choice of the point-coupling effective Lagrangian determines the mean-field energy (RMF) of a nuclear system in terms of local single-nucleon densities and currents:

$$
\begin{aligned}
E_{\mathrm{RMF}} &= \int d\boldsymbol{r}\ \mathcal{E}_{\mathrm{RMF}}(\boldsymbol{r}) \\
&= \sum_k \int d\boldsymbol{r}\ v_k^2\ \bar{\psi}_k(\boldsymbol{r})\left(-i\boldsymbol{\gamma}\boldsymbol{\nabla}+m\right)\psi_k(\boldsymbol{r}) \\
&\quad + \int d\boldsymbol{r}\ \left(\frac{\alpha_S}{2}\rho_S^2+\frac{\beta_S}{3}\rho_S^3+\frac{\gamma_S}{4}\rho_S^4+\frac{\delta_S}{2}\rho_S\Delta\rho_S+\frac{\alpha_V}{2}j_\mu j^\mu\right. \\
&\quad +\frac{\gamma_V}{4}(j_\mu j^\mu)^2+\frac{\delta_V}{2}j_\mu\Delta j^\mu+\frac{\alpha_{TV}}{2}j_{TV}^\mu(j_{TV})_\mu+\frac{\delta_{TV}}{2}j_{TV}^\mu\Delta(j_{TV})_\mu \\
&\quad \left.+\frac{\alpha_{TS}}{2}\rho_{TS}^2+\frac{\delta_{TS}}{2}\rho_{TS}\Delta\rho_{TS}+\frac{e}{2}\rho_p A^0\right)\ ,
\end{aligned}
\tag{1}
$$

where ψ denotes the Dirac spinor field of a nucleon. The local isoscalar (S) and isovector scalar (TS) densities, and the corresponding isoscalar and isovector (TV) currents for a nucleus with A nucleons

$$\rho_S(\boldsymbol{r}) = \sum_k v_k^2\ \bar{\psi}_k(\boldsymbol{r})\psi_k(\boldsymbol{r})\ , \tag{2}$$

$$\rho_{TS}(\boldsymbol{r}) = \sum_k v_k^2\ \bar{\psi}_k(\boldsymbol{r})\tau_3\psi_k(\boldsymbol{r})\ , \tag{3}$$

$$j^\mu(\boldsymbol{r}) = \sum_k v_k^2\ \bar{\psi}_k(\boldsymbol{r})\gamma^\mu\psi_k(\boldsymbol{r})\ , \tag{4}$$

$$j_{TV}^\mu(\boldsymbol{r}) = \sum_k v_k^2\ \bar{\psi}_k(\boldsymbol{r})\gamma^\mu\tau_3\psi_k(\boldsymbol{r})\ , \tag{5}$$

are calculated in the *no-sea* approximation: the summation in Eqs. (2)–(5) runs over all occupied states in the Fermi sea, i.e. only occupied single-nucleon states with positive energy explicitly contribute to the nucleon densities and currents. v_k^2 denotes the occupation factors of single-nucleon states. In Eq. (1) ρ_p is the proton density, and A^0 denotes the Coulomb potential. α, β, γ and δ denote the 11 constants that parameterize the particle-hole channel of the relativistic energy density functional in the corresponding space-isospace channels.

The single-nucleon wave functions are obtained as self-consistent solutions of the corresponding Dirac equation:

$$\left\{\boldsymbol{\alpha}\cdot\left[-i\boldsymbol{\nabla}-\boldsymbol{V}(\boldsymbol{r})\right]+V(\boldsymbol{r})+\beta\big(m+S(\boldsymbol{r})\big)\right\}\psi_k(\boldsymbol{r})=\epsilon_k\psi_k(\boldsymbol{r})\ . \tag{6}$$

The scalar and vector potentials

$$S(\boldsymbol{r})=\Sigma_S(\boldsymbol{r})+\tau_3\Sigma_{TS}(\boldsymbol{r})\ , \tag{7}$$

$$V^\mu(\boldsymbol{r})=\Sigma^\mu(\boldsymbol{r})+\tau_3\Sigma^\mu_{TV}(\boldsymbol{r})\ , \tag{8}$$

contain the nucleon isoscalar-scalar, isovector-scalar, isoscalar-vector and isovector-vector self-energies:

$$\Sigma_S=\alpha_S\rho_S+\beta_S\rho_S^2+\gamma_S\rho_S^3+\delta_S\triangle\rho_S\ , \tag{9}$$

$$\Sigma_{TS}=\alpha_{TS}\rho_{TS}+\delta_{TS}\triangle\rho_{TS}\ , \tag{10}$$

$$\Sigma^\mu=\alpha_V j^\mu+\gamma_V(j_\nu j^\nu)j^\mu+\delta_V\triangle j^\mu+eA^\mu\frac{1-\tau_3}{2}\ , \tag{11}$$

$$\Sigma^\mu_{TV}=\alpha_{TV}j^\mu_{TV}+\delta_{TV}\triangle j^\mu_{TV}\ , \tag{12}$$

respectively. Because of charge conservation, only the 3rd component of the isovector densities and currents contribute to the nucleon self-energies. In the present review we only consider even-even nuclei, i.e. time-reversal invariance is assumed, which implies that the spatial components of the single-nucleon currents vanish in the nuclear ground state.

The Dirac equation (6) is solved by expanding the nucleon spinors in the basis of a three-dimensional harmonic oscillator in Cartesian coordinates. In this way both axial and triaxial nuclear shapes can be described. In addition, to reduce the computational task, it is assumed that the total densities are symmetric under reflections with respect to all three planes xy, xz and yz. When combined with time-reversal invariance, this also implies that parity is conserved. Under these restrictions we consider only even-multipole deformations, whereas solutions for odd multipoles vanish. The method of solution of the Dirac equation is described in more detail in Refs. 14, 15.

In addition to the self-consistent mean-field potentials, for open-shell nuclei pairing correlations are taken into account in the BCS approximation, with a smooth cutoff factor to mimic the effect of finite range.[16,17] Thus, to the functional of Eq. (1) one adds a pairing energy term that explicitly depends on the pairing tensor κ,

$$E_{pair}[\kappa,\kappa^*] = \sum_{k,k'>0} f_k f_{k'} \langle k\bar{k}|V^{pp}|k'\bar{k}'\rangle \kappa_k^* \kappa_{k'}, \tag{13}$$

with the smooth cutoff factor

$$f_k = \frac{1}{1+exp[(\epsilon_k - \epsilon_F - \Delta E_\tau)/\mu_\tau]}, \tag{14}$$

where ϵ_k is the eigenvalue of the self-consistent single-particle field, and ϵ_F is the chemical potential determined by the particle number subsidiary condition $\langle \hat{N}_\tau \rangle = N_\tau$, with N_τ the number of neutrons or protons. The cutoff parameters ΔE_τ and $\mu_\tau = \Delta E_\tau/10$ are chosen in such a way that $2\sum_{k>0} f_k = N_\tau + 1.65 N_\tau^{2/3}$.[17] A density-independent δ force is adopted in the pairing channel, thus the pairing energy is given by

$$E_{pair}[\kappa,\kappa^*] = -\sum_{\tau=n,p} \frac{V_\tau}{4} \int d^3r \kappa_\tau^*(r)\kappa_\tau(r), \tag{15}$$

where V_τ is the pairing strength and the pairing tensor $\kappa(r)$ reads

$$\kappa(r) = -2\sum_{k>0} f_k u_k v_k |\psi_k(r)|^2. \tag{16}$$

The pairing strength parameters V_p and V_n are usually determined by adjusting the average single-particle pairing gaps for protons and neutrons

$$\langle \Delta_{p,n} \rangle = \frac{\sum_{k\in p,n} f_k u_k v_k \Delta_k}{\sum_k f_k u_k v_k} \tag{17}$$

to reproduce the data obtained from experimental masses using a five-point formula.[17]

Since translational symmetry is broken in the mean-field approximation, a proper treatment of the spurious center-of-mass motion is very important. The center-of-mass correction is included by adding the expectation value

$$E_{cm} = -\frac{\langle \hat{P}_{cm}^2 \rangle}{2mA}, \tag{18}$$

to the total energy.[18] Finally, the expression for the total energy reads

$$E_{\text{tot}} = \int \left[\mathcal{E}_{\text{RMF}}(\boldsymbol{r}) + \mathcal{E}^p_{\text{pair}}(\boldsymbol{r}) + \mathcal{E}^n_{\text{pair}}(\boldsymbol{r})\right] d\boldsymbol{r} + E_{cm}\,. \tag{19}$$

The pairing energy density reads

$$\mathcal{E}^{\tau}_{\text{pair}} = -\frac{1}{4} V_{\tau} \kappa^{*}_{\tau}(r) \kappa_{\tau}(r). \tag{20}$$

In the case of deformed nuclei, the map of the energy surface as a function of deformation is obtained by imposing constraints on mass multipole moments such as, for instance, the axial and triaxial quadrupole moments. The method of quadratic constraints uses an unrestricted variation of the function

$$\langle H \rangle + \sum_{\mu=0,2} C_{2\mu} \left(\langle \hat{Q}_{2\mu} \rangle - q_{2\mu} \right)^2 , \tag{21}$$

where $\langle H \rangle$ is the total energy, and $\langle \hat{Q}_{2\mu} \rangle$ denotes the expectation value of the mass quadrupole operator:

$$\hat{Q}_{20} = 2z^2 - x^2 - y^2 \quad \text{and} \quad \hat{Q}_{22} = x^2 - y^2 \, . \tag{22}$$

$q_{2\mu}$ is the constrained value of the multipole moment, and $C_{2\mu}$ the corresponding stiffness constant.[19]

The single-nucleon wave functions, energies and occupation factors, generated from constrained self-consistent solutions of the RMF+BCS equations, provide the microscopic input for consistent *beyond mean-field* structure models: the generator coordinate method configuration mixing of angular-momentum projected triaxial wave functions, or the collective Hamiltonian for vibrations and rotations. The solution of the corresponding eigenvalue problem yields the excitation spectra and collective wave functions that are used in the calculation of electromagnetic transition probabilities. This approach will be illustrated in the following sections.

3. Configuration mixing of angular-momentum and particle-number projected triaxial relativistic mean-field wave functions

This section presents the formalism and a few selected illustrative calculations for the angular-momentum and particle-number projected triaxial generator coordinate method (PN3DAMP GCM). In a first step we introduce a model with approximate particle-number projection and examine its deficiencies. In the next step the full particle-number projection is considered. As an example, we employ the PN3DAMP GCM to calculate the low-lying spectrum of the nucleus ^{76}Kr.

3.1. *The 3DAMP GCM model*

The generator coordinate method (GCM)[19] is based on the assumption that, starting from a set of mean-field states $|\Phi(q)\rangle$ that depend on a collective coordinate q, one can build approximate eigenstates of the nuclear Hamiltonian:

$$|\Psi_\alpha\rangle = \int dq f_\alpha(q)\,|\Phi(q)\rangle\ . \tag{23}$$

Here the basis states $|\Phi(q)\rangle$ are Slater determinants of single-nucleon states generated by self-consistent solutions of constrained relativistic mean-field (RMF) + BCS equations. To be able to compare theoretical predictions with data, it is necessary to construct states with good angular momentum. Thus the trial angular-momentum projected GCM collective wave function $|\Psi_\alpha^{JM}\rangle$, an eigenfunction of $\hat{J}^2$ and $\hat{J}_z$, with eigenvalues $J(J+1)\hbar^2$ and $M\hbar$, respectively, reads

$$|\Psi_\alpha^{JM}\rangle = \int dq \sum_{K\geq 0} f_\alpha^{JK}(q)\frac{1}{(1+\delta_{K0})}|JMK,q\rangle \tag{24}$$

where $\alpha = 1,2,\cdots$ labels collective eigenstates for a given angular momentum J. The details of the 3D angular-momentum projection in the RMF framework are given in Ref. 8, and here we only outline the essential features. Because of the D_2 and time-reversal symmetries of a triaxially deformed even-even nucleus, the projection of the angular momentum J along the intrinsic z-axis (K in Eq. (24)) takes only non-negative even values:

$$K = \begin{cases} 0,2,\cdots,J, & \text{for } J \bmod 2 = 0, \\ 2,4,\cdots,J-1, & \text{for } J \bmod 2 = 1. \end{cases} \tag{25}$$

The basis states $|JMK,q\rangle$ are projected from the intrinsic wave functions $|\Phi(q)\rangle$:

$$|JMK,q\rangle = [\hat{P}^J_{MK} + (-1)^J\hat{P}^J_{M-K}]|\Phi(q)\rangle, \tag{26}$$

where $\hat{P}^J_{MK}$ is the angular-momentum projection operator:

$$\hat{P}^J_{MK} = \frac{2J+1}{8\pi^2}\int d\Omega D^{J*}_{MK}(\Omega)\hat{R}(\Omega)\ . \tag{27}$$

Ω denotes the three Euler angles: (ϕ,θ,ψ), and $d\Omega = d\phi\sin\theta d\theta d\psi$. $D^J_{MK}(\Omega)$ is the Wigner D-function, with the rotational operator chosen in the notation of Edmonds:[20] $\hat{R}(\Omega) = e^{i\phi\hat{J}_z}e^{i\theta\hat{J}_y}e^{i\psi\hat{J}_z}$. The set of intrinsic wave functions $|\Phi(q)\rangle$, with the generic notation for quadrupole deformation parameters $q \equiv (\beta, \gamma)$, is generated by imposing constraints on the axial q_{20} and triaxial q_{22} mass quadrupole moments in a self-consistent RMF+BCS calculation. These moments are related to the Hill-Wheeler[21] coordinates β ($\beta > 0$) and γ by the following relations:

$$q_{20} = \sqrt{\frac{5}{16\pi}}\langle 2z^2 - x^2 - y^2\rangle = \frac{3}{4\pi}AR_0^2\beta\cos\gamma, \tag{28a}$$

$$q_{22} = \sqrt{\frac{15}{32\pi}}\langle x^2 - y^2\rangle = \frac{3}{4\pi}AR_0^2\frac{1}{\sqrt{2}}\beta\sin\gamma, \tag{28b}$$

where $R_0 = 1.2A^{1/3}$ fm. The total mass quadrupole moment q_m reads:

$$q_m = \sqrt{\frac{16\pi}{5}}\sqrt{q_{20}^2 + 2q_{22}^2}. \tag{29}$$

The weight functions $f_\alpha^{JK}(q)$ in the collective wave function Eq. (24) are determined from the variational equation:

$$\delta E^J = \delta\frac{\langle\Psi_\alpha^{JM}|\,\hat{H}\,|\Psi_\alpha^{JM}\rangle}{\langle\Psi_\alpha^{JM}|\Psi_\alpha^{JM}\rangle} = 0\,, \tag{30}$$

that is, by requiring that the expectation value of the Hamiltonian is stationary with respect to an arbitrary variation δf_α^{JK}. This leads to the Hill–Wheeler–Griffin (HWG) integral equation:

$$\int dq' \sum_{K'\geq 0}\left[\mathcal{H}_{KK'}^J(q,q') - E_\alpha^J\mathcal{N}_{KK'}^J(q,q')\right] f_\alpha^{JK'}(q') = 0, \tag{31}$$

where $\mathcal{H}$ and $\mathcal{N}$ are the angular-momentum projected GCM kernel matrices of the Hamiltonian and the norm, respectively. With the generic notation $\mathcal{O} \equiv \mathcal{N}$ or $\mathcal{H}$, the expression for the kernel reads:

$$\begin{aligned}\mathcal{O}_{KK'}^J(q,q') = \Delta_{KK'}[&O_{KK'}^J(q,q') + (-1)^{2J}O_{-K-K'}^J(q,q')\\ &+ (-1)^J O_{K-K'}^J(q,q') + (-1)^J O_{-KK'}^J(q,q')],\end{aligned} \tag{32}$$

where for the operator $\hat{O} \equiv 1$ or $\hat{H}$:

$$O_{KK'}^J(q,q') = \langle\Phi(q)|\hat{O}\hat{P}_{KK'}^J|\Phi(q')\rangle, \tag{33}$$

and $\Delta_{KK'} = 1/[(1+\delta_{K0})(1+\delta_{K'0})]$.

The overlap $\langle\Phi(q)|\hat{H}\hat{R}|\Phi(q')\rangle$ can be evaluated in coordinate space, and thus we rewrite the Hamiltonian kernel $H^J_{KK'}(q,q')$ in the following form:

$$H^J_{KK'}(q,q') = \int d\boldsymbol{r} H^J_{KK'}(\boldsymbol{r};q,q') \,, \tag{34}$$

where

$$H^J_{KK'}(\boldsymbol{r};q,q') = \frac{2J+1}{8\pi^2}\int d\Omega D^{J*}_{KK'}\mathcal{H}(\boldsymbol{r};q,q';\Omega)n(q,q';\Omega) \tag{35}$$

The norm overlap $n(q,q';\Omega)$ is defined by:

$$n(q,q';\Omega) \equiv \langle\Phi(q)|\hat{R}(\Omega)|\Phi(q')\rangle \,. \tag{36}$$

The calculation of the overlap matrix elements $\mathcal{H}(\boldsymbol{r};q,q';\Omega)$ requires the explicit form of $\hat{H}$. So far we have implicitly assumed that the system is described by a Hamiltonian. However, for energy density functionals this is strictly valid only if the density dependence can be expressed as a polynomial of ρ. By using product wave functions, a density functional can formally be derived from a Hamiltonian that contains many-body interactions. A prescription based on the generalized Wick theorem[22] states that the Hamilton overlap matrix elements have the same form as the mean field functional, with the intrinsic single particle density matrix elements replaced by the corresponding transition density matrix elements.[23] In this work we employ the relativistic point-coupling energy density functionals PC-F1[11] and PC-PK1,[13] which include powers of the scalar density and vector current up to the fourth order and, therefore, the above prescription can be applied. For a detailed discussion of open issues related to breaking and restoring symmetries in the framework of nuclear energy density functional theory, we refer the reader to Ref. 24, and references cited therein.

Consequently, $\mathcal{H}(\boldsymbol{r};q,q';\Omega)$ has the same form as the mean-field functional $\mathcal{E}_{RMF}(\boldsymbol{r})$ in Eq. (1) provided the *intrinsic* densities and currents are replaced by *transition* densities and currents. Further details about the calculation of the norm overlap $n(q,q';\Omega)$ and transition EDF $\mathcal{H}(\boldsymbol{r};q,q';\Omega)$ can be found in Ref. 8.

In practice, the first step in the solution of the HWG matrix eigenvalue equation is the diagonalization of the norm overlap kernel $\mathcal{N}^J(i,j)$

$$\sum_j \mathcal{N}^J(i,j)u^J_k(j) = n^J_k u^J_k(i). \tag{37}$$

Since the basis functions are not linearly independent, many of the eigenvalues n^J_k are very close to zero. They correspond to high momentum collective components, i.e., the corresponding eigenfunctions $u_k^J(i)$ are rapidly oscillating in the q space but carry very little physical information. However, due to numerical uncertainties, their contribution to the matrix elements of the collective Hamiltonian can be large, and these states should be removed from the basis. Therefore, a small positive constant ζ is introduced so that states with $n_k^J/n_{max}^J < \zeta$ are excluded from the GCM basis, where n_{max}^J is the largest eigenvalue of the norm kernel. From the remaining states, also called "natural states", one builds the collective Hamiltonian

$$\mathcal{H}_{kl}^J = \frac{1}{\sqrt{n_k^J}\sqrt{n_l^J}} \sum_{i,j} u_k^J(i)\mathcal{H}^J(i,j)u_l^J(j), \tag{38}$$

which is subsequently diagonalized

$$\sum_l \mathcal{H}_{kl}^J g_l^{J\alpha} = E_\alpha^J g_k^{J\alpha}. \tag{39}$$

The solution of Eq. (39) determines both the energies and the amplitudes $f_\alpha^{JK}(q)$ of collective states

$$f_\alpha^{JK}(q) = \sum_k \frac{g_k^{J\alpha}}{\sqrt{n_k^J}} u_k^J(i). \tag{40}$$

The collective wave functions $g_\alpha^J(i)$ are calculated from the norm overlap eigenstates:

$$g_\alpha^J(i) = \sum_k g_k^{J\alpha} u_k^J(i), \tag{41}$$

$g_\alpha^J(i)$ are orthonormal and, therefore, $|g_\alpha^J(i)|^2$ can be interpreted as a probability amplitude.

The center-of-mass (c.m.) correction is defined by:

$$\begin{aligned}\langle E_{\rm cm}\rangle(J_\alpha) &= \langle \Psi_\alpha^{JM}|\frac{\hat{\mathbf{P}}_{\rm cm}^2}{2mA}|\Psi_\alpha^{JM}\rangle \\ &= \sum_{ij}\sum_{KK'} f_\alpha^{JK*}(i) f_\alpha^{JK'}(j)\frac{1}{2mA}\langle\Phi(q_i)|\hat{\mathbf{P}}_{\rm cm}^2 P_{KK'}^J|\Phi(q_j)\rangle\,. \end{aligned}\tag{42}$$

The projected overlap matrix elements $\langle\Phi(q_i)|\hat{\mathbf{P}}_{\rm cm}^2 P_{KK'}^J|\Phi(q_j)\rangle$ are treated in zeroth order of the Kamlah approximation, i.e. considering the fact that $\langle\Phi(q_i)|\Phi(q_j)\rangle$ is sharply peaked at $q_i = q_j$, the projected matrix elements are approximated by the unprojected ones,[19] and

$$\frac{1}{2mA}\langle\Phi(q_i)|\hat{\mathbf{P}}_{\rm cm}^2 \hat{P}_{KK'}^J|\Phi(q_j)\rangle \approx \mathcal{N}_{KK'}^J(q_i,q_j)E_{\rm cm}(q_i)\,, \tag{43}$$

where $E_{\rm cm}(q)$ is the c.m. correction evaluated for the intrinsic states $|\Phi(q)\rangle$

$$E_{\rm cm}(q) = \frac{1}{2mA}\langle\Phi(q)|\hat{\mathbf{P}}^2_{\rm cm}|\Phi(q)\rangle\ . \tag{44}$$

m is the nucleon mass, and A denotes the number of nucleons. $\hat{\mathbf{P}}_{\rm cm} = \sum_i^A \hat{\mathbf{p}}_i$ is the total momentum. The energy of the collective state $|\Psi^J_\alpha\rangle$ is, therefore, given by the relation

$$\mathbb{E}(J^+_\alpha) = E^J_\alpha + \langle E_{\rm cm}\rangle(J_\alpha)\ . \tag{45}$$

The basis states $|\Phi(q)\rangle$ are not eigenstates of the proton and neutron number operator $\hat{Z}$ and $\hat{N}$, respectively. The location of the Fermi energy in a BCS calculation is determined by the requirement that the expectation value of the nucleon number operator corresponds to the actual number of nucleons. It follows that the wave functions $\left|\Psi^{JM}_\alpha\right\rangle$ are generally not eigenstates of the nucleon number operator and, moreover, the average value of the nucleon number in these states is not necessarily equal to the number of nucleons in a given nucleus. This happens because the binding energy increases with the average number of nucleons and, therefore, an unconstrained variation of the weight functions in a GCM calculation will generate a ground state with the average number of protons and neutrons larger than the actual values in a given nucleus. In order to restore the correct mean value of the nucleon number, we follow the standard prescription,[25,26] and modify the HWG equation by replacing $\mathcal{H}(\boldsymbol{r};q,q';\Omega)$ with

$$\begin{aligned}\mathcal{H}'(\boldsymbol{r};q,q';\Omega) = \mathcal{H}(\boldsymbol{r};q,q';\Omega) &- \lambda_p[Z(\boldsymbol{r};q,q';\Omega) - Z_0]\\ &-\lambda_n[N(\boldsymbol{r};q,q';\Omega) - N_0],\end{aligned} \tag{46}$$

where Z_0 and N_0 are the desired proton and neutron numbers, respectively. $Z(\boldsymbol{r};q,q';\Omega)$ and $N(\boldsymbol{r};q,q';\Omega)$ are the transition vector densities in $\boldsymbol{r}$-space for protons and neutrons, respectively. The Lagrange parameters $\lambda_{\tau=p,n}$ are in principle determined in such a way that each AMP GCM collective state has the correct average particle number. In that case, however, the Lagrange parameters λ_τ will be state dependent and, as a consequence, the orthonormality of the states $|\Psi^{JM}_\alpha\rangle$ is no longer guaranteed. In Ref. 26 a simple ansatz was introduced for a state-independent value of the Lagrange parameter, that is, the value of $\lambda_{\tau=p,n}$ is chosen to be the mean BCS Fermi energy, determined by averaging over the collective variable q (further denoted as type 2). We use a slightly modified ansatz: for the diagonal terms ($q' = q$) λ_τ is equal to the value obtained in the mean-field calculation for the given constraint q, while for the off-diagonal terms λ_τ is calculated as the mean value $[\lambda_\tau(q)+\lambda_\tau(q')]/2$ (further denoted as type 1). Our method ensures that at least the diagonal matrix elements are exact, however it should be noted that the difference between the two methods is not large. In Table 1 we have compared the average number of particles in various 1DAMP GCM states calculated with both types of corrections and the differences are below 1%, except the 0_2^+ state.

Table 1. The average number of neutrons / protons for various states obtained with the 1DAMP+GCM in the ^{24}Mg isotope. Calculations are performed using the energy density functional PC-F1.[11]

state	without correction N/P	correction type 1 N/P	correction type 2 N/P
0_1^+	11. 9989 / 11.9987	11.9945 / 11.9942	11.9952 / 11.9949
0_2^+	12.2605 / 12.2716	11.7202 / 11.7106	11.7356 / 11.7251
0_3^+	11.9878 / 11.9868	12.0368 / 12.0507	11.9906 / 12.0018
2_1^+	12.0000 / 12.0000	11.9993 / 11.9993	11.9995 / 11.9994
2_2^+	12.2069 / 12.2102	11.9029 /11.8946	11.9208 /11.9125
2_3^+	12.0303 / 12.0394	12.0265 / 12.0381	12.0359 / 12.0472
4_1^+	12.0009 / 12.0009	12.0000 / 12.0000	12.0000 / 12.0000
4_2^+	12.3492 / 12.3529	11.9681 / 11.9725	11.9643 /11.9684
4_3^+	11.9909 / 11.9919	12.0270 / 12.0288	12.0265 /12.0287

To emphasize the importance of particle number correction, in Fig. 1 we display the Fermi energies λ_τ (upper panel) and pairing energies E_p (middle panel) of neutrons and protons in the intrinsic deformed states, together with the corresponding average particle number $\langle \hat{N}_\tau \rangle$ in the angular-momentum projected states as functions of the deformation parameter β for ^{24}Mg. We notice that the deviation of the average particle number from the actual number of nucleons can be as large as 0.4 particles, both for neutrons and protons. Furthermore, the deviation varies significantly for different values of β and is much more pronounced in the regions of β with strong pairing. A distinct dependence on the value of the angular momentum J follows from the decomposition of the intrinsic wave function (see Fig. 2), as a weighted sum over all J components has to yield the unprojected particle number. Having to correct for 0.4 missing or additional nucleons when the Fermi energy is about -10 MeV, one obtains

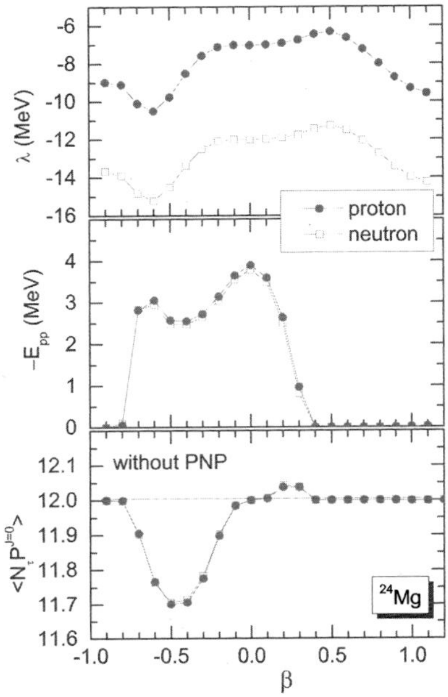

Fig. 1. (Color online) Fermi levels λ_τ and pairing energies E_p of neutrons (empty squares) and protons (filled circles) in the intrinsic deformed states, and the corresponding average particle numbers $\langle \hat{N}_\tau \rangle$ in the angular-momentum projected states, plotted as functions of the deformation parameter β for ^{24}Mg. The energy density functional PC-F1[11] has been used in this calculation.

a correction of the order of a few MeV, i.e., the correction is large in comparison to the spacing between curves with different values of the angular momentum. This effect can even cause the wrong ordering of the angular momentum projected energy curves, as shown in the left panel of Fig. 3. On the other hand, when the correction scheme of type 1 is included, the correct ordering of angular-momentum projected PECs is restored (see the right panel of Fig. 3).

3.2. *The model PN3DAMP*

To avoid problems with the non-conservation of particle number, we have developed the model further by including the full restoration of particle number in the wave functions of GCM states, that is, the model restores a symmetry that is broken on the mean-field level by the treatment of pairing correlations in the BCS approximation. A GCM configuration mixing of angular-momentum and particle-number projected relativistic wave func-

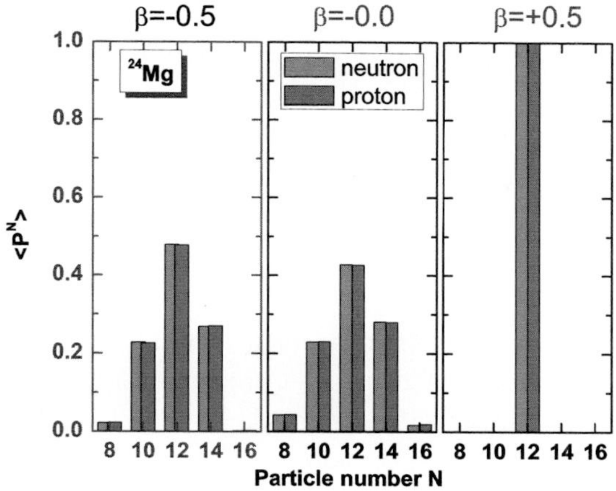

Fig. 2. (Color online) The decomposition of the intrinsic deformed wave functions into components of good particle numbers. The energy density functional PC-F1[11] is used in the calculation.

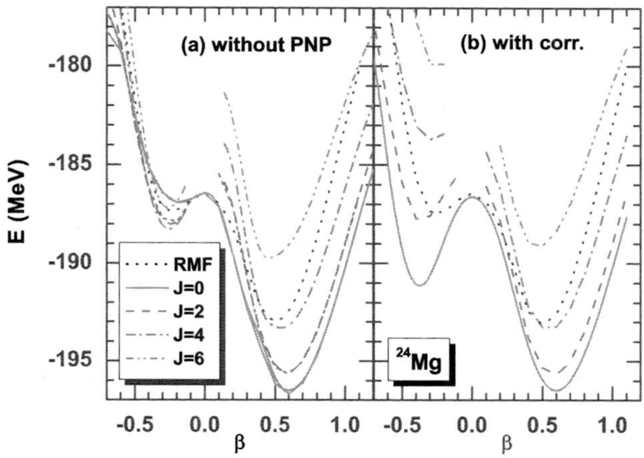

Fig. 3. (Color online) Mean-field and angular-momentum projected potential energy curves of ^{24}Mg calculated without (left panel) and with (right panel) particle-number correction. The energy density functional is PC-F1.[11]

tions is performed. The nuclear many-body wave function is expressed as a linear combination of projected mean-field configurations generated by the collective coordinates of quadrupole deformations,

$$|JNZ;\alpha\rangle = \int dq \sum_K f_\alpha^{JK} \hat{P}_{MK}^J \hat{P}^N \hat{P}^Z |q\rangle, \tag{47}$$

where $\alpha = 1, 2, \ldots$ denotes different collective states for a given angular momentum J, and $|q\rangle \equiv |\beta, \gamma\rangle$ denotes a set of RMF+BCS states with deformation parameters (β, γ). $\hat{P}^J_{MK}$ is the angular momentum projection operator. The operators $\hat{P}^N$ and $\hat{P}^Z$ project onto states with good neutron and proton number, respectively:

$$\hat{P}^N = \frac{1}{2\pi}\int_0^{2\pi} d\phi_n e^{i(\hat{N}-N)\phi_n}, \quad \hat{P}^Z = \frac{1}{2\pi}\int_0^{2\pi} d\phi_p e^{i(\hat{Z}-Z)\phi_p}, \tag{48}$$

where $\hat{N}(\hat{Z})$ is the number operator for neutrons (protons), and $N(Z)$ denotes the actual number of neutrons (protons) for a given nucleus. For an even number of particles the integration interval in Eq. (48) can be reduced to $[0, \pi]$. Furthermore, the integrals can be discretized by using the Fomenko expansion:[27]

$$\hat{P}^N = \frac{1}{L}\sum_{n=1}^{L} e^{i(\hat{N}-N)\phi_n}, \quad \phi_n = \frac{\pi}{L}n, \tag{49}$$

with L points in the expansion. To avoid numerical instabilities that might arise at $\phi = \pi/2$ when the occupation probability of a state is exactly $1/2$, an odd number of points must be used in the expansion.[27]

To illustrate the effect of non-conservation of particle number, in Fig. 2 we display the decomposition of the intrinsic deformed wave functions in ^{24}Mg into components of good particle numbers for three values of deformation: $\beta = -0.5$, $\beta = 0.0$ and $\beta = +0.5$. For configurations with large pairing energies a strong fragmentation into states with different nucleon number is observed, whereas the prolate deformed configuration with vanishing pairing is actually an eigenstate of the particle number operator.

3.3. *Multipole transition strength*

When the amplitudes $f^{JK}_\alpha(q)$ of the nuclear collective wave functions $|\Psi^{JM}_\alpha\rangle$ are known, it is straightforward to calculate all physical observables, such as the electromagnetic transition probability, spectroscopic quadrupole moments and the average particle number. The $B(E2)$ probability for a transition from an initial state (J_i, α_i) to a final state (J_f, α_f) is computed from:

$$B(E2; J_i, \alpha_i \to J_f, \alpha_f) = \frac{e^2}{2J_i+1}\left|\sum_{q_f, q_i}\langle J_f, q_f||\hat{Q}_2||J_i, q_i\rangle\right|^2. \tag{50}$$

Using the generalized Wigner–Eckart theorem for the spherical tensor operator $\hat{Q}_{\lambda\mu}$:

$$\hat{P}^J_{KM}\hat{Q}_{\lambda\mu}\hat{P}^{J'}_{M'K'} = C^{J\,M}_{J'M'\lambda\mu}\sum_{\bar{K}\mu'} C^{J\,K}_{J'\bar{K}\lambda\mu'}\hat{Q}_{\lambda\mu'}\hat{P}^{J'}_{\bar{K}K'}\,, \tag{51}$$

and the relation

$$\hat{P}^J_{MK}\hat{P}^{J'}_{M'K'} = \delta_{JJ'}\delta_{KM'}\hat{P}^J_{MK'} \tag{52}$$

for projection operators,[19] one obtains for the reduced matrix element $\langle J_f, q_f||\hat{Q}_2||J_i, q_i\rangle$:

$$\begin{aligned}\langle J_f, q_f||\hat{Q}_2||J_i, q_i\rangle = \hat{J}_f \sum_{K_i K_f} & f^{*J_f K_f}_{\alpha_f}(q_f) f^{J_i K_i}_{\alpha_i}(q_i) \\ & \times \sum_{\mu K'} (-1)^{J_f - K_f} \begin{pmatrix} J_f & 2 & J_i \\ -K_f & \mu & K' \end{pmatrix} Q_{2\mu}(K', K_i; q_f, q_i)\end{aligned} \tag{53}$$

with $\hat{J}_f = 2J_f + 1$, $f^{JK}_\alpha(q) = (-1)^J f^{J-K}_\alpha(q)$ for $K < 0$, and

$$Q_{2\mu}(K', K_i; q_f, q_i) \equiv \langle \Phi(q_f)|\hat{Q}_{2\mu}\hat{P}^{J_i}_{K'K_i}|\Phi(q_i)\rangle. \tag{54}$$

More details about the calculation of the reduced $E2$ matrix element are given in Appendix I of Ref. 9. The matrix elements of the charge quadrupole operator $\hat{Q}_{2\mu} = e\sum_p r_p^2 Y_{2\mu}(\Omega_p)$ are calculated in the full configuration space, that is, there are no effective charges and e is simply the bare value of the proton charge.

As an illustration, in Fig. 4 we display the energy spectrum of low-lying states in ^{24}Mg obtained from the 1DAMP+GCM calculations without and with the particle-number correction (type 1). The results are compared with the calculation that includes the full particle number projection.

3.4. *Illustrative calculation:* ^{76}*Kr*

As an illustrative example we present the PN3DAMP GCM analysis of low-energy collective states of ^{76}Kr, based on the energy density functional PC-PK1 and the density-independent delta force in the pairing channel.[13] The calculation is carried out in the model space of ten major oscillator shells. The Gauss–Legendre quadrature is used for the integrals over the three Euler angles (ϕ, θ, ψ) in the calculation of the projected kernels. The number of mesh points in the interval $[0, \pi]$ are chosen as follows:

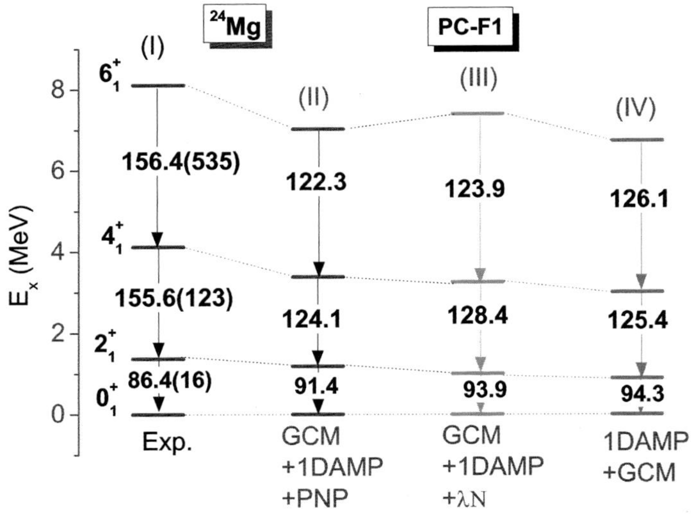

Fig. 4. (Color online) Low-lying states of ^{24}Mg calculated using the 1DAMP+GCM model based on the energy density functional PC-F1.[11] Results of calculations performed without (IV) and with particle-number correction (type 1) (III), are compared with the calculation that includes the full particle number projection (II), and with the experimental excitation spectrum (I).

$N_\phi = 10, N_\theta = 14, N_\psi = 12$. The number of gauge angles in the Fomenko expansion in Eq. (49) is $N_\phi = 9$. The Pfaffian method[28] is implemented to calculate the norm overlaps.

In Fig. 5 we display the constrained energy surfaces in the (β, γ) plane for ^{76}Kr. Figure 5(a) shows the energy calculated using the constrained RMF+BCS method. A spherical minimum is found that extends along oblate shapes and competes with the prolate deformed minimum. Figure 5(b) displays the corresponding energy surface with the inclusion of particle number projection and, finally, the energy surface shown in Fig. 5(c) includes also the angular momentum projection on $J = 0$. We note that particle number projection alone does not lead to significant modifications of the mean-field energy. 3DAMP, however, changes the topology in a considerable way. In particular, we note the onset of triaxial deformation, soft along the curve that connects the minimum with weak oblate deformation ($|\beta| \approx 0.2$) and the minimum with large prolate deformation ($\beta \approx 0.5$). As a result of angular-momentum projection both these minima become saddle points on the triaxial energy surface. Because of this effect, calculations restricted to axially symmetric configurations obviously cannot

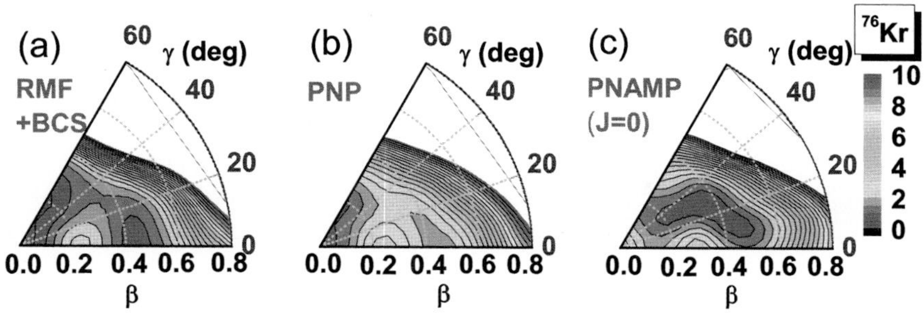

Fig. 5. (Color online) Constrained energy surfaces in the (β, γ) plane for ^{76}Kr: (a) RMF+BCS mean-field calculation, (b) RMF+BCS with particle-number projection (PNP), and (c) RMF+BCS with both PNP and angular-momentum projection (AMP) on $J = 0$. The energy density functional is PC-PK1,[13] and a density-independent delta force is used in the pairing channel.

provide a quantitative description of the structure of ^{76}Kr.[29–31] These calculations predict the equilibrium minimum at a small oblate deformation and, consequently, fail to reproduce the low-energy excitation spectrum of the isotope ^{76}Kr.

Figure 6 compares the low-lying states in ^{76}Kr, calculated both with the PN1DAMP GCM and PN3DAMP GCM models, to the experimental excitation spectrum. The comparison clearly shows that a restriction to

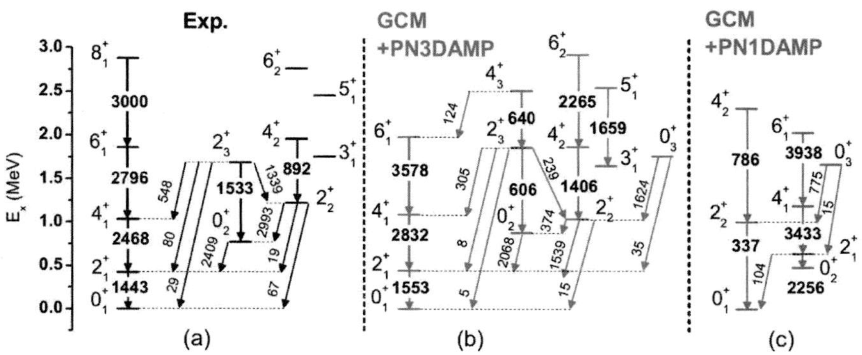

Fig. 6. (Color online) Low-lying excitation spectrum and $B(E2)$ values (in $e^2\mathrm{fm}^4$) of ^{76}Kr. Results obtained with the PN3DAMP GCM model (b), are compared to those of the corresponding PN1DAMP GCM calculation restricted to axially symmetric shapes (c), and to the experimental spectrum (a).[32]

axially symmetric configurations represents a poor approximation that fails to reproduce the experimental excitation energies and transition rates. The PN1DAMP GCM model predicts the coexistence of the ground-state band built on the oblate deformed minimum and the first excited band built on the prolate deformed minimum, consistent with previous calculations performed in Refs. 29, 30. The calculation that includes the triaxial degree of freedom, on the other hand, reproduces the details of the experimental spectrum with high accuracy. In particular, the strong electric quadrupole transition from the low-lying 0_2^+ state to the 2_1^+ state in the ground-state band observed in the data is also reproduced by the calculation. These results present strong evidence for the importance of the triaxial degree of freedom in the region of Kr isotopes, as already emphasized in Refs. 29, 30. The origin for the strong $0_2^+ \to 2_1^+$ transition can be traced back to the collective wave functions for the states 0_2^+ and 2_1^+, plotted in Fig. 7. The probability densities $\rho_{J\alpha}(q)$, calculated with the PN3DAMP GCM model, are defined by the following relation:

$$\rho_{J\alpha}(q) = \sum_K \left| \sum_{q'K'} \left[\mathcal{N}^J_{KK'}(q,q')\right]^{1/2} f_\alpha^{JK'}(q') \right|^2 . \qquad (55)$$

The distribution of the 2_1^+ collective wave function extends over the triaxial region, consistent with the topology of the PN3DAMP energy surface. This leads to a significant overlap with the probability density for the 0_2^+ state, and consequently to a strong $0_2^+ \to 2_1^+$ transition.

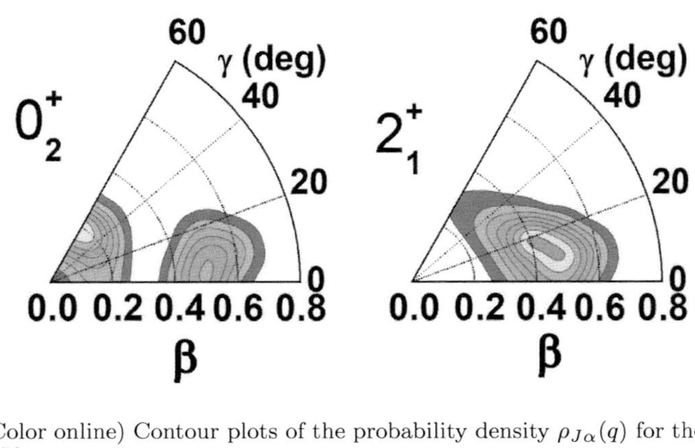

Fig. 7. (Color online) Contour plots of the probability density $\rho_{J\alpha}(q)$ for the states 0_2^+ and 2_1^+ in ^{76}Kr.

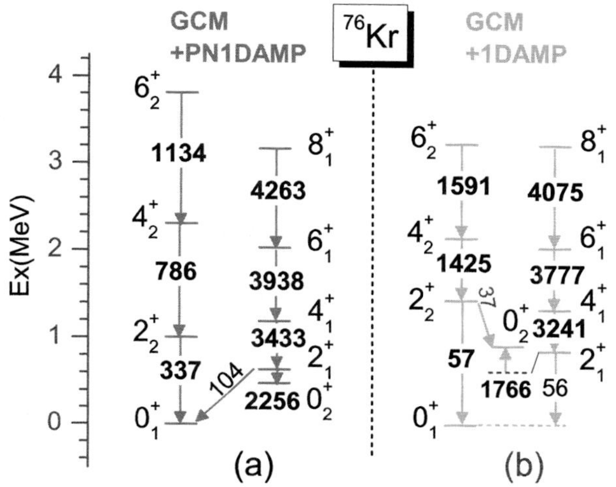

Fig. 8. (Color online) Low-lying energy states and $B(E2)$ values (in $e^2\text{fm}^4$) of ^{76}Kr. Results of the axially symmetric PN1DAMP GCM calculation (a), are compared to those obtained without particle-number projection (b).

Finally, to estimate the effect of particle-number projection (PNP), it is useful to compare the low-lying spectra calculated using the PN1DAMP GCM and 1DAMP GCM. This comparison is shown in Fig. 8, and one notices significant differences, especially for the position of the band-heads 0_1^+ and 0_2^+. This can be explained by the spurious mixing between the 0^+ states in the isotope ^{76}Kr and the 0^+ states in the neighbouring nuclei with $N \pm 2$ and/or $Z \pm 2$. The effect of spurious mixing can be particularly pronounced in transitional nuclei whose energy surfaces change rapidly with the number of nucleons.

4. Collective Hamiltonian in five dimensions

Collective quadrupole dynamics can also be described using an alternative method that includes rotational symmetry restoration and takes into account triaxial quadrupole fluctuations. In this approach the collective Bohr Hamiltonian is constructed, with deformation-dependent parameters determined from microscopic self-consistent mean-field calculations.[26,33] The collective Hamiltonian can be derived in the Gaussian overlap approximation (GOA)[19] to the full five-dimensional GCM. With the assumption that the GCM overlap kernels can be approximated by Gaussian functions, the local expansion of the kernels up to second order in the non-locality transforms the GCM Hill–Wheeler equation into a second-order differential equation — the Schrödinger equation for the collective Hamiltonian. The kinetic part of this Hamiltonian contains an inertia tensor,[34] and the potential energy is determined by the diagonal elements of the Hamiltonian kernel, and also includes zero-point energy (ZPE) corrections.[35] The adiabatic time-dependent Hartree–Fock (ATDHF) theory[36] provides an alternative way to derive a classical collective Hamiltonian and, after re-quantization, a Bohr Hamiltonian of the same structure is obtained, but with different microscopic expressions for the inertia parameters.[37] The choice of masses in the collective Bohr Hamiltonian has extensively been discussed in the literature:[38] the GCM-GOA expressions (the so-called Yoccoz masses[39]) and the ATDHF masses (the so-called Thouless–Valatin masses[40]). The latter have the advantage that they also include the time-odd components of the microscopic wave functions and, in this sense, the full dynamics of a nuclear system. In the GCM approach these components can only be included if, in addition to the coordinates q_i, the corresponding canonically conjugate momenta p_i are also taken into account. In many applications a further simplification is thus introduced in terms of cranking formulas,[35,41] i.e. the perturbative limit for the Thouless–Valatin masses, and the corresponding expressions for ZPE corrections. This approximation was applied in recent studies using models based both on the Gogny interaction,[42] and relativistic energy density functionals.[43]

4.1. *Theoretical framework*

Nuclear excitations determined by quadrupole vibrational and rotational degrees of freedom can be treated simultaneously by considering five quadrupole collective coordinates α_μ, $\mu = -2, -1, \ldots, 2$ that describe the surface of a deformed nucleus: $R = R_0[1 + \sum_\mu \alpha_\mu Y^*_{2\mu}]$. To separate rotational and vibrational motion, these coordinates are usually parameterized in terms of two deformation parameters β and γ, and three Euler angles $(\phi, \theta, \psi) \equiv \Omega$ which define the orientation of the intrinsic principal axes in the laboratory frame

$$\alpha_\mu = D^2_{\mu 0}(\Omega)\beta\cos\gamma + \frac{1}{\sqrt{2}}\left[D^2_{\mu 2}(\Omega) + D^2_{\mu -2}(\Omega)\right]\beta\sin\gamma\,, \tag{56}$$

where $D^\lambda_{\mu\nu}$ is the Wigner function.[44] The three terms of the classical collective Hamiltonian, expressed in terms of the intrinsic variables β, γ

and Euler angles

$$\mathcal{H}_{\text{coll}} = \mathcal{T}_{\text{vib}}(\beta,\gamma) + \mathcal{T}_{\text{rot}}(\beta,\gamma,\Omega) + \mathcal{V}_{\text{coll}}(\beta,\gamma)\ , \tag{57}$$

denote the contributions from the vibrational kinetic energy:

$$\mathcal{T}_{\text{vib}} = \frac{1}{2}B_{\beta\beta}\dot{\beta}^2 + \beta B_{\beta\gamma}\dot{\beta}\dot{\gamma} + \frac{1}{2}\beta^2 B_{\gamma\gamma}\dot{\gamma}^2\ , \tag{58}$$

the rotational kinetic energy:

$$\mathcal{T}_{\text{rot}} = \frac{1}{2}\sum_{k=1}^{3}\mathcal{I}_k\omega_k^2, \tag{59}$$

and the collective potential energy $\mathcal{V}_{\text{coll}}(\beta,\gamma)$. The mass parameters $B_{\beta\beta}$, $B_{\beta\gamma}$, $B_{\gamma\gamma}$, and the moments of inertia $\mathcal{I}_k$ depend on the quadrupole deformation variables β and γ.

The Hamiltonian Eq. (57) is quantized according to the general Pauli prescription:[45] for the classical kinetic energy

$$T = \frac{1}{2}\sum_{ij} B_{ij}(q)\dot{q}_i\dot{q}_j\ , \tag{60}$$

the corresponding quantized form reads:

$$\hat{H}_{\text{kin}} = -\frac{\hbar^2}{2}\frac{1}{\sqrt{\det B}}\sum_{ij}\frac{\partial}{\partial q_i}\sqrt{\det B}(B^{-1})_{ij}\frac{\partial}{\partial q_j}. \tag{61}$$

The kinetic energy tensor in Eq. (57) takes the block diagonal form:

$$B = \begin{pmatrix} B_{\text{vib}} & 0 \\ 0 & B_{\text{rot}} \end{pmatrix}, \tag{62}$$

with the vibrational part of the tensor

$$B_{\text{vib}} = \begin{pmatrix} B_{\beta\beta} & \beta B_{\beta\gamma} \\ \beta B_{\beta\gamma} & \beta^2 B_{\gamma\gamma} \end{pmatrix}. \tag{63}$$

In general the rotational part is a complicated function of the Euler angles but, using the quasi-coordinates related to the components of the angular momentum in the body-fixed frame, it takes a simple diagonal form

$$(B_{\text{rot}})_{ik} = \delta_{ik}\mathcal{I}_k, \quad k = 1,2,3\ , \tag{64}$$

with the moments of inertia expressed as

$$\mathcal{I}_k = 4B_k\beta^2\sin^2(\gamma - 2k\pi/3)\ . \tag{65}$$

This particular functional form is motivated by the fact that all three moments of inertia vanish for the spherical configuration ($\beta = 0$) and, additionally, $\mathcal{I}_z$ and $\mathcal{I}_y$ vanish for axially symmetric prolate ($\gamma = 0^0$) and oblate ($\gamma = 60^0$) configurations, respectively. The resulting determinant reads

$$\det B = \det B_{\text{vib}} \cdot \det B_{\text{rot}} = 4wr\beta^8 \sin^2 3\gamma \,, \tag{66}$$

where $w = B_{\beta\beta}B_{\gamma\gamma} - B_{\beta\gamma}^2$ and $r = B_1 B_2 B_3$. The quantized collective Hamiltonian can hence be written in the form:

$$\hat{H} = \hat{T}_{\text{vib}} + \hat{T}_{\text{rot}} + V_{\text{coll}} \,, \tag{67}$$

with

$$\begin{aligned}\hat{T}_{\text{vib}} = & -\frac{\hbar^2}{2\sqrt{wr}} \left\{ \frac{1}{\beta^4} \left[\frac{\partial}{\partial\beta} \sqrt{\frac{r}{w}} \beta^4 B_{\gamma\gamma} \frac{\partial}{\partial\beta} - \frac{\partial}{\partial\beta} \sqrt{\frac{r}{w}} \beta^3 B_{\beta\gamma} \frac{\partial}{\partial\gamma} \right] \right. \\ & \left. + \frac{1}{\beta \sin 3\gamma} \left[-\frac{\partial}{\partial\gamma} \sqrt{\frac{r}{w}} \sin 3\gamma B_{\beta\gamma} \frac{\partial}{\partial\beta} + \frac{1}{\beta} \frac{\partial}{\partial\gamma} \sqrt{\frac{r}{w}} \sin 3\gamma B_{\beta\beta} \frac{\partial}{\partial\gamma} \right] \right\} ,\end{aligned} \tag{68}$$

and

$$\hat{T}_{\text{rot}} = \frac{1}{2} \sum_{k=1}^{3} \frac{\hat{J}_k^2}{\mathcal{I}_k} \,, \tag{69}$$

where $\hat{J}_k$ denotes the components of the angular momentum in the body-fixed frame of a nucleus. V_{coll} is the collective potential. The Hamiltonian describes quadrupole vibrations, rotations, and the coupling of these collective modes. The determinant Eq. (66) specifies the volume element in the collective space:

$$\int d\tau_{coll} = \int d\Omega d\tau_0 \sqrt{wr} = \int_0^\infty d\beta \beta^4 \int_0^{2\pi} d\gamma |\sin 3\gamma| \int d\Omega \sqrt{wr} \,, \tag{70}$$

and the quantized Hamiltonian Eq. (67) is hermitian with respect to the collective measure Eq. (70).

The methods used to solve the eigenvalue problem of the general collective Hamiltonian Eq. (67) can be divided into two classes. The first is based on a direct numerical solution of a system of partial differential equations using finite-difference methods.[46–48] The second approach uses an expansion of eigenfunctions in terms of a complete set of basis functions, that depend on the deformation variables β and γ, and the Euler angles ϕ, θ and ψ.[49–52] The eigenvalue problem then reduces to a simple matrix diagonalization, and the main task is the construction of an appropriate basis for each value of the angular momentum quantum number.

Here we employ the second approach and construct basis states according to the method described in Refs. 42, 52–55. For each value of the angular momentum I, one chooses a complete set of square integrable functions

$$\phi_{Lmn}^{IM}(\beta,\gamma,\Omega) = e^{-\mu^2\beta^2/2}\beta^n \begin{Bmatrix} \cos m\gamma \\ \\ \sin m\gamma \end{Bmatrix} D_{ML}^{I*}(\Omega). \tag{71}$$

The projections M and L are determined by the angular momentum: $M, L = -I, \ldots, I$. In principle, the parameter n can take any non-negative integer value, but in actual calculations a certain cut-off value n_{max} has to be imposed. The allowed values of m are: $m = n, n-2, \ldots, 0$ or 1. The choice of the function $e^{-\mu^2\beta^2/2}$ ensures that the basis states generate wave functions that vanish at large deformations ($\beta \to \infty$). The basis parameter μ has to be adjusted for each nucleus individually, so that it minimizes the ground state energy of the nucleus. However, if the cut-off value n_{max} is large enough, a stable ground-state solution can be found for a broad range of values of the parameter μ .

The basis states have to fullfill certain symmetry conditions that originate from the fact that the choice of the body-fixed frame is not unique. For a given quadrupole tensor α_μ in the laboratory frame, there are 24 possible orientations of the body-fixed right-hand coordinate system, corresponding to different values of the variables β, γ, and Ω. The basis states in the body-fixed frame must be invariant with respect to the transformations that connect various choices of the body-fixed frame, and which form a finite group isomorphic to the octahedral point group O[46,56] (group of proper rotations which transform a cube or octahedron onto itself). This symmetry condition is fulfilled by linear combinations of the states (71)

$$\xi_{Lmn}^{IM}(\beta,\gamma,\Omega) = e^{-\mu^2\beta^2/2}\beta^n \sum_{K\in\Delta I} f_{LmK}^{I}(\gamma)\Phi_{MK}^{I}(\Omega)\,, \tag{72}$$

invariant under the transformations of the octahedral group. The angular part corresponds to linear combinations of the Wigner functions

$$\Phi_{MK}^{I}(\Omega) = \sqrt{\frac{2I+1}{16\pi^2(1+\delta_{K0})}}\left[D_{MK}^{I*}(\Omega) + (-1)^I D_{M-K}^{I*}(\Omega)\right]\,, \tag{73}$$

and the summation in Eq. (72) is over the allowed set of the K values:

$$\Delta I = \begin{cases} 0, 2, \ldots, I, & \text{for} \quad I \bmod 2 = 0, \\ \\ 2, 4, \ldots, I-1, & \text{for} \quad I \bmod 2 = 1. \end{cases} \tag{74}$$

In the next step linearly independent functions have to be selected from the overcomplete basis set Eq. (72). In addition, some of the basis states have to be discarded in order to enforce the correct behavior of solutions on the $\gamma = n\pi/3$ axes.[46] A simple and elegant solution of both problems is provided by group theoretical methods.[57] Finally, the basis states Eq. (72) are not orthogonal. Although the Hamiltonian could also be diagonalized directly in a non-orthogonal basis,[51] we choose to orthogonalize the basis states by applying the Cholesky–Banachiewicz method.[58]

The diagonalization of the collective Hamiltonian yields the energy spectrum E^I_α and the corresponding eigenfunctions

$$\Psi^{IM}_\alpha(\beta,\gamma,\Omega) = \sum_{K\in\Delta I} \psi^I_{\alpha K}(\beta,\gamma)\Phi^I_{MK}(\Omega). \tag{75}$$

Using the collective wave functions Eq. (75), various observables can be calculated and compared with experimental results. For instance, the quadrupole E2 reduced transition probability:

$$B(\mathrm{E2};\ \alpha I \to \alpha' I') = \frac{1}{2I+1}|\langle\alpha' I'||\hat{\mathcal{M}}(E2)||\alpha I\rangle|^2 , \tag{76}$$

and the spectroscopic quadrupole moment of the state $|\alpha I\rangle$:

$$Q_{\mathrm{spec},\alpha I} = \frac{1}{\sqrt{2I+1}} C^{II}_{II20}\langle\alpha I||\hat{\mathcal{M}}(E2)||\alpha I\rangle , \tag{77}$$

where $\hat{\mathcal{M}}(E2)$ denotes the electric quadrupole operator. Detailed expressions for the reduced matrix element $\langle\alpha' I'||\hat{\mathcal{M}}(E2)||\alpha I\rangle$ can be found in Ref. 46.

The shape of a nucleus can be characterized in a qualitative way by the expectation values of invariants β^2, $\beta^3\cos 3\gamma$, as well as their combinations. For example, the expectation value of the invariant β^2 in the state $|\alpha I\rangle$:

$$\langle\beta^2\rangle_{I\alpha} = \langle\Psi^I_\alpha|\beta^2|\Psi^I_\alpha\rangle = \sum_{K\in\Delta I}\int \beta^2|\psi^I_{\alpha,K}(\beta,\gamma)|^2 d\tau_0 , \tag{78}$$

and the expectation values of the deformation parameters β and γ in the state $|\alpha I\rangle$ are calculated from:

$$\langle\beta\rangle_{I\alpha} = \sqrt{\langle\beta^2\rangle_{I\alpha}}, \tag{79}$$

$$\langle\gamma\rangle_{I\alpha} = \frac{1}{3}\arccos\frac{\langle\beta^3\cos 3\gamma\rangle_{I\alpha}}{\sqrt{\langle\beta^2\rangle_{I\alpha}\langle\beta^4\rangle_{I\alpha}}}. \tag{80}$$

The mixing of different intrinsic configurations in the state $|\alpha I\rangle$ can be determined from the distribution of the projection K of the angular momentum I on the z axis in the body-fixed frame:

$$N_K = 6\int_0^{\pi/3}\int_0^{\infty} |\psi^I_{\alpha,K}(\beta,\gamma)|^2\beta^4|\sin 3\gamma| d\beta d\gamma, \tag{81}$$

where the components $\psi^I_{\alpha,K}(\beta,\gamma)$ are defined in Eq. (75). For large deformations the K quantum number is to a good approximation conserved. Consequently, only one of the integrals Eq. (81) will give a value close to one. A broader distribution of N_K values in the state $|\alpha I\rangle$ provides a measure of the mixing of intrinsic configurations.

4.2. *Parameters of the collective Hamiltonian*

The entire dynamics of the collective Hamiltonian is governed by the seven functions of the intrinsic deformations β and γ: the collective potential, the three mass parameters: $B_{\beta\beta}$, $B_{\beta\gamma}$, $B_{\gamma\gamma}$, and the three moments of inertia $\mathcal{I}_k$. These functions are determined by the choice of a particular microscopic nuclear energy density functional or effective interaction. The illustrative calculations presented in this section are based on the energy density functional PC-PK1.[13] The single-nucleon wave functions, energies and occupation factors, generated from constrained self-consistent solutions of the RMF+BCS equations, provide the microscopic input for the parameters of the collective Hamiltonian.

The moments of inertia are calculated according to the Inglis–Belyaev formula:[41,59]

$$\mathcal{I}_k = \sum_{i,j}\frac{(u_i v_j - v_i u_j)^2}{E_i + E_j}|\langle i|\hat{J}_k|j\rangle|^2, \quad k = 1,2,3, \tag{82}$$

where k denotes the axis of rotation, and the summation runs over the proton and neutron quasiparticle states. The quasiparticle energies E_i, occupation probabilities v_i, and single-nucleon wave functions ψ_i are determined by solutions of the constrained RMF+BCS equations. The mass parameters associated with the two quadrupole collective coordinates $q_0 = \langle\hat{Q}_{20}\rangle$ and $q_2 = \langle\hat{Q}_{22}\rangle$ are also calculated in the cranking approximation[35]

$$B_{\mu\nu}(q_0, q_2) = \frac{\hbar^2}{2}\left[\mathcal{M}^{-1}_{(1)}\mathcal{M}_{(3)}\mathcal{M}^{-1}_{(1)}\right]_{\mu\nu}, \tag{83}$$

with

$$\mathcal{M}_{(n),\mu\nu}(q_0,q_2) = \sum_{i,j} \frac{\langle i|\,\hat{Q}_{2\mu}\,|j\rangle\,\langle j|\,\hat{Q}_{2\nu}\,|i\rangle}{(E_i+E_j)^n}\,(u_i v_j + v_i u_j)^2\,. \tag{84}$$

The collective energy surface includes the energy of the zero-point motion, which has to be subtracted. The collective zero-point energy (ZPE) corresponds to a superposition of zero-point motion of individual nucleons in the single-nucleon potential. In the general case, the ZPE corrections on the potential energy surfaces depend on the deformation. The ZPE includes terms originating from the vibrational and rotational kinetic energy, and a contribution of potential energy

$$\Delta V(q_0,q_2) = \Delta V_{\rm vib}(q_0,q_2) + \Delta V_{\rm rot}(q_0,q_2) + \Delta V_{\rm pot}(q_0,q_2)\,. \tag{85}$$

The latter is much smaller than the contribution of kinetic energy, and is usually neglected.[42] Simple prescriptions for the calculation of vibrational and rotational ZPE were derived in Ref. 35. Both corrections are calculated in the cranking approximation, i.e. on the same level of approximation as the mass parameters and the moments of inertia. The vibrational ZPE is given by the expression:

$$\Delta V_{\rm vib}(q_0,q_2) = \frac{1}{4}{\rm Tr}\left[\mathcal{M}_{(3)}^{-1}\mathcal{M}_{(2)}\right]\,. \tag{86}$$

The rotational ZPE is a sum of three terms:

$$\Delta V_{\rm rot}(q_0,q_2) = \Delta V_{-2-2}(q_0,q_2) + \Delta V_{-1-1}(q_0,q_2) + \Delta V_{11}(q_0,q_2), \tag{87}$$

with

$$\Delta V_{\mu\nu}(q_0,q_2) = \frac{1}{4}\frac{\mathcal{M}_{(2),\mu\nu}(q_0,q_2)}{\mathcal{M}_{(3),\mu\nu}(q_0,q_2)}\,. \tag{88}$$

The individual terms are calculated from Eqs. (88) and (84), with the intrinsic components of the quadrupole operator defined by:

$$\hat{Q}_{21} = -2iyz\,,\quad \hat{Q}_{2-1} = -2xz\,,\quad \hat{Q}_{2-2} = 2ixy\,. \tag{89}$$

The potential $V_{\rm coll}$ in the collective Hamiltonian Eq. (67) is obtained by subtracting the ZPE corrections from the total mean-field energy defined in Eq. (19):

$$V_{\rm coll}(q_0,q_2) = E_{\rm tot}(q_0,q_2) - \Delta V_{\rm vib}(q_0,q_2) - \Delta V_{\rm rot}(q_0,q_2)\,. \tag{90}$$

Detailed expressions for the parameters of the collective Hamiltonian are given in the Appendices B and E of Ref. 7.

4.3. *Illustrative calculations:* ^{76}*Kr and* ^{240}*Pu*

To illustrate the application of the quadrupole collective Hamiltonian based on microscopic energy density functionals, we consider two examples of nuclei from different mass regions: ^{76}Kr and ^{240}Pu. Figure 9 displays the

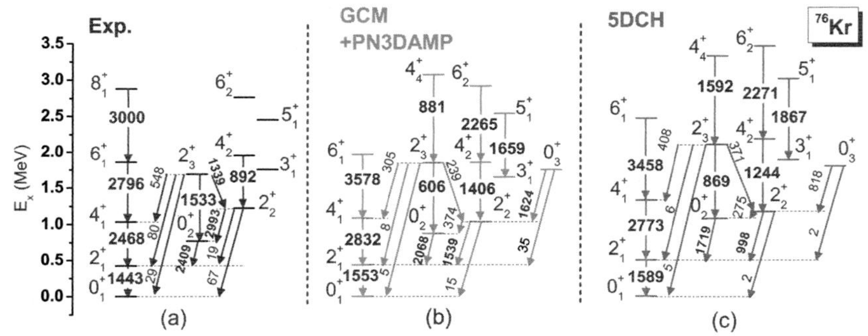

Fig. 9. (Color online) Low-lying spectra and $B(E2)$ values (in e^2fm^4) in ^{76}Kr. Available data (a)[32] are shown in comparison to results obtained with the full relativistic PN3DAMP GCM calculation (b), and with the collective Hamiltonian (5DCH) (c), both based on the energy density functional PC-PK1.

data on low-lying excitation spectra and $B(E2)$ values ^{76}Kr, in comparison with results obtained using the PN3DAMP GCM and with the collective Hamiltonian (5DCH), both based on the energy density functional PC-PK1 plus the density-independent delta-force interaction in the pairing channel. Overall the 5DCH model calculation yields a structure similar to that obtained with the PN3DAMP GCM: the ground-state rotational band, a quasi-β band based on the state $I^\pi = 0_2^+$, and a quasi-γ band with a band-head $I^\pi = 2_2^+$. We notice that the large electric quadrupole transition from the low-lying 0_2^+ state to the 2_1^+ state in the ground-state band is reproduced by both models, confirming once more the crucial role of triaxial degree of freedom in this particular isotope. Similar B(E2) values for the transitions within the ground-state band and the quasi-γ band are predicted by the two models. Significant differences are found for transitions within the quasi-β band and transitions from the quasi-β band to the ground-state band. The 5DCH model predicts somewhat stronger transitions within the quasi-β band and weaker rates from the quasi-β band or the quasi-γ band to the ground-state band. The differences arise from the

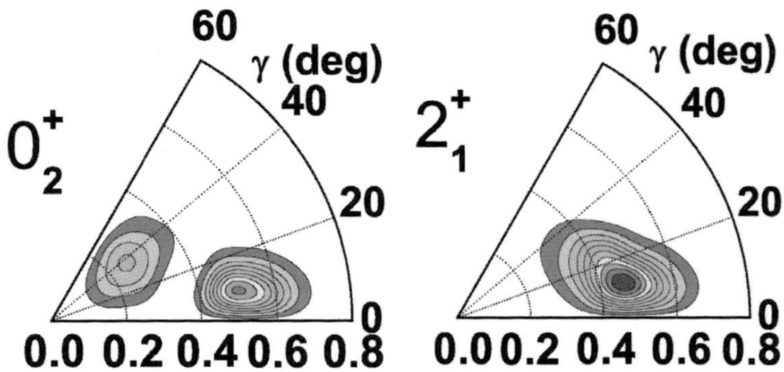

Fig. 10. (Color online) Contour plots of the probability densities $\rho_{J\alpha}(q)$ for the 0_2^+ and 2_1^+ states in ^{76}Kr, calculated with the 5DCH model.

corresponding collective functions. As an illustration, in Fig. 10 we plot the probability distributions in the $\beta-\gamma$ plane of the collective wave functions for the states 0_2^+ and 2_1^+, calculated with the 5DCH model. These should be compared with the ones obtained in the PN3DAMP GCM calculation, and shown in Fig. 7. One notices that the 5DCH model yields collective wave functions that are more concentrated close to the prolate axis. Similar differences are found also for other states, especially those that belong to the band based on the state 0_2^+.

As a second example, we consider the isotope ^{240}Pu. The structure of this nucleus and its double-humped fission barrier have become a standard benchmark for models based on the self-consistent mean-field approach and the corresponding effective interactions or density functionals. Numerous theoretical studies of this nucleus are reported in the literature. In their review of self-consistent mean-field models for nuclear structure,[1] which also contains an extensive list of references to previous studies of fission barriers using mean-field-based models, Bender *et al.* compared paths in the deformation energy landscape of ^{240}Pu obtained with various Skyrme, Gogny, and relativistic mean-field (RMF) interactions. In general, relaxing constraints on symmetries lowers the fission barriers. The predicted shapes are triaxial and reflection symmetric at the first barrier around $\beta \approx 0.6$, and they are axial and reflection asymmetric at the second barrier, located around $\beta \approx 1.3$. The systematics of axially symmetric fission barriers in Th, U, Pu, Cm, and Cf nuclei, as well as for superheavy elements $Z = 108-120$, using several Skyrme and RMF interactions, was investigated in Ref. 60. The fission barriers of 26 even-Z nuclei with $Z = 90-102$, up to and beyond

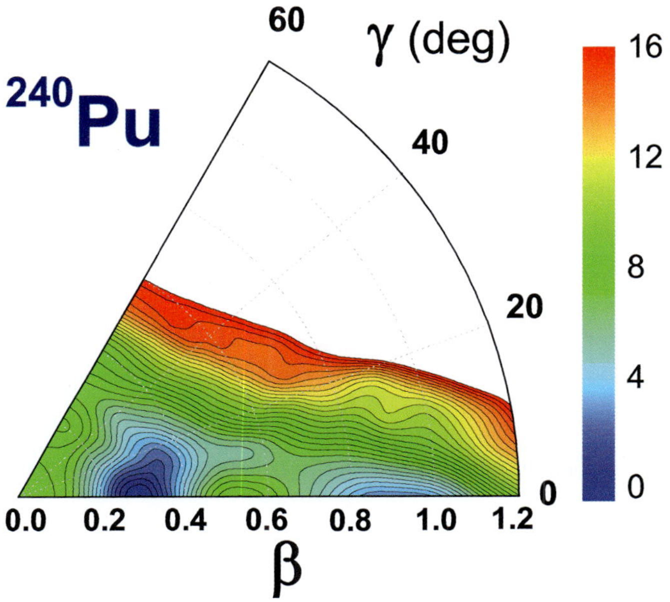

Fig. 11. Self-consistent RMF+BCS triaxial quadrupole binding energy maps of ^{240}Pu in the $\beta-\gamma$ plane ($0 \leq \gamma \leq 60°$). All energies are normalized with respect to the binding energy of the absolute minimum. The contours join points on the surface with the same energy (in MeV).

the second saddle point, were calculated in Ref. 61 with the constrained Hartree–Fock approach based on the Skyrme effective interaction SkM*. The fission barriers of ^{240}Pu beyond the second saddle point were also explored using the axially quadrupole constrained RMF model with the PK1 effective interaction.[62] A number of studies were also reported that include beyond mean-field methods (i.e., multireference calculations that explored the effects of symmetry restoration and configuration mixing).[63–65]

In Fig. 11 we display the RMF+BCS triaxial quadrupole binding energy map of ^{240}Pu in the $\beta - \gamma$ plane ($0 \leq \gamma \leq 60°$), calculated with the PC-PK1 energy density functional plus the density-independent delta-force interaction in the pairing channel. The calculation has been performed on a mesh of quadrupole deformation parameters with $\Delta\beta = 0.05$ and $\Delta\gamma = 6°$. All energies are normalized with respect to the binding energy of the absolute minimum, and the contours join points on the surface with the same energy (in MeV). The current implementation of the model does not include reflection-asymmetric shapes and, for this reason, the potential

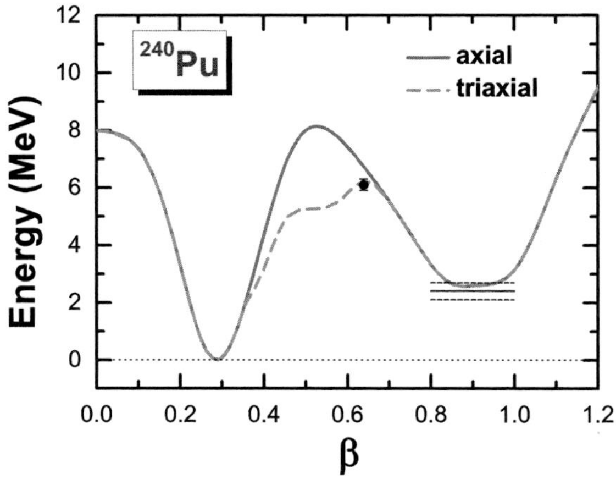

Fig. 12. (Color online) Deformation energy curves and the inner barrier of ^{240}Pu as functions of the axial deformation β. The two curves correspond to the axially symmetric RMF+BCS calculation (solid) and to the projection on the β-axis of the triaxial PES (dashed), calculated with the functional PC-PK1. The experimental values for the barrier height and the energy of the second minimum are indicated with a symbol with the error bar, and three lines indicating the value and its errors, respectively. The data are taken from Refs. 66–69.

energy surface (PES) is calculated only up to $\beta \leq 1.2$. For larger deformations (i.e., in the region of the second barrier), octupole deformations should also be taken into account. The absolute minimum is calculated at $\beta = 0.28, \gamma = 0°$, and a second (super deformed) valley is predicted around $\beta \approx 0.9$. The axially symmetric barrier at $\beta \approx 0.5$ is bypassed through the triaxial region, bringing the height of the barrier much closer to the empirical value. This is shown more clearly in Fig. 12, where we plot the deformation energy curves and the inner barrier of ^{240}Pu as functions of the axially symmetric RMF+BCS calculation (solid curve), and the projection on the β-axis of the triaxial PES (dashed curve). The empirical values for the barrier height and the energy of the second minimum are taken from Refs. 66–69. One notices a very good agreement between theory and available data. In particular, the inclusion of triaxial shapes lowers the inner barrier by ≈ 2 MeV. Similar results were also obtained in constrained self-consistent mean-field calculations using Skyrme functionals,[1] and in the HFB+Gogny analysis of the actinide region[65] it was shown that the inner barriers of the actinides were systematically lowered by up to 4 MeV when calculations included triaxial shapes.

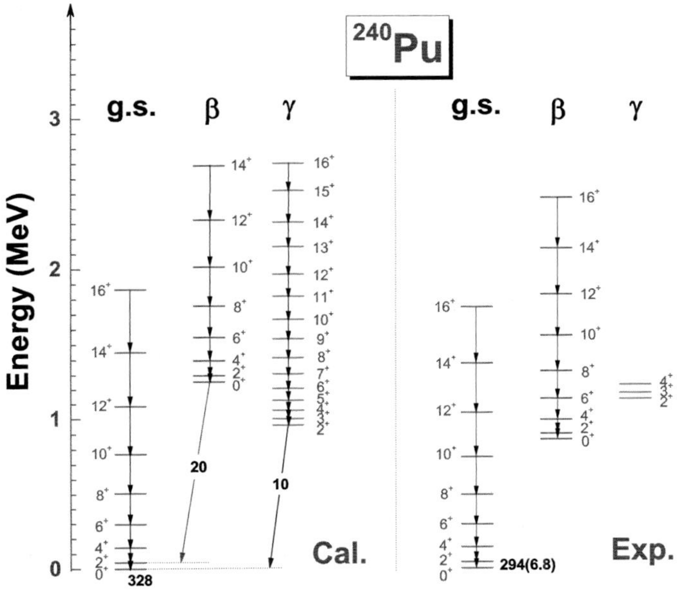

Fig. 13. (Color online) Low-energy spectrum of ^{240}Pu calculated with the relativistic density functional PC-PK1 plus the density-independent delta-force interaction in the pairing channel (left panel), compared to available data (right panel) for the three lowest positive-parity bands at normal deformation.

Starting from constrained self-consistent solutions, the parameters that determine the collective Hamiltonian are calculated as functions of the deformations β and γ. The excitation spectrum of collective states is obtained by diagonalization of the resulting Hamiltonian. In Fig. 13 the calculated low-energy spectrum of ^{240}Pu is compared to data for the three lowest positive-parity bands at normal deformation. The overall agreement with data is very good. Compared to the corresponding experimental sequence, the position of the β-band is predicted at somewhat higher excitation energy, whereas the lower position of the calculated γ-band indicates that the theoretical PES is probably too soft in γ.

4.4. *Extension of the collective Hamiltonian model to octupole degrees of freedom*

Nuclear excitations characterised by quadrupole and octupole vibrational and rotational degrees of freedom can be simultaneously described by considering quadrupole and octupole collective coordinates that specify the surface of a nucleus $R = R_0 \left[1 + \sum_\mu \alpha_{2\mu} Y^*_{2\mu} + \sum_\mu \alpha_{3\mu} Y^*_{3\mu}\right]$. In addition, when axial symmetry is imposed, the collective coordinates can be parameterized in terms of two deformation parameters β_2 and β_3, and three Euler angles $\Omega \equiv (\phi, \theta, \psi)$:

$$\alpha_{\lambda\mu} = \beta_\lambda D^\lambda_{0\mu}(\Omega), \quad \lambda = 2, 3. \tag{91}$$

The classical collective Hamiltonian is expressed as the sum of the vibrational kinetic energy, rotational kinetic energy, and the collective potential $\mathcal{V}_{coll}$. The vibrational and rotational kinetic energies read:

$$\mathcal{T}_{\text{vib}} = \frac{1}{2} B_{22} \dot{\beta}_2^2 + B_{23} \dot{\beta}_2 \dot{\beta}_3 + \frac{1}{2} B_{33} \dot{\beta}_3^2, \tag{92}$$

$$\mathcal{T}_{\text{rot}} = \frac{1}{2} \sum_{k=1}^{3} \mathcal{I}_k \omega_k^2, \tag{93}$$

respectively, where the mass parameters B_{22}, B_{23} and B_{33}, and the moments of inertia $\mathcal{I}_k$, are functions of the quadrupole and octupole deformations β_2 and β_3.

After quantization the collective Hamiltonian takes the form

$$\begin{aligned} \hat{H}_{\text{coll}} = & -\frac{\hbar^2}{2\sqrt{w\mathcal{I}}} \left[\frac{\partial}{\partial \beta_2} \sqrt{\frac{\mathcal{I}}{w}} B_{33} \frac{\partial}{\partial \beta_2} - \frac{\partial}{\partial \beta_2} \sqrt{\frac{\mathcal{I}}{w}} B_{23} \frac{\partial}{\partial \beta_3} \right. \\ & \left. - \frac{\partial}{\partial \beta_3} \sqrt{\frac{\mathcal{I}}{w}} B_{23} \frac{\partial}{\partial \beta_2} + \frac{\partial}{\partial \beta_3} \sqrt{\frac{\mathcal{I}}{w}} B_{22} \frac{\partial}{\partial \beta_3} \right] \\ & + \frac{\hat{J}^2}{2\mathcal{I}} + V(\beta_2, \beta_3), \end{aligned} \tag{94}$$

where $w = B_{22} B_{33} - B_{23}^2$, and the corresponding volume element in the collective space reads

$$\int d\tau_{\text{coll}} = \int \sqrt{w\mathcal{I}} d\beta_2 d\beta_3 d\Omega \,. \tag{95}$$

To solve the eigenvalue problem for the collective Hamiltonian Eq. (94), the eigenfunctions are expanded in terms of a complete set of basis functions. In particular, for each value of angular momentum I the basis is constructed from a complete set of square integrable functions

$$|n_2 n_3 I M K\rangle = (w\mathcal{I})^{-1/4} \phi_{n_2}(\beta_2) \phi_{n_3}(\beta_3) D^I_{MK}(\Omega), \tag{96}$$

where ϕ_{n_2} (ϕ_{n_3}) denotes the one-dimension harmonic-oscillator wave function of the deformation variable β_2 (β_3). For positive (negative) parity states, n_3 and I have to be even (odd). Since we only consider axially deformed nuclei, the projection of angular momentum on the symmetry axis $K = 0$. Finally, the collective wave function reads

$$\Psi_\alpha^{IM\pi}(\beta_2, \beta_3, \Omega) = \psi_\alpha^{I\pi}(\beta_2, \beta_3) D_{M0}^I(\Omega). \tag{97}$$

The reduced electromagnetic transition rates are calculated using the expression:

$$B(E\lambda, I_i \to I_f) = \langle I_i 0 \lambda 0 | I_f 0 \rangle^2 \left| \int d\beta_2 d\beta_3 \sqrt{w\mathcal{I}} \psi_i \mathcal{M}_{E\lambda}(\beta_2, \beta_3) \psi_f^* \right|^2, \tag{98}$$

where $\mathcal{M}_{E\lambda}(\beta_2, \beta_3)$ denotes the electric multipole moment of order λ. For the electric dipole, quadrupole, and octupole transitions, we use the following expressions for the operator $\hat{\mathcal{M}}(E\lambda)$:

$$D_1 = e\frac{N}{A}z_p - e\frac{Z}{A}z_n, \tag{99}$$

$$Q_2^p = e(2z_p^2 - x_p^2 - y_p^2), \tag{100}$$

$$Q_3^p = e[2z_p^3 - 3z_p(x_p^2 + y_p^2)]. \tag{101}$$

The moments of inertia are calculated according to the Inglis–Belyaev formula[41,59]

$$\mathcal{I} = \sum_{i,j} \frac{(u_i v_j - v_i u_j)^2}{E_i + E_j} |\langle i | \hat{I} | j \rangle|^2, \tag{102}$$

where $\hat{I}$ is the angular momentum along the axis perpendicular to the symmetric axis, and the summation runs over the proton and neutron quasiparticle states. The quasiparticle energies E_i, occupation probabilities v_i, and single-nucleon wave functions ψ_i are determined by solutions of the constrained RMF+BCS equations. The mass parameters associated with the collective coordinates $q_2 = \langle \hat{Q}_2 \rangle$ and $q_3 = \langle \hat{Q}_3 \rangle$ are also calculated in the cranking approximation[35]

$$B_{\lambda\lambda'}(q_2, q_3) = \frac{\hbar^2}{2} \left[\mathcal{M}_{(1)}^{-1} \mathcal{M}_{(3)} \mathcal{M}_{(1)}^{-1} \right]_{\lambda\lambda'}, \tag{103}$$

with

$$\mathcal{M}_{(n),\lambda\lambda'}(q_2, q_3) = \sum_{i,j} \frac{\langle i | \hat{Q}_\lambda | j \rangle \langle j | \hat{Q}_{\lambda'} | i \rangle}{(E_i + E_j)^n} (u_i v_j + v_i u_j)^2. \tag{104}$$

The collective energy surface includes the energy of the zero-point motion which has to be subtracted. In the present version of the model we subtract the vibrational and rotational ZPE, and both corrections are calculated in the cranking approximation. The vibrational ZPE is computed using the expression:

$$\Delta V_{\text{vib}}(\beta_2, \beta_3) = \frac{1}{4}\text{Tr}\left[\mathcal{M}_{(3)}^{-1}\mathcal{M}_{(2)}\right] , \tag{105}$$

and the rotational ZPE reads

$$\Delta V_{\text{rot}}(\beta_2, \beta_3) = \frac{\langle \hat{I}^2 \rangle}{2\mathcal{I}} . \tag{106}$$

4.4.1. *Illustrative calculation for* ^{224}*Ra*

As an illustration of the quadrupole-octupole collective Hamiltonian (QOCH) model, we present an analysis of collective excitations in ^{224}Ra. The calculation is carried out using the energy density functional PC-PK1 plus the density-independent delta-interaction in the pairing channel.

The self-consistent axially symmetric quadrupole-octupole energy surface of ^{224}Ra, calculated with constraints on the expectation values of the

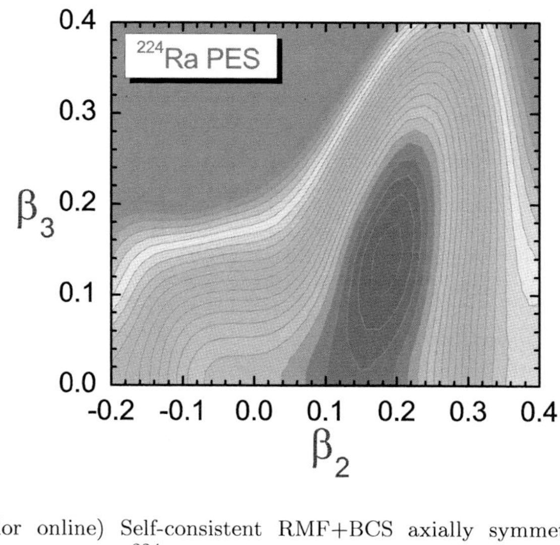

Fig. 14. (Color online) Self-consistent RMF+BCS axially symmetric quadrupole-octupole energy surface of ^{224}Ra in the $\beta_2 - \beta_3$ plane. The contours join points on the surface with the same energy, and the separation between neighbouring contours is 0.5 MeV.

quadrupole $\langle Q_{20}\rangle$ and octupole $\langle Q_{30}\rangle$ moments, is shown in Fig. 14. The contours join points on the surface with the same energy, and successive contours differ in energy by 0.5 MeV. We notice that already the mean-field calculation predicts a shallow octupole deformed minimum in the PES at the deformation: $\beta_2 \approx 0.2$ and $\beta_3 \approx 0.15$.

Starting from constrained self-consistent solutions, that is, using the single-quasiparticle energies and wave functions that correspond to each point on the energy surface shown in Fig. 14, the parameters that determine the collective Hamiltonian: mass parameters, the moment of inertia, as well as the zero-point energy corrections, are calculated as functions of the deformations β_2 and β_3. The spectrum of collective states is obtained by diagonalization of the resulting Hamiltonian. To illustrate in more detail the level of quantitative agreement between our microscopic model calculation and data, we analyze the low-lying energy spectra of positive- and negative-parity states, the $B(E2)$ values for in-band transitions, and the interband $B(E3)$ values (Fig. 15).

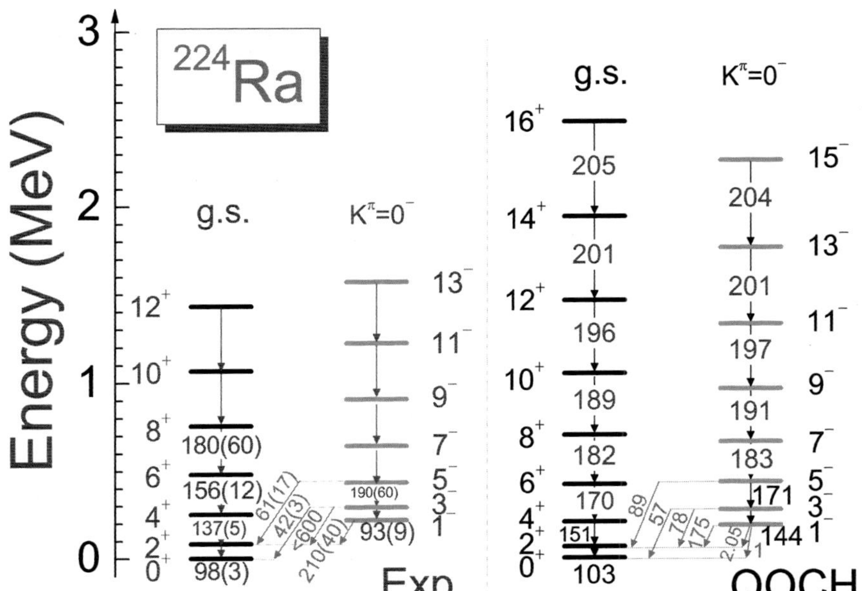

Fig. 15. (Color online) Partial level scheme of ^{224}Ra. The theoretical low-lying spectra, in-band $B(E2)$ values (in Weisskopf units), interband $B(E3)$ values (in Weisskopf unit), and the branching ratio $B(E1{:}1^- \to 2^+)/B(E1{:}1^- \to 0^+)$ are compared with the available empirical data.[70]

The level scheme of ^{224}Ra shows that the lowest negative-parity band is located close in energy to the ground-state positive-parity band. The lowest positive- and negative-parity bands form a single, alternating-parity band, starting with angular momentum J = 3. Overall, a very good agreement between theory and experiment is obtained for the excitation spectrum using the the QOCH model. Considering that the model is fully microscopic, that is, no effective charges are used in the calculation of transition probabilities, the agreement between the experimental and calculated values for the reduced E2 and E3 transition rates is remarkable.

5. Summary

The latest extensions of the relativistic mean-field framework that include the treatment of collective correlations have been reviewed. In recent years a number of RMF-based microscopic structure models have been developed and applied to detailed spectroscopic studies, often to regions of exotic nuclei far from stability, including predictions for excitation spectra and electromagnetic transitions. By restoring symmetries broken by the static mean-field and by considering fluctuations of collective deformation variables, this approach provides an accurate microscopic description of a variety of structure phenomena, with the only input being the relativistic energy density functional or effective interaction. Without including additional parameters specifically adjusted to spectroscopic data, these models predict results that can quantitatively be compared to data.

Many interesting phenomena related to shell evolution have been investigated by employing configuration mixing of particle-number and angular-momentum projected triaxial relativistic mean-field wave functions (PN3DAMP GCM). This is one of most effective microscopic approaches that take into account correlations that arise from symmetry restoration and fluctuations around the mean-field minimum, with multipole moments used as coordinates that generate the intrinsic wave functions. The model and basic approximations have been discussed in detail, and applications illustrated in a study of low-energy excitation spectrum of the nucleus ^{76}Kr.

Even though the model that uses the PN3DAMP GCM can predict results in excellent agreement with spectroscopic data, the applications have so far been somewhat limited to the region of light and medium-mass nuclei. The reason is that the numerical implementation of the model is very complex, and calculations of heavier nuclei are still computationally too demanding and time consuming. In addition, the use of general density functionals in GCM-type calculations often leads to discontinuities or even divergences of the energy kernels as a function of deformation. In an alternative approach to collective nuclear dynamics that includes rotational symmetry restoration and takes into account fluctuations, a collective Bohr Hamiltonian has been constructed, with deformation-dependent parameters determined from microscopic self-consistent mean-field calculations. The collective Hamiltonian can be derived in the Gaussian overlap approximation to the full GCM. With the assumption that the GCM overlap kernels can be approximated by Gaussian functions, the local expansion of the kernels up to second order in the non-locality transforms the GCM Hill–Wheeler equation into a second-order differential equation — the Schrödinger equation for the collective Hamiltonian.

In this review we have described two recent implementations of the collective Hamiltonian based on the self-consistent RMF framework: the five-dimensional collective Hamiltonian for quadrupole vibrational and rotational degrees of freedom, with parameters determined by constrained self-consistent relativistic mean-field calculations for triaxial shapes, and a quadrupole-octupole collective Hamitonian that describes excitations of nuclei with axially-symmetric and reflection-asymmetric shapes. The current implementations of these models have been illustrated in spectroscopic calculations of ^{76}Kr and ^{240}Pu (5D quadrupole Hamiltonian) and ^{224}Ra (quadrupole-octupole Hamiltonian). The microscopic approach to nuclear structure that uses collective Hamiltonians based on microscopic energy density functionals or effective interactions, is much less computationally demanding than the full GCM. Therefore, it is to a lesser extent constrained by the symmetries of the intrinsic shapes, or by the size of the nuclear system. A significant challenge, however, is the fully consistent microscopic calculation of the parameters that determine the dynamics of the Hamiltonian: the mass parameters and the moments of inertia, as well as the zero-point energy corrections.

References

1. M. Bender, P.-H. Heenen, and P.-G. Reinhard, Self-consistent mean-field models for nuclear structure, *Rev. Mod. Phys.* **75**(1), 121–180 (Jan, 2003). doi: 10.1103/RevModPhys.75.121.
2. D. Vretenar, A. V. Afanasjev, G. A. Lalazissis, and P. Ring, Relativistic Hartree Bogoliubov theory: static and dynamic aspects of exotic nuclear structure, *Phys. Rep.* **409**(3-4), 101–259 (2005). doi: 10.1016/j.physrep.2004.10.001.

3. J. Meng, H. Toki, S. Zhou, S. Zhang, W. Long, and L. Geng, Relativistic continuum Hartree Bogoliubov theory for ground-state properties of exotic nuclei. **57**(2), 470–563 (2006). ISSN 0146-6410. doi: http://dx.doi.org/10.1016/j.ppnp.2005.06.001. URL `http://www.sciencedirect.com/science/article/pii/S014664100500075X`.
4. N. Paar, D. Vretenar, E. Khan, and G. Colò, Exotic modes of excitation in atomic nuclei far from stability, *Rep. Prog. Phys.* **70**(5), 691–793 (2007). doi: 10.1088/0034-4885/70/5/R02.
5. T. Nikšić, D. Vretenar, and P. Ring, Beyond the relativistic mean-field approximation: Configuration mixing of angular-momentum-projected wave functions, *Phys. Rev. C.* **73**(3), 034308 (Mar, 2006). doi: 10.1103/PhysRevC.73.034308.
6. T. Nikšić, D. Vretenar, and P. Ring, Beyond the relativistic mean-field approximation. (ii): Configuration mixing of mean-field wave functions projected on angular momentum and particle number, *Phys. Rev. C.* **74**(6), 064309 (2006). doi: 10.1103/PhysRevC.74.064309.
7. T. Nikšić, Z. P. Li, D. Vretenar, L. Prochniak, J. Meng, and P. Ring, Beyond the relativistic mean-field approximation. iii. collective Hamiltonian in five dimensions, *Phys. Rev. C.* 79(3):034303 (2009). doi: 10.1103/PhysRevC.79.034303.
8. J. M. Yao, J. Meng, P. Ring, and D. Pena Arteaga, Three-dimensional angular momentum projection in relativistic mean-field theory, *Phys. Rev. C.* **79**(4), 044312 (Apr, 2009). doi: 10.1103/PhysRevC.79.044312.
9. J. M. Yao, J. Meng, P. Ring, and D. Vretenar, Configuration mixing of angular-momentum-projected triaxial relativistic mean-field wave functions, *Phys. Rev. C.* **81**(4), 044311 (2010). doi: 10.1103/PhysRevC.81.044311.
10. J. M. Yao, H. Mei, H. Chen, J. Meng, P. Ring, and D. Vretenar, Configuration mixing of angular-momentum-projected triaxial relativistic mean-field wave functions. ii. microscopic analysis of low-lying states in magnesium isotopes, *Phys. Rev. C.* **83**(1), 014308 (2011). doi: 10.1103/PhysRevC.83.014308.
11. T. Bürvenich, D. G. Madland, J. A. Maruhn, and P.-G. Reinhard, Nuclear ground state observables and QCD scaling in a refined relativistic point coupling model, *Phys. Rev. C.* **65**(4), 044308 (Mar, 2002). doi: 10.1103/PhysRevC.65.044308.
12. T. Nikšić, D. Vretenar, and P. Ring, Relativistic nuclear energy density functionals: Adjusting parameters to binding energies, *Phys. Rev. C.* 78(3): 034318 (2008). doi: 10.1103/PhysRevC.78.034318.
13. P. W. Zhao, Z. P. Li, J. M. Yao, and J. Meng, New parametrization for the nuclear covariant energy density functional with a point-coupling interaction, *Phys. Rev. C.* **82**(5), 054319 (Nov, 2010). doi: 10.1103/PhysRevC.82.054319.
14. J. Peng. *Nuclear triaxial deformation and magnetic rotation in relativistic mean field theory*. PhD thesis, Peking University (2006).
15. J. Peng, J. Meng, P. Ring, and S. Q. Zhang, Covariant density functional theory for magnetic rotation, *Phys. Rev. C.* **78**(2), 024313 (Aug, 2008). doi: 10.1103/PhysRevC.78.024313.
16. S. J. Krieger, P. Bonche, H. Flocard, P. Quentin, and M. S. Weiss, An improved pairing interaction for mean field calculations using Skyrme poten-

tials, *Nucl. Phys. A.* **517**(2), 275–284 (1990). doi: 10.1016/0375-9474(90)90035-K.
17. M. Bender, K. Rutz, P.-G. Reinhard, and J. A. Maruhn, Pairing gaps from nuclear mean-field models, *Eur. Phys. J. A.* **8**(1), 59–75 (July, 2000). doi: 10.1007/s10050-000-4504-z.
18. M. Bender, K. Rutz, P.-G. Reinhard, and J. A. Maruhn, Consequences of the center-of-mass correction in nuclear mean-field models, *Eur. Phys. J. A.* **7**(4), 467–478 (APR, 2000). doi: 10.1007/PL00013645.
19. P. Ring and P. Schuck, *The nuclear many-body problem*. Springer-Verlag, New York (1980).
20. A. R. Edmonds, *Angular Momentum in Quantum Mechanics*. Princeton University Press (1957).
21. D. L. Hill and J. A. Wheeler, Nuclear constitution and the interpretation of fission phenomena, *Phys. Rev.* **89**(5), 1102–1145 (1953). doi: 10.1103/PhysRev.89.1102.
22. R. Balian and E. Brezin, Nonunitary Bogoliubov tansformations and extension of Wick's theorem, *Il Nuovo Cimento.* **64**(1), 37–55 (1969). doi: 10.1007/BF02710281.
23. N. Onishi and S. Yoshida, Generator coordinate method applied to nuclei in the transition region, *Nuclear Physics.* **80**(2), 367–376 (May, 1966). ISSN 0029-5582. doi: 10.1016/0029-5582(66)90096-4. URL `http://dx.doi.org/10.1016/0029-5582(66)90096-4`.
24. T. Duguet and J. Sadoudi, Breaking and restoring symmetries within the nuclear energy density functional method, *J. Phys. G.* **37**(6), 064009 (June, 2010). doi: 10.1088/0954-3899/37/6/064009.
25. K. Hara, A. Hayashi, and P. Ring., Exact angular momentum projection of cranked Hartree–Fock–Bogoliubov wave functions, *Nucl. Phys. A.* **385**(1), 14–28 (1982). doi: 10.1016/0375-9474(82)90486-9.
26. P. Bonche, J. Dobaczewski, H. Flocard, P. H. Heenen, and J. Meyer, Analysis of the generator coordinate method in a study of shape isomerism in ^{194}Hg, *Nucl. Phys. A.* **510**(3), 466–502 (1990). doi: 10.1016/0375-9474(90)90062-Q.
27. V. N. Fomenko, Projection in the occupation-number space and the canonical transformation, *J. Phys. A: Gen. Phys.* **3**(1), 8 (Jan, 1970). ISSN 0022-3689. doi: 10.1088/0305-4470/3/1/002. URL `http://dx.doi.org/10.1088/0305-4470/3/1/002`.
28. L. M. Robledo, Technical aspects of the evaluation of the overlap of Hartree–Fock–Bogoliubov wave functions, *Phys. Rev. C.* **84**, 014307 (Jul, 2011). doi: 10.1103/PhysRevC.84.014307.
29. Y. Fu, H. Mei, J. Xiang, Z. P. Li, J. M. Yao, and J. Meng, Beyond relativistic mean-field studies of low-lying states in neutron-deficient krypton isotopes, *Phys. Rev. C.* **87**, 054305 (May, 2013). doi: 10.1103/PhysRevC.87.054305. URL `http://link.aps.org/doi/10.1103/PhysRevC.87.054305`.
30. M. Bender, P. Bonche, and P.-H. Heenen, Shape coexistence in neutron-deficient Kr isotopes: Constraints on the single-particle spectrum of self-consistent mean-field models from collective excitations, *Phys. Rev. C.* **74** (2), 024312 (2006). doi: 10.1103/PhysRevC.74.024312.

31. J. M. Yao, K. Hagino, Z. P. Li, J. Meng, and P. Ring, Microscopic benchmark study of triaxiality in low-lying states of ^{76}Kr, *Phys. Rev. C.* **89**(5) (May, 2014). ISSN 1089-490X. doi: 10.1103/physrevc.89.054306. URL `http://dx.doi.org/10.1103/PhysRevC.89.054306`.
32. E. Clément, A. Görgen, W. Korten, E. Bouchez, A. Chatillon, J.-P. Delaroche, M. Girod, H. Goutte, A. Hürstel, Y. Le Coz, A. Obertelli, P. S., C. Theisen, J. N. Wilson, M. Zielińska, C. Andreoiu, F. Becker, P. A. Butler, J. M. Casandjian, T. Catford, W. N.and Czosnyka, G. de France, J. Gerl, R.-D. Herzberg, J. Iwanicki, D. G. Jenkins, G. D. Jones, P. J. Napiorkowski, G. Sletten, and C. N. Timis, Shape coexistence in neutron-deficient krypton isotopes, *Phys. Rev. C.* **75**(5), 054313 (2007). doi: 10.1103/PhysRevC.75.054313.
33. P.-G. Reinhard and K. Goeke, The generator coordinate method and quantised collective motion in nuclear systems, *Rep. Prog. Phys.* **50**(1), 1–64 (1987). doi: 10.1088/0034-4885/50/1/001.
34. B. Giraud and B. Grammaticos, Collective potentials and inertia parameters (ii), *Nucl. Phys. A.* **255**(1), 141–156 (1975). doi: DOI:10.1016/0375-9474(75)90154-2.
35. M. Girod and B. Grammaticos, The zero-point energy correction and its effect on nuclear dynamics, *Nucl. Phys. A.* **330**(1), 40–52 (1979). doi: 10.1016/0375-9474(79)90535-9.
36. M. Baranger and M. Veneroni, An adiabatic time-dependent Hartree–Fock theory of collective motion in finite systems, *Ann. Phys.* **114**(1-2), 123–200 (1978). doi: 10.1016/0003-4916(78)90265-8.
37. F. Villars, Adiabatic time-dependent Hartree–Fock theory in nuclear physics, *Nucl. Phys. A.* **285**(2), 269–296 (1977). doi: 10.1016/0375-9474(77)90253-6.
38. K. Goeke and P.-G. Reinhard, The generator-coordinate-method with conjugate parameters and the unification of microscopic theories for large amplitude collective motion, *Ann. Phys.* **124**(2), 249–289 (1980). doi: 10.1016/0003-4916(80)90210-9.
39. R. E. Peierls and J. Yoccoz, The collective model of nuclear motion, *Proceedings of the Physical Society. Section A.* **70**(5), 381–387 (May, 1957). ISSN 0370-1298. doi: 10.1088/0370-1298/70/5/309. URL `http://dx.doi.org/10.1088/0370-1298/70/5/309`.
40. D. Thouless and J. Valatin, Time-dependent Hartree–Fock equations and rotational states of nuclei, *Nuclear Physics.* **31**, 211–230 (Mar, 1962). ISSN 0029-5582. doi: 10.1016/0029-5582(62)90741-1. URL `http://dx.doi.org/10.1016/0029-5582(62)90741-1`.
41. D. Inglis, Nuclear moments of inertia due to nucleon motion in a rotating well, *Phys. Rev.* **103**(6), 1786–1795 (Sep, 1956). ISSN 1536-6065. doi: 10.1103/physrev.103.1786. URL `http://dx.doi.org/10.1103/PhysRev.103.1786`.
42. J. Libert, M. Girod, and J.-P. Delaroche, Microscopic descriptions of superdeformed bands with the Gogny force: Configuration mixing calculations in the $A \sim 190$ mass region, *Phys. Rev. C.* **60**(5), 054301 (Sep, 1999). doi: 10.1103/PhysRevC.60.054301.
43. L. Próchniak and P. Ring, The relativistic mean field theory and low energy quadrupole collective excitations, *International Journal of*

Modern Physics E. **13**(01), 217–224 (Feb, 2004). ISSN 1793-6608. doi: 10.1142/s0218301304001977. URL `http://dx.doi.org/10.1142/S0218301304001977`.

44. D. A. Varshalovich, A. N. Moskalev, and V. K. Khersonskii, *Quantum theory of angular momentum.* World Scientific, Singapore (1988).
45. W. Pauli, *Handbuch der Physik, Vol. XXIV.* Springer Verlag, Berlin (1933).
46. K. Kumar and M. Baranger, Complete numerical solution of Bohr's collective Hamiltonian, *Nuclear Physics A.* **92**(3), 608–652 (Feb, 1967). ISSN 0375-9474. doi: 10.1016/0375-9474(67)90636-7. URL `http://dx.doi.org/10.1016/0375-9474(67)90636-7`.
47. S. G. Rohozinski, J. Dobaczewski, B. Nerlo-Pomorska, K. Pomorski, and J. Srebrny, Microscopic dynamic calculations of collective states in xenon and barium isotopes, *Nucl. Phys. A.* **292**(1-2), 66–87 (1977). doi: 10.1016/0375-9474(77)90358-X.
48. D. Troltenier, J. A. Maruhn, W. Greiner, and P. O. Hess, A general numerical solution of collective quadrupole surface motion applied to microscopically calculated potential energy surfaces, *Z. Phys. A.* **343**(1), 25–34 (1992). doi: 10.1007/BF01291593.
49. G. G. Dussel and D. R. Bès, Solution of Bohr's collective Hamiltonian, *Nucl. Phys. A.* **143**(3), 623–640 (1970). doi: 10.1016/0375-9474(70)90552-X.
50. G. Gneuss and W. Greiner, Collective potential energy surfaces and nuclear structure, *Nucl. Phys. A.* **171**(3), 449–479 (1971). doi: 10.1016/0375-9474(71)90596-3.
51. K. Kumar, Collective Hamiltonian derived from the pairing-plus-quadrupole model *1: Modification and application to the transitional nuclei, ^{150,152}Sm, *Nucl. Phys. A.* **231**(2), 189–232 (1974). doi: 10.1016/0375-9474(74)90518-1.
52. J. Libert and P. Quentin, Self-consistent description of heavy nuclei. i. static properties of some even nuclei, *Phys. Rev. C.* **25**(1), 571–585 (Jan, 1982). ISSN 0556-2813. doi: 10.1103/physrevc.25.571. URL `http://dx.doi.org/10.1103/PhysRevC.25.571`.
53. I. Deloncle. PhD thesis, Université de Paris 6 (1989).
54. I. Deloncle, J. Libert, L. Bennour, L. Bennour, and P. Quentin, The low energy spectra of heavy transitional even nuclei from an effective nucleon-nucleon interaction, *Phys. Lett. B.* **233**(1-2), 16–22 (1989). doi: 10.1016/0370-2693(89)90608-4.
55. L. Próhniak, K. Zajac, K. Pomorski, S. G. Rohozinski, and J. Srebrny, Collective quadrupole excitations in the $50 < Z, N < 82$ nuclei with the general Bohr Hamiltonian, *Nucl. Phys. A.* **648**(3-4), 181–202 (1999). doi: 10.1016/S0375-9474(99)00023-8.
56. M. Hammermesh, *Group Theory and Its Applications to Physical Problems.* Addison-Wesley, Reading, Massachusetts (1962).
57. L. Próchniak and S. G. Rohoziński, Quadrupole collective states within the Bohr collective Hamiltonian, *J. Phys. G: Nuclear and Particle Physics.* **36** (12), 123101 (2009). doi: 10.1088/0954-3899/36/12/123101.
58. W. Press, B. Flannery, S. Teukolsky, and W. Vetterling, *Numerical Recipes in C: The Art of Scientific Computing.* Cambridge University Press, Cambridge, England (1988).

59. S. T. Belyaev, Concerning the calculation of the nuclear moment of inertia, *Nucl. Phys.* **24**(2), 322–325 (1961). doi: 10.1016/0029-5582(61)90384-4.
60. T. Bürvenich, M. Bender, J. Maruhn, and P.-G. Reinhard, Systematics of fission barriers in superheavy elements, *Phys. Rev. C.* **69**(1), 014307 (Jan, 2004). ISSN 1089-490X. doi: 10.1103/physrevc.69.014307. URL `http://dx.doi.org/10.1103/PhysRevC.69.014307`.
61. L. Bonneau, P. Quentin, and D. Samsn, Fission barriers of heavy nuclei within a microscopic approach, *Eur. Phys. J. A.* **21**(3), 391–406 (Sep, 2004). ISSN 1434-601X. doi: 10.1140/epja/i2003-10224-x. URL `http://dx.doi.org/10.1140/epja/i2003-10224-x`.
62. L. Hong-Feng, G. Li-Sheng, and M. Jie, Fission barrier for 240 Pu in the quadrupole constrained relativistic mean field approach, *Chinese Phys. Lett.* **23**(11), 2940–2943 (Oct, 2006). ISSN 1741-3540. doi: 10.1088/0256-307x/23/11/016. URL `http://dx.doi.org/10.1088/0256-307X/23/11/016`.
63. M. Girod, J. Libert, J.-P. Delaroche, and P. Romain, Microscopic structure of superdeformed states in Th, U, Pu and Cm isotopes with Gogny force, *The European Physical Journal A - Hadrons and Nuclei.* **15**(1-2), 209–212 (Sep, 2002). ISSN 1434-601X. doi: 10.1140/epja/i2001-10256-2. URL `http://dx.doi.org/10.1140/epja/i2001-10256-2`.
64. M. Bender, P.-H. Heenen, and P. Bonche, Microscopic study of Pu-240: Mean field and beyond, *Phys. Rev. C.* **70**(5), 054403 (Nov, 2004). ISSN 1089-490X. doi: 10.1103/physrevc.70.054304. URL `http://dx.doi.org/10.1103/PhysRevC.70.054304`.
65. J.-P. Delaroche, M. Girod, H. Goutte, and J. Libert, Structure properties of even–even actinides at normal and super deformed shapes analysed using the Gogny force, *Nuclear Physics A.* **771**, 103–168 (May, 2006). ISSN 0375-9474. doi: 10.1016/j.nuclphysa.2006.03.004. URL `http://dx.doi.org/10.1016/j.nuclphysa.2006.03.004`.
66. S. Bjørnholm and J. E. Lynn, The double-humped fission barrier, *Rev. Mod. Phys.* **52**(4), 725–931 (Oct, 1980). doi: 10.1103/RevModPhys.52.725.
67. R. Firestone, V. Shirley, S. Chu, C. Baglin, and J. Zipkin, *Table of Isotopes.* Wiley, New York (1996).
68. K. Löbner, M. Vetter, and V. Hönig, *Nucl. Data Tables A.* **7** (1970).
69. C. Bemis, F. McGowan, J. Ford, W. Milner, P. Stelson, and R. Robinson, E2 and E4, transition moments and equilibrium deformations in the actinide nuclei, *Phys. Rev. C.* **8**(4), 1466–1480 (Oct, 1973). ISSN 0556-2813. doi: 10.1103/physrevc.8.1466. URL `http://dx.doi.org/10.1103/PhysRevC.8.1466`.
70. L. P. Gaffney, P. A. Butler, M. Scheck, A. B. Hayes, F. Wenander, M. Albers, B. Bastin, C. Bauer, A. Blazhev, S. Bnig et al., Studies of pear-shaped nuclei using accelerated radioactive beams, *Nature.* **497**(7448), 199–204 (May, 2013). ISSN 1476-4687. doi: 10.1038/nature12073. URL `http://dx.doi.org/10.1038/nature12073`.

Chapter 13

Heavy elements in astrophysical nucleosynthesis

Bao-Hua Sun
School of Physics and Nuclear Energy Engineering, Beihang University, Beijing 100191, China
bhsun@buaa.edu.cn

Zhong-Ming Niu
School of Physics and Material Science, Anhui University, Hefei 230039, China
zmniu@ahu.edu.cn

With the many successes of covariant density functional theory (CDFT) as seen in the previous chapters, there has been growing interest over the last years to examine directly their applicability in astrophysical nucleosynthesis simulations. This chapter thus concentrates on the very recent applications of CDFT in astrophysics nucleosynthesis, ranging from the calculations of nuclear physics inputs — masses and beta-decay half-lives — for rapid-neutron (r-) and rapid-proton (rp-) capture processes, to the nucleosynthesis studies that employed these inputs and to nuclear cosmochronology. The concepts of nucleosynthesis process and formulas on beta-decays are sketched briefly.

Contents

1. Introduction

Nuclear astrophysics has been an active branch of modern physics for over half a century and is still in the midst of an exciting revival in experimental and theoretical researches around the world. It refers generally to the basic physics of energy generation and element synthesis in stars. Its ultimate goals are to understand how nuclear processes generate the energy of stars over their lifetimes and, in doing so, synthesize heavier elements from the primordial hydrogen and helium produced in the big bang which leads to the expanding universe. Nuclear physics is thus indispensable for the transmutation of nuclei into new nuclei with lower mass. A small decrease in mass results in a relatively large amount of energy released as governed by the mass-energy equivalence formula.

This chapter concerns the recent applications of CDFT in nuclear astrophysics. We start with a brief introduction of the elemental abundance in our solar system and various processes proposed for revealing their origins. We then give a brief discussion of the rapid neutron-capture (r-) process, its modeling, and the nuclear physics inputs involved. Particular emphasis is placed on the nuclear physics inputs, i.e. masses and β-decay half-lives, which can be calculated in CDFT, as well as the r-process nucleosynthesis studies that employed these inputs. The scenario of r-process is out of the scope of this chapter, but we discuss that it is possible to constrain the astrophysical condition from solely nuclear masses within the so-called waiting-point approximation. We then discuss a specific application, i.e., by employing the CDFT predictions to determine the ages of ultra-metal-poor stars by nuclear chronometers. We close this chapter by showing an example of CDFT in the rapid-proton capture (rp-) process.

2. Origin of elements in Universe

Elements and their distribution in nature carry the signatures of astrophysical nucleosynthesis processes that occurred long before our sun was born. Indeed, the solar abundance distribution played a crucial role in studies on the origin of elements. The solar abundance distribution[1] (see Fig. 1) is frequently referred to as "universal" or "cosmic", but in reality it primarily represents the relative atomic abundances in the solar system. The abundance spans about 12 order-of-magnitude from the most enriched element hydrogen to the rarest isotope ^{183}W. The most important characteristics of this distribution are summarized here:

- Hydrogen accounts for about 75% of mass abundances and He of 23%: In other words, only roughly 2% of the mass in the solar system consists of all the heavier isotopes from lithium to the heaviest isotopes ^{238}U.
- Deuterium, lithium, beryllium, and boron are under-abundant in comparison with neighboring elements.
- Large abundances at carbon, oxygen, neon, magnesium, silicon: They differ by one alpha particle in mass unit. More specially, oxygen and carbon are the third and fourth most abundant, respectively.
- Exponential decrease up to iron in abundance.
- Abundance peak near iron.
- Almost constant abundances beyond iron peak.
- Double abundance peaks around $A = 130$ and $A = 190$.

Both theories and observations support that the elements observed in the Universe were created mainly in either of two ways. Light elements (mostly deuterium and helium, and partially lithium and beryllium) were produced in the first few minutes of the big bang, while elements heavier than boron are thought to have their origins in the interiors of stars which formed much later in the history of Universe. These two ways are called big-bang nucleosynthesis and stellar nucleosynthesis, respectively.

The big bang nucleosynthesis (BBN), known also as primordial nucleosynthesis, initially attempted to synthesize all elements during the big bang. It assumed that all elements formed from protons and neutrons by sequence of n-capture and β-decays, i.e., proceeded in steps of one mass unit at a time. BBN explained the origin of lightest isotopes in an excellent way, and results in mass abundances of ^{1}H(75%), ^{4}He(23%), ^{2}H(0.003%), ^{3}He(0.004%). However, this theory meets grave difficulties beyond ^{4}He since there is no way to bridge the gap at mass number $A = 5$ and 8 through a sequence of neutron captures, beta decays and gamma emissions.

Although a small amount of lithium and beryllium can be produced in the big bang, the remainder of the lithium and beryllium and all of the boron are thought to have been produced by the spallation of still heavier elements by the cosmic radiation in the interstellar medium between stars. These elements are in general very rare. This process is known as "cosmic ray".

The synthesis of the first carbon isotopes, the basis of all carbon-based life, however, had to wait until the formation of the first star via gravitation. Only in stars was it possible to induce charged particle reactions because matter coalesces to higher temperature and density. Nevertheless, the advance to heavier elements is a very slow process due to the Coulomb repulsion between charged particles. Depending on masses of stars, reactions from hydrogen fusion to silicon burning can subsequently take place. Almost 95% of all stars spend their lives burning the hydrogen in their core, including our sun. Helium burning eventually leads to carbon and oxygen formation bypassing the $A = 8$ gap, while ^{12}C formation sets the stage for the entire nucleosynthesis of heavy elements due to the hindrance of the famous triple-α reactions. The powerful isotope factory, the shining stars, will eventually have to cease the production along with the death of stars. No energy will be produced from burning lighter isotopes in the last phase of stars, or in other words, stars can no longer convert mass into energy via nuclear fusion. The heaviest element that was possibly produced inside stars is iron because fusion reactions beyond it are endothermic (see Fig. 1). Although there is probably no way to examine the burning directly, it is expected that a very massive star near the end of its lifetime has an "onion ring" structure as a result of quiescent stages of stellar evolution. Hydrogen, helium, up to the heaviest element iron, should be dominant in different rings from the outside to the inside of the star.

Now the question left is how, where and when the heavier elements beyond Fe originate. Because the transmission through the Coulomb barrier decreases drastically with increasing nuclear charges, charged-particle cross sections are far too small at stellar temperatures to explain the observed solar abundances of nuclides with mass number larger than about 60. At very high temperatures, on the other hand, charged-particle reactions will give rise to abundances that are described by the nuclear statistical equilibrium (NSE), either favoring nuclei of the iron peak group or lighter species (see Fig. 1).

This difficulty to advance to higher mass via charged particle reactions can be easily coped with once neutron-induced reactions are considered as the mechanism for the synthesis of the heavy nuclides. Without the hindrance of the Coulomb barrier, the neutron capture cross sections, even

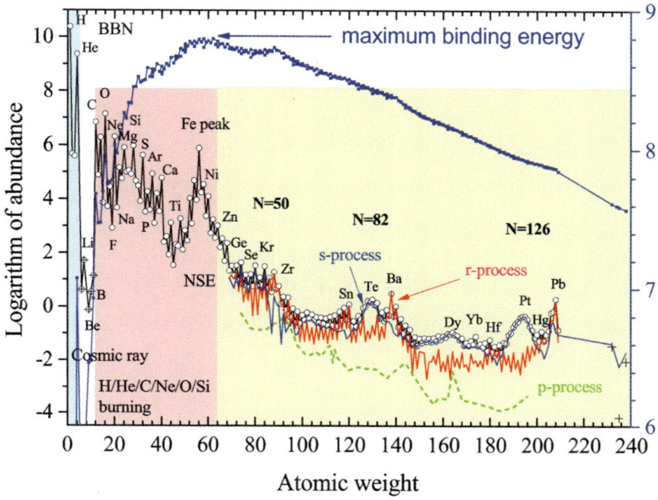

Fig. 1. Solar elemental abundances (with element of Si normalized to 10^6) and the curve of binding energy per nucleon. Indicated also are the different processes that account for the origin of elements in nature. The solar abundance data and the binding energies of stable nuclides are taken from Ref. 1 and AME2012,[2] respectively.

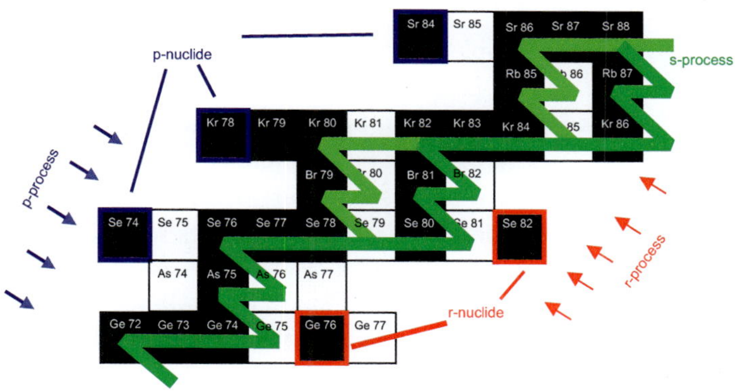

Fig. 2. A small portion of nuclear chart illustrating isotopes produced by the three basic processes, *s*-, *r*- and *p*-processes. Zigzag lines indicate possible *s*-process path. Black squares denote stable isotopes. *p*-nuclide and *r*-nuclide are marked especially, while most isotopes can be built by more than one process.

at moderate stellar energies, are often quite large. In fact, the cross sections for most neutron-induced reactions can even increase with decreasing incident neutron energies.[3]

As already mentioned before, there is an exponential decrease in abundances up to iron but an almost flat abundance beyond iron. This is naturally connected with the fact of exponential decrease in tunneling probability for charged-particle reactions, and non-charged-particle reactions for beyond iron. Moreover, a prominent feature of this solar distribution is the presence of three sets of double peaks in the region beyond the iron group nuclei. This was recognized as signatures of two distinct processes of neutron capture: a slow (s) one encountering the N =50, 82, and 126 closed neutron shells in the stable region and a rapid (r) one encountering the same in the extremely neutron-rich region of the nuclear chart.[4,5] Both processes are widely accepted to contribute in about equal parts to the total elemental abundances in the mass region between iron and bismuth. However, they differ in the amount of neutrons available (and thus astrophysical conditions), and accordingly the reaction paths and time scales. Specifically, the s-process runs along the path involving the mostly stable isotopes, indicating the slow time scale, and yields the peaks at mass numbers $A \sim 90$, 140, and 208. Most of the relevant nuclides involved in this process have been studied in worldwide nuclear physics laboratories. The r-process will be depicted in the next section. An illustration to distinguish these two n-capture processes is shown in Fig. 2.

Here one might already raise a question to the n-capture processes, namely, neutrons are unstable, with a half-life of about ten minutes. The interstellar medium cannot contain a significant concentration of free neutrons. Thus they must be and also, fortunately, can be produced in stars at stages of stellar evolution. There are numerous sources of neutrons in various stellar environments proposed so far. Here we would just like to point out that, unlike previously discussed processes inside stars, the neutron capture processes do not generate any significant amount of energy. This can be seen from the decline of the binding energy per nucleon beyond the iron peak in Fig. 1.

With all the mechanisms described above, however, there are still more than 30 stable neutron-deficient nuclides (p-nuclide) between ^{74}Se and ^{196}Hg that cannot be formed by the aforementioned neutron capture reactions. These nuclei cannot be produced in large (or any) amounts since they either lie outside the s-process flow or are shielded by stable isotopes from the r-process β-decay chains. As shown in Fig. 1, these p-nuclei are 10 to 100 times less abundant than the r- and s-nuclei in the same mass region, with exception of ^{92,94}Mo and ^{96,98}Ru. Historically, these nuclides are thought to have been produced in the so-called p-process, where proton-rich nuclei are made by sequences of photo-disintegrations on existing r- and s-seed nuclei and following β^+ decays. The astrophysical sites of the p-process are still under discussion.[6,7]

3. r-process nucleosynthesis

The r-process was proposed about 60 years ago to explain the solar abundances not created from the s-process.[4,5] The exact astrophysical site where the r-process proceeds has not been unambiguously identified, despite decades of work.[8,9] However, it is widely believed that the r-process is one of the major mechanisms for producing the stable and some long-lived radioactive (like ^{99}Tc) nuclides from about gallium, element thirty-one, up to long-lived radioactive uranium, element ninety-two. These isotopes are observed in stars of different metallicities as well as in the solar system.[1]

The r-process is somehow quite unique in the sense that it is the only known process that can produce elements heavier than Bi and reach thorium and uranium, and furthermore the only known process that can possibly synthesize superheavy elements in nature. A widely accepted picture is that the r-process occurs under conditions in which neutrons with densities of $10^{20} - 10^{30}$ cm^{-3} are captured on a very fast timescale of milliseconds. It starts with the reactions from hydrogen and is terminated when the heaviest nuclei created in the process become unstable to fission. The abundance flow runs up along the contour lines near the neutron drip line, and as it does so, highly unstable neutron-rich nuclei are created. A schematic view on r-process path is shown in Fig. 3. A few r-nuclides are also indicated in Fig. 2.

Understanding the r-process would definitely require knowledge of properties such as masses, β-decay lifetimes, and neutron-capture cross sections for several thousand extremely neutron-rich nuclei far from stability. Unfortunately, most of these nuclei are beyond the reach of current experiments[11] and therefore, these properties must be calculated with guidance from existing data and from measurements to be made at rare-isotope beam facilities such as FAIR, FRIB, HIRFL-CSR, and RIKEN.

In this section, we will first give a detailed description of the r-process modeling, from a dynamic to waiting-point approximation approach. Then we introduce a site-independent r-process method for the sensitivity studies of nuclear physics input. Finally we point out the nuclear data required for r-process nucleosynthesis studies.

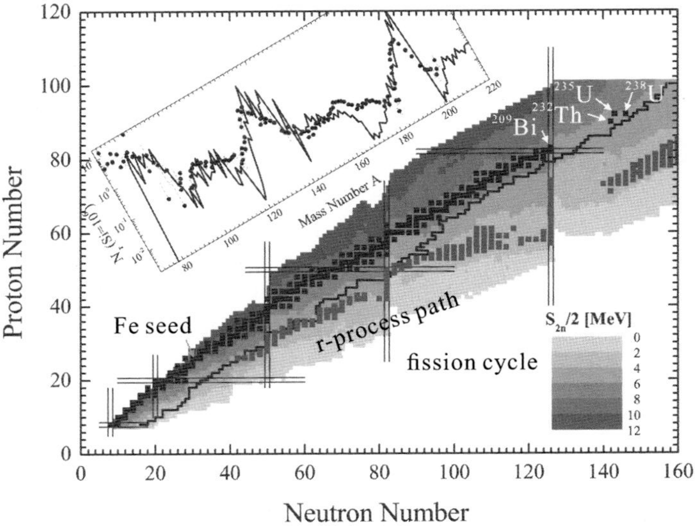

Fig. 3. Features of the r-process calculated using the RMF-BCS mass table. Black squares denote β-stable nuclei, and magic proton and neutron numbers are indicated by pairs of parallel lines. The region in the main graph shows the calculated average one neutron separation energy ($S_{2n}/2$). The solid line denotes the border of nuclides with known masses in the neutron-rich side. The dark grey squares show the r-process path when using the RMF mass predictions and the FRDM half-lives. The observed and calculated solar r-process abundance curves are plotted versus the mass number A in the inset, whose x-axis is curved slightly to follow the r-process path. Adaped from Ref. 10.

3.1. *r-process modeling*

During the r-process, the evolution of the abundances is mainly determined by neutron capture, photo-dissociation, and beta-decays. Fission and alpha decays also need to be included for heavy isotopes, while charged particle reactions such as (n,p) reactions need to be included for the light isotopes. This results in the following differential equation that determines the change of the abundance Y of a nucleus i with the relevant nuclei j, k, l:

$$\frac{dY_i}{dt} = \sum_j N_j^i \lambda_j Y_j + \sum_{j,k} \frac{N_{j,k}^i}{1+\delta_{jk}} \rho N_A \langle\sigma v\rangle_{j,k} Y_j Y_k + \sum_{j,k} \frac{N_{j,k,l}^i}{1+\Delta_{jkl}} \rho^2 N_A^2 \langle\sigma v\rangle_{j,k,l} Y_j Y_k Y_l \,, \tag{1}$$

here λ_i, $\langle\sigma v\rangle_{j,k}$, $\langle\sigma v\rangle_{j,k,l}$ are the "decay constant" for one-body reactions (e.g. β^-decays, electron capture, photo-disintegrations), reactivity for two-body reactions (e.g., n-capture reactions) and three-body reactions (e.g. triple-α reactions), respectively. The individual N_is are positive or negative integers and specify how many particles of species i are created (+) or destroyed (−) in a reaction. δ_{jk}, Δ_{jkl} are the factors that prevents double counting in two-body and three-body reactions, respectively. ρ and N_A are the mass density of the system and the Avogadro's number.

This set of equations is usually referred to as "reaction network", and can only be solved numerically. In a full dynamical r-process, the astrophysical conditions characterized by, e.g. mass density and temperature, are parameterized specially as functions of time or follow the evolution of astrophysical environments like supernovae explosion. On the other hand, the reaction rates for each nucleus are determined as a function of temperature, density, neutron number density etc. A full dynamical calculation would involve typically 5000–6000 nuclei. It starts with the reactions from hydrogen and is terminated when the heaviest nuclei created in the process become unstable to fission. Equation (1) can be further written explicitly when only neutron-capture reactions, photo-dissociations and beta-decays are involved. In this case, the abundance $Y(Z, A)$ of a nuclide with Z protons and A nucleons is determined by:

$$\begin{aligned}\frac{Y(Z,A)}{dt} &= \rho N_A \langle \sigma v \rangle_{Z,A-1} Y_n Y(Z,A-1) + \lambda_\gamma(Z,A+1)Y(Z,A+1) \\ &+ \sum_{j=0}^{J} \lambda_{\beta jn}(Z-1,A+j)Y(Z-1,A+j) \\ &- \rho N_A \langle \sigma v \rangle_{Z,A} Y_n Y(Z,A) \\ &- \left(\lambda_\gamma(Z,A) + \sum_{j=0}^{J} \lambda_{\beta jn}(Z,A) \right) Y(Z,A)\,. \end{aligned} \tag{2}$$

Here Y_n is the neutron abundance. For a nuclide (Z, A), $\langle \sigma v \rangle_{(Z,A)}$ is the thermal averaged neutron-capture rate, $\lambda_\gamma(Z, A)$ the photo-dissociation rate, and $\lambda_{\beta jn}(Z, A)$ is the corresponding β^- decay rate with emission of j delayed neutrons (up to a maximum of J).

The photo-dissociation rate is related to the neutron capture rate by detailed balance,

$$\begin{aligned}\lambda_\gamma(Z,A+1) = \langle \sigma v \rangle_{Z,A} \left(\frac{m_u kT}{2\pi\hbar^2} \right)^{3/2} \frac{2G(Z,A)}{G(Z,A+1)} \left(\frac{A}{A+1} \right)^{3/2} \\ \exp\left[-\frac{S_n(Z,A+1)}{kT} \right], \end{aligned} \tag{3}$$

where $\hbar$, m_u, k are the Planck constant, the atomic mass unit, and the Boltzmann constant, respectively, while G and S_n denote the partition function and one-neutron separation energy of the appropriate nucleus.

A commonly used assumption to the full network simulation is the $(n,\gamma) \rightleftarrows (\gamma,n)$ equilibrium, or "waiting-point approximation". This approximation is valid when the neutron density ($n_n \gtrsim 10^{20}$ cm^{-3}) and temperature ($T \gtrsim 1$ GK) are high enough to warrant that both the rate of neutron capture and photo-dissociation are much larger than the competing β-decay rate for all the nuclei participating in the network. Under this condition, the evolution of the system is mainly determined by the β decay rates as the abundances along an isotopic chain are immediately adjusted to an equilibrium between neutron captures and photo-dissociations, i.e. $n_n Y(Z,A)\langle\sigma v\rangle_{Z,A} = \lambda_\gamma(Z,A+1)Y(Z,A+1)$. Assuming furthermore that the neutron density $n_n = Y_n \rho N_A$ is constant over a time step, the ratio between the abundances of neighboring isotopes are then given by the following simple relation:

$$\frac{Y(Z,A+1)}{Y(Z,A)} = n_n \left(\frac{2\pi\hbar^2}{m_u kT}\right)^{3/2} \left(\frac{A+1}{A}\right)^{3/2} \frac{G(Z,A+1)}{2G(Z,A)} \exp\left[\frac{S_n(Z,A+1)}{kT}\right]. \tag{4}$$

As can be seen from the exponential dependence on the neutron separation energy in Eq. (4), nuclear masses are among the most important inputs for modeling the r-process.

For each isotopic chain, the above equation defines in fact a nuclide that has the maximum abundance. This nuclide is normally known as waiting point (WP) nuclide as the flow of neutron captures "waits" for this nucleus to beta-decay. The set of WP nuclei constitute the r-process path. The maximum of the abundance distribution can be determined by setting the left-hand side of Eq. (4) to 1. This results in a value of S_n,

$$\begin{aligned} S_n &\approx \kappa T \log\left[\frac{2}{n_n}\left(\frac{m_\mu \kappa T}{2\pi\hbar^2}\right)^{3/2}\right] \\ &= T_9\left\{2.79 + 0.198\left[\log\left(\frac{10^{20}}{n_n}\right) + \frac{3}{2}\log T_9\right]\right\}. \end{aligned} \tag{5}$$

Ideally, S_n is the same for all isotopic chains for a given neutron density n_n and temperature T_9, which is the temperature in 10^9 K. Equation (5) implies that the r-process proceeds along lines of constant neutron separation energies towards heavy nuclei. Higher temperature or lower neutron density will drive the r-process path towards the valley of stability. For typical r-process conditions this corresponds to $S_n \sim 2$–4 MeV. The separation energy is, however, not a continuous function of neutron number but shows large jumps particularly close to magic neutron numbers. This is also the reason that the r-process path moves closer to the stability where the relevant nuclei have larger S_n. Considering the pairing correlations the most abundant isotope always has an even neutron number N.

Under the $(n,\gamma) \rightleftarrows (\gamma,n)$ equilibrium, it is sufficient to consider only the time evolution of the total abundance of an isotopic chain $Y(Z) = \sum_A Y(Z,A)$, as the abundances of different isotopes are fully determined by Eq. (4). The abundance flow from one isotopic chain to the next is governed by β-decays and can be expressed by the following set of differential equations:

$$\frac{dY(Z)}{dt} = \lambda_\beta(Z-1)Y(Z-1) - \lambda_\beta(Z)Y(Z) \tag{6}$$

where $\lambda_\beta(Z) = \sum_A \lambda_\beta(Z,A)Y(Z,A)/Y(Z)$. In this way the r-process evolution can be computed independent of the neutron-capture rates, and only beta-decays are necessary for Eq. (6) and masses via S_n in Eq. (4). After the neutrons freeze out, all the isotopes will proceed to the corresponding stable isotopes via β-decays or α-decays. The matter flow towards heavier mass systems by following the procedure in Eqs. (4, 6) is known as the classical r-process.[12,13]

A few recent discussions (see, e.g. Ref. 14) argued that r-process may proceed in conditions with low temperatures and neutron densities. For these cases, one has to resort to the full r-process network calculations. However, in the following discussion we will always refer to the "traditional" r-process, in which the waiting point approximation holds well.

3.2. *Site-independent r-process approach*

Because the specific astrophysical conditions among the different scenarios may change, solar r-process abundances have been used in the past to constrain the astrophysical conditions using a site-independent approach.[13] In this approach seed-nuclei (usually the iron group) are irradiated by neutron sources of high and continuous neutron densities n_n ranging from 10^{20} to 10^{28} cm^{-3} over a timescale τ in a high temperature environment ($T \sim 1$GK). A superposition of r-process components (n_n,τ) is needed to reproduce the overall shapes and positions of the solar r-process abundances and it is equivalent to the exponential neutron exposures in the s-process.[15] The configuration of many r-process components seems to be also a reasonable approximation to the real r-process event. For instance, one can think of it as the "onion ring" structure of neutron sources with different densities, where the seed-nuclei capture neutrons while moving through different zones with different thicknesses. In section 4.1 we will introduce the first application of the mass model rooted in relativistic mean field theory (RMF) in an r-process calculation.

To simulate a "realistic" r-process condition, we will apply sixteen components with neutron densities in the range of 10^{20}–3×10^{27} cm^{-3} in our calculation. The temperature is fixed to $T = 1.5$ GK. We further assume that the irradiation time τ and the corresponding weight w follow the exponential dependent of neutron density n_n, i.e.,

$$w(n_n) = n_n^a, \; \tau(n_n) = b \times n_n^c \; , \tag{7}$$

where a, b, c are parameters to be optimized. These parameters can be obtained from a least-square fit to the solar r-process abundances.

The classical r-process approach leaves out many important details, such as the time evolution of temperature and neutron density, the finite duration of the freeze-out, and the breakdown of $(n, \gamma) \rightleftharpoons (\gamma, n)$ equilibrium during the freeze-out. However, so long as $(n, \gamma) \rightleftharpoons (\gamma, n)$ equilibrium can be achieved in an r-process environment, the conditions immediately before the freeze-out in that environment should be close to those derived here. Nevertheless, it is essential in the future to carry out parametric studies of the r-process based on more detailed and more realistic astrophysical models, thus to explore the effects of various nuclear inputs on the model such as CDFT.

3.3. *Nuclear physics challenges*

As already mentioned, the r-process runs along the path far away from the β-stability line, where the majority of those nuclei cannot be reached experimentally even in the near future. Therefore, we have to rely on the extrapolation of theoretical nuclear models for these "terra incognita". Following the discussions on the r-process modeling, we can easily summarize some of the most important nuclear properties for r-process in Table 13.1.

As seen from the classical r-process, mass and half-lives play the most important role in matter flow from light to heavy nuclei. To first order, they determine the reaction path, the abundance distribution and the time scale of r-process. Besides those quantities tabulated in the above table, neutrino related properties are also important for simulations with neutrino-driven r-process mechanism.

Table 13.1. Selected nuclear physics data required for the r-process. Q is the reaction Q-value, and P_n is the branching ratio of beta-delayed neutron-emission.

Property	Effect in r-process
S_n, Q	path of r-process
$T_{1/2}$	time scale of r-process
P_n	smoothing of the r-process abundance
G	abundance pattern (weakly dependent)
Fission branchings and products	r-process endpoint, fission cycling
$N_A\langle\sigma v\rangle$	abundance pattern during freeze out
Isomers, excited states	branch of r-process path, time scale

It should be pointed out that half-lives and cross sections here do not refer to the terrestrial data measured in laboratories, but the stellar data in conditions characterized by high temperature and high density, in which the nuclei are typically highly charged or even fully ionized. The influence of bound electrons may lead to a significant difference from the corresponding terrestrial data. For instance, to derive astrophysical rates, one has to take into account the additional transitions that are not included in the laboratory cross-section measurements.[16] Meanwhile, the nuclei inside stars are immersed in a dense electron gas, the charge shielding for charged particle reactions exists due to the Coulomb repulsion between projectile and target. The corresponding charged particle reaction rates can thus be enhanced in a stellar plasma, compared to reaction rates for bare nuclei. Therefore, measured reaction rates may need to be corrected somehow to obtain bare reaction rates depending on the stellar conditions.

Furthermore, new decay modes — strongly suppressed or disabled in neutral atoms — can open up, or in contrast some channels are strongly forbidden.[17] A well known example in s-process studies is the bound beta decay (β_b^-) of $^{187}Re^{75+}$ nuclei.[18] Neutral ^{187}Re atoms decay with $T_{1/2} = 42 \times 10^9$ years to ^{187}Os atoms. The decay energy $Q_{\beta_b^-} = 2.7$ keV is the smallest known $Q_{\beta_b^-}$ value. However, the bare ^{187}Re nuclei decay in merely $T_{1/2}(^{187}Re^{75+}) = 33(2)$ years, which is by more than 9 orders of magnitude shorter than the half-life of neutral atoms. Thus the ^{187}Re–^{187}Os galactic chronometer calibration should be corrected, by taking account of the β_b^- decay enhancement in stellar interiors.

In astrophysical conditions, each nuclide can be thermally populated in both ground state and excited states, and their ratio is given by the Saha equation. In typical r-process conditions, $kT < 300$ keV is smaller than the typical level spacing in nuclei at low energies of about 1 MeV. However, in special cases, e.g. when low lying excited states exist or the populated state has a very different rate in comparison with the ground state, the partition function G needs to be treated carefully.

When a low-lying isomer with excitation energy E_x less than 100 keV exists along the r-process path, a new branching may be open in the matter flow. However, the isomeric effect so far has not been put into any r-process calculations self consistently. This is due to the difficulties in both the unknown nuclear properties and the complicity in numerical modeling. Fission properties, which determine the endpoint of r-process, can now be calculated either in empirical formulas or in nuclear models, and have been included in the full network recently.

4. Nuclear physics inputs

To start this section, we would like to emphasize that it is important to have a consistent nuclear data set, i.e., to calculate every relevant property within the same framework. Otherwise, artificial effects may be introduced. Triggered by the many successes of CDFT as discussed in previous chapers, interests have grown over the last years to provide the nuclear physics input for nuclear astrophysical application. Indeed, many pilot works have been done, e.g. on the nuclear mass tables and applications of these data in r-process simulations. More efforts and breakthroughs are, however, definitely needed towards a fully consistent nuclear database for astrophysical interest, especially considering the fact that key properties such as neutron-capture reaction rates and partition functions are still missing. Therefore, in the following discussion most of the r-process simulations are in fact not "fully" consistent. We will concentrate on the progress in computing nuclear masses, lifetimes in CDFT and their tentative applications in r-process applications.

4.1. *Mass*

We first review briefly the progress in large-scale mass predictions based on CDFT, and then introduce the first application of the CDFT mass input in r-process simulations and the relevant conclusions. Considering the mass model uncertainties, several selected nuclear mass models are used for comparisons. These models range from the macroscopic-microscopic kind, represented by the finite-range droplet model (FRDM),[19] extended Thomas-Fermi plus Strutinsky integral ETFSI-Q,[20] the Duflo-Zuker model (DZ)[21] and a more recent Weizsäcker-Skyrme (WS*) model,[22] to the microscopic kind, represented by the Skyrme–Hartree–Fock–Bogolyubov mean-field HFB-13,[23] HFB-17 model.[24]

4.1.1. *Overview of mass database in CDFT*

The application of CDFT to the large-scale mass predictions can be traced back to the 1990's. Limited mainly by the computation power, those calculations had to be simplified by either neglecting the pairing correlations[25] or limiting to even-even nuclei.[26] Only in 2005 was the first global mass database[27] possible in reality. Masses of all nuclei from the proton drip line to the neutron drip line with $Z, N \geqslant 8$ and $Z \leqslant 100$ were calculated based on the RMF model with the TMA effective interaction, and the state dependent BCS method with a zero-range δ force was used. Hereafter, we refer to this mass database as RMF+BCS otherwise specified. Globally, the root-mean square (rms) deviation with respect to known masses in AME2003[28] is about 2 MeV. This large rms deviation may be partially attributed to the TMA effective interaction, since it was only optimized with the properties of a few selected nuclei.[29]

Although a relatively large rms deviation for the absolute mass value is found for the RMF+BCS calculation in comparison with other popular mass models such as FRDM, the finite differences in binding energies such as the practical used one-neutron separation energies S_n in r-process simulations are well predicted due to the cancellation of systematic error. The rms deviation of S_n for the RMF+BCS model with respect to experimental values are 654 keV. This value is about 400 keV for FRDM. Here the comparisons include nuclei with $Z, N \geq 8$. Comparing the predictions of the RMF+BCS model to the known values results in discrepancies between -1.4 MeV to 1 MeV, while the difference between the FRDM and the experimental data is in the range of -1.3 MeV to 0.5 MeV. It shows that the microscopic model such as the RMF approach can almost achieve the same level of accuracy for known S_n as the phenomenological FRDM. For each isobaric chain with mass A, the distance between the nuclide (Z, A) and the nuclide (Z_0, A) in the β-stability line is defined by $\varepsilon = Z_0 - Z$ with

$$Z_0 = \frac{A}{1.98 + 0.0155A^{2/3}} \,. \tag{8}$$

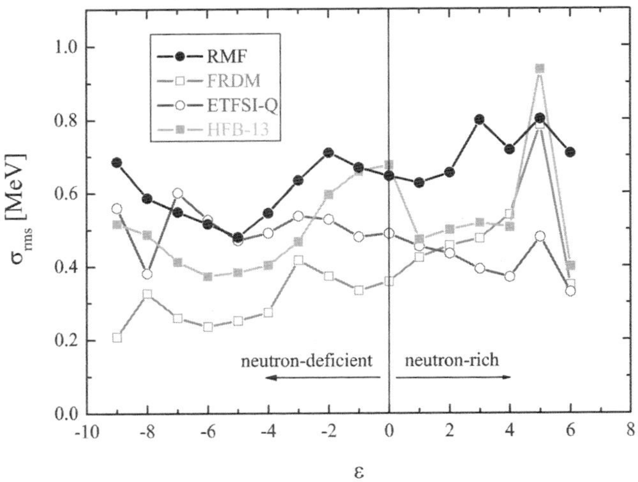

Fig. 4. (Color online) Rms deviation σ_{rms} of one-neutron separation energy S_n with respect to AME2003 as a function of the distance from the β-stability line $\varepsilon = Z_0 - Z$ for different mass models, where Z_0 stands for the proton number of the most stable isotope in the isobaric chain with mass number A. Adapted from Ref. 13.

$\varepsilon = 0$ stands for the most stable nuclei and $\varepsilon > 0$ the neutron-rich nuclei. The rms deviation σ_{rms} of S_n as a function of $\varepsilon = Z_0 - Z$ for RMF is shown in Fig. 4. Three other widely used mass models, FRDM, ETFSI-Q and HFB-13, are also plotted for comparison. It is remarkable that almost the same order of prediction power of S_n from the neutron-deficient side to the neutron-rich side is achieved for all the models, even though models like FRDM and HFB-13 have about 10 more free parameters than the RMF model and were optimized for all the known masses. The good prediction power of S_n in the RMF+BCS make its first application in r-process simulations possible.

The evolution of the nuclear structure around the shell closures $N = 82$ and 126 is critical in understanding the r-process abundance distribution around the $A = 130$ and $A = 195$ abundance peaks. In Fig. 5(a), the predicated two-neutron separation energies S_{2n} in the RMF+BCS model are displayed as a function of mass number together with the available experimental data for nuclei with $50 \leqslant Z \leqslant 90$. In general, the RMF+BCS model reproduces well the experimental data, but shows a very subtle variance relative to experiments. We would like to draw attention especially to the shell structure at $N = 82, 126$ (corresponds to the abrupt drop at $A \sim 140, 210$) and shape transition near $N = 90$ (corresponds to the small

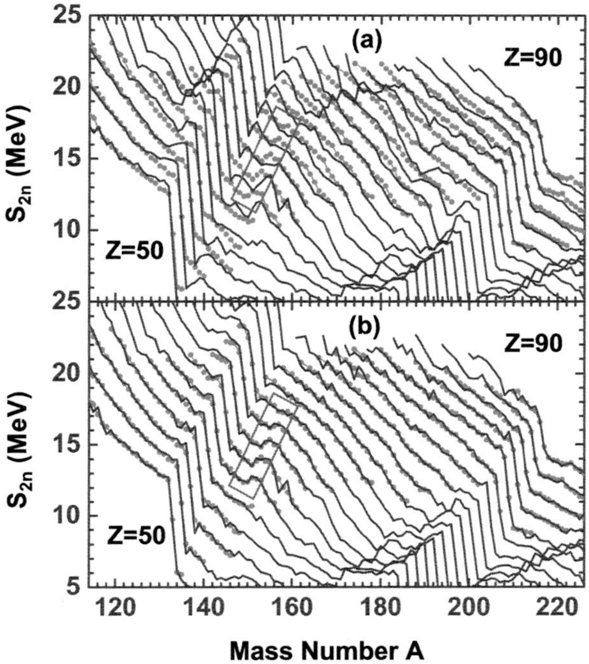

Fig. 5. (Color online) Two neutron separation energies S_{2n} for nuclei with $50 \leqslant Z \leqslant 90$. For clarity, only nuclei with even Z are shown. The black lines in the panels (a) and (b) denote those predicted by the RMF+BCS model and the RMF+RBF model, respectively. The experimental data of S_{2n} and shape transition regions are indicated by dots and rectangles. Adapted from Ref. 30.

dip at $A \sim 150$). By comparing with the experimental data, one can see that: (a) The predictions can reproduce well the position of shell structure at $N = 82, 126$, however, the strength of shell closure is generally overestimated, especially in the region near the doubly-magic nuclide ^{132}Sn. This can be more clearly seen in Fig. 6. (b) Different from the known data, pronounced dips before the magic numbers $N = 82, 126$ are observed in the predictions. (c) A considerable deviation from known data is seen for the region of shape transitions near $N = 90$, including the exact location and also the strength. These will affect the simulated abundance distribution in r-process nucleosynthesis as will be discussed in Section 4.1.3.

The neutron shell gaps, defined as $\Delta_n(Z, A) = S_{2n}(Z, A) - S_{2n}(Z, A+2)$, can be more clearly seen from Fig. 6, which shows the shell gaps for $N = 82$ and 126 in the RMF+BCS approach in comparison with the data available

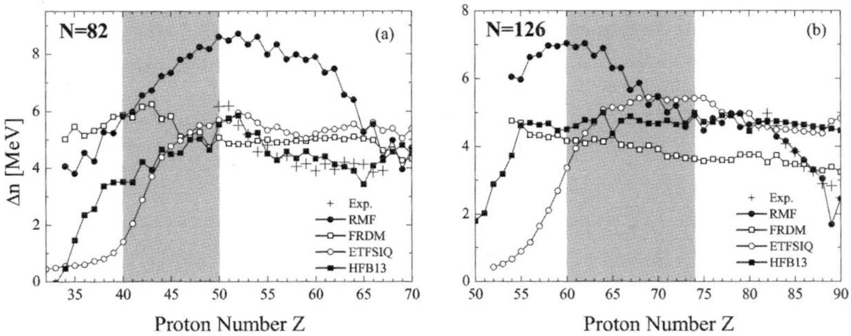

Fig. 6. Neutron shell gaps $\Delta_n(Z, A) = S_{2n}(Z, A) - S_{2n}(Z, A+2)$ for $N = 82$ and 126 in the RMF approach compared with those in the FRDM, ETFSI-Q and HFB-13 models together with the data available. The nuclei in the shadowed areas are involved in the r-process paths based on our calculations. Adapted from Ref. 13.

and those in the FRDM, ETFSI-Q and HFB-13 models. The nuclei in the shadowed area are in the r-process path. Experimentalists have been trying to make a breakthrough on the critical waiting point nuclei ^{130}Cd for more than 20 years, but experimental data are still in debate.

Considering the theoretical predictions, all the mass models except the FRDM model show a strong quenching effect (i.e., the shell gap drops) towards the neutron-rich side at the $N = 82$ shell. The RMF+BCS shell gap is overestimated compared with the data available and it is around 2 MeV larger than other models for $45 \leq Z \leq 60$. Regardless of this, the RMF+BCS model succeeds in predicting the enhanced double-magic effect at $Z = 50$ together with the HFB-13 model. Concerning the $N = 126$ shell, the relevant nuclei are so neutron-rich that we are unlikely to measure them in the foreseeable future. Theoretically, there is no sign of shell quenching observed in the r-process region for all the models. A unique feature of the RMF+BCS model is that it fully coincides with the available data and it is also the only model to reproduce the enhanced double-magic effect at $Z = 82$. In comparison, the other models fail to reproduce the trend of the known $N = 126$ shell. Towards the neutron drip line, the RMF+BCS prediction tends to enhance the shell until the maximum is reached around $Z = 60$ while the other models have a roughly constant shell gap.

To further improve the accuracies of mass models, it is certainly consequential to optimize the effective interactions of CDFT by using more observables as constraints, such as binding energies, charge radii and single particle levels. This procedure has been well demonstrated already in the non-relativistic approaches. A recent attempt in this direction was made by fitting to properties of 60 selected spherical nuclei, and a new effective interaction PC-PK1 was proposed.[31] With the PC-PK1, the masses of nuclei with proton number $Z \geqslant 8$ and neutron number $N \geqslant 8$ in AME2003 are calculated and the rms deviation with respect to the known masses in AME2003 was successfully reduced to 1.4 MeV for even-even nuclides. This effective interaction particularly improves the description for isospin dependence of binding energies, and results in excellent agreement with the new experimental data from Sn to Pa.[32] Now the global mass set with the improved interaction are still in progress. Some pilot results can be found in Ref. 33.

An alternative way to improve the predictions of CDFT is to examine directly the nuclear masses computed and then to recognize the pattern of how the calculations differ from experiments. Efforts in this directions will help not only to isolate the problems in CDFT, but also to suggest lessons to progress. For such purpose, the radial basis function (RBF) has been borrowed from the field of surface reconstruction, while the improvement is quite encouraging. This is witnessed by the much better accuracy of RMF when training the RBF with thousands of known nuclear masses. The rms deviation of the mass predictions in the RMF+BCS model is reduced from the 2.217 MeV to 0.488 MeV. The RBF approach also shows more reliable extrapolative power along the distance from the β-stability line. Moreover, it improves the description of separation energies, resulting in a much smoother evolution of S_{2n} as a function of A. The resulting S_{2n} surface moves more closely to reality as seen in Fig. 5(b). Meanwhile, the shell strength at $N = 82$ is reduced and now agrees well with experimental data, although the deviations are not fully smoothed out. The dips before $N = 82, 126$ are washed out to a large extent and the phase transition near $N = 90$ are reproduced nicely. All these results demonstrate the feasibility of RBF although it is only qualified as a local approach.

4.1.2. *r-process conditions under the WP approximation*

While there are substantial uncertainties in both the r-process sites and the relevant nuclear physics inputs, an essential feature of the r-process is considered robust: the observed abundance peaks at $A \sim 80$, 130, and 195 correspond to the intrinsic properties of extremely neutron-rich nuclei with $N = 50$, 82, and 126 closed neutron shells that are produced in the r-process. As discussed already, when $(n, \gamma) \rightleftharpoons (\gamma, n)$ equilibrium is achieved, the total abundance of an isotopic chain is concentrated in the

Table 13.2. Critical waiting-point (CWP) nuclei.

N	CWP nuclei
50	^{80}Zn, ^{79}Cu, ^{78}Ni
82	^{130}Cd, ^{129}Ag, ^{128}Pd, ^{127}Rh, ^{126}Ru
126	^{195}Tm, ^{194}Er, ^{193}Ho, ^{192}Dy, ^{191}Tb

corresponding WP nucleus. The β-decay lifetimes of the WP nuclei then regulate the abundance pattern resulting from an r-process episode. In particular, the much longer β-decay lifetimes of extremely neutron-rich nuclei with closed neutron shells than those without produce peaks in r-patterns. Consequently, in order to produce the observed peaks in r-patterns under the WP approximation, nuclei with $N = 50$, 82, and 126 closed neutron shells and with $A \sim 80$, 130, and 195, respectively, must be among the WP nuclei. The critical waiting-point (CWP) nuclei selected are listed in Table 13.2 based on similar considerations to these studies. Below we aim to derive the conditions required for r-process nucleosynthesis under the WP approximation by considering the properties of only those CWP nuclei. To do so, precise binding energies of the CWP are essential.

To be quantitative, we define a WP nucleus (Z, A_{WP}) as the one that has an abundance

$$Y(Z, A_{\text{WP}}) \geq 0.5 \sum_{A} Y(Z, A), \tag{9}$$

where the sum over A gives the total abundance of the corresponding isotopic chain. For specific T and n_n, we can use the above criterion and the relative abundance $Y(Z, A+1)/Y(Z, A)$ given by Eq. (4) to determine (Z, A_{WP}) from the nuclear partition functions and neutron separation energies provided by a model. Conversely, one can also determine the T and n_n conditions required by a specific WP nucleus. As can be seen from Eq. (4), the predominant dependence of $Y(Z, A+1)/Y(Z, A)$ is on the neutron separation energy, which can be calculated from a nuclear mass model. Eq. (4) is used in the calculations below.

As an example, we adopt the RMF+BCS mass model to calculate the sets of T_9 and n_n within the ranges $1 \leq T_9 \leq 3$ and $10^{20} \leq n_n \leq 10^{30}$ cm^{-3} that are required by the $N = 82$ CWP nuclei. The results are shown in Fig. 7. For a specific T_9, the values of n_n between two identical symbols in this figure would allow the corresponding nucleus to have $\geq 50\%$ of the total abundance of its isotopic chain. In order to accommodate all the

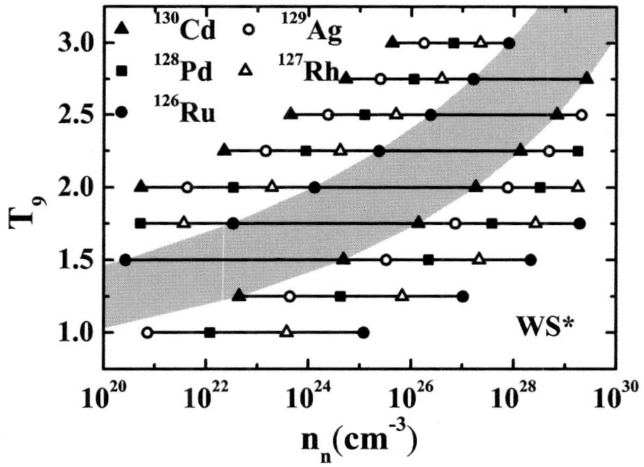

Fig. 7. T_9−n_n conditions required by the $N = 82$ CWP nuclei based on the RMF+BCS mass model. For a specific T_9, the values of n_n between two identical symbols would allow the corresponding nucleus to have $\geq 50\%$ of the total abundance of its isotopic chain. The conditions indicated by the shaded band are required to accommodate all the $N = 82$ CWP nuclei. Adapted from Ref. 34.

$N = 82$ CWP nuclei, the common range of n_n for a specific T_9 is bounded from below by ^{126}Ru (filled circle) and from above by ^{130}Cd (filled triangle). This range of n_n changes with T_9 and is shown as the shaded band in Fig. 7. This band represents the T_9−n_n conditions required by the $N = 82$ CWP nuclei.

The calculations above can be generalized to determine the T_9−n_n conditions required by the $N = 50$, 82, and 126 CWP nuclei, respectively, for any specific nuclear mass model. To show the nuclear model uncertainties, the results for the $N = 50$ CWP nuclei are presented for the FRDM, WS*, and RMF models in Fig. 8(a), which clearly show that the required conditions change with models. Similar to the case of the $N = 82$ CWP nuclei, the upper curve for each model in Fig. 8(a) is effectively determined by the two-neutron separation energy of the lightest $N = 50$ CWP nucleus ^{78}Ni and the lower curve by that of the $N = 52$ isotope ^{82}Zn of the heaviest $N = 50$ CWP nucleus ^{80}Zn. Therefore, the large differences among the conditions required by the $N = 50$ CWP nuclei for different models can be traced to the differences in S_{2n} of ^{78}Ni and ^{82}Zn provided by these models. In particular, the differences for ^{82}Zn among the models appear to be substantially larger than those for ^{78}Ni. It is also found that no conditions can accommodate all the $N = 50$ CWP nuclei for the HFB-17 model, for

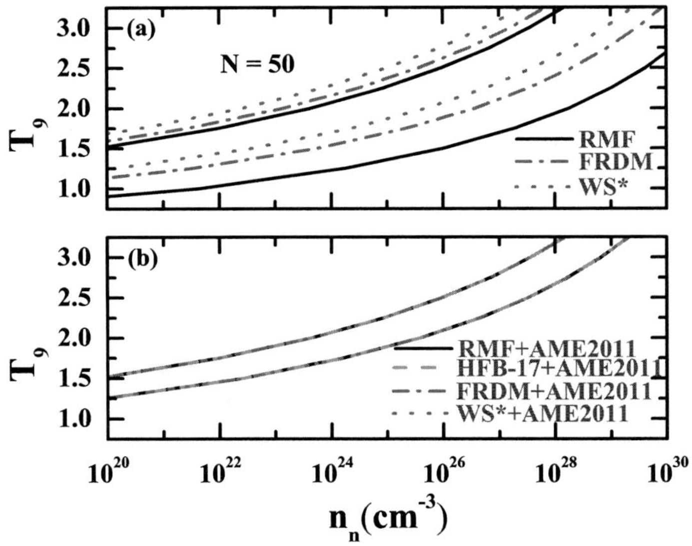

Fig. 8. (Color online) T_9–n_n conditions required by the $N = 50$ CWP nuclei. (a) The band between two curves of the same kind represents the required conditions based on the corresponding mass model. Note the large differences among the results for the three indicated models. Note also that no conditions can be found to accommodate all the $N = 50$ CWP nuclei based on the HFB-17 model. (b) Same as (a), but for those nuclei the masses of which are tabulated in the AME2011-preview, the model predictions are replaced by the tabulated values. All four models now give the same conditions required by the $N = 50$ CWP nuclei. Adapted from Ref. 34.

which the odd-even effects in the neutron separation energy for Ni, Cu, and Zn isotopes around $N = 50$ are larger by $\sim$ 1–1.5 MeV than those for the FRDM, WS*, and RMF+BCS models.

Noting that the two-neutron separation energies of ^{78}Ni and ^{82}Zn can be calculated from the masses tabulated in AME2011-preview, we augment the nuclear mass models by using the tabulated values in AME2011-preview when they are available to replace the corresponding model predictions. Remarkably, all four models, including the HFB-17 model, now give the same conditions required by the $N = 50$ CWP nuclei as shown in Fig. 8(b). It is found that the changes between Figs. 8(a) and 8(b) are caused dominantly by the use of the tabulated masses of ^{76}Ni to ^{78}Ni and ^{78}Zn to ^{82}Zn, which confirms the crucial roles of the two-neutron separation energies of ^{78}Ni and ^{82}Zn in determining the conditions required by the $N = 50$ CWP nuclei. In the calculations below, we use the FRDM, WS*, HBF-17, and RMF predictions augmented by AME2011-preview.

We calculate the conditions required by the $N = 82$ and 126 CWP nuclei, respectively, as in the case of the $N = 50$ CWP nuclei. The results are summarized in Fig. 9. Again three other models are used to indicate the nuclear model uncertainty. It can be seen that the conditions required by the $N = 82$ CWP nuclei (shaded band) are essentially converged for the four augmented nuclear mass models just like those required by the $N = 50$ CWP nuclei (horizontally hatched band between solid curves). In contrast, the conditions required by the $N = 126$ CWP nuclei (vertically hatched band between dashed curves) are still strongly dependent on models. This is because these nuclei and the majority of those in the nearby region of the nuclear chart are still out of the reach of experiments while theoretical predictions for their masses involve dramatic extrapolations with large uncertainties.

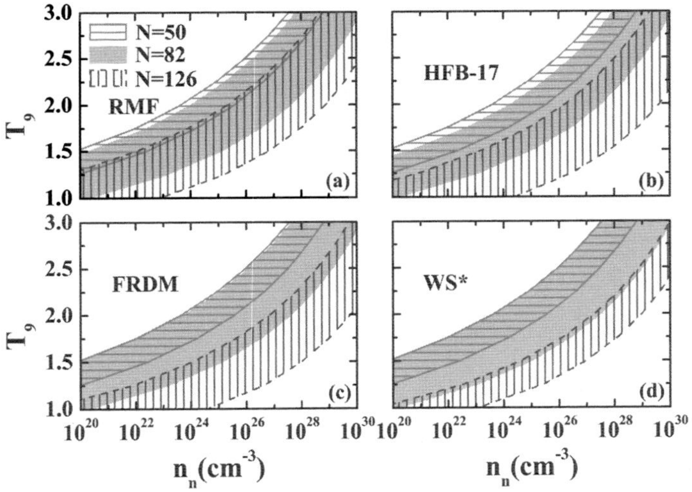

Fig. 9. (Color online) T_9−n_n conditions required by the $N = 50$ (horizontally hatched band between solid curves), 82 (shaded band), and 126 (vertically hatched band between dashed curves) CWP nuclei, respectively, for four nuclear mass models: (a) RMF, (b) HFB-17, (c) FRDM, and (d) WS*. See text for details. Adapted from Ref. 34.

Figure 9 resembles a phase diagram in terms of three bands for the production of the peaks at $A \sim 80$, 130, and 195 in r-patterns that correspond to the $N = 50$, 82, and 126 CWP nuclei. For the T_9−n_n conditions inside the non-overlap region of a band, only a single peak can be produced. For those conditions inside the overlap region of two bands, it is possible to produce two peaks simultaneously. For the RMF+BCS model only, there is a very thin sliver where three bands overlap. Consequently, we consider it very unlikely that three peaks can be produced simultaneously. A close examination of Fig. 9 shows that the $T_9 - n_n$ conditions required by the $N = 50$ CWP nuclei (horizontally hatched band between solid curves) are distinct from those required by the $N = 126$ CWP nuclei (vertically hatched band between dashed curves) for the FRDM, HFB-17, and WS* models. These two sets of conditions overlap only slightly for the RMF+BCS model. This suggests that the peaks at $A \sim 80$ and 195 in r-patterns are not produced simultaneously. Their production may differ in the time of occurrence within the same astrophysical site or in the astrophysical site itself. In contrast, there is large overlap between the conditions required by the $N = 50$ and 82 CWP nuclei (shaded band) for the four models considered. In addition, there is slight to significant overlap between the conditions required by the $N = 82$ and 126 CWP nuclei for all the models. Therefore, it is possible to produce the peaks at $A \sim 80$ and 130 or those at $A \sim 130$ and 195 simultaneously. These conclusions drawn from the waiting point approximation are very instructive for more comprehensive r-process simulations. Eventually, the above discussion can be used to identified some key nuclei including ^{76}Ni to ^{78}Ni, ^{82}Zn, ^{131}Cd, and ^{132}Cd for precise mass measurements at rare-isotope beam facilities.

4.1.3. *Solar r-process abundances within WP approximation*

Now we will explore to what extent the solar r-process abundance can be reproduced using the RMF+BCS mass table. As mentioned already, this is the first application of CDFT in r-process simulations. In the calculations, unknown neutron separation energies were calculated from the RMF+BCS and β-decay properties were taken from Refs. 35, 36. Experimental data[28,37] was used when available. The site-independent approach discussed in section 3.2 is used.

Calculated solar r-process abundances are displayed in Fig. 10. Shadowed areas show the regions with underproduced abundances before the neutrons freeze out. After β-decays to the stability line, these gaps are too large to be completely filled in by β-delayed neutron emissions. Although it is thought that the r-process is responsible for the majority of those isotopic abundances with $Z \geq 56$, its contribution to the lighter elements is still debatable. It is possible that some of the discrepancies in the reproduction of the low mass abundances may be due to an additional nucleosynthesis component creating some of those abundances. However,

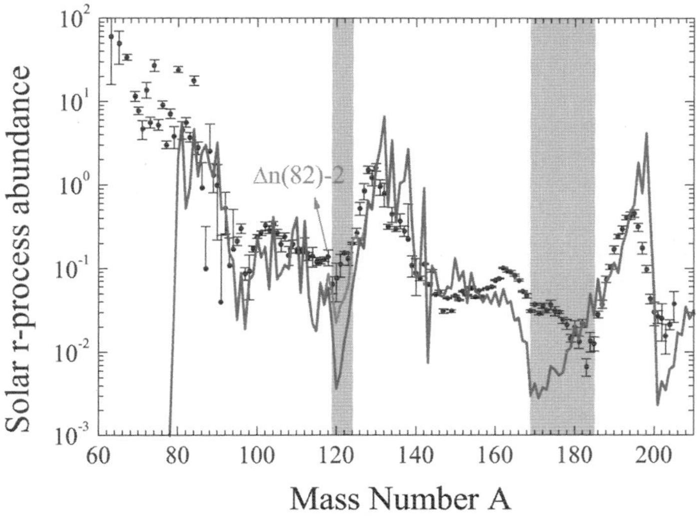

Fig. 10. (Color online) Our best fits to the solar r-process abundances using the nuclear mass inputs from the RMF+BCS. The β-decay properties are taken from the FRDM model. The best fits are displayed as solid lines. The dashed curve is the same as the solid curve but with a shell closure at $N = 82$ that is 2 MeV smaller. The shadowed areas correspond to the range where the abundances of these isotopes are largely underestimated before neutrons freeze out. Adapted from Ref. 13.

since astrophysical conditions and nuclear properties both affect the resulting r-process abundances, one needs to determine or at least understand the uncertainties in the nuclear physics properties in future to disentangle both effects.[10]

Traditionally, the underestimation of the isotopic abundances before $A \sim 130$ peak has been attributed to the overestimated strength of the $N = 82$ shell closure in the theoretical nuclear physics model. Since it is not possible to do a complete study of the shell-quenching effect at the single particle level which should affect more nuclei than the one with $N = 82$, we only study the effect of a reduced shell closure by artificially decreasing the shell gap energies at $N = 82$ in the RMF+BCS model by 2 MeV. In such a way, the shell gaps relevant to the r-process would roughly have the same values as those in the quenched models ETFSI-Q and HFB-13. Eventually, a better agreement with the observation at $A \sim 120$ is obtained as shown in Fig. 10. This can be easily understood as follows. A reduction of shell gap leads to a nuclear matter repopulation in the isotopic chain according to Eq. 4. R-process waiting points located at $N = 82$ move closer to the valley of stability and thus some of the underabundance can be filled.

However, new experimental evidence tends to show no shell quenching effect at $N = 82$.

The abundance trough at $A \sim 115$ for the RMF+BCS model can be related to the additional bump of S_n at $A = 110-120$ in Fig. 5, and thus the associated nuclear shape transition. Nuclear shape changes from prolate to oblate and then to spherical nuclei with $N = 82$. This transition leads to a deviation from the approximate relationship between neutron separation energies and mass number for each isotope, and can be clearly recognized in Fig. 5 by the sudden increase of the separation energies. Similar conclusion can be drawn for the trough around $A \sim 170$. This suggests that the potentially wrongly assigned location of the shape transition before the neutron magic number in the theoretical predictions can lead to the troughs before the abundance peaks.

Finally we wish to note without details that there is also work to apply the RMF+BCS data in simulating non-solar-like r-process, e.g. Ref. 34.

4.1.4. *Accuracy of nuclear masses*

The nuclear mass precision needed for r-process studies has been stressed in the literature (e.g. Ref. 38). Generally, a precision of $\delta m/m$ on the order of around 10^{-6} is needed, while 10^{-7} or even lower is needed for some specific cases like waiting point nuclei in the r-process path. However, it should be noted that in the foreseeable future we still have to resort to nuclear theories for modeling r- process. Unfortunately, even the most accurate mass models have rms mass errors near 500 keV for known nuclei,[11,38] but it is in general comparable to the known experimental data for the most exotic nuclei. Here we would like to illustrate the importance of accuracy of nuclear masses in r-process studies with a specific example.

We have explored the effects of four nuclear mass models (FRDM, WS*, HBF-17, and RMF+BCS) on the conditions required by r-process nucleosynthesis under the WP approximation. As discussed in Sec. 4.1.2, the required T_9-n_n conditions are mostly determined by the two-neutron separation energies of the CWP nuclei with $N = 50$, 82, and 126 and of those nuclei around them. Figure 8 shows the dramatic effect of using the tabulated values in the AME2011-preview when they are available to replace the masses predicted by models. As noted before, the tabulated masses of ^{76}Ni to ^{78}Ni and ^{78}Zn to ^{82}Zn play crucial roles in determining the conditions required by the $N = 50$ CWP nuclei. However, the tabulated masses of ^{76}Ni to ^{78}Ni and ^{82}Zn are extrapolated rather than measured.

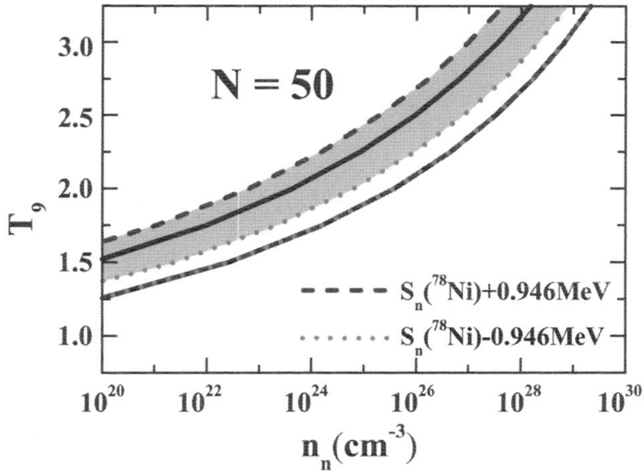

Fig. 11. (Color online) Effects of the uncertainty in the neutron separation energy S_n for ^{78}Ni on the required T_9−n_n conditions. The solid curves are the same as those in Fig. 8(b). The shaded regions in each panel show the effects on the required T_9−n_n conditions when the corresponding S_n values are varied within the estimated uncertainties. See text for details.

To emphasize the effects of these masses on the conditions required by the $N = 50$ CWP nuclei, we first repeat the calculations of Sec. 4.1.2 by varying the neutron separation energy of ^{78}Ni within the estimated uncertainty of 0.946 MeV while keeping the other input the same as for Fig. 8(b). The results are shown in Fig. 11. Increasing the neutron separation energy of ^{78}Ni by 0.946 MeV raises the upper bound from the solid curve (upper bound in Fig. 11) to the dashed curve and decreasing this quantity by the same amount lowers it to the dotted curve. In any case, the significant effects of uncertainties in neutron separation energies on the required T_9−n_n conditions shown in Fig. 11 clearly demonstrate the importance of precise mass measurements for ^{76}Ni to ^{78}Ni. Other key nuclei including ^{82}Zn, ^{131}Cd, and ^{132}Cd can be identified for precise mass measurements at rare-isotope beam facilities. These nuclei have the largest impact and are important candidates for precise mass measurements at rare-isotope beam facilities with high priority.

4.2. *Beta Decay*

The nuclear β decay is a type of nuclear process in which an atomic nucleus can emit a β particle or capture an atomic electron. The β particle denotes an electron or a positron. There are three basic β-decay processes: β^-

decay, in which an atomic nucleus emits an electron; β^+ decay, in which an atomic nucleus emits a positron; and electron capture (EC), in which an atomic electron is captured by a nucleus. In this part, we will first briefly introduce the Fermi theory of β decay, and then review the recent progress on β-decay calculations with CDFT. Finally, the impact of new β-decay half-lives predicted with CDFT on the r-process simulations is given as well.

4.2.1. *Fermi theory of β decay*

In the 1930's, the quantum theory of radiation was well established. It was known that the interaction between the electromagnetic field and orbit electron induces the transition of electron between different atomic states, in which photons are emitted. Analogously to the process of photon emission, in 1934, Fermi proposed that β decay is induced by the interaction between the electron-neutrino field and nucleus, in which the electron and antineutrino were created.[39]

With this idea, the total β-decay probability λ can be derived based on the time-dependent perturbation theory, which is:

$$\lambda = \int_0^{p_m} \frac{g^2|\bar{M}_{fi}|^2}{2\pi^3\hbar^7c^3} F_0(Z_D, A, p_e) p_e^2 (E_m - E_e)^2 dp_e. \tag{10}$$

where p_e and E_e are the momentum and energy of emitted electron, p_m and E_m is the maximum value of p_e and E_e. $F_0(Z_D, A, p_e)$ is the Fermi function, in which Z_D denotes the proton number of the daughter nucleus. $\bar{M}_{fi}$ denotes the reduced matrix element between the initial and final states. By transforming the momentum (p_e) and energy (E_e) into dimensionless variables, i.e.,

$$\eta = p_e/m_e c, \quad \omega = E_e/m_e c^2, \tag{11}$$

the relativistic energy-mass relation then becomes

$$\omega^2 = \eta^2 + 1. \tag{12}$$

The transition probability then becomes

$$\lambda = \frac{g^2 m_e^5 c^4}{2\pi^3\hbar^7} \int_1^{\omega_m} |\bar{M}_{fi}|^2 F_0(Z_D, A, \eta)\eta\omega(\omega_m - \omega)^2 d\omega. \tag{13}$$

If the dependence of matrix element $\bar{M}_{fi}$ on the electron energy ω can be neglected, one gets

$$\lambda = \frac{g^2 m_e^5 c^4}{2\pi^3\hbar^7} |\bar{M}_{fi}|^2 f(Z_D, A, \omega_m) \tag{14}$$

with

$$f(Z_D, A, \omega_m) = \int_1^{\omega_m} F_0(Z_D, A, \eta)\eta\omega(\omega_m - \omega)^2 d\omega. \tag{15}$$

Then the half-life $T_{1/2}$ can be predicted with

$$T_{1/2} = \frac{\ln 2}{\lambda} = \frac{D}{|\bar{M}_{fi}|^2 f(Z_D, A, \omega_m)}, \tag{16}$$

where

$$D = \frac{2\pi^3\hbar^7 \ln 2}{m_e^5 c^4 g^2} = 6163.4 \pm 3.8 \text{ s}. \tag{17}$$

If electron-capture process occurs simultaneously, the decay function of electron capture f^{EC} should be added to that in Eq. (15), which is

$$f^{\mathrm{EC}} = \frac{\pi}{2}\sum_x [E_\nu^2 \psi_e^2(R)]_x B_x, \tag{18}$$

where x denotes the atomic subshell from which the electron is captured, E_ν is the neutrino energy, $\psi_e(R)$ is the radial component of the bound-state electron wave function at the nuclear surface, and B stands for other exchange and overlap corrections.

4.2.2. *Recent progress on β-decay calculations with CDFT*

Nuclear β decays play important roles in many subjects of nuclear physics. Specifically, the investigation of β decay provides information on the spin and isospin dependence of the effective nuclear interaction, as well as on nuclear properties such as masses,[38] shapes,[40] and energy levels.[41] Moreover, nuclear β decays are also important in nuclear astrophysics, because they set the time scale of the r-process,[5,9,42] which is a major mechanism for producing the elements heavier than iron. In addition, nuclear β decays can provide tests for the electroweak standard model.[43,44] Therefore, the nuclear β decay is an important topic and has achieved wide attention in the physics studies.

With the development of radioactive ion beam facilities, the measurement of nuclear β-decay half-lives has achieved great progress in recent years.[45–48] On the theoretical side, apart from the macroscopic gross theory,[49] two different microscopic approaches have been widely used to describe and predict the nuclear β-decay rates. They are the shell model[42] and the proton-neutron quasiparticle random phase approximation (QRPA).[35,50] While the shell model takes into account the detailed structure of the β-strength function, the proton-neutron QRPA approach provides a systematic description of β-decay properties of arbitrarily heavy nuclei. In order to reliably predict properties of thousands of unknown nuclei relevant to the r-process, the self-consistent QRPA approach has become a current trend in nuclear structure study, including those based on the Skyrme HFB theory[51] and the relativistic density functional theory.[52–54]

In the standard relativistic Hartree–Bogoliubov (RHB) model, a low effective nucleon mass m^* is essential to reproduce correctly the empirical spin-orbit splittings in finite nuclei and the binding energy at the saturation density in nuclear matter, while this results in a low density of states around the Fermi surface. Therefore self-consistent QRPA calculations based on the RHB model predict low transition energies and hence long β-decay half-lives. For reliably predicting β-decay half-lives, the standard effective Lagrangian has been extended by including an isoscalar tensor-coupling term or the couplings between the isoscalar meson fields and the derivatives of the nucleon fields. In these two extended models, it is possible to increase the effective nucleon mass m^*, while keeping a reasonable description of the empirical spin-orbit splittings.[52,55]

In Ref. 52, the self-consistent QRPA based on the effective Lagrangian including an isoscalar tensor-coupling term was developed and employed to calculate the nuclear β-decay half-lives. By employing the DD-ME1 interaction of standard RHB model, the calculation predicts a half-life $T_{1/2} = 7$ s for ^{78}Ni, which is an order of magnitude longer than the experimental value $T_{1/2} = 140 \pm 80$ ms.[48] By including an isoscalar tensor-coupling term, a new density-dependent effective interaction DD-ME1*, with an enhanced value of the nucleon effective mass, was adjusted simultaneously to properties of nuclear matter and finite nuclei. Thus, the calculations with DD-ME1* effective interaction produce a higher density of single-nucleon states at the Fermi surface. Consequently, the self-consistent QRPA calculation with DD-ME1* predicts a shorter half-life of $T_{1/2} = 0.9$ s for ^{78}Ni. This reduction of half-life presents a significant improvement over the result obtained with DD-ME1, although it still overestimates the empirical half-life of ^{78}Ni.

Furthermore, the QRPA approach was employed to calculate the β-decay half-lives of neutron-rich nuclei in the $N \approx 50$ and $N \approx 82$ regions with DD-ME1*. However, the calculations are still significantly longer than the known half-lives of the Fe, Zn, Cd, Sn, and Te isotopes. Similar to that in the non-relativistic framework, it was found that the $T = 0$ pairing plays an important role in describing the half-lives of these open shell nuclei. Since the calculated half-lives with a simple δ pairing force diverge steadily as canonical single quasiparticle states are added to the basis used in the QRPA, a finite-range pairing force is employed:[51]

$$V_{T=0}(1,2) = -V_0 \sum_{j=1}^{2} g_j e^{-[(\boldsymbol{r}_1-\boldsymbol{r}_2)/\mu_j]^2} \hat{\prod}_{S=1,T=0}, \tag{19}$$

where $\mu_1 = 1.2$ fm, $\mu_2 = 0.7$ fm, $g_1 = 1$, $g_2 = -2$. The operator $\hat{\prod}_{S=1,T=0}$ projects onto states with $S = 1$ and $T = 0$. V_0 is the overall strength of the $T = 0$ proton-neutron pairing. By adjusting the strength V_0 of $T = 0$ pairing interaction to the known half-lives of selected nuclei in the corresponding isotopic chain, the calculations with DD-ME1* reproduce the experimental data well. However, very different values of V_0 are used, e.g. 115 MeV for Fe isotopes and 330 MeV for Zn isotopes.

In Ref. 53, a new parametrization D^{3}C* with the couplings of the meson fields to derivative nucleon densities was adjusted for the purpose of calculating β-decay half-lives, whose effective mass is even larger than that in DD-ME1*. With this effective interaction D^{3}C*, the self-consistent QRPA has been applied in the calculation of β-decay half-lives of neutron-rich nuclei in the $Z \approx 28$ and $Z \approx 50$ regions. The results for Fe, Ni, and Zn isotopes are shown in Fig. 12. For comparison, the experimental values and calculations with the DD-ME1 and D^{3}C effective interactions are also displayed. Clearly, the new effective interaction D^{3}C* significantly reduces the nuclear half-lives compared with the traditional effective interaction DD-ME1. By including the $T = 0$ pairing in Eq. (19), the calculated values are in qualitative agreement with the data. However, there still exist quantitative deviations, in particular for Zn isotopes. In this framework, the additional enhancement of the effective nucleon mass is due to the coupling of single-nucleon levels to low-energy collective vibrational states, an effect which goes entirely beyond the mean-field approximation and is not included in the present model. In principle, the effect of two- and three-phonon states on the weak-interaction rates could be taken into account by explicitly considering the coupling of single-quasiparticle states to phonons. In the non-relativistic framework, it has been found that particle-vibration coupling (PVC) plays a crucial role in reproducing the β-decay half-lives of magic nuclei.[56] Therefore, it is interesting to investigate the PVC effect on the calculations of β-decay half-lives based on the CDFT in the future.

To reliably predict properties of thousands of unknown nuclei relevant to the r-process, the self-consistency of the QRPA approach is essential.

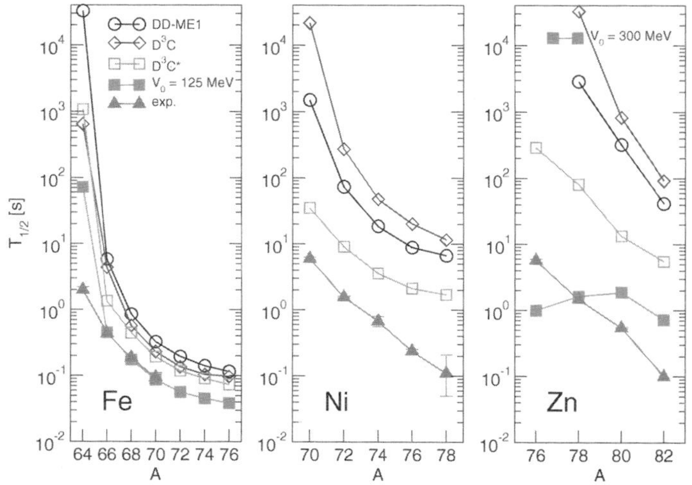

Fig. 12. (Color online) β-decay half-lives of Fe (left panel), Ni (middle panel), and Zn (right panel) nuclei, calculated with the DD-ME1, D^3C, and D^3C^* effective interactions, compared with the experimental values. Open symbols correspond to QRPA values calculated without the inclusion of the $T = 0$ pairing interaction. The filled squares are half-lives calculated with the D^3C^* interaction and $T = 0$ pairing, with the strength parameter $V_0 = 125$ MeV for Fe, and $V_0 = 300$ MeV for Zn isotopes. Taken from Ref. 53.

Only recently has the fully self-consistent relativistic RPA been formulated based on the relativistic Hartree–Fock (RHF) theory.[57] The RHF+RPA model produces results in excellent agreement with data on the Gamow-Teller (GTR) and spin-dipole resonances in doubly magic nuclei, without any readjustment of the parameters of the covariant energy density functional.[57,58] Recently, the self-consistent QRPA based on the relativistic Hartree–Fock–Bogoliubov (RHFB) approach[59] was developed and a systematic study on the β-decay half-lives of neutron-rich even-even nuclei with $20 \leqslant Z \leqslant 50$ has been performed.[54]

Nuclear β-decay half-lives are very sensitive to the $T = 0$ proton-neutron pairing interaction, and its strength V_0 is determined by adjusting QRPA results to empirical half-lives.[51–53] For Ni, Zn, Ge, and Sn isotopes with N before the corresponding magic number ($N = 50$ for Ni, Zn, and Ge, $N = 82$ for Sn), the β-decay is dominated by the transitions with the main configurations of particle-hole (ph) type, so the contribution from the $T = 0$ pairing becomes negligible. In recent studies based on the non-relativistic framework, it was found that the overestimation of half-lives of ^{78}Ni and ^{132}Sn can be improved by taking into account PVC[56] or including the tensor force.[61] For other nuclei, especially the Fe and Cd isotopes with N just before the corresponding magic number, the β decay is dominated by the transitions with the main configurations of particle-particle (pp) type, so it is reasonable to extract the strength of $T = 0$ pairing with experimental half-lives of these nuclei.

Taking ^{70}Fe and ^{130}Cd as reference nuclei for the two mass regions, the value of V_0 is determined as 155 MeV and 255 MeV, respectively. Using these two values, the calculated half-lives of Fe and Cd isotopes are shown in Fig. 13. For comparison, the experimental values and the results of a calculation without the $T = 0$ pairing are also displayed. One notices that the β-decay half-lives calculated without the inclusion of $T = 0$ pairing are systematically much longer than the experimental half-lives, both for Fe and Cd isotopes. A constant value of V_0 was usually taken for one isotopic chain, and it was determined by adjusting to the known half-lives of selected nuclei in the isotopic chain.[52,53] This procedure limits the predictive power of the model. Moreover, as shown in Fig. 13, when V_0 is determined by the β-decay half-life of ^{130}Cd, the calculated results underestimate the half-lives of ^{118,120,122}Cd. This indicates that the half-lives of an isotopic chain cannot always be reproduced using a constant value V_0, and points to a

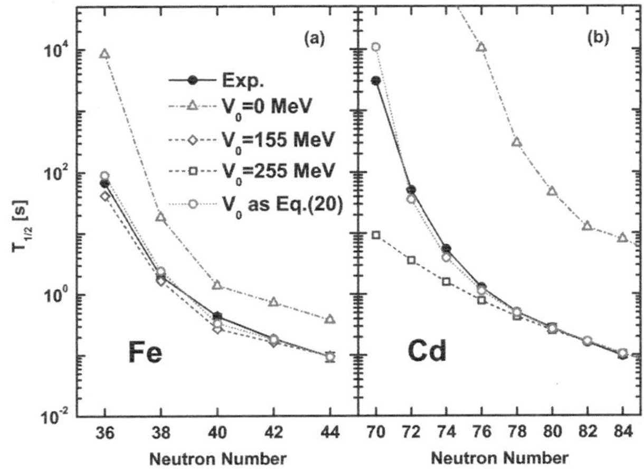

Fig. 13. (Color online) Nuclear β-decay half-lives of Fe (left) and Cd (right) isotopes, calculated with the PKO1 effective interaction,[60] compared to the experimental values.[48] Open triangles, diamonds, and squares denote values obtained using the RHFB+QRPA with the strength parameter of the $T = 0$ pairing: $V_0 = 0, 155$, and 255 MeV, respectively. The RHFB+QRPA values obtained with the V_0 of Eq. (20) are denoted by open circles. Taken from Ref. 54.

possible isospin-dependence of V_0. Therefore, the following ansatz for an isospin-dependent pairing strength

$$V_0 = V_L + \frac{V_D}{1 + e^{a+b(N-Z)}} \tag{20}$$

was employed in Ref. 54. The parameters in Eq. (20) were adjusted to reproduce the known half-lives of even-even nuclei with $20 \leqslant Z \leqslant 50$ except Ni, Zn, Ge, and Sn isotopes with N before the corresponding magic number. The resulting values: $V_L = 134.0$ MeV, $V_D = 121.1$ MeV, $a = 8.5$, and $b = -0.4$ are used in the calculation of β-decay half-lives for nuclei in the interval $8 \leqslant N - Z \leqslant 50$. With the isospin-dependent strength Eq. (20) of the proton-neutron pairing interaction, the calculated β-decay half-lives of both the Fe and Cd isotopic chains are in excellent agreement with data. For the half-lives of even-even nuclei with $20 \leqslant Z \leqslant 50$, calculations with the RHFB+QRPA model are compared with data in Fig. 14. Clearly, The RHFB+QRPA calculation well reproduces the experimental half-lives of these neutron-rich nuclei except for Ni, Zn, Ge, and Sn isotopes with N before the corresponding neutron magic $N = 50$ and $N = 82$.

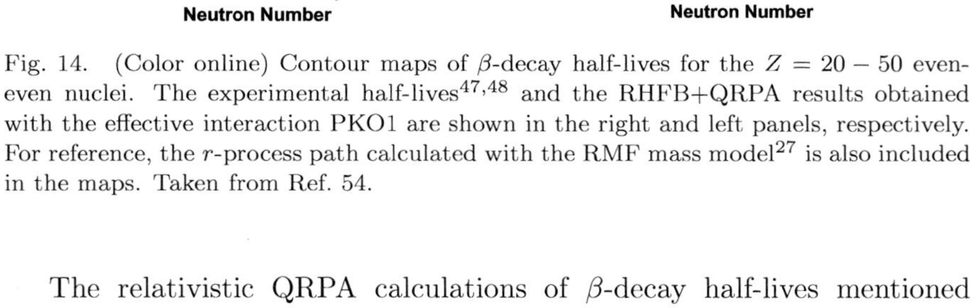

Fig. 14. (Color online) Contour maps of β-decay half-lives for the $Z = 20 - 50$ even-even nuclei. The experimental half-lives[47,48] and the RHFB+QRPA results obtained with the effective interaction PKO1 are shown in the right and left panels, respectively. For reference, the r-process path calculated with the RMF mass model[27] is also included in the maps. Taken from Ref. 54.

The relativistic QRPA calculations of β-decay half-lives mentioned above are based on meson-exchange version of CDFT framework. Recently, the relativistic functionals based on the zero-range point-coupling interactions has attracted more and more attention due to its simplicity and several other advantages.[62] For example, it is even possible to include the effects of Fock terms in a local RHF equivalent scheme,[63,64] which can make the QRPA approach self-consistent even in the Hartree approximation.

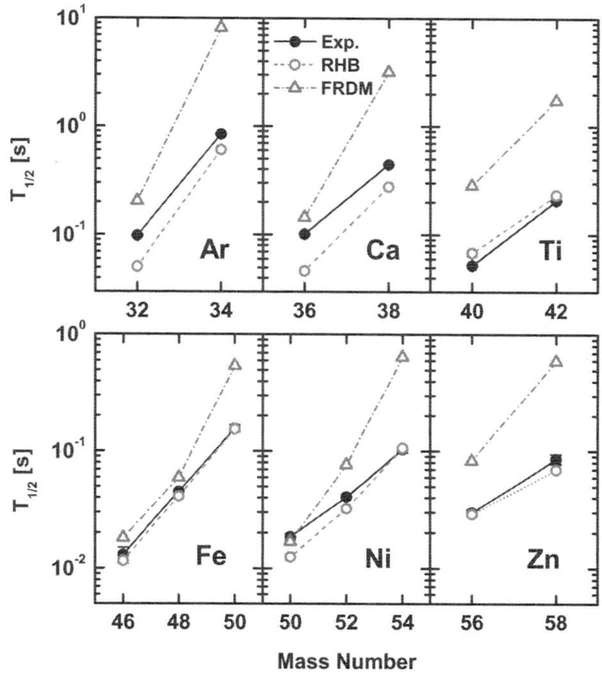

Fig. 15. (Color online) Nuclear β^+/EC-decay half-lives for Ar, Ca, Ti, Fe, Ni, and Zn isotopes calculated by RHB + QRPA approach with the effective interaction PC-PK1 and $V_0 = 175$ MeV. For comparison, the experimental data[48] (filled circles), as well as theoretical results obtained from FRDM + QRPA[35] (open upward triangles) approach are also shown. Taken from Ref. 65

Self-consistent proton-neutron quasiparticle random phase approximation based on the spherical point-coupling relativistic Hartree–Bogoliubov theory has been established and used to investigate the β^+/EC-decay half-lives of neutron-deficient nuclei with the effective interaction PC-PK1. The $T = 0$ pairing is found to play an important role in reducing the decay half-lives, which is consistent with the same mechanism in the β decays of neutron-rich nuclei. From Eq. (20), one finds that isospin-dependent pairing strength V_0 is nearly constant for nuclei with $N - Z < 5$, which is exactly the case for the neutron-deficient nuclei with $20 \lesssim Z \lesssim 50$. Therefore, the half-lives of Ar, Ca, Ti, Fe, Ni, and Zn isotopes are calculated with the same V_0 ($V_0 = 175$ MeV) used in calculations of half-lives of Cd and Sn neutron-deficient isotopes. The results are shown in Fig. 15. For comparison, the calculated results obtained from the FRDM + QRPA[35] are also shown. It is found that the self-consistent RHB + QRPA calculations well reproduce the experimental data with a universal pairing strength. For the FRDM + QRPA approach, it almost systematically overestimates the experimental half-lives. It has been pointed out that the overestimation of half-lives in the FRDM + QRPA approach can be attributed partially to the neglect of the $T = 0$ pairing.[51,54,66] This was supported by this investigation on the β^+/EC decays in neutron-deficient nuclei, which further implies the importance of $T = 0$ pairing to β-decay calculations.

To predict the nuclear properties of thousands of mostly unknown nuclei reliably, the microscopic QRPA model should employ a universal spin-isospin NN-interaction and be based on the self-consistent ground state description. It is found that an isospin-dependent $T = 0$ proton-neutron pairing interaction and the tensor coupling in the isoscalar vector channel should be self-consistently included in the β-decay calculations. Recent β-decay calculations based on deformed QRPA found that deformation plays an important role in the description of β-decay properties for nuclei with large deformation.[67] Since many nuclei in the nuclear chart are deformed, the deformed QRPA approach should be employed in the β-decay calculations to quantitatively reproduce the half-lives of deformed nuclei. The inclusion of deformation degrees of freedom can make the QRPA calculations of β-decay half-lives more complicated and time-consuming. Thus a high computational efficiency of QRPA approach is necessary for the β-decay calculations of all nuclei in nuclear chart. The relativistic point-coupling model is a realistic candidate due to the zero-range point-coupling interactions.[68] Therefore, in the future, a deformed self-consistent QRPA based on the relativistic point-coupling model should be developed to reliably predict the nuclear β-decay half-lives with a universal spin-isospin NN-interaction, in which an isospin-dependent $T = 0$ proton-neutron pairing interaction and the tensor coupling in the isoscalar vector channel are included.

4.2.3. *r-process simulations with β-decay half-lives in CDFT*

To analyze the impact of β-decay half-lives predicted with RHFB+QRPA approach[54] on r-process abundances, a classical r-process calculation is performed with neutron density $n_n = 10^{22} - 10^{24}$ cm^{-3} and temperature $T = 1.5 \times 10^9$ K. Figure 16 displays four snapshots of r-process abundances at different neutron irradiation times τ_r. One notices that the half-lives calculated with RHFB+QRPA model (solid curves) produce a faster r-matter flow in the $N = 82$ region, and thus yield higher r-process abundances of

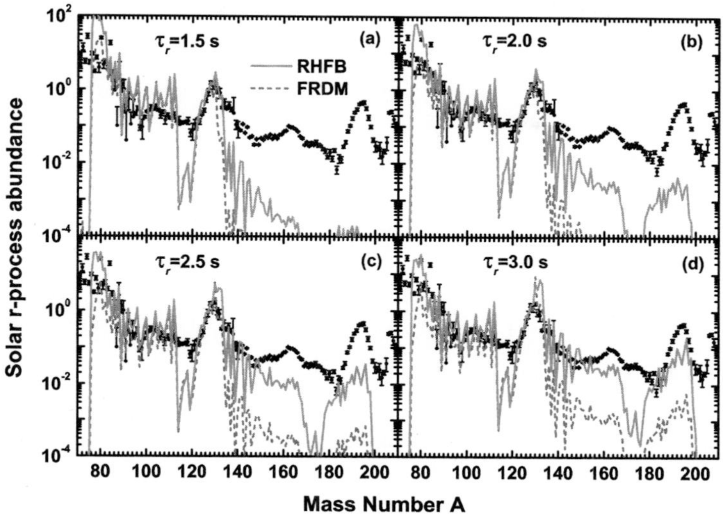

Fig. 16. (Color online) The impact of nuclear β-decay half-lives on the calculated r-process abundances. The solid (dashed) curves correspond to r-process abundances calculated with the RHFB+QRPA (FRDM+QRPA) β-decay half-lives in comparison to the data denoted by the points. In all calculations, nuclear masses are taken from the mass evaluation[28] if available, otherwise predictions of RMF mass model[27] are employed. Panels (a)–(d) correspond to the neutron irradiation times $\tau_r = 1.5, 2.0, 2.5$, and 3.0 s, respectively. Taken from Ref. 54

elements with $A \gtrsim 140$. By summing up the half-lives of r-path nuclei at $N = 82$, one can roughly estimate the time when the r-process passes the $N = 82$ shell. Based on the RHFB+QRPA results this time is speeded up to 0.25 s, from the 1.29 s predicted by the FRDM+QRPA calculation. This is an important result for the estimate of the duration of the r-process, and hence the origin of heavy elements in the universe.

5. Nuclear cosmochronology

Nuclear cosmochronology is an important topic in nuclear astrophysics. In this section, we will first introduce the methods used to estimate the age of our Universe. More attention will be focused on the nuclear cosmochronology and its progress in recent years.

5.1. *Age determination of our Universe*

The cosmic age is an important parameter in cosmology. According to the Big Bang cosmology, the age of the universe usually refers to the time elapsed since the Big Bang itself. It can be directly determined from the cosmological method. This is usually carried out in the Lambda Cold Dark Matter (ΛCDM) model, where the constituents of the universe are dominated by cold dark matter (abbreviated with CDM) and a cosmological constant associated with dark energy (denoted by Λ) at late times. By using the Friedmann equation, the cosmic age is determined to be

$$t = \frac{1}{H_0} F(\Omega_m, \Omega_\Lambda, ...), \tag{21}$$

where H_0 denotes the Hubble constant. The function F depends on the density parameters of various constituents of the universe, in which the matter density parameter Ω_m and the dark energy density parameter Ω_Λ are the most important.

With the cosmic microwave background (CMB) temperature spectra, these density parameters can be determined, and then the cosmic age is directly derived. Based on the nine-year Wilkinson microwave anisotropy probe (WMAP) observations and the Planck 2013 measurements, the cosmic age is determined to be 13.772 ± 0.059 Gyr[69] and 13.813 ± 0.058 Gyr,[70] respectively. We should keep in mind that the cosmic age determined here is based on the ΛCDM model, in which the universe is assumed to be spatially-flat. If we relax this assumption within the allowed range, the uncertainty of cosmic age inevitably increases.

The age of the galaxy is usually used to set a stringent lower limit on the cosmic age. There are three independent methods to reliably infer the age of the galaxy: the ages of the globular clusters from main sequence turnoffs,[71–73] the ages of white dwarfs from cooling calculations,[74,75] and the age of the elements by radioactive dating. These derived ages can then be used as a consistency check on the cosmic age determined from cosmological method.

The time when a star evolves to the main sequence turnoff is well-known, which can be determined from the luminosity of the main sequence turnoff. A recent study based on the main-sequence turnoff found that the age of the oldest globular clusters in our galaxy is $12.6^{+3.4}_{-2.2}$ Gyr with a 95% confidence level.[73] To constrain cosmological parameters, a time of 0.8 Gyr that corresponds to the time between the Big Bang and the formation of globular clusters in our galaxy is added, and then the lower limit on the cosmic age is 11.2 Gyr with a 95% confidence level.[73] This results are still consistent with that determined from the current cosmological model.

There is no nuclear energy generation and gravitational contraction in white dwarfs, so their rate of cooling with time is relatively simple. By comparing the observed luminosities with predictions of theoretical cooling models, their ages can be estimated. However, only the ages of the oldest (faintest) white dwarfs can be used to effectively constrain the cosmic age. The globular clusters are good candidates for finding these oldest white dwarfs. Recently, the age estimate for the globular cluster M4 was determined to be 12.7 ± 0.7 Gyr,[74] in which the quoted error only takes into account the observational uncertainties. After taking into account the various uncertainties from both observations and theoretical calculations, a best-fit age of 12.1 Gyr with a 95% confidence level lower limit of 10.3 Gyr was obtained.[75] This is remarkably similar to the age estimate for the cluster system using the main-sequence turnoff, which is 12.6 Gyr best fit with a 95% confidence level lower limit of 10.4 Gyr.[73]

5.2. *Nuclear cosmochronology*

The age of the elements by radioactive dating is another method to constrain the cosmic age,[76–79] which is also called nuclear cosmochronology. The radioactive dating method is independent of cosmological models and the stellar evolution models. This method can be traced back to the early twentieth century, when Rutherford outlined the essential features of this method by analyzing the decay in radioactive mineral bröggerite.[80] In this method, the age is determined by comparing the current abundances of radioactive nuclei with the initial abundances at their production sites. The abundance evolution of a radioactive nuclide i follows the exponential decay, i.e.

$$N_i(t) = N_i(0)e^{-\lambda_i t}, \tag{22}$$

where the $N_i(t), N_i(0)$, and λ_i are the abundance observed at present, the initial abundance at its production, and the decay constant of nuclide i, respectively. According to Eq. (22), the time elapsed t since the production of the radioactive element can be determined from:

$$t = \frac{1}{\lambda_i} \ln \frac{N_i(0)}{N_i(t)}. \tag{23}$$

The decay constant $\lambda_i = \ln 2/\tau_i$ (τ_i is the half-life of the nuclide) and the current abundance can be easily measured from experiments. Thus the most difficult part of the radioactive dating method comes from the determination of initial abundance.

For determining the age of an object, the half-life of the radioactive nuclide should be comparable to age itself. If the half-life of the radioac-

Table 13.3. The nuclides with appropriate half-lives for the nuclear cosmochronology. The half-lives are taken from NUBASE2012.[48]

Radioactive Nuclide	Half-life (Gyr)
^{147}Sm	106.6 ± 0.7
^{138}La	102 ± 1
^{87}Rb	49.23 ± 0.22
^{187}Re	43.3 ± 0.07
^{176}Lu	37.6 ± 0.7
^{232}Th	14.0 ± 0.1
^{238}U	4.468 ± 0.003
^{40}K	1.248 ± 0.003
^{235}U	0.704 ± 0.001

tive nuclide is much longer, the amount of decay will be very small, and the error in the derived age will be comparable to the age itself, hence a fairly useless result. If the half-life is much shorter, then the nuclide will decay to undetectable levels due to the exponential decay. The nuclides with appropriate half-lives for nuclear cosmochronology are listed in the Table 13.3. In practice, ^{232}Th, ^{238}U, and ^{235}U are usually employed in nuclear cosmochronology. For ^{147}Sm, ^{138}La, ^{87}Rb, ^{176}Lu, and ^{40}K, the use of these nuclides in nuclear cosmochronology is constrained by the lack of an adequate knowledge of their nucleosynthesis histories. For ^{187}Re, a major difficulty is the fact that its decay product ^{187}Os is produced in the s-process. The uncertainties introduced by the subtraction of the s-process contribution to isolate the cosmoradiogenic component are significant. Further complications are associated with the fact that the β decay rate of ^{187}Re in stellar environments is sensitive to temperature.

By far the most accurate radioactive dates still come from samples from Earth or elsewhere in the Solar System. Since these samples can be analyzed in the laboratory, the amount of radioactive nuclide can be measured to extremely high precision. Furthermore, the initial abundance of radioactive nuclide can be well determined by measuring the amount of its decay products, which are trapped in the samples. Using $N_p(t)$ and $N_p(0)$ to denote the abundances of radioactive parent nuclide at time t and zero-decay time, Eq. (22) leads to

$$N_p(t) = N_p(0)e^{-\lambda t}. \tag{24}$$

Assuming the daughter nuclide is totally produced by the decay of radioactive nuclide, then one has

$$N_d(t) = N_p(0) - N_p(t). \tag{25}$$

Combining Eqs. (24) and (25), then

$$t = \frac{1}{\lambda} \ln \left[1 + \frac{N_d(t)}{N_p(t)} \right]. \tag{26}$$

Therefore, with the ratio of the number of daughter and parent nuclides, the age of the sample can be well determined.

For the radioactive nuclides used in nuclear cosmochronology, such as ^{235}U and ^{238}U, they respectively decay to the stable nuclides ^{207}Pb and ^{206}Pb by a series of α and β transformations of very short lifetime compared to that of parent nuclides. In principle, the intermediate products should be considered in the left side of Eq. (24). However, the number of these intermediate products is much less than the number of daughter nuclides, so the age of the sample can also be estimated with Eq. (26). For ^{235}U and ^{238}U, the corresponding equations for age determination are

$$t = \frac{1}{\lambda_{^{235}\mathrm{U}}} \ln \left[1 + \frac{N_{^{207}\mathrm{Pb}}(t)}{N_{^{235}\mathrm{U}}(t)} \right], \tag{27}$$

and

$$t = \frac{1}{\lambda_{^{238}\mathrm{U}}} \ln \left[1 + \frac{N_{^{206}\mathrm{Pb}}(t)}{N_{^{238}\mathrm{U}}(t)} \right]. \tag{28}$$

For a given sample, the ages determined with Eqs. (27) and (28) should be equal to each other. However, these two ages may show some deviations due to the large systematic uncertainties. To reduce the influence of systematic uncertainties, the age is usually determined with the ratio of $N_{^{207}\mathrm{Pb}}(t)$ and $N_{^{206}\mathrm{Pb}}(t)$. Combing Eqs. (27) and (28), one derives

$$\frac{N_{^{207}\mathrm{Pb}}(t)}{N_{^{206}\mathrm{Pb}}(t)} = \frac{N_{^{235}\mathrm{U}}(t)}{N_{^{238}\mathrm{U}}(t)} \cdot \frac{e^{\lambda_{^{235}\mathrm{U}} t} - 1}{e^{\lambda_{^{238}\mathrm{U}} t} - 1}. \tag{29}$$

By measuring the ratios $\frac{N_{^{207}\mathrm{Pb}}(t)}{N_{^{206}\mathrm{Pb}}(t)}$ and $\frac{N_{^{235}\mathrm{U}}(t)}{N_{^{238}\mathrm{U}}(t)}$, then the age of the sample can be easily obtained from Eq. (29).

It is known that the nuclear process occurring in the sun only transform hydrogen into helium, so the heavy elements found in the Solar System must be synthesized at a time before the formation of the Solar System. Therefore, the ages of these heavy elements can give a relatively strong constraint on the cosmic age. However, their abundances may change due to galactic chemical evolution, which make age determinations of these elements very complicated. Then the equation governing the abundance evolution of a radioactive nuclide becomes

$$\frac{dN_i(t)}{dt} = -\lambda_i N_i(t) + f(t, t_n, P_i, T_R), \quad (30)$$

where the function f describes the galactic chemical evolution. The paramters t_n, P_i, and T_R represent the onset time of nucleosynthesis, the primary production rate of nuclide i, and the characteristic time for element synthesis. To be useful for nucleocosmochronology, Eq. (30) must be simplified by means of a galactic evolution model. If one assumes that the history of material going into the solar system was spatially homogeneous and that the instantaneous recycling approximation holds,[81] Eq. (30) then simplifies to

$$\frac{dN_i(t)}{dt} = -\lambda_i N_i(t) - \omega(t) N_i(t) + P_i \psi(t), \quad (31)$$

where $\omega(t)$ is a time-dependent parameter representing the rate of movement of metals into and out of the interstellar medium for reasons other than decay, and $\psi(t)$ is the rate of conversion of mass into stars.

Assuming no gain or loss of matter ($\omega(t) = 0$), one can obtain two extreme assumptions of the galactic chemical evolution models: the sudden synthesis and the uniform synthesis, whereby the former can provide a lower boundary of the galactic age. The sudden synthesis means a single event nucleosynthesis history, thus the production function f is described by a δ-function and Eq. (30) becomes

$$\frac{dN_i(t)}{dt} = -\lambda_i N_i(t) + P_i \delta(t_n - t). \quad (32)$$

With the boundary condition $N_i(t_n) = P_i$, the solution of Eq. (32) is

$$N_i(t) = P_i e^{-\lambda_i t}. \quad (33)$$

Taking the radioactive nuclides ^{232}Th and ^{238}U as a pair, one obtains

$$\left(\frac{N_{232}}{N_{238}}\right)_{\rm SS} = \left(\frac{P_{232}}{P_{238}}\right) e^{-(\lambda_{232} - \lambda_{238})T}. \quad (34)$$

The uniform synthesis means there are numerous nucleosynthesis events before the condensation of solar nebula, and the nucleosynthesis rate is assumed to be uniform. Then Eq. (30) becomes

$$\frac{dN_i(t)}{dt} = -\lambda_i N_i(t) + P_i. \quad (35)$$

With the boundary condition $N_i(t_n) = 0$, the solution of Eq. (35) is

$$N_i(t) = \frac{P_i}{\lambda_i}(1 - e^{-\lambda_i t}). \tag{36}$$

For ^{232}Th/^{238}U pair, one obtains

$$\left(\frac{N_{232}}{N_{238}}\right)_{\rm SS} = \left(\frac{P_{232}}{P_{238}}\right)\frac{\lambda_{238}}{\lambda_{232}}\frac{1 - e^{-\lambda_{232}T}}{1 - e^{-\lambda_{238}T}}. \tag{37}$$

The evolution of abundance ratio of ^{232}Th and ^{238}U is shown in Fig. 17. It is known that ^{232}Th and ^{238}U are synthesized by the r-process. Therefore, the production abundance ratio of ^{232}Th and ^{238}U can be predicted with the r-process simulations. Taking the ratio predicted using CDFT mass model, the lower limits of the galactic age are 7.1 Gyr and 10.1 Gyr based on the sudden and uniform nucleosynthesis assumptions, respectively. Apart from these two extreme assumptions, a number of investigations[82,83] concerning chemical evolution effects on the age dating have been made in literatures, including considerations of varied prescriptions for the star formation history, and of the consequences of infall and outflow of gas from the star-forming regions.

Recent observations have found that the element abundance pattern in these r-process-rich metal-poor halo stars is consistent with the solar r-process abundance pattern.[1,85] This strongly supports the hypothesis that the r-process was responsible for the formation of the elements in these stars in the early history of the Galaxy. The very metal-poor stars were formed at the early epoch of the Universe and the elemental abundances in these stars are usually thought to be not influenced by the galactic chemical evolution, so the radioactive dating technique based on the metal-poor halo stars can be used as a relatively reliable dating technique for the universe.

The radioactive element Th in the r-process enhanced metal-poor halo star CS 22892-052 was detected in the 1990's,[86] and later the element Th was also observed in many other metal-poor stars, e.g., HD 115444[87] and HD 221170.[88] For the element U, it was first detected in CS 31802-001.[78] However, due to the weakness of U lines and severe blending issues, so far U has only been observed in two other metal-poor stars, namely, BD $+17^\circ 3248$[89] and HE 1523-0901.[90] With these abundance observations, the ages of these metal-poor stars can be estimated from Th/X, U/X, or Th/U chronometers (X represents a certain element). Due to the relatively short half-life of ^{235}U, the present abundances of ^{235}U in metal-poor stars are negligible, so we use Th and U respectively to represent ^{232}Th and ^{238}U for simplicity hereafter.

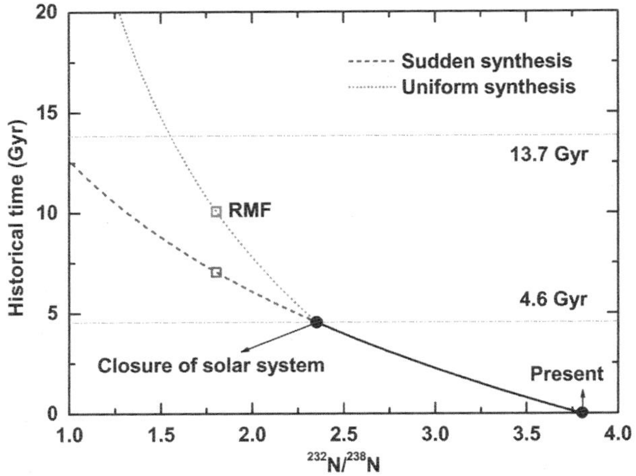

Fig. 17. (Color online) Historical time and the corresponding abundance ratio of ^{232}Th and ^{238}U. The solid line between the present and the time when the solar system became a closed system is deduced from the radioactive decay of ^{232}Th and ^{238}U. The predictions from the assumptions of sudden and uniform synthesis are shown by the dashed and dotted curves, respectively. The predicted production ratio $\frac{P_{232}}{P_{238}}$ using CDFT mass model is taken from Ref. 30, and is marked in the corresponding curves. The dash-dotted lines represent the age of the solar system (4.6 Gyr[3]) and the age of the universe (13.7 Gyr[84]).

Since the radioactive dating method is independent on the galactic chemical evolution model when it is applied to metal-poor halo stars, the abundance evolution of a radioactive element follows the normal exponential decay, i.e. Eq. (22). According to Eq. (22), the time elapsed t since the production of the radioactive element can be determined from:

$$t = 46.7[\log_{10}(\mathrm{Th/X})_0 - \log_{10}(\mathrm{Th/X})_{\mathrm{obs}}] \text{ Gyr}, \tag{38}$$

$$t = 14.8[\log_{10}(\mathrm{U/X})_0 - \log_{10}(\mathrm{U/X})_{\mathrm{obs}}] \text{ Gyr}, \tag{39}$$

$$t = 21.8[\log_{10}(\mathrm{U/Th})_0 - \log_{10}(\mathrm{U/Th})_{\mathrm{obs}}] \text{ Gyr}, \tag{40}$$

where $(\mathrm{Th/X})_0$ and $(\mathrm{Th/X})_{\mathrm{obs}}$ denote the initial (production) abundance ratio and the observed abundance ratio at present. Since the Th and U are totally synthesized in the r-process,[5] the initial abundances can be predicted with the r-process model. During the past two decades, a number of investigations on the age of metal-poor stars has been made.[12,30,76,77,91,92] For example, the average age of very metal-poor stars HD 115444 and CS 22892-052 was determined to be 15.6 ± 4.6 Gyr with the Th/Eu chronometer,[76] and the age of CS 31082-001 was determined to be 15.5 ± 3.2 Gyr with the Th/U chronometer.[91] The obtained ages also agree well with cosmological theory.

Together with the CDFT mass model, the influence of nuclear physics inputs and astrophysical conditions on the Th/U chronometer were carefully investigated based on the classical r-process approach recently.[30] For studying the influence of nuclear masses on the age determinations, eight widely used nuclear mass models (FRDM,[19] ETFSI-Q,[20] DZ10, DZ28,[21] HFB-17,[24] RMF,[27] KTUY,[93] and ETFSI-2[94] models), ranging from macroscopic-microscopic to self-consistent microscopic models, were employed. To estimate the astrophysical uncertainties, three different fits to the observations were performed to get a set of potential astrophysical conditions. The three fits correspond to the fit using the average scaled abundances of elements with $38 \leqslant Z \leqslant 82$ (labeled as fit I), the fit using the average scaled abundances of elements with $56 \leqslant Z \leqslant 82$ (labeled as fit II), and the fit using the solar r-process isotopic abundances with $125 \leqslant A \leqslant 209$ (labeled as fit III). The average scaled abundances of elements are obtained by averaging those from metal-poor halo stars CS 22892-052 and CS 31082-001.

With the initial r-process abundance ratio Th/U and their present observed values, one can easily deduce the age of low metallicity stars with Eq. (40). The estimated ages of the metal-poor stars HE 1523-0901 and CS 31082-001 using different combinations of various mass models and fits are given in Fig. 18. From this figure, it is found that the ages determined using different fits are in good agreements with each other for a given mass model. The discrepancy is within 2.7 Gyr except for the ETFSI-2 or FRDM models, whose production ratio of Th/U is not consistent with the constraint from the Th/U ratio measured in the solar system.[30] This implies that the age estimate is not sensitive to different fits, which correspond to the possible astrophysical conditions of r-process calculation. On the other hand, for a fixed fit, the deviation using different mass models is rather large. The corresponding difference can be as large as 5.9, 5.0, and 6.3 Gyr for the fit I, II, and III, respectively. Since the DZ10, DZ28, and KTUY simulations better reproduce the stable element abundances than the others and their predictions for Th/U production ratio are also consistent with the constraint from the Th/U ratio measured in the solar system,[30] therefore these mass models might be more credible and are selected to estimate the age of the metal-poor star.

In order to estimate the age of the star more reliably, one could adopt the average values for different simulations. The corresponding uncertainty

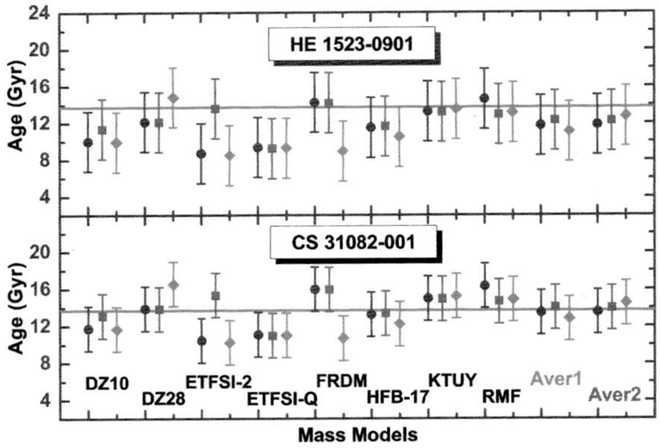

Fig. 18. (Color online) The ages of the metal-poor stars HE 1523-0901 and CS 31082-001 determined using different combinations of various mass models and fits. The circles (squares, diamonds) denote ages determined using the fit I (II, III). The average ages with all mass models and the three mass models DZ10, DZ28, KTUY are labeled as Aver1 and Aver2, respectively. The errors are attributed to uncertainties from observations of the corresponding metal-poor stars. The thick horizontal line corresponds to the cosmic age determined from WMAP data.[84] Taken from Ref. 30

is their rms deviation. In this way, the discrepancy between the average values of ages estimated using the three selected mass models for different fits decreases to 0.9 Gyr. For the fit I, the age of the metal-poor star HE 1523-0901 is obtained as 11.8 ± 3.7 Gyr which agrees with the value of 13.2 Gyr in Ref. 90. The uncertainty of age determined here includes that of 1.6 Gyr from nuclear mass models and 3.3 Gyr from observation. Similarly, ages of CS 31082-001 and BD $+17^{\circ}3248$ are deduced as 13.5 ± 2.9 Gyr and 10.9 ± 2.9 Gyr, respectively. These results can serve as the lower limit of the cosmic age, and are consistent with results derived from the WAMP data.[84] For comparison, the corresponding average ages for all the mass models using the fit I are determined as 11.7 ± 4.0, 13.5 ± 3.3, and 10.9 ± 3.3 Gyr for HE 1523-0901, CS 31082-001, and BD $+17^{\circ}3248$, respectively, with uncertainties of 2.2 Gyr from nuclear mass models.

The radiometric method based on metal-poor halo stars can avoid the uncertainties of galactic chemical evolution model, while uncertainties in initial abundances still exist. Because the r-process site is still in debate and large amounts of neutron-rich isotopes which are involved in r-process are still out of the reach of laboratory experiments, the initial elemental abundances which are obtained from theoretical r-process calculations have large uncertainties. These lead to large uncertainties in the age estimates.

Recently, a way to determine the Galactic age independent of the r-process calculations of initial Th/U ratio was proposed. In this approach, the age of the Galaxy was precisely determined based on the observed abundance ratio Th/U in metal-poor halo star CS 31082-001 and that in meteorites in conjunction with the nonlinear infall Galactic chemical evolution model.

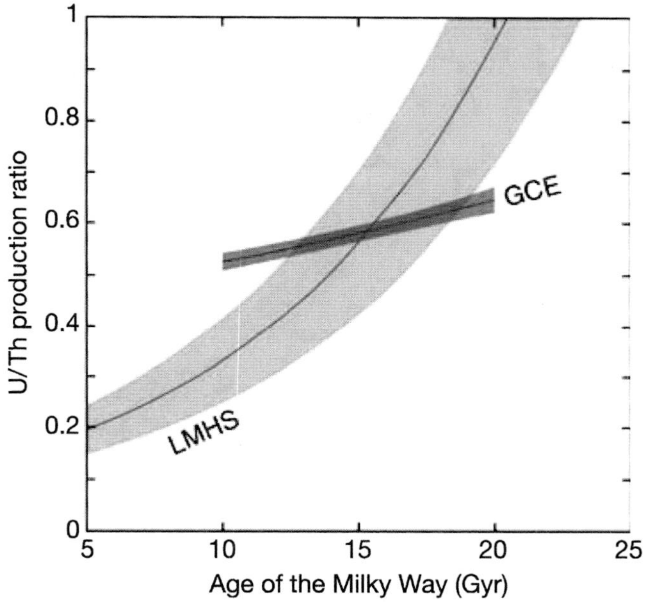

Fig. 19. (Color online) Determination of the age of the Galaxy. The light gray curve labeled LMHS is derived from the determination of the U/Th abundance ratio in a low metallicity halo star CS 31082-C001. The dark gray curve labeled GCE is derived from the solar U/Th ratio in conjunction with the Galactic chemical evolution model of a gaussian rate of infall. Taken from Ref. 95.

The principle of this approach to the age determination is shown in Fig. 19. The light gray curve in this figure denotes the age determination with the observed Th/U ratio in metal-poor halo star CS 31082-001, which is described by Eq. (40) since the initial Th/U ratio trapped in CS 31082-001 was only modified by free decay during the life of the star. The dark gray curve in this figure denotes the age determination with the solar Th/U ratio in meteorites in conjunction with the nonlinear infall Galactic chemical evolution model, which can be approximately described by a simple formula $(\mathrm{U/Th})_{\mathrm{SS}}/(\mathrm{U/Th})_0 = aT_G + b$ ($a = -1.576 \times 10^{-2}$ and $b = 0.9946$ for a gaussian rate of infall, and $a = -1.968 \times 10^{-2}$ and $b = 1.0114$ for an exponential rate of infall) when the Galactic age T_G is in the range from 10 Gyr to 20 Gyr. By assuming the initial Th/U ratio did not vary from one star to another and the age determinations with the Th/U ratios in CS 31082-001 and meteorites predict the same Galactic age, then it is possible to determine the initial Th/U ratio and the age of the galaxy by the intersection between the two curves. By a careful analyse of uncertainties, the Galactic age was precisely determined to be $14.5^{+2.8}_{-2.2}$ Gyr for a gaussian rate of infall or $15.1^{+2.8}_{-2.3}$ Gyr for an exponential rate of infall. Clearly, this result avoids the calculations of initial Th/U ratio with the r-process model and is thus independent of the uncertainties of nuclear physics inputs and other uncertainties of r-process sites, while the dependence on the Galactic chemical evolution model remains.

The age determinations with the observed abundances of radioactive elements in metal-poor stars are independent of the Galactic chemical evolution model, while the uncertainties in the r-process calculations of initial abundances are significant. The approach in Ref. 95 avoids the r-process calculations of initial abundances by employing the observed abundances both in metal-poor halo stars and in meteorites, while the uncertainty in the Galactic chemical evolution model remains. A way to avoid the uncertainties both in the theoretical r-process calculations and Galactic chemical evolution model is to employ solar r-process abundances at the time when the Solar System became a closed system to directly approximate the initial r-process abundances. This provides a lower limit to the initial r-process abundances, because the radioactive elements have partially decayed although they can be constantly produced and ejected into the interstellar medium. Furthermore, it is found that the solar r-process abundances at that time are close to the initial r-process elemental abundances,[30] so the age determination with the solar r-process abundances as the initial r-process abundances can give a relatively strong and reliable constraint to the lower limit of the cosmic age.

By subtracting the solar s-process abundances from the observed total solar abundances, the solar r-process abundances have been well determined based on the s-process nucleosynthesis model in low-mass asymptotic giant branch (AGB) stars.[96,97] Recently, the accurate abundances of rare earth (RE) elements in five metal-poor stars CS 22829-052, CS 31082-001, HD 115444, HD 221170, and BD $+17^\circ 3248$ have been derived,[85] so their scaled average (first scaled to Eu data and then averaged) abundances can be used as the observed abundances of RE elements in metal-poor halo stars. Together with the Th abundances in metal-poor halo star CS 22892-052[98]

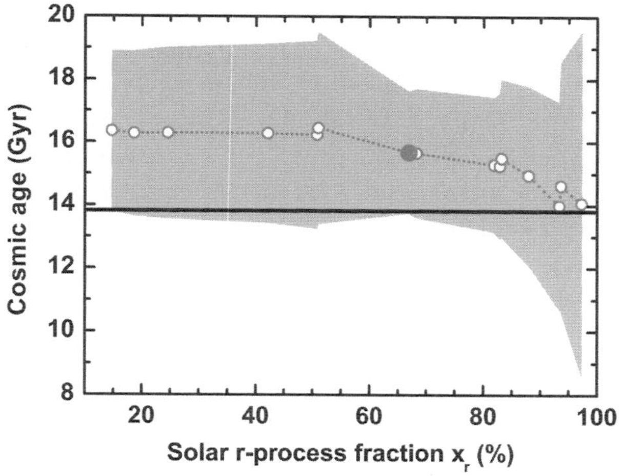

Fig. 20. (Color online) The average cosmic age calculated with different groups of Th/X chronometers. The circle at the r-process fraction $x_r = x_{r0}$ represents the average age of that group of Th/X chronometers with $x_r \geqslant x_{r0}$. For comparison, the cosmic age from Planck 2013 results[70] is shown by the thick solid line. Taken from Ref. 100.

and the solar system,[99] the galactic age can be determined with Eq. (38) using different Th/X chronometers.

To estimate the stellar age more reliably, one could adopt the average value from various Th/X chronometers. In Fig. 20, the average ages are shown for 15 groups of Th/X chronometers. The circle at the r-process fraction (the ratio of the solar r-process abundance to its total abundance) $x_r = x_{r0}$ represents the average age of that group of Th/X chronometers with $x_r \geqslant x_{r0}$. Clearly, these 15 average ages agree with each other and also agree with the latest cosmic age 13.813 ± 0.058 Gyr determined from Planck 2013 results.[70] However, the uncertainties are relatively smaller for the average age shown by the filled circle in Fig. 20. For average ages on the left side of filled circle, the relatively larger uncertainties come from the relatively larger rms deviations among the ages of different Th/X chronometers. These larger rms deviations may originate from the uncertainties of the s-process abundances, since the s-process gives larger contributions for the abundances of these elements. For average ages on the right side of filled circle, the number of Th/X chronometers is smaller and hence the large uncertainty from observation of Th abundance is mostly transferred to the average age. By taking the Th/X ages with $x_r > 60\%$ as a group (denoted by the filled circle in Fig. 20), the average cosmic age and the related error are 15.68 ± 1.94 Gyr. Although the average age of this group has a relatively small uncertainty, it should be emphasized that 1.94 Gyr is still a relatively large value and guarantees the agreement with the cosmic age determined from Planck 2013 results. To test whether the two values might be discrepant, the uncertainty of 1.94 Gyr should be further reduced. Since this uncertainty of determined cosmic age mainly originates from the error on thorium abundance observed in metal-poor star CS 22892-052, future high-precision abundance observations on CS 22892-052 are needed to further test the consistency of these two results.

6. *rp*-process

The astrophysical rapid proton capture process (*rp*-process)[101] is a reaction sequence involving very neutron-deficient nuclides near the proton drip line, possibly up to mass number 100. It is thought to cause astronomically observed x-ray bursts by repeated thermonuclear explosions in a thin, proton-rich fuel layer on the surface of a mass accreting neutron star. The *rp*-process may also produce the neutron-deficient ^{92,94}Mo and ^{96,98}Ru isotopes, whose origin are still open in the framework of standard nucleosynthesis. The nuclear energy generation in a burst and the composition of the burst ashes need to be reliably predicted to understand the properties of bursts and their impact on the structure of the neutron star crust. In this section, we present an example to apply CDFT in the *rp*-process simulations.

6.1. *Absolute mass vs. Coulomb displacement energy*

If the nuclear forces are charge symmetric, the Coulomb displacement energy CDE(A,T) between a mirror nucleus pair with the isospin T is given by:

$$\mathrm{CDE}(A,T) = M(A,T_{Z>}) - M(A,T_{Z<}) + 2T_{Z<} \cdot \Delta_{np}, \tag{41}$$

where $M(A,T_{Z>})$ and $M(A,T_{Z<})$ represent the masses of a pair of mirror nucleus with the larger and smaller charge, respectively. Δ_{np} is the proton-neutron atomic mass difference 782.3466(5) keV.[2]

It is found that the displacement energies can be calculated much more accurately than the absolute mass values in theoretical models. This is because the CDE depends mainly on the Coulomb interaction and moreover most nuclear interactions, which can hardly be treated in modeling with an

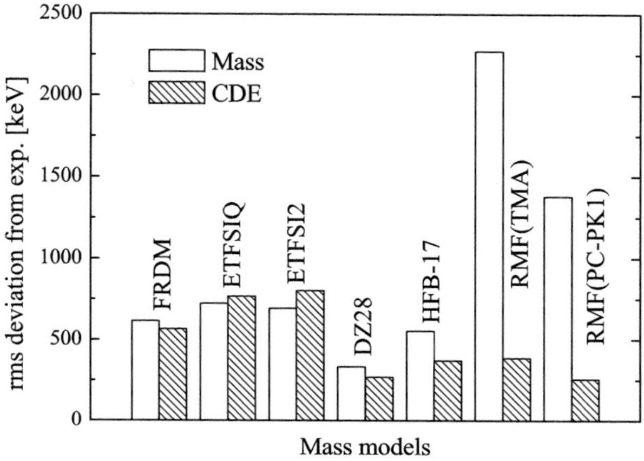

Fig. 21. rms deviations of absolute masses and the CDEs in various mass models with respect to the experimental data. In this comparison only experimental data with $Z \geq 8$ and accuracy less than 100 keV are included. Adapted from Ref. 10.

accuracy better than a few hundred keV, are canceled somehow by subtraction. On condition that the mass of the corresponding neutron-rich nuclide is known, absolute mass value for proton-rich nuclide can be deduced with Eq. (41) by combining the calculated CDE and experimentally measured value for the neutron-rich mirror. This offers an alternative method in the estimates of proton-rich nuclear masses. This is indeed very helpful considering the fact that the masses of many neutron-rich nuclei above ruthenium ($Z = 44$) are unknown.

Shown in Fig. 21 are the rms deviations of the calculated CDEs in various mass models from the available experimental data. The merits of the CDE method are clearly demonstrated in the HFB17, the RMF with meson-exchange interaction (RMF-ME),[27] and point-coupling interaction (RMF-PC),[31] in particular the latter two versions of CDFT. The best rms deviation with respect to all the available CDEs with $Z \geq 8$ is obtained in the RMF-PC model and amounts to about 250 keV. This is more than a factor of 5 better than the corresponding rms value for absolute masses. On the other hand, the models such as the FRDM, ETFSI-2,[94] and ETFSI-Q, and the Duflo–Zuker model (DZ) show the same level of predictive power for both the CDEs and the absolute masses.

The dependence of calculated displacement energies on mass numbers and isospins in the RMF-PC model can be found in Fig. 22. It shows that the calculations are in excellent agreement with the experimental values

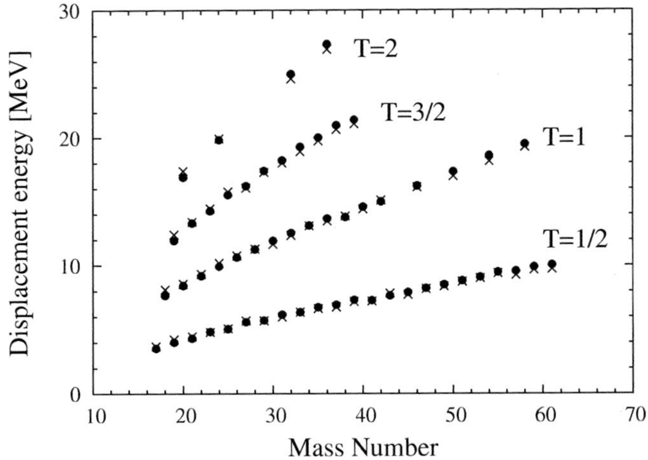

Fig. 22. Calculated (filled circles) vs. experimental (crosses) coulomb displacement energies (crosses) as a function of mass number. The experimental uncertainties are within the symbols. The isospin number is also labeled. Adapted from Ref. 10.

with a mean difference of only about 50 keV. Systematic underestimation of the displacement energies, namely, the Nolen–Schiffer anomaly,[102] is not observed, but it is shown in the case of the RMF-ME calculations. The improvement of the CDEs in the RMF-PC is partially due to a proper treatment of deformation and center-of-mass corrections.

6.2. *Tentative application in the rp-process*

Similar to the r-process, accurate reaction Q-values are essential for reliable estimates of reaction rates in the rp-process. The masses and the exact location of proton drip line determine the possible path. For example, whether ^{64}Ge is a "waiting point" or not in the path largely depends on the stability of ^{65}As. A similar remark is valid as well for the nuclides ^{68}Se and ^{72}Kr. Here we employ the prediction for the key nuclide ^{65}As deduced from the CDE value in the rp-process, and further estimate the effective stellar half-life of ^{64}Ge.

At temperatures of 1–2$\times10^9$ K in the explosion environment such as X-ray burst, the proton-capture and photodisintegration rates are predicted to be greater than the β-decay rates of the nuclides involved. Thus, if ^{65}As were significantly proton unbound, ^{64}Ge would then become a "waiting point" in the rp-path because of the long beta-decay half-life of ^{64}Ge, and

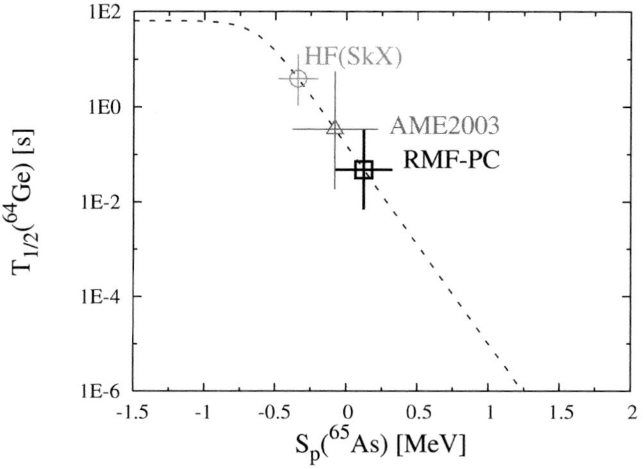

Fig. 23. Color online) Effective stellar half-life of ^{64}Ge as a function of $S_p(^{65}\text{As})$. The results based on the predicted $S_p(^{65}\text{As})$ in the RMF-PC, HF(SkX) and AME2003 are illustrated. Adapted from Ref. 10.

thereby inhibit the production of heavier elements. The total half-life of the waiting points, on the other hand, can be significantly reduced by two-proton capture reactions towards heavier masses.

Here a simple estimation is given to illustrate the sensitivity of the "waiting-point" and the relevant separation energy. At a temperature T of 1.2 GK, the decay constant λ of the ^{64}Ge is determined by its β-decay rate λ_β and the two-proton capture rate, namely,

$$\lambda = \lambda_\beta + Y_p^2\rho^2 N_A^2 \left(\frac{2\pi\hbar^2}{kT}\right)^{3/2} \frac{G(^{65}As)}{2G(^{64}Ge)} exp\left(\frac{S_p(^{64}Ge)}{kT}\right) < p\gamma >_{^{65}As} , \tag{42}$$

where the partition function G of 1.0 is used for both ^{64}Ge and ^{64}As. The abundance of proton Y_p of 0.771 mol/g and the density of 10^6 g/cm^3 are chosen as the stellar condition. The $< p\gamma >_{^{65}As}$ rate is taken from Ref. 101.

The effective stellar half-life of ^{64}Ge as a function of $S_p(^{64}\text{As})$ is presented as a dashed line in Fig. 23. Indicated also are the relevant predictions when using various S_p values. In the RMF-PC calculation, ^{65}As is found to be loosely bound to one-proton emission by 0.12 MeV. The uncertainty of present calculations is considered as 0.2 MeV. This prediction agrees well with the latest experimental data of $-0.090(85)$ MeV.[103] The RMF-PC prediction suggests that the effective stellar half-life of ^{64}Ge is between

7 ms and 330 ms, which is smaller than the results based on AME2003 and HF-SkX predictions. Thus contrary to the HF-SkX result no significant delay at ^{64}Ge is concluded in the present calculation.

7. Conclusion remark

In this chapter, we gave a brief overview of the efforts in computing masses and β-decay half-lives of exotic nuclei, two of the most important nuclear properties in nucleosynthesis applications, in CDFT. We furthermore introduced several applications of these results to discuss the nuclear model uncertainties in r- and rp-process and to determine the age of Universe by nuclear cosmochronology method. There are already some interesting and instructive results by combining with the augmented knowledge of both experimental and theoretical progress. However, there are still many challenges and big problems that the CDFT community has to face, before one can provide a self-consistent database for astrophysical application. On top of all that, one question has to be addressed with highest priority, i.e., how can we get a better accuracy in computing nuclear masses and β-decay half-lives while keeping the merit of covariance?

This review is entirely based on the common effort of our colleagues working with us over many years in different collaborations. Especially, we would like to thank Li-Sheng Geng, Jie Meng, Fernando A. Montes, Yong-Zhong Qian, Xiao-Dong Xu, Peng-Wei Zhao for fruitful discussions. This work was supported partially by the National Natural Science Foundation of China (Nos. 11235002, 11128510, 11475014, 11205004).

References

1. C. Sneden, J. J. Cowan, and R. Gallino, Neutron-capture elements in the early galaxy, *Annu. Rev. Astron. Astrophys.* **46**, 241 (2008).
2. M. Wang, G. Audi, A. H. Wapstra, F. G. Kondev, M. MacCormick, X. X., and B. Pfeiffer, The NUBASE2012 evaluation of nuclear properties (II). tables, graphs and references, *Chin. Phys. C.* **36**(12), 1603–2014 (Dec., 2012). URL `http://amdc.impcas.ac.cn/`.
3. C. Rolfs and W. Rodney, *Cauldrons in the Cosmos: Nuclear Astrophysics*. The University of Chicago Press (1988).
4. A. G. W. Cameron, Nuclear reactions in stars and nucleogenesis, *Pub. Astron. Soc. Pac.* **69**, 201–222 (1957).
5. E. M. Burbidge, G. R. Burbidge, W. A. Fowler, and F. Hoyle, Synthesis of the elements in stars, *Rev. Mod. Phys.* **29**(4), 547–650 (Oct, 1957). doi: 10.1103/RevModPhys.29.547.

6. M. Arnould and S. Goriely, The p-process of stellar nucleosynthesis: astrophysics and nuclear physics status, *Phys. Rep.* **384**, 1–84 (2003). ISSN 0370–1573. doi: http://dx.doi.org/10.1016/S0370-1573(03)00242-4. URL `http://www.sciencedirect.com/science/article/pii/S0370157303002424`.
7. T. Rauscher, N. Dauphas, I. Dillmann, C. Fröhlich, Z. Fülöp, and G. Gyürky, Constraining the astrophysical origin of the p-nuclei through nuclear physics and meteoritic data, *Rep. Prog. Phys.* **76**(6), 066201 (2013). URL `http://stacks.iop.org/0034-4885/76/i=6/a=066201`.
8. M. Arnould, S. Goriely, and K. Takahashi, The r-process of stellar nucleosynthesis: Astrophysics and nuclear physics achievements and mysteries, *Phys. Rep.* **450**(4–6), 97–213 (2007). ISSN 0370–1573. doi: http://dx.doi.org/10.1016/j.physrep.2007.06.002. URL `http://www.sciencedirect.com/science/article/pii/S0370157307002438`.
9. Y.-Z. Qian and G. Wasserburg, Where, oh where has the r-process gone?, *Phys. Rep.* **442**(1-6), 237–268 (2007). doi: DOI:10.1016/j.physrep.2007.02.006. The Hans Bethe Centennial Volume 1906-2006.
10. B. Sun and J. Meng, Challenge on the astrophysical r-process calculation with nuclear mass models, *Chin. Phys. Lett.* **25**(7), 2429–2431 (2008). doi: 10.1088/0256-307X/25/7/027.
11. B. H. Sun, Y. A. Litvinov, I. Tanihata, and Y. H. Zhang, Toward precision mass measurements of neutron-rich nuclei relevant to r-process nucleosynthesis, *Front. Phys.* **10**, 102102 (2015). doi: 10.1007/s11467-015-0503-z.
12. K.-L. Kratz, K. Farouqi, B. Pfeiffer, J. W. Truran, C. Sneden, and J. J. Cowan, Explorations of the r-processes: Comparisons between calculations and observations of low-metallicity stars, *Astrophys. J.* **662**(1), 39–52 (2007). doi: 10.1086/517495.
13. B. Sun, F. Montes, L. S. Geng, H. Geissel, Y. A. Litvinov, and J. Meng, Application of the relativistic mean-field mass model to the r-process and the influence of mass uncertainties, *Phys. Rev. C.* **78**(2), 025806 (Aug, 2008). doi: 10.1103/PhysRevC.78.025806.
14. P. Banerjee, W. C. Haxton, and Y.-Z. Qian, Long, cold, early r process? neutrino-induced nucleosynthesis in He shells revisited, *Phys. Rev. Lett.* **106**, 201104 (May, 2011). doi: 10.1103/PhysRevLett.106.201104. URL `http://link.aps.org/doi/10.1103/PhysRevLett.106.201104`.
15. F. Käppeler, H. Beer, and K. Wisshak, s-process nucleosynthesis-nuclear physics and the classical model, *Rep. Prog. Phys.* **52**, 945–1013 (1989).
16. T. Rauscher, The path to improved reaction rates for astrophysics, *Int. J. Mod. Phys. E.* **20**(2), 1071–1168 (2011).
17. K. Takahashi and K. Yokoi, Beta-decay rates of highly ionized heavy atoms in stellar interiors, *At. Data Nucl. Data Tables.* **36**, 375–409 (1987).
18. F. Bosch, T. Faestermann, J. Friese, F. Heine, P. Kienle, E. Wefers, K. Zeitelhack, K. Beckert, B. Franzke, O. Klepper, C. Kozhuharov, G. Menzel, R. Moshammer, F. Nolden, H. Reich, B. Schlitt, M. Steck, T. Stöhlker, T. Winkler, and K. Takahashi, Observation of bound-state β^- decay of fully ionized ^{187}Re: ^{187}Re–^{187}Os cosmochronometry, *Phys. Rev. Lett.* **77**,

5190–5193 (Dec, 1996). doi: 10.1103/PhysRevLett.77.5190. URL `http://link.aps.org/doi/10.1103/PhysRevLett.77.5190`.
19. P. Möller, J. R. Nix, W. D. Myers, and W. J. Swiatecki, Nuclear ground-state masses and deformations, *At. Data Nucl. Data Tables.* **59**(2), 185–381 (1995). ISSN 0092-640X. doi: 10.1006/adnd.1995.1002.
20. J. M. Pearson, R. C. Nayak, and S. Goriely, Nuclear mass formula with Bogolyubov-enhanced shell-quenching: application to r-process, *Phys. Lett. B.* **387**(3), 455–459 (1996). doi: 10.1016/0370-2693(96)01071-4.
21. J. Duflo and A. P. Zuker, Microscopic mass formulas, *Phys. Rev. C.* **52**(1), R23–R27 (Jul, 1995). doi: 10.1103/PhysRevC.52.R23.
22. N. Wang, Z. Liang, M. Liu, and X. Wu, Mirror nuclei constraint in nuclear mass formula, *Phys. Rev. C.* **82**, 044304 (Oct, 2010). doi: 10.1103/PhysRevC.82.044304. URL `http://link.aps.org/doi/10.1103/PhysRevC.82.044304`.
23. S. Goriely, M. Samyn, and J. Pearson, Further explorations of Skyrme–Hartree–Fock–Bogoliubov mass formulas; vi: Weakened pairing, *Nucl. Phys. A.* **773**(3-4), 279–299 (2006). ISSN 0375-9474. doi: 10.1016/j.nuclphysa.2006.05.002.
24. S. Goriely, N. Chamel, and J. M. Pearson, Skyrme–Hartree–Fock–Bogoliubov nuclear mass formulas: Crossing the 0.6 MeV accuracy threshold with microscopically deduced pairing, *Phys. Rev. Lett.* 102(15):152503 (2009). doi: 10.1103/PhysRevLett.102.152503.
25. D. Hirata, K. Sumiyoshi, I. Tanihata, Y. Sugahara, T. Tachibana, and H. Toki, A systematic study of even-even nuclei up to the drip lines within the relativistic mean field framework, *Nucl. Phys. A.* **616**, 438c–445c (1997). doi: 10.1051/0004-6361:20020434.
26. G. A. Lalazissis, S. Raman, and P. Ring, Ground-state properties of even-even nuclei in the relativistic mean-field theory, *At. Data Nucl. Data Tables.* **71**(1), 1–40 (Jan., 1999). doi: 10.1006/adnd.1998.0795.
27. L. S. Geng, H. Toki, and J. Meng, Masses, deformations and charge radii–nuclear ground-state properties in the relativistic mean field model, *Prog. Theor. Phys.* **113**(4), 785–800 (2005). doi: 10.1143/PTP.113.785.
28. G. Audi, A. H. Wapstra, and C. Thibault, The 2003 atomic mass evaluation: (II). tables, graphs and references, *Nucl. Phys. A.* **729**(1), 337–676 (2003). doi: 10.1016/j.nuclphysa.2003.11.003.
29. Y. Sugahara and H. Toki, Relativistic mean-field theory for unstable nuclei with non-linear [sigma] and [omega] terms, *Nucl. Phys. A.* **579**(3-4), 557–572 (1994). ISSN 0375-9474. doi: 10.1016/0375-9474(94)90923-7.
30. Z. M. Niu, B. H. Sun, and J. Meng, Influence of nuclear physics inputs and astrophysical conditions on the Th/U chronometer, *Phys. Rev. C.* **80**(6), 065806 (Dec, 2009). doi: 10.1103/PhysRevC.80.065806.
31. P. W. Zhao, Z. P. Li, J. M. Yao, and J. Meng, New parametrization for the nuclear covariant energy density functional with a point-coupling interaction, *Phys. Rev. C.* **82**(5), 054319 (Nov, 2010). doi: 10.1103/PhysRevC.82.054319.
32. P. W. Zhao, L. S. Song, B. Sun, H. Geissel, and J. Meng, Crucial test

for covariant density functional theory with new and accurate mass measurements from Sn to Pa, *Phys. Rev. C.* **86**, 064324 (Dec, 2012). doi: 10.1103/PhysRevC.86.064324.
33. X. Y. Qu, Y. Chen, S. Q. Zhang, P. W. Zhao, I. J. Shin, Y. Lim, Y. Kim, and J. Meng, Extending the nuclear chart by continuum: From oxygen to titanium, *Sci. China Phys. Mech. Astron.* **56**, 2031–2036 (2013). doi: 10.1007/s11433-013-5329-5.
34. X. D. Xu, B. Sun, Z. M. Niu, Z. Li, Y.-Z. Qian, and J. Meng, Reexamining the temperature and neutron density conditions for r-process nucleosynthesis with augmented nuclear mass models, *Phys. Rev. C.* **87**, 015805 (Jan, 2013). doi: 10.1103/PhysRevC.87.015805. URL `http://link.aps.org/doi/10.1103/PhysRevC.87.015805`.
35. P. Möller, J. R. Nix, and K. L. Kratz, Nuclear properties for astrophysical and radioactive-ion-beam applications, *At. Data Nucl. Data Tables.* **66**(2), 131–343 (1997). ISSN 0092-640X. doi: 10.1006/adnd.1997.0746.
36. P. Möller, B. Pfeiffer, and K.-L. Kratz, New calculations of gross β-decay properties for astrophysical applications: Speeding-up the classical r process, *Phys. Rev. C.* **67**(5), 055802 (May, 2003). doi: 10.1103/PhysRevC.67.055802.
37. NNDC National Nuclear Data Center, Brookhaven National Laboratory (http://www.nndc.bnl.gov/).
38. D. Lunney, J. M. Pearson, and C. Thibault, Recent trends in the determination of nuclear masses, *Rev. Mod. Phys.* **75**(3), 1021–1082 (Aug, 2003). doi: 10.1103/RevModPhys.75.1021.
39. E. Fermi, Versuch einer theorie der β-strahlen. i, *Z. Phys.* **88**(3-4), 161–177 (1934). ISSN 0044-3328. doi: 10.1007/BF01351864. URL `http://dx.doi.org/10.1007/BF01351864`.
40. E. Nácher, A. Algora, B. Rubio, J. L. Taín, D. Cano-Ott, S. Courtin, P. Dessagne, F. Maréchal, C. Miehé, E. Poirier, M. J. G. Borge, D. Escrig, A. Jungclaus, P. Sarriguren, O. Tengblad, W. Gelletly, L. M. Fraile, and G. L. Scornet, Deformation of the $N = Z$ nucleus ^{76}Sr using β-decay studies, *Phys. Rev. Lett.* **92**, 232501 (Jun, 2004). doi: 10.1103/PhysRevLett.92.232501. URL `http://link.aps.org/doi/10.1103/PhysRevLett.92.232501`.
41. V. Tripathi, S. L. Tabor, P. F. Mantica, Y. Utsuno, P. Bender, J. Cook, C. R. Hoffman, S. Lee, T. Otsuka, J. Pereira, M. Perry, K. Pepper, J. S. Pinter, J. Stoker, A. Volya, and D. Weisshaar, Intruder configurations in the $A = 33$ isobars: ^{33}Mg and ^{33}Al, *Phys. Rev. Lett.* **101**, 142504 (Oct, 2008). doi: 10.1103/PhysRevLett.101.142504. URL `http://link.aps.org/doi/10.1103/PhysRevLett.101.142504`.
42. K. Langanke and G. Martínez-Pinedo, Nuclear weak-interaction process in stars, *Rev. Mod. Phys.* **75**, 819 (2003).
43. N. Severijns, M. Beck, and O. Naviliat-Cuncic, Tests of the standard electroweak model in nuclear beta decay, *Rev. Mod. Phys.* **78**, 991–1040 (Sep, 2006). doi: 10.1103/RevModPhys.78.991. URL `http://link.aps.org/doi/10.1103/RevModPhys.78.991`.

44. I. S. Towner and J. C. Hardy, The evaluation of Vud and its impact on the unitarity of the Cabibbo–Kobayashi–Maskawa quark-mixing matrix, *Rep. Prog. Phys.* **73**(4), 046301 (2010). URL http://stacks.iop.org/0034-4885/73/i=4/a=046301.
45. S. Grévy, J. Angélique, P. Baumann, C. Borcea, A. Buta, G. Canchel, W. Catford, S. Courtin, J. Daugas, F. de Oliveira, P. Dessagne, Z. Dlouhy, A. Knipper, K. Kratz, F. Lecolley, J. Lecouey, G. Lehrsenneau, M. Lewitowicz, E. Liénard, S. Lukyanov, F. Maréchal, C. Miehé, J. Mrazek, F. Negoita, N. Orr, D. Pantelica, Y. Penionzhkevich, J. Peter, B. Pfeiffer, S. Pietri, E. Poirier, O. Sorlin, M. Stanoiu, I. Stefan, C. Stodel, and C. Timis, Beta-decay half-lives at the $N = 28$ shell closure, *Phys. Lett. B.* **594**(3-4), 252–259 (2004). ISSN 0370-2693. doi: http://dx.doi.org/10.1016/j.physletb.2004.06.005. URL http://www.sciencedirect.com/science/article/pii/S0370269304007981.
46. P. T. Hosmer, H. Schatz, A. Aprahamian, O. Arndt, R. R. C. Clement, A. Estrade, K.-L. Kratz, S. N. Liddick, P. F. Mantica, W. F. Mueller, F. Montes, A. C. Morton, M. Ouellette, E. Pellegrini, B. Pfeiffer, P. Reeder, P. Santi, M. Steiner, A. Stolz, B. E. Tomlin, W. B. Walters, and A. Wöhr, Half-life of the doubly magic r-process nucleus ^{78}Ni, *Phys. Rev. Lett.* **94**, 112501 (Mar, 2005). doi: 10.1103/PhysRevLett.94.112501. URL http://link.aps.org/doi/10.1103/PhysRevLett.94.112501.
47. S. Nishimura, Z. Li, H. Watanabe, K. Yoshinaga, T. Sumikama, T. Tachibana, K. Yamaguchi, M. Kurata-Nishimura, G. Lorusso, Y. Miyashita, A. Odahara, H. Baba, J. S. Berryman, N. Blasi, A. Bracco, F. Camera, J. Chiba, P. Doornenbal, S. Go, T. Hashimoto, S. Hayakawa, C. Hinke, E. Ideguchi, T. Isobe, Y. Ito, D. G. Jenkins, Y. Kawada, N. Kobayashi, Y. Kondo, R. Krücken, S. Kubono, T. Nakano, H. J. Ong, S. Ota, Z. Podolyák, H. Sakurai, H. Scheit, K. Steiger, D. Steppenbeck, K. Sugimoto, S. Takano, A. Takashima, K. Tajiri, T. Teranishi, Y. Wakabayashi, P. M. Walker, O. Wieland, and H. Yamaguchi, β-decay half-lives of very neutron-rich Kr to Tc isotopes on the boundary of the r-process path: an indication of fast r-matter flow, *Phys. Rev. Lett.* **106**, 052502 (Feb, 2011). doi: 10.1103/PhysRevLett.106.052502. URL http://link.aps.org/doi/10.1103/PhysRevLett.106.052502.
48. G. Audi, F. Kondev, M. Wang, B. Pfeiffer, X. Sun, J. Blachot, and M. MacCormick, The NUBASE2012 evaluation of nuclear properties, *Chin. Phys. C.* **36**(12), 1157 (2012). URL http://stacks.iop.org/1674-1137/36/i=12/a=001.
49. K. Takahashi, M. Yamada, and T. Kondoh, Beta-decay half-lives calculated on the gross theory, *At. Data Nucl. Data Tables.* **12** (1), 101–142 (1973). ISSN 0092-640X. doi: http://dx.doi.org/10.1016/0092-640X(73)90015-6. URL http://www.sciencedirect.com/science/article/pii/0092640X73900156.
50. M. Hirsch, A. Staudt, K. Muto, and H. Klapdorkleingrothaus, Microscopic predictions of β^+/ec-decay half-lives, *At. Data Nucl. Data Tables.* **53**(2), 165–193 (1993). ISSN 0092-640X. doi: http://dx.doi.org/10.1006/

adnd.1993.1004. URL http://www.sciencedirect.com/science/article/pii/S0092640X83710041.

51. J. Engel, M. Bender, J. Dobaczewski, W. Nazarewicz, and R. Surman, β decay rates of r-process waiting-point nuclei in a self-consistent approach, *Phys. Rev. C.* **60**(1), 014302 (1999). doi: 10.1103/PhysRevC.60.014302.
52. T. Nikšić, T. Marketin, D. Vretenar, N. Paar, and P. Ring, β-decay rates of r-process nuclei in the relativistic quasiparticle random phase approximation, *Phys. Rev. C.* **71**(1), 014308 (2005). doi: 10.1103/PhysRevC.71.014308.
53. T. Marketin, D. Vretenar, and P. Ring, Calculation of beta-decay rates in a relativistic model with momentum-dependent self-energies, *Phys. Rev. C.* 75(2):024304 (2007). doi: 10.1103/PhysRevC.75.024304.
54. Z. Niu, Y. Niu, H. Liang, W. Long, T. Nikšić, D. Vretenar, and J. Meng, β-decay half-lives of neutron-rich nuclei and matter flow in the r-process, *Phys. Lett. B.* **723**(1–3), 172–176 (2013). ISSN 0370-2693. doi: http://dx.doi.org/10.1016/j.physletb.2013.04.048.
55. S. Typel, Relativistic model for nuclear matter and atomic nuclei with momentum-dependent self-energies, *Phys. Rev. C.* **71**, 064301 (Jun, 2005). doi: 10.1103/PhysRevC.71.064301. URL http://link.aps.org/doi/10.1103/PhysRevC.71.064301.
56. Y. F. Niu, Z. M. Niu, G. Colò, and E. Vigezzi, Particle-vibration coupling effect on the β decay of magic nuclei, *Phys. Rev. Lett.* **114**, 142501 (Apr, 2015). doi: 10.1103/PhysRevLett.114.142501. URL http://link.aps.org/doi/10.1103/PhysRevLett.114.142501.
57. H. Liang, N. Van Giai, and J. Meng, Spin-isospin resonances: A self-consistent covariant description, *Phys. Rev. Lett.* **101**(12), 122502 (2008). doi: 10.1103/PhysRevLett.101.122502.
58. H. Liang, P. Zhao, and J. Meng, Fine structure of charge-exchange spin-dipole excitations in ^{16}O, *Phys. Rev. C.* **85**, 064302 (Jun, 2012). doi: 10.1103/PhysRevC.85.064302. URL http://link.aps.org/doi/10.1103/PhysRevC.85.064302.
59. W. H. Long, P. Ring, N. Van Giai, and J. Meng, Relativistic Hartree–Fock–Bogoliubov theory with density dependent meson-nucleon couplings, *Phys. Rev. C.* **81**(2), 024308 (2010). doi: 10.1103/PhysRevC.81.024308.
60. W. H. Long, N. Van Giai, and J. Meng, Density-dependent relativistic Hartree–Fock approach, *Phys. Lett. B.* **640**(4), 150–154 (2006). doi: 10.1016/j.physletb.2006.07.064.
61. F. Minato and C. L. Bai, Impact of tensor force on β decay of magic and semimagic nuclei, *Phys. Rev. Lett.* **110**, 122501 (Mar, 2013). doi: 10.1103/PhysRevLett.110.122501. URL http://link.aps.org/doi/10.1103/PhysRevLett.110.122501.
62. T. Nikšić, D. Vretenar, and P. Ring, Relativistic nuclear energy density functionals: Mean-field and beyond, *Prog. Part. Nucl. Phys.* **66**(3), 519–548 (2011). ISSN 0146-6410. doi: 10.1016/j.ppnp.2011.01.055.
63. H. Liang, P. Zhao, P. Ring, X. Roca-Maza, and J. Meng, Localized form

of Fock terms in nuclear covariant density functional theory, *Phys. Rev. C.* **86**, 021302 (Aug, 2012). doi: 10.1103/PhysRevC.86.021302.

64. H.-Q. Gu, H. Liang, W. H. Long, N. Van Giai, and J. Meng, Slater approximation for Coulomb exchange effects in nuclear covariant density functional theory, *Phys. Rev. C.* **87**, 041301 (Apr, 2013). doi: 10.1103/PhysRevC.87.041301. URL `http://link.aps.org/doi/10.1103/PhysRevC.87.041301`.
65. Z. M. Niu, Y. F. Niu, Q. Liu, H. Z. Liang, and J. Y. Guo, Nuclear β^+/ec decays in covariant density functional theory and the impact of isoscalar proton-neutron pairing, *Phys. Rev. C.* **87**, 051303 (May, 2013). doi: 10.1103/PhysRevC.87.051303. URL `http://link.aps.org/doi/10.1103/PhysRevC.87.051303`.
66. I. N. Borzov and S. Goriely, Weak interaction rates of neutron-rich nuclei and the r-process nucleosynthesis, *Phys. Rev. C.* **62**, 035501 (Aug, 2000). doi: 10.1103/PhysRevC.62.035501. URL `http://link.aps.org/doi/10.1103/PhysRevC.62.035501`.
67. P. Sarriguren and J. Pereira, β-decay properties of neutron-rich Zr and Mo isotopes, *Phys. Rev. C.* **81**, 064314 (Jun, 2010). doi: 10.1103/PhysRevC.81.064314. URL `http://link.aps.org/doi/10.1103/PhysRevC.81.064314`.
68. B. A. Nikolaus, T. Hoch, and D. G. Madland, Nuclear ground state properties in a relativistic point coupling model, *Phys. Rev. C.* **46**(5), 1757–1781 (Nov, 1992). doi: 10.1103/PhysRevC.46.1757.
69. C. L. Bennett, D. Larson, J. L. Weiland, N. Jarosik, G. Hinshaw, N. Odegard, K. M. Smith, R. S. Hill, B. Gold, M. Halpern, E. Komatsu, M. R. Nolta, L. Page, D. N. Spergel, E. Wollack, J. Dunkley, A. Kogut, M. Limon, S. S. Meyer, G. S. Tucker, and E. L. Wright, Nine-year Wilkinson Microwave Anisotropy Probe (WMAP) observations: Final maps and results, *Astrophys. J. Suppl. Ser.* **208**(2), 20 (2013). URL `http://stacks.iop.org/0067-0049/208/i=2/a=20`.
70. P. Collaboration, Planck 2013 results. xvi. cosmological parameters, *Astron. Astrophys.* **571**, A16 (2014). doi: 10.1051/0004-6361/201321591. URL `http://www.aanda.org/articles/aa/abs/2014/11/aa21591-13/aa21591-13.html`.
71. B. Chaboyer, P. Demarque, P. J. Kernan, and L. M. Krauss, A lower limit on the age of the universe, *Science.* **271**(5251), 957–961 (1996). doi: 10.1126/science.271.5251.957. URL `http://www.sciencemag.org/content/271/5251/957.abstract`.
72. R. G. Gratton, F. F. Pecci, E. Carretta, G. Clementini, C. E. Corsi, and M. Lattanzi, Ages of globular clusters from Hipparcos parallaxes of local subdwarfs, *Astrophys. J.* **491**(2), 749 (1997). URL `http://stacks.iop.org/0004-637X/491/i=2/a=749`.
73. L. M. Krauss and B. Chaboyer, Age estimates of globular clusters in the Milky Way: Constraints on cosmology, *Science.* **299**, 65–69 (2003). doi: 10.1126/science.1075631.
74. B. M. S. Hansen, J. Brewer, G. G. Fahlman, B. K. Gibson, R. Ibata, M. Limongi, R. M. Rich, H. B. Richer, M. M. Shara, and P. B. Stetson, The white dwarf cooling sequence of the globular cluster Messier 4, *Astrophys. J.*

Lett. **574**(2), L155 (2002). URL `http://stacks.iop.org/1538-4357/574/i=2/a=L155`.

75. B. M. S. Hansen, H. B. Richer, G. G. Fahlman, P. B. Stetson, J. Brewer, T. Currie, B. K. Gibson, R. Ibata, R. M. Rich, and M. M. Shara, Hubble space telescope observations of the white dwarf cooling sequence of M4, *Astrophys. J. Suppl. Ser.* **155**(2), 551–576 (2004). doi: 10.1086/424832.
76. J. J. Cowan, B. Pfeiffer, K.-L. Kratz, F.-K. Thielemann, C. Sneden, S. Burles, D. Tytler, and T. C. Beers, r-process abundances and chronometers in metal-poor stars, *Astrophys. J.* **521**(1), 194–205 (1999). doi: 10.1086/307512.
77. S. Goriely and M. Arnould, Actinides: How well do we know their stellar production?, *Astron. Astrophys.* **379**(3), 1113–1122 (2001). doi: 10.1051/0004-6361:20011368.
78. R. Cayrel, V. Hill, T. C. Beers, B. Barbuy, M. Spite, F. Spite, B. Plez, J. Andersen, P. Bonifacio, P. Francois, P. Molaro, B. Nordstrom, and F. Primas, Measurement of stellar age from uranium decay, *Nature.* **409**, 691–692 (Feb., 2001). doi: 10.1038/35055507.
79. S. Wanajo, N. Itoh, Y. Ishimaru, S. Nozawa, and T. C. Beers, The r-process in the neutrino winds of core-collapse supernovae and U-Th cosmochronology, *Astrophys. J.* **577**(2), 853 (2002). URL `http://stacks.iop.org/0004-637X/577/i=2/a=853`.
80. E. Rutherford, Origin of actinium and age of the earth, *Nature.* **123**, 313–314 (1929). doi: 10.1038/123313b0. URL `http://www.nature.com/nature/journal/v123/n3096/pdf/123313b0.pdf`.
81. E. M. D. Symbalisty and D. N. Schramm, Nucleocosmochronology, *Rep. Prog. Phys.* **44**(3), 293 (1981). URL `http://stacks.iop.org/0034-4885/44/i=3/a=002`.
82. J. J. Cowan, F.-K. Thielemann, and J. W. Truran, The r-process and nucleochronology, *Phys. Rep.* **208**(4-5), 267–394 (1991). ISSN 0370-1573. doi: http://dx.doi.org/10.1016/0370-1573(91)90070-3. URL `http://www.sciencedirect.com/science/article/pii/0370157391900703`.
83. J. J. Cowan, F.-K. Thielemann, and J. W. Truran, Radioactive dating of the elements, *Annu. Rev. Astron. Astrophys.* **29**(1), 447–497 (1991). doi: 10.1146/annurev.aa.29.090191.002311. URL `http://dx.doi.org/10.1146/annurev.aa.29.090191.002311`.
84. E. Komatsu, J. Dunkley, M. R. Nolta, C. L. Bennett, B. Gold, G. Hinshaw, N. Jarosik, D. Larson, M. Limon, L. Page, D. N. Spergel, M. Halpern, R. S. Hill, A. Kogut, S. S. Meyer, G. S. Tucker, J. L. Weiland, E. Wollack, and E. L. Wright, Five-year Wilkinson Microwave Anisotropy Probe observations: Cosmological interpretation, *Astrophys. J. Suppl. Ser.* **180**(2), 330–376 (2009). doi: 10.1088/0067-0049/180/2/330.
85. C. Sneden, J. E. Lawler, J. J. Cowan, I. I. Ivans, and E. A. D. Hartog, New rare earth element abundance distributions for the sun and five r-process-rich very metal-poor stars, *Astrophys. J. Suppl. Ser.* **182**(1), 80 (2009). URL `http://stacks.iop.org/0067-0049/182/i=1/a=80`.
86. C. Sneden, A. McWilliam, G. W. Preston, J. J. Cowan, D. L. Burris, and

B. J. Armosky, The ultra-metal-poor, neutron-capture-rich giant star CS 22892-052, *Astrophys. J.* **467**(2), 819–840 (1996). doi: 10.1086/177656.

87. J. Westin, C. Sneden, B. Gustafsson, and J. J. Cowan, The r-process-enriched low-metallicity giant HD 115444, *Astrophys. J.* **530**(2), 783–799 (2000). doi: 10.1086/308407.
88. I. I. Ivans, J. Simmerer, C. Sneden, J. E. Lawler, J. J. Cowan, R. Gallino, and S. Bisterzo, Near-ultraviolet observations of HD 221170: New insights into the nature of r-process-rich stars, *Astrophys. J.* **645**(1), 613–633 (2006). doi: 10.1086/504069.
89. J. J. Cowan, C. Sneden, S. Burles, I. I. Ivans, T. C. Beers, J. W. Truran, J. E. Lawler, F. Primas, G. M. Fuller, B. Pfeiffer, and K.-L. Kratz, The chemical composition and age of the metal-poor halo star BD $+17°3248$, *Astrophys. J.* **572**(2), 861–879 (2002). doi: 10.1086/340347.
90. A. Frebel, N. Christlieb, J. E. Norris, C. Thom, T. C. Beers, and J. Rhee, Discovery of he 1523-0901, a strongly r-process-enhanced metal-poor star with detected uranium, *Astrophys. J.* **660**(2), L117–L120 (2007). doi: 10.1086/518122.
91. H. Schatz, R. Toenjes, B. Pfeiffer, T. C. Beers, J. J. Cowan, V. Hill, and K.-L. Kratz, Thorium and uranium chronometers applied to CS 31082-001, *Astrophys. J.* **579**(2), 626–638 (2002). doi: 10.1086/342939.
92. W.-H. Zhang, Z.-M. Niu, F. Wang, X.-B. Gong, and B.-H. Sun, Uncertainties of nucleo-chronometers from nuclear physics inputs, *Acta Phys. Sin.* **61**, 112601 (2012). doi: 10.7498/aps.61.112601. URL `http://wulixb.iphy.ac.cn/CN/abstract/article_47477.shtml`.
93. H. Koura, T. Tachibana, M. Uno, and M. Yamada, Nuclidic mass formula on a spherical basis with an improved even-odd term, *Prog. Theor. Phys.* **113**(2), 305–325 (2005). doi: 10.1143/PTP.113.305.
94. S. Goriely, Nuclear inputs for astrophysics applications, *AIP Conf. Proc.* **529**, 287–294 (2000). doi: 10.1063/1.1361389.
95. N. Dauphas, The U/Th production ratio and the age of the Milky Way from meteorites and galactic halo stars, *Nature.* **435**, 1203–1205 (June, 2005). doi: 10.1038/nature03645.
96. C. Arlandini, F. Käppeler, K. Wisshak, R. Gallino, M. Lugaro, M. Busso, and O. Straniero, Neutron capture in low-mass asymptotic giant branch stars: Cross sections and abundance signatures, *Astrophys. J.* **525**(2), 886 (1999). URL `http://stacks.iop.org/0004-637X/525/i=2/a=886`.
97. J. Simmerer, C. Sneden, J. J. Cowan, J. Collier, V. M. Woolf, and J. E. Lawler, The rise of the s-process in the galaxy, *Astrophys. J.* **617**(2), 1091 (2004). URL `http://stacks.iop.org/0004-637X/617/i=2/a=1091`.
98. C. Sneden, J. J. Cowan, J. E. Lawler, I. I. Ivans, S. Burles, T. C. Beers, F. Primas, V. Hill, J. W. Truran, G. M. Fuller, B. Pfeiffer, and K.-L. Kratz, The extremely metal-poor, neutron capture-rich star CS 22892-052: A comprehensive abundance analysis, *Astrophys. J.* **591**(2), 936–953 (2003). doi: 10.1086/375491.
99. K. Lodders, Solar system abundances and condensation temperatures of the elements, *Astrophys. J.* **591**(2), 1220–1247 (2003). doi: 10.1086/375492.

100. T. H. Heng, X. D. Xu, Z. M. Niu, B. H. Sun, and J. Y. Guo, Constraint on the cosmic age from the solar r -process abundances, *J. Phys. G: Nucl. Part. Phys.* **41**(10), 105202 (2014). URL `http://stacks.iop.org/0954-3899/41/i=10/a=105202`.
101. H. Schatz, A. Aprahamian, J. Görres, M. Wiescher, T. Rauscher, J. Rembges, F.-K. Thielemann, B. Pfeiffer, P. Möller, K.-L. Kratz, H. Herndl, B. Brown, and H. Rebel, rp-process nucleosynthesis at extreme temperature and density conditions, *Phys. Rep.* **294**, 167–263 (1998).
102. J. A. Nolen and J. P. Schiffer, Coulomb energies, *Annu. Rev. Nucl. Sci.* **19**, 471–526 (1969). CDE.
103. Y. H. Zhang, H. S. Xu, Y. A. Litvinov, X. L. Tu, X. L. Yan, S. Typel, K. Blaum, M. Wang, X. H. Zhou, Y. Sun, B. A. Brown, Y. J. Yuan, J. W. Xia, J. C. Yang, G. Audi, X. C. Chen, G. B. Jia, Z. G. Hu, X. W. Ma, R. S. Mao, B. Mei, P. Shuai, Z. Y. Sun, S. T. Wang, G. Q. Xiao, X. Xu, T. Yamaguchi, Y. Yamaguchi, Y. D. Zang, H. W. Zhao, T. C. Zhao, W. Zhang, and W. L. Zhan, Mass measurements of the neutron-deficient ^{41}Ti, ^{45}Cr, ^{49}Fe, and ^{53}Ni nuclides: First test of the isobaric multiplet mass equation in fp-shell nuclei, *Phys. Rev. Lett.* **109**, 102501 (Sep, 2012). doi: 10.1103/PhysRevLett.109.102501. URL `http://link.aps.org/doi/10.1103/PhysRevLett.109.102501`.

Chapter 14

Relativistic density functional theory for finite nuclei and neutron stars

Jorge Piekarewicz

Department of Physics; Florida State University,
Tallahassee, FL 32306-4350, USA
jpiekarewicz@fsu.edu

In 1939 Oppenheimer and Volkoff demonstrated using Einstein's theory of general relativity that a neutron star supported exclusively by neutron degeneracy pressure will collapse into a black hole if its mass exceeds seven tenths of a solar mass. Seventy five years after such a pioneering prediction the existence of neutron stars with masses as large as two solar masses has been firmly established. This fact alone highlights the critical role that nuclear interactions play in explaining the structure of neutron stars. Indeed, a neutron star is a gold mine for the study of nuclear phenomena that span an enormous range of densities and neutron-proton asymmetries. Physical phenomena over such diverse scales are best described by a formalism based on Relativistic Density Functional Theory. In this contribution I focus on the synergy between theory, experiment, and observation that is needed to elucidate the myriad of exotic states of matter that are believed to exist in a neutron star.

Contents

1. Introduction

The birth of a star is marked by the conversion of hydrogen into helium nuclei (α particles) in their hot dense cores. This thermonuclear reaction is the main source of energy generation during the main stage of stellar evolution and provides the pressure support against gravitational collapse. Once the hydrogen in the stellar core is exhausted, thermonuclear fusion stops and the star contracts. As a result of the gravitational contraction, the temperature in the stellar core increases to about 100 million K allowing the heavier helium ashes to overcome their electrostatic repulsion and fuse into heavier elements. However, the absence of stable nuclei containing either five or eight nucleons hinders the production of heavy elements. Remarkably, the conditions of density and temperature in the stellar interior are such that a minute equilibrium concentration of ^{8}Be develops; the concentration of ^{8}Be relative to that of ^{4}He is about 1 parts per billion! Yet this minute concentration is sufficient for another α-particle to be captured leading to the formation of a ^{12}C nucleus. The physics of the "triple-alpha" reaction, including the prediction of the resonant Hoyle state, is one of the most fascinating chapters in the story of stellar nucleosynthesis.[1,2] Although in stars as our Sun the formation of heavier elements is hindered by the degeneracy pressure of the electrons, the conditions in the core of more massive stars are conducive to the formation of heavier elements, such as ^{16}O, ^{24}Mg, ^{28}Si, ^{32}S. However, abruptly and unavoidably, the fusion of light nuclei into ever increasing heavier elements terminates with the synthesis of the iron-group elements (Fe, Co, and Ni) that are characterized by having the largest binding energy per nucleon. That is, once iron-group elements are produced in the stellar interior, it is no longer feasible to generate energy by thermonuclear fusion. Indeed, if the iron core exceeds the "Chandrasekhar limit" of about 1.4 solar masses, neither thermonuclear fusion nor electron degeneracy pressure can prevent the rapid collapse of the stellar core. The collapse of the core, with the ensuing shock wave that disseminates the chemical elements crafted during the lifetime of the star, produces one of the most remarkable events in the Universe: a *Supernova Explosion* (see Fig. 1). Besides creating an ejecta that contains some of the essential elements necessary for life, core-collapse supernovae leave behind exotic compact remnants in the form of either black holes or *neutron stars*. Neutron stars are the central theme of the present contribution.

Indirectly and inadvertently, Subramanyan Chandrasekhar ("Chandra") may have been the discoverer of neutron stars. In a pioneering paper

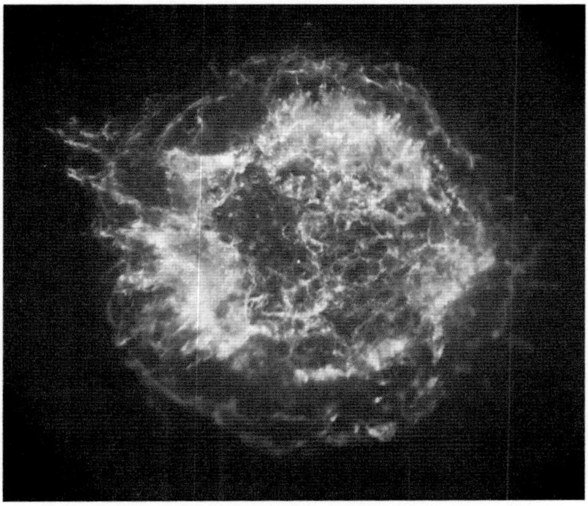

Fig. 1. (Color online) Cassiopeia A is the remnant of a supernovae explosion that was witnessed in the northern sky about 300 years ago. The supernova remanent is located about 10,000 light years away from Earth and the small "dot" near the center of the image represents the dense neutron star. The image was created using NASA's Chandra x-ray observatory, an observatory named in honor of Subramanyan Chandrasekhar.

published in 1931, Chandra re-examined the role of electron degeneracy pressure in supporting a white-dwarf star against gravitational collapse, a fact that was already well known at the time. Chandra realized, however, that as the electrons become relativistic, the pressure support weakens and a white-dwarf star with a mass in excess of about 1.4 solar masses (the so-called "Chandrasekhar mass limit") will collapse under its own weight.[3] Chandra summarized eloquently this critical finding: *For a star of small mass the white-dwarf stage is an initial step towards complete extinction. A star of large mass cannot pass into the white-dwarf stage and one is left speculating on other possibilities.* One may ask why Chandra never speculated that neutron stars may be among the "other possibilities". As luck will have it, the neutron was not yet discovered in 1931; it would take Chadwick another year to announce the discovery.[4] However, soon after Chadwick's announcement, the term *neutron star* appears in writing for the first time in the 1933 proceedings of the the American Physical Society by Baade and Zwicky who wrote: *With all reserve we advance the view that supernovae represent the transition from ordinary stars into "neutron stars", which in their final stages consist of extremely closed packed neutrons.*[5]

It appears, however, that speculations on the possible existence of both the neutron and neutron stars may have an earlier origin. In the case of the neutron, the story starts with the "boys of Via Panisperna" who included such luminaries as Enrico Fermi, Franco Rasetti, and Ettore Majorana, among others. During a brief stay at Caltech (in the 1928–29 period) working in Millikan's laboratory, Rasetti measured the ground-state angular momentum of ^{14}N to be $J=1$. At the time, the widespread belief was that the nucleus of ^{14}N must contain 14 protons (to account for its mass) and 7 electrons (to account for its charge). However, Majorana was the first one to realize that 21 spin-1/2 fermions cannot account for the measured spin of ^{14}N, thereby postulating the existence of an electrically neutral spin-1/2 particle having the same mass as the proton.[6] Moreover, the requirement for such a particle fitted correctly Majorana's interpretation of some experiments carried out in 1932 by Irène Joliot-Curie and Frédéric Joliot. Apparently Fermi pleaded with Majorana to write an article on the neutron, but Majorana did not find it worthy.

Fig. 2. Subrahmanyan Chandrasekhar (left) and Lev Landau (right) who wrote seminal papers on the theory of stars at the "ripe" age of 19 and 23, respectively.

Insofar as Landau's role on the history of neutron stars is concerned, the first recollection appears on a paper entitled "On the theory of stars" that Landau submitted for publication in early 1932 at the age of 23.[7]

In that paper Landau calculates—independently of Chandrasekhar—the maximum mass of a white dwarf star. Moreover, unlike Chandrasekhar, Landau did speculate on "other possibilities", namely, the existence of dense stars that look like giant atomic nuclei. For an in-depth and fascinating tale on Landau's role on the possible existence of neutron stars see Ref. 8. Note that in Fig. 2 we display pictures of both Chandra and Landau as very young men.

Perhaps the last great theoretical landmark of that time involves the 1939 work by Oppenheimer and Volkoff on the structure of neutron stars.[9] By then, Einstein's general theory of relativity was firmly established as was Tolman's framework to compute solutions appropriate to spherical systems in hydrostatic equilibrium.[10] In what it is now referred to as the Tolman–Volkoff–Oppenheimer (TOV) equations—effectively the generalization of Newtonian gravity to the domain of general relativity—Oppenheimer and Volkoff concluded that a neutron star supported exclusively by the pressure from its degenerate neutrons will collapse into (what we now know as) a black hole for masses in excess of about 0.7 solar masses. This critical finding, together with our present knowledge of neutron-star masses, has made nuclear physics and astrophysics intimately intertwined.

Although firmly established theoretically, it would take almost three decades for the discovery of neutron stars. This momentous discovery started with a young graduate student by the name of Jocelyn Bell (see Fig. 3)—now *Dame* Jocelyn Bell Burnell—who detected a "bit of scruff" in the data arriving into her radio telescope, a telescope originally designed to study distant quasars. The arriving signal was "pulsing" with such an enormous regularity, indeed once every 1.337 302 088 331 seconds, that both Bell and her research advisor, Anthony Hewish, were so mesmerized by the observation that they were convinced that the signal was a beacon from an extraterrestrial civilization. Initially dubbed as "Little Green Man 1" the source, now known as radio pulsar "PSR B1919+21", was shortly identified as a rapidly rotating neutron star.[11] Although Hewish was recognized with the Physics Nobel Prize in 1974 for "his decisive role in the discovery of pulsars", Jocelyn Bell did not share the award. The exclusion of Jocelyn Bell as co-recipient of the Nobel Prize was both controversial and roundly condemned by the astrophysics community. Still, Bell has always displayed enormous grace and humility in the face of this controversy. Indeed, she has stated: *I believe it would demean Nobel Prizes if they were awarded to research students, except in very exceptional cases, and I do not believe this is one of them.* It appears that Dr. Iosif Shklovsky, as well as many

Fig. 3. Jocelyn Bell as a young graduate student with her radio telescope designed to search for quasars. Instead, and according to Dr. Iosif Shklovsky, she made the greatest astronomical discovery of the twentieth century.

others, did not share her views. Dr. Shklovsky—recipient of the 1972 Bruce Medal for outstanding lifetime contributions to astronomy—told Jocelyn Bell: *Miss Bell, you have made the greatest astronomical discovery of the twentieth century.*

The finding by Oppenheimer and Volkoff of a maximum neutron star mass of about 0.7 solar masses[9] will eventually place nuclear physics at the forefront of neutron-star structure. This is because the masses of various neutron stars have been determined very accurately and they exceed, often by a significant margin, the $0.7\,M_{\odot}$ limit (see Ref. 12 and references contained therein). Indeed, the present "record" stands at $2\,M_{\odot}$.[13,14] Given that the Oppenheimer–Volkoff limit was obtained by assuming that the support against gravitational collapse is provided by a degenerate gas

Table 1. Approximate characteristics of a "canonical" neutron star, such as the 960 year old Crab pulsar.

Name: PSR B0531+21	Constellation: Taurus
Distance: 2.2 kpc	Age: 960 years
Mass: $1.4\,M_{\odot}$	Radius: 10 km
Density: 10^{15}g/cm^3	Pressure: 10^{29} atm
Surface Temperature: 10^6 K	Escape velocity: 0.6 c
Period: 33 ms	Magnetic Field: 10^{12} G

of neutrons, the large deficit must then be supplied by nuclear interactions. Thus, neutron stars are enormously valuable in constraining the largely undetermined equation of state of neutron-rich matter at high densities. However, the reach of neutron stars is not limited to the nuclear physics domain. Indeed, neutron stars are unique laboratories for the study of matter under extreme conditions of density and isospin asymmetry. In particular, their extreme compactness has been used to test the basic tenets of general relativity.[15] Moreover, by spanning many orders of magnitude in density, neutron stars display exotic phases that cannot be realized under normal laboratory conditions. Yet, some of these phases have direct counterparts in both atomic physics and condensed-matter physics. Finally, their very dense stellar cores may harbor novel states of matter, such as color superconductors, that are a direct prediction of Quantum Chromodynamics.[16] To appreciate some of the unique properties of neutron stars, such as their density, pressure, spin period, and magnetic fields, we have listed some of these properties in Table 1 for the case of the well-known Crab pulsar.

We have organized this chapter on neutron stars as follows. After this Introduction, we present the formalism that will be used to compute some structural properties of neutron stars, such as their mass-vs-radius relation. Once the formalism is in place, we embark on a journey into a neutron star that involves a detailed discussion of the outer crust, the inner crust, and the outer core. For all these three cases we present results that highlight those stellar observables that are particularly sensitive to the choice of the nuclear density functional. We finish with a summary of our results and an outlook for the future.

2. Formalism

The ultimate goal of a properly constructed nuclear energy density functional is to provide a unified description of a wide variety of physical phenomena ranging from the properties of finite nuclei to the structure and dynamics of neutron stars. This ambitious goal involves physical systems that differ in mass and size by about 55 and 18 orders of magnitude, respectively. Moreover, given that extrapolations into regions that are inaccessible in laboratory experiments are unavoidable, the predictions of such microscopic theory should always be accompanied by well-quantified theoretical uncertainties.[17,18]

Historically, relativistic models of nuclear structure were limited to renormalizable field theories.[19,20] The appeal of renormalizability was evident: with only a handful of model parameters calibrated to well-known physical observables one could then extrapolate to unknown physical regions without the need for introducing additional parameters. However, the modern viewpoint suggests that any relativistic model, although often inspired in Quantum Chromodynamics (QCD), should be treated as an *effective field theory* (EFT) where the demand for renormalizability is no longer required. An effective field theory is designed to describe low-energy physics without any attempt to account for its detailed short-distance behavior.[21] Although in principle the empirical parameters of the EFT may be calculable from QCD, in practice this becomes enormously challenging in the non-perturbative regime of relevance to nuclear systems. Hence, the parameters of the model are directly calibrated (i.e., fitted) to physical observables. By doing so, the short-distance structure of the theory as well as many other complicated many-body effects get implicitly encoded in the parameters of the model. In this regard, density functional theory (DFT), a powerful and highly successful theoretical framework pioneered by Kohn and collaborators,[22–24] provides a unified approach for the construction of an EFT that may be used to compute phenomena ranging over many distance scales.

2.1. *Relativistic Density Functional Theory*

In the framework of the relativistic density functional theory, the effective degrees of freedom include nucleons (protons and neutrons), three "mesons", and the photon. The interactions among the particles can be described as a generalization of the original Lagrangian density of Serot and Walecka.[19,25–28] That is,

$$\mathscr{L}_{\rm int} = \bar{\psi}\left[g_{\rm s}\phi - \left(g_{\rm v}V_\mu + \frac{g_\rho}{2}\boldsymbol{\tau}\cdot\mathbf{b}_\mu + \frac{e}{2}(1+\tau_3)A_\mu\right)\gamma^\mu\right]\psi$$
$$- \frac{\kappa}{3!}(g_{\rm s}\phi)^3 - \frac{\lambda}{4!}(g_{\rm s}\phi)^4 + \frac{\zeta}{4!}g_{\rm v}^4(V_\mu V^\mu)^2$$
$$+ \Lambda_{\rm v}\left(g_\rho^2\,\mathbf{b}_\mu\cdot\mathbf{b}^\mu\right)\left(g_{\rm v}^2 V_\nu V^\nu\right), \qquad (1)$$

where ψ is the isodoublet nucleon field, A_μ is the photon field, and ϕ, V_μ, and $\mathbf{b}_\mu$ represent the isoscalar-scalar σ-, isoscalar-vector ω-, and isovector-vector ρ-meson field, respectively. We note that the pion is not explicitly included in the Lagrangian density as it does not contribute to the nuclear dynamics at the mean field level. The Lagrangian density incorporates the conventional Yukawa terms between the nucleon and the various mesons and the photon. However, in order to improve the quality of the model it is critical to supplement the dynamics with nonlinear interaction terms between the various mesons. In the spirit of an effective field theory, one should incorporate all possible meson interactions that are allowed by symmetry considerations to a given order in a power-counting scheme. Moreover, once the *dimensionful* meson fields have been properly scaled using strong-interaction mass scales, the remaining dimensionless coefficients of the effective Lagrangian should all be "natural", namely, neither too small nor too large.[29–33] However, given the limited experimental database of nuclear observables, certain empirical coefficients—or linear combinations of them—may remain poorly determined even after the optimization procedure. This results in "unnatural" coefficients that deviate significantly from unity. Therefore, in an effort to avoid this problem, only those non-linear meson interactions with a clear physical interpretation are retained. For instance, for the Lagrangian density depicted in Eq. (1), we have only kept the four non-linear meson interactions that are denoted by the coefficients: κ, λ, ζ, and $\Lambda_{\rm v}$. Two of the isoscalar parameters, κ and λ, were introduced by Boguta and Bodmer[34] to soften the equation of state of symmetric nuclear matter, primarily the incompressibility coefficient,[19,25] in an effort to make the theory consistent with measurements of giant monopole resonances in finite nuclei. In turn, ζ may be used to efficiently tune the maximum neutron star mass without sacrificing the agreement with other well reproduced observables.[26] Finally, $\Lambda_{\rm v}$ is highly sensitive to the density dependence of symmetry energy—and in particular to its slope at saturation density—which has important implications in the structure and dynamics of neutron stars.[28,35–37]

With the Lagrangian density given in Eq. (1), one can derive the equations of motion for each of the constituent particles in the mean-field approximation.[38] In particular, the nucleons satisfy a Dirac equation in the presence of mean-field potentials of Lorentz scalar and vector character. In turn, the various meson fields satisfy both nonlinear and inhomogeneous Klein-Gordon equations with the various nuclear densities acting as source terms. Given that the nuclear densities act as sources for the meson fields and, in turn, the meson fields determine the mean-field potentials for the nucleons, the set of equations must be solved self-consistently. Once solved, these equations determine the ground-state properties of the nucleus of interest—such as its total binding energy, the single-nucleon energies and Dirac orbitals, the distribution of meson fields, and the various density profiles.

2.2. *Nuclear Matter Equation of State*

The solution of the mean-field equations is simplified considerably in the case of infinite nuclear matter, which is assumed to be spatially uniform. Although the solution may be found at finite temperature, our main goal is to solve the mean-field equations at zero temperature because of their relevance to the structure and dynamics of neutron stars. Indeed, if the validity of Einstein's theory of General Relativity is assumed, then the equation of state of asymmetric nuclear matter represents the sole ingredient required to compute the properties of neutron stars (see next section).

In the simplified case of infinite nuclear matter, the meson fields are uniform (i.e., constant throughout space) and the nucleon orbitals are plane-wave Dirac spinors with medium-modified effective masses and energies that must be determined self-consistently. By constructing the energy-momentum tensor in the mean-field approximation,[25] one obtains the equation of state of asymmetric nuclear matter, namely, the energy density and pressure of the system as a function of both the conserved baryon density $\rho = \rho_n + \rho_p$ and the neutron-proton asymmetry $\alpha \equiv (\rho_n - \rho_p)/(\rho_n + \rho_p)$. A particularly insightful view of the EOS is obtained by expanding the energy per nucleon in powers of the neutron-proton asymmetry. That is,

$$\frac{E}{A}(\rho, \alpha) - M \equiv \mathcal{E}(\rho, \alpha) = \mathcal{E}_{\rm SNM}(\rho) + \alpha^2 \mathcal{S}(\rho) + \mathcal{O}(\alpha^4) \,. \tag{2}$$

Here $\mathcal{E}_{\rm SNM}(\rho) = \mathcal{E}(\rho, \alpha \equiv 0)$ is the energy per nucleon of symmetric nuclear matter (SNM) and the symmetry energy $\mathcal{S}(\rho)$ represents the first-order correction to the symmetric limit. Note that no odd powers of α appear as the nuclear force is assumed to be isospin symmetric and long-range electromagnetic effects have been "turned off". Also note that to a very good approximation the symmetry energy represents the energy cost required to convert symmetric nuclear matter into pure neutron matter (PNM). That is,

$$\mathcal{S}(\rho) \approx \mathcal{E}(\rho, \alpha = 1) - \mathcal{E}(\rho, \alpha = 0)\,. \tag{3}$$

Such a separation is useful because symmetric nuclear matter is sensitive to the isoscalar sector of the density functional which is well constrained by the properties of stable nuclei. In contrast, the symmetry energy probes the isovector sector of the density functional which at present is poorly constrained. However, this problem will be mitigated with the commissioning of radioactive beam facilities throughout the world.

Besides the separation of the EOS into symmetric and asymmetric components, it is also useful to characterize the behavior of the equation of state in terms of a few bulk parameters. To do so one performs a Taylor series expansion around nuclear matter saturation density ρ_0. That is,[39]

$$\mathcal{E}_{\rm SNM}(\rho) = \varepsilon_0 + \frac{1}{2}K_0 x^2 + \ldots, \tag{4a}$$

$$\mathcal{S}(\rho) = J + Lx + \frac{1}{2}K_{\rm sym}x^2 + \ldots, \tag{4b}$$

where $x = (\rho - \rho_0)/3\rho_0$ is a dimensionless parameter that quantifies the deviations of the density from its value at saturation. Here ε_0 and K_0 represent the energy per nucleon and the incompressibility coefficient of SNM; J and $K_{\rm sym}$ are the corresponding quantities for the symmetry energy. However, unlike symmetric nuclear matter whose pressure vanishes at ρ_0, the slope of the symmetry energy L does not vanish at saturation density. Indeed, assuming the validity of Eq. (3), L is directly proportional to the pressure of PNM (P_0) at saturation density, namely,

$$P_0 \approx \frac{1}{3}\rho_0 L\,. \tag{5}$$

In computing various neutron-star observables in the next few sections, we will rely on several nuclear density functionals that while successful in reproducing a myriad of laboratory observables, predict significant differences in the properties of neutron stars.

2.3. *Tolman–Oppenheimer–Volkoff Equations*

With masses comparable to that of our Sun but with radii that are almost five orders of magnitude smaller (i.e., of the order of 10 km) neutron stars are highly compact objects that must be described using Einstein's theory of General Relativity. The generalization of Newtonian gravity to the realm of general relativity is expressed in the *Tolman–Oppenheimer–Volkoff* (TOV) equations, which are usually presented as a coupled set of first-order differential equations of the following form:

$$\frac{dP}{dr} = -G\,\frac{\mathcal{E}(r)M(r)}{r^2}\left[1+\frac{P(r)}{\mathcal{E}(r)}\right]\left[1+\frac{4\pi r^3 P(r)}{M(r)}\right]\left[1-\frac{2GM(r)}{r}\right]^{-1}, \tag{6}$$

$$\frac{dM}{dr} = 4\pi r^2 \mathcal{E}(r)\,, \tag{7}$$

where G is Newton's gravitational constant and $P(r)$, $\mathcal{E}(r)$, and $M(r)$ represent the pressure, energy density, and enclosed-mass profiles of the star, respectively. Note that the three terms enclosed in square brackets in Eq. (6) are of general-relativistic origin. As already alluded earlier and particularly interesting, the only input that neutron stars are sensitive to is the equation of state of neutron-rich matter. This fact alone creates a unique synergy between nuclear physics and astrophysics. In essence, by specifying the central pressure and enclosed mass, i.e., $P_c = P(r=0)$ and $M(r=0) = 0$—together with a suitable EOS—the TOV equations may be solved using a standard numerical algorithm, such as the Runge–Kutta method.

3. Anatomy of a Neutron Star

In the next few sections we embark on a journey through a neutron star. According to Baade and Zwicky, the most common perception of a neutron star is that of a uniform assembly of extremely closely packed neutrons.[5] We will now show, however, how the reality is very different and much more interesting. In particular, as we journey through the neutron star we will discover a myriad of exotic states of matter and will discuss the critical role that laboratory experiments can play in elucidating their fascinating nature. Because of their enormous relevance to nuclear physics, we focus our attention on three components of the neutron star: (a) the outer crust, (b) the inner crust, and (c) the outer core. For two recent reviews on the exotic nature of the stellar crust see Refs. 40, 41 and references therein. The three regions are clearly highlighted in Fig. 4, which includes two physically accurate renditions of a neutron star. In particular, the question mark at the center of the left-hand illustration denotes the possibility that the inner stellar core harbors exotic states of matter, such as hyperons, meson

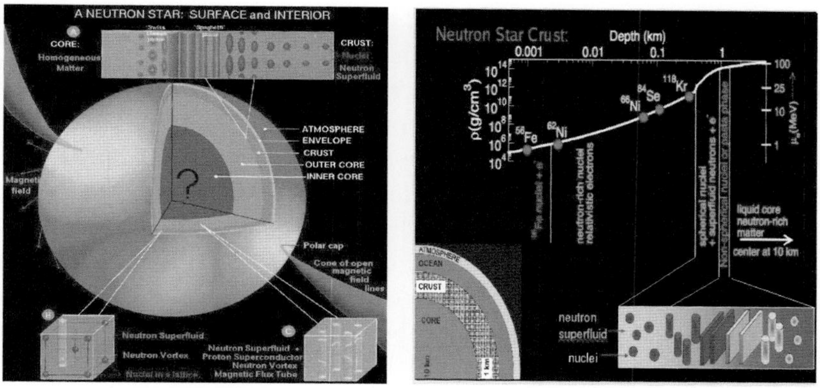

Fig. 4. (Color online) Two physically accurate renditions of the structure, composition, and possible phases of a neutron star. Courtesy of Dany Page (left panel) and Sanjay Reddy (right panel).

condensates, and strange quark matter. Although this idea is enormously provocative, at present there is simply not enough experimental information to properly constrain the dynamics of the inner core. Thus, the possible existence of such exotic states of matter will be ignored hereafter. Although the stellar atmosphere and the envelope will also be ignored, we briefly discuss now some of its most relevant features.

Because of the enormous gravitational fields around a neutron star, the atmosphere is believed to be about a mere 10 cm thick. However, the atmosphere shapes the thermal radiation from the photosphere which is customarily assumed to be that of a black body. Hence, detailed knowledge of the atmosphere is critical for the reliable extraction of, for example, stellar radii. In turn, the 100 m envelope acts as a blanket that modulates the huge temperature gradient between the core and the crust. For more information about the role of the stellar atmosphere and envelope see Ref. 42 and references contained therein.

3.1. *The Outer Crust*

The outer crust of a neutron star comprises a region that spans almost 7 orders of magnitude in baryon density; from about $10^4 \mathrm{g/cm^3}$ to $4 \times 10^{11} \mathrm{g/cm^3}$.[43] Recall that nuclear-matter saturation density is $\rho_0 \simeq 2.4 \times 10^{14} \mathrm{g/cm^3}$, which corresponds to a baryon density of $n_0 \simeq 0.15\,\mathrm{fm}^{-3}$. Thus, these densities are significantly lower that those encountered in the interior of the atomic nucleus. However, at these densities the electrons—

which represent a critical component of the star in order to maintain the overall charge neutrality of the system—have been pressure ionized. Thus, they can be accurately described by a relativistic free Fermi gas. Moreover, given that at these densities the average inter-nucleon distance is significantly larger than the range of the strong nuclear force, the uniform ground state becomes unstable against cluster formation. That is, under these conditions of density it is energetically favorably for translational invariance to be broken and for the individual nucleons to cluster into "normal" nuclei. Thus, the outer stellar crust consists of isolated nuclei embedded in a uniform electron gas. Moreover, because the short-range nuclear force saturates within the individual clusters, nuclei interact exclusively via the long-range Coulomb interaction. This promotes the formation of a Coulomb crystal of nuclei arranged in a body-centered-cubic (bcc) lattice that itself is embedded in a neutralized uniform electron gas.[43] In the particular case of the top layers of the crust where the density is at its lowest, the energetically preferred nucleus is ^{56}Fe; see the right-hand panel of Fig. 4. Recall that ^{56}Fe is the nucleus with the lowest mass per nucleon.

However, as one moves inward towards the center of the star, the density increases, and so does the electronic contribution to the total energy. Thus, it becomes energetically advantegeous to remove a fraction of the electrons (through electron capture) albeit at the expense of an increase in the neutron-proton asymmetry. In this manner, the energetically most favorable nucleus emerges from a competition between the electronic contribution, which favors a small electron (and proton) fraction, and the nuclear symmetry energy which, in turn, favors symmetric nuclei. Hence, the nuclear contribution to the composition of the outer crust appears in the form of a nuclear mass table that is generated from a combination of experimental data and theoretical predictions. Indeed, in full thermodynamic equilibrium, one determines the crustal composition by minimizing the chemical potential of the system (μ) at zero temperature and fixed pressure. That is,[44]

$$\mu(A,Z;P) = \frac{M(N,Z)}{A} + \frac{Z}{A}\mu_e - \frac{4}{3}C_\ell \frac{Z^2}{A^{4/3}} p_{\rm F}\,. \tag{8}$$

The total chemical potential consists of nuclear, electronic, and lattice contributions. As already mentioned, computing the nuclear contribution requires "only" a reliable nuclear mass table. Moreover, given that at the relevant densities the electrons can be accurately modeled by a Fermi-gas distribution, its contribution to the chemical potential is simply given by

$$\mu_e = \sqrt{(p_{\rm F}^e)^2 + m_e^2} = \sqrt{(yp_{\rm F})^2 + m_e^2}\,, \tag{9}$$

where $y = Z/A$ is the electron fraction and the Fermi momentum $p_{\rm F}$ is related to the baryon density n by

$$p_{\rm F} = \left(3\pi^2 n\right)^{1/3}. \tag{10}$$

Finally, the last term in Eq. (8) represents the complicated lattice contribution. This contribution is complex because it involves the long-range nature of the Coulomb interaction. Nevertheless, the overall charge neutrality of the system ensures that the contribution is finite.[45,46] In the particular case of the energetically preferred bcc lattice, one obtains[43]

$$\varepsilon_\ell(A,Z;n) = -C_\ell \frac{Z^2}{A^{4/3}} p_{\rm F}\,, \tag{11}$$

where $C_\ell = 3.40665 \times 10^{-3}$ is a dimensionless constant.[44] Note that although the expression for the chemical potential is written in terms of the density rather than the pressure, chemical equilibrium demands that the minimization of $\mu(A,Z;P)$ be carried out at a fixed pressure rather than at a fixed density. Thus, one needs an equation of state to properly relate them. However, given that for the outer crust the pressure is dominated by the degenerate electrons (with only a small lattice contribution) to a very good approximation the relevant equation of state is that of a relativistic Fermi gas of electrons.

To assess the sensitivity of the crustal composition to the mass model we display in Fig. 5 predictions using two accurately calibrated relativistic EDFs, i.e., FSUGold[47] and NL3[48,49] as well as two highly successful microscopic-macroscopic (mic-mac) models; one due to Möller and collaborators[50,51] and the other one due to Duflo and Zuker.[52–54] As already anticipated, the top (i.e., most dilute) layers of the outer crust consist of a crystalline lattice of ^{56}Fe nuclei embedded in a uniform electron gas; for a more aesthetic view see the right-hand panel of Fig. 4. However, as the density increases—and with it the electron chemical potential—it becomes energetically favorable to replace ^{56}Fe with a slightly more neutron-rich nucleus. That is, at slightly higher densities ^{56}Fe ceases to be the most stable nucleus. Rather, the slightly more isospin asymmetric (even-even) Nickel isotopes become energetically favored. As the density continues to increase further, the system must decide whether to reduce Z at neutron number $N \approx 34$ or to increase both N and Z by jumping to the next magic

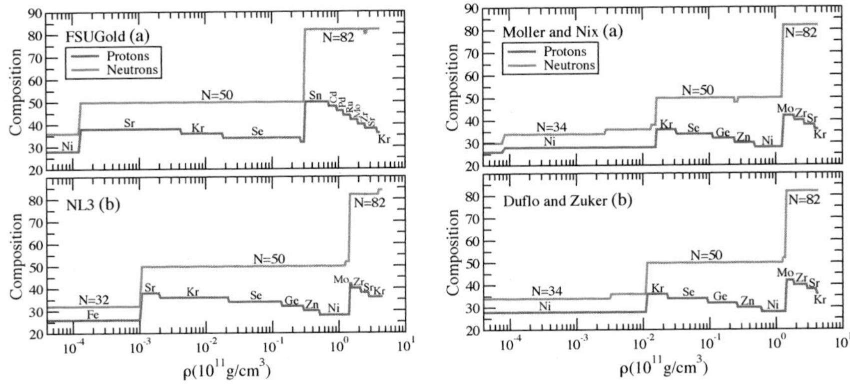

Fig. 5. (Color online) Composition of the outer crust as predicted by two accurately-calibrated relativistic mean field models (left) and by two microscopic-macroscopic models (right). Protons are depicted with the blue (lower) line whereas neutrons with the green (upper) line.

shell at $N=50$. Although it is seen that all four models favor an eventual jump to the $N=50$ shell, the predictions for the density at which the jump occurs is highly model dependent. Indeed, while FSUGold predicts the transition at a density of about 10^7g/cm^3, both mic-mac models suggest a density that is almost two orders of magnitude larger. Our results also indicate that in all cases the proton fraction decreases systematically with increasing density in an effort to reduce the electronic contribution to the chemical potential. Yet, at one point reducing the electron fraction even further becomes too expensive for the symmetry energy to sustain and the system jumps to the next plateau at magic number $N=82$. We want to underscore that both the transition density as well as the crustal composition depend sensitively on the symmetry energy at sub-saturation density.

In the particular case of the two microscopic models (FSUGold and NL3) the symmetry energy may be computed at all densities, so a study of the trends displayed in Fig. 5 are illuminating. At a density of relevance to finite nuclei, which consists of an average between the nuclear interior and the nuclear surface, the symmetry energy is known to be larger for FSUGold than for NL3. This implies that a neutron-proton mismatch at those densities is more costly for FSUGold than for NL3.[38] Thus, the tolerance to a larger neutron-proton asymmetry is responsible for delaying the transition to the next higher plateau for NL3 relative to FSUGold, a fact that is clearly evident in the figure. By the same token, NL3 predicts a more exotic crustal composition than FSUGold. Indeed, whereas FSUGold suggests the formation of ^{132}Sn at the transition to the $N=82$ shell, NL3 predicts the formation of the significantly more neutron-rich isotope ^{124}Mo. Note that as the density continues to increase even further, the neutron-proton asymmetry will become so large that the neutron drip line will be reached. For all models, the drip line is predicted to occur at a density of about 4×10^{11}g/cm^3 and with the formation of the highly exotic ^{118}Kr isotope.

We close this section with a brief comment on the assumptions and extrapolations required to predict the composition of the outer crust. As just described, there are three main regions in the nuclear chart that are of direct relevance to the outer crust: (a) The Fe-Ni region, (b) the $N=50$ isotones from Ni ($Z=28$) to Sr ($Z=38$), and (c) the $N=82$-isotone region from Kr ($Z = 36$) to Sn ($Z = 50$). In regards to (a), all the masses in this region have been measured with great precision.[55] In the case of (b), precise mass values exist, but only for the cases of ^{88}Sr, ^{86}Kr, ^{84}Se, ^{82}Ge, and ^{80}Zn, and not for either ^{78}Ni or ^{82}Zn.[55,56] However, in a pioneering Penning trap experiment with the ISOLTRAP setup at the ISOLDE-CERN facility, the mass of ^{82}Zn has been recently determined.[57] This new mass determination has ruled out the presence of ^{82}Zn in the outer crust and provides the most stringent constraint to date on its composition profile. Finally, the $N=82$-isotone region remains largely unexplored and it is likely to remain so even after the construction of a new generation of rare isotope facilities. Thus, the only hope to elucidate the composition of the bottom layers of the outer crust is through theoretical modeling. In this regard, measuring a large number of as yet unknown masses of exotic nuclei—even if of no direct relevance to the composition of the outer crust—will still be instrumental in guiding the calibration of future nuclear density functionals.

3.2. *The Inner Crust*

The inner crust of a neutron star comprises the region from neutron-drip density up to the density at which uniformity in the system is restored; about one third to one half of normal nuclear density. However, the transition from the highly ordered Coulomb crystal to the uniform liquid is both interesting and complex. This is because distance scales that are well separated in both the crystalline phase, where the long-range Coulomb interaction dominates, and in the uniform phase, where the short-range strong interaction dominates, become comparable in the inner stellar crust. This unique situation involving competing distance scales gives rise to *Coulomb frustration*. Frustration, a universal phenomenon characterized by the existence of a very large number of low-energy configurations, emerges from the impossibility to simultaneously minimize all elementary interactions in the system. Ultimately, the competition between the short-range nuclear interaction and the long-range Coulomb repulsion results in the formation of complex topological structures collectively referred to as *nuclear pasta*. Given that these complex structures are very close in energy, it has been speculated that the transition from the highly ordered crystal to the uniform phase must proceed through a series of changes in the dimensionality and topology of these structures.[58,59] Moreover, due to the preponderance of low-energy states, frustrated systems display an interesting and unique low-energy dynamics. We note that a seemingly unrelated condensed-matter problem, namely, the strongly-correlated electron gas, also displays Coulomb frustration. In the case of the electron gas, one aims to characterize the transition from the low-density Wigner crystal, where the long-range Coulomb potential dominates, to the uniform Fermi liquid, where the kinetic energy dominates.[60] It has been shown that such a transition must be mediated by the emergence of "microemulsions", namely, exotic pasta-like structures with interesting topologies. Indeed, it has been proven that in two-spatial dimensions a direct first-order phase transition is forbidden in the presence of long-range (e.g., Coulomb) forces.[61]

To illustrate the complexity of the pasta phase we display in Fig. 6 two snapshots obtained from Monte-Carlo and Molecular-Dynamics simulations of a nuclear system at densities of relevance to the inner stellar crust.[37,62] The left-hand panel displays how at a density of about one sixth of normal nuclear density and a proton fraction of $Z/A = 0.2$, the system organizes itself into neutron-rich clusters (i.e., "nuclei") of complex topologies that are surrounded by a dilute vapor of likely superfluid neutrons. In turn, the right-hand panel displays the complex topology of a proton iso-surface (i.e., a surface of constant proton density) for a system of 100,000 nucleons. Such complex pasta structures may have a significant impact on various transport properties, such as neutrino propagation and electron conductivity. We should underscore that the emergence of such complex structures is a true dynamical effect associated with Coulomb frustration, as no *a-priori* shapes (such as spheres, rods, slabs, etc.) are ever assumed.

A great advantage of molecular-dynamics (MD) simulations is that many-body correlations are properly accounted for within the formalism.[37,62–66] However, given their classical nature, MD simulations fail

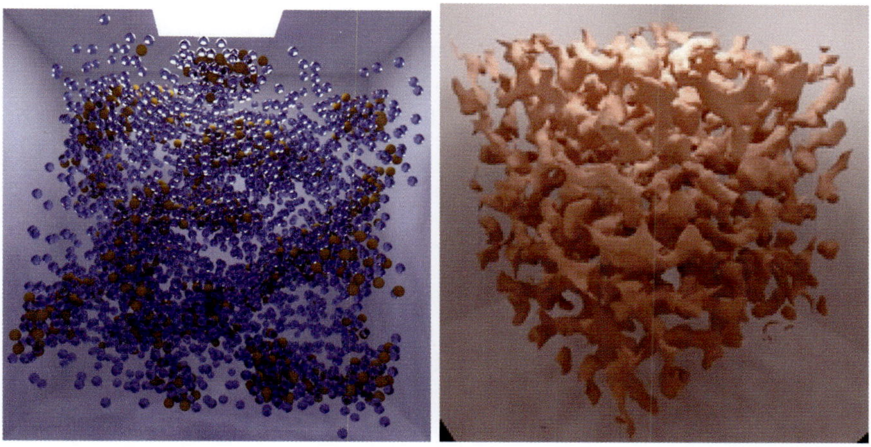

Fig. 6. (Left panel) A snapshot of a Monte Carlo simulation for one configuration of 4,000 nucleons at a baryon density of $0.025\,\mathrm{fm}^{-3}$, a proton fraction of $Z/A = 0.2$, and a temperature of 1 MeV. (Right panel) The $0.03\,\mathrm{fm}^{-3}$ proton density iso-surface for one configuration of 100,000 nucleons at a density of $0.05\,\mathrm{fm}^{-3}$, a proton fraction of $Z/A = 0.2$, and a temperature of 1 MeV. Note that the simulation volume in this latter case is fairly large, i.e., $V = (126\,\mathrm{fm})^3$.

to capture any quantum-mechanical detail that the system may be sensitive to; for example, the superfluid nature of the dilute neutron vapor. In contrast, mean-field approximations incorporate quantum-mechanical effects (at least on average) but fail to describe the important and complicated clustering correlations.[67–71] Nevertheless, because the robustness of Coulomb frustration, both sets of theoretical approaches seem to reach similar conclusions. For example, for relatively large proton fractions in the 0.3–0.5 range (such as in core-collapse supernovae) there appears to be general agreement that the transition from the ordered Coulomb crystal to the uniform phase must proceed via intermediate pasta phases. What is unclear, however, is whether such exotic pasta shapes can develop in the proton-poor environment characteristic of the inner stellar crust.[72] Note that mean-field models that impose β-equilibrium predict proton fractions at densities of relevance to the inner crust of only a few percent.[67–69,71]

Another enormous challenge associated with the purported pasta phases is the identification of a set of astrophysical observables that are sensitive to their formation. Colloquially, we refer to this challenge as "how to smell the pasta?" A possible manifestation of the nuclear pasta on the dynamics of neutron stars has been suggested recently by Pons, Viganò, and Rea.[73]

These authors have identified a special class of rotation-powered pulsars—*the isolated x-ray pulsars*—that appear to display spin periods shorter than about 12 seconds. Although enormously stable, rotation-powered pulsars are known to slow down, albeit very slowly, due to the emission of magnetic dipole radiation. The lack of isolated x-ray pulsars with spin periods longer than 12 seconds seems to suggest magnetic field decay due to the existence of a highly resistive layer in the inner crust; such a layer has been speculated to be the exotic nuclear pasta phase.[73] Very recently, Horowitz and collaborators have carried out large molecular dynamics simulations to explore a possible magnetic field decay in the inner crust due to the existence of a nuclear pasta phase.[74] In particular, it was concluded that the formation of *topological defects* in the nuclear pasta could reduce both its electrical and thermal conductivity. Hence, the formation of this "highly resistive layer" could promote magnetic field decay and may ultimately explain the lack of x-ray pulsars with long spin periods.[73]

3.3. *The Outer Core*

Structurally, the stellar core is by far the most critical component of the star. Because of the enormous interior densities, practically all the mass and most of the size reside in the stellar core. At densities of about 10^{14}g/cm^3, the pasta phase "melts" and uniformity in the system is restored. It is in the stellar core where the original perception of Baade and Zwicky[5] is finally realized, namely, a neutron star as a uniform assembly of extremely closely packed neutrons. However, in order to maintain both chemical equilibrium and charge neutrality, a small fraction (of about 10%) of protons and leptons is also required. Remarkably, given that the densities in the stellar core are so large, the leptonic component consists of both electrons and *muons*; indeed, a neutron star typically contains about 10^{56} muons! Although exotic, the presence of muons is a model-independent consequence of chemical equilibrium. Instead, the physics of some of the more exotic states of matter that have been speculated to exist in the stellar core—such as hyperons, meson condensates, and quark matter—are much more uncertain. Thus, we limit ourselves to model the stellar core exclusively in terms of non-exotic constituents, namely, neutrons, protons, electrons, and muons. Moreover, this approach enables us to test and improve relativistic density functionals that were calibrated using only the properties of finite nuclei. This is highly desirable as the equation of state of dense neutron-rich matter is poorly constrained by laboratory observables.

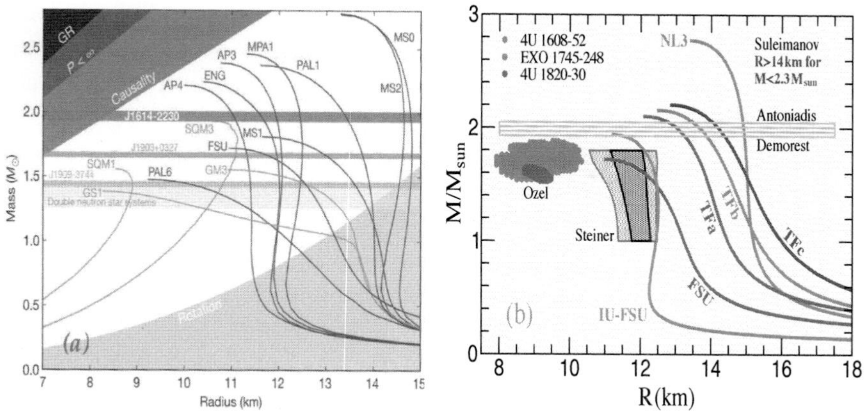

Fig. 7. (Color online) (Left panel) Predictions for the mass-vs-radius relation for a variety of models of the EOS using both exotic and non-exotic cores. The recent discovery of the massive neutron star PSR J1614-2230 by Demorest and collaborators is clearly indicated in the figure.[13] (b) (Right panel) Constraints on both stellar masses and radii extracted from various analyses of x-ray bursts.[75–77] Also shown are constraints obtained from the measurement of two massive neutron stars by Demorest[13] and Antoniadis.[14]

The cleanest constraint on the EOS at high density emerges from the accurate measurement of massive neutron stars. In this regard, enormous progress has been made by the recent observations of two massive neutron stars by Demorest[13] and Antoniadis[14] at the Green Bank Telescope. In particular, the accurate mass measurement of PSR J1614-2230 (with a mass of $M_{\star}=1.97\pm0.04M_{\odot}$) was carried out using "Shapiro delay", a beautiful and powerful technique that relies on the tiny delay in the arrival of the pulsar signal as it "dips down" into the gravitational well of its white-dwarf companion.[13] This observation provides a clear example of the synergy between nuclear physics, astrophysics, and general relativity.

The impact of such mass measurements on the predictions of the *mass-vs-radius* relation of neutron stars is displayed in Fig. 7a. Note that the constraint from PSR J1614-2230 alone can rule out models with equations of state that are too soft to support a $2M_{\odot}$ neutron star, such as those that predict the existence of exotic cores. Undoubtedly, the quest for even more massive neutron stars will continue and this will provide unique and valuable insights into the behavior of ultra-dense matter.

Although enormously valuable, the accurate measurement of massive neutron stars does not impose any significant constraint on stellar radii. Indeed, Fig. 7a indicates that predictions for the radius of a "canonical"

$1.4M_\odot$ neutron star can vary by almost a factor of two; between 8 km and 15 km. Whereas observations of various spectroscopic phenomena in x-ray bursters are promising in the quest to measure stellar radii, the approach currently suffers from large systematic uncertainties. For example, in one of the first analyses of this kind by Özel, Baym, and Güver,[75] it was suggested that neutron stars with masses of about $1.4M_\odot$ have very small radii of about 8–10 km; see Fig. 7b. This conclusion seems to favor models with exotic cores[78] which, in turn, are disfavored by the observation of $2M_\odot$ neutron stars. However, it was recognized soon after, first by Steiner, Lattimer, and Brown[76] and shortly after by Suleimanov,[77] that if one corrects for systematic uncertainties in the analysis by Özel, then it is possible to obtain larger (in the case of the former) or even much larger (in the case of the latter) stellar radii. Indeed, the results by Steiner[76] depicted in Fig. 7b by the two shaded areas that indicate 1σ and 2σ contours, suggest stellar radii in the 10–13 km range. However, even this more conservative limit has been challenged by Suleimanov, who has proposed a *lower limit* on the stellar radius of 14 km for neutron stars with masses below $2.3\mathrm{M}_\odot$.[77] This conclusion suggests that the neutron-star matter EOS must be stiff, in agreement with the recent measurement of massive stars.

While at present a reliable technique based on thermal emissions from x-ray bursts has not yet been realized, a recent analysis limited to the observation of accreting neutron stars during quiescence, the so-called *quiescent low-mass x-ray binaries* (qLMXBs), has challenged our current understanding of the equation of state of dense matter. qLMXBs are particularly attractive as they appear to be free from some of the uncertainties that plague x-ray bursters. By analyzing the thermal spectra of five qLMXBs inside globular clusters, Guillot and collaborators have reported a *common radius* for all five sources of $R_{NS} = 9.1^{+1.3}_{-1.5}$ km at a 90% confidence level.[79] One should mention that whereas the approach developed in Ref. 79 provides a careful accounting of all uncertainties, some of the assumptions—such as the single common radius—and some of the adopted uncertainties have been called into question.[80] Regardless, it is interesting to note that few, if any, nuclear density functionals—whether relativistic or non-relativistic—can accommodate both a small stellar radius and a large limiting mass.[79] However, we are confident that with new space missions—such as GAIA—that will provide unprecedented positional measurements, many of the current problems plaguing such challenging analyses will be mitigated.

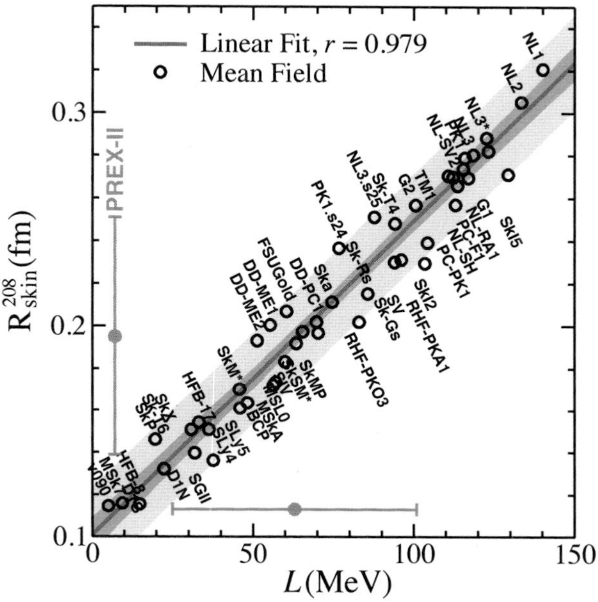

Fig. 8. (Color online) Predictions from a large number of nuclear density functionals for the neutron-skin thickness of ^{208}Pb and the slope of the symmetry energy L.[81] Constraints from an anticipated upgraded PREX measurement ("PREX-II") have been incorporated into the plot.

Whereas laboratory experiments are of marginal utility in constraining the limiting mass of a neutron star, they play an essential role in constraining stellar radii. This is because the radius of a neutron star is sensitive to the density dependence of the symmetry energy in the immediate vicinity of nuclear-matter saturation density.[82] A fundamental property of the EOS that has received considerable attention over the last decade is the slope of the symmetry energy at saturation density.[39] As already shown in Eq. (5), the slope of the symmetry energy L is proportional to the pressure of pure neutron matter at saturation density. In turn, the slope of the symmetry energy is also strongly correlated to a myriad of neutron-star observables.[28,35,83,84] Remarkably, L is also strongly correlated to the thickness of the neutron skin of heavy nuclei,[85,86] which is defined as the difference between the neutron and proton root-mean-square radii. The physical reason behind this correlation is particularly insightful. Heavy nuclei favor a neutron excess as a result of the repulsive Coulomb interaction between the protons. Energetically, it is advantageous—to both the surface tension and to the symmetry energy—to form an isospin symmetric ($N = Z$) core. So the basic question to be answered is *where do the extra neutrons go?* Placing them in the core reduces the surface tension but increases the symmetry energy. Moving the excess neutrons to the surface increases the surface tension but reduces the symmetry energy, which is lower in the dilute surface than in the dense core. So the thickness of the neutron skin emerges from a competition between the surface tension and the *difference* between the value of the symmetry energy at the surface relative to that at the center; this difference is encoded in the slope of the symmetry energy L.[87] Hence, if L is large, then it is energetically advantegeous to move the neutrons to the surface, which results in a thick neutron skin. This suggests a powerful correlation: *the larger the value of L the thicker the neutron skin.*[35] The strong correlation between L and the neutron-skin thickness of ^{208}Pb ($R_{\rm skin}^{208}$) has been verified using a large and representative set of density functionals—both relativistic and non-relativistic—and is displayed in Fig. 8.[81] The strong correlation coefficient of $r = 0.979$ suggests how a laboratory observable such as $R_{\rm skin}^{208}$ may serve to determine a fundamental parameter of the equation of state.

Recently, the Lead Radius Experiment ("PREX") at the Jefferson Laboratory has provided the first model-independent evidence in favor of a neutron-rich skin in ^{208}Pb.[88,89] Given that the Z^0 boson couples strongly to the neutron, parity violating electron scattering provides a clean probe of neutron densities that is free from strong-interaction uncertainties. As the proton radius of ^{208}Pb is known extremely accurately from conventional (parity conserving) electron scattering, PREX effectively determined the neutron-skin thickness of ^{208}Pb to be: $R_{\rm skin}^{208} = 0.33^{+0.16}_{-0.18}\,{\rm fm}$.[88] In a much anticipated and already approved follow-up experiment ("PREX-II") the neutron-skin thickness of ^{208}Pb will be measured with a significantly increased accuracy. The impact of such a measurement in constraining the value of L is displayed with the green error bars in Fig. 8.

Perhaps surprisingly, the same pressure that is responsible for creating a neutron-rich skin in heavy nuclei is also responsible for supporting a neutron star against gravitational collapse. Thus, models that predict thicker neutron skins often produce neutron stars with larger radii.[28,35] This makes it possible to establish a powerful "data-to-data" relation between the neutron-rich skin of a heavy nucleus and the radius of a neutron star; a relation involving physical observables that differ by 18 orders of magnitude! This powerful connection is nicely illustrated in Fig. 9 which displays a strong correlation between $R_{\rm skin}^{208}$ and the stellar radius of a $0.8M_{\odot}$

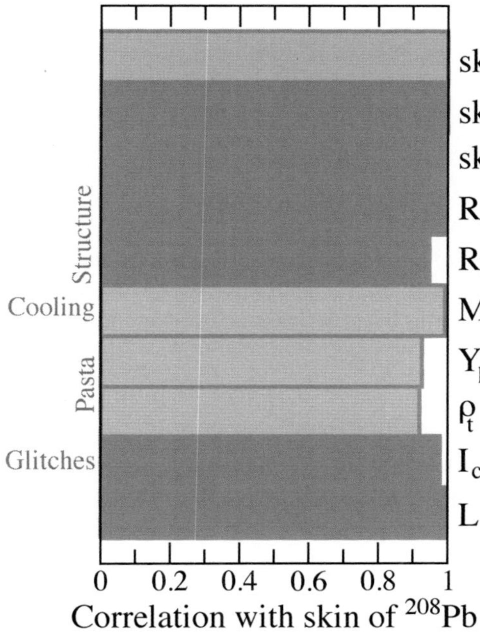

Fig. 9. (Color online) Correlation coefficients between $R_{\rm skin}^{208}$ and a variety of neutron-star observables as predicted by the FSUGold density functional.[47]

neutron star. In the case of a $1.4M_{\odot}$ neutron star, the correlation weakens slightly because, unlike low-mass neutron stars, medium-to-heavy neutron stars are also sensitive to the symmetry energies at densities above saturation density.[36] Moreover, Fig. 9 also displays the enormous reach that an accurate measurement of $R_{\rm skin}^{208}$ could have on several neutron-star properties;[84] note that red bands imply a direct correlation whereas the blue bands an inverse correlation (i.e., an anticorrelation). Indeed, an accurate measurement of $R_{\rm skin}^{208}$ could have a strong impact on a variety of neutron-star observables related to their structure, cooling, and glitch mechanism.[90] We underscore that all properties displayed in Fig. 9—whether pertaining to finite nuclei or neutron stars—were computed with a unique relativistic energy density functional.

4. Summary and Outlook

Neutron stars are truly gold mines for the study of physical phenomena across a variety of disciplines ranging from elementary particle physics to general relativity. Indeed, binary pulsars have been used to conduct some of the most stringent tests of general relativity and their merger may provide both a powerful source of gravitational waves as well as a plausible site for r-process nucleosynthesis. Moreover, some of the fascinating phases predicted to exist in the neutron-star crust, such as Coulomb crystals of neutron-rich nuclei and nuclear pasta, have counterparts in condensed-matter physics. Finally, the stellar core may harbor new states of matter that may consist of deconfined quark matter, such as color superconductors.

Of course, from the perspective of nuclear physics, neutron stars hold the answer to one of the most fundamental questions in the field: *How does subatomic matter organize itself and what phenomena emerge?* Indeed, this question figures prominently in the recent community report entitled "Nuclear Physics: Exploring the Heart of Matter". In this contribution we challenged the original view of Baade and Zwicky that neutron stars consist of extremely closely packed neutrons.[5] Although undoubtedly such is the most common perception of a neutron star, we showed that the reality is far different and much more interesting. In particular, during our journey through a neutron star we uncovered a myriad of exotic states of matter that are speculated to exist in a neutron star. Moreover, we underscored the critical role that laboratory experiments will play in elucidating the fascinating nature of these exotic states. By the same token, we highlighted the fundamental role that observation of neutron stars using telescopes operating over an enormous range of frequencies will have in constraining the equation of state of neutron-rich matter. As such, neutron stars provide a powerful intellectual bridge between Nuclear Physics and Astrophysics.

Whereas the main goal of the present contribution was an introduction to the fascinating world of neutron stars, the powerful tool used for their study, namely, *relativistic density functional theory* represents the overarching theme of this whole volume. Density functional theory provides a powerful—and perhaps unique—framework for the accurate calculation of nuclear properties. Based on the seminal work by Kohn and collaborators, DFT shifts the focus from the complicated many-body wave function to the much simpler one-body density. By doing so, the formidable challenge of deducing the exact ground-state energy and one-body density from the many-body wave function "reduces" to the minimization of a suitable functional of the density. Whereas the implementation of DFT to the understanding of electronic properties of materials (work for which Walter Kohn was recognized in 1998 with the Nobel prize in chemistry) is firmly based on the Coulomb interaction, the situation is significantly more difficult in the nuclear physics domain. This is because the parameters of the nuclear density functional cannot be computed directly from QCD. By necessity then, the parameters of the model must be directly calibrated (i.e., fitted) to physical observables. However, by directly fitting to data, the short-distance structure of the theory as well as many other complicated many-body effects get implicitly encoded in the parameters of the model. In this regard, DFT provides a unified and powerful approach that may be used to compute physical phenomena ranging over many distance scales, such as the properties of finite nuclei and neutron stars; systems that differ in size by 18 orders of magnitude! Moreover, by implementing the calibration of the model parameters via a standard optimization procedure, DFT becomes a powerful microscopic theory that both predicts and provides well-quantified theoretical uncertainties.[18] Documenting theoretical uncertainties is particularly critical as one extrapolates into uncharted regions of the observable landscape.

In this contribution we relied on relativistic DFT to explore three regions of the neutron star: the outer and inner crust as well as the outer core. We determined that the outer stellar crust consists of a Coulomb crystal of neutron-rich nuclei embedded in a uniform electron gas. Remarkably, the dynamics of the outer crust is only sensitive to the masses of both stable and unstable nuclei. Thus, mass measurements of exotic nuclei at rare isotope facilities will be instrumental in both constraining theoretical models as well as guiding the extrapolations into regions of the nuclear chart where mass measurement are unlikely to occur. Moving into the inner crust, we established that at the top layers the Coulomb crystal of neutron-rich nuclei is now in equilibrium with a dilute, and likely superfluid, neutron vapor. Hence, in this region one is sensitive to the equation of state of dilute neutron matter, a topic of enormous interest and one that holds an intimate connection to cold atomic Fermi gases in the unitary regime. Given that the EOS of neutron-star matter must span an enormous range of densities, a study of dilute neutron matter can be used to further constrain the form of the relativistic density functional. As one moves even deeper into the inner crust, one discovers a complex pasta phase that displays unique and fascinating dynamical features. Although finding clear signatures of its existence has proved elusive, the lack of x-ray pulsars with long spin periods has recently been suggested as the first observable manifestation of the nuclear pasta phase. Recent molecular dynamics simulations seem to support such an assertion.

Finally, we explored the deep stellar core which consists of a uniform quantum liquid of neutrons, protons, electrons and muons in chemical equilibrium. In this region the relativistic character of the density functional is particularly critical, as it provides a Lorentz covariant extrapolation to the high-density regime. That is, the relativistic nature of the functional guarantees that the speed of sound in matter will always remain below the speed of light at all densities, a feature that is often not displayed by *non-relativistic* density functionals. Whereas one of the main goals of a relativistic density functional is to provide a unified description of phenomena that happens at many distance scales, we particularly highlighted the strong correlation between a fundamental property of finite nuclei—namely, the thickness of the neutron skin—and the radius of a neutron star. Thus, a laboratory measurement of the neutron-skin thickness of ^{208}Pb (PREX) can constrain the radius of a neutron star. Conversely, accurate measurements of stellar radii can help refine the functional. In particular, we discussed how a recent analysis of quiescent low-mass x-ray binaries seem to favor neutron stars with small stellar radii. Such a finding, which seems to favor a soft EOS, poses serious challenges to theoretical models that must simultaneously account for massive neutron stars of at least 2 solar masses. Moreover, small stellar radii also appear at odds with the original PREX report of a fairly large neutron-skin thickness in ^{208}Pb, albeit with large error bars. However, if future laboratory experiments and astronomical observations confirm that both $R_{\rm skin}^{208}$ is thick and stellar radii are small, this would strongly suggest a softening of the EOS due to the onset of a phase transition. Regardless, the EOS must eventually stiffen again to account for the existence of massive neutron stars. Such extraordinary behavior will confirm the fundamental role of neutron stars as unique laboratories for the study of dense nucleonic matter. Independent of the surprises that may lie ahead, relativistic density functional theory provides the only tractable microscopic theory that can describe nuclear phenomena ranging from the physics of finite nuclei to the structure of neutron stars.

Acknowledgement

This material is based upon work supported by the United States Department of Energy Office of Science, Office of Nuclear Physics under Award Number DE-FD05-92ER40750.

References

1. D. D. Clayton, *Principles of Stellar Evolution and Nucleosynthesis.* University of Chicago Press, Chicago (1983).

2. C. Iliadis, *Nuclear Physics of Stars*. Wiley-VCH, Weinheim (2007).
3. S. Chandrasekhar, The maximum mass of ideal white dwarfs, *Astrophys. J.* **74**, 81 (1931).
4. J. Chadwick, Possible Existence of a Neutron, *Nature*. **129**, 312 (1932). doi: doi:10.1038/129312a0.
5. W. Baade and F. Zwicky, (in minutes of the Stanford meeting, proceedings of the American Physical Society, December 15–16, 1933), *Phys. Rev.* **45**, 138 (Jan, 1934). doi: 10.1103/PhysRev.45.130.
6. J. a. Magueijo, *A Brilliant Darkness: The Extraordinary Life and Mysterious Disappearance of Ettore Majorana*. Perseus Books Group, New York, NY 10016 (2009).
7. L. D. Landau, On the theory of stars, *Phys. Z. Sowjetunion.* **1**, 285 (1932). Translated into Russian: in Landau L D Sobranie Trudov (Collected Works) Vol. 1 (Moscow: Nauka, 1969) p. 86.
8. D. G. Yakovlev, P. Haensel, G. Baym, and C. J. Pethick, Lev Landau and the conception of neutron stars, *Phys. Usp.* **56**, 289–295 (2013). doi: 10.3367/UFNe.0183.201303f.0307.
9. J. R. Oppenheimer and G. M. Volkoff, On massive neutron cores, *Phys. Rev.* **55**, 374 (1939).
10. R. C. Tolman, Static solutions of Einstein's field equations for spheres of fluid, *Phys. Rev.* **55**, 364 (1939).
11. A. Hewish, S. Bell, J. Pilkington, P. Scott, and R. Collins, Observation of a Rapidly Pulsating Radio Source, *Nature*. **217**, 709–713 (1968). doi: 10.1038/217709a0.
12. J. M. Lattimer, The nuclear equation of state and neutron star masses, *Ann. Rev. Nucl. Part. Sci.* **62**, 485–515 (2012). doi: 10.1146/annurev-nucl-102711-095018.
13. P. Demorest, T. Pennucci, S. Ransom, M. Roberts, and J. Hessels, Shapiro delay measurement of a two solar mass neutron star, *Nature*. **467**, 1081–1083 (2010). doi: 10.1038/nature09466.
14. J. Antoniadis, P. C. Freire, N. Wex, T. M. Tauris, R. S. Lynch, et al., A Massive Pulsar in a Compact Relativistic Binary, *Science*. **340**, 6131 (2013). doi: 10.1126/science.1233232.
15. R. A. Hulse and J. H. Taylor, Discovery of a pulsar in a binary system, *Astrophys. J.* **195**, L51–L53 (1975).
16. M. G. Alford, K. Rajagopal, and F. Wilczek, Color-flavor locking and chiral symmetry breaking in high density QCD, *Nucl. Phys.* **B537**, 443–458 (1999). doi: 10.1016/S0550-3213(98)00668-3.
17. M. Kortelainen, T. Lesinski, J. More, W. Nazarewicz, J. Sarich, N. Schunck, M. V. Stoitsov, and S. Wild, Nuclear Energy Density Optimization, *Phys. Rev. C.* **82**, 024313 (2010).
18. PRA-Editors, Editorial: Uncertainty estimates, *Phys. Rev. A.* **83**(4), 040001 (Apr, 2011). doi: 10.1103/PhysRevA.83.040001.
19. J. D. Walecka, A theory of highly condensed matter, *Annals Phys.* **83**, 491 (1974).

20. B. Serot, A relativistic nuclear field theory with pi and rho mesons, *Phys. Lett.* **B86**, 146–150 (1979). doi: 10.1016/0370-2693(79)90804-9.
21. R. J. Furnstahl and B. D. Serot, Quantum hadrodynamics: Evolution and revolution, *Comments Nucl. Part. Phys.* **2**, A23–A45 (2000).
22. P. Hohenberg and W. Kohn, Inhomogeneous Electron Gas, *Phys. Rev.* **136**, B864–B871 (1964). doi: 10.1103/PhysRev.136.B864.
23. W. Kohn and L. J. Sham, Self-consistent equations including exchange and correlation effects, *Phys. Rev.* **140**, A1133–A1138 (Nov, 1965). doi: 10.1103/PhysRev.140.A1133. URL `http://link.aps.org/doi/10.1103/PhysRev.140.A1133`.
24. W. Kohn, Nobel lecture: Electronic structure of matter¯wave functions and density functionals, *Rev. Mod. Phys.* **71**(5), 1253–1266 (Oct, 1999). doi: 10.1103/RevModPhys.71.1253.
25. B. D. Serot and J. D. Walecka, The relativistic nuclear many body problem, *Adv. Nucl. Phys.* **16**, 1–327 (1986).
26. H. Mueller and B. D. Serot, Relativistic mean-field theory and the high-density nuclear equation of state, *Nucl. Phys.* **A606**, 508–537 (1996).
27. B. D. Serot and J. D. Walecka, Recent progress in quantum hadrodynamics, *Int. J. Mod. Phys.* **E6**, 515–631 (1997).
28. C. J. Horowitz and J. Piekarewicz, Neutron star structure and the neutron radius of 208pb, *Phys. Rev. Lett.* **86**, 5647 (2001).
29. R. Furnstahl, B. D. Serot, and H.-B. Tang, A Chiral effective Lagrangian for nuclei, *Nucl.Phys.* **A615**, 441–482 (1997). doi: 10.1016/S0375-9474(96)00472-1.
30. R. J. Furnstahl, B. D. Serot, and H.-B. Tang, Vacuum nucleon loops and naturalness, *Nucl. Phys.* **A618**, 446–454 (1997).
31. J. J. Rusnak and R. Furnstahl, Relativistic point coupling models as effective theories of nuclei, *Nucl. Phys.* **A627**, 495–521 (1997). doi: 10.1016/S0375-9474(97)00598-8.
32. R. Furnstahl and J. C. Hackworth, The Skyrme energy functional and naturalness, *Phys. Rev.* **C56**, 2875–2878 (1997). doi: 10.1103/PhysRevC.56.2875.
33. M. Kortelainen, R. Furnstahl, W. Nazarewicz, and M. Stoitsov, Natural Units For Nuclear Energy Density Functional Theory, *Phys. Rev.* **C82**, 011304 (2010). doi: 10.1103/PhysRevC.82.011304.
34. J. Boguta and A. R. Bodmer, Relativistic calculation of nuclear matter and the nuclear surface, *Nucl. Phys.* **A292**, 413 (1977).
35. C. J. Horowitz and J. Piekarewicz, The neutron radii of lead and neutron stars, *Phys. Rev.* **C64**, 062802 (2001).
36. J. Carriere, C. J. Horowitz, and J. Piekarewicz, Low mass neutron stars and the equation of state of dense matter, *Astrophys. J.* **593**, 463 (2003).
37. C. J. Horowitz, M. A. Perez-Garcia, and J. Piekarewicz, Neutrino-pasta scattering: The opacity of nonuniform neutron-rich matter, *Phys. Rev.* **C69**, 045804 (2004).
38. B. G. Todd and J. Piekarewicz, Relativistic mean-field study of neutron-rich nuclei, *Phys. Rev.* **C67**, 044317 (2003).

39. J. Piekarewicz and M. Centelles, Incompressibility of neutron-rich matter, *Phys. Rev.* **C79**, 054311 (2009). doi: 10.1103/PhysRevC.79.054311.
40. N. Chamel and P. Haensel, Physics of Neutron Star Crusts, *Living Rev. Rel.* **11**, 10 (2008).
41. C. Bertulani and J. Piekarewicz, *Neutron Star Crust*. Nova Science Publishers, Hauppauge New York (2012).
42. D. Page and S. Reddy, Dense Matter in Compact Stars: Theoretical Developments and Observational Constraints, *Ann. Rev. Nucl. Part. Sci.* **56**, 327–374 (2006). doi: 10.1146/annurev.nucl.56.080805.140600.
43. G. Baym, C. Pethick, and P. Sutherland, The ground state of matter at high densities: Equation of state and stellar models, *Astrophys. J.* **170**, 299–317 (1971).
44. X. Roca-Maza and J. Piekarewicz, Impact of the symmetry energy on the outer crust of non-accreting neutron stars, *Phys. Rev.* **C78**, 025807 (2008). doi: 10.1103/PhysRevC.78.025807.
45. R. A. Coldwell-Horsfall and A. A. Maradudin, Zero-point energy of an electron lattice, *Journal of Mathematical Physics*. **1**(5), 395–404 (1960). doi: 10.1063/1.1703670.
46. C. A. Sholl, The calculation of electrostatic energies of metals by plane-wise summation, *Proceedings of the Physical Society*. **92**(2), 434–445 (1967).
47. B. G. Todd-Rutel and J. Piekarewicz, Neutron-rich nuclei and neutron stars: A new accurately calibrated interaction for the study of neutron-rich matter, *Phys. Rev. Lett.* **95**, 122501 (2005).
48. G. A. Lalazissis, J. Konig, and P. Ring, A new parameterization for the lagrangian density of relativistic mean field theory, *Phys. Rev.* **C55**, 540–543 (1997).
49. G. A. Lalazissis, S. Raman, and P. Ring, Ground-state properties of even-even nuclei in the relativistic mean-field theory, *At. Data Nucl. Data Tables*. **71**, 1 (1999).
50. P. Möller, J. R. Nix, W. D. Myers, and W. J. Swiatecki, Nuclear ground state masses and deformations, *Atom. Data Nucl. Data Tabl.* **59**, 185–381 (1995).
51. P. Möller, J. R. Nix, and K. L. Kratz, Nuclear properties for astrophysical and radioactive-ion- beam applications, *Atom. Data Nucl. Data Tabl.* **66**, 131–343 (1996).
52. J. Duflo, Phenomenological calculation for nuclear masses and charge radii, *Nucl. Phys.* **A576**(1), 29–64 (August, 1994). doi: 10.1016/0375-9474(94) 90737-4.
53. A. Zuker, On the microscopic derivation of a mass formula, *Nucl. Phys.* **A576** (1), 65–108 (August, 1994). doi: doi:10.1016/0375-9474(94)90737-4.
54. J. Duflo and A. Zuker, Microscopic mass formulas, *Phys. Rev. C.* **52**(1), R23–R27 (Jul, 1995). doi: 10.1103/PhysRevC.52.R23.
55. M. Wang, G. Audi, A. Wapstra, F. Kondev, M. MacCormick, X. Xu, and B. Pfeiffer, The AME2012 atomic mass evaluation (II). Tables, graphs and references, *Chinese Phys. C.* **36**, 1603–2014 (2012). doi: 10.1088/1674-1137/ 36/12/003.
56. S. Goriely, N. Chamel, and J. Pearson, Further explorations of Skyrme–

Hartree–Fock–Bogoliubov mass formulas. XII: Stiffness and stability of neutron-star matter, *Phys. Rev.* **C82**, 035804 (2010). doi: 10.1103/PhysRevC.82.035804.

57. R. Wolf et al., Plumbing Neutron Stars to New Depths with the Binding Energy of the Exotic Nuclide Zn-82, *Phys. Rev. Lett.* **110**, 041101 (2013). doi: 10.1103/PhysRevLett.110.041101.
58. D. G. Ravenhall, C. J. Pethick, and J. R. Wilson, Structure of matter below nuclear saturation density, *Phys. Rev. Lett.* **50**, 2066–2069 (1983).
59. M. Hashimoto, H. Seki, and M. Yamada, Shape of nuclei in the crust of neutron star, *Prog. Theor. Phys.* **71**, 320–326 (1984).
60. A. L. Fetter and J. D. Walecka, *Quantum Theory of Many Particle Systems.* McGraw-Hill, New York (1971).
61. R. Jamei, S. Kivelson, and B. Spivak, Universal aspects of Coulomb-frustrated phase separation, *Phys. Rev. Lett.* 94(5):056805 (2005).
62. C. J. Horowitz, M. A. Perez-Garcia, J. Carriere, D. K. Berry, and J. Piekarewicz, Nonuniform neutron-rich matter and coherent neutrino scattering, *Phys. Rev.* **C70**, 065806 (2004).
63. C. J. Horowitz, M. A. Perez-Garcia, D. K. Berry, and J. Piekarewicz, Dynamical response of the nuclear pasta in neutron star crusts, *Phys. Rev.* **C72**, 035801 (2005).
64. G. Watanabe, K. Sato, K. Yasuoka, and T. Ebisuzaki, Structure of cold nuclear matter at subnuclear densities by quantum molecular dynamics, *Phys. Rev.* **C68**, 035806 (2003). doi: 10.1103/PhysRevC.68.035806.
65. G. Watanabe, T. Maruyama, K. Sato, K. Yasuoka, and T. Ebisuzaki, Simulation of transitions between 'Pasta' phases in dense matter, *Phys. Rev. Lett.* **94**, 031101 (2005). doi: 10.1103/PhysRevLett.94.031101.
66. G. Watanabe, H. Sonoda, T. Maruyama, K. Sato, K. Yasuoka, et al., Formation of Nuclear 'Pasta' in Supernovae, *Phys. Rev. Lett.* **103**, 121101 (2009). doi: 10.1103/PhysRevLett.103.121101.
67. T. Maruyama, T. Tatsumi, D. N. Voskresensky, T. Tanigawa, and S. Chiba, Nuclear pasta structures and the charge screening effect, *Phys. Rev.* **C72**, 015802 (2005). doi: 10.1103/PhysRevC.72.015802.
68. S. Avancini, D. Menezes, M. Alloy, J. Marinelli, M. Moraes, et al., Warm and cold pasta phase in relativistic mean field theory, *Phys. Rev.* **C78**, 015802 (2008). doi: 10.1103/PhysRevC.78.015802.
69. S. Avancini, L. Brito, J. Marinelli, D. Menezes, M. de Moraes, et al., Nuclear pasta phase within density dependent hadronic models, *Phys. Rev.* **C79**, 035804 (2009). doi: 10.1103/PhysRevC.79.035804.
70. W. Newton and J. Stone, Modeling nuclear 'pasta' and the transition to uniform nuclear matter with the D-3 Skyrme–Hartree–Fock method at finite temperature: Core-collapse supernovae, *Phys. Rev.* **C79**, 055801 (2009). doi: 10.1103/PhysRevC.79.055801.
71. G. Shen, C. Horowitz, and S. Teige, A New Equation of State for Astrophysical Simulations, *Phys. Rev.* **C83**, 035802 (2011). doi: 10.1103/PhysRevC.83.035802.
72. J. Piekarewicz and G. Toledo Sanchez, Proton fraction in the inner neutron-

star crust, *Phys. Rev.* **C85**, 015807 (2012). doi: 10.1103/PhysRevC.85.015807.
73. J. A. Pons, D. Vigano; and N. Rea, A highly resistive layer within the crust of X-ray pulsars limits their spin periods, *Nature Physics, 9,.* **431-434** (2013). doi: 10.1038/nphys2640.
74. C. Horowitz, D. Berry, C. Briggs, M. Caplan, A. Cumming, et al., Disordered nuclear pasta, magnetic field decay, and crust cooling in neutron stars (2014).
75. F. Ozel, G. Baym, and T. Guver, Astrophysical Measurement of the Equation of State of Neutron Star Matter, *Phys. Rev.* **D82**, 101301 (2010). doi: 10.1103/PhysRevD.82.101301.
76. A. W. Steiner, J. M. Lattimer, and E. F. Brown, The Equation of State from Observed Masses and Radii of Neutron Stars, *Astrophys. J.* **722**, 33–54 (2010). doi: 10.1088/0004-637X/722/1/33.
77. V. Suleimanov, J. Poutanen, M. Revnivtsev, and K. Werner, Neutron star stiff equation of state derived from cooling phases of the X-ray burster 4U 1724-307, *Astrophys. J.* **742**, 122 (2011). doi: 10.1088/0004-637X/742/2/122.
78. F. J. Fattoyev and J. Piekarewicz, Relativistic models of the neutron-star matter equation of state, *Phys. Rev.* **C82**, 025805 (2010). doi: 10.1103/PhysRevC.82.025805.
79. S. Guillot, M. Servillat, N. A. Webb, and R. E. Rutledge, Measurement of the radius of neutron stars with high signal-to-noise quiescent low-mass x-ray binaries in globular clusters, *Astrophys. J.* **772**(1), 7 (2013).
80. J. M. Lattimer and A. W. Steiner, Neutron Star Masses and Radii from Quiescent Low-Mass X-ray Binaries (2013).
81. X. Roca-Maza, M. Centelles, X. Vinas, and M. Warda, Neutron skin of ^{208}Pb, nuclear symmetry energy, and the parity radius experiment, *Phys. Rev. Lett.* **106**, 252501 (2011). doi: 10.1103/PhysRevLett.106.252501.
82. J. M. Lattimer and M. Prakash, Neutron Star Observations: Prognosis for Equation of State Constraints, *Phys. Rept.* **442**, 109–165 (2007). doi: 10.1016/j.physrep.2007.02.003.
83. C. J. Horowitz and J. Piekarewicz, Constraining URCA cooling of neutron stars from the neutron radius of ^{208}Pb, *Phys. Rev.* **C66**, 055803 (2002).
84. F. Fattoyev and J. Piekarewicz, Neutron skins and neutron stars, *Phys. Rev.* **C88**, 015802 (2012).
85. B. A. Brown, Neutron radii in nuclei and the neutron equation of state, *Phys. Rev. Lett.* **85**, 5296 (2000).
86. R. J. Furnstahl, Neutron radii in mean-field models, *Nucl. Phys.* **A706**, 85–110 (2002).
87. C. Horowitz, E. Brown, Y. Kim, W. Lynch, R. Michaels, et al., A way forward in the study of the symmetry energy: experiment, theory, and observation, *J. Phys.* **G41**, 093001 (2014). doi: 10.1088/0954-3899/41/9/093001.
88. S. Abrahamyan, Z. Ahmed, H. Albataineh, K. Aniol, D. Armstrong, et al., Measurement of the Neutron Radius of 208Pb Through Parity-Violation in Electron Scattering, *Phys. Rev. Lett.* **108**, 112502 (2012). doi: 10.1103/PhysRevLett.108.112502.

89. C. Horowitz, Z. Ahmed, C. Jen, A. Rakhman, P. Souder, et al., Weak charge form factor and radius of 208Pb through parity violation in electron scattering, *Phys. Rev.* **C85**, 032501 (2012). doi: 10.1103/PhysRevC.85.032501.
90. F. J. Fattoyev and J. Piekarewicz, Sensitivity of the Moment of Inertia of Neutron Stars to the Equation of State of Neutron-Rich Matter, *Phys. Rev.* **C82**, 025810 (2010). doi: 10.1103/PhysRevC.82.025810.

Chapter 15

Relativistic versus non-relativistic mean field

Paul-Gerhard Reinhard

Institut für Theoretische Physik,
Universität Erlangen/Nürnberg,
91077 Erlangen/Germany
paul-gerhard.reinhard@fau.de

Three variants of the relativistic mean-field model (RMF) and the non-relativistic Skyrme–Hartree–Fock model (SHF) are compared. Overall quality, predictive power, and correlations between observables are addressed using statistical analysis on the basis of least squares fits. Appropriate density dependence is a crucial ingredient for good performance of RMF. However, SHF shows still more flexibility particularly in the isovector channel.

Contents

1. Introduction

As we have seen in many chapters along this book, RMF has grown over decades to a versatile and reliable tool for the description of nuclear structure and dynamics. Much in parallel went the development of non-relativistic mean-field models such that both versions of self-consistent mean-field models are strong competitors those days. A basic presentation of RMF and non-relativistic mean-field models side by side was given in the review.[1] This chapter here concentrates on a close comparison of relativistic and non-relativistic mean field models. To the latter family belong the presently widely used Skyrme-Hartree–Fock (SHF) model and the Gogny force. To keep the comparison overseeable, we will consider at the non-relativistic side only SHF and will also select only three sub-cases of the reach variety of RMF models. All three have the same structure in the basic couplings, but differ in the way density-dependence and meson finite-range are treated. There are many more ways to introduce a more versatile density dependence into RMF models as, e.g., more terms in non-linear meson coupling. Such variants are found elsewhere in the book, but are not considered here. The description of the models as such is kept to a minimum as it is well documented elsewhere, for RMF in this book and for SHF in various reviews.[1–3] We will briefly sketch the relations between RMF and SHF from a formal perspective and mainly concentrate on the empirical aspect of performance of either model.

2. Synopsis of the functionals

The basis of any mean-field approach is a set of single-nucleon canonical (Kohn–Sham) orbitals, Dirac four-spinor wave functions $\psi_\alpha(\mathbf{r})$ in the RMF framework and two-component-spinor wave functions $\varphi_\alpha(\mathbf{r})$ in non-relativistic SHF. An occupation amplitude v_α, determined by the pairing interaction, is associated to each orbital. Starting point of each mean-field model is an energy-density functional (EDF) expressed in terms of single particle (s.p.) orbitals and the local densities derived therefrom. The static and dynamic mean-field equations are derived from the EDF variationally in standard manner and will not be discussed here. Relativistic EDF are discussed in great detail in Chapter 2. The EDF for SHF reads

$$
\begin{aligned}
E &= \int d^3r\ (\mathcal{E}_{\text{kin}} + \mathcal{E}_{\text{pot}}) + E_{\text{Coul}} + E_{\text{pair}} + E_{\text{cm}}, \\
\mathcal{E}_{\text{kin}} &= \frac{\hbar^2}{2m_{\text{p}}}\tau_{\text{p}}(\mathbf{r}) + \frac{\hbar^2}{2m_{\text{n}}}\tau_{\text{n}}(\mathbf{r}) \\
E_{\text{cm}} &= -\frac{1}{2mA}\langle(\hat{P}_{\text{cm}})^2\rangle.
\end{aligned} \tag{1}
$$

The kinetic energy $\mathcal{E}_{\text{kin}}$ is expressed in terms of the local kinetic-energy densities $\tau_{p/n}$, see upper panel of Table 1. The Coulomb energy E_{Coul} consists of the direct Coulomb term, and the Coulomb exchange that is usually taken into account at the level of the Slater approximation. In most applications the center-of-mass correction E_{cm} is applied *a posteriori* because its variation would considerably complicate the mean-field equations. The pairing functional E_{pair} will be detailed later.

Table 1. Upper: The basic isoscalar ($T = 0$) and isovector ($T = 1$) local densities of SHF (density ρ_T, kinetic density τ_T, spin-orbit density $\mathbf{J}_T$. Lower: The basic terms in the potential-energy densities for SHF. Terms are sorted into isoscalar and isovector couplings. The density dependence, applying only to the $\rho\rho$-coupling, is summarized in the right column.

SHF densities	
$T = 0$	$T = 1$
$\rho_0(\mathbf{r}) = \sum_\alpha v_\alpha^2\, \varphi_\alpha^\dagger \varphi_\alpha$	$\rho_1(\mathbf{r}) = \sum_\alpha v_\alpha^2\, \varphi_\alpha^\dagger \hat{\tau}_3 \varphi_\alpha$
$\tau_0(\mathbf{r}) = \sum_\alpha v_\alpha^2\, \boldsymbol{\nabla}\varphi_\alpha^\dagger \boldsymbol{\nabla}\varphi_\alpha$	$\tau_1(\mathbf{r}) = \sum_\alpha v_\alpha^2\, \boldsymbol{\nabla}\varphi_\alpha^\dagger \hat{\tau}_3 \boldsymbol{\nabla}\varphi_\alpha$
$\mathbf{J}_0(\mathbf{r}) = -\mathrm{i}\sum_\alpha v_\alpha^2\, \varphi_\alpha^\dagger \boldsymbol{\nabla}\times\hat{\boldsymbol{\sigma}}\varphi_\alpha$	$\mathbf{J}_1(\mathbf{r}) = -\mathrm{i}\sum_\alpha v_\alpha^2\, \varphi_\alpha^\dagger \hat{\tau}_3 \boldsymbol{\nabla}\times\hat{\boldsymbol{\sigma}}\varphi_\alpha$

SHF potential-energy density				
	basic couplings		**density dependence**	
	$T = 0$	$T = 1$	$T = 0$	$T = 1$
$\rho\rho$	$C_0^\rho \rho_0^2$	$C_1^\rho \rho_1^2$	$C_0^\rho = c_0^\rho + d_0^\rho \rho_0^a$	$C_1^\rho = c_1^\rho + d_1^\rho \rho_0^a$
mass	$C_0^\tau \rho_0 \tau_0$	$C_1^\tau \rho_1 \tau_1$		
$\boldsymbol{\ell}\cdot\boldsymbol{s}$	$C_0^{\boldsymbol{\nabla}\mathbf{J}} \rho_0 \boldsymbol{\nabla}\cdot\mathbf{J}_0$	$C_1^{\boldsymbol{\nabla}\mathbf{J}} \rho_1 \boldsymbol{\nabla}\cdot\mathbf{J}_1$		
grad.	$C_0^{\Delta\rho} (\boldsymbol{\nabla}\rho_0)^2$	$C_1^{\Delta\rho} (\boldsymbol{\nabla}\rho_1)^2$		

The essence of the EDF, the SHF functional, is contained in the interaction part with the potential-energy density $\mathcal{E}_{\rm pot}$. It is summarized in Table 1. There we show for simplicity only the couplings between time even densities which suffices for comparison with RMF functionals. The time odd terms (employing local current $\mathbf{j}(\mathbf{r})$ and spin density $\hat{\boldsymbol{\sigma}}(\mathbf{r})$) are found in the above mentioned reviews. They play no role for the ground state observables discussed below.

The basic building blocks of an EDF are local densities built from single-nucleon wave functions. These are summarized in the upper part of Table 1. The non-relativistic EDF employ, besides the local density ρ_T, the kinetic-energy density τ_T, the spin-orbit density $\mathbf{J}_T$, and the time odd quantities current and spin density (not shown here). All densities appear in two flavors: isoscalar ($T = 0$), or total density (sum of proton and neutron densities), and isovector ($T = 1$) density (difference between neutron and proton densities). The isovector density can be conveniently expressed using the isospin operator $\hat{\tau}_3$.

The lower part of Table 1 displays the various terms of the SHF potential-energy density $\mathcal{E}_{\rm pot}$. The underlying idea is to take all bi-linear isoscalar combinations of the local densities and to associate a coupling constant with each term. SHF confines the combinations to have at most second order of derivatives (this would include also the tensor spin-orbit term $\mathbf{J}^2$ which is dropped here). This form of the EDF can be deduced formally from a low-momentum expansion of the T-matrix (effective interaction) in the nuclear medium.[4–6] Note that the expansion proceeds analogously in the isovector channel as in the isoscalar channel thus filling all entries also for $T = 1$. Each term is given a coupling constant $C_T^{(\rm typ)}$ which, in principle, is density dependent. Neither empirical data (see section 5) nor *ab-initio* calculations supply sufficient information to determine each density dependence for all $C_T^{(\rm typ)}$ unambiguously. Thus SHF attributes a (simple) density dependence only to the leading terms $\propto \rho_0^2$ and ρ_1^2, i.e. to C_0^ρ and C_0^ρ, as indicated in Table 1 while all terms containing derivatives remain in simple bi-linear form. The chosen density dependence $\propto \rho^\alpha$ turns out to be extremely efficient. But its non-analytic structure raises problems in symmetry restoration.[7] There are attempts with more regular forms of density dependence which could resolve the projection problem.[8] For the present survey which does not require symmetry restoration, we stay with the widely used conventional form of SHF as given in Table 1.

Table 2 summarizes the basic structure of RMF in similar fashion as in Table 1 for SHF. As seen in Chapter 2, there are, in fact, many variants of

Table 2. Upper: The basic isoscalar ($T = 0$) and isovector ($T = 1$) local densities of RMF. Lower: The basic terms in the potential-energy densities for the three RMF models to be compared with SHF. Terms are sorted into isoscalar and isovector couplings. The density dependence, different for each model, is summarized in the right column. The free meson propagator is defined as $\mathcal{F}_i = [-\nabla^2 + m_i^2]^{-1}$, $i \in \{\sigma, \omega, \rho\}$ and the non-linear coupling of the scalar meson read $U'_\sigma = m_\sigma^2 + b_3\sigma^3 + b_4\sigma^4$.

RMF densities	
$T=0$	$T=1$
$\rho_0(\mathbf{r}) = \sum_\alpha v_\alpha^2\,\psi_\alpha^\dagger\psi_\alpha$	$\rho_1(\mathbf{r}) = \sum_\alpha v_\alpha^2\,\psi_\alpha^\dagger\hat{\tau}_3\psi_\alpha$
$\rho_S(\mathbf{r}) = \sum_\alpha v_\alpha^2\,\psi_\alpha^\dagger\hat{\gamma}_0\psi_\alpha$	

DD-PC potential-energy density			
	basic couplings $T=0$	$T=1$	**density dependence** T=0 & 1
$\rho\rho$	$\alpha_\omega\rho_0^2$	$\alpha_\rho\rho_1^2$	$\alpha_i = a_i + (b_i + c_i x)e^{-d_i x}$
mass	$\alpha_\sigma\rho_S^2$		$i \in \{\sigma,\omega,\rho\}\,,\ x = \dfrac{\rho_0}{\rho_{\rm sat}}$
grad.	$f_S(\boldsymbol{\nabla}\rho_S)^2$		

DD-ME potential-energy density			
	basic couplings $T=0$	$T=1$	**density dependence**
$\rho\rho$	$g_\omega^2\rho_0\mathcal{F}_\omega\rho_0$	$g_\rho^2\rho_1\mathcal{F}_\rho\rho_1$	$g_i = a_i\frac{1+b_i(x+d_i)^2}{1+c_i(x+d_i)^2}\,,\ i \in \{\sigma,\omega\}$
mass	$g_\sigma^2\rho_S\mathcal{F}_\sigma\rho_S$		$g_\rho = g_\rho e^{-a_\rho(x-1)}$

NL-ME potential-energy density			
	basic couplings $T=0$	$T=1$	**density dependence**
$\rho\rho$	$g_\omega^2\rho_0\mathcal{F}_\omega\rho_0$	$g_\rho^2\rho_1\mathcal{F}_\rho\rho_1$	
mass	$g_\sigma^2\rho_S\sigma$		$-\nabla^2\sigma + U'_\sigma(\sigma) = g_\sigma\rho_S$

RMF around which can differ substantially in their formal structure. We have chosen here three representatives from that large family. The first one is the traditional RMF based on a coupling of meson exchange (ME) terms[9] augmented with a non-linear self-coupling of the scalar-isoscalar meson.[10] A first quality fit, parametrization NL1,[11] has proven that the non-linear ME model can compete in performance with SHF and it has been used widely since, for early reviews see Refs. 12, 13. Meanwhile, there have been a great variety of developments to improve RMF. As the other two candidates in our comparison, we include recent models with educated density dependencies. One is based on the ME form with density dependent meson coupling strengths, abbreviated as DD-ME. The other is based on a point coupling (PC) model where mesons are eliminated in favor of terms bi-linear in nucleon densities,[14] used here also with density dependent couplings, thus abbreviated as DD-PC.[15–17]

The upper panel of Table 2 shows the densities which enter all RMF. As in SHF, we show only the time even parts used in stationary calculations, thus the zeroth components of the vector density, isoscalar ρ_0 as well as isovector ρ_1, and the isoscalar scalar density ρ_S. This looks at first glance very different than in SHF with its kinetic-energy and spin-orbit densities. But one can show that kinetic density τ and spin-orbit density $\mathbf{J}$ emerge in the non-relativistic limit from ρ_S, for details see section 3. The principal difference between SHF and RMF is that the two densities τ and $\mathbf{J}$ are independent quantities in SHF, whereas they are tightly related through ρ_S in the non-relativistic limit of RMF. Moreover, the RMF does not invoke an isovector counterpart of ρ_S thus being more restricted in the isovector channel.

The lower panels of Table 2 display the main components of the potential-energy density $\mathcal{E}_{\mathrm{pot}}$ for the three variants of RMF under consideration. Structurally closest to SHF is the case of DD-PC.[17,18] It is built from bi-linear terms (point couplings) which form a Lorentz scalar. A term with gradients of density is added to account for finite-range effects. This is done only in the scalar channel where the range of the corresponding scalar meson field is largest in the ME models. Note that, unlike SHF, the model is rather sparse in the isovector channel. On the other hand, the density dependence is richer, attributing density dependent coupling constants to each non-derivative term.

The DD-ME model has the same structure of couplings as DD-PC, two vector terms and one isoscalar-scalar term. But here, the densities are connected by the meson-field propagators $\mathcal{F}_i$ giving each coupling a finite range.[18–20] This allows us to skip the gradient term. As in DD-PC, the DD-ME model is sparse in the isovector channel but rich in density dependence augmenting each coupling constant with density dependence. (Note that there are boundary conditions in the density dependencies such that not every parameter therein is a free parameter).

The NL-ME model has the same nucleon couplings, mediated by meson fields, as DD-ME. It differs in the way the density dependence is parametrized. Here the coupling strengths are constants independent of density. The necessary non-linearity is brought in by a non-linear self coupling of the scalar meson field. This has only two free parameters and acts only in the scalar-isoscalar channel. Thus it is less flexible than both DD models.

The four self-consistent mean-field models thus allow us to probe the influences from different structures, relativistic versus non-relativistic kinematics, point coupling versus finite range, and meson self-coupling versus density dependent nucleon couplings. The four models differ in the number of free parameters for the potential-energy density $\mathcal{E}_{\rm pot}$. There are 11 adjustable parameters for SHF, 10 for DD-PC, 8 for DD-ME, and 6 for NL-ME (the masses of the two vector meson fields, ω and ρ meson, are frozen at their empirical values).

All models are augmented by a pairing functional to allow for the proper description of open-shell nuclei. There is a great variety of pairing recipes around, for the RMF see Chapter 2. For SHF, we use the pairing functional derived from a density-dependent zero-range force:

$$E_{\rm pair} = \frac{1}{4}\sum_{q\in\{p,n\}} V_{{\rm pair},q}\int d^3r\tilde{\rho}_q^2\left[1-\frac{\rho(\mathbf{r})}{\rho_{\rm pair}}\right], \tag{2a}$$

$$\tilde{\rho}_q(\mathbf{r}) = \sum_{\alpha\in q} u_\alpha v_\alpha \left|\psi_\alpha(\mathbf{r})\right|^2, \tag{2b}$$

where q runs over protons and neutrons. It involves the pair-density $\tilde{\rho}_q$ and is usually augmented by some density dependence, switched by the parameter $\rho_{\rm pair}$. We consider here $V_{{\rm pair},p}$, $V_{{\rm pair},n}$, and $\rho_{\rm pair}$ as free parameters of the pairing functional in SHF. Note that we do not recouple to isoscalar and isovector terms because pairing is considered independently for protons and neutrons. Actually, the zero-range pairing force (2) works only together with a limited phase space for pairing. We use here a soft cut-off in the space of single-nucleon energies[21] in the dynamical formulation of Ref. 22. One RMF model, namely NL-ME, uses the same pairing prescription as SHF. For DD-PC and DD-ME, we use the recently developed separable pairing force.[23,24] It is separable in momentum space, and is completely determined by two parameters corresponding to the pairing strengths $V_{\mathrm{pair},p}$ and $V_{\mathrm{pair},n}$ above. All these modern pairing recipes yield very similar results and most observables we will look at are rather insensitive to it anyway.

3. Non-relativistic functional as limit from RMF

The Schrödinger equation can be derived as the non-relativistic limit of the Dirac equation by means of a Foldy–Wouthuysen transformation.[25] With the same techniques, one can map the Lagrangian of RMF to a Skyrme-like EDF.[12,26] This rather simple derivation works nicely for stationary states. However, it fails to recover Galilean invariance of the non-relativistic EDF as non-relativistic limit of Lorentzean invariance of RMF. To obtain proper Galilean invariance, one needs to drive the expansion to second order and to handle normalization with great care as was worked out carefully in Ref. 27. For simplicity, we review here briefly the simpler version of the non-relativistic limit which suffices for stationary theories. This complies with our presentation of the EDF in section 2 where we also ignored all time-odd contributions.

Dirac four-spinor wavefunctions ψ_α separate into upper and lower two-spinor components as

$$\psi_\alpha = \begin{pmatrix} \varphi_\alpha^{(\mathrm{u})} \\ \varphi_\alpha^{(\mathrm{d})} \end{pmatrix} . \tag{3}$$

Its normalization is

$$1 = \int \mathrm{d}^3 r \bar{\psi}\gamma_0\psi = \int \mathrm{d}^3 r \left((\varphi_\alpha^{(\mathrm{u})})^\dagger \varphi_\alpha^{(\mathrm{u})} + (\varphi_\alpha^{(\mathrm{d})})^\dagger \varphi_\alpha^{(\mathrm{d})} \right) . \tag{4}$$

The non-relativistic two-spinor wavefunction $\varphi_\alpha^{(\mathrm{cl})}$ is closely related to $\varphi_\alpha^{(\mathrm{u})}$, but not identical to it because it should be normalized as

$$\int \mathrm{d}^3 r \left|\varphi^{(\mathrm{cl})}\right|^2 = 1 . \tag{5}$$

The way from ψ_α to $\varphi_\alpha^{(\mathrm{cl})}$ is textbook standard since long.[25] The task here is more involved. We aim at deriving a non-relativistic EDF from a relativistic one which requires us to extend the derivation also to the densities. For simplicity, we reduce the EDF to the essential and consider the simplest covariant EDF built from point couplings. We ignore isovector coupling, density dependence, and finite range as being irrelevant for what we want to demonstrate. The starting EDF for potential energy thus reads

$$\mathcal{E}_{\rm pot} = \frac{\alpha_\sigma}{2}\rho_S^2 + \frac{\alpha_\omega}{2}\rho_0^2 \tag{6}$$

with

$$\rho_S = \sum_\alpha \bar{\psi}_\alpha \psi_\alpha = \sum_\alpha \left[\varphi_\alpha^{(\rm u)\dagger}\varphi_\alpha^{(\rm u)} - \varphi_\alpha^{(\rm d)\dagger}\varphi_\alpha^{(\rm d)}\right] , \tag{7a}$$

$$\rho_0 = \sum_\alpha \bar{\psi}_\alpha \gamma_0 \psi_\alpha = \sum_\alpha \left[\varphi_\alpha^{(\rm u)\dagger}\varphi_\alpha^{(\rm u)} + \varphi_\alpha^{(\rm d)\dagger}\varphi_\alpha^{(\rm d)}\right] , \tag{7b}$$

with γ_0 being the zeroth component of the four-vector of Dirac matrices.[25] Note that the index 0 in ρ_0 has now a double meaning. It stands for the $T = 0$ component as we are discussing here a simple model with isoscalar couplings only, and it stands for the zeroth component of the four-vector density ρ_μ. But this ambiguity exists only in this formal section and should not cause confusion. Furthermore, we simplify in this section notation by dropping the state label α, identifying, e.g., the scalar density with $\rho_S = \bar{\psi}\psi$ and similarly for all other densities and currents. The goal is now to obtain from the relativistic parent functional (6) a non-relativistic functional based on the non-relativistic densities ρ, τ, $\mathbf{J}$, see Table 1. Solutions in the positive-energy branch (particle-like) are distinguished by a dominance of $\varphi^{(\rm u)}$ over $\varphi^{(\rm d)}$. The strategy is thus to eliminate $\varphi^{(\rm d)}$ and identify the upper component with the classical two-spinor wavefunction, $\varphi^{(\rm u)} \longleftrightarrow \varphi^{(\rm cl)}$. We finally express ψ in terms of $\varphi^{(\rm cl)}$. Inserting that into the EDF (6) will generate the desired limit.

Starting point are the relativistic equations of motion derived from the EDF (6)

$$0 = \left(\mathrm{i}\gamma^\mu\partial_\mu - m + S + \gamma_0 V_0\right)\psi_\alpha , \tag{8a}$$

$$S = -c_s\rho_S, \quad V_0 = c_v\rho_0 , \tag{8b}$$

with the self-consistent scalar and vector potentials S and V_μ. Note that we consider a stationary case where only the component ρ_0 of the vector density ρ_μ survives and similarly only component V_0 of the vector field V_μ.

We insert the decomposition (3) into the stationary Dirac equation according to the EDF (6) which yields the coupled equations

$$\begin{aligned}(-\epsilon + S + V_0)\varphi_\alpha^{(\rm u)} + \hat{\boldsymbol{\sigma}}\cdot\hat{\mathbf{p}}\varphi_\alpha^{(\rm d)} &= 0 ,\\ (2m + \epsilon + S - V_0)\varphi_\alpha^{(\rm d)} - \hat{\boldsymbol{\sigma}}\cdot\hat{\mathbf{p}}\varphi_\alpha^{(\rm u)} &= 0 ,\end{aligned} \tag{9a}$$

where $\hat{\mathbf{p}} = -\mathrm{i}\nabla$. We resolve the second equation for $\varphi^{(\mathrm{d})}$. Keeping only terms up to order p/m yields

$$\varphi^{(\mathrm{d})} = B_0 \hat{\boldsymbol{\sigma}}\cdot\hat{\mathbf{p}}\varphi^{(\mathrm{u})} \ , \ B_0 = \frac{1}{2m+S-V_0} \approx \frac{1}{2m} . \tag{10}$$

The approximation $B_0 = 1/m$ ignores the density dependence in B_0. It suffices for the present derivation. The form (10) violates the normalization (4) to second order in p/m. The procedure of Ref. 12 aims to restore (ortho-)normalization at operator level by imposing the relativistic normalization (4) up to $(p/m)^2$, namely

$$1 = \int \mathrm{d}^3 r\, \varphi^{(\mathrm{u})\dagger} \left[1 + \hat{\mathcal{T}}\right] \varphi^{(\mathrm{u})} \ , \ \hat{\mathcal{T}} = B_0^2 \left(\hat{\boldsymbol{\sigma}}\cdot\hat{\mathbf{p}}\right)^2 \ . \tag{11}$$

The non-relativistic wavefunction $\varphi^{(\mathrm{cl})}$ should recover non-relativistic normalization (5) which is achieved by regrouping the normalization integral as

$$1 = \int \mathrm{d}^3 r\, \underbrace{\varphi^{(\mathrm{u})\dagger} \left[1 + \hat{\mathcal{T}}\right]^{1/2}}_{\varphi^{((\mathrm{cl})\dagger}} \underbrace{\left[1 + \hat{\mathcal{T}}\right]^{1/2} \varphi^{(\mathrm{u})}}_{\varphi^{((\mathrm{cl})}} .$$

We expand in second order of p/m and obtain

$$\varphi^{(\mathrm{u})} = \left[1 + \frac{1}{2}\hat{\mathcal{T}}\right]^{-1/2} \varphi^{(\mathrm{cl})} \approx \left[1 - \frac{1}{2}\hat{\mathcal{T}}\right] \varphi^{(\mathrm{cl})} \tag{12}$$

which, together with relation (10), provides a complete description of ψ in terms of $\varphi^{(\mathrm{cl})}$ up to order p/m. Using $(\hat{\boldsymbol{\sigma}}\cdot\mathbf{A})(\hat{\boldsymbol{\sigma}}\cdot\mathbf{B}) = \mathbf{A}\cdot\mathbf{B} - \mathrm{i}\mathbf{A}\cdot(\hat{\boldsymbol{\sigma}}\times\mathbf{B})$ yields

$$\frac{\hat{\mathcal{T}}}{2B_0^2} = \hat{\mathbf{p}}^2 + \mathrm{i}\hat{\mathbf{p}}\cdot\hat{\boldsymbol{\sigma}}\times\hat{\mathbf{p}} \ . \tag{13}$$

We insert eqs. (12) with (13) into expressions (7) for the relativistic densities and finally obtain (of course, up to order $(p/m)^2$)

$$\rho_S = \rho^{(\mathrm{cl})} - 2B_0^2 \Big[\tau - \nabla\cdot\mathbf{J}\Big] \ , \ \rho_0 = \rho^{(\mathrm{cl})} \ , \tag{14}$$

where we introduce for this moment the non-relativistic density $\rho^{(\mathrm{cl})}$ to avoid ambiguities in notation. The $\rho^{(\mathrm{cl})}$ is the density according to Table 1. The same relations are obtained when going through the Foldy–Wouthuysen transformation up to second order.

$$\mathcal{E}_{\mathrm{pot}}^{(\mathrm{class})} = \underbrace{\frac{\alpha_\sigma + \alpha_\omega}{2}}_{c_0^\rho} \rho^2 - \underbrace{\alpha_\sigma B_0^2}_{C_0^\tau = C_0^{\nabla \mathbf{J}}} \rho \left[\tau - \nabla\cdot\mathbf{J}\right] \ . \tag{15}$$

This is the non-relativistic limit of the EDF. It has indeed the structure of the isoscalar channel of the SHF functional from Table 1. The kinetic and spin-orbit terms emerge from the non-relativistic expansion of the scalar density ρ_S. It is particularly noteworthy that these two terms are closely linked in RMF because the limit delivers the same parameter for both. This means that an appropriate description of the spin-orbit splitting requires a certain value for the effective mass (kinetic term). We will see later on that RMF models all have rather low effective mass.

The strict relation between effective mass and spin-orbit splitting holds only for the models with the simple couplings as the ones which we are considering here (see Table 2). There are variants of RMF which include tensor coupling. These render, in principle, the spin-orbit strength independent of the effective mass. But empirical fits so far led to rather small tensor terms such that the tight connection between m^*/m and spin-orbit terms persists in practice.[12]

The non-relativistic functional (15) shows the spin-orbit and kinetic term as bi-linear point couplings. This is the outcome of the approximation $B_0 = 1/(2m)$ in eq. (10). Actually, $S - V_0$ is not small and we deal in practice with $B_0 = 1/(2m + S - V_0)$. The scalar field S and vector field V_0 depend on density. Thus kinetic and spin-orbit terms acquire some density dependence. This differs from SHF models which ignore density dependence in these terms. It seems that this difference is a detail which has not so much effect in the practical performance as we will see in section 6.

We remind once more that the above result is correct for stationary states where all time-odd densities and fields vanish, i.e. where $\mathbf{j} = 0$ and $\boldsymbol{\sigma} = 0$. A proper derivation which includes also time-odd densities requires a more elaborate discussion of the normalization condition because non-vanishing velocities enter the volume element of the normalization integral, for details see Ref. 27. We also warn that the limit up to $(p/m)^2$ is not sufficient for a quantitative approximation to the Dirac equation. This requires higher orders.

4. Observables

In this section, we introduce the observables which will be used for the comparison of models in section 6. These are energies and radii in a few selected nuclei as outlined in section 4.1. Moreover, we include a couple of nuclear matter properties (NMP) as outlined in section 4.2. NMP are, in fact, not accessible by experiment. But they serve as extremely useful pseudo-observables which allow us to characterize the basic physical properties of a model. Most EDFs allow a one-to-one mapping of the basic NMP to model parameters such that NMP are a more intuitive way to describe the model parameters and compare it across models.

4.1. *Observables from finite nuclei*

Calculations based on an EDF by definition deliver binding energy E_B and local density $\rho_{p/n}(\mathbf{r})$ as observables. Binding energy is the most crucial observable and we will discuss it for a great variety of nuclei. A more detailed view is often obtained by considering differences of binding energies which can reveal other features than total binding alone. We will consider in the following the two-neutron separation energy S_{2n}, the two-proton separation energy S_{2p}, and the α-particle energy Q_α, which are defined in detail as

$$S_{2n} = E_B(Z,N-2) - E_B(Z,N) \ , \tag{16a}$$

$$S_{2p} = E_B(Z-2,N) - E_B(Z,N) \ , \tag{16b}$$

$$Q_\alpha = E_B(Z-2,N-2) + E_B(2,2) - E_B(Z,N) \ . \tag{16c}$$

For the nuclear density distribution, the best accessible signal is the charge density $\rho_C(\mathbf{r})$ which is measured in elastic electron scattering. The scattering cross sections are analyzed with distorted-wave Born approximation which allows a more or less model-free experimental determination of the nuclear charge formfactor $F_\mathrm{C}(\mathbf{k})$.[28] It is related to the density by a simple Fourier transformation $F_\mathrm{C}(\mathbf{k}) = \int \mathrm{d}\mathbf{r}\, e^{\mathrm{i}\mathbf{k}\cdot\mathbf{r}}\rho_C(\mathbf{r})$. For spherical systems, it depends only on $k = |\mathbf{k}|$, which we will assume for the following explanations. Mean-field calculations yield first proton and neutron formfactors $F_p(k)$ and $F_n(k)$. The charge formfactor F_C is obtained by multiplying $F_p(k)$ and $F_n(k)$ with the intrinsic proton and neutron formfactors G_E and G_M (equivalent to folding of the densities). This is indicated for the case of SHF in the first line of Table 3. Note that this also includes the spin-orbit formfactors $F_{ls,q}$ which are Fourier transforms of $\nabla \cdot \mathbf{J}_q$ augmented by a factor $\mu_q/4m^2$ where μ_q is the magnetic moment of the nucleon. The $G_{E,q}$ is the electric formfactor of the nucleon and G_M its magnetic formfactor (treated as equal for both species). The exponential factor accounts for the center-of-mass correction. It employs the same variance of the total momentum $\langle\hat{\mathbf{P}}^2_\mathrm{cm}\rangle$ as the energy correction E_cm in eq. (1). The nucleon formfactors $G_{E,q}$ and G_M are taken from nucleon scattering data, for details see Ref. 1, 28. For RMF, the composition of F_C reads a bit different

Table 3. Definition of the charge formfactor $F_C(k)$, the r.m.s. radius r_C, diffraction radius R_C and the surface thickness σ_C.

charge formfactor:	$F_C(k)$	=	$\sum_{q\in\{p,n\}}[F_q G_{E,q} + F_{ls,q} G_M] \exp\left(\frac{\hbar^2 k^2}{8\langle\hat{\mathbf{P}}^2_{\rm cm}\rangle}\right)$
r.m.s. radius:	$r_{\rm rms,C}$	=	$\frac{3}{F_C(0)}\frac{d^2}{dk^2}F_C(k)\Big\vert_{k=0}$
diffraction radius:	$R_{\rm diffr,C}$	=	$\frac{4.493}{k_0^{(1)}}$, $F_C(k_0^{(1)}) = 0$
surface thickness:	σ_C	=	$\frac{2}{k_m}\log\left(\frac{F_{\rm box}(k_m)}{F_C(k_m)}\right)$
			$F_{\rm box}(k) = 3\frac{j_1(kR_C)}{kR_C}$, $k_m = 5.6/R_C$.

in that the spin-orbit formfactor is to be replaced by the formfactor of the tensor density.[11,12]

Exact EDFs should, in principle, provide density distributions $\rho(\mathbf{r})$, or formfactors $F(\mathbf{k})$ respectively, in all details.[29] However, practicable functionals as SHF and RMF employ analytically simple forms which are smooth functions of the densities. It has been shown that this limits the predictive value for formfactors to the regime $k < 2k_F$ where k_F is the Fermi momentum.[30] Fortunately, the most crucial features are determined in that range of low k. Table 3 shows the definition of these bulk parameters of the charge density: r.m.s. radius $r_{\rm rms,C}$, diffraction radius $R_{\rm diffr,C}$, and surface thickness σ_C.[28] The R_C is called diffraction radius because it parametrizes the overall diffraction pattern which resembles those of a filled sphere of radius R_C. The actual F_C decreases faster than the box formfactor $F_{\rm box}$ due to the finite surface thickness σ_C of nuclei which is thus determined by comparing the height of the first maximum of the box equivalent formfactor and of the mean-field result F_C.

The charge formfactor is predominantly a measure of the the proton distribution. Useful complementing information is contained in the neutron distribution which was not well accessible in the past. Recent scattering experiments with high-energy protons[31] or electrons[32] together with elaborate scattering theory start to deliver reliable information on the neutron formfactor, in particular the neutron r.m.s. radius r_n. We will use this observable in terms the neutron skin $r_{\rm skin} = r_{\rm n} - r_{\rm p}$. This difference of radii amplifies effects and is thus often better suited for a comparison of models.

Table 4. Definitions of NMP. All derivatives are to be taken at the density of symmetric matter at equilibrium.

incompressibility:	K	$=$	$9\,\rho_0^2 \frac{d^2}{d\rho_0^2}\frac{E}{A}\Big\vert_{\rm eq}$
symmetry energy:	$a_{\rm sym}$	$=$	$\frac{1}{2}\rho_0^2\frac{d^2}{d\rho_1^2}\frac{E}{A}\Big\vert_{\rm eq}$
slope of $a_{\rm sym}$:	L	$=$	$3\rho_0\frac{da_{\rm sym}}{d\rho_0}\Big\vert_{\rm eq}$
effective mass – SHF:	$\frac{\hbar^2}{2m^*}$	$=$	$\frac{\hbar^2}{2m}+\frac{\partial}{\partial\tau_0}\frac{E_{\rm EDF}}{A}\Big\vert_{\rm eq}$
effective mass – RMF:	m^*	$=$	$m+G_\sigma\rho_S$
TRK enhancement:	$\kappa_{\rm TRK}$	$=$	$\frac{m}{m^*}-1+\frac{1}{2}\rho_0\frac{d}{d\rho_1}\frac{m}{m_n^*}$

4.2. *Nuclear matter properties (NMP)*

Starting point for the definition of NMP is the energy per nucleon in the nuclear matter $E_{\rm EDF}/A(\rho_0,\rho_1,\tau_0,\tau_1)$. At the level of an energy functional, kinetic-energy densities τ_T are independent of densities ρ_T. Energy minimization establishes a relation $\tau_i=\tau_i(\rho_0,\rho_1)$. This yields the commonly used binding energy of nuclear matter

$$\frac{E}{A}(\rho_0,\rho_1)=\frac{E_{\rm EDF}}{A}(\rho_0,\rho_1,\tau_0(\rho_0,\rho_1),\tau_1(\rho_0,\rho_1)) \tag{17}$$

as a function of the densities ρ_T alone. Table 4 lists the NMP discussed in this work. We consider τ_T as independent variables for the purpose of a formally compact definition of the effective mass in SHF. This is indicated by the partial derivatives with respect to τ_T for m^*/m in SHF and $\kappa_{\rm TRK}$. Static properties are deduced from the binding energy of nuclear matter E/A which depends on ρ_T only. This is indicated by using the total derivatives for K, $a_{\rm sym}$, and L. The slope of the symmetry energy L parametrizes the density dependence of $a_{\rm sym}$. The kinetic properties, effective mass m^*/m and Thomas–Reiche–Kuhn (TRK) sum-rule enhancement factor $\kappa_{\rm TRK}$, require access to the kinetic densities τ_T which still exists in $E_{\rm EDF}/A$. The notion of effective mass is ambiguous in RMF. In this survey, we have decided to use what is called the Dirac mass.[16] (Alternatively, one may talk about the effective mass at the Fermi surface, which is typically 10% larger than the Dirac mass.) The $\kappa_{\rm TRK}$[33] characterizes the isovector effective mass. Again, this quantity is ambiguous in RMF. We use here $\kappa_{\rm TRK}$ based on Dirac masses. (The alternative value deduced from the mass at the Fermi momentum is typically 20% smaller.) The quantity m_n^* is the neutron effective mass. It is defined in the same way as the isoscalar effective mass m^* but using the derivative with respect to the neutron kinetic density τ_n.

Each one of these NMP characterize one specific part of the EDF. The incompressibility K stands for the static isoscalar response while the symmetry energy $a_{\rm sym}$ quantifies the static isovector response, closely related to dipole polarizability.[34] The slope of symmetry energy L plays, together with $a_{\rm sym}$, a role in static isovector response for finite nuclei and neutron stars.[35] Dynamic response is characterized by effective mass m^* for the isoscalar channel and by the TRK sum-rule enhancement $\kappa_{\rm TRK}$ for the isovector one; it is crucial for determining the peak frequencies of giant resonances.[36]

5. Least squares fits and statistical analysis

5.1. *Optimization, variances and covariances*

It is typical for nuclear self-consistent mean-field models that one can motivate their formal structure by microscopic considerations as, e.g., low-momentum expansion.[1,4,5] But the model parameters remain undetermined. We denote them throughout this section in a model-independent way as $\boldsymbol{p} = (p_1 ... p_{N_p})$. We have seen above that all self-consistent mean-field models remain with a couple of free parameters: 11 for SHF, 10 for DD-PC, 8 for DD-ME, and for NL-ME. Additionally we have 2–3 parameters for the pairing functionals. Ab-initio calculations are not yet precise enough to use them for a high-quality calibration of the free parameters. The present day standard is to optimize them by a least squares fit to a well chosen pool of experimental data. To that end, one selects a representative set of observables $\{\hat{\mathcal{O}}_i, i = 1...N_d\}$, typically gross properties of the nuclear ground state as binding energies and radii. Then one proceeds along the standard scheme of statistical χ^2 analysis.[37] First straightforward χ^2 fits for SHF were done in Ref. 38 and for RMF in Ref. 11. Meanwhile, the scheme has been developed steadily concerning data base and numerical procedures, see e.g. Ref. 39. The recent paper[40] contains a very detailed explanation of least-squares (χ^2) fits in nuclear context and related error analysis. We repeat here briefly the basic formula. It begins with defining a quality function as

$$\chi^2(\boldsymbol{p}) = \sum_{i=1}^{N_d} \frac{(\mathcal{O}_i(\boldsymbol{p}) - \mathcal{O}_i^{\rm exp})^2}{\Delta\mathcal{O}_i^2} , \tag{18}$$

where $\mathcal{O}_i(\boldsymbol{p})$ stands for the calculated values, $\mathcal{O}_i^{\rm exp}$ for experimental data, and $\Delta\mathcal{O}_i$ for adopted errors. The best-fit parameters $\boldsymbol{p}_0$ are those for which χ^2 becomes the minimum, i.e. $\chi_0^2 = \chi^2(\boldsymbol{p}_0) = \chi^2_{\rm min}$. The adopted errors $\Delta\mathcal{O}_i$ should be chosen such that $\chi_0^2 = N_d - N_p$ which is the number of degrees of freedom in the fit, see Ref. 40.

Least squares fits can deliver more than just an optimized parametrization $\mathbf{p}_0$. This idea is based on the observation that not only the minimum, but also some vicinity of it represents a reasonable reproduction of data. Assuming a statistical distribution of errors, one can deduce a probability distribution $W(\boldsymbol{p}) \propto \exp(-\chi^2(\boldsymbol{p}))$ of reasonable model parameters. Their domain is characterized by $\chi^2(\boldsymbol{p}) \leq \chi_0^2 + 1$. This vicinity is usually very narrow and we can apply Taylor expansion up to second order, i.e.[37,40]

$$\chi^2(\boldsymbol{p}) \approx \chi_0^2 + \sum_{\alpha,\beta=1}^{N_p} (p_\alpha - p_{0,\alpha})(\mathcal{C}^{-1})_{\alpha\beta}(p_\beta - p_{0,\beta}), \tag{19}$$

$$(\mathcal{C}^{-1})_{\alpha\beta} = \tfrac{1}{2}\partial_{p_\alpha}\partial_{p_\beta}\chi^2\Big|_{\boldsymbol{p}_0} = \sum_i J_{i\alpha}J_{i\beta} , \tag{20}$$

$$J_{i\alpha} = \frac{\partial_{p_\alpha}\mathcal{O}_i\Big|_{\boldsymbol{p}_0}}{\Delta\mathcal{O}_i} , \tag{21}$$

where $\hat{J}$ is the Jacobean matrix and $\mathcal{C}$ the covariance matrix. The matrix $\mathcal{C}$ carries information about the allowed variations in each direction and thus plays the key role in covariance analysis. Inserting approximation (19) into the probability distribution yields the simple Gaussian profile

$$W(\boldsymbol{p}) \propto \exp(-\frac{1}{2}\boldsymbol{p}\cdot\hat{\mathcal{C}}^{-1}\cdot\boldsymbol{p}) . \tag{22}$$

The domain of reasonable (most probable) parameters is thus given by $\boldsymbol{p}\cdot\hat{\mathcal{C}}^{-1}\cdot\boldsymbol{p} \leq 1$ which defines a confidence ellipsoid in the space of model parameters. The variance of $\boldsymbol{p}$ thus obtained can be carried forth to estimate the variances, in other words the extrapolation errors, of an observable.

Statistical expectation values of a function $F(\boldsymbol{p})$ are defined as

$$\overline{F} = \int d\boldsymbol{p}\, W(\boldsymbol{p})\, F(\boldsymbol{p}) . \tag{23}$$

Any observable A is a function of model parameters $A = A(\boldsymbol{p})$. Within the Gaussian approximation (22), the statistical average then recovers the value at the optimal parametrization $\overline{A} = A_0 = A(\boldsymbol{p}_0)$. The values of A vary within the confidence ellipsoid, and this results in some uncertainty ΔA around the best-fit value A_0. Usually, one can assume that A varies weakly such that one can linearize

$$A(\boldsymbol{p}) \simeq A_0 + \boldsymbol{G}^A \cdot (\boldsymbol{p} - \boldsymbol{p}_0) \ , \ \boldsymbol{G}^A = \partial_{\boldsymbol{p}} A\Big|_{\boldsymbol{p}_0} \quad . \tag{24}$$

Its variance is obtained as average

$$\Delta A = \sqrt{\overline{\Delta A^2}} \quad , \quad \overline{\Delta A^2} = \overline{A^2} - (\overline{A})^2 = \boldsymbol{G}^A \hat{\mathcal{C}} \boldsymbol{G}^A \ . \tag{25}$$

In section 6, we will show the results for expectation values $\overline{A}$ together with variances ΔA where available. Alternative names for variances are r.m.s. deviations or extrapolation errors and will be used in the following.

One can go one step further and consider correlated averages $\overline{AB}$. This yields the correlation, or covariance, between two observables A and B as

$$c_{AB} = \frac{|\overline{\Delta A \, \Delta B}|}{\Delta A_0 \, \Delta B_0} \quad . \tag{26}$$

A value $c_{AB} = 1$ means fully correlated where knowledge of $A(\boldsymbol{p})$ fully determines $B(\boldsymbol{p})$. A value $c_{AB} = 0$ means uncorrelated, i.e. $A(\boldsymbol{p})$ and $B(\boldsymbol{p})$ are statistically independent. We will discuss insights from covariance analysis in section 6.6.

5.2. *Fitting strategy and choice of data*

The key to success in parameter calibration is an appropriate choice of the data basis for the fit. A great variety of different fit data pools has been used in the literature producing accordingly a variety of parametrizations.[1] For a fair comparison of different EDF at the same footing, one must fit them to the same data pool. This is what we will do in the following. We calibrate the four models under consideration, SHF, DD-PC, DD-ME, and NL-ME, to the recently developed large data set of Ref. 36 which was selected to include only nuclei which have small correlation effects[41] and can thus be computed safely within a pure self-consistent mean-field model. The selection of data is listed in Table 5. For three models, this amounts to the already published parametrizations, the non-relativistic SV-min from Ref. 36 and the best-fit DD-PC as well as DD-ME parametrizations from Ref. 34 which were obtained using the published code of Ref. 18.

Table 5. Compilation of phenomenological input for fitting the four newly optimized parametrizations SV-min, DD-PC, DD-ME, and NL-ME. Abbreviations mean: E = binding energy, $r_{\rm rms,C}$ = charge r.m.s. radius, $R_{\rm diffr,C}$ = charge diffraction radius, $\sigma_{\rm C}$ = charge surface thickness, pairing gaps Δ_p for protons and Δ_n for neutrons deduced from even-odd staggering, isotope shift of charge r.m.s. radii $\delta r^2(^{214-208}{\rm Pb})$, and $l*s$ = spin-orbit splitting of selected single-particle states.

E:	$^{36-52}$Ca, ^{68}Ni, $^{100,126-134}$Sn, $^{204-214}$Pb, ^{34}Si, ^{36}S, ^{38}Ar, ^{50}Ti, ^{86}Kr, ^{88}Sr, ^{90}Zr, ^{92}Mo, ^{94}Ru, ^{96}Pd, ^{98}Cd, ^{134}Te, ^{136}Xe, ^{138}Ba, ^{140}Ce, ^{142}Nd, ^{144}Sm, ^{146}Gd, ^{148}Dy, ^{150}Er, ^{152}Yb, ^{206}Hg, ^{210}Po, ^{212}Rn, ^{214}Ra, ^{216}Th, ^{218}U
$R_{\rm diffr,C}$:	^{16}O, $^{40-44,48}$Ca, $^{58-64}$Ni, $^{118-124}$Sn, $^{204-208}$Pb, ^{50}Ti, 52Cr, ^{54}Fe, ^{86}Kr, ^{88}Sr, ^{90}Zr, ^{92}Mo, ^{138}Ba, ^{142}Nd,
$\sigma_{\rm C}$:	^{16}O, $^{40-44,48}$Ca, $^{60-64}$Ni, $^{118,122-124}$Sn, $^{204-208}$Pb, ^{50}Ti, ^{86}Kr, ^{88}Sr, ^{90}Zr, ^{92}Mo, ^{138}Ba, ^{142}Nd
$r_{\rm rms,C}$:	^{16}O, $^{40-48}$Ca, $^{108,118-124}$Sn, $^{200-214}$Pb, ^{36}S, ^{38}Ar, ^{50}Ti, 52Cr, ^{54}Fe, ^{86}Kr, ^{88}Sr, ^{90}Zr, ^{92}Mo, ^{136}Xe, ^{138}Ba, ^{140}Ce, ^{142}Nd, ^{144}Sm, ^{146}Gd, ^{148}Dy, ^{150}Er, ^{206}Hg, ^{210}Po, ^{212}Rn, ^{214}Ra
Δ_p	^{134}Te, ^{136}Xe, ^{138}Ba, ^{140}Ce, ^{142}Nd, ^{144}Sm, ^{146}Gd, ^{212}Rn, ^{214}Ra
Δ_n	$^{112-128}$Sn, ^{204}Pb
δr^2	$r(^{214}{\rm Pb}) - r(^{208}{\rm Pb})$
$l*s$	$^{16}{\rm O}(1p_n, 1p_p)$ $^{132}{\rm Sn}(2p_p, 2d_n)$, $^{208}{\rm Pb}(2d_p, 1f_n, 3p_n)$, only SV-min

A parametrization for the traditional EDF of the NL-ME model was freshly adjusted using the same fit data and strategy. To indicate the great variety of parametrizations, we also consider a few parametrizations taken from literature which were tuned with different selections off it data. These are from SHF the force SLy6[42] and SkI3[43] which both have also high quality but differ from SV-min in having a much lower effective mass. SLy6 was fitted with a bias to neutron matter and neutron rich nuclei while SkI3 used a slightly extended SHF functional which allows for a separate isovector spin-orbit term and takes care of the charge formfactor (as SV-min does). For RMF we chose two not too old parametrizations from the traditional

NL-ME family, namely NL3 and NL-Z2. NL3 was fitted with a large data set from exotic nuclei and with a bias on large incompressibility K.[44] NL-Z2 was fitted to the data set of NL1 extended by new data from exotic nuclei.[45] Although the data sets of the different parametrizations differ by the selection of nuclei, they all include already a great deal of exotic nuclei which renders that difference less important. A major difference lies in the fact that NL3 and SLy6 use only the charge r.m.s. radius as observable for the nuclear density profile while the five other parametrizations follow the strategy initiated in Ref. 38 and use more information from the nuclear charge formfactor, namely the diffraction radius $R_{\mathrm{diffr,C}}$ and surface thickness σ_{C}, see section 4.

6. Results

Throughout this section, we are going to use the parametrizations as explained in the section 5.2. At the core of the comparison are the four parametrizations SV-min, DD-PC, DD-ME, and NL-ME which we summarize in the following under the name the "four reference parametrizations". Additionally, we also consider results from four older parametrizations SLy6, SkI3, NL3, and NL-Z2, all being developed in the 1990 decade. These latter parametrizations are taken as published. No statistical analysis (variances and covariances) is possible for them in the present context.

6.1. *An introductory example*

Comparisons between different models have a long tradition. For example, the review[1] discusses RMF and SHF in parallel. A particularly interesting figure therein compares binding energies for all (at that time) known super-heavy elements. We have taken the data from this old figure and complemented it with data from more recent parametrizations. As all these nuclei are well deformed we did include also a correction for the rotational zero-point energy in addition to the center-of-mass correction.[36,41] The summary view is shown in figure 1. All SHF parametrizations tend to underbinding (positive deviation), sometimes a rather dramatic one as for SLy6, while the RMF models of NL-ME type (NL-Z2, NL3) are rather at the side of overbinding with NL-Z2 showing least deviations of all cases. The underbinding in SHF seems to be incurable with the present terms of the functional.[8,46] The exciting new aspect is now to have a look at the RMF variants with elaborate density dependence, DD-PC and DD-ME.

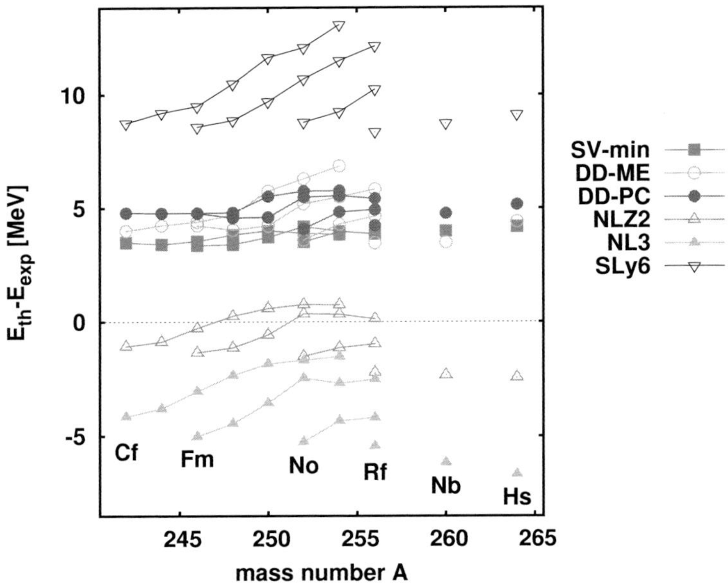

Fig. 1. (Color online) Deviations $E_B^{(\mathrm{th})} - E_B^{(\mathrm{exp})}$ of binding energies for known super-heavy elements and for a couple of relativistic as well as non-relativistic parametrizations as indicated.

These do not follow the trends of NL-ME, but share the underbinding with the SHF models. This indicates that the under- or over-binding is not related to (non-)relativistic kinematics. It is a consequence of the modeling of density dependence. This little example demonstrates how instructive such comparisons between models can be. This motivates us for the nonce following extensive survey. Among the many observables, we will also come back to binding of super-heavy elements. As we learn from figure 1 that the mismatches are very similar throughout the trans-actinides, we will confine later on the discussion to the heaviest ones which are found here to be the most critical.

6.2. *Average quality*

A first characterization of the quality of a parametrization is given by the multitude of deviations $\mathcal{O}_i(\boldsymbol{p}) - \mathcal{O}_i^{\mathrm{exp}}$ for all data $\mathcal{O}_i$ in the fit. This amounts to a lot of detailed information which is thus hard to visualize. A more compact measure of the overall quality is obtained by averaging the deviations for a given observable over all nuclei. For this averaging, we consider the selection of data as listed in table 5. Figure 2 shows average and r.m.s.

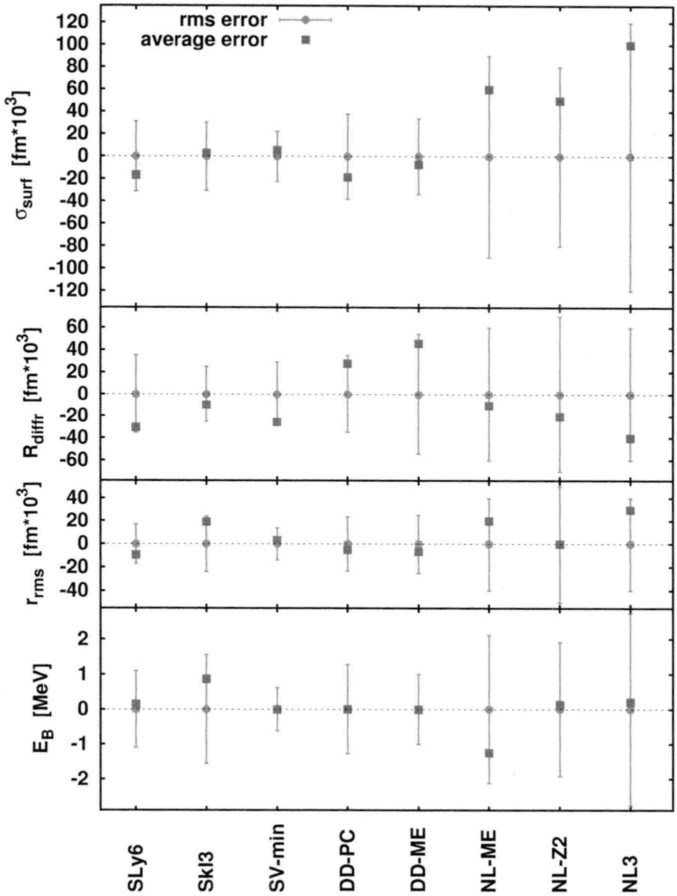

Fig. 2. (Color online) Average quality of the four newly fitted parametrizations for the four basic bulk properties (binding energy and charge distribution), each one displayed in a separate panel. Shown are the deviations $\mathcal{O}_i(\boldsymbol{p}) - \mathcal{O}_i^{\mathrm{exp}}$ averaged for a given observable over all fit nuclei. The mere average is shown by boxes and the r.m.s. averaged deviation by error bars centered around zero for better comparison.

deviation for the observables binding energy, r.m.s. radius, diffraction radius, and surface thickness. Let us first look at the r.m.s. deviations. We see large differences between between the parametrizations. The SHF parametrizations have generally smaller errors than RMF. Of course, within each model family we find that the most recent fits perform best. This is in two ways a consequence of the fit data: first, they constitute a large pool of data which thus gives the fit more grip, and second, the average is done over the same data as were used for the fit. The true comparison of models as such takes place between the four reference parametrizations and here we see that SHF with SV-min has significantly lowest deviations, the RMF models with educated density dependence, DD-PC and DD-ME perform with comparable, although lower, quality, and the traditional model NL-ME has the largest error. The smaller deviations for SV-min relative to all RMF has two sources: SHF has greater versatility concerning effective mass and concerning the isovector channel (each isoscalar coupling term has an isovector cousin) while RMF is inevitably tied to very low effective mass (see section 3) and reduces isovector coupling to vector density only. The larger effect seems to be the deteriorating influence of effective mass as we can read off from the comparison of SV-min (high effective mass) with SLy6 (lower effective mass) and SkI3 (even lower effective mass). There are also noteworthy differences within the RMF family. DD-PC/DD-ME perform better than NL-ME which shows clearly the impact of a better shaping of density dependence. Finally, we learn from the close competition between DD-PC and DD-ME that point couplings (compensated by a gradient term) and finite range meson fields have comparable capability. Comparing r.m.s. deviations between the four observables in table 5 shows that the fits successfully managed to find a good compromise with equilibrating the errors among these fit observables.

Further useful information can be gained from looking at the average deviations (boxes). Ideally they should all be near zero. Larger average deviations indicate stress within the model. Such a situation arises in each parametrization. NL-ME has a large average error in binding energy E_B and surface thickness $\sigma_{\rm C}$, a non-negligible one also for the r.m.s. radius $r_{\rm rms,C}$. This is the price for the compromise between different observables. One may improve, e.g., on the energy but will then drive the formfactor observables, particularly surface thickness, farther off as seen from the case of NL3. It seems that the modeling of density dependence by non-linear meson self-coupling is not well suited to shape the nuclear surface profile properly. DD-PC/DD-ME have perfect average energy and r.m.s. radius. But they have enormous stress with the diffraction radius as the average error is just as large as the r.m.s. error. At a lower level of error, the situation is similar for SV-min. However, the average errors on $R_{\rm diffr}$ go in opposite direction for SV-min as compared to DD-PC/DD-ME. This is most probably again an effect of modeling of density dependence which is equally versatile in these three models, but much different in functional form.

Already at that stage of investigations, we can summarize a few crucial observations: (1) SHF is more versatile than RMF concerning effective mass and isovector channel which leads to a higher overall quality of the description. (2) NL-ME is inferior to DD-PC/DD-ME in r.m.s. deviations and in the trends of average errors which indicates the enormous importance of a proper modeling of density dependence.

6.3. *Nuclear matter properties (NMP)*

As said above, basic physical properties of a nuclear EDF can be conveniently characterized by NMP. These are summarized in figure 3 for our selection of parametrizations. Uncertainties computed with Eq. (25) are given for the four reference parametrizations. Density ρ_{eq} (panel a) and energy per particle E/A (panel b) are the ground state properties of symmetric matter. They determine also the leading contributions to binding of (super-)heavy nuclei. The equilibrium densities ρ_{eq} separate into two groups, SHF, DD-PC, and DD-ME on the one side versus NL-ME models on the other. This shows that the modeling of density dependence has a large impact on the extrapolation of density, and with it radii, to superheavy nuclei. We have seen this already in figure 2 where the ME models behave different, generally worse, concerning radii and surface thickness.

Binding energies E/A (panel b) can also be separated into two groups, although different ones, namely SHF versus RMF. SHF yields generally less binding (see also figure 1). But the variations within the RMF family are large. There is no particular trend among these differences visible and thus no guess for a mechanism explaining it.

Incompressibilities K (panel c) vary widely. For formal reasons, it is obvious that modeling of density dependence plays a crucial role for K, but finite range versus zero range matters as well. Even the fit strategy seems to play a role. The three SHF forces and the three NL-ME forces were fitted to different data sets and this leads to different K for the same model. That variance in predictions has a positive aspect. Reliable experimental information on K provides highly selective information for calibrating a model. Unfortunately its value in matter is not experimentally accessible. But K in finite nuclei is closely related to the giant monopole resonance[47,48] and adding information about it can indeed confine the parametrization.[36]

Effective masses m^*/m (panel d) show particularly interesting trends. All RMF models agree in producing about the same, very low m^*/m. This is clear from the structure of all RMF version considered here. Binding

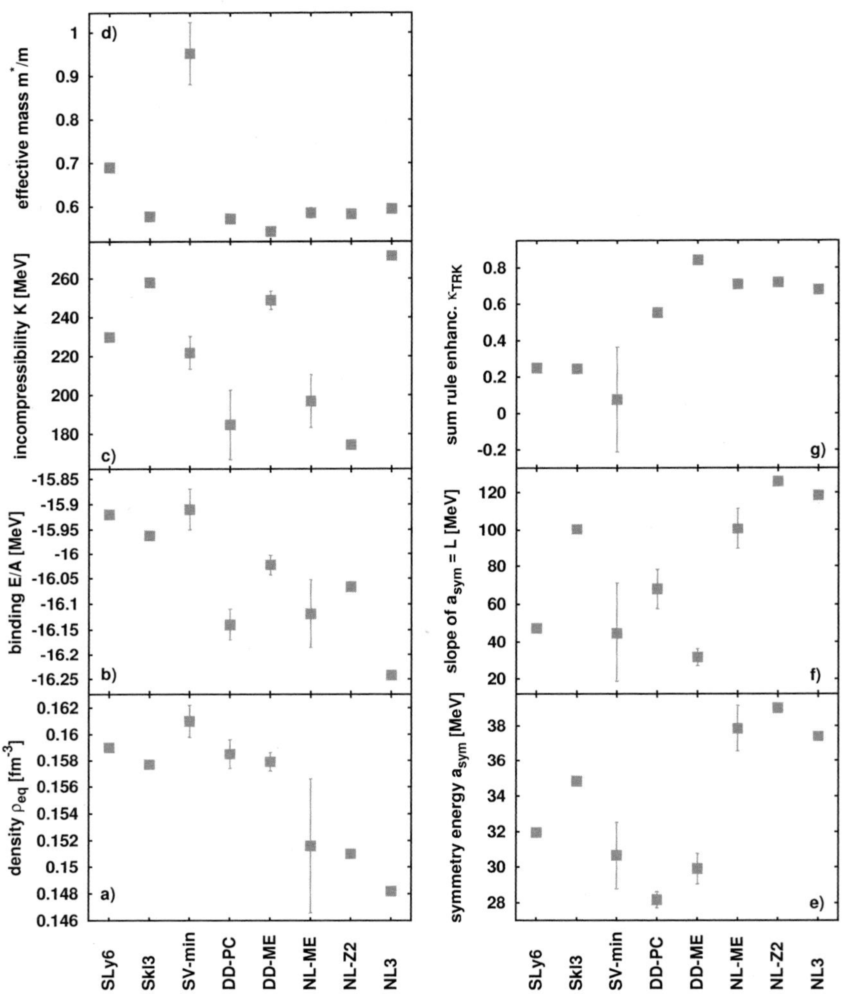

Fig. 3. (Color online) Isoscalar NMP (left panel) and isovector NMP (right panel) for a selection of parametrizations. Extrapolation uncertainties are available for the four fitted parametrizations (SV-min, DD-PC, DD-ME, NL-ME) and are indicated by error bars.

results from near compensation of large scalar and vector fields (see also Chapter 6) and the non-relativistic limit in section 3 shows that this inevitably leads to low m^*/m, and reasonable spin-orbit splitting as we will see in section 6.5. SHF, on the other hand, allows independent calibration of the terms $\propto \tau_{p/n}$ which determine m^*/m. This, in turn, yields a broader span of m^*/m within the pool of reasonable parametrizations. The outcome for m^*/m then depends on the fit strategy and data. SkI3, for example, was fitted to stay close to RMF and indeed has equally low m^*/m. SLy6 did not include R_{diffr} and σ in the fit which leads to an intermediate value of m^*/m. Its inclusion in SV-min drives m^*/m to the larger value which is also required by the giant monopole resonance.[36]

The right column of figure 3 shows isovector response properties in symmetric nuclear matter. The symmetry energy a_{sym} (panel e) shows, once more, a clear distinction of groups. All parametrizations within the traditional NL models produce an unusually large a_{sym} while SHF and the DD-RMF models have smaller a_{sym} in the range of the commonly accepted values around 30–32 MeV.[49] Only SkI3 is a bit off, due to its particular bias on RMF properties in fitting.[43] The fact that DD-PC/DD-ME yield reasonably low a_{sym} indicates that density dependence is here more important than kinematics (relativistic versus non-relativistic).

The slope of symmetry energy L (panel f) shows basically the same trends as a_{sym}. We find indeed that L and a_{sym} are more or less strongly tied to each other. This will be discussed quantitatively in section 6.6 in the framework of covariance analysis. For the discussion in the next subsection, we remark that there are slightly different trends in detail. While a_{sym} has a minimum for DD-PC, the value of L for this force lies above those of SV-min and DD-ME. This is obviously another (small) effect of density dependence.

The TRK sum rule enhancement κ_{TRK} (panel g) is equivalent to the isovector effective mass and consequently it shows the same large distinction between SHF and RMF. All RMF models produce notoriously very large values for κ_{TRK} while SHF results stay in a moderate range. The explanation is the same as for the isoscalar effective mass m^*/m.

We have not yet discussed the uncertainties on the NMP which are available and shown for the four reference parametrizations. In all cases, the uncertainties are smaller than the variations of the NMP. If the uncertainties were purely statistical errors, the variation of results should stay within the error bars. This indicates that we are plagued here by systematic errors of the modeling for which systematic estimates do not exist.[40] What we are doing here, namely to compare different forms of EDFs is one strategy to get an idea about the systematic errors lurking underneath any modeling. We have seen this way the impact of density dependence, meson range, and kinematics. However, the variances of results seen at several places of figure 3 are not directly the systematic errors. These could be smaller. Recall that the quality of all NL models is much worse than for any other model, see figure 2. These models should, in fact, be discarded. This removes unusually large a_{sym} and L as well as surprisingly small densities ρ_0, and so reduces dramatically the variations among the results.

Even though the statistical errors from χ^2 analysis are not covering all possible uncertainties, they are useful quantities. Comparison of them between models can indicate weaknesses and strengths of a model. For example, uncertainties on all isovector properties (right column of figure 3) are systematically larger for SHF as compared to all RMF models. This is understandable as SHF has more freedom (free parameters) in the isovector channel. The low uncertainties in RMF are not necessarily good news. They indicate, in fact, a rigidity of the modeling concerning isovector terms. There is also a large effect for the effective mass m^*/m where SHF, again, allows larger span of value while RMF models show practically no uncertainties. This is a consequence of the strict limitation of the effective mass within RMF as discussed above.

6.4. *Prediction/reproduction of other observables*

Figure 4 shows results for a couple of observables in finite nuclei which were not included in the fit data. This means that we are here exploring the extrapolation properties of the models. Neutron skin (panel a) and neutron radius (panel b) are considered to be observables which are closely tied to isovector response.[50,51] Indeed, we see that its trends behave similarly to the trends of a_{sym} and L in figure 3. On a closer look, we realize that the detailed trends around DD-PC resemble those of L more rather than a_{sym}. This lets us expect a close correlation between L and neutron skin or r_n. This will be confirmed unambiguously by the covariance analysis in section 6.6.

The proton isotope shift $r_p^2(^{214}Pb) - r_p^2(^{208}Pb)$ (panel c) is generally large for the NL RMF models. This was taken as a strong argument in favor of RMF (at this time at the level of NL-ME).[52] However, we see here that DD-PC/DD-ME yield somewhat lower values close to those from SHF. The isotope shift is related to the position of the lowest unoccupied neutron level in ^{208}Pb.[43,52] However, this level is not only dictated by spin-orbit splitting, but also by the nuclear surface profile which determines the position of shells with high-angular momentum. Thus an effect of density dependence is very plausible. After all, this isotope shift remains a subtle observable which is influenced by many ingredients.

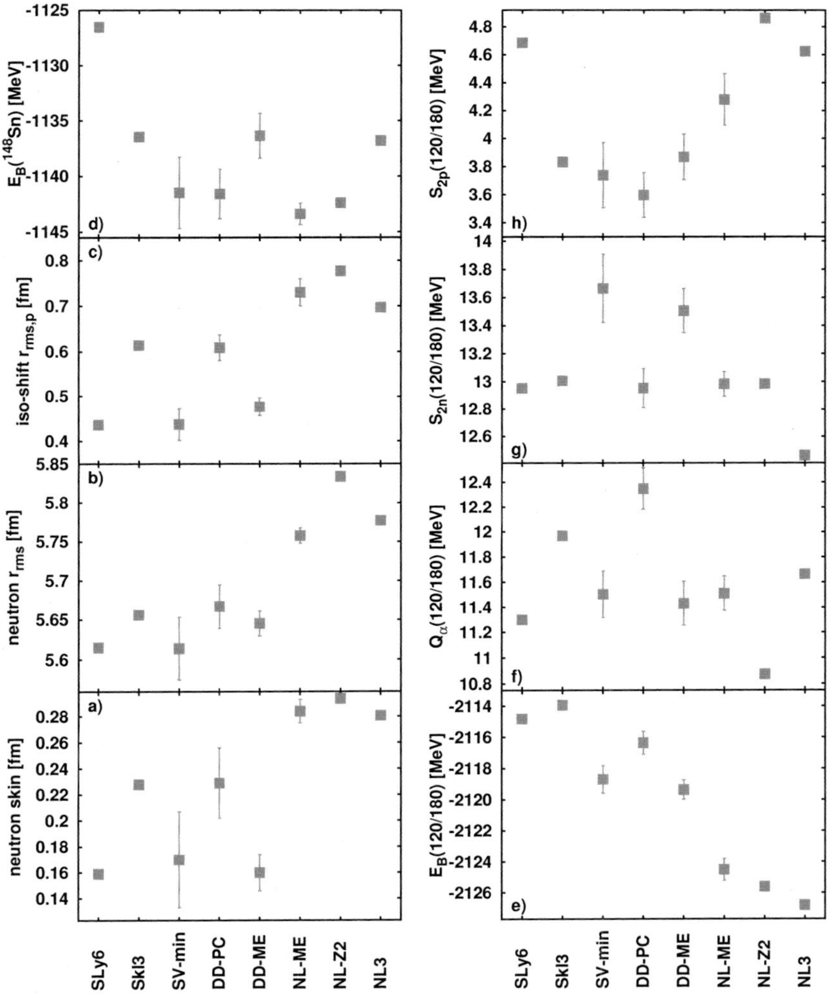

Fig. 4. (Color online) Observables for finite nuclei for a selection of parametrizations. Extrapolation uncertainties are available for the four fitted parametrizations (SV-min, DD-PC, DD-ME, NL-ME) and indicated by error bars.

Panel d shows the binding energy of the very neutron rich hypothetical nucleus ^{148}Sn. This should probe the extrapolation to neutron rich nuclei in the astrophysical r-process region. One expects an influence from isovector properties. But this is not clearly visible from the trends here. Again, we encounter an observable with mixed influences. Moreover, we will see in covariance analysis that the connection to isovector properties is not the same for all forces.

Properties of a hypothetical super-heavy nucleus with $Z = 120$ and $N = 182$ are shown in the right column of figure 4. The binding energy E_B (panel e) follows the same trends as the binding energy E/A in symmetric matter, see figure 3. This means, quite plausibly, that the nuclear matter binding is closely related to bulk binding in super-heavy elements. Particularly pronounced here is the discrepancy between small statistical extrapolation errors and the large variances among the results which remains a large discrepancy even when skipping the entries from the three NL models. It seems that (uncontrollable) systematic errors are becoming increasingly important with increasing distance to the fit nuclei.

Energy differences are shown in the further panels, the Q_α value for α decay in panel f, the two-neutron separation energy S_{2n} in panel g, and the two-proton separation energy S_{2p} in panel h. All three have different trends than E_B as such and each one has another one. This demonstrates nicely that differences of observables allow us to focus on other properties of a model. It is hard to read off correlation with particular NMP from the trends of these difference observables. We will evaluate these correlation later on with covariance analysis in section 6.6.

6.5. *Single particle energies*

Single particle (s.p.) energies do not belong to the observables for which density functional theory gives a guarantee.[29] Nonetheless, what comes out from mean-field calculations looks quite reasonable. Thus one should check the performance also for this, less safe, observable. In fact, spin-orbit splittings which are a particular combination of s.p. energies are part of the fit data in all SHF models to better confine the free parameters for the spin-orbit term. Comparison with experiment has to be taken with care because the experimental s.p. energies are taken from the energy difference to neighboring odd nuclei. A large survey within SHF has shown that the effect of the extra nucleon (or hole) is not negligible.[53] It can be of order up to 0.3 MeV for the levels as such. But this means also that an inspection of s.p. energies is legitimate as long as one does not scrutinize too deeply into the details. There is, however, one severe limitation. One should only look at states a few MeV below the highest occupied and above the lowest unoccupied state. Farther away from the Fermi surface, one may encounter large corrections from coupling to surface vibrations.[54] The

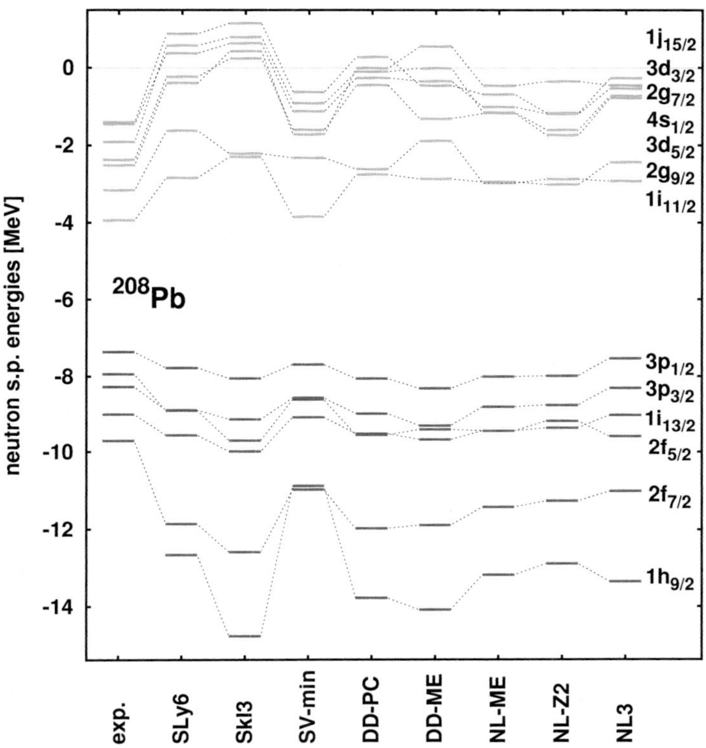

Fig. 5. (Color online) Neutron s.p. energies near the Fermi energy in ^{208}Pb for the standard selection of parametrizations.

difference of s.p. energies as built in spin-orbit splittings is less affected by the polarization effects and thus we can use data on spin-orbit splittings sufficiently close to the Fermi surface rather safely.

Figure 5 shows neutron s.p. energies in ^{208}Pb near the Fermi energy for the chosen sample of parametrizations. First of all, we emphasize that all parametrizations provide at once reasonable level schemes, in particular great agreement in the ordering of levels. Note that no information on s.p. energies was used in the fits (except for the spin-orbit splitting which, however, do not refer to the absolute position of a level). Nonetheless, proper level schemes emerge. This happens because the systematics of total binding energies reflects shell closures as well as sub-shell closures and these are only well reproduced if the shell sequence is sort of correct. Information on shell structure is thus also contained in gross trends of energies, not only in the excitation energies of the nearest odd neighbors.

On a closer look, one sees, of course, differences between the models. SV-min looks the most different from all others. Its spectrum is more compressed which is an obvious consequence of the larger effective mass m^*/m. The RMF models and SkI3 all have about the same level density which is plausible because they all have about the same m^*/m. The spectrum of SLy6 with its m^*/m between RMF values and SV-min lies also between the extremely compressed SV-min spectrum and extremely stretched other ones. The experimental levels are rather dense such that SV-min comes closest to it. Thus the level density near the Fermi surface speaks very much in favor of high effective masses as, e.g., SV-min has. This does not fully rule out low effective masses. Investigations using more microscopic diagrammatic analysis suggest that one unique value for m^*/m is inappropriate. It is very well conceivable that m^*/m in the nuclear interior is low and switches gently to large m^*/m at the surface.[55] States near the Fermi surface are most sensitive to the nuclear surface region and thus the empirical data favor high m^*/m. The problem with all EDFs considered here is that they know only one value for m^*/m. A solution of the conflict at the level of mean-field models is to develop a more flexible modeling of the density dependence of the kinetic terms. It is, of course, easy to invent such functionals. The challenge is to find experimental data which allow one to calibrate new terms unambiguously. This is a task for future development of nuclear EDF. Another option is to go beyond mean field and include correlations from low-lying collective states (often called nuclear phonons). These are known to modify the s.p. energies and produce a higher level density near the Fermi surface.[54,56] This, however, is the less convenient approach and one can still try to incorporate also these correlations effects into the EDF.

Useful further aspects of s.p. energies can be uncovered by recoupling the energies of the spin-orbit partners to spin-orbit splitting and centroid energy. This reads for a pair with orbital angular momentum l and total angular momentum $j_\pm = l \pm 1/2$

$$\delta_{ls}\varepsilon_l = \varepsilon_{lj_-} - \varepsilon_{lj_+} \ , \ \overline{\varepsilon_l} = \frac{(2j_- + 1)\varepsilon_{lj_-} + (2j_+ + 1)\varepsilon_{lj_+}}{4l+2} \ . \qquad (27)$$

These splittings and centroids for neutron states in ^{208}Pb are shown in figure 6. The situation is very similar for the proton states such that it suffices to discuss the neutron case, shown here. The spin-orbit splittings (lower panel) are very similar for all models. The $2f_{3/2} - 2f_{5/2}$ splitting tends to overestimation while the $3p_{1/2} - 3p_{1/2}$ splitting to underestimation.

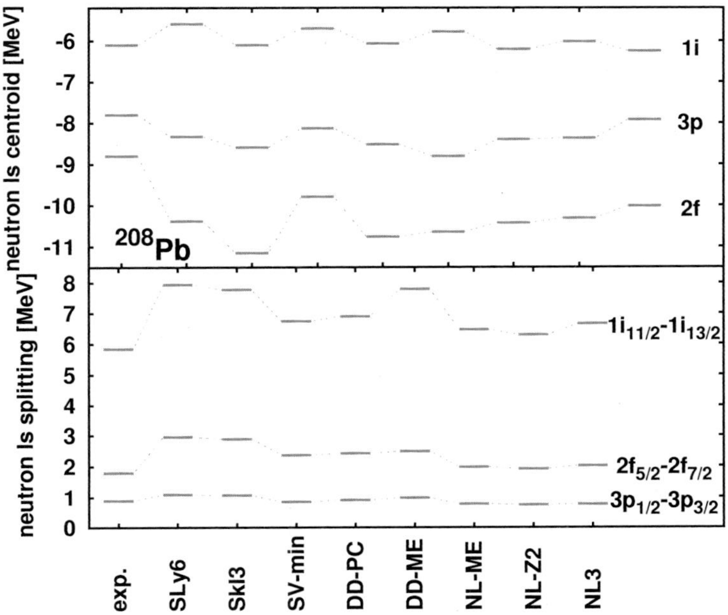

Fig. 6. (Color online) Spin-orbit splittings (lower panels) and centroids (upper panel) as defined in Eq. (27) for neutron states in ^{208}Pb near the Fermi energy for the standard selection of parametrizations.

For RMF, this is just the inevitable outcome of the model as spin-orbit properties are automatically included in the model. For SHF this is result of a compromise over the many spin-orbit splittings in doubly magic nuclei. It is very interesting that the resulting splittings are so similar for SHF and RMF although they have somewhat different modeling. The trend is the same in both models: the splittings are more overestimated the larger l is. The reason for this (moderate) mismatch may be similar as for the conflict between low and high effective masses, namely lacking density dependence in the modeling of the spin-orbit term (which is the same as the mass term for RMF).

The upper panel of figure 6 shows the centroids. These characterize the positioning due to orbital angular momentum. The two states closest to the Fermi surface, $3p$ and $1i$, behave rather well while the $2f$ states shows larger deviations from the data. Recall that the $1f_{7/2}$ state is already far away (about 4 MeV) from the Fermi surface. Quadrupole and octupole vibrational states in ^{208}Pb lie in that energy range. Thus we most probably encounter already effects from particle-core coupling.[54]

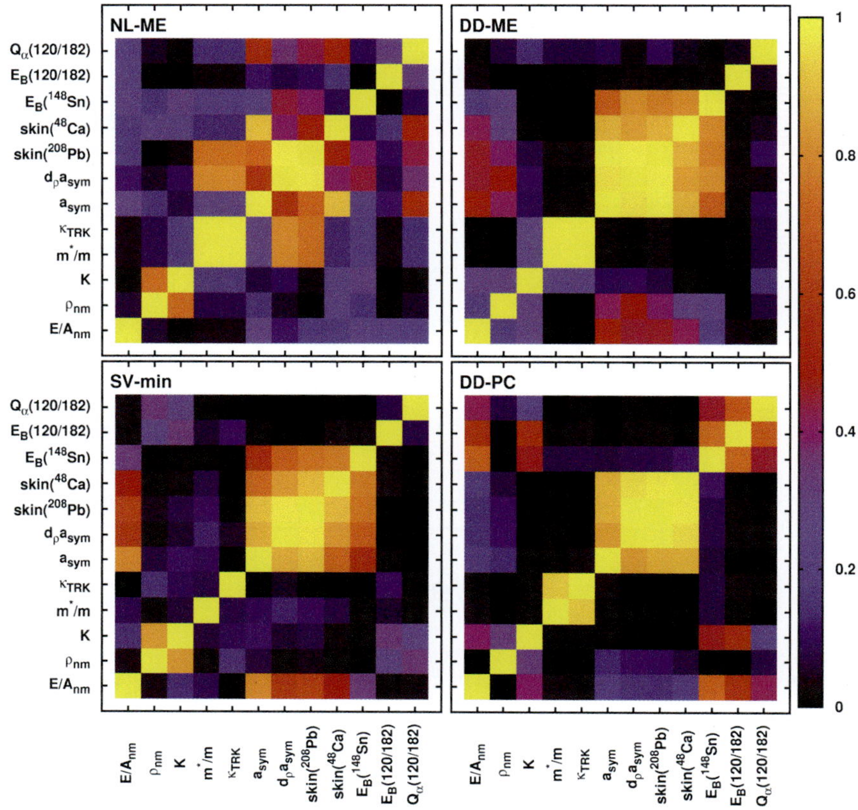

Fig. 7. Matrix of covariances between selected observables for the four parametrizations fitted to the same data set, namely SV-min, DD-PC, DD-ME, and NL-ME. The color scale is shown to the right of the panels. Yellow indicates $c_{AB} = 1$ and thus high correlation while black stands for no correlation.

6.6. *Covariances*

We have seen in the discussion of results above that uncertainties (25) of observables carry much useful information. Even more powerful are the correlations, or covariances (26) between observables. They carry a great multitude of information which is obvious already from the fact that one can square a large set of observables to a giant amount of pairs. A most compact, though not always easy to comprehend, way to represent the mass of information is to plot it as a matrix of covariances. This is done for the four recently fitted parametrizations in figure 7. A rather small set of observables was chosen in order to keep the matrix small and to allow the different parametrizations to be presented side by side. The set of observables includes first the NMP which can also be viewed as model parameters and additionally five observables in finite nuclei. The two entries for neutron skin are known to be isovector observables. The binding energies of two far exotic nuclei and the Q_α values of a super-heavy element are expected to explore mixed influences as we deduced from the above discussion of results.

Let us first have a look at the correlations between NMP (left lower block in each panel). The patterns differ dramatically between the models. The two dynamic responses, m^*/m and κ_{TRK}, are totally isolated quantities for SV-min, but very strongly correlated in all three RMF models. This tells us from the complementing perspective of covariances that m^*/m and κ_{TRK} are independent (and free) parameters in SHF. In RMF, on the other hand, they are rather frozen for the same reason, namely the interplay between strong vector and strong scalar field which is available only in one channel, the isoscalar one. It may be possible to decouple m^*/m from κ_{TRK} also in RMF by allowing additionally an isovector-scalar field (δ meson). This is worth exploring.

What the other NMP is concerned, we see in most cases E/A as independent (i.e. isolated) observable and some correlations between K and ρ_0. These correlations can be understood from the fact that both observables are influenced by the modeling of density dependence.

Always well separated from the other NMP are the two static isovector response observables a_{sym} and L. We see in all four cases more or less good correlation between these two NMP. This is found for all present parametrizations and fit data sets. It would require for each model some extensions to achieve independence between a_{sym} and L. At present, there is no pressing need to do that because we have not yet encountered so far conflicting data which would ask for independent a_{sym} and L to resolve a mismatch.

The two static isovector NMP, a_{sym} and L, are in all cases correlated with the neutron skins which confirms that these are indeed good isovector observables. A distinction remains between neutron skins in ^{48}Ca and ^{208}Pb. The skin in ^{208}Pb is the cleaner isovector observable because it always strongly correlated with L. Different results among the different models are obtained for the skin in ^{48}Ca where NL-ME shows significantly less correlations with L. Remember that ^{208}Pb is closer to bulk and thus more easily dominated by bulk properties whereas ^{48}Ca may have much stronger influences from shell structure for which no unique observable could be included in the covariance matrices of figure 7. It has yet to be worked out why some models correlate the skins in ^{48}Ca and ^{208}Pb more strongly than others. Understanding that would very much help to better understand the models and hopefully improve on them.

The three observables in very exotic nuclei (upper right block) shows only weak correlations with NMP and very different ones for the different parametrizations. This indicates that the prediction of properties of very exotic nuclei is an extremely subtle task where each ingredient of a model contributes.

Although this was a very compressed presentation of covariance matrices, the examples demonstrate how covariance analysis can be used and how powerful it is. One is tempted to apply it to all observables discussed in the previous subsections. But that would quickly exceed the bounds of this book and it would unfold a huge lot of information in which one is quickly lost. Patience in looking at many data and guidance in concentrating on the most telling observables is required for efficient use of the powerful covariance analysis.

7. Summary

In this chapter, we have compared three variants of RMF with the closest non-relativistic EDF, the Skyrme–Hartree–Fock (SHF) approach. We have done this in two ways, first, at a formal level along the non-relativistic limit of point-coupling RMF, and second, from an empirical perspective comparing the performance of the models concerning reproduction and prediction of several observables. We try to summarize the main observations:

(1) The non-relativistic limit allowed us to recover the structure of SHF from a point-coupling version of RMF. Unlike SHF, spin-orbit and kinetic term are in RMF closely related and dictated by the same model parameter. This means the effective mass m^*/m and spin-orbit splittings are intimately connected.
(2) Empirical fits produce notoriously very low effective masses m^*/m and very high TRK sum-rule factors κ_{TRK}. This is at variance with SHF models which have generally higher m^*/m and κ_{TRK}. Proper reproduction of giant resonances (not discussed here) calls for high m^*/m.
(3) RMF models have much fewer free parameters for isovector properties than SHF. As a consequence, extrapolation errors for isovector observables (e.g. neutron radius) are smaller for RMF. This looks deceivingly

like a higher predictive power, but it means higher rigidity in the isovector channel and so possibly higher systematic uncertainties.

(4) Many of the observed differences are due to modeling of density dependence. The predictions of more recent versions of RMF with explicit density dependence of the coupling parameters, DD-PC and DD-ME, stay closer to SHF than to the traditional RMF with non-linear meson self-coupling.

These points set also the directions for future development of the EDFs. At the side of RMF, there is a need to introduce more isovector flexibility. The simplest step in this direction is to introduce into RMF an isovector-scalar coupling term. However, one needs to include sufficiently many isovector data into the fit pool in order to calibrate the new couplings sufficiently well. This has probably to go beyond pure ground state properties and to include response properties as, e.g., the isovector dipole polarizability. For both branches, relativistic as well as non-relativistic EDF, there exists a large demand and space for development concerning density dependence, in general and particularly on the kinetic terms. Again, the major task is here to find out data which are most sensitive to density dependence such that they can be used for a proper determination of an extended model.

References

1. M. Bender, P.-H. Heenen, and P.-G. Reinhard, Self-consistent mean-field models for nuclear structure, *Rev. Mod. Phys.* **75**(1), 121–180 (Jan, 2003). doi: 10.1103/RevModPhys.75.121.
2. J. Stone and P.-G. Reinhard, The Skyrme interaction in finite nuclei and nuclear matter, *Prog. Part. Nucl. Phys.* **58**, 587 (2007).
3. J. Erler, P. Klüpfel, and P.-G. Reinhard, Self-consistent nuclear mean-field models: example Skyrme–Hartree–Fock, *J. Phys. G.* **38**, 033101 (2011).
4. J. W. Negele and D. Vautherin, Density–matrix expansion for an effective nuclear hamiltonian, *Phys. Rev. C.* **5**, 1472 (1972).
5. P.-G. Reinhard and C. Toepffer, Correlations in nuclei and nuclear dynamics, *Int. J. Mod. Phys. E.* **3**, 435 (1994).
6. B. G. Carlsson and J. Dobaczewski, Convergence of density-matrix expansions for nuclear interactions, *Phys. Rev. Lett.* **105**, 122501 (2010). doi: 10.1103/PhysRevLett.105.122501.
7. J. Dobaczewski, M. Stoitsov, W. Nazarewicz, and P.-G. Reinhard, Particle-number projection and the density functional theory, *Phys. Rev. C.* **76**, 054315 (2007). URL `http://link.aps.org/doi/10.1103/PhysRevC.76.054315`. http://www.arxiv.org/abs/0708.0441.

8. J. Erler, P. Klüpfel, and P.-G. Reinhard, Exploration of a modified density dependence in the Skyrme functional, *Phys. Rev. C.* **82**, 044307 (2010). URL `http://link.aps.org/doi/10.1103/PhysRevC.82.044307`. http://arxiv.org/abs/1009.0624.
9. J. D. Walecka, A theory of highly condensed matter, *Ann. Phys.* **83**(2), 491–529 (1974). doi: 10.1016/0003-4916(74)90208-5.
10. J. Boguta and A. R. Bodmer, Relativistic calculation of nuclear matter and the nuclear surface, *Nucl. Phys. A.* **292**(3), 413–428 (1977). doi: 10.1016/0375-9474(77)90626-1.
11. M. Rufa, J. Maruhn, W. Greiner, P.-G. Reinhard, and J. Friedrich, Nuclear ground-state properties in a meson-mean-field theory, *Z. Phys. A.* **323**, 323 (1986).
12. P.-G. Reinhard, The relativistic mean-field description of nuclei and nuclear-dynamics, *Rep. Prog. Phys.* **52**(4), 439–514 (1989). doi: 10.1088/0034-4885/52/4/002.
13. P. Ring, Relativistic mean field theory in finite nuclei, *Prog. Part. Nucl. Phys.* **37**, 193–263 (1996). doi: 10.1016/0146-6410(96)00054-3.
14. T. Bürvenich, D. G. Madland, J. A. Maruhn, and P.-G. Reinhard, Nuclear ground state observables and QCD scaling in a refined relativistic point coupling model, *Phys. Rev. C.* **65**(4), 044308 (Mar, 2002). doi: 10.1103/PhysRevC.65.044308.
15. W. Long, J. Meng, N. Van Giai, and S.-G. Zhou, New effective interactions in relativistic mean field theory with nonlinear terms and density-dependent meson-nucleon coupling, *Phys. Rev. C.* **69**(3), 034319 (Mar, 2004). doi: 10.1103/PhysRevC.69.034319.
16. W. H. Long, N. Van Giai, and J. Meng, Density-dependent relativistic Hartree–Fock approach, *Phys. Lett. B.* **640**(4), 150–154 (2006). doi: 10.1016/j.physletb.2006.07.064.
17. T. Nikšić, D. Vretenar, and P. Ring, Relativistic nuclear energy density functionals: Adjusting parameters to binding energies, *Phys. Rev. C.* 78(3): 034318 (2008). doi: 10.1103/PhysRevC.78.034318.
18. T. Nikšić, N. Paar, D. Vretenar, and P. Ring, DIRHB — a relativistic self-consistent mean-field framework for atomic nuclei, *Comput. Phys. Commun.* **185**, 1808–1821 (2014). doi: 10.1016/j.cpc.2014.02.027.
19. T. Nikšić, D. Vretenar, P. Finelli, and P. Ring, Relativistic Hartree–Bogoliubov model with density-dependent meson-nucleon couplings, *Phys. Rev. C.* **66**(2), 024306 (2002). doi: 10.1103/PhysRevC.66.024306.
20. G. A. Lalazissis, T. Nikšić, D. Vretenar, and P. Ring, New relativistic mean-field interaction with density-dependent meson-nucleon couplings, *Phys. Rev. C.* **71**(2), 024312 (Feb, 2005). doi: 10.1103/PhysRevC.71.024312.
21. P. Bonche, H. Flocard, P. H. Heenen, S. J. Krieger, and M. S. Weiss, Self-consistent study of triaxial deformations: Application to the isotopes of Kr, Sr, Mo, *Nucl. Phys. A.* **443**(1), 39–63 (1985). doi: 10.1016/0375-9474(85)90320-3.
22. M. Bender, K. Rutz, P.-G. Reinhard, and J. A. Maruhn, Pairing gaps from

nuclear mean-field models, *Eur. Phys. J. A.* **8**(1), 59–75 (July, 2000). doi: 10.1007/s10050-000-4504-z.
23. Y. Tian, Z. Ma, and P. Ring, A finite range pairing force for density functional theory in superfluid nuclei, *Phys. Lett. B.* **676**(1–3), 44–50 (2009). ISSN 0370-2693. doi: 10.1016/j.physletb.2009.04.067.
24. Y. Tian, Z.-y. Ma, and P. Ring, Separable pairing force for relativistic quasi-particle random-phase approximation, *Phys. Rev. C.* **79**(6) (Jun, 2009). ISSN 1089-490X. doi: 10.1103/physrevc.79.064301.
25. J. D. Bjorken and S. D. Drell, *Relativistic Quantum Mechanics.* McGraw–Hill, New York (1964).
26. M. Thies, *Nucl. Phys. A.* **460**, 636 (1986).
27. A. Sulaksono, P. G. Reinhard, T. J. Bürvenich, P. O. Hess, and J. A. Maruhn, From self-consistent covariant effective field theories to their Galilean-invariant counterparts, *Phys. Rev. Lett.* **98**, 262501 (2007). URL `http://link.aps.org/doi/10.1103/PhysRevLett.98.262501`. http://www.arxiv.org/abs/0705.3192.
28. J. Friedrich and N. Vögler, The salient features of charge density distribution of medium and heavy even–even nuclei determined from a systematic analysis of elastic electron scattering formfactors, *Nucl. Phys. A.* **373**, 192–224 (1982).
29. R. M. Dreizler and E. K. U. Gross, *Density Functional Theory: An Approach to the Quantum Many-Body Problem.* Springer, Berlin (1990).
30. P.-G. Reinhard, Correlations and local-density approximation, *Phys. Lett. A.* **169**, 281 (1992).
31. B. C. Clark and L. J. Kerr, Neutron densities from a global analysis of medium-energy proton-nucleus elastic scattering, *Phys. Rev. E.* **68**, 031204 (2003).
32. S. Abrahamyan et al., Measurement of the neutron radius of ^{208}Pb through parity violation in electron scattering, *Phys. Rev. Lett.* **108**, 112502 (2012). doi: 10.1103/PhysRevLett.108.112502.
33. P. Ring and P. Schuck, *The nuclear many-body problem.* Springer-Verlag, New York (1980).
34. W. Nazarewicz, P.-G. Reinhard, W. Satuła, and D. Vretenar, Symmetry energy in nuclear density functional theory, *Eur. Phys. J. A.* **50**, 20 (2014). URL `http://dx.doi.org/10.1140/epja/i2014-14020-3`. arXiv:1307.5782.
35. J. M. Lattimer, The nuclear equation of state and neutron star masses, *Annu. Rev. Nucl. Part. Sci.* **62**, 485–515 (2012). doi: 10.1146/annurev-nucl-102711-095018.
36. P. Klüpfel, P.-G. Reinhard, T. J. Bürvenich, and J. A. Maruhn, Variations on a theme by Skyrme: A systematic study of adjustments of model parameters, *Phys. Rev. C.* **79**, 034310 (2009). doi: 10.1103/PhysRevC.79.034310.
37. P. R. Bevington and D. K. Robinson, *Data Reduction and Error Analysis for the Physical Sciences.* McGraw-Hill (2003).
38. J. Friedrich and P.-G. Reinhard, The Skyrme force parametrisation and nuclear ground state properties, *Phys. Rev. C.* **33**, 355 (1986). URL `http://link.aps.org/doi/10.1103/PhysRevC.33.335`.

39. M. Kortelainen, T. Lesinski, J. Moré, W. Nazarewicz, J. Sarich, N. Schunck, M. V. Stoitsov, and S. Wild, Nuclear energy density optimization, *Phys. Rev. C.* **82**, 024313 (2010). doi: 10.1103/PhysRevC.82.024313.
40. J. Dobaczewski, W. Nazarewicz, and P.-G. Reinhard, Error estimates of theoretical models: a guide, *J. Phys. G.* **41**, 074001 (2014). doi: 10.1088/0954-3899/41/7/074001.
41. P. Klüpfel, J. Erler, P.-G. Reinhard, and J. A. Maruhn, Systematics of collective correlation energies from self-consistent mean-field calculations, *Eur. Phys. J A.* **37**, 343 (2008). URL `http://dx.doi.org/10.1140/epja/i2008-10633-3`.
42. E. Chabanat, P. Bonche, P. Haensel, J. Meyer, and R. Schaeffer, A Skyrme parametrization from subnuclear to neutron star densities part II. Nuclei far from stabilities, *Nucl. Phys. A.* **635**(1-2), 231–256 (1998). doi: 10.1016/S0375-9474(98)00180-8.
43. P.-G. Reinhard and H. Flocard, Nuclear effective forces and isotope shifts, *Nucl. Phys. A.* **584**, 467 (1995).
44. G. A. Lalazissis, J. König, and P. Ring, New parametrization for the Lagrangian density of relativistic mean field theory, *Phys. Rev. C.* **55**(1), 540–543 (Jan, 1997). doi: 10.1103/PhysRevC.55.540.
45. M. Bender, K. Rutz, P.-G. Reinhard, J. A. Maruhn, and W. Greiner, Shell structure of superheavy nuclei in self–consistent mean–field models, *Phys. Rev. C.* **60**, 034304 (1999).
46. J. Erler, P. Klüpfel, and P.-G. Reinhard, Misfits in Skyrme–Hartree–Fock, *J. Phys. G.* **37**, 064001 (2010). URL `http://dx.doi.org/10.1088/0954-3899/37/6/064001`. http://www.arxiv.org/abs/1002.0027.
47. J. P. Blaizot, Nuclear compressibilities, *Phys. Rep.* **64**, 171–248 (1980).
48. D. H. Youngblood, H. L. Clark, and Y.-W. Lui, Incompressibility of nuclear matter from the giant monopole resonance, *Phys. Rev. Lett.* **82**(4), 691–694 (Jan, 1999). doi: 10.1103/PhysRevLett.82.691.
49. P. Möller, J. R. Nix, W. D. Myers, and W. J. Swiatecki, Nuclear ground-state masses and deformations, *At. Data Nucl. Data Tables.* **59**(2), 185–381 (1995). ISSN 0092-640X. doi: 10.1006/adnd.1995.1002.
50. W. Nazarewicz and P.-G. Reinhard, Information content of a new observable: The case of the nuclear neutron skin, *Phys. Rev. C.* **81**, 051303 (2010). URL `http://link.aps.org/doi/10.1103/PhysRevC.81.051303`. http://arxiv.org/abs/1002.4140.
51. P.-G. Reinhard, J. Piekarewicz, W. Nazarewicz, B. K. Agrawal, N. Paar, and X. Roca-Maza, Information content of the weak-charge formfactor, *Phys. Rev. C.* **88**, 034325 (2013). arXiv:1308.1659.
52. M. M. Sharma, G. A. Lalazissis, and P. Ring, Anomaly in the charge radii of Pb isotopes, *Phys. Lett. B.* **317**, 9–13 (1993).
53. K. J. Pototzky, J. Erler, P.-G. Reinhard, and V. O. Nesterenko, Properties of odd nuclei and the impact of time-odd mean fields: A systematic Skyrme–Hartree–Fock analysis, *Eur. Phys. J. A.* **46**, 299 (2010). doi: 10.1140/epja/i2010-11045-6.

54. P. F. Bortignon, R. A. Broglia, C. H. Dasso, and C. Mahaux, Level dependence of the self-energy correction to the Hartree–Fock single-particle energies in ^{208}Pb, *Phys. Lett. B.* **140**, 163 (1984).
55. C. Mahaux, P. F. Bortignon, R. A. Broglia, and C. H. Dasso, Dynamics of the shell model, *Phys. Rep.* **120**, 1–274 (1985).
56. N. V. Mau and D. Vautherin, Effect of residual interactions on nuclear level densities, *Nucl. Phys. A.* **445**, 245 (1985).

Author Index